AQUATIC MESOCOSM STUDIES in ECOLOGICAL RISK ASSESSMENT

Edited by

Robert L. Graney, Miles, Inc.
James H. Kennedy, University of North Texas
John H. Rodgers, University of Mississippi

SETAC Special Publications Series

Series Editors

Dr. C. H. Ward
Department of Environmental Science and Engineering, Rice University
Dr. B. T. Walton
Environmental Sciences Division Oak Ridge National Laboratory
Dr. T. W. La Point
The Institute of Wildlife and Environmental Toxicology
Clemson University

Publication sponsored by the Society of Environmental Toxicology
and Chemistry (SETAC) and the SETAC Foundation for Environmental Education, Inc.

 **LEWIS PUBLISHERS**
Boca Raton   Ann Arbor   London   Tokyo

Library of Congress Cataloging-in-Publication Data

Aquatic mesocosm studies in ecological risk assessment / edited by
    Robert L. Graney, James H. Kennedy, John H. Rodgers, Jr.
        p.   cm.
    "A symposium . . . held at the 11th Annual Meeting of the Society of
Environmental Toxicology and Chemistry, in Arlington, Virginia in
November of 1990"—Chapter 1, introd.
    Includes bibliographical references and index.
    ISBN 0-87371-592-6
    1. Ecological risk assessment — Congresses.   2. Aquatic ecology —
Congresses.   3. Pesticides — Environmental aspects — Congresses.
4. Water — Pollution — Congresses.   I. Graney, Robert L.
II. Kennedy, James H., 1947–   .   III. Rodgers, John H. (John Hasford), 1950–   .
QH541.15.R57A68   1993
574.5'263 — dc20   93-3144
                                                                                      CIP

The SETAC Special Publication Series

The SETAC Special Publications series was established by the Society of Environmental Toxicology and Chemistry to provide in-depth reviews and critical appraisals on scientific subjects relevant to understanding the impacts of chemicals and technology on the environment. The series consists of single- and multiple-authored/edited books on topics selected by the SETAC Board of Directors for their importance, timeliness, and their contribution to multidisciplinary approaches to solving environmental problems. The diversity and breadth of subjects covered in the series will reflect the wide range of disciplines encompassed by environmental toxicology, environmental chemistry, and hazard/risk assessment. Despite this diversity, the goals of these volumes are similar; they are to present the reader with authoritative coverage of the literature, as well as paradigms, methodologies, controversies, research needs, and new development specific to the featured topics. All books in the series are peer reviewed for SETAC by acknowledged experts.

The SETAC Special Publications will be useful to environmental scientists in research, research management, chemical manufacturing, regulation, and education, as well as to students considering careers in these areas. The series will provide information for keeping abreast of recent developments in familiar areas and for rapid introduction to principles and approaches in new subject areas.

Aquatic Mesocosm Studies in Ecological Risk Assessment is the eighth volume to be published in this series. The book presents the collected papers stemming from a symposium on ''Utilization of Simulated Field Studies in Aquatic Ecological Risk Assessment,'' held at the 11th Annual meeting of the Society of Environmental Toxicology and Chemistry on November 11–15, 1990, in Arlington, Virginia. The symposium focuses on simulated field tests designed to understand basic mechanisms by which toxic chemicals affect ecosystem structure and function. Regulatory use of such systems stems from the need to effectively evaluate the ecological risk of agrichemical application. The papers contained herein critically evaluate endpoints, methods, and analyses in simulated field tests.

Thomas W. La Point
Editor, SETAC Special Publications
The Institute of Wildlife and Environmental Toxicology
Clemson University
September 1993

Robert L. Graney is Manager of the Ecological Effects Group at Miles Incorporated. He received a B.S. degree in biology (1977) and an M.S. degree in zoology (1980) from Virginia Tech and in 1986 obtained his Ph.D. from Michigan State University in Environmental Toxicology and Fisheries and Wildlife.

Dr. Graney has been involved in the research and development of ecological risk assessment procedures for the past 15 years. During the last five years, he has been manager of a research group responsible for developing the laboratory toxicity information and field ecotoxicological and exposure assessment data necessary for performing pesticide ecological risk assessments. Dr. Graney has been involved in the design, conduct, and interpretation of six aquatic mesocosm studies performed on various classes of pesticide chemistry. He has authored over 20 publications covering various aspects of aquatic and ecological toxicology, including laboratory and field toxicity testing procedures, biomarkers, and simulated field studies. He is an active member of the Society of Environmental Toxicology and Chemistry. (SETAC).

James H. Kennedy is an associate professor of biology and director of the Water Research Field Station at the University of North Texas. Dr. Kennedy received his Ph.D. in zoology from Virginia Polytechnic Institute and State University in 1981. Before joining the faculty at the University of North Texas, he was a private environmental consultant. In a collaborative effort, he and researchers from ZENECA agricultural chemical division designed and performed the first mesocosm study in support of agricultural chemical registration (1986). Dr. Kennedy has continued to be involved with the use of surrogate ecosystems for evaluating response of aquatic communities to chemicals. He has participated in a number of national and international workshops dealing with the use of surrogate aquatic ecosystems in the ecological risk assessment process.

Dr. Kennedy's research interests include benthic macroinvertebrate ecology, environmental impact assessment, and the use of surrogate ecosystems for evaluating ecological risk assessment. He has authored or co-authored a number of book chapters, papers, research reports, and presentations.

John H. Rodgers, Jr. is professor of biology, adjunct professor of the Research Institute for Pharmaceutical Sciences, and Director of the Biological Field Station at the University of Mississippi.

Dr. Rodgers received B.S. and M.S. degrees in botany at Clemson University and a Ph.D. in biology/aquatic ecology at Virginia Polytechnic Institute and State University (VPI & SU). Prior to joining the University of Mississippi faculty in 1989, he was an associate professor at the University of North Texas, founding director of the Water Research Field Station, and conducted research in the Institute for Applied Sciences. He began his academic career as an assistant professor at East Tennessee State University after a post-doctoral fellowship at VPI & SU.

For almost two decades, he has authored or co-authored over 80 research articles based on research that has focused on aquatic ecotoxicology, scaling in aquatic studies, sediment toxicology, and aquatic and wetland plants. He and his colleagues have studied the fate and effects of pesticides and other materials in aquatic systems (including microcosms, mesocosms, and field studies) for ecological risk evaluations. He is a member of the Society of Environmental Toxicology and Chemistry (SETAC) and was a board member from 1990–1992.

Contents

1. **Introduction** 1
    Robert L. Graney

SECTION I. INTRODUCTION AND REGULATORY BACKGROUND

Introduction 5
Robert L. Graney

2. **Mesocosms in Risk Assessment** 7
    Douglas J. Urban

3. **EPA Mesocosm Study Review** 17
    Candace A. Brassard, Arthur L. Buikema, Jr., Tom A. Bailey,
    Mary J. Frankenberry, Richard M. Lee, Ann M. Stavola,
    Clayton L. Stunkard, and Leslie W. Touart

4. **Regulatory Endpoints and the Experimental Design of
    Aquatic Mesocosm Tests** 25
    Leslie W. Touart

5. **Simulated or Actual Field Testing: A Comparison** 35
    Reinhard Fischer

6. **Contaminant Effects in Marine/Estuarine Systems:
    Field Studies and Scaled Simulations** 47
    James R. Clark and John L. Noles

7. **Summary and Discussion** 61
    Robert L. Graney

SECTION II. DESIGN OF MESOCOSM STUDIES AND STATISTICAL ANALYSIS OF DATA

Introduction 69
John H. Rodgers, Jr.

8. **Tests of Proportional Means for Mesocosm Studies** 71
    Clayton L. Stunkard

9. **Design and Statistical Analysis of Field Aquatic Mesocosm Studies** 85
    Jennifer L. Shaw, Malcolm Moore, James H. Kennedy, and Ian R. Hill

10. **Ecological Development and Biometry of Untreated Pond Mesocosms** 105
    Van D. Christman, J. Reese Voshell, Jr., David G. Jenkins, Michael S. Rosenzweig, Raymond J. Layton, and Arthur L. Buikema, Jr.

11. **Applying Concentration-Response Theory to Experimental Design of Aquatic Enclosure Studies** 129
    Dean G. Thompson, Stephen B. Holmes, Douglas G. Pitt, Keith R. Solomon, and Kerrie L. Wainio-Keizer

12. **Optimal Design of Aquatic Field Studies** 157
    Ronald R. Regal and Stephen J. Lozano

13. **Summary and Discussion** 173
    John H. Rodgers, Jr.

SECTION III. MANAGEMENT AND TREATMENT OF MESOCOSMS

Introduction 177
Robert L. Graney

14. **Sediment Transfers and Representativeness of Mesocosm Test Fauna** 179
    Leonard C. Ferrington, Jr., Mary Anne Blackwood, Christopher A. Wright, Tracey M. Anderson, and David S. Goldhammer

15. **Spray Drift and Runoff Simulations of Foliar-Applied Pyrethroids to Aquatic Mesocosms: Rates, Frequencies, and Methods** 201
    Ian R. Hill, Kim Z. Travis, and Paul Ekoniak

16. **Use of Mesocosm Data to Predict Effects in Aquatic Ecosystems: Limits to Interpretation** 241
    Thomas W. La Point and James F. Fairchild

17. **Impact of 2,3,4,6-Tetrachlorophenol (DIATOX®) on Plankton Communities in Limnocorrals** 257
    Karsten Liber, Keith R. Solomon, Narinder K. Kaushik, and John H. Carey

18. **Summary and Discussion** 295
    Robert L. Graney

SECTION IV. SCALING

Introduction 301
John H. Rodgers, Jr.

19. **Fate and Biological Effects of an Herbicide on Two**
    Artificial Pond Ecosystems of Different Size.........................303
    Fred Heimbach, Juergen Berndt, and Wolfgang Pflueger

20. **Earthen Ponds vs. Fiberglass Tanks as Venues For**
    Assessing the Impact of Pesticides on Aquatic
    Environments: A Parallel Study With Sulprofos 321
    Gregory L. Howick, Frank deNoyelles, Jr., Jeffrey M. Giddings,
    and Robert L. Graney

21. **Fate and Effects of Cyfluthrin (Pyrethroid Insecticide) in**
    Pond Mesocosms and Concrete Microcosms 337
    Philip C. Johnson, James H. Kennedy, R. Gregg Morris,
    and Faithann E. Hambleton

22. **Pyrethroid Insecticide Effects on Bluegill Sunfish in**
    Microcosms and Mesocosms and Bluegill Impact on
    Microcosm Fauna 373
    R. Gregg Morris, James H. Kennedy, Philip C. Johnson,
    and Faithann E. Hambleton

23. **Summary and Discussion** 397
    John H. Rodgers, Jr.

SECTION V. CASE HISTORIES

Introduction 401
James H. Kennedy

24. **Lambda-Cyhalothrin: A Mesocosm Study of Its Effects on**
    Aquatic Organisms...403
    Ian R. Hill, Jill K. Runnalls, James H. Kennedy, and Paul Ekoniak

25. **The Fate and Effects of Guthion (Azinphos Methyl)**
    in Mesocosms..469
    Jeffrey M. Giddings, Ronald C. Biever, Raymond L. Helm,
    Gregory L. Howick, and Frank J. deNoyelles, Jr.

26. **Evaluation of the Ecological and Biological Effects of
Tralomethrin Utilizing an Experimental Pond System** 497
Joseph M. Mayasich, James H. Kennedy, and Joseph S. O'Grodnick

27. **Response of Zooplankton to Dursban® 4E Insecticide
in a Pond Experiment**...517
Willem G.H. Lucassen and Peter Leeuwangh

28. **Algal Periphyton Structure and Function in Response
to Consumer Chemicals in Stream Mesocosms** 535
Scott E. Belanger, James B. Barnum, Daniel M. Woltering,
John W. Bowling, Roy M. Ventullo, Scott D. Schermerhorn,
and Rex L. Lowe

29. **Summary and Discussion** 569
James H. Kennedy

SECTION VI. ECOSYSTEM ANALYSIS

Introduction 575
James H. Kennedy

30. **Aquatic Mesocosms in Ecological Effects Testing:
Detecting Direct and Indirect Effects of Pesticides** 577
Frank deNoyelles, Jr., Sharon L. Dewey, Donald G. Huggins, and
W. Dean Kettle

31. **On the Use of Ecosystem Stability Measurements in
Ecological Effects Testing**...605
Sharon L. Dewey and Frank deNoyelles, Jr.

32. **Structural Equation Modeling and Ecosystem Analysis** 627
Michael L. Johnson, Donald G. Huggins, and Frank deNoyelles, Jr.

33. **The Ecotoxic Effects of Atrazine on Aquatic Ecosystems:
An Assessment of Direct and Indirect Effects Using
Structural Equation Modeling**...653
Donald G. Huggins, Michael L. Johnson, and Frank deNoyelles, Jr.

34. **Summary and Discussion** 693
James H. Kennedy

Index..701

CHAPTER 1

Introduction

Robert L. Graney

A symposium on the use of aquatic mesocosm studies in the ecological risk assessment of pesticides was held at the 11th annual meeting of the Society of Environmental Toxicology and Chemistry, in Arlington, VA in November of 1990. The objective of the symposium was to review the "state of the art" in the area of aquatic mesocosm testing, with special emphasis on the use of these studies in the pesticide regulatory arena. Confirmation of laboratory-based risk assessments via mesocosm studies is an area of active research, and this symposium served to bring together experts for discussion in areas such as study design, methodology, interpretation, and regulatory evaluation.

Classically, ecological effects evaluations have been based on results from single species toxicity tests. Results from standardization laboratory test procedures have provided an efficient and cost effective means for assessing potential adverse effects of chemicals (Rand and Petrocelli, 1985). Exposure-response information developed under controlled laboratory conditions can be compared with predicted or measured environmental concentrations and provide estimates of relative risk. The accuracy of such predictions is partially based on the reliability of the exposure estimates and the ability to extrapolate data on laboratory effects to the "real world". Laboratory toxicity tests are designed to be reproducible; however, they are not necessarily designed to be reflective of single species effects which may occur during exposure in the field. It has been assumed that by testing the "most sensitive species" in the laboratory and applying appropriate safety factors that structural

0-87371-592-6/94/$0.00 + $.50

and functional components of aquatic ecosystems would be protected. While there is some evidence that this approach can work and will afford adequate environmental protection, many questions remain to be addressed through multispecies ecosystem tests such as "mesocosms" or full field studies. Single species tests do not provide direct information on potential indirect effects which may occur in complex ecosystems. In a general sense, the types of indirect effects which may occur are often predictable; however, the details associated with specific changes are difficult to predict with any degree of resolution.

Field testing may be needed to address uncertainty associated with extrapolating results from single species toxicity tests to the "real world". Field testing can serve a number of basic functions, including (1) "validation" of the risk assessment process which is currently based on single species toxicity tests, (2) evaluation of potential fate and effects of a specific chemical, and (3) basic research on the influence of stressors on the structure and function of aquatic ecosystems, including the development and validation of effects models.

"Field" studies can be subdivided into two areas: natural field studies and simulated field studies. Natural field studies can be defined as those studies in which both the test system and exposure to the stressor are "naturally" derived. Examples of such studies may include (1) a pond in an agricultural setting in which the influence of pesticide applications to the surrounding watershed is evaluated, (2) an evaluation of the impact of acid rain on acidified and nonacidified lakes, and (3) a lotic field survey of specific organisms located upstream and downstream of a power plant or chemical plant effluents. All natural field studies are site specific and designed to evaluate the impact of chemicals from identified sources on specific ecosystems. These studies are not designed to be predictive, making extrapolation to other sites or situations difficult.

Simulated field studies are composed of either an isolated subsection of the natural environment or a man-made physical model of a lotic or lentic ecosystem. Such test systems are often referred to as mesocosms, although they may be designed to represent a wide range of physical structures simulating various components of aquatic ecosystems. In these studies, the test systems are manually treated with the test chemical at predetermined concentrations. Various test systems utilized by researchers will be discussed in this book, including:

- Large Ponds Systems — artificially constructed earthen ponds ranging in size from 0.01 to 0.1 hectares and in volume from 100 to 1000 cubic meters. Ponds are allowed to colonize for a predetermined period and fish are stocked prior to treatment. These studies are currently required by the Environmental Protection Agency (EPA) for the registration of some pesticides.
- Outdoor "Microcosms" — fabricated tanks large enough to be representative of lentic ecosystems and not greatly influenced by ambient environmental conditions (i.e., temperature, light, wind, etc.). These generally vary in size from 2000 to 20,000 l.
- Limnocorrals — artificial enclosures placed in the pelagic region of ponds, lakes, or marine environments. These systems have varied in size from less than 100 l

to greater than 100,000 l and may or may not be in contact with the profundal region. Fish are generally excluded from these test systems.

- Littoral Enclosures — plastic dividers are used in this test system to isolate the littoral region of ponds. These test systems, developed by the EPA Duluth Laboratory, have a volume of approximately 50,000 l and a maximum depth of 2 m.
- Lotic Systems — artificial streams of various sizes have been used to evaluate the effects of chemicals. Unlike the lentic systems, no standard design has been developed for flowing water test systems.

This symposium only addressed simulated field studies (mesocosms). These multispecies tests can generate data on a plethora of additional species not tested, or unsuitable for testing, under laboratory conditions. Mesocosms continue to provide a data base for evaluating the relative sensitivity of laboratory test organisms and are useful in identifying species that have an unusual sensitivity to a particular chemical or class of chemicals. Since the test systems are complex functioning ecosystems, they also provide information on indirect effects that may result from direct toxicity to one or more components of the ecosystem. In addition, more natural exposure regimes are provided in simulated field studies. Laboratory studies are normally conducted under continuous test chemical exposure regimes which, except for more persistent chemicals or for point sources with a constant input, do not occur in the field. Mitigating water quality or environmental conditions which can alter the exposure regime are accounted for in mesocosm studies.

However, ecosystems are by nature variable and somewhat uncontrollable. Identification of ecosystem level effects requires the ability to separate treatment or chemical related changes in the system from natural or background variability. The greater the variability in a particular endpoint, the more difficult to identify or measure chemical-induced alterations. Although natural field or simulated field studies provide a test system which evaluates ecosystem level effects, interpretation of such effects can be difficult. This difficulty arises due to the variability problem discussed above and due to the regulatory dilemma often associated with making decisions on "minor" or secondary effects which may be observed in such systems.

Mesocosm studies can be extremely expensive to conduct. The advantages and disadvantages of single species tests are known. They have been the mainstay of ecological risk assessment and have effectively served their purpose by providing a cost effective means for regulating toxic chemicals. It is relatively easy to criticize single species tests; however, resources are limited and must be applied in a manner which provides the most useful information. For the purpose of evaluating potential effects of toxic chemicals, amelioration of the environmental effects of toxic chemicals may be better realized by evaluating numerous chemicals in rather simple microcosm system vs. conducting a large field study for a single chemical. Resources must be allocated to maximize the amount of useful information obtained without adversely interfering with continued scientific and economic development.

This symposium tried to address many aspects of simulated field testing. Due to time constraints, the focus of the symposium was freshwater lentic systems; however, this is not to minimize the importance of freshwater lotic systems or estuarine/marine systems or the research conducted in these systems. Many of the papers on test design and interpretation are generic and are applicable to many ecosystems not specifically discussed. Numerous case histories are presented and many of them deal with pesticide-related studies. This reflects recent trends in the pesticide registration arena and the increasing number of requirements to conduct mesocosm studies for pesticide registration.

Proceedings from the symposium have been divided into six sections. These sections do not reflect the actual format of the symposium, but rather reflect a logical organization of the manuscripts based on their content and relationship to other topics. A general introduction is provided for each section which presents, in very brief format, the manuscripts included in that section. At the end of each section is a brief overview provided by the editors. The purpose of the overview is to try to capture the main points of the papers in that section. As preparation for reading each section, it may be worthwhile to read the summary section first. This may provide some useful insight and perspective for reading the manuscripts. Each summary section also addresses the questions/discussion sessions which were held at the symposium. The exact transactions from the symposium are not provided; however, relevant questions/discussions are reviewed and discussed by the editors.

REFERENCE

Rand G.M. and S.R. Petrocelli. 1985. *Fundamentals of Aquatic Toxicology: Methods and Applications*. Hemisphere Publishing Corporation. 666 pp.

Introduction and Regulatory Background

Recent regulatory initiatives have intensified the amount of research on contaminant impacts on the structure and function of aquatic ecosystems. The central role of legislative mandates in requiring multispecies field testing makes it important to understand the basis for the requirements and how they influence the conduct and interpretation of such studies. The first three chapters of this section were all written by personnel from the Ecological Effects Branch (EEB) of the Office of Pesticide Programs (OPP) of the Environmental Protection Agency (EPA). These papers provide an overview of the current status and use of mesocosm studies as they apply to pesticide registration under the Federal Insecticide, Fungicide, and Rodenticide Act (FIFRA). Under this statute, the Agency must determine whether or not to register a pesticide and, as stated by all three of the authors, the registrant must provide data to establish that "when used in accordance with widespread and commonly accepted practice, the pesticide will not generally cause unreasonable adverse effects on the environment".

These papers do not address the use of mesocosm studies in the context of other regulatory statutes (i.e., TSCA). Currently such studies are not required under any other statutes; however, the concept of ecosystem level testing and the need for test systems which extend beyond the limitations of single species tests continues to be debated in other regulatory arenas. The EPA Risk Assessment Forum, which was established to develop Agency-wide risk assessment guidelines, is critically evaluating the need to include community and ecosystem level parameters in the standard risk assessment guidelines for the Agency (EPA, 1991).

The first paper (Urban) in this section, entitled "Mesocosms in Risk Assessment", reviews the risk assessment process which precedes the requirement to conduct mesocosm studies and briefly evaluates how the data from such studies are used in making regulatory decisions. The second manuscript (Brassard et al.), authored by a group of reviewers from within the Ecological Effects Branch of EPA, provides an overview of the process which the Agency goes through during

5

the review of a mesocosm study. These authors touch upon specific design and measurement aspects of the studies as they relate to study review and interpretation. The third manuscript (Touart), entitled "Regulatory Endpoints and Experimental Design of Aquatic Mesocosm Tests", provides an overview of the current Agency philosophy on endpoints of importance and the advantages and disadvantages of different experimental designs. All of the papers provide unique information and, taken collectively, provide a good overview on the regulatory use of mesocosms.

The final two chapters in this section provide an overview of the use and importance of "natural" or "simulated" field studies for freshwater and marine ecosystems. The paper by R. Fischer compares and contrasts actual field testing and simulated field testing, using an example of an actual field study on endosulfan to highlight some of the disadvantages of such studies. The final paper on marine field studies and simulated field studies by Clark and Noles provides an overview of the importance of field testing for evaluating impacts on marine systems.

CHAPTER 2

Mesocosms in Risk Assessment

Douglas J. Urban

Abstract: Since 1975, the Office of Pesticide Programs (OPP) has included aquatic field testing in its aquatic tier testing scheme. Beginning in 1988, OPP provided additional guidance concerning such testing and focused on mesocosms as the preferred method for aquatic field testing. Using endpoints as described in the regulations, and considering the weight-of-evidence approach to ecological risk assessment, we will describe why mesocosms are important and how mesocosms are used in OPP's ecological risk assessment process.

INTRODUCTION

Under the Federal Insecticide, Fungicide, and Rodenticide Act (FIFRA), the United States Environmental Protection Agency (EPA) has been conducting ecological risk assessments using tiered data requirements since 1975. Guidance for conducting ecotoxicity tests in the laboratory and field have been periodically issued and updated to the present time. This guidance has specifically covered aquatic field testing (Tier 4, the highest for Aquatic Organisms), both actual field tests and simulated field tests. Over the years, the Agency has required these tests to support pesticide registration. The purpose of this data requirement has always been to negate the presumption of risk to aquatic organisms indicated by lower tiered data. The Office of Pesticide Programs (OPP) has reviewed and validated these studies and has made regulatory decisions based upon their results. While the total number of these studies is small, and the early aquatic field studies were wanting in many areas including design and methods, OPP was able to use the results as part of weight-of-evidence risk conclusions.

Generally, the adverse effect of greatest concern and most easily measured in all the tiers of testing over the years has been direct mortality. Mortality as well as reproductive impairment and growth are effects on single species which have been used as regulatory endpoints under FIFRA. They have been used for specifying label precautions, for classifying pesticide use, and for making decisions concerning registration, reregistration, and special review.

Recently,[1] OPP has issued specific technical guidance concerning the conduct of a particular kind of simulated aquatic field study, the mesocosm. An aquatic mesocosm is described as "an intermediate-sized system, such as a dug-out pond or *in situ* enclosure, that can be replicated and manipulated to test both structural and functional parameters as a representative aquatic ecosystem".[2] These types of test systems provide results which OPP can use to establish endpoints other than direct mortality and reproductive impairment in its ecological risk assessments and regulatory decisions.

The history of the development of the mesocosm has been covered elsewhere.[1,3] Over the last few years, this test system has received numerous reviews at workshops and in the literature. It is currently the preferred test system for satisfying OPP's Tier 4 aquatic field testing data requirement. The test system has numerous advantages over actual field tests.[1] While direct effects upon single species or multiple species within the system can be measured, it also provides effects data on many structural and functional parameters of aquatic populations and communities for ecological risk assessments. In addition, it provides better replication under field-like conditions.

Under FIFRA, the Agency must determine whether a pesticide can be registered for a particular use. The test for whether a pesticide can be registered is that it *will not* "generally cause unreasonable adverse effects on the environment" (P.L.95396, Sec. 3(c)(5)(D)). The environment "includes water, air, land, and all plants and man and other animals living therein, and the interrelationships which exist among these." (ibid, Sec. 2(j)). With this test system, OPP can consider both direct and indirect adverse effects on single species, populations, and their communities in its risk-benefit decisions.

The remainder of this paper will cover the history (Figure 1) and importance of aquatic field testing in pesticide guidelines and regulations, and the use of aquatic field tests, mesocosms in particular, in pesticide ecological risk assessment.

AQUATIC FIELD TESTING: HISTORY IN PESTICIDE REGULATION AND GUIDELINES

The Agency requires a pesticide registrant to submit Tier 4 aquatic-simulated field or actual field testing on a case-by-case basis. Generally, this testing is required when the results of previous tier testing data (laboratory acute and chronic), after

Guidelines - 1975

Guidelines - 1978

Guidelines - 1982

Regulations - 1984

Technical Guidance - 1988

Figure 1.   History of aquatic field testing.

being compared to exposure levels of the pesticide in the environment, have indicated a potential risk to aquatic organisms. Thus, the Agency requires Tier 4 data to negate the presumption of adverse effects and thus serious risk of pesticides to aquatic organisms.

In 1975, the Agency issued "Guidelines for Registering Pesticides in the United States"[4] which discussed aquatic field testing. The advantages and disadvantages of various methods and techniques that were available at that time were weighed and the lack of standardization lamented. These included field surveys, monitoring programs, models of laboratory streams, and microcosms. The authors of the document concluded that

"at this point in time it appears that the only possible approaches to aquatic field testing that might be standardized are the cage-type and confined-area exposure. Since the confined area type of study requires a high degree of knowledge of the life history of various species, the ecology of the area, etc., this would be very difficult to reduce to a routine."

Thus, at that time, the field testing method of choice to measure adverse effects from pesticides in the field was the caged-aquatic studies. Further, it was clear from the discussion of parameters, that the focus of the aquatic field studies at that time was on single-species effects, especially survival:

"Parameters that should be studied which would relate to individual species include: mortality, growth, reproductive success, and gross behavior, and other observable changes within such fields as physiology, biochemistry, pathology, etc. which relate directly to the survival of the species."

Finally, no guidance was provided concerning when this study may be required, and there was no reference to mesocosm.

The Agency issued updated guidance concerning aquatic field testing in 1978. This document[5] provided the first guidance concerning when stimulated or actual field testing for aquatic organisms would be required. Under Part 163.72-6 of this document,

> "a short-term simulated field test (where confined populations are observed, or an actual short-term field test (where natural populations are observed), or both are required if laboratory data indicate adverse short-term or acute effects may result from intended use."

And,

> "a long-term simulated field test (e.g., where reproduction and growth of confined populations are observed) or an actual field test (e.g., where reproduction and growth of natural populations are observed), or both are required if laboratory data indicate adverse long-term, cumulative, or life-cycle effects may result from intended use."

The 1975 guidelines were referenced for protocol development. No other guidance was given, and no reference was made to mesocosms.

In 1982, the Agency published the "Pesticide Assessment Guidelines, Subdivision E, Hazard Evaluation: Wildlife and Aquatic Organisms".[6] This document expanded upon previous guidance concerning when these data would be required. First, emphasis was placed upon the case-by-case nature of the requirement. Second, simulated field tests were generally given precedence over actual fields. For example,

> "The short-term simulated field test (where confined populations are observed) should be selected if it can yield data useful in assessing such [adverse short-term or acute effects] risks. An actual short-term field test (where natural populations are observed) may be needed if a simulated test would not suffice."

The document also provided the first guidance for general testing standards for simulated and actual field tests. Specific test standards covered test substance, concentration analysis, test conditions, endangered species, and residue levels. In addition, an updated list of references was provided, and the reference to the 1975 guidance document was dropped. While no specific reference was made to mesocosms by name, one of the references in the 1982 guidelines[7] was an experimental pond study with a design similar to current mesocosm designs.

In 1984, the Agency issued data requirements for pesticide registration.[8] Part 158.145 of the document included a condensed version of the required guidance for simulated or actual field testing — Aquatic Organisms:

> "Tests are required on a case-by-case basis depending upon the results of lower tier studies such as acute and subacute testing, intended use pattern, and pertinent environmental fate characteristics."

In 1988, OPP published a Technical Guidance Document, "Aquatic Mesocosm Tests to Support Pesticide Registrations".[1] Based upon an Agency Workshop in 1987 and after review by the FIFRA Scientific Advisory Panel, this document provided detailed guidance on the objectives of the test system, proposed design criteria, measured parameters, interpretation of results, and current literature references. Mesocosms are currently the preferred test systems for satisfying the Agency's Tier 4 data requirement for pesticides — simulated or actual field testing for aquatic organisms.

AQUATIC SIMULATED (MESOCOSMS) AND ACTUAL FIELD STUDIES: THEIR PLACE IN THE PESTICIDE RISK ASSESSMENT PROCESS

The Agency may register a pesticide when it determines among other findings that the pesticide will perform its intended function without unreasonable adverse effects on the environment, and, when in accordance with widespread and commonly accepted practice, the pesticide will not generally cause unreasonable adverse effects on the environment. FIFRA has always placed the burden of proof for this finding upon the pesticide registrant,[9] whether it be to support new registration or continued registration (reregistration). This burden of proof involves developing and submitting data to support this finding. Concerning ecological effects data required for pesticides, OPP establishes ecological effects testing guidelines and data requirements, reviews the data when submitted and evaluates it (see Reference 10), and writes a risk assessment. Generally, if the assessment concludes little or no risk, and other conclusions such as human health and groundwater are similar, then a decision for registration is likely. If, however, the risk assessment concludes risk, and the benefits case is not strong, then decisions such as to deny, suspend, or cancel an existing registration, or initiate a special review, are more likely.

ECOLOGICAL RISK ASSESSMENT: THE PROCESS

The process (see Figure 2) involved in ecological risk assessment for pesticides includes the following components: hazard (toxicity), exposure, quotient, weight-of-evidence, confirmatory data (field testing and incident reports), and ecological risk conclusion. Briefly, the tiered ecotoxicological data, including aquatic single-species laboratory toxicity tests and simulated or actual aquatic field tests, are reviewed, evaluated, and the most appropriate ecotoxicity values are chosen for expressing the hazard of the pesticide to aquatic organisms based upon the proposed/registered labeled use. These values are compared to the estimated or actual exposure levels of the pesticide likely to be found in the environment. This exposure is determined from information on the fate and transport of the pesticide, labeled use pattern including area of use, application rates, frequency of application, application methods, etc., as well as the type, abundance, distribution, natural history, etc. of the organisms likely to be exposed. Based on these data, the pesticide concentrations to which the aquatic organisms are likely to be exposed are estimated (often by standard fate and transport models such as SWRRB [simulator for water resources on rural basins] and EXAMS [exposure analysis modeling system]). When available, actual measured concentrations are used.

OPP then uses the quotient method to express likely ecological risk. The quotient method is a common sense method whereby the hazard value(s) of the pesticide for the most sensitive species tested are compared to the estimated or actual exposure value(s) to provide an estimate of ecological risk. This can be expressed as follows:

$$\text{Quotient} = \frac{\text{Exposure}}{\text{Hazard}}$$

Hazard is defined as the intrinsic quality of a pesticide to cause an adverse effect under a particular set of circumstances. Exposure has two components, the first is the estimated or measured amount of pesticide residue that will be in the environment and available to aquatic organisms. The second consists of the numbers, types, distribution, abundance, dynamics, and natural history of the aquatic organisms which will be in contact with these residues.[11]

If the quotients are equal to or greater than one, then they indicate a high risk to aquatic organisms.[12] Quotients considerably less than one indicate a low risk. And quotients approaching one indicate an uncertain risk. If the hazard component of the quotient was based on single-species laboratory tests, i.e., aquatic Tiers 1,

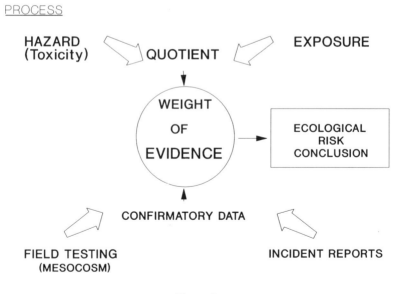

Figure 2.

2, and 3, and the quotient indicates a high or uncertain risk, then Tier 4 aquatic-simulated or actual field testing (usually a mesocosm) is required on a case-by-case basis in order to negate or confirm the risk indicated by the quotient. The results of the Tier 4 testing are considered confirmatory data, that is, data that confirm or negate the ecological risk estimates indicated by the quotient. These data also include field incident reports, such as fish kills.

While the results of the confirmatory data may confirm or negate the risk estimates from the quotient, they are not considered pass-fail conclusions. The final step in the ecological risk assessment is the balancing of the weight of all the evidence. This includes balancing the results from ecotoxicity data, exposure estimates, simulated or actual field studies, and available incident reports, as shown in Figure 2. This weight-of-evidence review can be complex considering the extent of the use area of the pesticide, and the number of species likely to be exposed. During this review, consideration is given to the quality and adequacy of the data, as well as the frequency and magnitude of the observed or estimated adverse effect(s). The conclusions drawn are considered the ecological risk component of the FIFRA mandated risk-benefit analysis.

ECOLOGICAL RISK ASSESSMENT: IMPORTANT REGULATORY DECISIONS

Ecological risk assessments are conducted for FIFRA Section 3 (registration of pesticides), Section 4 (reregistration of registered pesticides), Section 5 (experimental use permits), Section 18 (exemptions of federal and state agencies), and Section 24(c) (authority of states — addition uses) actions. During the ecological risk assessment process, OPP makes certain important regulatory decisions for which the results of certain aquatic tests are crucial. First, aquatic single-species acute laboratory test results are used to establish environmental label precautions. For example,[13]

"This pesticide is toxic to fish" is based on the results of the fish 96-hour LC50 test. The statement is required if the fish LC50 is <1 ppm for the most sensitive fish species tested.

The results from the same testing are used to determine if a pesticide product intended for outdoor use will be considered for restricted use classification.[14] Generally, a pesticide will be considered for such when, among other factors, its use according to label directions

"results in residues of the pesticide, its metabolites or degradation products, in water that equal or exceed one-tenth of the acute LC50 for non-target aquatic organisms likely to be exposed."

Classification for Restricted Use means that the application of the product is restricted to EPA certified applicators.

Acute, chronic, reproductive, simulated or actual field data, or another appropriate data such as field incident reports are used to determine if a Special Review should be initiated. Special Review is a process to "help the Agency determine whether to initiate procedures to cancel, deny, or reclassify registration of a pesticide product because uses of that product may cause unreasonable adverse effects on the environment . . . ."[15] Specifically,[16]

"The Administrator may conduct a Special Review of a pesticide use if he determines, based on a validated test or other significant evidence, that the use of the pesticide . . . (3) May result in residues in the environment of nontarget organisms at levels which equal or exceed concentrations acutely or chronically toxic to such organisms, or at levels which produce adverse reproductive effects in such organisms, as determined from tests conducted on representative species or other appropriate data. (4) May pose a risk to the continued existence of any endangered or threatened species . . . (6) May otherwise pose a risk to humans or the environment which is of sufficient magnitude to merit a determination whether the use of the pesticide product offers offsetting social, economic, and environmental benefits that justify initial or continued registration."

While applicable to all actions under FIFRA, the results of mesocosm-simulated field tests most directly apply to actions and decisions made under Sections 3, 4, and the Special Review process. OPP has focused its ecological risk assessment on single-species effects endpoints, especially mortality. Mesocosms provide data that can expand the endpoints of concern to also include adverse effects from pesticides on community metabolism and populations of phytoplankton, periphyton, filamentous algae, macrophytes, zooplankton, macroinvertebrates, and fish. Since finfish occupy the higher trophic levels of aquatic organisms and therefore are the summation of much of the activity at lower levels, the interpretation of the results from mesocosms will generally focus on the adverse effects to the finfish populations.[1] Notable exceptions include results indicating adverse effects on commercial shellfish, endangered or threatened species, food sources for terrestrial organisms, such as breeding waterfowl, etc.

CONCLUSIONS

Mesocosm test systems should enable the Agency to better meet the demands and intent of FIFRA, i.e., to determine whether a pesticide can be registered or continued to be registered for a particular use conditioned upon the determination that it will not "generally cause unreasonable adverse effects on the environment". The environment includes all plants and other animals living therein, and the interrelationships which exist among these. With this test system, OPP will have data to more confidently consider endpoints of concern including indirect adverse effects on single species, populations, and their communities. Focus of concern will continue to be on the effects of pesticides on finfish and their populations as integrators in the aquatic ecosystem. Further, there are a growing number of circumstances where pesticide effects on lower trophic level aquatic organisms, their populations, and communities may be considered in regulatory decisions.

The results from this test system challenge the Agency to decide how best to incorporate endpoints other than direct mortality, reproductive impairment, and growth for single species in regulatory decision making for pesticides. This will certainly engender a lively debate, and the Agency will ensure that this debate takes place in the public arena.

REFERENCES/LITERATURE CITED

1. Touart, L.W. 1988. Hazard evaluation division, technical guidance document: aquatic mesocosm tests to support pesticide registrations. EPA-540/09-88-035. U.S. Environmental Protection Agency. Washington, D.C.
2. Touart, L.W. 1991. Regulatory endpoints and the experimental design of aquatic mesocosms. In *Utilization of Simulated Field Studies in Ecological Risk,* a symposium sponsored by society of environmental toxicology and chemistry, at the 11th annual meeting, November 11–15, 1990, Arlington, VA.
3. Touart, L.W. and M.W. Slimak. 1989. Mesocosm approach for assessing the ecological risk of pesticides. In *Using Mesocosms to Assess the Aquatic Ecological Risk of Pesticides: Theory and Practice,* a symposium sponsored by section E, extension and regulatory entomology, at the 34th annual meeting of the entomological society of America, November 29–December 3, 1987, Boston, MA. Ed.: J. Reese Voshell, Jr.
4. U.S. Environmental Protection Agency. 1975. Guidelines for registering pesticides in the United States. *Fed. Regist.* 40 CFR (123). Part 162.
5. U.S. Environmental Protection Agency. 1978. Proposed guidelines for registering pesticides in the United States. *Fed. Regist.* 40 CFR (132). Part 163.
6. U.S. Environmental Protection Agency. 1982. Pesticide assessment guidelines, subdivision E, hazard evaluation: wildlife and aquatic organisms. EPA-540/09-82-024. Washington, D.C.
7. Macek, K.J., D.F. Walsh, J.W. Hogan and D.D. Holz. 1972. Toxicity of the insecticide Dursban® to fish and aquatic invertebrates in ponds. *Trans. Am. Fish. Soc.* 101(3):420–427.
8. U.S. Environmental Protection Agency. 1984. Data requirements for pesticide registration; final rule. *Fed. Regist.* 40 CFR (207). Part 158.
9. U.S. Environmental Protection Agency. 1985. Burden of persuasion in determinations under this part. *Fed. Regist.* 40 (CFR (229). Part 154.5.
10. Brassard, C.A., A.L. Buikema, M. Frankenberry, R. Lee, A. Stavola, C. Stunkard, and L.W. Touart. 1991. EPA mesocosm study review. In *Utilization of Simulated Field Studies in Ecological Risk,* a symposium sponsored by society of environmental toxicology and chemistry, at the 11th annual meeting, November 11–15, 1990, Arlington, VA.
11. Urban, D.J. and N.J. Cook. 1986. Hazard evaluation division, standard evaluation procedure: ecological risk assessment. EPA-540/09-86-167. U.S. Environmental Protection Agency, Washington, D.C.

12. Rodier, D. and D.J. Urban. 1991. Ecological risk assessment in the office of pesticides and toxic substances. A paper given at the society of environmental toxicology and chemistry, at the 11th annual meeting, November 11–15, 1990. Arlington, VA.

13. U.S. Environmental Protection Agency. 1989. Labeling requirements for pesticides and devices. 40 CFR Part 156.10(h)(2)(ii)(B).

14. U.S. Environmental Protection Agency. 1989. Criteria for restriction by use by certified applicators. 40 CFR Part 152.170(c)(1)(iii).

15. U.S. Environmental Protection Agency. 1985. Special reviews of pesticides; criteria and procedures; final rule. *Fed. Regist.* 40 CFR (229). Part 154.1(a).

16. U.S. Environmental Protection Agency. 1985. Special reviews of pesticides; criteria and procedures; final rule. *Fed. Regist.* 40 CFR (229). Part 154.7(a), (3), (4), (6).

CHAPTER 3

EPA Aquatic Mesocosm Study Review

Candace A. Brassard, Arthur L. Buikema, Jr., Tom A. Bailey,
Mary J. Frankenberry, Richard M. Lee, Ann M. Stavola,
Clayton L. Stunkard, and Leslie W. Touart

Abstract: The results of a mesocosm study are team-reviewed by staff scientists and statisticians to ensure that data used in an ecological risk assessment are scientifically acceptable. Structural and functional variables of plankton communities, macroinvertebrate communities, and fish populations are measured. These variables are analyzed for statistical significance, and biological trends are reviewed to determine if the results have demonstrated safety.

INTRODUCTION

Since the mid 1970s the Environmental Protection Agency has required toxicity data to evaluate the potential ecological hazards associated with the use of pesticides. A variety of ecological effects data are required to support a new chemical registration, conditional registration, and reregistration of a pesticide. These data are obtained through a tiered system of data requirements which may culminate in field testing. Urban and Cook (1985) and Jenkins et al. (1989) summarize the tiered testing requirements and their use in an ecological risk assessment.

The initial tier of testing consists of acute toxicity studies to support registration of a pesticide. Intermediate tiers of aquatic testing include fish early-life stage studies, invertebrate life cycle studies, and fish full-life cycle studies.

17

The final tier of testing, simulated or actual field testing, may be required by the EPA to support a pesticide registration depending on the results of lower tiered studies, intended use pattern, and pertinent environmental fate characteristics. A variety of field study types have been employed to address the aquatic field testing data requirement. A field test may involve (1) only the monitoring of pesticide residues in water and sediment associated with a particular use, (2) an aquatic mesocosm test where a series of replicated ponds or enclosures are treated at levels representative of expected pesticide loadings, or (3) an "actual" field test where biological effects are assessed in an aquatic environment (e.g., farm pond) receiving residues from a treated field under actual use conditions. The pesticide registrant will generally have an option of which of the two latter types of field study to pursue, but EPA prefers aquatic mesocosm tests because of their improved experimental control. The registrant often prefers the aquatic mesocosm test since the requirements for a farm pond study are logistically difficult to achieve, i.e., replication.

The aquatic mesocosm study or actual field study may be required for major use pesticides when information, including the lower tier data, indicates the use of a pesticide may cause adverse effects to the environment. When a detrimental effect level for nontarget aquatic organisms is exceeded by the expected exposure concentrations of a pesticide, there is potential for widespread impact, and an intensive field study is warranted. This initial risk assessment is made by comparing biological response data, such as acute toxicity values, to aquatic estimated environmental concentrations (EEC). Adverse effects are presumed if the EEC exceeds 1/2 the acute LC_{50} (test concentration which kills half of the test population) or is equal to or greater than a level which may result in chronic adverse effects (greater than the no observable effect concentration).

The aquatic mesocosm test serves a primary regulatory objective by providing a pesticide registrant supportable means for negating presumptions of unacceptable risks to aquatic organisms associated with their product. The question of whether pesticide concentrations which adversely affect aquatic organisms in the laboratory fail to adversely impact aquatic organisms under field conditions is addressed. Since 1986, EPA has encouraged the use of mesocosms (experimental ponds and *in situ* enclosures) in lieu of the nonreplicated pond study. If the actual field (pond) study design is chosen, replication of the control and treated ponds is required.

Experimental pond systems have been used since the 1930s (Swingle, 1950). The Agency describes a mesocosm study as a ecosystem study, consisting of at least 12 ponds approximately 0.1 to 0.25 surface acres each (Touart, 1988). Each mesocosm pond simulates a pond containing a natural assemblage of organisms ranging from plankton to fish. Each mesocosm pond, except control ponds, is treated with an amount of pesticide equivalent to the loading expected from surface runoff and spray drift consistent with label application rates. The purpose of this paper is to describe the Agency's review process employed for studies addressing the aquatic mesocosm test requirement. This paper does not address the risk assessment process after the evaluation is completed.

PROTOCOL REVIEW

Prior to conducting aquatic field testing to support a pesticide registration, the Agency recommends that a protocol be submitted for review. The protocol is reviewed by a staff scientist and then forwarded to the Aquatic Field Studies Team (AFST). The AFST was established in 1985 and consists of OPP scientists, specifically for reviewing aquatic field study protocols and final reports. The protocol is reviewed initially for consistency with the Agency's Technical Guidance Document (Touart, 1988).

The Agency reviews the proposed or existing registrations for the pesticide of concern. Chemical/physical characteristics, number of pesticide applications, application rate, use pattern, area of use, and toxicity data are all considered. The Agency then performs a comprehensive review of all salient aspects of the proposed mesocosm study. These include the test system, pesticide application rates and methods, residue analyses, biological organisms, quality assurance, data formatting, and proposed statistical analysis. Information required for the protocol review is similar to the required data included in the final report. The Agency reviews all factors to assure risk concerns will be adequately addressed.

The use of mesocosm tests in support of pesticide registration is an evolving process. Variables or test methods not addressed in previous studies may need to be considered in future studies. For instance, the first few mesocosm studies submitted to the Agency did not include fish tagging. However, on the basis of the data recently submitted for review, the Agency now requires fish tagging since it allows us to monitor impacts on growth of adult fish. In other cases the Agency may determine that some of the biological samples may be archived and analyzed only if concerns are identified during the review of the final study results. After completing the protocol review, the registrant may submit a counterproposal suggesting design changes before final acceptance.

CONDUCT OF STUDY

The mesocosm study is expected to be performed according to the accepted protocol. The study, which includes the collection of pretreatment data as well as preparation of the final report for EPA review, requires a considerable investment of time and resources. This can typically take 2 to 3 years to complete.

It is advantageous for study authors to discuss any anticipated difficulties in adhering to the accepted protocol prior to conducting the study. Ultimately, the Agency identifies the information which may be necessary for completing a risk assessment. Depending on the chemical/physical characteristics of the pesticide and the label, the Agency will specify when residue analyses are required or when samples can be archived. EPA understands that unanticipated complications can arise during the course of a complex study.

If the study authors significantly deviate from recommendations in the protocol review, these deviations may affect the scientific soundness of the study or its adequacy in negating the presumption of risk. Early communication with the Agency is encouraged to resolve difficulties.

Generally, the Agency receives the final mesocosm study report within 6 to 9 months after study completion.

FINAL MESOCOSM STUDY REVIEW

Scientists with a variety of expertise are utilized in the review of the final mesocosm study report. Specifically, fisheries biologists, aquatic ecologists, environmental fate modelers, statisticians, agronomists, and chemists, with one scientist acting as a team leader, are needed to complete the review. The areas discussed below are reviewed in depth.

Test System

The reviewers first establish whether the test system is consistent with the Technical Guidance Document (Touart, 1988) and is appropriate to negate the identified presumption of risk.

The test system can be defined as 12 or more ponds, each being at least 0.1 acre (Touart 1988). It is to the advantage of the registrant to include as many replicates as possible within each treatment level and the control within the design of the study. The treatment regime depends on the label the registrant wants to support. These ponds are used to maintain test organisms during the study. The Agency reviews the baseline data which include: pond construction, characteristics of test water and hydrosoil (ions, metals, pesticides, etc.), and uniformity of biota among mesocosm ponds. The Agency also compares biota of mesocosms to those in established natural ponds. If test ponds have been used for other studies, data must be submitted to verify that the test ponds are not contaminated with other chemicals. Baseline control data from previous studies are also useful.

The Agency then reviews the test system data after pesticide application to the ponds. The report must include dissolved oxygen (DO), acidity (pH), temperature, rainfall, wind direction, evaporation, and other factors identified in the protocol. Depending on the variable, measurements may have been required on a daily, weekly, or monthly basis.

The study authors must indicate the methods used to prevent the invasion of extrinsic fish or biota from the stock pond. The mesh size used to cover outflow pipes from the stock pond should be indicated. The Agency has reviewed data in the past which indicated there was an invasion of extraneous fish into mesocosm ponds from the reservoir.

The study authors must specify how federal and state requirements for handling waste water from the study will be met. This is of special interest when the ponds are drained at the end of the study to evaluate fish production.

Pesticide Application

The method of application as well as the loading rate must represent the pesticide label being supported by the registrant. If a pesticide is aerially applied (according to the label), the method of application and loading to each of the treated ponds must represent spray drift and runoff loadings. However, if the chemical is applied with ground equipment (depending on type of equipment specified on label), the exposure may result only from runoff or from runoff and drift.

Applications must be conducted to achieve target loading and timing stated on the pesticide label. Methods used to prevent cross contamination between ponds while applying the chemical must be reported. In most cases the chemical is applied as a typical end-use product. Depending on the recommended application method of the label that the registrant is supporting, application may be made across the surface of each pond to simulate spray drift (generally aerial application) and/or as a water soluble or a slurry mixture to simulate runoff (aerial and/or ground application). Slurry mixtures generally contain soil, water, and chemical for runoff simulation and are applied across the surface of the pond or at discrete entry points.

The loading rates must be based on actual residue monitoring data or upon an agreed estimated loading. The total loading must be consistent with the expected loading range to encompass typical and maximum expected concentrations. The concentrations of the chemical in the mesocosms must be measured and the loadings used for each treatment level must be verified by analyzing aliquots of the introduced spray and/or slurry.

Residue Analyses

Residue analyses are conducted on samples obtained from the water and sediment (for spatial-temporal heterogeneity), drift cards, and biological organisms. Methods of collecting the samples for residue analyses are also reviewed. The sampling program must be sufficient to track intensity and duration of exposures.

The water/sediment analysis should include various depths of the water column and sediment of both the littoral and pelagic zones. For example, the residues in the sediment should be measured at the top 1 cm layer if the chemical is not expected to leach. Residues within the various levels of the water column are useful in measuring the rate of mixing and understanding reported effects on various organisms.

Analysis of water and sediment residues must be conducted throughout the study. The Agency has determined that with some chemicals (e.g., synthetic pyrethroids) residues are so persistent in sediment that sampling may be required to continue well after the biological sampling has terminated. In some cases, sediment samples may be analyzed for residues for an additional year after the fish have been harvested. In other words, the sampling strategy must correspond to the application strategy as well as take into account the physical/chemical characteristics of the chemical.

The analysis should also delineate both the temporal and spatial-heterogeneity of the chemical in a particular substrate. This information is useful in interpreting macroinvertebrate distribution data. Additionally, analysis of residues on drift cards from all the ponds are required to confirm that no cross contamination occurred during application. If cross contamination is detected, the study may be invalidated.

Results from residue analysis harvested of dead and living fish must be reported. The bioconcentration factor (BCF) should also be estimated, based on the residues in the fish and the water column. This BCF will be compared to the BCF obtained under laboratory conditions.

The methods of collection, preservation, and analyses of the residue samples are also reviewed. Quality assurance steps used to assure that the samples were adequately preserved, stored, and measured must be reported.

The Agency reviews and audits residue data from all substrates. The dissipation rate of the chemical is determined within the water column to provide information on the chemical properties of the chemical under field conditions. The residues in the sediment are determined, as well as the persistence of the compound in sediment. The methods of residue analyses must be validated at relevant levels of detection.

Biological Organisms

The Agency determines which biological organisms are to be measured based on the ecological effects and environmental fate data available to the Agency. The conclusions drawn from the measured biological variables must be sufficient to negate the presumption of risk. Variables generally considered useful in the mesocosm study include the following: community metabolism, phytoplankton, periphyton, filamentous algae, macrophytes, zooplankton, macroinvertebrates, and fish.

The sampling regime (timing and number of samples) must consider the application strategy (numbers of applications and application interval) as well as physical/chemical characteristics of the pesticide. The methods used to collect, preserve, and analyze the various biological variables should be specified. The lowest level of taxonomic identification that can be practically determined for each of these organisms must be reported (i.e., genus, species).

Once all the data are reported for all fish, the numbers are tabulated and verified. The Agency reviews fish length and weight data and estimates the growth rate when possible. All stocked fish must be accounted for, and all measurements on fish must be reported, regardless of whether the individual fish survived to harvest. All dead fish prior to harvest must be measured at time of retrieval. Abundance and biomass data on phytoplankton, zooplankton, macroinvertebrates, and macrophytes are reviewed. The Agency compares data within a certain time frame between control and treated ponds including whether or not recovery was observed for a particular organism. Taxa richness, species diversity, and community similarities are also considered.

Quality Assurance

The study authors must provide documentation for quality assurance for all aspects of the study. The provisions for the field/laboratory Good Laboratory Practice Standards (USEPA 1989) must be satisfied.

Data Formatting

The Agency requests that all data be provided in an ASCII file on an IBM compatible diskette, to expedite the review process. In the past, data have been formatted for spreadsheets. The results should be presented in graphic form and all data must be submitted for statistical analyses.

Statistical Analyses

The study authors must indicate which statistical methods were used to analyze the data. The Agency emphasizes hypothesis testing for analyzing the results of mesocosm studies. In cases where experience has shown that typical variability in the data may render traditional hypothesis tests of limited power for analyzing results, the Agency has worked with registrants in the design of studies to specify hypotheses that will both increase the power of the tests as well as conform to the directives of Federal Insecticide, Fungicide, and Rodenticide Act. According to FIFRA, section 3 (c)(5), as amended in October 1988, "the Administrator shall register a pesticide if he determines that . . . it will perform its intended function without unreasonable adverse effects on the environment; and when used in accordance with widespread and commonly recognized practice it will not generally cause unreasonable adverse effects on the environment." The registrant must demonstrate to the Agency that the use of the chemical will not cause unreasonable adverse effects. Therefore, in these cases, the hypotheses specified assume that the treatment will result in deleterious effects (Stunkard, 1990). The Agency examines the data and determines whether or not the results refute the presumption of risk. In all cases, presentation of the summary data should correspond to the statistical tests performed. For example, combined replicate means for each treatment and control must be included.

In addition to the statistical analyses, biological trends are evaluated. Studies submitted in the past have demonstrated that the biological trends support what has been statistically observed.

CONCLUSIONS

Review of a mesocosm study involves exhaustive efforts of several scientists with many talents and areas of expertise. Once the review process is initiated, it may require 6 months or longer to complete. The Agency interprets observed effects at the various treatment levels. These data are then coupled with the available toxicity data and compared to the exposure information. Ultimately, the Agency reviews all available data pertaining to the proposed or existing registration being supported by the company and determines whether the presumption of unreasonable adverse effects to the aquatic environment will be negated. This completes a risk assessment. This risk assessment is then used for implementing future regulatory decisions.

REFERENCES

1. Jenkins, D.G., R.J. Layton, and A.L. Buikema, Jr. 1989. State of the art in aquatic ecological risk assessment, MPPEAL 75:18–32. In: J.R. Voshell, Jr. [ed.], *Using Mesocosms to Assess the Aquatic Ecological Risk of Pesticides: Theory and Practice,* Entomological Society of America, Lanham, MD, pp. 18–32.
2. Stunkard, C. 1990. Tests of proportional means for mesocosm studies. In: R. Graney, J. Kennedy, and J. Rodgers [Cochairs], *Utilization of Simulated Field Studies in Aquatic Ecological Risk Assessment,* presented at the 11th Annual Meeting of the Society of Environmental Toxicology and Chemistry, Washington, D.C.
3. Swingle, H.S. 1950. Relationships and dynamics of balanced and unbalanced fish populations. Bulletin No. 274, Alabama Polytechnical Institute, Agricultural Experiment Station.
4. Touart, L.W. 1988. Hazard Evaluation Division, Technical Guidance Document: Aquatic mesocosm tests to support pesticide registrations. EPA-540/09-88-035. U.S. Environmental Protection Agency, Washington, D.C., pp. 1–35.
5. Urban, D.J. and N.J. Cook. 1985. Hazard Evaluation Division, Standard Evaluation Procedure, Ecological Risk Assessment. EPA 540/9-85-001. U.S. Environmental Protection Agency, Washington, D.C., pp. 1–96.
6. U.S. Environmental Protection Agency. 1989. Federal Insecticide, Fungicide and Rodenticide Act (FIFRA); Good Laboratory Practice Standards. Final Rule. *Fed. Regist.* Vol. 54, No. 158. 40 CFR Part 160. U.S. Government Printing Office.

CHAPTER 4

Regulatory Endpoints and the Experimental Design of Aquatic Mesocosm Tests

Leslie W. Touart

To make an end is to make a beginning.
The end is where we start from.

T. S. Eliot

Abstract: Mesocosm tests provide important data for evaluating ecological risk in the aquatic environment. In 1988, the Office of Pesticide Programs provided guidance concerning the conduct of such tests. Endpoints measured in such tests include both structural and functional elements. The experimental design is mostly independent of measurement endpoints and mostly dependent on regulatory endpoints. Endpoints of regulatory concern (e.g., biomass change) can effectively be addressed only from an appropriate experimental design. Hybrid test designs which include both multiple replicates and multiple treatment levels offer, in general, the most appropriate design.

INTRODUCTION

Ecological risk assessment for aquatic organisms under the Federal Insecticide, Fungicide, and Rodenticide Act (FIFRA) may include aquatic field data, developed on a case-by-case basis, in addition to single-species laboratory toxicity tests. These data with information on the fate of the pesticide, its use pattern, and the potential for its exposure are assessed in a process which includes weight-of-the-evidence

0-87371-592-6/94/$0.00 + $.50
© 1994 by CRC Press, Inc.

review and risk-benefit analysis. Recently, a guidance document (Touart, 1988) was published to describe an ecosystem-level test advocated by EPA to satisfy an aquatic field testing requirement for those products with potential risks sufficient to warrant such testing. This test is referred to as an aquatic mesocosm test and represents the highest tier of testing in support of a pesticide registration.

An aquatic mesocosm can be defined as an intermediate-sized system, such as a dug-out pond or *in situ* enclosure, that can be replicated and manipulated to test both structural and functional parameters as a representative aquatic ecosystem. The dual objectives of an aquatic mesocosm test are, first and foremost, to provide data to negate a presumption of serious risk indicated in the lower tiered testing and, next, to estimate the intensity and duration of any adverse effects identified. An understanding of the study objectives and endpoints are essential to adequately design these studies.

The purpose of this paper is to discuss endpoints of an aquatic mesocosm test and the experimental design of the test from a regulatory context. In effect, it is the end result of the test which determines the beginning of the test.

ENDPOINTS

Endpoints can be described in many ways, but here they are simply defined as conclusions drawn from the results of an experimental study, in this case a mesocosm test. Endpoints can be separated by type into two distinct categories and, by relevance, into three roughly related groups. The first category may be termed "observational endpoints" and includes those endpoints which are derived directly from measurements or calculations of specific parameters or phenomena embodied within the test. Good examples would be a reduction in numbers of a class of macroinvertebrates or in total biomass of finfish. The other category, "interpretational endpoints", is determined indirectly through assessment or judgment of an observational endpoint. In other words, an interpretational endpoint is explanatory of an observational endpoint. For example, a reduction in finfish biomass could be interpreted to indicate poor growth and foretell poor reproductive performance in subsequent seasons. The three endpoint groups are ecological, socioeconomic, and regulatory and, though conceptually distinct, they are operationally related. Regulatory endpoints, to be effective, should embrace through legislative language or interpretation those endpoints which are both ecologically and socioeconomically relevant. Sound public policy must be attuned to both the perception of and actual best interests of the public.

Ecological

Ecological endpoints are those which are directly related to observable changes in the abiotic and biotic components of an aquatic ecosystem. Kelly and Harwell (1989) suggest that ecological endpoints be limited to only those with "relevance to issues of concern to humans'', so that the myriad of potential measures may be narrowed to a manageable level. This suggestion implies that ecological impacts observed without apparent human relevance would not constitute an *endpoint*. Another perspective could be that if an impact is relevant to the health of an ecosystem, then it is necessarily relevant to humankind.

Harte et al. (1980, 1981) identified direct chemical threats to the quality of drinking water, impairment of sport fish populations, aesthetic loss from increasing turbidity or eutrophication, enhanced odor-producing biological activity, and increased likelihood of disease-bearing vectors and pathogens as areas of human concern. Any observed effects which degrade or threaten to alter the ecosystem of an aquatic mesocosm should be considered a relevant observational endpoint. Ecosystem stress is manifested through changes in nutrient cycling, productivity, the size of dominant species, species diversity, and/or a shift in species dominance to opportunistic shorter-lived forms (Rapport et al., 1985).

Levels of biological organization range from that of biochemical structures and activity to that of the ecosphere, which embodies all living systems. The organizational levels most relevant for consideration within an aquatic mesocosm test are those of individual species, populations, communities, and ecosystem. Emergent properties of individuals, populations, and communities have been discussed by Meyerhoff (1990) that are relevant to this discussion and have been adapted in what follows.

Survival is the fundamental emergent property of an individual; one is either alive or dead. Without life, one cannot possess any other property at the individual level or contribute to a property at another organizational level. Other major properties at the individual level include behavior, growth, and reproduction. Behavior measures are somewhat species specific in that one may measure swimming speed for a finfish or siphon rate for a sedentary bivalve, and the relevance of these measures also varies by their importance to the survival, growth, or reproduction of the given individual. Growth is usually determined as a change in size, length, and/or weight over a set period of time or as a change in morphological development. Measures of reproduction (i.e., fecundity, number of offspring, etc.) determine the success or adaptive fitness of the individual.

Populations are defined by their persistence, production, and structure. Measures of abundance and distribution over time are used in determining the persistence of a population. Change in biomass and density (population number per unit area) determine population production. The structure of a population is characterized by size/age class distribution, recruitment, and spatial distribution.

Emergent properties of communities are expressed in their composition, organization, and productivity. Community composition is expressed in species diversity, species dominance, and succession (change in species dominance). Organization is measured by functional groups or guild structure and complexity of the trophic structure (linkages). The productivity of a community is expressed by measures of energy or material flow (trophic dynamics) or production capacity of the functional groups.

Jenkins et al. (1989) identified several methods sufficiently developed for measuring aquatic ecosystem functioning. These include primary production, secondary production of zooplankton and benthic macroinvertebrates, respiration rates, and leaf degradation rates.

Socioeconomic

Socioeconomic endpoints relate directly to an aesthetic, human health, or economic condition resulting from a change in an aquatic ecosystem. Although seemingly defined in much the same way as an ecological endpoint of human concern, the distinction is in the direct nature or influence of an ecosystem change on the human condition. Examples include reductions in abundance and production of commercial or sport (game) fish populations or development of algal populations that detract from water use (Barnthouse et al., 1986).

Regulatory

A regulatory endpoint has been defined as "a regulatory decision-making norm which translates fundamental legislative purposes into regulatory action" (Harwell and Harwell, 1989). FIFRA oversees three types of regulatory action: (1) labeling, (2) classification, and (3) registration.

Regulatory endpoints associated with labeling are those needed to establish appropriate label precautions. For example, the precautionary statement — "This pesticide is toxic to fish" — is based on lethality and is required if a laboratory derived LC_{50} is less than 1 ppm for the most sensitive fish surrogate tested. More relevant to the mesocosm test, the statement — "This pesticide is extremely toxic to fish" — is based on the laboratory data as above and, also, on lethality confirmed in a simulated or actual field study or under normal use. That is, if direct mortality to finfish is confirmed (as opposed to expected) under normal use of a pesticide, the adverb "extremely" is used to accentuate the concern. Therefore, endpoints are directed at label precautions to reduce risks to the environment and/or warn the user of expected or possible adverse effects to the environment.

Under FIFRA, pesticides are classified for *general use* or *restricted use*. A general use product is one available for use to the general public, while a restricted use product is available for use only by or under the supervision of an applicator certified to have completed pesticide safety training. Criteria have been established to identify those pesticides which pose sufficient risk to warrant a restricted use classification. A pesticide product intended for outdoor use will be considered for restricted use classification, on ecological grounds, if it causes discernible adverse effects on nontarget organisms, such as significant mortality or effects on the physiology, growth, population levels, or reproduction rates of such organisms, resulting from direct or indirect exposure to the pesticide, its metabolites, or its degradation products (40 CFR Part 152.170(c)(iv)). The classification for restricted use is also contingent on the condition that restricting the use will result in the associated risks being acceptable after risk/benefit analysis.

For regulatory action on registration, there are three decisions — register, deny registration, or suspend/cancel an existing registration. The Agency under FIFRA (Section 3(c)(5)) may register a pesticide when it determines, among other findings, the pesticide will perform its intended function when used in accordance with widespread and commonly recognized practice and will not generally cause *unreasonable* adverse effects on the environment.

Criteria have been established for identifying potentially unreasonable adverse effects (40 CFR Part 154.7), albeit a determination that adverse effects are indeed *unreasonable* is made only after risk/benefit analysis in a process termed "Special Review". Briefly, a pesticide use exceeds the criteria for unreasonable adverse effects and could be considered for Special Review if environmental concentrations equal or exceed concentrations toxic to nontarget organisms or which impair reproduction, if it poses a risk to the continued existence of any endangered or threatened species, or if it otherwise poses a risk to the environment of sufficient magnitude to warrant risk/benefit analysis.

Implicit within FIFRA and explicit in the regulations (40 CFR Part 154.5) "the burden of persuasion that a pesticide product is entitled to registration or continued registration for any particular use or under any particular set of terms and conditions of registration is always on the proponent(s) of registration." Therefore, a registration decision hinges on the ability of the registrant (proponent of registration) to demonstrate that a given product will not result in *unreasonable* adverse effects.

EXPERIMENTAL DESIGN

Graney et al. (1989) have discussed several of the experimental design strategies which are conceptually possible under the minimum design constraints of the Agency's guidance (Touart, 1988). Basically, the designs fall into one of three categories: hypothesis tests, point estimate tests, or hybrid tests which incorporate elements of hypothesis and point estimate tests. Certain ones of these design strategies are more appropriate or relevant than others in addressing the regulatory endpoints of concern.

Hypothesis tests are used for investigating whether the response of an experimental unit is different from that of a control unit. Hypothesis tests generally consist of a null hypothesis and an alternate hypothesis, such that rejection of the null requires acceptance of the alternate. Examples of null (Ho) and alternate (Ha) hypotheses follow:

1. Ho: $\mu_c = \mu_t$ Ha: $\mu_c \neq \mu_t$
2. Ho: $\mu_c \leq \mu_t$ Ha: $\mu_c > \mu_t$
3. Ho: $\mu_t \leq b\mu_c$ Ha: $\mu_t > b\mu_c$

(for b = specified level of unacceptable effect; μ_c = control response; μ_t = treatment response)

A test of significance provides an avenue for rejecting a null hypothesis and, therefore, providing evidence in support of the alternate hypothesis. At an α =

0.05 for example, we are 95% confident that a true null hypothesis will be retained. In rejecting a null hypothesis, we have evidence which suggests that the alternate hypothesis is true. However, failure to reject a null does not mean we are 95% confident that *it* is true, nor that we are 95% confident that the *alternate* is not true. In other words, failure to reject a null hypothesis provides us with no confidence in differentiating the true relationship between control and treatment means. Scientists generally want to base new scientific discoveries on positive evidence (rejecting the null), whereas prudent risk managers do not want to err by accepting a false hypothesis that there is no effect due to treatment.

Hypothesis tests are used for comparing means and are characterized by having multiple replicates in control and treatment groups. The greater number of replicates, the more accurately is the group variability defined and the greater the power of the test for resolving differences. Hypothesis tests are best for objectively determining if an effect (identified differences between control and treatment groups) is real.

Point estimate tests are designed to evaluate regression relationships and, in this instance, estimate an exposure level which will not cause an adverse effect (NOEL or threshold level) or predict the intensity of an effect at a given exposure level. Regression analysis is used to iteratively fit observed data to a theoretical equation. This requires multiple treatments at various exposure concentrations with an objective of defining an exposure (dose) response relationship. Once a dose-response has been determined, then a specified point, such as an effect threshold level, may be estimated. The greater number of treatment levels along the response gradient, the greater the confidence in the fitted dose-response line.

Hybrid tests incorporate features of both hypothesis and point estimate tests. Employing multiple replicates and multiple doses, one can determine if a given treatment level significantly differs from control and may estimate how different another treatment level will be above or below the given treatment level. The difficulty is, with a limited number of test units (mesocosms), one reduces the number of replicates to add dose levels and reduces the number of dose levels to add replicates. Reduced replicates mean reduced power in resolving significant effects and reduced dose levels mean reduced confidence in estimating the fit of the regression line. The Agency has recommended a balanced experimental design consisting of "a control which receives no test compound, an X treatment concentration representing expected exposures, an X+ treatment level representing an upper bound and an X− treatment level representing a lower bound", with a minimum of three replicates and four treatment levels (Touart, 1988). Graney et al. (1989) argue for increased treatment levels and Stunkard (1990) argues for increased replicates to "best" meet the stated objectives of the test. Limited resources may prevent adding sufficient test units to satisfy both design camps.

DISCUSSION

The mesocosm test strategy was developed as an ecosystem-level test to address adverse ecological effects which are inappropriate for single-species laboratory studies (Touart and Slimak, 1989). Ecologists and aquatic toxicologists have long and well recognized the limitations of single-species laboratory toxicity tests alone for assessing potential ecosystem impacts (Cairns, 1981, 1984; Pimentel and Edwards, 1982; Levin et al., 1984; Odum, 1984; Kimball and Levin, 1985; Kelly and Harwell, 1989). FIFRA was amended in 1972, among other reasons, to prevent the marketing of pesticides which pose unreasonable adverse effects to the environment, and defines the environment as including water, air, land, and all plants and man and other animals living therein, and the interrelationships which exist among these (Sec. 2(j)). It is this language which drives the regulatory concern for ecosystem-level impacts.

The experimental design of choice must permit the resolution of effects (changes) in the endpoints selected. For this reason, the design chosen is mostly independent of ecological or socioeconomic endpoints per se, but is dependent on the regulatory endpoints. A hybrid design is generally preferred by the Agency to adequately address the regulatory endpoints. This does not imply that the hybrid design is considered the "best" design, only that it is preferred for addressing the multiple questions usually confronted by risk scientists. Repeating, the regulatory objectives of a mesocosm study are to negate a presumption of serious risk indicated in the lower tiered testing and to define the intensity and duration of any adverse effects identified. FIFRA is not a "no-effect" statute, but rather a "no-*unreasonable*-effect" statute. A presumption of risk may be easily negated by demonstration of no effect at worst-case exposure conditions where detrimental effects, if observed, must be weighed carefully in assessing risk acceptability and could involve risk-benefit analysis under special review. Defining the effect may, therefore, be necessary to negate presumption of serious risks which may effectively connotate with *unreasonable* effects.

The results of a mesocosm are then subject to adaptive interpretation by both statistical and nonstatistical (e.g., biological) analysis. Statistical analysis is used to assist in the objective evaluation of studies and to verify the sensitivity of the test to detect an undesirable effect if one truly occurs. Nonstatistical analysis is performed by professional judgment of qualitative information to ascertain relevance of the statistically quantifiable effects. Together, these analyses allow the risk scientist to complete an aquatic risk assessment to be employed in risk management decisions.

Pass/fail conclusions are not drawn directly from a mesocosm study. Instead a weight-of-evidence review is used to place any adverse effects in an appropriate perspective for use with risk-benefit analysis and subsequent risk management decisions.

CONCLUSIONS

Observational and interpretational endpoints can be grouped into regulatory, ecological, and socioeconomic categories. Though these categories can be conceptually separated, they are in practice related. To be effective, a regulatory endpoint must be both ecologically and socioeconomically relevant.

Experimental designs can be separated into hypothesis tests, point estimate tests, and hybrid tests. The choice of experimental design is not driven by ecological or socioeconomic endpoint categories, but are driven by the regulatory endpoint category. Hybrid tests which include both multiple replicates and multiple treatment levels offer, in compromise, what the Agency considers the optimal design for addressing regulatory concerns. The Agency does recognize that hypothesis or point estimate designs may be most appropriate in specific circumstances, but favor the hybrid design in the absence of case-specific justification warranting an alternative design.

LITERATURE CITED

Barnthouse, L.W., G.W. Suter II, S.M. Bartell, J.J. Beauchamp, R.H. Gardner, E. Linder, R.V. O'Neill, and A.E. Rosen. 1986. *User's Manual for Ecological Risk Assessment.* ORNL-6251. Environmental Sciences Division Publication No. 2679, Oak Ridge National Laboratory, Oak Ridge, TN, 215 pp.

Cairns, J., Jr. 1981. Committee to review methods for ecotoxicology, National Research Council, testing for effects of chemicals on ecosystems. National Academy Press, Washington, D.C.

Cairns, J., Jr. 1984. Are single species toxicity tests alone adequate for estimating environmental hazard? *Environ. Monit. Assess.* 4:259–273.

Graney, R.L., J.P. Giesy, Jr., and D. DiToro. 1989. Mesocosm experimental design strategies: advantages and disadvantages in ecological risk assessment, MPPEAL 75:74–88. In: J.R. Voshell, Jr. (Ed.), *Using Mesocosms to Assess the Aquatic Ecological Risk of Pesticides: Theory and Practice.* Entomological Society of America, Lanham, MD.

Harte, J., D. Levy, J. Rees, and E. Sagebarth. 1980. Making microcosms an effective assessment tool, pp. 105–137. In: J.P. Giesy, Jr. (Ed.), *Microcosms in Ecological Research.* Symposium Series 52, Conference 781101. U.S. Department of Energy, Washington, D.C.

Harte, J., D. Levy, J. Rees, and E. Sagebarth. 1981. Assessment of optimum aquatic microcosm design for pollution impact studies. Rep. EA-1989. Electric Power Research Institute, Palo Alto, CA.

Harwell, M.A. and C.C. Harwell. 1989. Environmental decision making in the presence of uncertainty, Chapter 18, pp. 515–540. In: S.A. Levin, M.A. Harwell, J.R. Kelly, and K.D. Kimball (Eds.), *Ecotoxicology: Problems and Approaches.* Springer-Verlag, New York.

Jenkins, D.G., R.J. Layton, and A.L. Buikema, Jr. 1989. State of the art in aquatic ecological risk assessment, MPPEAL 75:18–32. In: J.R. Voshell, Jr. (Ed.), *Using Mesocosms to Assess the Aquatic Ecological Risk of Pesticides: Theory and Practice.* Entomological Society of America, Lanham, MD.

Kelly, J.R. and M.A. Harwell. 1989. Indicators of ecosystem response and recovery, Chapter 2, pp. 9–35. In: S.A. Levin, M.A. Harwell, J.R. Kelly, and K.D. Kimball (Eds.), *Ecotoxicology: Problems and Approaches.* Springer-Verlag, New York.

Kimball, K.D. and S.A. Levin. 1985. Limitations of laboratory bioassays: the need for ecosystem-level testing. *BioScience* 35:165–171.

Levin, S.A., K.D. Kimball, W. H. McDowell, and S.F. Kimball (Eds.). 1984. New perspectives in ecotoxicology. *Environ. Manage.* 8:375–442.

Meyerhoff, R. 1990. Biological systems of concern. Presentation to the Aquatic Effects Dialogue Group, May 15, 1990, The Conservation Foundation, Washington, D.C.

Odum, E.P. 1984. The mesocosm. *BioScience* 34:558–562.

Pimentel, D. and C.A. Edwards. 1982. Pesticides and ecosystems. *BioScience* 32:595–600.

Rapport, D.J., H.A. Regier, and T.C. Hutchinson. 1985. Ecosystem behavior under stress. *Am. Nat.* 125:617–640.

Stunkard, C.L. 1990. Tests of proportional means for mesocosm studies. In: R. Graney, J. Kennedy, and J. Rodgers (Eds.) *Utilization of Simulated Field Studies in Aquatic Ecological Risk Assessment,* presented at the 11th Annual Meeting of the Society of Environmental Toxicology and Chemistry, Washington, D.C.

Touart, L.W. 1988. Hazard Evaluation Division, Technical Guidance Document: Aquatic mesocosm tests to support pesticide registrations. EPA-540/09-88-035. U.S. Environmental Protection Agency, Washington, D.C.

Touart, L.W. and M.W. Slimak. 1989. Mesocosm approach for assessing the ecological risk of pesticides. MPPEAL 75:33–40. In: J.R. Voshell, Jr. (Ed.), *Using Mesocosms to Assess the Aquatic Ecological Risk of Pesticides: Theory and Practice.* Entomological Society of America, Lanham, MD.

CHAPTER 5

Simulated or Actual Field Testing: A Comparison

Reinhard Fischer

Abstract: Under certain conditions EPA requires simulated or actual field testing for the registration of pesticides. Simulated field tests are generally known as mesocosm studies, while farm pond studies are field tests performed under actual pesticide use conditions. Both approaches have advantages and disadvantages. Pesticide application to mesocosms is based on computer simulations. In farm pond studies, chemicals are applied to fields surrounding ponds and follow natural routes of entry into the aquatic environment. In mesocosm studies, simple pond-like systems at an early stage of ecological succession are replicated at one location enabling a statistical evaluation. Farm ponds are often located over a larger area making true replication difficult. However, as farm ponds are relatively stable, changes in ecosystem composition or function can be observed even within one system. Biological sampling is similar in both methods, but the agricultural field portion of a farm pond study adds efforts in sampling and chemical analyses. Depending on the specific objective of the study, both methods can yield valuable results. As ecosystem studies require considerable efforts the problem should be defined clearly, so the results can be used for a firm decision.

INTRODUCTION

For the registration of pesticides EPA requires after careful evaluation of results of laboratory tests and theoretical concentrations in the environment under certain conditions the conduct of "simulated or actual" aquatic field testing. The requirement is contained in the Pesticide Assessment Guideline, Subdivision E, Hazard Evaluation: Wildlife and Aquatic Organisms from October 1982 under Section 72-7. The guideline specifies: "The (short-term or long-term) simulated field test (where growth and reproduction of confined populations are observed) should be selected, if it can yield data useful in assessing such risks" while "an actual (short-term or long-term) field test (where growth and reproduction of natural populations are observed) may be needed, if the simulated test does not suffice." The present guideline distinguishes between "short-term" or "long-term" studies. A drafted revision of this guideline from 1988 changes the expressions "short-term" and "long-term" into "qualitative" and "quantitative" and adds a preference for mesocosm tests.

Basically, either mesocosm studies or actual field tests are possible to fulfill a requirement for ecosystem level testing. Presently there is a very strong preference for mesocosm studies, but actual field studies should still be acceptable. There can be reasons for testing a particular pesticide in a field test scenario and not in a mesocosm and vice versa.

Both types of studies offer advantages and disadvantages. This paper will compare the two strategies. For the purpose of comparison, only the major aspects of a field study will be given.

MESOCOSM CONCEPT

Increased interest in field tests has resulted in quite a few conferences between regulators, academia, industries, and testing facilities (e.g., a mesocosm workshop in April 1986, a workshop on aquatic risk assessment in October 1986, a workshop on aquatic field testing in September 1987). The objective of these workshops, as well as of many unofficial meetings between regulators and industry, was to establish scientifically recognized methods for assessing pesticidal effects in aquatic ecosystems.

Some of the theories developed during these workshops were subject to a confirmation process in the past 3 to 4 years by a variety of ecosystem tests, mostly mesocosm trials. As a result of this large "method try-out", considerable practical experience has been gained in mesocosm trials. Therefore it will not be necessary to go into a detailed description of the mesocosm test method itself. Numerous other papers in this symposium outline the procedure. However, a short summary of the mesocosm method should be useful.

Mesocosms are artificial pond systems. Approximately 12 ponds are used in each mesocosm study. Each treatment or control group is replicated at least three times for statistical reasons. Dosing occurs after calculating theoretical loadings by runoff and drift in water and sediment or estimating theoretical concentrations. Different approaches in dosing have been practiced. Originally three dosing levels plus a control were suggested: x (which is the computed environmental concentration or a theoretical loading scenario), n * x, and x/n. For evaluation purposes, stepwise (e.g., logarithmic scaled) dilutions have also been used. As a statistically oriented evaluation still proved to be difficult, it was proposed in the meantime to replicate one treatment group and the control at least six times. After application, biological as well as chemical parameters are monitored.

FARM POND CONCEPT

The various approaches to a mesocosm design demonstrate that there are still many unanswered questions in performing this type of study. This was one of the main reasons that a farm pond study was selected to study the effects of endosulfan in the field, when a field test was required for this insecticidal active ingredient. Another methodological reason to be discussed later was the unrealistic exposure of mesocosm systems.

Endosulfan is very toxic to aquatic animals, particularly fish (Figure 1). Therefore a special concern for assessing effects of endosulfan on fish populations was raised. Exposing only one species and only one trophic level of fish, as would have been the case in a mesocosm study, was regarded as insufficient for endosulfan. This further lends support to the farm pond approach for endosulfan.

The basic concept behind a farm pond study is the same as in the mesocosm — investigate the possible impact of a pesticide on an aquatic pond-like ecosystem. There are, however, some major differences between the mesocosm and the farm pond strategy. In order to achieve a better understanding of the farm pond approach, I will further use the example of our endosulfan experiment.

MATERIALS AND METHODS

One of the first problems encountered when planning a farm pond study is the question of location. There are numerous conditions to be met before the actual experiment may start (Table 1). In our case, after a laborious and time-consuming evaluation process, during which over 400 ponds were inspected, six relatively similar ponds were selected. These ponds were 3 to 15 mi from each other in a rural area of Georgia. Each pond had a surface area of approximately 3 acres, and was surrounded by fields of at least ten times that acreage draining into the pond.

In mesocosm tests, observations from treatment ponds are compared to observations in the control ponds. In a farm pond experiment such a comparison can be difficult. Based solely on acreage requirements for farm pond studies, treatment and control ponds cannot be established side-by-side. Therefore a control in the strict sense is not possible. Different locations have different climatic conditions, different histories, etc. To address the need for a control system, two approaches were taken. First, since farm ponds are relatively established aquatic systems, one should expect a comparable behavior of the system every year. Therefore during a baseline year, the ponds were biologically monitored for the desired parameters. The result of the monitoring led to exclusion of two ponds, and to generation of

Table 1. Criteria For the Selection of Farm Pond Systems

• Pond surrounded by fields	• Representative fish populations
• Ratio of field to pond surface area 10:1	• Stable, active, healthy pond-ecosystems
• Pond size 3–5 acres	• Cooperative owners
• Slope of fields 3–8%	• Additional irrigation source

reference or background data from four ponds for the treatment year. From these four ponds, two systems were treated, two served as references.

The fields surrounding the ponds were planted with tomatoes following normal agricultural practices. Thiodan 3EC™, an emulsifiable concentrate containing 35% endosulfan, was applied by ground application using the highest rate permitted on the label and the highest frequency (1 lb a.i./a applied three times at intervals of 14 days).

As indicated above, treatment was to occur "naturally" by drift during applications, and runoff after rainstorms. In order to measure these impacts appropriately,

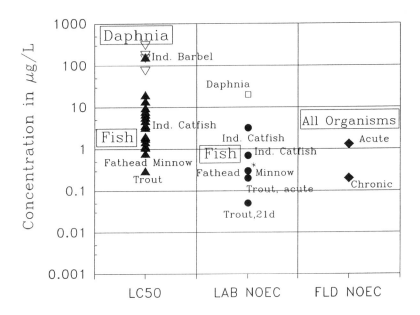

▲ LC₅₀ – values of fish in the laboratory
▽ LC₅₀ – values of daphnia in the laboratory
● NOEC – values of fish in the laboratory (*=chronic)
□ NOEC – values of daphnia in the laboratory
◆ NOEC observed in field

Figure 1.   Aquatic toxicity of endosulfan.

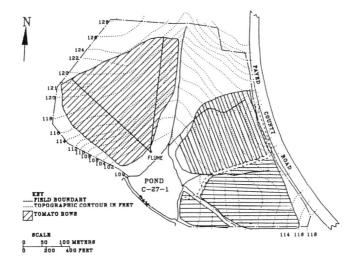

TOMATO ACREAGE AT C-27-1. 14.2 hectares (35 acres) of
tomatoes were planted in the two fields. Location of the
flume in this drainage area is also shown.

Figure 2. Layout of endosulfan farm pond study.

application cards (filter paper with a defined surface area) within the field verified
the amounts applied. Drift cards around the field, around the pond, and above the
water surface monitored the drift portions. Flumes supplied with automatic samplers
measured the runoff (Figure 2).

After application, drift and application cards as well as samples from soil, water,
sediment, and leaves were collected and subsequently analyzed for residues. At
regular intervals, biological samples were taken from all trophic levels to investigate
possible changes in community structure. Additionally, measurements for assessing
community functions were performed (Table 2). Relevant abiotic factors such as
temperature, pH, oxygen, etc., as well as weather data, were also collected for the
different systems.

The rainfall during this study period was atypical for Georgia. Although historically Georgia had heavy rainstorms during the summer season, 1988 was one of
the driest years in decades. Only insufficient runoff was achieved by natural rainfall.
Therefore artificial irrigation was set up to create runoff. After installing the irrigation system, natural rainfall delivering 1.5 in. produced considerable surface
runoff into one of the two treatment ponds. At the other pond site, irrigation guns
delivering 1.5 in. of water achieved the same runoff. Both runoff events occurred
within 2 days after the last application. Until the termination of the study in the
fall, a total of 17 runoff events were observed. Various measurements (Table 2)
were taken for 25 weeks after the last application.

Table 2. Measurements in Farm Pond Study

Chemical-Analytical
  • Concentration in water
  • Concentration in sediment
  • Residues in fish
  • Water quality (hardness, alkalinity, TOC, oxygen, total P, nitrate, nitrite, ammonium, suspended solids)
  • Distribution of pesticide in field (plants, soil, runoff, drift)

Biological
  • Phytoplankton (pigments and biomass)
  • Zooplankton (taxa and biomass)
  • Periphyton (pigments and biomass)
  • Macroinvertebrates (taxa, diversity, biomass, emergence)
  • Macrophytes (percent cover)
  • Fish (species, abundance, mortality, growth, condition, reproduction)
  • Ecosystem metabolism

RESULTS AND DISCUSSION

Monitoring concentrations of endosulfan in runoff water, pond water, and sediment supported exposure of the pond-ecosystems (Table 3).

The surface runoff led to the only effects directly attributable to endosulfan. At points where runoff entered the ponds, many of the small fish and fish fry died within 2 days. All other results were much more difficult to understand and to interpret.

Statistical comparisons of the single parameters were performed within each pond as well as between treatment pond and that reference pond most similar for the respective parameter before the treatment period. The comparison resulted in a large number of statistically significant differences (Table 4). Each of these differences had to be scrutinized carefully for its relevance. Was it caused directly or indirectly by the application of endosulfan or were there other factors responsible? Detailed

Table 3. Maximum Mean Measured Concentrations and Overall Ecological Effects in Endosulfan Farm Pond Study

	Pond 1 C 27-1	Pond 2 M 55-8
Maximum Mean Measured Concentrations		
Drift Cards Pond	218 μg/m^2	99.3 μg/m^2
Runoff Water	203 μg/L	75 μg/L
Pond Water	1.3 μg/L	0.58 μg/L
Sediment	49 μg/kg	99 μg/kg

Ecological Effects
  • Fish-kill in shallow areas
  • No ecologically relevant effects on fish populations
  • No ecologically relevant effects on benthic communities
  • No ecologically relevant effects on phyto- or zooplankton

Table 4. Overview of Possible Effects at Endosulfan Pond Study

Parameter	Comparison of Ponds M-55-8/M-55-4 Pre	Post	End	Comparison of Ponds M-55-8/T-4-1 Pre	Post	End	Comparison of Ponds C-27-1/M-55-4 Pre	Post	End	Comparison of Ponds C-27/1/T-4-1 Pre	Post	End
Phytoplankton diversity	*	*	*	—	—	—	*	*	*	—	—	—
Zooplankton diversity	NS	*	*	—	—	—	*	NS	*	—	—	—
Chironomid emergence	—	—	—	NS	NS	NS	NS	*	NS	NS	*	NS
Chironomid density (Ekman)	—	—	—	NS	NS	NS	NS	*	NS	NS	*	*
Chaoborus density (Ekman)	NS	NS	NS	—	—	—	NS	—	NS	NS	*	*
Oligochaeta density	—	—	—	NS	NS	NS	—	—	—	NS	—	—
Chironomid density (S-sampler)	*	NS	NS	—	—	—	—	—	—	NS	NS	NS
Chaoborus density (S-sampler)	*	NS	NS	—	—	—	—	—	—	NS	NS	NS
Oligochaeta biomass	NS	NS	NS	—	—	—	—	—	—	*	*	*
Insect biomass	*	*	*	—	—	—	—	—	—	*	*	*
Non-insect biomass	—	—	—	*	*	*	—	*	*	*	*	NS
Net metabolism	*	*	*	—	—	—	NS	*	*	—	—	—
Gross metabolism	NS	*	*	—	—	—	NS	*	*	—	—	—
Autotrophic index	NS	*	*	—	—	—	—	—	—	*	*	NS
Significant differences	5	6	6	1	1	1	2	5	4	4	6	2

Note: Treatment pond codes: M-55-8 and C-27-1. — = Pair not selected for comparison due to large differences in first year, NS = no significant difference, * = significant difference, Pre = preapplication, Post = postapplication, End = end of study.

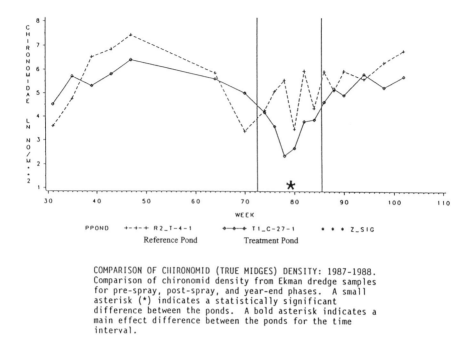

COMPARISON OF CHIRONOMID (TRUE MIDGES) DENSITY: 1987-1988.
Comparison of chironomid density from Ekman dredge samples
for pre-spray, post-spray, and year-end phases. A small
asterisk (*) indicates a statistically significant
difference between the ponds. A bold asterisk indicates a
main effect difference between the ponds for the time
interval.

Figure 3.   Comparison of chironomid density in farm pond treated with endosulfan.

evaluations of all results would contradict the purpose of this paper as outlined above. Only some examples will be discussed therefore.

Chironomid density evaluated by dredge samples was statistically significantly lower during the application period in the pond that received the higher endosulfan input (Figure 3). This could be interpreted as an impact of endosulfan. However, a closer look shows a similar, but earlier decline in the respective reference pond. Additionally, emergence of chironomids during that time period was unaffected. Other aquatic macroinvertebrates like chaoborids were not reduced, but oligochaetes were. In the other pond, endosulfan residues in the sediment were higher than in the first one. However, chironomids were not affected. Therefore, sediment toxicity was excluded. Inspection of the data revealed a localized heavy thunderstorm over the first pond system dumping loads of sediment into the pond. It is probable that animals living on the bottom were covered by the settling sediment affecting chironomid density.

Another example for the necessity of a very thorough evaluation are the abundance measurements of spike rush (*Eleocharis* sp.), an aquatic weed. In one of the treatment ponds, the abundance was significantly lower than in all other ponds at the time after application (Figure 4). As an insecticide, endosulfan was not expected to inhibit growth of an aquatic weed. A careful comparison of all data disclosed the most probable explanation. The reduction of plants was caused by drying of a particularly shallow area within that pond.

A comparable scrutiny had to be applied to all of the statistically significant differences, compiling all available information. As an overall summary from the endosulfan pond study, we conclude that the only major effect clearly attributed to endosulfan was localized fish kill around shallow areas at runoff locations. All other statistical differences were attributable to either natural sources or "typical" biological variations (Tables 3 and 4).

The examples above demonstrate the particular emphasis of a very thorough evaluation of farm pond data for each of the locations. Contrary to mesocosms, ponds are normally spread over a larger area. Therefore, localized events might occur leading to localized effects. This statement does not imply, however, that evaluations of mesocosms are easy and do not require careful evaluation. Examples in the literature demonstrate the difficulty in doing so (a few selected examples are given in the references). Some of the variations typical of the farm pond approach are excluded in mesocosms.

When comparing the two strategies, the major differences are as in Table 5.

In mesocosms, replication is possible; a dose-response may be obtained, if the concentrations are chosen appropriately. Relatively true control groups can be incorporated into a test. The experiment is designed for a statistical evaluation. Effects in fish are exactly assessable. As mesocosm replicates are all concentrated at one location, logistics for sampling do not pose a problem.

The following disadvantages, on the other hand, must also be considered. Dosing of mesocosms depends on theoretical predictions of environmental situations. The dosing of these predicted levels occurs evenly over the entire surface, unlike the natural situation. There is only a limited possibility for a repopulation. Only one

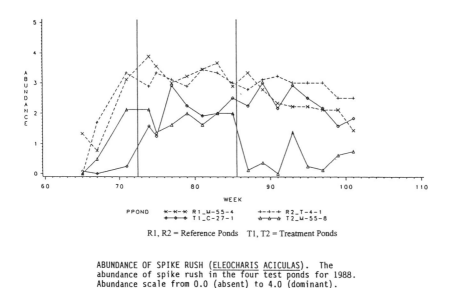

R1, R2 = Reference Ponds   T1, T2 = Treatment Ponds

ABUNDANCE OF SPIKE RUSH (ELEOCHARIS ACICULAS). The abundance of spike rush in the four test ponds for 1988. Abundance scale from 0.0 (absent) to 4.0 (dominant).

Figure 4.   Abundance of spike rush in farm ponds treated with endosulfan.

Table 5. Advantages and Disadvantages of Mesocosm or Farm Pond Study Designs

Advantages Mesocosm Study	Advantages Farm Pond Study
• Replication possible	• Reality (but only one case?)
• Dose-response obtainable	• Dosing natural
• Statistical analysis relatively easy	• Natural repopulation possible
• Effects in fish exactly assessable	• Representative food chain
	• Established ecosystem
Disadvantages Mesocosm Study	• All pond elements present
• Dependence from computer modeling	
• Dosing artificial	Disadvantages Farm Pond Study
• Natural repopulation limited	• Location of appropriate sites difficult
• Extrapolation to "reality" necessary	• True replication impossible
• Only one top predator species	• No dose-response obtainable
• Ecosystem not fully established	• Disturbances by public access
• Not all pond elements present	• Extrapolation to other conditions?

top predator fish species is seeded in the systems. The mesocosm ecosystem has not reached a climax, but is in an establishing succession. Not all elements of the system are present, e.g., aquatic weeds are eliminated on purpose from the ponds.

In farm pond testing, the system is exposed by realistic paths. The farm pond ecosystem is relatively stable, and fully established, as it is used for testing many years after its initial construction. All elements to be expected in a pond-ecosystem are present, including more than one top predator species.

In farm pond studies, the main problems are the difficulties in locating appropriate sites, the impossibility of having true replications or true control sites, and a high need for ecological and biological interpretation. The experiment may be disturbed by the public, as access cannot be totally excluded.

Both strategies have common problems of interpretation, representativeness, and extrapolation. Although many aquatic ecosystem field tests have been performed, there are still different perceptions in interpreting effects (La Point, 1989b,c). How should one interpret declines occurring earlier or later than in other (control-reference) systems? Are temporary differences ecologically adverse effects under all circumstances? Are the causes of the observed effects unequivocally direct or indirect results of the exposure to the pesticide? How representative are the results for other types of at least lentic ecosystems? How do we extrapolate the results to other environmental or agricultural conditions?

During the 1989 SETAC meeting, examples were presented for the comparableness of different types of mesocosm approaches under various conditions for esfenvalerate (Lozano et al., 1989; La Point et al., 1989a). Although comparison between nature and various ecosystem studies have been made, to my knowledge, similar comparisons for farm-pond studies as described above are still missing.

If field ecosystem studies can reveal effects unnoticeable with other test approaches, both strategies can be practiced successfully. It will depend on the question asked, the effort involved, and on the certainty required, if either of the choices will suffice.

ACKNOWLEDGMENTS

The study was carried out by *Battelle Columbus* and *Hickey's Agri-Services*. The principal investigators were B.W. Cornaby, A.F. Maciorowski, M.G. Griffith, J.E. Navarro, S.E. Pomeroy, N.G. Reichenbach, J.S. Shuey, and M.F. Yancey, all of Battelle, and Steve Hickey of Hickey's Agri-Services. During the planning stage, R. Graney, and during the course and in the evaluation, J. Mayasich, also contributed to this study.

REFERENCES

Boyle, T., S.E. Finger, R.L. Paulson, C.F. Rabeni, 1985. Comparison of Laboratory and Field Assessment of Fluorene-Part II: Effects on the Ecological Structure and Function of Experimental Pond Ecosystems. In *Validation and Predictability of Laboratory Methods for Assessing the Fate and Effects of Contaminants in Aquatic Ecosystems,* ASTM STP 865, T.P. Boyle (Ed.), American Society for Testing and Materials, Philadelphia, 134–151.

Crossland, N.O., 1982. Aquatic Toxicology of Cypermethrin. II. Fate and Biological Effects in Pond Experiments, *Aquat. Toxicol.,* 2, 205–222.

Environmental Protection Agency, 1982. Pesticide Assessment Guidelines. Subdivision E, Hazard Evaluation, Wildlife and Aquatic Organisms, EPA 540/9-82-024, Washington, D.C.

Environmental Protection Agency, 1984. 40 CFR Part 158 Data Requirement for Pesticide Registration; Final Rule, *Fed. Regist.* 49, No. 207, 42856-42905.

Environmental Protection Agency, 1988. 40 CFR Parts 152, 153, 156,158, and 162 Pesticide Registration Procedure, Pesticide Data Requirement; Final Rule, *Fed. Regist.* 53, No. 86, 15652–15990.

Giddings, J.M., P.J. Franco, S.M. Bartell, R.M. Cushman, S.E. Herbes, L.A. Hook, J.D. Newbold, G.R. Southworth, A.J. Stewart, 1985. Effects of Contaminants on Aquatic Ecosystems: Experiments with Microcosms and Outdoor Ponds, Environmental Science Division Publication No. 2381, Oak Ridge National Laboratory, 104 S.

Hill, I.R., S.T. Hadfield, J.H. Kennedy, P. Ekoniak, 1988. Assessment of the impact of PP321 on aquatic ecosystems using tenth-acre experimental ponds, British Crop Protection Conference, Pest and Diseases, 4B-1, 309–318.

Hurlbert, S.H., 1975. Secondary effects of pesticides on aquatic ecosystems, *Resid. Rev.,* 57, 81–148.

La Point, T.W., J.C. Brazner, E.C. Webber, J.J. Dulka, A.L. Hosmer, 1989a. A geographic comparison of aquatic ecosystem responses to the pyrethroid insecticide, esfenvalerate, Asana, Paper #419, given at the 10th Annual Meeting of the Society of Environmental Toxicology and Chemistry (SETAC), Oct. 28–Nov. 2, 1989, Toronto.

La Point, T.W., J.F. Fairchild, E.E. Little, S.E. Finger, 1989b. Laboratory and field techniques in ecotoxicological research: Strengths and limitations in: Boudou, A. and Ribeyre, F., (ed): *Aquatic Ecotoxicology: Fundamental Concepts and Methodologies,* pp. 240–255, CRC Press, Boca Raton, FL.

La Point, T.W., J.A. Perry, 1989c. Use of experimental ecosystems in regulatory decision making, *Environ. Manage.,* 13, No. 5, 539–544.

Lozano, S., L. Heinis, J. Brazner, M., Knuth, D. Tanner, R. Siefert, 1989. The effects of esfenvalerate in littoral enclosures: The merits of mesocosm vs natural pond studies, Paper #417, given at the 10th Annual Meeting of the Society of Environmental Toxicology and Chemistry (SETAC), Oct. 28–Nov. 2, 1989, Toronto.

Mulla, M.S., H.A. Darwazeh, M.S. Dhillon, 1981. Impact and joint action of Decamethrin and Permethrin and freshwater fishes on mosquitoes, *Bull. Environ. Contam. Toxicol.,* 26, 689–695.

Neugebaur, K., F.-J. Zieris, W. Huber, 1990. Ecological effects of atrazine on two outdoor artificial freshwater ecosystems, *Z. Wasser-, Abwasser-Forsch.,* 23, 11–17.

Touart, L.W., 1987. Aquatic Mesocosm Testing To Support Pesticide Registrations, Issue Paper, U.S. Environmental Protection Agency, Washington, D.C.

CHAPTER 6

Contaminant Effects in Marine/Estuarine Systems: Field Studies and Scaled Simulations

James R. Clark and John L. Noles

Abstract: Attempts to obtain field data for risk assessment of contaminants released into marine/estuarine systems can be complicated by a number of interrelated factors, such as complex circulation and mixing patterns, diverse stratification forces, dynamic short-term changes as well as seasonal movements of biota, and the ecosystem's physical scale. Tests conducted in simulated ecosystems are subject to constraints that restrict the effect of physical forces, limit physical scale of the test, and introduce biases from chemical partitioning and processing along the walls of the test system. These constraints restrict the broad application of test results as a model of dynamic marine systems. Through selected examples from literature and ongoing studies, we provide illustrations of how contaminant effects are studied at the individual, population, and community level in the field and/or in simulated ecosystems, such as mesocosms. We discuss marine-environment field studies and simulated field studies that measure contaminant effects with respect to exposure-response relationships, food-web interactions, competition/colonization studies, and selected aspects of nutrient cycling. Based on results to date, we conclude that: (1) successful field studies must focus on selected endpoints fundamental to our understanding of contaminant effects, and (2) endpoints studied in simulated ecosystems must be representative of key structural and/or functional factors of the system of interest.

INTRODUCTION

Marine and estuarine systems may receive direct pesticide inputs from coastal agricultural activities, from chemical control of pest species, such as mosquitoes or aquatic weeds in or near coastal habitats, or from chemical use in coastal resorts or residential developments. Freshwater streams and rivers also can affect marine environments by introducing soluble or particulate-sorbed pesticides transported from inland habitats. A combination of environmental fate and toxicological effects data is used to evaluate the potential for ecological effects in marine environments as part of the pesticide registration process in the U.S.[1] Because there are uncertainties in the way we extrapolate from simple laboratory data to predict effects in marine ecosystems, additional studies in the field or in controlled, large-scale simulated systems (mesocosms) can be used to obtain more definitive answers and to reduce uncertainties.[1,2]

Marine systems are open, dynamic ecosystems with regular tidal exchange of media and planktonic biota. Direct extrapolation of exposure-response data from simple laboratory systems or from closed, scaled test systems may overestimate chemical exposures and ecological effects likely to occur in the field where pesticides are introduced in pulses and concentrations are diminished by tidal exchange. On the other hand, both tidal exchange with concomitant transport of plankton and the seasonal movements of motile biota through marine systems can bring a variety of species into habitats where direct pesticide exposures are likely, thus enhancing the potential for exposures. We have reviewed published field and mesocosm studies of pesticides to survey the kinds of experimental and system designs that have been used to study the effects of pesticides in marine ecosystems and to quantify uncertainties in extrapolating toxicological effects observed in the laboratory to the field. We also considered mesocosm studies used to evaluate ecological effects of other types of contaminants and offered suggestions about how these systems might be utilized for future pesticide studies. Readers are referred to reviews by Grice and Reeve,[3] Lalli,[4] and Gearing[5] for comprehensive discussions on use of mesocosms for fundamental and applied ecological study.

The issues surrounding estimations of releases and evaluations of ecological effects of pesticides have become increasingly complex with the commercial advent of new pesticides that are toxic at lower thresholds and have more specific modes of action. In order to improve our understanding of ecological effects of pesticides in marine environments, historical and contemporary approaches must be evaluated for their utility as either screening-level studies or as comprehensive assessments of ecological effects for pest control chemicals. As part of the comprehensive ecological risk assessment process for pesticides, we examined the applicability of generalized environmental assessments and the ability to extrapolate from ecosystem to ecosystem. We also reviewed some comparative aspects of freshwater and marine systems, and posed research questions to evaluate the need for developing ecological response information for marine systems as a database separate from ecological studies in freshwater. Finally, we summarized the state of field and mesocosm studies with pesticides in marine environments and recommended directions for future research and ecological risk assessment procedures.

LITERATURE REVIEW OF MARINE ECOSYSTEM LEVEL PESTICIDE STUDIES

Field Studies of Pesticide Effects

Several environmental and experimental design factors influence field studies of pesticide effects in marine systems. Clark[6] reviewed approaches for assessing contaminant effects at various levels of ecological organization in estuarine systems. The interactions of hydrodynamic, morphometric, and anthropogenic factors specific to estuarine study sites were discussed and their influence on the assumptions and types of comparisons applicable to exposure-response assessments at several levels of ecological organization was examined. Attempts to follow basic scientific principles, i.e., to select an appropriate control site in the field or to achieve true replication in the experimental design, can present major challenges.[7] Site-to-site differences in water quality and background contaminant inputs within and between systems can greatly affect the distribution of estuarine biota.[6] Field assessments in marine systems require that sampling activities be conducted over time frames of weeks to months, if not years, to elicit trends associated with pesticide exposure distinct from short-term trends in population and community dynamics.[8] With considerable forethought and coordination effort, designs employed for monitoring the distribution of contaminants and associated ecological effects in estuarine systems can be adapted to experimental, exposure-response comparisons and hypothesis testing,[9] thus allowing evaluations of the ecological effects of pesticide inputs. The studies discussed in the remainder of this section provide examples of field assessments of pesticide effects in estuarine environments which were focused on specific exposure-response issues. We found no references that addressed holistic assessments of system-level effects of pesticides for estuarine systems.

Mirex® is an organochlorine insecticide that was used to control fire ants in the southeastern U.S. until 1978, when the EPA agreed to a plan resulting in cancellation of all uses of Mirex® in order to protect aquatic biota and wildlife. As a follow-up to laboratory studies that demonstrated toxicity of Mirex® to estuarine biota at aqueous concentrations ≥ 0.50 mg/l or from residues in food,[10-13] the U.S. EPA Environmental Research Laboratory at Gulf Breeze monitored the movement of Mirex® to coastal habitats following large-scale applications of 1.7 g Mirex® per acre to control fire ants. Collections focused on animals representative of various levels in the food web during all seasons of the year to quantify residues accumulated and to allow detection of any resulting mass mortalities among nontarget, estuarine species. Although the fieldwork was not designed to quantify subtle or indirect effects on populations or communities of estuarine biota, residue data supported the predicted persistence and movement of the compound from treated fields to coastal habitats and bioaccumulation of Mirex® by a variety of estuarine species.[14,15] These data and other laboratory tests supported the ban of Mirex® in 1978.

Kepone® is another persistent organochlorine insecticide that has been shown to affect marine and freshwater biota following acute or chronic exposures at low aqueous concentrations or via the food web. Kepone® is of special interest and relevant to current pesticide issues because it is primarily a sediment-source and sediment-transport problem, an area of increasing concern with ecological risk assessments of contemporary and new insecticides. The environmental fate and effects of Kepone® have been investigated in a variety of laboratory studies, characterizing its toxicity, bioaccumulation potential, and persistence.[16-20] Bender and Huggett[21] reviewed existing field data on exposures and effects of Kepone® and discussed studies from the James River estuary in Virginia. Although no single study has addressed the Kepone® problem in the James River, through a systematic, ecological assessment, Bender and Huggett placed the laboratory and field data into perspective relative to quantified exposure-effects relationships for individuals of sensitive species. Bioaccumulation of Kepone® in the James River food chain and reductions in the abundance of crustaceans in contaminated areas support the laboratory assessments of ecological effects.

Field studies on the acute effects of insecticide releases into estuarine habitats have been conducted with the organophosphate insecticides malathion and fenthion following mosquito control applications. Ground applications of malathion were studied by Tagatz et al.[22] to evaluate lethal effects on caged and local biota as well as quantifying environmental exposures and residues in saltmarsh habitats. Clark et al.[23] summarized results of ground and aerial applications of saltmarsh mosquito control applications of fenthion and effects on several test species held in cages at field sites. These studies demonstrated that laboratory toxicity tests can provide reliable estimates of acutely toxic pesticide concentrations, but only when laboratory and field exposure regimes were similar. Characterizing and quantifying exposure-response relationships for pesticides and estuarine biota in the field are among the most challenging aspects of estuarine field work with pesticides.

Studies of herbicide effects also have focused on laboratory and field comparisons of short-term, exposure-response relationships rather than system-level assessments. Assessments of herbicide runoff effects in Chesapeake Bay focused on experimental manipulations of small-scale laboratory test systems and on field measurements of exposure and plant abundances in areas where submerged aquatic vegetation was declining.[24,25] Short-term toxic effects and bioaccumulation of residues following atrazine additions to field plots in a Georgia marsh were assessed relative to toxicological and fate studies by using small-scale laboratory test systems.[26] Both studies attempted to relate assessments of individual plant responses to changes in standing crop (plant populations).

Comprehensive ecological assessments from 4 years of field research of agricultural pesticide runoff into tidal creeks in North Carolina and South Carolina are only now becoming available. Initial findings on environmental concentrations of pesticides (alachlor, permethrin, terbufos) and short-term and long-term effects on estuarine biota in the South River Estuary near Beaufort, NC have been discussed.[27-29] These studies have demonstrated that complex mixing patterns and particulate transport characteristics of tidal creeks govern the exposure of estuarine biota during pesticide runoff events, and may obscure direct exposure-response relationships characterized from caged-animal studies in the field or laboratory toxicity tests.

Similarly, Scott et al.[30,31] have reported on short-term effects of endosulfan and fenvalerate runoff into tidal creeks of South Carolina. Long-term effects on populations and communities of pelagic and benthic biota currently are being assessed as part of their ongoing field research. Estuarine systems under study openly mix with coastal water masses and are continually recolonized by mobile and pelagic life forms. Only by evaluating long-term trends can we obtain an accurate assessment of pesticide effects on resident populations and communities in the estuary.

Large-Scale Simulated Marine Systems (Mesocosm Studies)

Many researchers have studied selected components of estuaries in small-scale systems with noted success in establishing exposure-response relationships under complex testing conditions, but with little success in convincing extrapolation of results to estuarine ecosystems. Few large-scale marine systems (>100 l volume) have been tested with pesticides. Tagatz et al.[32] tested Mirex® in systems containing plants, sediment, and contaminated water to study effects on invertebrates and fishes as well as to compare fate and transformations. Grass shrimp were more vulnerable to fish predation during exposure to Mirex®, and Mirex® was bioaccumulated by plants and animals. Effects of pentachlorophenol on benthos of soft bottom communities and predator-prey relationships were studied by Diaz and Livingston[33,34] using a large experimental system in the field, as well as a number of small-scale laboratory test systems. Again, prey species appeared to be more vulnerable when chemically stressed. Functional assessments of benthic infaunal assemblages were used to demonstrate changes in the relative abundance of several species assigned to various guilds based on feeding strategies and mobility capabilities that influenced the group's exposure to pentachlorophenol-contaminated sediments. The utilization of guilds was a useful approach for determining animal life stages and activity traits that contribute to enhancing exposure to sediment-associated contaminants. While these intensive and comprehensive studies provide insight into population- and community-level ecological effects of pesticides in estuarine environments, exposure-response assessments at the ecosystem level will require an order-of-magnitude greater level of effort and complexity to characterize the direct and indirect pathways through which pesticides might change ecological systems.

Additional Mesocosm Designs Applicable For Pesticide Studies

A variety of mesocosm-scale test systems have been used to enclose masses of seawater or benthic habitats and to study effects of oil, metals, toxic chemicals, or nutrient additions on the structure and function of planktonic communities, including heterotrophic, autotrophic, microfaunal, and macrofaunal components (Table 1). The scale of pelagic test systems used to enclose seawater to study planktonic communities has ranged from 1.5 m^3 to 300 m^3, with various configurations for flotation and support as well as for collecting settled particles or biota. Systems

Table 1. Summary of Marine Mesocosms, Previously Used To Study Contaminant Effects, That Could Be Used for Pesticide Studies

System Design Features	Approximate Volume and Dimensions	Contaminants Studied	References
In Situ Pelagic Enclosures			
CEPEX — moored pelagic enclosures deployed in embayments to study plankton dynamics	Cone-shaped bags, 68 m³, 2.5 m diam top, 0.2 m diam bottom, 16 m depth; Second version 1300 m³, 10 m diam top, 0.2 m diam bottom, 24 m depth	Heavy metals, nutrients, crude oil, refined petroleum products	Menzel and Case,[35] Grice and Menzel[36] (See Gearing, 1989 for comprehensive listing)
Pelagic enclosures moored in lagoons or embayments to study plankton dynamics	Cylindrical bags, 3 to 5 m diam at top, depth varied, ≦18 m	Crude oil, refined petroleum products	Lacaze,[37] Grice and Reeve,[3] Skjoldal et al.[38]
Loch Ewe enclosures — pelagic bags moored in Scotland to study plankton dynamics	Cylindrical tubes, 3 m or 10 m diam, 17 m depth	Oil, copper, mercury	Davies et al.,[39] Gamble et al.,[40] Davies and Gamble[41] (See Gearing, 1989 for comprehensive listing)
Pelagic enclosures deployed in harbors along the North Sea to study plankton dynamics	Cylindrical tubes, 1.5 m³ or 0.75 m diam, 3.5 m depth; Second version, 16 m³, 0.75 m diam, 20 m depth	Mercury, cadmium, phenols	Kuiper,[42] Kuiper[43]
Hamburg enclosures — pelagic bags deployed to study plankton dynamics	Cylindrical tubes, 30 m³, 1 m diam, 40 m depth	Dissolved organic substances affecting plankton growth	Grice and Reeve[3]
In Situ Pelagic and Sediment Enclosures			
Plankton Tower — fixed structure in embayment to study plankton and sediment interactions	Rectangular enclosure, 30 m³, 10 m × 10 m, 16.5 m depth	Nutrient release from sediments	von Bodurgen et al.,[44] Smetacek et al.[45]
Constructed Pelagic and Sediment Systems			
MERL — fiberglass tanks constructed on shore, mechanical circulation, scaled to model Narragansett Bay	Cylindrical tanks, 13 m³, 2 m diam, 6 m depth, 0.76 m² sediment surface	Refined petroleum products, sewage, heavy metals	Pilson et al.[46] (See Gearing, 1989 for comprehensive listing)
In Situ Benthic Enclosures			
Bremerhaven Caisson — partitions inserted into mud flat or intertidal area to enclose habitat, controlled water exchange	Rectangular area, 13 m², 5.6 m × 2.4 m, wall height 2 m	Lead, chromium, oil, dispersants	Farke et al.,[47] Schulz-Blades et al.[48]

with a sediment component may provide a more realistic exposure scenario for long-term studies of pesticide effects because they allow sediment to serve as both a source and sink for pesticides. There have been fewer tests with mesocosm-scale benthic communities than with planktonic enclosures. Benthic mesocosms that are modular and portable could be deployed in a variety of intertidal habitats such as mudflats, marshes, or rocky shorelines.

It is likely that the mesocosm test systems listed in Table 1, or ones of similar design, could be adapted to study pesticide effects in marine ecosystems after some consideration of potentially confounding effects. The fate of pesticides in meso-cosms must be evaluated in light of system design problems, such as sorption of the soluble portion of the chemical to sidewalls or incorporation of the chemical into a sidewall fouling community, especially for chemicals with high organic partition coefficients. Static systems such as CEPEX could be modified to allow some degree of tidal exchange for dilution and recruitment, although treatment of contaminated water leaving a test system must be considered in the design.

Because many environmental concerns center around the fate and effects of pesticides that sorb to particulate matter and may be incorporated into estuarine sediments, test systems that address impacts on benthic biota are needed. Current research focusing on the toxic effects of sediment-associated contaminants utilize an assortment of chemical and biological measurements in the laboratory and field to develop and quantify exposure-response relationships for benthic species and/or benthic communities.[49-51] Mesocosm-scale benthic test systems such as the Bremerhaven Cassions[47,48] could be utilized for controlled studies with single or multiple pesticide exposures for benthic communities.

Incorporation of sediment components with benthic communities into a balanced ecosystem with pelagic and planktonic components would be a significant advancement for marine mesocosm testing programs. Systems with sufficient scale to assess interactions of pelagic and benthic communities are needed. However, such test systems must be designed with some reference system as a model. While freshwater mesocosms have farm ponds as a real-world model upon which their design is based, no standard for marine systems has yet been agreed upon. There is no consensus on the suitability of, or preference for, physical models of mudflats, tidal creeks, or rocky shorelines as mesocosms for pesticide risk assessment for marine ecosystems. Before extensive testing for regulatory assessments is begun in marine mesocosms, some scientific evaluation of the merits of various approaches and test systems must be completed.

Expense in construction and design as well as operational costs have limited the numbers of test systems used. Desired numbers of treatments and replication within treatments may prove to be a limiting factor in attempts to adapt existing mesocosm test systems to meet the needs required for pesticide studies. Design considerations discussed at this symposium as well as those presented elsewhere[7,52,53] will greatly influence the type of system to be modeled and the cost of conducting tests with marine mesocosms.

COMPARATIVE NATURE OF FRESHWATER AND MARINE ECOSYSTEMS EXPOSURE

The use of farm ponds as a model for freshwater mesocosm studies to test agricultural pesticides was based chiefly on several factors: (1) farm ponds represent worst-case exposure scenarios among lakes, rivers, and streams as a result of the maximum pesticide loadings and minimum dilution potential characteristic of a pond, (2) we have a comprehensive conceptual and empirical understanding of ponds as ecosystems, and (3) the physical and biological characteristics of a pond can be modeled from a quantitative database.[2] Marine systems have no small-scale surrogate system upon which a physical model can be based. Maximum pesticide inputs to marine systems may occur in tidal creeks or marshes located near coastal agricultural activities, but the wide diversity in physical scale and processes present in marine systems provide for no single conceptual or physical model as being representative. All marine systems are subject to some degree of tidal exchange, resulting in rapidly decreasing aqueous concentrations of pesticides over time. Tidal ranges vary by more than an order of magnitude for sites along the U.S. coastline, providing a variety of conceptual opportunities for modeling a test system.

In marshes and tidal creeks, input and resuspension of particulate matter plays a dynamic role in controlling the fate and transport of contaminants, directly altering the bioavailability of soluble pesticides with time. A comprehensive evaluation of pesticide exposure and fate processes in marine habitats would likely show that predictions of pesticide exposure and fate in freshwater ponds will not be directly applicable to marine systems, nor provide confidence in developing safe use guidelines near estuaries. Such an evaluation must be based upon further, focused research on comparative aspects of endpoints used in quantifying pesticide fate and bioavailability.

Ecosystem Sensitivity

Comparative ecological toxicology is still in its infancy, although we are obtaining substantial knowledge of the rates of response to stress and recovery rates relating to issues such as atmospheric deposition, eutrophication, hazardous waste disposal, waste clean-up, and oil spill issues. The ultimate goal of aquatic field testing or mesocosm studies for environmental risk assessments of pesticides is to ensure that wide-scale pesticide uses will not result in inputs to aquatic systems that will significantly change ecological structure or inhibit ecosystem functions. If there was sufficient evidence to convince scientists that one ecosystem type (i.e., freshwater or marine ecosystems) was more sensitive than another, then, under worst-case assessment scenarios, only one type would have to be tested. Decisions on use restrictions of a chemical that would protect the more sensitive ecosystem would be protective of the others.

Scientist are actively debating which are the best bioindicators of stress or health for assessments at the species, population, community, and ecosystem level.[54,55]

Our ability to extrapolate from one system to another is highly limited, because of deficiencies in our understanding of comparative ecotoxicology. Extrapolating direct toxic effects measured on one species from system A to the same effects on another species in system B can be discussed and debated because we have a more comprehensive understanding of comparative toxicology at the species level.[56,57] Indirect effects of toxicants (such as alterations in competition, food sources, or behavior) on the survival and growth of individuals cannot be extrapolated as readily to effects at higher levels of ecosystem organization because of our less comprehensive understanding of the dynamics of individual and population interactions within ecosystems.[58-60]

Extrapolating population changes observed in one system to effects in another is extremely difficult, because of the natural dynamic aspects of populations and our limited understanding of how toxicants can affect controlling functions. Contaminant-induced changes at the community and ecosystem level of organization perhaps could be extrapolated more readily from system to system because of the redundant functional properties embedded among the individuals and species comprising certain aquatic communities. Because many species can contribute to community functions, such as primary productivity, detritus processing, nitrogen fixation, and others, contaminant-induced effects on a community function may have affected such a broad array of species that similar effects may be expected to occur in another system if exposure concentrations and durations are matched. However, our ability to explain how direct or indirect effects of pesticide inputs have altered ecosystem structure and function are extremely limited, and examples of single species effects altering system functions can be found.[60] Confidence in any extrapolations will be developed only as comparative ecotoxicology research provides effective examples.

Nixon[61] edited a collection of papers addressing the comparative ecology of freshwater and marine ecosystems and demonstrating that there are strong, fundamental similarities between system-level structure and function of microbial, algal, plant, planktonic, benthic, and fisheries components of freshwater and marine ecosystems. Neither ecosystem has specific, key components or links that could be forwarded as the most representative, sensitive, or vulnerable aspect (its "Achilles' heel") upon which risk assessments from a single approach or ecosystem could be based so that it would be protective of all ecosystems. Ecological risk assessment must focus on all levels of organization in all types of systems to evaluate potential for ecological effects of chemical perturbation. The comparative aspects of freshwater and marine ecology will remain an interesting area of future ecological research.

CONCLUSIONS AND FUTURE DIRECTIONS

Appropriately organized studies in the field or studies in suitably scaled test systems are needed to develop our understanding of exposure-response relationships for individual, population, and community levels of assessment in marine ecosystems. Historical focus has been on understanding the relationship of laboratory-derived toxicity data to direct toxic effects occurring in the field. Indirect effects and population- or community-level assessments have not been studied to the same extent. Our understanding today does not allow us to determine with confidence which types of ecosystems or which ecosystem traits are more sensitive than others, allowing us to focus on key indicators or crucial links essential to preserving a healthy ecosystem, and thus minimizing the cost of studying chemical effects on marine ecosystems. Indeed, a concerted effort is needed to characterize the roles played by the diverse biota that contribute to the breadth of ecosystem structure and function to ensure their protection from adverse impacts of intentional chemical releases into the environment. Endpoints studied in marine systems must focus on key structural or functional factors that sustain the ecosystem.

ACKNOWLEDGMENT

Ms. E.J. Pinnell provided valuable assistance in literature searches to support this effort. This manuscript is contribution #731 of the U.S. Environmental Protection Agency, Environmental Research Laboratory, Gulf Breeze, FL 32561.

REFERENCES

1. Urban, D.J. and J.J. Cook. 1986. *Hazard Evaluation Division Standard Evaluation Procedure, Ecological Risk Assessment.* U.S. EPA 540/9-85-001, Office of Pesticide Programs, Washington, D.C.
2. Touart, L.W. 1988. *Hazard Evaluation Division: Technical Guidance Document, Aquatic Mesocosm Tests to Support Pesticide Registration.* U.S. EPA 540/09-88-035, Office of Pesticide Programs, Washington, D.C.
3. Grice, G.D. and R.R. Reeve. 1982. *Marine Mesocosms: Biological and Chemical Research in Experimental Ecosystems.* Springer-Verlag, New York.
4. Lalli, C.M. 1990. *Enclosed Experimental Marine Ecosystems: A Review and Recommendations.* Springer-Verlag, New York.
5. Gearing, J.N. 1989. The Role of Aquatic Microcosms in Ecotoxicologic Research as Illustrated by Large Marine Systems. In S.A. Levin, M.A. Harwell, J.R. Kelly, and K.D. Kimball, Eds., *Ecotoxicology: Problems and Approaches.* Springer-Verlag, New York, pp. 411–470.
6. Clark, J.R. 1989. Field Studies in Estuarine Ecosystems: A Review of Approaches for Assessing Contaminant Effects. In U.M. Cowgill and L.R. Williams, Eds., *Aquatic Toxicology and Hazard Assessment: 12th Volume,* ASTM STP 1027, American Society for Testing and Materials, Philadelphia, pp. 120–133.
7. Hurlbert, S. 1984. Pseudoreplication and the design of ecological fish experiments. *Ecol. Monogr.* 54:187–211.

8. Livingston, R.J., N.P. Thompson, and D.A. Meeter. 1978. Long-term variation of organochlorine residues and assemblages of epibenthic organisms in a shallow north Florida (USA) estuary. *Mar. Biol.* 46:355–372.

9. O'Connor, J.S. and D.A. Flemer. 1987. Monitoring, Research, and Management: Integration for Decisionmaking in Coastal Maine Environments. In T.P. Boyle, Ed., *New Approaches to Monitoring Aquatic Ecosystems*, ASTM STP 940, American Society for Testing and Materials, Philadelphia, pp. 70–90.

10. Lowe, J.I., P.R. Parrish, A.J. Wilson, Jr., P.D. Wilson, and T.W. Duke. 1971. Effects of Mirex® on selected estuarine organisms. In J.B. Trefethen, Ed., *Transactions of the 36th North American Wildlife and Natural Resources Conference,* Wildlife Management Institute, Washington, D.C., pp. 171–186.

11. Tagatz, M.E. 1976. Effect of Mirex® on predatory-prey interaction in an experimental estuarine ecosystem. *Trans. Am. Fish. Soc.,* 105:546–549.

12. Bookhout, C.G., A.J. Wilson, Jr., T.W. Duke, and J.I. Lowe. 1972. Effects of Mirex® on the larval development of two crabs. *Water, Air, Soil Pollut.* 1:165–180.

13. Bookhout, C.G. and J.D. Costlow, Jr. 1975. Effects of Mirex® on the larval development of blue crab. *Water, Air, Soil Pollut.* 4:113–126.

14. Borthwick, P.W., T.W. Duke, A.J. Wilson, Jr., J.I. Lowe, J.M. Patrick, Jr., and J.C. Oberheu. 1973. Accumulation and movement of Mirex® in selected estuaries of South Carolina, 1969–1971. *Pest. Monit. J.* 7:6–26.

15. Borthwick, P.W., G.H. Cook, and J.M. Patrick, Jr. 1974. Mirex® residues in selected estuaries of South Carolina — June 1972. *Pest. Monit. J.* 7:144–145.

16. Bahner, L.H., A.J. Wilson, Jr., J.M. Sheppard, J.M. Patrick, Jr., L.R. Goodman, and G.E. Walsh. 1977. Kepone® bioconcentration, accumulation, loss, and transfer through estuarine food chains. *Chesapeake Sci.* 18:299–308.

17. Schimmel, S.C. and A.J. Wilson. 1977. Acute toxicity of Kepone® to four estuarine animals. *Chesapeake Sci.* 18:224–227.

18. U.S. Environmental Protection Agency. 1978. Proceedings of the Kepone® Seminar II. U.S. EPA-903/9-78-011, Region III, Philadelphia.

19. Goodman, L.R., D.J. Hansen, C.S. Manning, and L.F. Faas. 1982. Effects of Kepone® on the sheepshead minnow in an entire life-cycle toxicity test. *Arch. Environ. Contam. Toxicol.* 11:335–342.

20. O'Neill, E.J., C.A. Monti, P.W. Pritchard, A.W. Bourquin, and D.G. Ahearn. 1985. Effects of lugworms and seagrass on Kepone® (chlordecone) distribution in sediment/water laboratory systems. *Environ. Toxicol. Chem.* 4:453–458.

21. Bender, M.E. and R.J. Huggett. 1984. Fate and Effects of Kepone® in the James River. In E. Hojdgson, Ed., *Reviews in Environmental Toxicology I,* Elsevier Science, New York, pp. 5–50.

22. Tagatz, M.E., P.W. Borthwick, G.H. Cook, and D.L. Coppage. 1974. Effects of ground applications of malathion on salt marsh environments in northwestern Florida. *Mosquito News* 34:309–315.

23. Clark, J.R., P.W. Borthwick, L.R. Goodman, J.M. Patrick, Jr., E.M. Lores, and J.C. Moore. 1987. Comparison of laboratory toxicity test results with responses of caged estuarine animals exposed to fenthion in the field. *Environ. Toxicol. Chem.* 6:151–160.

24. Kemp, W.M., J.C. Means, T.W. Jones, and J.C. Stevenson. 1982. Herbicides in Chesapeake Bay and their effects on submerged aquatic vegetation: A synthesis of research supported by U.S. EPA Chesapeake Bay Program. In E.G. Macalaster, D.A. Barker, and N.E. Kasper, Eds., *Chesapeake Bay Program Technical Studies: A synthesis,* Washington, D.C., pp. 503–567.

25. Flemer, D.A., R.B. Biggs, V.K. Tippie, W. Nehlsen, G.B. Mackierman, and K.S. Price. 1987. Characterizing the Chesapeake Bay ecosystem and lessons learned. *Proceedings of the Tenth National Conference of the Coastal Society,* The Coastal Society, Washington, D.C., pp. 153–177.

26. Davis, D.E., J.D. Weete, C.G.P. Pillai, F.G. Plumley, J.T. McEnerney, J.W. Everest, B. Truelove, and A.M. Diner. 1979. Atrazine Fate and Effects in a Salt Marsh. U.S. EPA 600/3-79-111, Washington, D.C.

27. Kirby-Smith, W., S. Thompson, and R.B. Forward, Jr. 1989. Use of grass shrimp (*Palaemonetes pugio*) larvae in field bioassays for the effects of agricultural runoff into estuaries. In D.L. Weigman, Ed., *Pesticides in Terrestrial and Aquatic Environments.* Virginia Water Resources Research Center, Blacksburg, VA, pp. 29–36.

28. Kirby-Smith, W. and S. Thompson. The effects of agricultural runoff on the benthic and nektonic communities in a tributary of the Pamlico sound estuarine system. In review.

29. Howe, J., M.W. Sandstrom, and S.J. Eisenreich. Alachlor and permethrin from agricultural runoff in South River estuary, North Carolina. *Environ. Sci. Technol.,* in review.

30. Scott, G.I., D.S. Baughman, A.H. Trim, and J.C. Dee. 1987. Lethal and sublethal effects of insecticides commonly found in agricultural runoff to estuarine fish and shellfish. In W.B. Vernberg, A. Calabreese, F.P. Thurberg, and F.J. Vernberg, Eds., *Pollution Physiology of Estuarine Organisms, Baruch Series #17,* Univ. South Carolina Press, Columbia, pp. 251–273.

31. Baughman, D.S., D.W. Moore, and G.I. Scott. 1989. A comparison of field and laboratory toxicity tests with fenvalerate on an estuarine crustacean. *Environ. Toxicol. Chem.* 8:417–429.

32. Tagatz, M.E., P.W. Borthwick, J.M. Ivey, and J. Knight. 1976. Effects of leached Mirex® on experimental communities of estuarine animals. *Arch. Environ. Contam. Toxicol.* 4:435–442.

33. Livingston, R.J., R.J. Diaz, and D.C. White. 1985. Field Validation of Laboratory-Derived Multispecies Aquatic Test Systems. EPA 600/4-85/039, NTIS, Springfield, VA.

34. Diaz, R.J., M. Luckenbach, S. Thornton, M.H. Roberts, Jr., R.J. Livingston, C.C. Koenig, G.L. Ray, and L.E. Wolfe. 1987. Field Validation of Multi-Species Laboratory Test Systems for Estuarine Benthic Communities. EPA 600/3-87/016, NTIS, Springfield, VA.

35. Menzel, D.W. and J. Case. 1977. Concept and design: controlled ecosystem pollution experiment. *Bull. Mar. Sci.* 27:1–7.

36. Grice, G.D. and D.W. Menzel. 1978. Controlled ecosystem pollution experiment: effect of mercury on enclosed water columns. VIII. Summary of results. *Mar. Sci. Comm.* 4:23–31.

37. Lacaze, J. 1974. Ecotoxicology of crude oils and the use of experimental marine ecosystems. *Mar. Poll. Bull.* 5:153–156.

38. Skjoldal, H.R., T. Dale, H. Haldorsen, B. Pengerud, T.F. Thingstad, K. Tjessem, and A. Aaberg. 1982. Oil pollution and plankton dynamics. I. Controlled ecosystem experiment during the 1980 spring bloom in Lindaspollene, Norway. *Neth. J. Sea Resh.* 16:511–523.

39. Davies, J.M., I.E. Baird, L.C. Massie, S.J. Hay, and A.P. Ward. 1980. Some effects of oil derived hydrocarbons on a pelagic food web from observations in an enclosed ecosystem and a consideration of their implications for monitoring. *Rapp. PV. Reun. Cons. Int. Explor. Mer.* 179:201–211.

40. Gamble, J.C., J.M. Davies, and J.H. Steele. 1977. Loch Ewe bag experiment, 1974. *Bull. Mar. Sci.* 27:146–175.

41. Davies, J.M. and J.C. Gamble. 1979. Experiments with large enclosed ecosystems. *Philas. Trans. R. Soc. Lond. B.* 286:523–544.

42. Kuiper, J. 1977. Development of North Sea coastal plankton communities in separate plastic bags under identical conditions. *Mar. Biol.* 44:97–107.

43. Kuiper, J. 1982. Ecotoxicological experiments with marine plankton communities in plastic bags. In G.D. Grice and M.R. Reeve, Eds., *Marine Mesocosms, Biological and Chemical Research in Experimental Ecosystems.* Springer-Verlag, New York, pp. 181–193.

44. von Bodungen, B., K. von Grockel, V. Smetacek, and B. Zeitzschel. 1976. The plankton tower. I. A structure to study water/sediment interactions in enclosed water columns. *Mar. Biol.* 34:369–372.

45. Smetacek, V., B. von Bodungen, B. Knoppers, F. Pollehne, and B. Zeitzschel. 1982. The plankton tower. IV. Interactions between water column and sediment in enclosure experiments in Kiel Bight. In G.D. Grice and M.R. Reeve, Eds., *Marine Mesocosms, Biological and Chemical Research in Experimental Ecosystems.* Springer-Verlag, New York, pp. 205–216.

46. Pilson, M.E.Q., C.A. Oviatt, G.A. Vargo, and S.L. Vargo. 1979. Replicability of MERL microcosms: Initial observations. In F.S. Jacoff, Ed., *Advances in Marine Environmental Research.* EPA 600/9-79-035, Narragansett, RI, pp. 359–381.

47. Farke, J., M. Schulz-Baldes, K. Ohm, and S.A. Gerlach. 1984. Bremerhaven caisson for intertidal field studies. *Mar. Ecol. Prog. Ser.* 16:193–197.

48. Schulz-Baldes, M., E. Rehm, and H. Farke. 1983. Field experiments on the fate of lead and chromium in an intertidal benthic mesocosm, the Bremerhaven caisson. *Mar. Biol.* 75:307–318.

49. Chapman, P.M. 1986. Sediment quality criteria from the sediment quality triad: an example. *Environ. Toxicol. Chem.* 5:957–964.

50. Chapman, P.M., R.C. Barrick, J.M. Neff, and R.C. Swartz. 1987. Four independent approaches to developing sediment quality criteria yield similar values for model contaminants. *Environ. Toxicol. Chem.* 6:723–725.

51. Long, E.R., M.F. Buchman, S.M. Bay, R.J. Breteler, R.S. Carr, P.M. Chapman, J.E. Hose, A.L. Lissner, J. Scott, and D.A. Wolfe. 1990. Comparative evaluation of five toxicity tests with sediments from San Francisco Bay and Tomales Bay, California. *Environ. Toxicol. Chem.* 9:1193–1214.

52. Gamble, J.C. 1990. Mesocosms: Statistical and experimental design considerations. In C.M. Lalli, Ed., *Enclosed Experimental Marine Ecosystems: A Review and Recommendations.* Springer-Verlag, New York, pp. 188–196.

53. Greene, R. 1979. *Sampling Design and Statistical Methods for Environmental Biologists.* Wiley, New York.

54. National Research Council. 1981. *Testing for Effects of Chemicals on Ecosystems.* National Academy Press, Washington, D.C.

55. Segar, D.A., D.J.H. Phillips, and E. Stamman. 1987. Strategies for long-term pollution monitoring of the coastal oceans. In T.P. Boyle, Ed., *New Approaches to Monitoring Aquatic Ecosystems,* ASTM STP 940, American Society for Testing and Materials, Philadelphia, pp. 12–27.

56. Mayer, F.L., C.H. Deans, and A.G. Smith. 1987. Inter-taxa correlations for toxicity to aquatic organisms. U.S. EPA/600/X-87/332. Gulf Breeze, FL.

57. Mayer, F.L. and M.R. Ellersieck. 1988. Experiences with single-species tests for acute toxic effects on freshwater animals. *Ambio* 17:367–375.

58. Slooff, W., J.A.M. van Oers, and D. deZwart. 1986. Margins of uncertainty in ecotoxicological hazard assessment. *Environ. Toxicol. Chem.* 5:841–852.

59. Perry, J.A. and N.H. Troelstrup, Jr. 1988. Whole ecosystem manipulation: a productive avenue for test system research? *Environ. Toxicol. Chem.* 7:941–951.

60. Kelly, J.R. 1988. Ecotoxicology beyond sensitivity: a case study involving "unreasonableness" of environmental change. In S.A. Levin, M.A. Harwell, J.R. Kelly, and K.D. Kimball, eds., *Ecotoxicology: Problems and Approaches,* Springer-Verlag, New York, pp. 473–496.

61. Nixon, S.W., editor. 1988. Comparative ecology of freshwater and marine ecosystems. *Limnol. Oceanogr.* 33:649–1025.

CHAPTER 7

Introduction and Regulatory Background: Summary and Discussion

Robert L. Graney

The manuscripts in this section provided an overview of the current status and use of mesocosm studies as they apply to pesticide registration under FIFRA. Overall, there were a number of basic concepts stressed by all three authors which should be reiterated. First, mesocosm studies represent the final tier in the risk assessment process and are only required if the laboratory-based risk assessment indicates that the use of the compound may present an unacceptable risk. The primary objective of the mesocosm study is to refute this presumption of risk and the burden of proof is on the registrant. Second, mesocosm studies may provide the "opportunity" to go beyond single species tests and evaluate direct and indirect effects on community and ecosystem level parameters. Third, the final regulatory decision on the "safety" of the compound will be based on "weight-of-the-evidence", and not solely on the result of the mesocosm study or laboratory studies. Beyond these three common themes, each of the papers evaluated different regulatory components of simulated field studies.

In the first paper of this section, Urban addressed the use of mesocosm studies in OPP/EPA's risk assessment process. A good overview was provided on historical development of aquatic field studies and their use under FIFRA. Over time, there has been a shift from "natural field studies" to mesocosms as the preferred test system for fulfilling the higher tier requirements. Although not discussed directly

0-87371-592-6/94/$0.00 + $.50
© 1994 by CRC Press, Inc.

by Urban, the primary reason for this shift appears to be the greater experimental control obtained in mesocosm studies vs. "natural field studies". The primary problem with "natural field studies" is variability among aquatic habitats and the resulting difficulty in obtaining statistically meaningful data for evaluating potential impacts. These factors are also discussed in Chapters 3, 5, and 30 of these proceedings (Brassard, 1992; Fischer, 1992; deNoyelles, 1992).

Urban also provided a good overview of the risk assessment process and how the results from both laboratory and field data fit within regulatory decision making. Laboratory data can be used to conclude that a product is "safe" and/or that specific label restrictions are required. However, if laboratory data indicates that there may be unacceptable risk associated with the use of the product, then "confirmatory" data are needed to refute this risk. The weight of all the evidence (i.e., laboratory and field data) is used in the final evaluation and, if the potential risk is not refuted, then Special Review is initiated. The Special Review process is used to help the Agency decide if a pesticide's registration should be canceled, denied, or reclassified based on the potential for adverse effects on aquatic organisms. It is during this process that the benefits associated with the use of the product are considered.

Endpoints were briefly discussed by Urban. He stated that during the Special Review process the interpretation of the mesocosm results "will generally focus on the adverse effects on the finfish populations". However, he also states that "there are growing number of circumstances where pesticide effects on lower trophic level aquatic organisms, their populations and communities may be considered in regulatory decisions". Unfortunately, no examples of the types of circumstances were provided. As discussed in other manuscripts, to maximize experimental design, it is important to identify endpoints of concern prior to test initiation.

The second paper of this section (Brassard et al.) primarily addressed the process for reviewing mesocosm studies submitted to the Agency in support of pesticide registration. The authors state that "mesocosm tests in support of pesticide registration is an evolving process" and essentially allude to the fact that methods and approaches need to be continually reevaluated and improved. I strongly support this statement and would like to emphasize the need for continued improvement and modification in test methods. Although a number of mesocosm studies have been performed, problems in methodology still exist (i.e., fish stocking rates and the assessment of fish reproductive performance). It should be realized that improvement does not necessarily equate with expansion of endpoints. Endpoints tested to date should be evaluated and as experience is gained in conducting these studies, efforts should be made to better focus the studies to obtain results most useful to the regulators.

Brassard et al. provide an overview of the major factors which are considered by the Agency when reviewing a mesocosm study. The final Agency review includes a detailed evaluation of the test system, pesticide dose levels and application procedures, residue analysis, biological assessments, quality assurance, and statistical analysis. The authors state that the "biological organisms to be measured (are) based on the ecological effects and environmental fate information'' and ''must be sufficient to negate the presumption of risk''. No attempt was made to address interpretational aspects of endpoints measured in mesocosm studies. Presently, the question ''what does it all mean'' represents the greatest unknown relative to the usefulness of these test systems in making regulatory decisions. As recently submitted studies are reviewed and used in regulatory decision making, a greater understanding of the interpretational aspects of the studies should emerge.

A particular topic which Brassard et al. discussed in considerable detail was the residue analysis required for mesocosm studies. Since the need for the extensive residue requirements outlined by the authors is debatable, it is appropriate to mention some concerns relative to escalations in residue requirements for these studies. As outlined by all three papers in this section, the primary objective of the study is to refute the presumption of risk identified in the laboratory based risk assessment. To meet this objective, measurement of the chemical exposure profile in the test system is useful in understanding the effects observed. However, these studies are not triggered based on environmental fate data and it is this editor's opinion that the objectives should not be expanded to include detailed mass balance information for the test compound.

Residue data from simulated field studies should be collected in support of the biological effects data. Data on the distribution of residues in the water column and/or sediment are only useful if the biological data are collected from the same spatial compartments. For example, Brassard et al. state that water column analysis should include various depths because this information is ''useful in measuring the rate of mixing and understanding reported effects on various organisms''. This may be the case if biological samples were collected as discrete samples at the same sample depths as the residue samples. However, water column biological samples are often collected as depth integrated samples. Depth-specific residue samples will not improve our ability to interpret effects on integrated biological samples. In addition, relative to the need to measure the mixing rate, it should be noted that the mixing rate within the water column is site specific (i.e., influenced by pond configuration and depth), time specific (wind direction relative to greatest fetch), and provides no predictive information. Such measurements are expensive, time consuming, and provide minimal information relative to the stated objectives of the study (i.e., refuting the presumption of risk).

Overall, given the expense and difficulty associated with residue analyses, it is important to be sure that the residue sampling regime fits within the objectives of the study. Extensive documentation of residue levels within a pond system may add little to interpretation of results relative to refuting the presumption of risk.

In the final paper of this section, Touart discussed ''Regulatory Endpoints and the Experimental Design of Aquatic Mesocosm Tests''. In this manuscript, the author identified two objectives of aquatic mesocosm tests. The first objective is to ''provide data to negate a presumption of serious risk indicated in the lower tier testing''. This objective was also identified by the other authors. The lower tier tests mentioned by the author consist of standard laboratory toxicity tests which, when compared to the "worst case" Estimated Environmental Concentration (EEC), indicate that adverse effects on aquatic organisms may occur during the normal use of the product. The second objective stated by Touart is "to estimate the intensity and duration of any adverse effects identified".

The two objectives identified by Touart are basically the same as those outlined in the Mesocosm Guidance Document (Touart, 1988). However, relative to the second objective, a difference exists in the definitions between the Touart manuscript and the Guidance Document. In the Guidance Document, the objective includes greater detail on the proposed use of the information generated in mesocosm studies, specifically: "it (mesocosm data) will provide risk managers descriptive information on the extent of adverse impacts, both in duration and magnitude, likely to occur in aquatic ecosystems which can then be evaluated in risk-benefit analyses". It is important to retain this as part of the second objective because it identifies risk managers as one of the primary "users" of the information being generated in mesocosm studies. The fact that the information generated in a mesocosm study will be used in risk-benefit analysis should be considered when designing the study. What type of information is most useful to risk managers? What are their "needs" for making risk-benefit decisions? In his manuscript, Touart alludes to the type of information needed, although the specific information required is not identified. As discussed in more detail below, it is important to design a study to not only "refute a presumption of risk", but also to provide risk managers with information that they need to make risk-benefit decisions for *all* use patterns and conditions.

Touart defines two types of endpoints in his manuscript: "observational endpoints and interpretational endpoints". Observational endpoints are those parameters measured directly in the test and interpretational endpoints represent the assessment, judgement, or explanation of the observational endpoint. The author further separates endpoints in three groups: ecological, socioeconomic, and regulatory. He states that regulatory endpoints should be both ecologically and socioeconomically relevant. The author discusses ecologically relevant endpoints in terms of ecosystem emergent properties; however, he never states exactly what should be measured in these test systems. He specifically states that "any observed effects which degrade or threaten to alter the ecosystem of an aquatic mesocosm should be considered a relevant observation endpoint". Obviously such a statement can lead to measurement of a vast array of both structural and functional endpoints. In the Summary Report on Issues in Ecological Risk Assessment (EPA, 1991), the authors conclude that "the identification of both assessment endpoints (impacts of interest) and measurement endpoints (parameters to be measured) is best done at the initial stages of a risk assessment". Although discussed in generalities, this still has not been clearly defined for mesocosm studies conducted for registration of pesticides under FIFRA.

Relative to the question of endpoints of ecological concern which should be measured in these test systems, there are still misunderstandings concerning the concept of endpoints and the need to strictly identify these prior to test initiation. Two points of view often expressed are provided below.

1. If the specific observational endpoints of regulatory concern are not clearly stated up front, how can an appropriate experiment be designed? For example, if it is known that a decline of specific invertebrate species is of regulatory concern, then, knowing the typical variation associated with the distribution of those species, an appropriate experimental design can be adopted to evaluate those endpoints or species. However, if certain species are not of regulatory consequence (i.e., because they are "minor" and/or redundant within the system), then this should be known before test initiation. Why design a study to have the statistical power to differentiate specific treatment related effects for a non-regulatory endpoint? At this point in time, regulatory endpoints are not clearly defined. This poses considerable difficulty in experimental design. Studies must become more focused and should not rely on "shot-gun ecology" in selection of measurement endpoints.

2. The risk-benefit analysis will ultimately determine whether a particular effect is of regulatory concern. Therefore, data must be collected to adequately evaluate the impact of the chemical on all components of the ecosystem. All components should be evaluated because only a limited number of species were tested in the laboratory. Additional effects may be identified in the mesocosm due to the large number of species present in the test systems and the potential for indirect effects due to predator/prey interactions and behavioral effects not measured in the laboratory. Laboratory toxicological data may help focus the efforts on endpoints most likely to be impacted; however, there should be some consistency in the endpoints measured among studies for different compounds so comparative assessments can be performed. Also, as stated by Touart "in effect, it is the end result of the test which determines the beginning of the test". This statement can be interpreted in a number of ways. Relative to risk-benefit analysis, this may mean that until you evaluate the benefits, it is difficult to establish which endpoints should be used for regulation. For a compound shown to have few benefits to agriculture or society, "minor" adverse effects may be unacceptable. However, for a compound shown to be "irreplaceable", greater potential for adverse effects or risks may be tolerated.

There are valid points raised in both positions outlined above and there is likely no right or wrong position. In designing a study, all factors listed above should be considered and the most appropriate endpoints and design developed. As alluded to by Brassard et al., the final overall study design is usually the result of negotiations between the Agency and registrant.

Touart outlines three categories of experimental design: (1) hypothesis tests, (2) point estimate tests, and (3) hybrid tests, which incorporate elements of hypothesis and point estimate tests. The basic rationale, strengths, and weaknesses for these different approaches are discussed. The overall conclusion provided is that the "hybrid design is generally preferred by the Agency to adequately address the regulatory endpoints". Although not specifically stated, we can assume the reason is that this design offers the most flexibility. It provides replication so that specific hypotheses can be tested and it provides multiple dose levels which enable the evaluation, in some manner, of the dose-response relationship for the test substance and the organism(s) of interest. Obviously, the disadvantage of this compromise is that the full advantages of either approach are not realized. Due to a limitation on the number of mesocosms which can be realistically tested, replication may not be sufficient to permit identification of effects on variable endpoints. Similarly, due to the varying sensitivity of organisms and the occurrence of nonlinear indirect effects, it is doubtful that the range of concentrations necessary to obtain solid dose-response information can be obtained with the limited number of experimental units available. As with most areas which deal with expensive studies conducted under variable field conditions, a compromise is required. The final experimental design should be determined by the specific objectives of the study and endpoints of interest. This is why it is extremely important to identify, if possible, the endpoints of concern prior to test initiation.

Relative to the most appropriate experimental design, the second study objective outlined in the Guidance Document should be remembered. It is important to provide results for a variety of potential exposure scenarios so risk managers are not restricted in their ability to evaluate the potential impacts under different use scenarios. Studies should not be designed to "refute the presumption of risk" at a single exposure level. Rather, the risk manager needs information on the potential for effects at a variety of exposure levels. This will provide the risk manager the data necessary to determine how changes in the exposure (i.e., via exposure reduction programs or alternative use patterns) will influence the potential for unacceptable adverse effects.

The final two chapters in this section provide specific perspectives on the importance of simulated (i.e., mesocosm) and "natural" field studies. The authors concur that the results obtained from ecosystem level tests can provide a considerable amount of information on both direct and indirect effects of a stressor. Although the authors addressed the underlying theme and importance of ecosystem level testing, they did not specifically address the regulatory interpretation of results from field studies. However, this is covered in greater detail in other chapters in this book. The "what does it all mean" question has been difficult to answer for single species tests because of the unknown relationship between the "unnatural" laboratory exposure scenario and the real world. In large scale ecosystem tests such as mesocosms, "reality" is not the difficult question, but rather the ability to separate acceptable or minor alterations from background variability and to identify impacts which truly alter the functioning of the ecosystem.

In Chapter 5, Fischer discussed the two options available for fulfilling the final tier of the pesticide hazard assessment process. As currently written, when ecosystem level testing is required for pesticide registration or reregistration, FIFRA guidelines allow the conduct of either an actual field study or a simulated field study. Fischer discusses the advantages and disadvantages of each of these study designs. The main advantages of the actual field study is reality. The entry of the chemical into the aquatic habitat occurs via natural processes. The ecosystem is generally well established and responds and adapts to the chemical insult as a natural ecosystem. In mesocosm studies, treatment of ponds may not be representative of natural exposure scenarios and the systems are in early stages of successional development. However, simulated systems can be replicated and controlled, allowing greater power and ability to identify treatment-related effects. Neither approach is perfect. The strengths and weaknesses of the different approaches need to be considered relative to the objectives of the study. Only after study objectives are identified can test system advantages and disadvantages be considered and the most appropriate approach be determined.

After discussing the different approaches, Fischer reviews an actual field study performed for endosulfan. He uses the results to emphasize one of the primary weaknesses of actual field studies, which is interpretation of the overall results. The lack of a control and variability within and between systems requires that the results be interpreted with a careful eye on cause and effect and a thorough understanding of the biology of the organisms involved. Interpretation of such studies can become extremely subjective and are often open to considerable debate. This is a major reason why the EPA is currently recommending mesocosm studies, where replication and statistical power can help prevent the occurrence of such debates. Overall, an actual field study is most useful in assessing the exposure under real world conditions; the mesocosm most useful in interpretation of cause-effect relationships.

In the last chapter of this section, Clark and Noles provide an overview of actual and simulated marine/estuarine field studies. Currently, the EPA is only requiring freshwater field testing for pesticide registration. The assumption is that a lentic ecosystem, as modeled in the current mesocosm design, represents a reasonable worst case test system and decisions made based on such studies should be protective of marine environments. Based on the discussions provided by Clark and Noles, there is little empirical data to support such an assumption. Species sensitivities may be similar; however, there is not enough information to conclude that indirect effects observed in freshwater systems will be reflective of effects likely to occur in marine systems; in fact, there will likely be differences. At higher levels of organization (community and ecosystem level), extrapolation among systems may be easier due to the functional redundancy imparted by the individuals comprising aquatic communities. Another uncertainty is the differing exposure regimes in marine vs. freshwater environments. Given a similar exposure regime, effects on the structure or function of marine and freshwater systems may be similar; however, tidal fluxes and dilution effects make the exposure regimes in marine/estuarine systems considerably different.

Clark also discussed the various simulated test systems and/or experimental designs which have been used and are available for testing the effects of pesticides on marine/estuarine ecosystems. Unfortunately, no consensus has been reached on the suitability or preference for simulated marine test systems. The difficulty in working in marine habitats due to tidal flow, etc. and the expense associated with construction and operation of such test systems have limited the number of test systems used. Clark and Noles conclude that ''before extensive testing for regulatory assessments is begun in marine mesocosms, some scientific evaluation of the merits of various approaches and test systems must be completed''.

REFERENCE

Touart, L.W. 1988. Hazard Evaluation Division, Technical Guidance Document: Aquatic Mesocosm Tests to Support Pesticide Registrations. EPA-540/0988-035. U.S. Environmental Protection Agency, Washington, D.C.

Design of Mesocosm Studies and Statistical Analysis of Data

INTRODUCTION

The following five chapters present a spectrum of approaches for evaluation of data collected in aquatic mesocosm studies for ecological risk assessment of pesticides. You may wish to recall that a mesocosm study is usually undertaken (1) to negate a presumption of adverse ecological effects or unreasonable risks based upon laboratory derived information, or (2) to define the intensity and duration of adverse effects of pesticides in aquatic systems. These objectives may impose special constraints upon the chosen experimental design (e.g., allocation of replicates and the number and concentrations of treatments) since "proof" of safety may be much more difficult than "proof" of adverse effects (Brass, 1985). It is also important to remember that statistical testing can only identify *statistically* significant or nonsignificant differences or relationships. We are still left with considerable interpretation after the statistical drums have sounded.

Regal and Lozano (Chapter 12) emphasize the importance of clearly identifying the goals as mesocosm study in order to develop the optimal experimental design. Stunkard (Chapter 8) proposed that multiple treatment mesocosm study designs with minimal replication would not permit rejection of the null hypothesis of no treatment effects with acceptable frequency, even when large treatment effects exist. He presents an alternative design that involves a two-sample *t*-test of a null hypothesis involving proportional means. This approach relies on a biologist or exo-toxicologist to identify population mean disparities that are considered to have substantive meaning or ecological significance relative to adverse effects. Shaw et al. (Chapter 9) also emphasize the importance of designating ecologically important effects *a priori* for proper study design. They propose ANOVA with hypothesis testing as the preferred approach for statistical evaluation of data from mesocosm studies. In an alternative approach, Thompson et al. (Chapter 11) offer a somewhat novel design for mesocosm studies based on experience with limnocorrals and concentration-response theory. Nonlinear modeling provided insight that may be useful for regulatory decisions in ecological risk assessment. To provide some guidance for future studies, Christman et al. (Chapter 10) evaluated colonization of phytoplankton, zooplankton, and macroinvertebrate assemblages in new, un-managed and untreated pond mesocosms. Based on the precision of various parameters, they found that some measures would be more useful than others for detection of treatment effects. One could also extract the magnitude of treatment effects that would be necessary to discern statistical differences based on the variance of these parameters in this scenario.

LITERATURE CITED

Brass, *Biometrics,* 41, 785, 1985.
Midlard, *Biometrics,* 43, 719, 1987.

CHAPTER 8

Tests of Proportional Means for Mesocosm Studies

Clayton L. Stunkard

Abstract: The genesis and reasons for testing hypotheses of proportional means rather than of equal means in aquatic mesocosm studies are presented in this paper. Symbolically, the hypotheses of proportional and equal means may be expressed, respectively, as $\mu_T = b\mu_C$ and $\mu_T = \mu_C$, with μ_T and μ_C standing for the treatment and control means. Special consideration is given to the minimum sample sizes permitted for such studies in the U.S. EPA's technical guidance document *Aquatic Mesocosm Tests to Support Pesticide Registrations*. A measure of effect size is defined as a function of b, the coefficient of proportionality, and the within treatments standard deviation. Characteristics of theoretical sampling distributions are summarized for two designs with multiple treatments and for a two-treatment design, all with a minimal total sample size and each with tests for equality of means. Characteristics of empirical sampling distributions are summarized for the two treatment design with tests for proportional means.

INTRODUCTION

The U.S. Environmental Protection Agency recently issued a document[1] setting forth guidance for the conduct of mesocosm studies. The document indicated that an acceptable design could consist of "a minimum of four (4) experimental treatments", including a control, with "at least three replicates per treatment". Since

0-87371-592-6/94/$0.00 + $.50

71

the document did not specify any statistical hypotheses to be tested, an omnibus F-test of a null hypothesis of equality of all four treatment means would be considered by some persons as an appropriate first test. This would then be followed by pairwise comparisons of the control with other treatment means only when the omnibus test is rejected. Since the guidance document indicated that an aquatic mesocosm test should "provide a pesticide registrant supportable means for negating presumptions of unacceptable risks to aquatic organisms for their product", such an approach is inappropriate.

The procedure employing the omnibus F-test and pairwise comparisons is not consistent with the objective of the guidance document, because *failure to reject the null hypothesis should not be considered as support* for its assertion. Nonrejection of a null hypothesis is a very common and expected result of inadequate replication, and certainly cannot "provide a pesticide registrant supportable means for negating presumptions of unacceptable risks to aquatic organisms for their product".

This paper provides documentation of the inability of multiple treatment designs with minimal replication to allow rejection of the omnibus F-test with acceptable frequency even when large effects exist. This is shown to be the case for conventional levels of significance, as well as for values which are much larger. Utilization of a design consisting of only six replicates in each of two treatments, one of which is the control, with a directional *t*-test for equality of means at conventional levels of significance also does not provide an acceptable frequency of rejection of the null hypothesis except for what may best be described as "gross" differences.

An approach consistent with the objective of the guidance document, but employing the minimum total number of replicates, is proposed. The approach equates rejection of a null hypothesis with negation of a presumption of an "unacceptable adverse effect". It requires rejection of a null hypothesis that asserts that the difference in treatment and control means is at least as great as some specified proportion of the control mean, which may be considered as a critical value for classifying an adverse effect of a toxic substance on the environment as "acceptable" or "unacceptable".

MINIMAL "ACCEPTABLE" DESIGNS

This section examines characteristics of 3 basic designs, each employing the minimum total of 12 replicates. First, a design consisting of four treatments, each with the minimal number of replicates of three. Next, a design having three treatments, each with four replicates, and lastly, a design with only two treatments, each having six replicates. These three basic designs will be labeled as the "k = 4/n = 3," "k = 3/n = 4," and "k = 2/n = 6" designs, respectively. Versions with maximum and minimum variation among means will be considered for the first two basic designs, and will be referred to as MAX and MIN models.

Notation

First, we symbolize some relevant quantities for the general case of $k \geq 2$:

k the number of populations (treatments)

μ_h the mean of the hth population; $h = 1, \ldots, k$

μ the mean of the combined populations

σ_W^2 the common variance within each population (within treatments variance)

σ_B^2 the variance of the population means (between treatments variance)

σ_T^2 the total variance (variance of measures with respect to μ)

Note: $\sigma_T^2 = \sigma_W^2 + \sigma_B^2$. (See Technical Note 1.)

The magnitude of differences among the population means most generally can be expressed by the correlation ratio, also known as eta-square, which is defined as

$$\eta^2 = \frac{\sigma_B^2}{\sigma_T^2} \tag{1}$$

Eta-square is restricted to values between zero and unity, inclusive. When all population means are equal, then $\eta^2 = 0$. When the means are unequal, but $\sigma_W^2 = 0$, then $\eta^2 = 1$. It is known that for k populations, for which the means fall within a fixed range for the minimum and maximum values, that σ_B^2 is a maximum when half of the means have the minimum value and half the maximum value. The minimum for σ_B^2 is attained when exactly one mean has the minimum value, exactly one mean has the maximum value, and the remaining means have a common value at the average of these extremes. Any other arrangement results in a value for σ_B^2 intermediate to those specified for the minimum and maximum.

In all models, μ_1 anchors a position at one end of the range, and μ_k a position at the other end of the range. The value of μ_1 is taken as a proportion (b) of μ_k, i.e., $\mu_1 = b\mu_k$, so that the range is fixed as $|1 - b|\mu_k$. Table 1 gives the location of all means, including that of the combined populations, in relation to the value anchored by μ_k for the $k = 4/n = 3$ models, and Table 2 gives locations of means for the $k = 3/n = 4$ models. For the $k = 2/n = 6$ design, there is only one model — that of $\mu_1 = b\mu_2$, with the mean of the combined populations as $\mu = [(1+b)/2]\mu_2$. For example, if μ_k is fixed at a value of 100 and b is selected as either 0.8 or 1.2, then μ_1 is either 80 or 120 and the range of means is fixed at 20.

Before further specification can be made for the models, we need to decide what values of b can be considered as potentially meaningful in establishing the range of means given by $|1 - b|\mu_k$. It is helpful to digress momentarily by defining a concept known as *effect size*.

Table 1. Algebraic Location of Population Means For k=4/n=3 Models

Model	μ_1	μ_2	μ_3	μ_4	μ
MAX	$b\mu_4$	$b\mu_4$	μ_4	μ_4	$\dfrac{1+b}{2}\mu_4$
MIN	$b\mu_4$	$\dfrac{1+b}{2}\mu_4$	$\dfrac{1+b}{2}\mu_4$	μ_4	$\dfrac{1+b}{2}\mu_4$

Effect Size

Cohen[2] has defined several different measures which are related to eta-square. For the two-population problem, he standardizes the mean difference as

$$ES = \frac{\mu_1 - \mu_2}{\sigma_w} \tag{2}$$

assuming equal variances, and labels three classes of effect size. He considers an absolute value of 0.2 as "small", that of 0.5 as "medium", and that of 0.8 as "large". For the $k > 2$ population problem, Cohen[2] defines effect size as

$$f = \frac{\sigma_B}{\sigma_w} \tag{3}$$

and assigns adjectives of "small", "medium", and "large" to values of f to be consistent with those assigned to ES for the two-population problem. When $k = 2$, $|ES| = 2f$.

Cohen's three classes of effect size are listed below with their equivalent values of eta and eta-square:

Class	ES	f	η	η^2
"Small"	0.2	0.10	0.0995	0.0099
"Medium"	0.5	0.25	0.2425	0.0588
"Large"	0.8	0.40	0.3714	0.1379

In summary, Cohen considers "small", "medium", and "large" effect sizes among means as those accounting for 1, 6, and 14 percent of the total variance, respectively. (See Technical Notes 2 and 3.)

For the two population situation with means whose separation is equivalent to an effect size of unity, approximately 16 percent of the members of the population with the smaller mean (say, population 1) have values greater than the mean for the other population (population 2); and 16 percent of the members of the second population have values less than the mean of the first population.

Suppose that a random sample of size n is selected from two normal populations with unknown means and variances. Also suppose the unknown variances are equal in the two populations. Designate the populations a Population 1 and Population

Table 2. Algebraic Location of Population Means For k=3/n=4 Models

Model	μ_1	μ_2	μ_3	μ
MAX	$b\mu_3$	$b\mu_3$	μ_3	$\dfrac{1+3b}{3}\mu_3$
MIN	$b\mu_3$	$\dfrac{1+b}{2}\mu_3$	μ_3	$\dfrac{1+b}{2}\mu_3$

2, and the means of their respective samples as Mean 1 and Mean 2. The probability that Mean 1 is less than Mean 2 may be determined by reference to normal distribution theory, provided the disparity of population means is expressed as a measure of effect size as defined by Equation 2, above. (See Technical Note 4.) This probability is 0.50 if the two population means are equal, the condition for which the effect size (ES) is zero. As the population means become disparate, the probability converges toward either zero or unity, depending upon the direction of the difference of the two population means and their relative sizes. Similar, but complementary, conditions exist for the probability that Mean 1 is greater than Mean 2.

Table 3 displays the probability that Mean 1 is less than Mean 2 for disparities in population means expressed as effect sizes ranging from -0.25 to -2.00 in increments of -0.25, and for sample sizes from 1 to 6 in unit increments. Values for the sample size of 1 are representative of random pairings of single elements from each of the two populations. It is noteworthy that the probability is slightly greater than the midpoint of the probability scale of observable ESs less than zero for an effect size of -1.00. For this reason, effect sizes with absolute value equal to or greater than unity are considered hereafter to reflect "large" disparities between two population means.

We shall examine our models for each of five values for the range of means, by considering the two-population measure of effect size as equal to the standardized difference between the maximum and minimum mean values. Specifically, we shall use the following absolute effect size values: 0.50, 0.75, 1.00, 1.25, and 1.50.

In order to calculate the theoretical power of the omnibus F-test of the null hypothesis,

$$H_0 = \mu_1 = \mu_2 \ldots = \mu_k \tag{4}$$

Table 3. Probabilities Associated With Effect and Sample Sizes

Effect Size	Sample Size (n)					
	1	2	3	4	5	6
-0.25	.570	.599	.620	.638	.654	.667
-0.50	.638	.691	.730	.760	.785	.807
-0.75	.702	.773	.821	.856	.882	.903
-1.00	.760	.841	.890	.921	.943	.958
-1.25	.812	.894	.937	.961	.976	.985
-1.50	.856	.933	.967	.983	.991	.995
-1.75	.892	.960	.984	.993	.997	.999
-2.00	.921	.977	.993	.998	.999	***

Note: *** indicates a value greater than .999.

Table 4. Theoretical Power of Tests For the $k=4/n=3$ Models With Equal Variances

Model	|ES|	λ	.01	.05	.10	.15	.20	.25
					Level of Significance (α)			
MAX	0.50	0.750	.018	.080	.149	.213	.274	.332
	0.75	1.688	.030	.122	.213	.292	.362	.426
	1.00	3.000	.052	.185	.303	.396	.474	.541
	1.25	4.688	.086	.271	.415	.518	.598	.662
	1.50	6.750	.135	.377	.538	.642	.717	.773
MIN	0.50	0.375	.014	.065	.124	.182	.237	.291
	0.75	0.844	.019	.084	.156	.221	.283	.341
	1.00	1.500	.028	.113	.200	.276	.345	.408
	1.25	2.344	.040	.153	.258	.345	.420	.486
	1.50	3.375	.059	.204	.329	.425	.504	.571

when each of our models hold, it is necessary to also specify the level of significance (α) and to determine the noncentrality parameter (λ) associated with the appropriate noncentral F-distribution.[3-5] Since a directional test is desired for the $k=2/n=6$ design, it is necessary to determine the noncentrality parameter (δ) for the appropriate noncentral t-distribution with 10 degrees of freedom. We shall consider each of six levels of significance, namely: 0.01, 0.05, 0.10, 0.15, 0.20, and 0.25. (See Technical Note 5.)

Because of the assumption of normality, it is necessary to restrict the within populations variation so that the smallest mean is at least two and one-half standard deviations above the zero of the scale of measures, assuming that the ratio scale of measurement[6] is involved. An apparent anomaly results in the fact that the coefficient of variation for populations with larger means must therefore be less than that of the population with the smallest mean. This restriction is equivalent to restricting the common standard deviation to be less than or equal to 40% of the smallest mean for the k populations. In other words, the coefficient of variation of measures in the population with the smallest mean is restricted to a value less than or equal to 0.40. (See Technical Note 6.)

Table 4 presents the theoretical power values for the various conditions associated with the basic $k=4/n=3$ models. These values correspond to the proportion of times the null hypothesis of equal means is rejected when in fact the model holds. The models here assume the populations means specified in Table 1, and a common variance within each population. If one considers a power of 0.80 or better as acceptable, then none of the models attains this criterion even with a level of significance of 0.25 for the maximum range of means associated with the largest ES considered. Clearly, the four treatment/three replicates design must be considered inadequate from this viewpoint. For the ES value of ±1.00, the MAX model has a power of 0.185 with an alpha of 0.05, and a power of 0.541 with an alpha of 0.25. For the same ES value, the MIN model has power values of 0.113 and 0.408 with alpha values of 0.05 and 0.25, respectively.

Table 5. Theoretical Power of Tests For the k = 3/n = 4 Models With Equal Variances

Model	|ES|	λ	.01	.05	.10	.15	.20	.25
			Level of Significance (α)					
MAX	0.50	0.667	.021	.088	.161	.226	.287	.345
	0.75	1.500	.038	.140	.238	.318	.388	.451
	1.00	2.667	.067	.218	.342	.436	.512	.575
	1.25	4.167	.112	.321	.467	.566	.641	.700
	1.50	6.000	.178	.441	.598	.693	.759	.808
MIN	0.50	0.500	.018	.078	.146	.208	.266	.322
	0.75	1.125	.030	.117	.203	.277	.344	.405
	1.00	2.000	.049	.173	.283	.370	.444	.507
	1.25	3.125	.080	.249	.382	.478	.555	.618
	1.50	4.500	.124	.343	.492	.592	.666	.723

Table 5 presents theoretical power values for the various conditions associated with the $k = 3/n = 4$ models. Again, these values represent the proportion of times the null hypothesis of equality of means is rejected when, in fact, the model holds. The models here assume that the populations means are as specified in Table 2, and that there is a common variance within each population. The power values here are somewhat larger than those for the $k = 4/n = 3$ models, but can hardly be considered as acceptable, except for the largest ES of ± 1.50 for the MAX model with an alpha of 0.25. For the ES value of ± 1.00, the MAX model has a power of 0.218 with an alpha of 0.05, and a power of 0.575 with an alpha of 0.25. For the same ES value, the MIN model has power values of 0.173 and 0.507 with alpha values of 0.05 and 0.25, respectively.

The theoretical power values for directional tests of the equal means null hypothesis for the $k = 2/n = 6$ design are displayed in Table 6. These values are the proportion of times the null hypothesis is rejected when, in fact, the model holds. The model here assumes that the mean of Population 1 is a proportion (b) of the mean of Population 2, and that there is a common variance within each population. An acceptable power may be considered as attained for directional tests for an ES value of ± 1.00 with an alpha of 0.20 or higher. For this ES value, a power of 0.488 is observed with an alpha of 0.05. Acceptable power values are also attained for an ES of ± 1.25 with a level significance of 0.15 or higher, and for an ES of ± 1.50 with a level of significance of 0.10 or higher.

Table 6. Theoretical Power of Directional Tests For the k = 2/n = 6 Design With Equal Variances

|ES|	|δ|	.01	.05	.10	.15	.20	.25
		Level of Significance (α)					
0.50	0.866	.058	.201	.326	.423	.504	.572
0.75	1.299	.116	.332	.485	.590	.668	.729
1.00	1.732	.206	.488	.647	.741	.805	.850
1.25	2.165	.327	.645	.786	.857	.900	.929
1.50	2.598	.469	.780	.886	.932	.956	.971

AN ENVIRONMENTAL HYPOTHESIS

Suppose that Populations 1 and 2 correspond to sets of measurements following treatment with and without a pesticide, respectively. The usual simple null hypothesis is H_0: $\mu_1 = \mu_2$. Directional or nondirectional tests of this null hypothesis are not logically consistent with the EPA's goal for aquatic mesocosm studies — namely, negation of a presumption of *unacceptable* adverse risks.

For example, let a directional test of this standard null hypothesis be performed with the alternative as $\mu_1 < \mu_2$, which is assumed to be in the direction of adverse effects. If the alternative is true, an effect of -1.00 has a low power of being detected with a sample size of six (6), except for an alpha of 0.20 or larger. (See Table 6.) The power with a level of significance of 0.20 could be considered a fair result, but the procedure is inconsistent with the regulatory objective because it corresponds to "proving the existence of an adverse effect" rather than negation of a presumption of such an effect.

Next, let a directional test of the standard null hypothesis be performed with the alternative as $\mu_1 > \mu_2$, but assume that the true state of affairs is, again, $\mu_1 < \mu_2$ and in the direction of adverse effects. Here, rejection of the null hypothesis is consistent with the negation of a presumption of *"no-effect"* rather than negation of a presumption of an *"unacceptable adverse effect"*.

A procedure, which resolves the logical inconsistency of employing the above hypothesis, is to test a null hypothesis which is consonant with a presumption of an *unacceptable* adverse effect when a pesticide is applied. A threshold for an *unacceptable* adverse ecological effect may be defined in terms of the difference between treatment and control means as a proportion of the control mean. For some variables, the difference may be defined in a negative direction, with the control mean larger than the treatment mean. Another variable may have the difference defined in a positive direction, with the control mean smaller than the treatment mean. We assume that μ_1 and μ_2 are both nonzero quantities. Any observed difference equal to or larger than the defined difference and in the direction consistent with the definition shall be considered as an outcome expected when the null hypothesis holds. The simple form of the null hypothesis may be written as

$$H_0: \mu_1 = b\mu_2 \tag{5}$$

Here b must be a specified positive scalar value. In effect, the hypothesis states that for b less than unity ($b < 1$), the mean of Population 1 is less than that of Population 2 ($\mu_1 < \mu_2$), and for b greater than unity ($b > 1$), the mean of Population 1 is greater than that of Population 2 ($\mu_1 > \mu_2$). For $b = 1$, the hypothesis is readily recognized as the standard hypothesis of equal means. The complex form depends upon whether the proportionality constant b is less than or greater than unity. These two cases are listed below with the appropriate forms for the alternative hypotheses:

Case 1: b < 1	Case 2: b > 1
$H_0: \mu_1 \leq b\mu_2$	$H_0: \mu_1 \geq b\mu_2$
$H_1: \mu_1 > b\mu_2$	$H_1: \mu_1 < b\mu_2$

It is of particular interest to note that $\mu_1 = \mu_2$ is an alternative to the complex null hypothesis of both Case 1 and Case 2.

The Test Statistic

For pairs of samples from the two populations, we define a statistic

$$t = \frac{\bar{y}_1 - b\bar{y}_2}{s \sqrt{\dfrac{1}{n_1} + \dfrac{1}{n_2}}} \tag{6}$$

with

$$s = \sqrt{\frac{(n_1 - 1)\, s_1^2 + (n_2 - 1)\, b^2 s_2^2}{n_1 + n_2 - 2}} \tag{7}$$

(See Technical Notes 7 and 8.)

When the null hypothesis specified by Equation 5 is true, the statistic given by Equation 6 is distributed as "Student's" t with $v = n_1 + n_2 - 2$ degrees of freedom, provided that the measures are distributed normally in each population with the following assumption regarding equality of variances:

$$\sigma_1^2 = b^2 \sigma_2^2 \tag{8}$$

This test statistic is algebraically equivalent to the standard two-independent samples t-ratio for testing equality of means performed after transforming the measures by multiplying each value in sample 2 by b. The transformed measures are assumed to have equal variances within their respective populations.

Decision Rules

Employment of the test statistic defined by Equation 6 requires the following decision rules for the two composite hypotheses in which the level of significance is symbolized by the Greek lower-case alpha (α); $t_{\alpha,v}$ refers to the α-percentile (left tail), and $t_{1-\alpha,v}$ refers to the $(1 - \alpha)$-percentile of the t-distribution with $v = n_1 + n_2 - 2$ degrees of freedom:

Case 1: b < 1. Reject H_0 if $t \geq t_{1-\alpha,v}$, else retain H_0 if $t < t_{1-\alpha,v}$.

Case 2: b > 1. Reject H_0 if $t \leq t_{\alpha,v}$, else retain H_0 if $t > t_{\alpha,v}$.

Table 7. Power of Directional Tests of Proportional Means at Two Effect Sizes, Two Coefficients of Variation, Four Sample Sizes, and Six Levels of Significance

				Level of Significance (α)					
ES	V	b	n	.01	.05	.10	.15	.20	.25
− 1.00	.2	0.8	6	.250	.544	.700	.781	.835	.868
			7	.310	.607	.757	.826	.876	.905
			8	.364	.673	.797	.852	.889	.922
			9	.444	.733	.852	.909	.933	.949
	.4	0.6	6	.313	.615	.756	.831	.877	.914
			7	.386	.681	.807	.867	.912	.941
			8	.444	.742	.843	.894	.920	.948
			9	.513	.809	.894	.940	.954	.965
1.00	.2	1.2	6	.185	.435	.581	.678	.756	.811
			7	.226	.474	.632	.728	.792	.843
			8	.268	.526	.689	.791	.842	.883
			9	.288	.551	.714	.806	.863	.905
	.4	1.4	6	.159	.392	.529	.632	.704	.770
			7	.189	.419	.571	.674	.751	.807
			8	.231	.469	.634	.728	.803	.846
			9	.237	.487	.644	.752	.809	.863

Note: ES is the two-population measure of effect size. V is the coefficient of variation. b is the coefficient of proportionality of means.

Power Characteristics

Table 7 presents empirical power values for the alternative $\mu_1 = \mu_2$, when testing the hypothesis given by Equation 5 for the two-population effect size of − 1.00 produced by combinations of proportionality constants of 0.8 and 0.6 with coefficients of variation of 0.2 and 0.4, and for an effect size of 1.00 produced by combinations of proportionality constants of 1.2 and 1.4 with the same coefficients of variation. Because of the assumption of equal coefficients of variation, a redefinition of effect size was necessitated. Here, the denominator of Equation 2, the common within populations standard deviation, is replaced by the standard deviation of population 2. Power values are given for sample sizes of six (6) through nine (9), inclusive. Values were obtained from simulations based on 1000 sample sets for each sample size. The sets used assumed equal means and variances for the two populations sampled. Four simulations were required. Hence these results represent the proportion of time the null hypothesis of specified proportional means is rejected with the accompanying coefficient of variation. An acceptable power value of 0.80 or larger is obtained with an alpha of 0.05 for only one of the 16 results at this significance level — that for a sample size of nine (9). All illustrated sample sizes have acceptable power values for the negative effect size with an alpha of 0.20 or larger. Sample sizes of eight (8) and nine (9) provide acceptable power values for the positive effect size with an alpha of 0.20 or larger.

DISCUSSION

This paper has documented inadequacies of employing minimal replication in multiple treatment mesocosm designs. Minimal replication is defined as a total of 12 replicates in all treatments, the number indicated as acceptable by the guidance document[1] for the conduct of mesocosm studies.

An examination of the nature of mean differences for random pairing of samples of size 1 from two normal populations lead to a labeling of a "large" effect in this paper. A disparity between two population means as great as or greater than one standard deviation was considered to operationally define the meaning for such a label.

Samples of size three (3) in the $k = 4/n = 3$ design, and those of size four (4) in the $k = 3/n = 4$ design were shown to provide unacceptable power for rejection of the omnibus F-test when large effect sizes exist, i.e., when there is an effect size greater than or equal to $|\pm 1.00|$. The reader should note that the MIN model for the $k = 3/n = 4$ design and a model intermediate to the MAX and MIN models for the $k = 4/n = 3$ design have means monotonically ordered with respect to the treatment variable. Such an arrangement is consonant with a design for which a regression explanation may be sought.

More important, the procedure involving the omnibus F-test may be considered to be flawed in that it is logically inconsistent with the regulatory objective of mesocosm studies — to "provide a pesticide registrant supportable means for negating a presumption of unacceptable risks".

Examination of the $k = 2/n = 6$ design for tests of the standard null hypothesis of equal means indicated unacceptable power for traditional levels of significance. This design attains marginally acceptable power to detect an effect size in the neighborhood of ± 1.00 with a level of significance of 0.20. Directional alternatives for a null hypothesis of two equal means were examined with respect to their implications when performing a t-test. Alternatives in either direction were found wanting with respect to the objective, mentioned above, for aquatic mesocosm studies.

A two-sample t-test of a null hypothesis of proportional means was presented. Equal coefficients of variation are assumed for the test. The hypothesis testing procedure for this test is consistent with the regulatory objective discussed earlier in this section. Power characteristics of the procedure may be considered as marginally acceptable for sample sizes of six (6) with a level of significance of 0.20.

The value of b, and its associated effect size, must in all circumstances be determined by the biologist rather than by the statistician. This decision needs to reflect what population mean disparities are considered by the biologist to have substantive meaning with respect to environmental impacts. The variables involved should require different b values, depending upon their importance within the aquatic environment. Selection of the level of significance is also a decision for the biologist to make. Adequacy of the design of a mesocosm study *cannot* definitively be assessed before the biologist has made such decisions.

REFERENCES

1. Touart, L.W. 1988. *Aquatic Mesocosm Tests to Support Pesticide Registrations.* Hazard Evaluation Division Technical Guidance Document, Office of Pesticide Programs, U.S. Environmental Protection Agency, EPA 540/09-88-035. National Technical Information Service, Springfield, VA.
2. Cohen, Jacob. 1988. *Statistical Power Analysis for the Behavioral Sciences* (2nd Ed.). Laurence Erlbaum, Hillsdale, NJ.
3. Hogg, R.V. and Craig, A.T. 1978. *Introduction to Mathematical Statistics* (4th Ed.). Macmillan, New York.
4. Johnson, N.L. and Kotz, S. 1970. *Distributions in Statistics: Continuous Univariate Distributions — 2.* Wiley, New York.
5. Kendall, M.G. and Stuart, A. 1979. *The Advanced Theory of Statistics: Volume 2, Inference and Relationship* (4th Ed.). Macmillan, New York.
6. Stevens, S.S. 1946. On the Theory of Scales of Measurement. *Science* 103:677–680.

TECHNICAL NOTES

Note 1. $\sigma_B^2 = \dfrac{1}{k} \sum\limits_{h=1}^{k} \tau_h^2$, with $\tau_h = \mu_h - \mu$, and $\mu = \dfrac{1}{k} \sum\limits_{h=1}^{k} \mu_h$.

Note 2. The two population measure of effect size may be translated to values of eta-square (or eta) through the relationship

$$\eta^2 = \frac{(ES)^2}{4 + (ES)^2} \tag{N2.1}$$

Note 3. The k populations measure of effect size may be translated to values of eta-square (or eta) through the relationship

$$\eta^2 = \frac{f^2}{1 + f^2} \tag{N3.1}$$

Note 4. A value of zero for the difference between sample means (Mean 1 − Mean 2) equates to a unit normal deviate of

$$-\frac{(\mu_1 - \mu_2)}{\sigma_w \sqrt{\dfrac{2}{n}}} = -\frac{ES\sqrt{n}}{\sqrt{2}} \tag{N4.1}$$

Hence, the probability that Mean 1 is less than Mean 2 is the cumulative probability associated with this quantity.

Note 5. The noncentrality parameter for each of the models of the $k = 4/n = 3$ and $k = 3/n = 4$ designs for the different range values may be determined by

$$\lambda = \frac{kn\sigma_B^2}{\sigma_W^2} \tag{N5.1}$$

or

$$\lambda = knf^2 \tag{N5.2}$$

Since we are interested in minimal number of replicates designs, our values for the degrees of freedom for the appropriate noncentral F-distributions are 3 and 8, 2 and 9, and 1 and 10, respectively, for the $k = 4/n = 3$, $k = 3/n = 4$, and $k = 2/n = 6$ designs. Again, eta and eta-square are related to the noncentrality parameter by

$$\eta^2 = \frac{\lambda}{kn + \lambda} \tag{N5.3}$$

The noncentrality parameter (δ) associated with directional tests for the $k = 2/n = 6$ design has a value equal to the appropriate square root of lambda (λ).

Note 6. For ES < 0 ($b < 1$), $b = (1 + |ES|V_1)^{-1}$, and for ES > 0 ($b > 1$), $b = 1 + (ES)(V_k)$.

Note 7. The sample means and variances are defined below, with n_h being the sample size associated with the hth population for $h = 1$ and 2:

$$\bar{y}_h = \frac{\sum_{i=1}^{n_h} y_{hi}}{n_h} \tag{N7.1}$$

and

$$s_h^2 = \frac{\sum_{i=1}^{n_h} (y_{hi} - \bar{y}_h)^2}{n_h - 1} \tag{N7.2}$$

Note 8. If equal size samples are employed, with $n = n_1 = n_2$, we have

$$s = \sqrt{(s_1^2 + b^2 s_2^2)/2} \tag{N8.1}$$

and the test statistic becomes

$$t = \frac{\bar{y}_1 - b\bar{y}_2}{\sqrt{(s_1^2 + b^2 s_2^2)/n}} \tag{N8.2}$$

CHAPTER 9

Design and Statistical Analysis of Field Aquatic Mesocosm Studies

Jennifer L. Shaw, Malcolm Moore, James H. Kennedy, and Ian R. Hill

Abstract: Over the past five years many field aquatic mesocosm studies have been conducted or are under way to meet regulatory requirements of the U.S. Environmental Protection Agency (EPA). Mesocosm studies as an ecosystem-level test are still developing, and, consequently, considerable variations are evident in study design and data analyses. In addition, experimental design often does not properly address crucial points at the study outset, e.g., defining estimated entry rates (EER) for spray drift and/or runoff into aquatic environments and what are the ecologically important effects. Study endpoints need to be carefully selected relative to the test chemical, to avoid pointless and tedious measurements of a generic list of parameters that are not pertinent to the study objectives. This paper delineates criteria to be considered in mesocosm study design and then discusses the uses of one-way analysis of variance (ANOVA), where specific hypotheses are tested, and non-hypothesis testing regression methods. Conventionally in hypothesis testing, data have been tested against the null hypothesis of "no difference between treatment and control means" using Student's t-test. A new method has been proposed by the EPA that reverses this approach and tests a null hypothesis of "a difference between treatment and control means which is greater than a specific amount". These two methods of statistical analyses are compared.

0-87371-592-6/94/$0.00 + $.50

INTRODUCTION

Large-scale aquatic experimental field studies are required by the U.S. Environmental Protection Agency (EPA) as part of an ecological risk assessment for pesticide registration. A guideline for mesocosm study design has been published by the EPA (Touart, 1988). These ecosystem-level mesocosm studies are the highest tier in aquatic toxicity testing of agricultural pesticides. In these studies, chemical application to mesocosm ponds simulate natural field entry by runoff and/or spray drift. Mesocosm studies are primarily intended to evaluate effects on organisms; however, in addition, physicochemical parameters are monitored and pesticide residues are determined.

In the mid 1980s, mesocosm studies replaced farm pond studies as the final stage aquatic test for risk assessment. Natural pond systems are variable and difficult to replicate. Therefore, statistically inconclusive data had been obtained from the farm pond studies and were difficult to interpret. The current large mesocosm studies use ≥ 0.1 acre (0.04 ha) experimental ponds which are replicated in the study design. Each pond is still a complex and potentially variable system, and, therefore, the experimental design and statistical analyses of data are crucial if the study objectives are to be achieved, i.e., assess ecologically important risk.

This paper outlines essential points to be considered in designing a mesocosm study. Regression using dose responses, and ANOVA where comparisons were made between treated and control groups, have both previously been used within mesocosm study designs. The advantages and disadvantages of these methods are discussed. The appropriate use of *t*-tests and multiple comparison tests are described. Finally, the effectiveness of Student's *t*-test and the EPA inverted *t*-test (reversed null hypothesis), in the context of the relationship between the magnitude of the detectable effect, the level of experimental variability, the number of replicates, the significance level, and the power of the test, is evaluated.

MESOCOSM STUDY DESIGN CRITERIA

The enthusiasm of scientists to perfect study techniques in an experiment often detracts their attention from the overall study design, which is the basis of the experiment. Consequently, experimental design and statistical analyses are often not as sensitive as they could be at addressing the objectives of the study. Without careful design of a mesocosm study, the resulting data may not be suitable for deciding whether or not an ecologically important effect is present.

Mesocosm design considerations are delineated sequentially in Figure 1. The "list" approach to "where are the toxicological effects" will have been covered by laboratory studies. These will have identified the areas of concern which triggered the necessity for a final tier study. The field mesocosm study should therefore be designed to demonstrate that these selective anticipated effects are not excessive. At the start of a mesocosm study, if additional toxicological information is necessary,

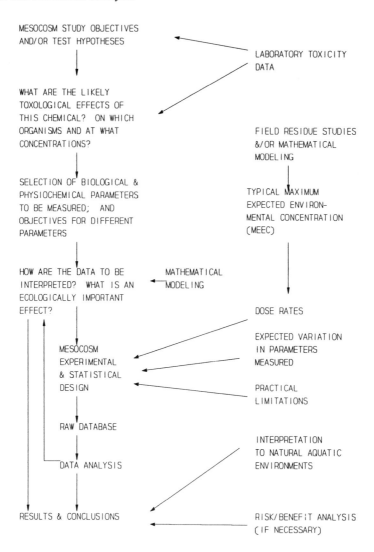

Figure 1. Mesocosm study design criteria.

e.g., for organisms which have not triggered a risk concern, additional laboratory studies should be conducted and not field study measurements. Consequently, the mesocosm study endpoints should be limited and tailored to match the potential problems associated with the test chemical, and may include both structural (e.g., population sizes) and functional (related to energy flow or nutrient cycling) parameters.

Toxicological effects of interest are those that occur at or below the typical maximum estimated environmental concentration (MEEC). The MEEC is dependent upon the typical maximum estimated entry rate (MEER) for spray drift and/or runoff.

This exposure component is normally determined either by mathematically modeling "typical worst case" scenarios of field runoff and/or spray drift studies, or from agricultural "in-use" studies or incidents. For example, agricultural chemical runoff has been modeled using the "Pesticide Run-off Simulator" computer program, SWRRB (Computer Sciences Corporation, 1980). The MEER value is at present mostly obtained from a combination of worst case scenarios and, hence, encompasses a very substantial, and often excessive, safety margin (Hill et al., 1991b).

A robust study design should be decided *a priori*. The biologically important effect for each measured variate should be determined at this stage. If this important question is addressed at the field study outset, the level of detection which the methods of statistical analyses should ideally achieve will be known. This will, in turn, affect the test power required to detect the required effect and the number of replicates in the study design. Several different mathematical models have been proposed to evaluate the environmental impact of agricultural chemicals (e.g., Patten, 1975; Swartzman and Rose, 1984; Brown and Sharpe, 1988; Barnthouse et al., 1990). Many aquatic organisms are "r-strategic" opportunistic species, with some resilience to adverse changes in their habitat. Therefore risk assessment should not be determined as a simple, direct effect (e.g., 20% reduction) on a variate at one moment in time. In addition, many effects may be influenced by density-dependence. Consequently, aquatic ecosystem risk assessment models need to incorporate the dynamics of the system.

The experimental design and subsequent statistical analyses should be sensitive enough to detect a biologically important effect if one is present. Coefficients of variation expected in biological measurements can be based on those calculated from previous pond or other field aquatic studies. These values often exceed 100% (Hill et al., 1988, 1991a). These will help determine the number of replicates necessary for ANOVA to detect, with a given level of statistical confidence, a biologically important effect. The dose rates included in the study design normally have to be set around the MEER or the EC_{50} values of variates to be measured. Finally, practical limitations have to be considered in the study design. Large 0.1 acre mesocosms are logistically difficult to manage and require substantial manpower and finances; therefore, most studies have been restricted to a 12 pond experimental design, the minimum recommended by the EPA study guidelines (Touart, 1988).

In conventional biological experiments using a null hypothesis of "treatment and control group means do not differ", the experimenter attempts to reject the hypothesis and conclude with a high degree of statistical confidence ($\alpha = 0.05$) that "there is a difference". Conversely, in some toxicological studies where the same null hypothesis of "no difference between treatment and control groups" is tested, the experimenter wishes to have statistical confidence in failing to reject the hypothesis. This emphasizes how controversial and crucial the design and statistical analyses of a toxicological experimental study are, as they set the power (probability of detecting a false null hypothesis) to locate effects, and the associated confidences that these are true differences between treatment and control groups. Focusing on both this issue and the other criteria to be recognized in the design of a mesocosm study, this paper assesses several alternative methods of data analyses for mesocosm studies. These techniques are ANOVA, and regression utilizing dose responses, and the appropriate use of *t*-tests and multiple range tests.

COMPARISON OF ANOVA WITH THE REGRESSION APPROACH

The one-way analysis of variance (ANOVA) has been used to test hypotheses about treatment means in mesocosm studies; the associated experimental design is based on treatment groups with replicates corresponding to different dose rates to be tested, and a control. Statistical confidence (probability of type I and II errors) is attained by having adequate replication, which, in turn, is based on the variability expected in a given measurement. The response of individual variates to a treatment rate in mesocosm studies has been tested against the control using Student's *t*-test or Dunnett's test (often referred to as a multiple comparison test, but not truly so).

ANOVA tests for variation in mean response, and it requires accurate application of the chemical to the ponds at the predetermined rate to reduce variability between replicates within a treatment group. In large-scale mesocosm studies to date, three treatment rates (and an untreated control) have normally been used, with the median mesocosm rate based upon the MEER (Hill et al., 1991b). The effectiveness of this test in risk assessment depends on the mesocosm design criteria being followed (see Figure 1). The ANOVA must be able to demonstrate, statistically, biologically important effects if they occur. Risk assessment is then in part based on the observed effects and the probability of occurrence (following normal field usage) of the concentrations which produce them.

The regression approach has not normally been used to test specific hypotheses in mesocosm studies. Instead a series of dose rates are tested, with the objective to obtain dose-response curves with no-observed-effect-concentrations (NOECs). Therefore, the regression approach outwardly appears to offer more flexibility than ANOVA, a seemingly attractive alternative considering that pesticide labels often describe a number of use patterns and application rates with different MEEC calculations. Essentially, if dose responses and NOELs are obtained, these can then be compared with different MEEC values and used in the estimation of risk for labeled and future use patterns. In addition, the pesticide concentration causing a specific effect can be determined (by interpolation), and also extrapolations have been made to untested concentrations.

Since ANOVA was the original method of statistical analysis suggested by the EPA when mesocosm studies began (Touart, 1988), it was the first method to be utilized by agricultural chemical companies (e.g., Hill et al., 1988, 1991a). As MEER values are not always available at a study outset, and as ANOVA tests the response to a specific treatment rather than measuring the degree of the response, the regression approach because of its flexibility is an attractive alternative.

More recently, mesocosm studies which have adopted the regression approach have neared completion. Unfortunately in large mesocosm studies a wide range of doses cannot be tested. This is due to the practical limitations (number of ponds and management logistics) of testing a range of different dose rates with a minimum of two mesocosm replicates per dose. Effects of individual chemicals on different aquatic organisms are often varied, as is seen with the toxicological potency of pyrethroids to nontarget organisms (Hill, 1985, 1989; Hill et al., 1991a). Therefore, a dose-response with a NOEL for a particular effect on an organism may encompass only a very small dose range, which may not overlap with dose responses for other organisms. Consequently, the regression approach in practice results in a series of threshold effects with few, if any, dose-response relationships. In theory, to obtain log logistic dose-response curves for the parameters typically measured in a meso-cosm study, a set of doses tested around the specific EC_{50} value and encompassing the NOEC for each variable would be essential. As many variables of varied toxicological sensitivity are tested in a mesocosm study, this would generally result in a study design of massive size and complexity.

Graney et al. (1989) compared the statistical power associated with different designs. They concluded that the probability of detecting a significant effect at or below the EEC is greater for the regression approach than for ANOVA. However, this depended on good dose-response relationships in the regression experiment. If dose responses are not obtained, the ANOVA method has superior statistical certainty over the regression approach. Furthermore, although the regression approach does offer more flexibility, there is a degree of risk associated with extrapolating to untested concentrations. The EEC or EER still has to be defined and, similarly, what constitutes a biologically and ecologically important effect must be established. In ANOVA, such questions cannot be put aside and need to be addressed early in the study design.

THE APPROPRIATE USE OF MULTIPLE COMPARISON TESTS

Multiple comparison tests have been used in mesocosm studies. The objective of multiple comparison tests is to reduce the probability of committing a type I error (i.e., it protects from concluding that there is a difference between treatment and control means when in reality they do not differ). However, multiple comparison tests, such as Duncan's Multiple Range test and Tukey's test, are designed for use when the comparisons to be performed are all possible pairs of means. In this situation, generally it would be invalid to employ a series of independent *t*-tests. This could occur during an epidemiological field survey when the experimenter cannot predict the direction of the results. Such an occurrence can be referred to as a "fishing expedition" because we are dredging the data to identify trends, and the use of multiple range tests is appropriate.

In contrast to the above, a mesocosm study is a designed experiment where an ordered set of quantitative treatment rates of the test chemical are set, and it is stated at the study outset that the responses of variates to treatment rates are to be compared to the untreated control group. In this situation, Student's independent *t*-tests and Dunnett's test are appropriate to compare the control mean to treatment group means. However, it may be necessary to use the Bonferroni correction with Student's *t*-test to avoid inflation of the type I error. The appropriate use of these analytical procedures will enable the experimenter to have maximum confidence in the test conclusions.

In summary, for designed experiments where we have specific objectives that are declared at the outset of the experiment, Student's *t*-test or Dunnett's test are the preferred hypotheses tests.

COMPARISON OF STUDENT'S *T*-TEST WITH THE EPA INVERTED *T*-TEST

Student's *t*-test is a well recognized statistical procedure that has been widely used for analysis of biological data for many years. Recently, the EPA has proposed that this should be replaced by the inverted *t*-test (Stunkard 1989, 1991) in the analysis of mesocosm data.

The use of Student's *t*-test and the EPA inverted *t*-test to test specific hypotheses are considered here. For the purpose of illustration and comparison, as presented below, it will be assumed that treatment will cause a *reduction* in the response variable (i.e., one-tailed test).

As mesocosm data do not follow a normal distribution, data are transformed using, e.g., a log (x + 1) transformation (e.g., Hill et al., 1991a), so that variances within groups are equal.

Student's *t*-Test

Accordingly, the null hypothesis for Student's *t*-test is that the treatment mean is greater than or equal to the control mean, and this is tested against the alternative hypothesis that treatment reduces the magnitude of the variate under investigation. Algebraically these hypotheses can be expressed as follows:

H_0: $\mu_t - \mu_c \geq 0$ (null hypothesis; the difference between the treatment and control means is greater than or equal to zero; 1-tailed test)
H_1: $\mu_t - \mu_c < 0$ (the treatment mean is less than the control mean)

where μ_t and μ_c are the means of the treatment and control group, respectively. The test statistic is:

$$t = \frac{\bar{y}_t - \bar{y}_c}{s \sqrt{\dfrac{1}{n_t} + \dfrac{1}{n_c}}}$$

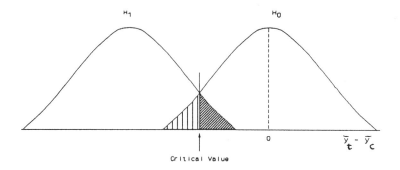

Figure 2. Student's *t*-test.

where $\bar{y}_t$ is the sample mean in the treated group; $\bar{y}_c$ is the sample mean in the control group; s is the pooled estimate of the standard deviation which is obtained from the analysis of variance; n_t is the number of replicates in the treated group; and n_c is the number of replicates in the control group.

This follows a Student's *t*-distribution with degrees of freedom (df) equal to the error degrees of freedom from analysis of variance. If $t < -t_{\alpha;df}$ (where α = significance level, 1-tailed test, df = degrees of freedom), then the null hypothesis is rejected and it is concluded that the difference between treatment and control means is unlikely to have occurred by chance (here the critical value takes a negative value because we are interested in detecting a reduction in the treated group).

Student's *t*-test is demonstrated pictorially in Figure 2.

The frequency distribution of $\bar{y}_t - \bar{y}_c$ under the null hypothesis (H_0) is depicted by the curve on the right (Figure 2); the frequency distribution of $\bar{y}_t - \bar{y}_c$ under the alternative hypothesis (H_1) is illustrated by the curve on the left.

The critical value is given by:

$$-t_{\alpha;df} \times s \sqrt{\frac{1}{n_t} + \frac{1}{n_c}}$$

If $\bar{y}_t - \bar{y}_c$ is less than this critical value then we conclude that $\mu_t - \mu_c < 0$, i.e., the treatment mean is less than the control mean (H_0 rejected). Otherwise we conclude that $\mu_t - \mu_c \geq 0$, i.e., the difference between the treatment and control means is greater than or equal to zero (H_0 is not rejected). The two shaded regions indicate the two types of error that can occur with this test:

The type I error or the probability of concluding that treatment and control means differ (H_0 rejected) when there is no real difference. The type I error is less than or equal to the significance level (α; alpha) and is selected by the experimenter (usually 5%).

The type II error (β; beta), or the probability of not rejecting the null hypothesis when it is false. The power of the test is [1-(probability of a type II error)], or the probability of rejecting the null hypothesis when it is not true. Type II error and power are both functions of the real difference.

(Note: The type I and type II errors in Figure 2 are displayed as symmetrical for illustrative purposes, although this may not always be the case).

When choosing the number of replicates for a mesocosm study, we wish to have both a small type I error and a large power. Thus we should select a sample size which allows us to detect the specified effect (e.g., a 20% reduction, if this is biologically important) with a small type I error (α) and a large power (small β). It is important that the design objective is to detect only differences which are biologically important. If we wish to detect smaller differences (or look for differences where variability is greater, as for example at species rather than family levels), this will be at the expense of increasing the error rates and thus having less confidence in the test conclusions. Unfortunately in practice, in order to detect a specified difference, it is difficult to have both a small type I error and a large power. This is because lowering one error type is usually at the expense of increasing the other; assuming that further replication is not effective or practically possible. However, failure to pay attention to the power of the statistical test when designing a mesocosm study may result in a high risk of failing to detect a difference between treatment and control means when one is present. This is undesirable when the concern is to protect the environment. Therefore in a toxicological study with a null hypothesis of "no difference between treatment and control means" the experimenter should concentrate on having a small type II error (β) in order to have a safety margin.

Inverted *t*-Test

The EPA inverted *t*-test proposed by Stunkard (1989) is a variant of Student's *t*-test. It tests whether the treatment group mean is within $(100 \times b)\%$ of the mean of the control group (where "b" represents a proportion of the control mean which is defined as a biologically important effect). The null hypothesis of this test is:

$H_0: \mu_t - b\mu_c \leq 0$ (the difference between treatment and control means which is $\geq 100(1 - b)\%$ of the control mean, e.g., for a 20% response, $b = 0.8$)

This is tested against the alternative hypothesis:

$H_1: \mu_t - b\mu_c > 0$ (the difference between treatment and control means is less than $100(1 - b)\%$ of the control mean)

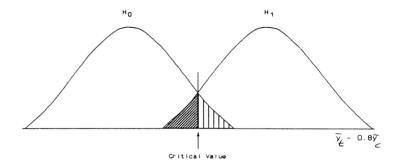

Figure 3.　　The inverted t-test, when b = 0.8.

The test statistic is

$$t_i = \frac{\bar{y}_t - b\bar{y}_c}{s\sqrt{\dfrac{1}{n_t} + \dfrac{b^2}{n_c}}}$$

(N.B. Here, it is assumed that variances within groups are equal [e.g., log transformed data]. If the assumption is that variances are proportional to treatment means, then "b" would be replaced by "b^2" in the test statistic [as described by Stunkard, 1991]. However, the result of the test comparisons are similar irrespective of whether "b" or "b^2" is used in the test statistic.)

This follows a Student's t-distribution with degrees of freedom equal to the error degrees of freedom from analysis of variance. If $t_i > t_{\alpha;df}$ then we reject the null hypothesis and conclude that the difference between treatment and control means is less than $100(1 - b)\%$ of the control mean. Algebraically, the inverted t-test is similar to Student's t-test, the difference being the presence of "b" in both the numerator and the denominator of the inverted t-test.

The inverted t-test is illustrated by Figure 3 for the situation when b = 0.8.

The frequency distribution of $\bar{y}_t - 0.8\bar{y}_c$ under the null hypothesis (H_0) is depicted by the curve on the left (Figure 3). The frequency distribution of $\bar{y}_t - 0.8\bar{y}_c$ under the alternative hypothesis (H_1) is illustrated by the curve on the right. Generally the null hypothesis of the inverted t-test corresponds to the alternative hypothesis of Student's t-test. Similarly, the alternative hypothesis of the inverted t-test corresponds to the null hypothesis of Student's t-test, hence giving the test its name. The critical value in Figure 3 is given by:

$$t_{\alpha;df} \times s\sqrt{\frac{1}{n_t} + \frac{0.64}{n_c}}$$

(e.g., if $\alpha = 0.05$ and df = 12 then $t_{0.05;12} = 1.782$)

If $\bar{y}_t - 0.8\bar{y}_c$ is greater than the critical value, then we conclude that $\mu_t - 0.8\mu_c$ > 0, i.e., the difference between treatment and control means is $<20\%$ of the control (H_0 rejected). Otherwise we conclude that $\mu_t - 0.8\mu_c \leq 0$, i.e., the difference between treatment and control means is $\geq 20\%$ of the control (H_0 not rejected). The two shaded regions are as follows:

 The type I error (α) or the probability of rejecting H_0 when in reality there is a difference between treatment and control means which is $\geq 20\%$ of the control mean.

 The type II error (β) or the probability of failing to reject H_0 when in reality the difference between treatment and control means is $<20\%$ of the control mean.

(Note: The type I and type II errors in Figure 3 are displayed as symmetrical for illustrative purposes, although this may not always be the case.)

Comparison of Tests

The one-tailed Student's t-test has a null hypothesis of "the difference between treatment and control means is greater than or equal to zero"; therefore, a small type II error (β) is more important than a small type I error (α) to ensure a safety margin. In contrast, for the EPA test it is desirable not to reject the null hypothesis when biologically meaningful effects exist, and, therefore, to have a safety margin, a small type I error (α) is more important than a small type II error (β). The EPA proposed the inverted t-test to be used with a type I error of 20% for a specified number of variates. Student's t-test has previously been used in mesocosm studies with a type I error of 5%. Therefore the comparison of the two tests (Figures 4a through f) adopts these conditions rather than those recommended above for a safety margin. It also assumes that the treatment will cause a reduction in the variate measured and is set up as follows:

 The type I error of Student's t-test is set at 5% ($\alpha = 0.05$) by convention and thus the type II error of the inverted t-test is also 5% ($\beta = 0.05$).

 The type I error of the inverted t-test is set at 20% ($\alpha = 0.2$) and thus the type II error of Student's t-test is also 20% ($\beta = 0.2$).

Figures 4a through f show the effect of replication on the magnitude of the difference between treatment and control means that are capable of being detected with 80% probability by the EPA inverted t-test and 95% probability by Student's t-test. The differences are expressed as a percentage of the control mean [$100(\mu_t - \mu_c)/\mu_c$] as described in Appendix I. This is presented for coefficients of variation (CV) ranging from 10 to 100%. The number of replicates and the level of experimental variability were determined from Armitage and Berry (1988) (using Equation 6.8, page 182 for Student's t-test and an appropriately modified version for the inverted t-test; see Appendix I).

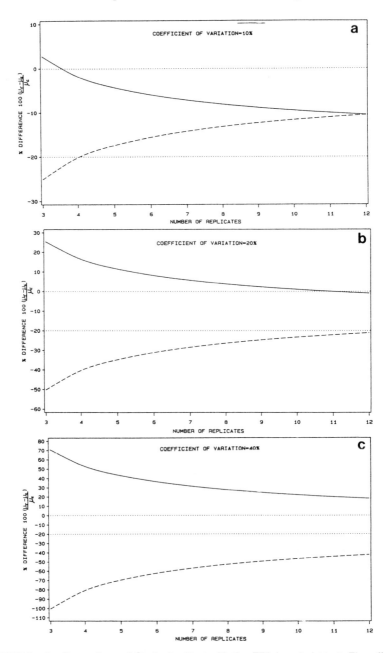

FIGURES 4a–f. Comparison of Student's *t*-test with the EPA inverted *t*-test. The effects of increasing replication (x-axis) on the percentage difference between treatment and control means (y-axis) for the two tests are shown for a range of coefficients of variation (CV). The inverted *t*-test (with b = 0.8) is expressed in terms of percentage of the control mean

$$\frac{\mu_t - \mu_c}{\mu_c}$$

ACKNOWLEDGMENTS

The authors gratefully acknowledge the valued advice and experience of many colleagues at their respective establishments and elsewhere. In particular we thank Dr. T. Springer, Wildlife International; Dr. T. Beitinger, University of North Texas; Dr. P. Chapman, ICI Agrochemicals; and Dr. J. Shelton, ICI Americas.

REFERENCES

Armitage, P. and G. Berry. 1988. *Statistical Methods in Medical Research,* 2nd ed. Blackwell Scientific, Boston.

Barnthouse, L.W., G.W. Suter II, and A.E. Rosen. 1990. Risks of toxic contaminants to exploited fish populations: influence of life history, data uncertainty and exploitation intensity. *Environmental Toxicology and Chemistry,* 9:297–311.

Brown, R.A. and J.E. Sharpe. 1988. Interpreting the results of terrestrial impact studies with reference to modelling an ecologically significant effect. In M.P. Greaves, B.D. Smith, and P.W. Greig-Smith, eds., *Environmental Effect of Pesticides.* British Crop Protection Council, Monograph No. 40, pp. 283–291. Environmental Effect of Pesticides.

Computer Sciences Corporation. 1980. *Pesticide Run-off Simulator (SWRRB) — User's Manual,* prepared for U.S. Environmental Protection Agency, Falls Church, Virginia.

Graney, R.L., J.P. Giesy, Jr., and D. DiToro. 1989. Mesocosm experimental design strategies: advantages and disadvantages in ecological risk assessment. In J. Reese Voshell ed., *Using Mesocosms to Assess the Aquatic Ecological Risk of Pesticides: Theory and Practice,* Ento. Soc. Am. Misc. Publicn., MPPEAL No. 75, pp. 74–88.

Hill, I.R. 1985. Effects on non-target organisms in terrestrial and aquatic environments. In J.P. Leahey, ed., *The Pyrethroid Insecticides.* Taylor & Francis, London, pp. 151–262.

Hill, I.R. 1989. Aquatic organisms and pyrethroids. *Pestic. Sci.,* 27:429–465.

Hill, I.R., S.T. Hadfield, J.H. Kennedy, and P. Ekoniak. 1988. Assessment of the impact of PP321 on aquatic ecosystems using tenth acre experimental ponds. *Proc. Brit. Crop Protect. Conf., Pests and Dis.,* 1988 1:309–318.

Hill, I.R., J.K. Sadler, J.H. Kennedy, P. Ekoniak. 1993a. Lambda-cyhalothrin: a mesocosm study of its effects on aquatic organisms, this volume, Ch. 24.

Hill, I.R., K.Z. Travis, and P. Ekoniak. 1993b. Spray-drift and runoff applications of foliar-applied pyrethroids to aquatic mesocosms: rates, frequencies & methods, this volume, Ch. 15.

Patten, B.C. 1975. A reservoir core ecosystem model. *Trans. Am. Fish Soc.,* 104:596–619.

Southwood, T.R.E. 1978. *Ecological Effects,* 2nd ed. Chapman & Hall, New York.

Stunkard, C.L. 1989. *Statistical Hypothesis for EPA.* Statistical Policy Branch, OPPE, Technical Note. U.S. Environmental Protection Agency, Washington, D.C.

Stunkard, C.L. 1993. Tests of proportional means for mesocosm studies, this volume, Ch. 8.

Swartzman, G. and K.A. Rose. 1984. Simulating the biological effects of toxicants in aquatic microcosm systems. *Ecol. Model,* 22:122–134.

Touart, L.W. 1988. *Hazard Evaluation Division, Technical Guidance Document: Aquatic Mesocosm Tests to Support Pesticide Registrations.* Report No. EPA-540/09-88-035. U.S. Environmental Protection Agency, Washington, D.C.

APPENDIX

1. Reassessment of the inverted t-test into units of percentage of control means (the dependent variable in Figures 4a through f), so that it is comparable to Student's *t*-test. If

$$\mu_t - 0.8\mu_c > t_{\alpha;df} \times s \sqrt{\frac{1}{n_t} + \frac{0.64}{n_c}}$$

then we reject the null hypothesis and conclude that the effect of treatment is to reduce the response by less than 20% of the control. If we divide both sides by μ_c this becomes

$$\frac{\mu_t - 0.8\mu_c}{\mu_c} > t_{\alpha;df} \times \frac{s}{\mu_c} \sqrt{\frac{1}{n_t} + \frac{0.64}{n_c}}$$

multiplying both sides by 100 we get

$$\frac{100(\mu_t - 0.8\mu_c)}{\mu_c} > t_{\alpha;df} \times \frac{100 \times s}{\mu_c} \sqrt{\frac{1}{n_t} + \frac{0.64}{n_c}}$$

and since

$$\frac{100 \times s}{\mu_c}$$

is the coefficient of variation (CV) this becomes

$$\frac{100(\mu_t - 0.8\mu_c)}{\mu_c} > t_{\alpha;df} \times CV \sqrt{\frac{1}{n_t} + \frac{0.64}{n_c}}$$

by subtracting 20% from both sides of this equation we get

$$\frac{100(\mu_t - 0.8\mu_c)}{\mu_c} - 20 > t_{\alpha;df} \times CV \sqrt{\frac{1}{n_t} + \frac{0.64}{n_c}} - 20$$

or

$$100 \left[\frac{(\mu_t - 0.8\mu_c)}{\mu_c} - 0.2 \right] > t_{\alpha;df} \times CV \sqrt{\frac{1}{n_t} + \frac{0.64}{n_c}} - 20$$

$$100 \frac{(\mu_t - 0.8\mu_c - 0.2\mu_c)}{\mu_c} > t_{\alpha;df} \times CV \sqrt{\frac{1}{n_t} + \frac{0.64}{n_c}} - 20$$

therefore

$$\frac{100(\mu_t - \mu_c)}{\mu_c} > t_{\alpha;df} \times CV \sqrt{\frac{1}{n_t} + \frac{0.64}{n_c}} - 20$$

Thus we now have the inverted t-test expressed in terms of percent of the control mean. It is this quantity (from Armitage and Berry, 1988) that is plotted in Figures 4a through f.

2. Similarly, the quantity expressed in Figures 4a through f for Student's t-test is as follows:

$$\frac{100(\mu_t - \mu_c)}{\mu_c} < -t_{\alpha;df} \times \frac{100 \times s}{\mu_c} \sqrt{\frac{1}{n_t} + \frac{1}{n_c}}$$

and since

$$\frac{100 \times s}{\mu_c}$$

is the CV this becomes

$$\frac{100(\mu_t - \mu_c)}{\mu_c} < -t_{\alpha;df} \times CV \sqrt{\frac{1}{n_t} + \frac{1}{n_c}}$$

CHAPTER 10

Ecological Development and Biometry of Untreated Pond Mesocosms

Van D. Christman, J. Reese Voshell, Jr., David G. Jenkins,
Michael S. Rosenzweig, Raymond J. Layton, and Arthur L. Buikema, Jr.

Abstract: The objectives of this study were to: (1) analyze colonization of 12 new, unmanaged, untreated pond mesocosms to determine if they acquired a biota typical of shallow lentic environments within 1 year and (2) analyze the precision of various biometrics in terms of the percent change that must occur to distinguish treatment effects from natural variability. We analyzed colonization of phytoplankton, zooplankton, and macroinvertebrate communities in conjunction with various environmental characteristics. Phytoplankton, zooplankton, and macroinvertebrate metrics were analyzed by a statistical procedure that indicates the percent change that must occur to detect true differences between two means. Biometrics evaluated included various measures of community structure and function. For the overall mesocosm facility, communities followed expected seasonal patterns of succession and resembled the communities that would be expected in shallow lentic environments; however, there were considerable differences among individual mesocosms. With some biological management (introduction of certain macroinvertebrates and macrophytes), the mesocosms might be ready for conducting pesticide registration tests within 1 year. Biometrics showed

0-87371-592-6/94/$0.00 + $.50

a wide range of detection limits with number of taxa metrics ranging from 2 to 40%, proportion of numerical abundance metrics ranging from 2 to 60%, and density metrics ranging from 10 to 100%. Of the 20 metrics evaluated, 7 allowed detection of impact with a 20% change from the mean. Four of these were number of taxa metrics and the other three were proportion of numerical abundance metrics. Density metrics would not allow detection of treatment effects at the 20% level. For all metrics, using six replicate mesocosms, instead of three, would usually improve the ability to detect differences between treatments. With a management program (mixing water and sediments, planting macrophytes), phytoplankton and zooplankton might be more evenly distributed among the mesocosms. Macroinvertebrates were more evenly distributed among the mesocosms and might only require seeding of a few nonflying taxa. Based on the results of this study, statistically realistic detection limits would be <20% for metrics involving number of taxa, <40% for metrics involving proportions of numerical abundance, and <60% for density and biomass metrics. These criteria should provide adequate environmental protection because changes of this magnitude occur naturally in unperturbed ecosystems. We recommend analyzing pretreatment data by this approach to establish biological criteria that are realistic for a particular study.

INTRODUCTION

The U.S. Environmental Protection Agency (EPA) now requires simulated field studies in pond mesocosms as part of the registration requirements for some new pesticides.[1,2] This is a large step in environmental toxicology, and there are many unanswered questions about mesocosms as test systems.[3] Mesocosms are being used because of two major characteristics:[4] mesocosms are similar to natural ecosystems, and they can be experimentally manipulated with sufficient replication for statistical testing. The underlying assumption that the EPA recommended mesocosm test achieves these characteristics has not been proven.

Mesocosm tests for pesticide registration in the U.S. are being conducted in small excavated ponds (0.04 to 0.10 ha). According to EPA recommendations, the naturalness of pond mesocosms can be obtained by lining them with sediment from an established pond and waiting 6 months for "seasoning" before beginning a test, or by lining them with topsoil and waiting 12 months. Howick et al.[5] and Ferrington et al. (Chapter 14) demonstrated that mesocosms can be colonized quickly by lining them with pond sediments, but there have been no long-term studies that followed the colonization of new pond mesocosms lined with topsoil to serve as comparisons. Although the colonization of various new lentic environments has been investigated,[6-10] most results are not applicable to pond mesocosms because the studies were conducted in larger environments in which colonizing organisms could arrive from incoming surface waters. Thus, it is not known how long it takes for new pond mesocosms that are lined with topsoil to develop a biota that is representative of shallow lentic environments.

Stunkard[11] (Chapter 8) recommended that the EPA use an hypothesis-testing design in which registration applicants are required to refute with statistical tests the presumption of adverse effects when the pesticide is applied. This approach differs from what is usually done in toxicological studies in that the pesticide is assumed to cause adverse effects at the estimated environmental concentration, and the results of the mesocosm test must reject the hypothesis to refute the presumption of adverse effects. Geisy[12] called this approach, which is intended to minimize type II errors instead of type I errors, the "guilty until proven innocent policy". Stunkard[11] (Chapter 8) also recommends that the hypothesis be expressed in terms of "proportional means". The EPA suggested that effects are >15% reduction in number of taxa or >20% change in other biometrics. Henceforth in this paper, we will refer to suggested percent change in biota and the hypothesis testing design that Stunkard[11] (Chapter 8) has recommended for the EPA to use in evaluating the results of mesocosm tests as the inverted proportional means hypothesis.

Insufficient attention has been given to distinguishing changes attributable statistically to treatment effects from changes that occur naturally in mesocosms. Because mesocosms are reasonably natural systems, variability can be expected to be greater than in laboratory test systems,[4,13] and it may not be possible to achieve sufficient replication for the stated purpose of the tests. High natural variability would cause the inverted proportional means hypothesis to be accepted (presumption of adverse effects not refuted) regardless of pesticide treatment effects. It is generally recognized that the number of replicate samples and the desired level of precision (within practicality for a given study) should be determined from preliminary data obtained before treatment effects are studied.[14-17]

We conducted a 2-year study of a set of new pond mesocosms that were lined with topsoil, filled with water containing no aquatic life, then were unmanaged except to maintain constant water level. The objectives of the study were to: (1) determine if the mesocosms acquired a diverse biota typical of shallow lentic environments within 1 year, and (2) analyze the precision of various biometrics in terms of the percent of change that must occur to distinguish treatment effects from natural variability.

STUDY SITE

This study was conducted at the experimental pond facility at the Southern Piedmont Agricultural Experiment Station near Blackstone, VA (longitude 77°57'30''W latitude 37°5'30''N). The facility is located within the Piedmont Physiographic Province in a transition zone between northern and southern climates characterized by a long (206 d), humid growing season.[18] Mean annual temperature is 14.4°C, mean annual precipitation is 105.8 cm, and elevation is 128 m (above mean sea level). Several impoundments are located near the site, which range in size from 0.04 to 2 ha. One of the largest impoundments (1.5 ha) is located on the experiment station 0.5 km northeast of the experimental pond facility.

The experimental pond facility consists of 12 square 0.04-ha pond mesocosms and a 0.36-ha reservoir (see diagram in [19]). Each mesocosm is 2.1 m deep with all four sides sloping at 2.5:1 and a volume of approximately 520 m³. The mesocosms were excavated during 1987, lined with a 15-cm compacted clay layer, and backfilled with a 15-cm layer of topsoil obtained at the site. Water can be supplied by a series of wells, the reservoir, or a temporary connection to the local municipal water supply. Each mesocosm can be filled or emptied independently.

The 12 pond mesocosms were first filled January 25 through 31, 1988. Only the municipal water supply was used for initial filling because production of the wells was inadequate, the reservoir was not constructed until summer 1988, and it was certain that the chlorinated tap water would not introduce any organisms. Chlorinated water had no impact on colonization after filling because 5 zooplankton taxa (copepod adults and nauplii and rotifers) were collected in the mesocosms after 1 week, and the number of taxa and density increased steadily thereafter.[20] The only management of the mesocosms was to add water to compensate for evaporation losses. Water levels were kept the same in all mesocosms by adding only well water during the first year. During the second year reservoir water was used; however, water was drawn from immediately below the surface of the reservoir during daytime and would not have transported macroinvertebrates to the mesocosms. The reservoir was also filled originally from the municipal water supply. There were no fish in the mesocosms during the study.

METHODS

Sampling Design

We sampled phytoplankton, zooplankton, and macroinvertebrates in 12 mesocosms the first year of the study and only macroinvertebrates in 6 mesocosms the second year. We measured an array of physicochemical parameters at the same time as biological sampling in the first year, but only measured temperature in the second year. Although some biological components were sampled more frequently, only data from monthly intervals are included in this paper for consistency in reporting results from the different biological groups. We took samples from an inflatable boat so that all areas of the mesocosms could be reached with minimal disturbance of sediment. The mesocosms were sampled in a different random sequence on each occasion to eliminate the possibility of causing colonization patterns by inadvertent transfer of organisms.

Environmental Characteristics

Physicochemical parameters measured in the mesocosms were Secchi depth, temperature, dissolved oxygen, pH, hardness, conductivity, alkalinity, ammonia, nitrate, nitrite, and soluble reactive phosphate. Standard methods were used for all analyses.[21] Additional details of the physicochemical methods can be found in Jenkins.[20]

We qualitatively observed macrophyte colonization and growth in the ponds throughout the 2-year study. As new taxa of macrophytes appeared in the meso-cosms, representative specimens were collected by hand and taken to the laboratory for identification.

Phytoplankton

An integrated depth sample of water, consisting of three subsamples, was col-lected with a 2.00 × 0.04-m transparent acrylic plastic tube sampler.[22] A 200-ml aliquot was preserved with 2% acid Lugol's solution for subsequent identification and enumeration of phytoplankton. Two 400-ml aliquots were filtered through Whatman GF/C filters to collect phytoplankton for chlorophyll *a* determinations. The filters were placed in sealed test tubes and stored on dry ice for transport to the laboratory.

In the laboratory, each 200-ml aliquot was reduced to a final volume of 5 ml through a series of settling procedures. From the final 5 ml, three 0.2-ml aliquots were examined at 500× magnification in a Palmer-Maloney phytoplankton counting cell for taxonomic identification. For taxa density, an entire Whipple counting grid field was counted. Chlorophyll *a* samples were first extracted by grinding the filters in a 9:1 acetone:$MgCO_3$ mixture and then analyzed on a Perkin-Elmer spectropho-tometer. See Rosenzweig[23] for a complete explanation of phytoplankton methods.

Zooplankton

The same tube sampler as used for phytoplankton was used to collect rotifers. Four liters of pond water were sieved through 35-μm plankton netting, and the organisms were rinsed into a jar and preserved with 4% buffered formalin. A Wisconsin-style plankton net (80-μm mesh) was used to collect crustaceans by vertical tows. The organisms were rinsed into a second jar and preserved with 4% buffered formalin.

In the laboratory, zooplankton were identified and counted with a Sedgwick-Rafter chamber and a compound microscope at 40×. Samples with high densities were subsampled before counting. Productivity was calculated using the increment-summation technique. Respiration was determined according to the method of Owens and King.[24] Additional details of zooplankton methods can be found in Jenkins.[20]

Macroinvertebrates

Artificial substrate samplers were used to reduce variation and shorten processing time in the laboratory. The samplers consisted of small plastic buckets (10.8 cm tall, 12.2 cm diameter) with 3.8-cm holes drilled in the sides and tops, a 1.5-cm layer of topsoil in the bottom, and the remaining space filled with 5-cm spherical TRI-PACK® units which are sold for biological treatment of waste water (Jaeger Products Inc., P.O. Box 16117, Houston, TX). Before filling the mesocosms with water, sufficient samplers were placed on the bottom of each mesocosm to remove one sampler from two depths on each date for the first year of study. In this paper, we report the results from only one depth (1.0 m). All samplers were replaced at the end of the first year. Samplers were allowed to colonize for at least 4 weeks before the first ones were retrieved at the beginning of the first and second year. Samplers to be retrieved on each date were chosen randomly. To reduce the loss of organisms during retrieval, an inverted funnel was dropped over the samplers before they were pulled up, and a 100-μm mesh net was placed under the samplers as soon as they were visible underwater. We also took qualitative samples with a D-frame dip net in shallow water around the edge of the mesocosms. All samples were preserved with 5% formalin.

In the laboratory, all surfaces of the samplers were scrubbed with a soft brush and the contents rinsed in a 106-μm sieve to remove fine sediment. The material on the sieve was placed in a jar with 70% ethyl alcohol, and the macroinvertebrates were sorted later with a stereomicroscope at low magnification (4 to 10$\times$). Organisms were identified to the lowest possible taxonomic level, using either Brigham et al.[25] or Merritt and Cummins,[26] then counted. Taxa were assigned to functional feeding groups using Merritt and Cummins.[26] See Layton[27] for a complete explanation of macroinvertebrate methods.

Biometrics

We analyzed various measures of community structure and function, but in this paper we concentrated on the metrics most commonly reported in macroinvertebrate studies. For phytoplankton these included: taxa richness, total density, density of one dominant taxon, total biovolume, and chlorophyll *a* biomass. Zooplankton metrics were taxa richness, percent contribution of dominant taxon, total density, density of one dominant taxon, total biomass, community productivity, and community respiration. Macroinvertebrate metrics were taxa richness, EOT index (number of taxa in pollution sensitive orders Ephemeroptera, Odonata, Trichoptera), percent contribution of dominant taxon, percent collector-gatherers, percent predators, total density, and density of two dominant taxa. Several of these metrics, or slight modifications, were chosen because the EPA has recommended that they be used in rapid bioassessment procedures for streams (percent contribution of dominant taxon, taxa richness, number of taxa in pollution sensitive orders, functional groups; see Reference 28).

We evaluated all biometrics in terms of the percent change that must occur to be attributed to a treatment effect (detection limit) by the following formula from Sokal and Rohlf[16]:

$$\delta \geq ((2/n)^{0.5} * CV * (t_{\alpha(v)} + t_{2(1-p)(v)}))$$

where δ is the amount of change that must occur to be attributed to a treatment effect; n is the number of mesocosms used for each treatment; CV is the coefficient of variation; α is the confidence of the test; p is the power of the test; and v is the degrees of freedom.

A pesticide treatment would have to cause a change greater than δ to be distinguished from natural variability. Resh and McElravy[17] suggested this approach for designing studies in which the objective is to detect differences in biotic variables over space or time.

The alpha level was set at 0.2 and the power was set at 0.8 to be consistent with the recommendations of Stunkard[11] (Chapter 8) for how the EPA should determine the effect of a treatment. The detection limit was calculated for six and three replicates. This was done by randomly selecting the desired number of replicates three times. The three iterations were then combined to get a grand mean and an average standard deviation. From this, the above formula was used to calculate the detection limit.

RESULTS

Environmental Characteristics

Three parameters were relatively unaffected by developmental changes during the first year the mesocosms were in existence. Water temperature and dissolved oxygen followed typical seasonal curves for temperate regions, ranging from 4 to 30°C and 6.0 to 13.0 mg/l, respectively. Because the mesocosms were shallow and exposed to prevailing winds, stratification was slight and transient, if present at all. In addition, water hardness was generally steady at about 75 mg/l as $CaCO_3$.

All other parameters exhibited temporal trends that were related to developmental changes in the mesocosms. There was an increase in pH over the first year, from initial values of about 6.7 to later values of about 7.7. Alkalinity also exhibited a general increase, roughly corresponding to trends in pH. The mesocosms were never well buffered — year-end alkalinities were about 45 mg/l as $CaCO_3$. Conductivity values declined steadily over the first year: initial values were about 230 μmhos/cm and final values about 160 μmhos/cm.

Nutrients exhibited variable patterns with little apparent relationship among nutrients during the first year. In addition, there was more variation among mesocosms for nutrients than other physicochemical parameters. Ammonia was variable but tended to be highest early in the study (0.06 to 0.09 mg/l, February through May 1988), then fluctuated at or below detection limits (0.02 mg/l) thereafter. Nitrate exhibited a general decline from about 1.0 mg/l during February and March 1988 to low levels (<0.05 mg/l) after June 1988. Nitrite levels were very low except during April through June 1988 and December 1988 through January 1989. Nitrite levels during these peaks approached 1.0 mg/l, and peaks occurred in all mesocosms.

Soluble reactive phosphate was highly variable among mesocosms and through time, but levels were generally very low or undetectable (<0.02 mg/l).

Secchi depth varied widely among mesocosms and was not closely related to chlorophyll *a* levels for much of the first year. Secchi depths were fairly shallow (about 0.75 to 1.0 m) until summer 1988, when average depths tended to increase to about 1.5 m. Shallow Secchi depths were apparently due to suspended sediments from initial filling of the mesocosms, storms, and winds. Secchi depths declined again in November 1988 after a storm and averaged about 1.3 m through January 1989.

Macrophytes were first observed in the ponds in June 1988. By late summer, emergent macrophytes included species of *Carex, Cyperus, Eleocharis, Hypericum, Juncus, Ludwigia,* and *Typha.*[23] One submerged macrophyte, *Potamogeton,* was first observed in June 1989. By the end of the study in April 1990, *Potamogeton* had spread to about half of the mesocosms.

Colonization

Phytoplankton

Of the 40 phytoplankton taxa observed during the first year, 17 were Chlorophyceae. The next most diverse group was Desmidaceae (11 taxa). Other groups included Cyanophyceae (5 taxa), Dinophyceae (3 taxa), Bacillariophyceae (2 taxa), and Chrysophyceae (1 taxon). A complete taxonomic list can be found in Rosenzweig.[23]

The first phytoplankton taxa to colonize the mesocosms were three genera of Cyanophyceae (*Anabaena, Chroococcus, Oscillatoria*) in March 1988. Other groups appeared by April and May. Colonization occurred more rapidly during summer, and the accrual of taxa peaked by early fall.

Numerically dominant taxa during the first year were *Dinobryon* (Chrysophyceae), *Peridinium,* and *Glenodinium* (Dinophyceae). Densities reached 1000 to 4000 cells/ml on some sampling dates. Biomass, as indicated by chlorophyll *a* concentration, exhibited a general increase from 0 to 5 mg/m^3 during spring 1988 when phytoplankton populations developed. Chlorophyll *a* exhibited low but roughly level values (about 3 mg/m^3) during June through July and a sharp decline after water temperatures reached 30°C in August 1988. Chlorophyll *a* concentrations increased to spring 1988 values during fall and winter 1988.

The collective phytoplankton community (all mesocosms considered together) was similar to that expected in shallow lentic environments. In addition, seasonal periodicity of these taxa was also similar to expected trends. Chlorophyceae were dominant during summer and Desmidaceae increased during the cooler fall months.

Phytoplankton taxa, density, and biomass were not uniform among mesocosms during the first year. Only 7 of 40 taxa (18%) occurred in all mesocosms. Densities and biomass varied as much as 2000- and 24-fold, respectively, among mesocosms on some dates.

Zooplankton

Rotifera, Copepoda, Cladocera, Ostracoda, and larvae of the phantom midge, *Chaoborus* (Diptera), were evaluated in this study. Of the 61 zooplankton species observed from February 1988 to February 1989, 47 were rotifers. The next most diverse taxonomic group was Cladocera, with seven species observed. In addition, three copepod species and three *Chaoborus* species were observed in the mesocosms. Ostracoda were not identified to species but were considered one taxon. A complete taxonomic list can be found in Jenkins.[20]

Several rotifer species and the copepod *Eucyclops agilis* were present in some mesocosms 1 week after the mesocosms were filled. These species probably originated from forest soil before the site was cleared and became established in puddles on the bottoms of the mesocosms prior to filling. Colonization of the mesocosms continued thereafter and was more rapid during warm months. Most of the observed species were present in some mesocosms by fall 1988. *Chaoborus* appeared almost simultaneously in all mesocosms in spring 1988. Most cladoceran species did not appear until late summer 1988.

Rotifers were numerically dominant during the first year, with densities >1000 organisms per l on many sampling dates. Copepods approached densities of 700 organisms per l in a few mesocosms in spring 1988, and other taxa had lesser densities. Collectively, rotifers also maintained greater biomass than other taxonomic groups, because there were greater densities and more species of rotifers. However, some individual crustacean species attained greater biomass than individual rotifer species.

Collectively, the mesocosms contained a zooplankton community typical of shallow lentic environments by the end of the first year. Many species observed in the mesocosms are known to inhabit benthic or littoral areas, reflecting the shallow design of the mesocosms. Dominant rotifers included the cosmopolitan genera *Keratella, Polyarthra, Anuraeopsis,* and *Trichocerca.* The major copepods were the herbivorous *Eucyclops agilis* and *Tropocyclops prasinus.* The predatory copepod *Macrocyclops albidus* also appeared later in 1988. Cladocera included species typically found in ponds, such as *Chydorus sphaericus, Bosmina longirostris,* and *Simocephalus serrulatus.*

Many zooplankton species were not uniformly distributed among mesocosms by the end of the first year (i.e., colonization did not occur evenly). Of the 61 zooplankton taxa, only 14 (23%) were observed in all 12 mesocosms. In addition, 29 species (48%) occurred in ≤6 mesocosms. Cladocera were most poorly distributed among mesocosms. Population peaks also differed among mesocosms in terms of timing and magnitude. For example, the relatively synchronous and ubiquitous *Chaoborus* exhibited a seven-fold range among mesocosms in peak densities.

Macroinvertebrates

A total of 42 macroinvertebrate taxa were collected with the artificial substrate samplers during the 2-year study (Table 1). Of these taxa, 35 were collected during

Table 1. Macroinvertebrates Collected in Mesocosms from May 1988 to April 1990

Taxa	Functional Feeding Group	Year 1	Year 2
Nematoda	CG	X	X
Annelida			
Oligochaeta	CG	X	X
Hirudinea	PR	X	X
Crustacea			
Amphipoda			
Talitridae			
Hyalella azteca	CG	X	
Insecta			
Ephemeroptera			
Baetidae			
Callibaetis	CG	X	X
Leptophlebiidae			
Leptophlebia	CG	X	X*
Caenidae			
Caenis simulans	CG	X	X
Ephemeridae			
Hexagenia	CG	X	X
Odonata			
Aeshnidae			
Anax junius	PR	X	X
Gomphidae			
Gomphus	PR	X	X
Corduliidae			
Tetragoneuria	PR		X
Libellulidae			
Celithemis	PR	X	X
Erythemis simplicicollis	PR		X
Erythrodiplax	PR	X	X
Ladona	PR	X	X
Libellula	PR	X	X
Pantala	PR	X	X
Perithemis	PR	X	X
Sympetrum	PR	X	X
Lestidae			
Lestes	PR	X	X
Coenagrionidae			
Anomalagrion hastatum	PR	X	X
Enallagma	PR	X	X
Ischnura	PR	X	X
Hemiptera			
Hydrometridae			
Hydrometra martini	PR	X*	X*
Gerridae	PR	X*	X*
Corixidae			
Hesperocorixa	PH	X*	X*
Notonectidae			
Buenoa margaritacea	PR	X*	X
Notonecta irrorata	PR		X*
Nepidae			
Ranatra drakei	PR	X*	
Belastomatidae			
Belastoma lutarium	PR		X*
Megaloptera			
Sialidae			
Sialis	PR		X*

Table 1. (*continued*)

Taxa	Functional Feeding Group	Year 1	Year 2
Trichoptera			
Polycentropodidae			
Cernotina	PR	X	X
Hydroptilidae			
Oxyethira	PH	X	X
Orthotrichia	CG	X	X
Leptoceridae			
Oecetis	PR	X	X
Phryganeidae			
Ptilostomis	SH	X	X
Coleoptera			
Haliplidae			
Peltodytes	PH		X*
Gyrinidae			
Dineutus carolinus	PR	X	X
Noteridae			
Notomicrus	PR		X*
Dytiscidae			
Agabus	PR	X	X
Bidessonotus	PR	X	X
Hydroporus	PR		X*
Laccophilus	PR	X*	X
Hydrophilidae			
Berosus	PH	X	X
Paracymus	PR	X*	X
Tropisternis	CG		X*
Elmidae			
Dubiraphia	CG		X
Diptera			
Tipulidae	SH	X	
Chaoboridae	PR	X	X
Psychodidae	CG	X	
Ceratopogonidae	PR	X	X
Chironominae/Orthocladiinae	CG	X	X
Tanypodinae	PR	X	X
Culicidae	CF		X
Tabanidae	PR		X
Ephydridae	CG		X*
Gastropoda			
Limnophila			
Ancylidae	SC		X

Note: CG = collector-gatherers; PR = predators; PH = piercer-herbivores; SH = shredders; SC = scrapers; X = present; * = collected only from qualitative dip net samples.

the first year and 38 were collected during the second year. In the second year, 4 taxa previously present did not reappear and 7 taxa appeared for the first time. Numerically dominant macroinvertebrate taxa in both years were *Caenis* (Ephemeroptera: Caenidae), *Anax* (Odonata: Aeshnidae), Libellulidae (Odonata), *Enallagma* (Odonata: Coenagrionidae), Ceratopogonidae (Diptera), Chironominae/Orthocladiinae (Diptera: Chironomidae), and Tanypodinae (Diptera: Chironomidae). None of the taxa lost or gained in the second year were numerically dominant.

Colonization began immediately after the mesocosms were filled with water, even though it was midwinter. The earliest colonizers were members of the Coleoptera and Diptera. These taxa colonized quickly because their adults are active throughout the year at the latitude where the mesocosms are located. We observed Chironomidae laying eggs in the mesocosms on calm, warm (>15°C) days in February. The arrival of other taxa was coincident with the time of adult emergence from surrounding environments.[19] Populations increased rapidly in the spring and summer of 1988, when adults of other taxa began to lay eggs in the mesocosms.

Chironomidae were the most abundant organisms in both years, reaching densities as high as 1200 organisms per sampler in October and November. On average, Chironomidae compromised 85% of total density during both years. All other taxa usually occurred in comparatively low numbers (1 to 50 organisms per sampler). Some of the other taxa, such as the Odonata, had a very large biomass as compared to Chironomidae.

The macroinvertebrate community in the mesocosms was typical of shallow lentic environments by the end of the first year. Most expected taxonomic groups were well represented in the mesocosms (Table 1). Hemiptera and Coleoptera, which spend most of their time swimming in very shallow areas, were not collected very well with the artificial substrates, but members of these groups were common in dip net samples. The only expected group not present in the mesocosms during the first year was Gastropoda. Two other nonflying taxonomic groups that would be expected to be abundant in small ponds, Oligochaeta and Amphipoda, were present only in very low densities. Two functional feeding groups that would be expected to dominate in shallow lentic environments, collector-gatherers and predators, were well represented by a variety of taxa (Table 1).

Distribution of individual macroinvertebrate taxa was not completely uniform among the mesocosms but was much more uniform than phytoplankton or zooplankton. Of the 35 macroinvertebrate taxa collected during the first year, 20 (57%) occurred in all 12 mesocosms, and 27 of the 38 taxa (71%) collected during the second year were found in all 6 mesocosms that were studied. In addition, only 8 taxa (23%) were found in less than half of the mesocosms the first year, and only 5 taxa (13%) were collected in less than half of the mesocosms in the second year. In general, the taxa that colonized the mesocosms earliest were also those that were found in the most mesocosms. There were considerable differences (up to tenfold) in the numerical abundances of various taxa in individual mesocosms on a given date.

Biometrics

We calculated the detection limits that would be obtained with three and six replicates for each metric on each date (Figures 1 through 4). The graphs include data for phytoplankton and zooplankton from May 1988 to February 1989 and for macroinvertebrates from May 1988 to April 1990. Data from February to April 1988 were not included because very few organisms were collected during the 3 months immediately following filling of the mesocosms in January. A pesticide treatment would have to cause a percent change greater than values plotted in the graphs in order to be distinguished from natural variability. Horizontal lines on the graphs are the percent changes that have been suggested by EPA as biologically significant ($\geq$15% reduction in taxa richness, $\geq$20% change in all other metrics). Whenever plotted values lie above the horizontal lines on the graphs, it means that the inverted proportional means hypothesis would be accepted, even though the cause is natural variability.

Phytoplankton

Detection limits for the various phytoplankton metrics ranged from 5 to 120% (Figure 1). Of the 5 metrics evaluated in this study, only taxa richness (Figure 1A) had detection limits that were almost always within the criterion (<15% reduction) suggested by EPA as biologically significant. A 10 to 15% change in taxa richness could usually have been detected using either six or three replicate mesocosms on most dates. Taxa richness for phytoplankton was calculated at the lowest possible taxonomic level (usually genus).

Detection limits of no other phytoplankton metrics on any date during the first year were within the criterion (<20% change) suggested by EPA as biologically significant. Detection limits for total density and density of an individual dominant taxon (*Peridinium*) were similar, usually ranging from 40 to 80% with 6 replicates and 60 to 100% with 3 replicates (Figures 1B and C). Biovolume had detection limits between 60 to 70% with 6 replicates and 90 to 100% with 3 replicates on most dates (Figure 1D). Chlorophyll *a* biomass had somewhat lower detection limits than biovolume, ranging from 40 to 50% with 6 replicates and being about 70% with 3 replicates (Figure 1E).

Zooplankton

Figures 2 and 3 show that different zooplankton metrics had detection limits ranging from 2 to 127%. Of the 7 metrics evaluated, only two had detection limits that were usually within the suggested criteria for rejecting the inverted proportional means hypothesis. With 6 replicates, taxa richness had detection limits at or below 15% on most dates, but with 3 replicates detection limits were only 20 to 25% (Figure 2A). Taxa richness was usually evaluated at the species level for zooplankton. The proportion of rotifers metric represents the proportion of the numerically dominant taxon. Detection limits for proportion of rotifers with either 6 or 3 replicates were <20% on most dates, except the first 3 months when relative densities of rotifers were low due to copepod population peaks, natural seasonal cycles, and a brief colonization period (Figure 2B).

Detection limits of other zooplankton metrics were not within the suggested criteria for rejecting the inverted proportional means hypothesis on any dates during the first year. Total density and rotifer density usually had detection limits of 60

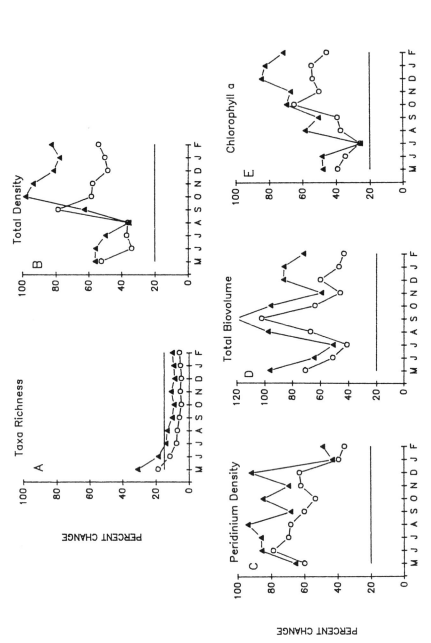

Figure 1.   Percent change that must occur to detect a treatment effect for phytoplankton metrics, May 1988 through February 1989. Circle = 6 replicate mesocosms; triangle = 3 replicate mesocosms; horizontal line = current EPA recommended criterion for biological significance.

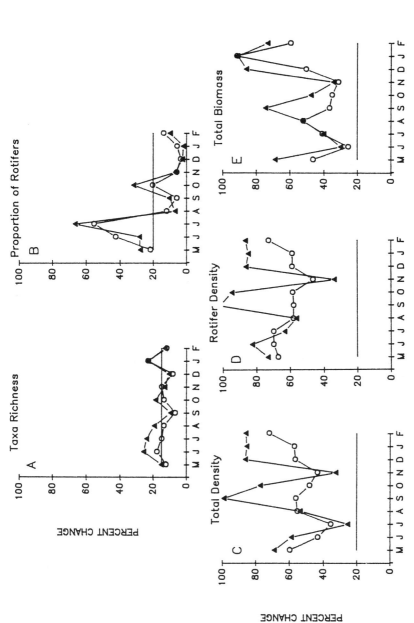

Figure 2. Percent change that must occur to detect a treatment effect for zooplankton community structural metrics, May 1988 through February 1989. Circle = 6 replicate mesocosms; triangle = 3 replicate mesocosms; horizontal line = current EPA recommended criterion for biological significance.

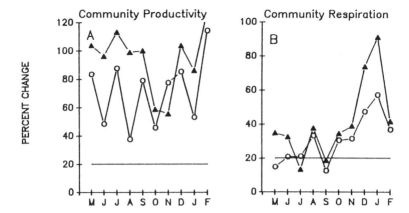

Figure 3.   Percent change that must occur to detect a treatment effect for zooplankton functional metrics, May 1988 through February 1989. Circle = 6 replicate mesocosms; triangle = 3 replicate mesocosms; horizontal line = current EPA recommended criterion for biological significance.

to 70% with 6 replicates and about 90% with 3 replicates (Figures 2C and D). Detection limits were not calculated for individual zooplankton genera or species, because variability was even greater for lower taxonomic levels than for higher taxonomic groupings, such as rotifers. Total biomass had detection limits similar to the density metrics (usually 40 to 50% with 6 replicates, 60 to 70% with 3 replicates; Figure 2E). The highest detection limits of any biometrics were for zooplankton community productivity (usually about 80% with 6 replicates, 100 to 110% with 3 replicates; Figure 3A). Community respiration, which is closely related to productivity, had detection limits that were considerably lower than productivity during part of the first year (Figure 3B). If only the warm months from May to November are considered, which corresponds to the period when greatest zooplankton respiration occurs, detection limits were often about 30% with 6 replicates and 40% with 3 replicates. Community productivity and respiration are functional metrics and may provide information that is useful for interpreting interactions between trophic levels.

Macroinvertebrates

Detection limits for the various macroinvertebrates ranged from 2 to 100% (Figure 4). Of the eight metrics evaluated, four had detection limits that were usually within the suggested criteria for rejecting the inverted proportional means hypothesis. Two of these were metrics that involve number of taxa. With 6 replicates, overall taxa richness had detection limits of 10 to 15% on most dates, but with 3 replicates detection limits were 15 to 25% (Figure 4A). The EOT Index, a subset of overall taxa richness, usually had the same range of detection limits with both six and three

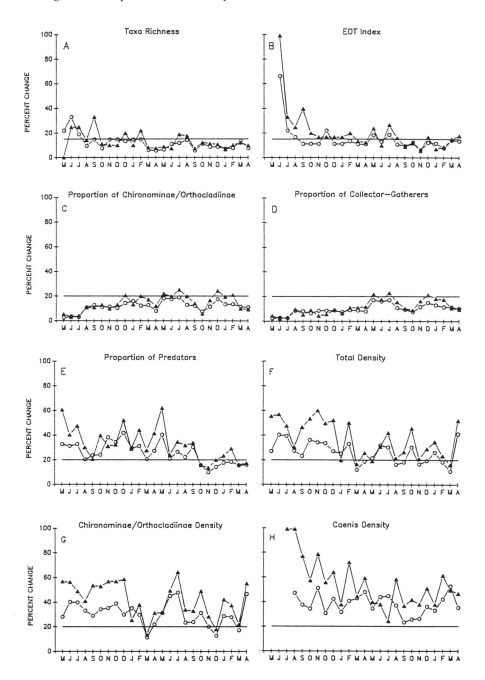

Figure 4. Percent change that must occur to detect a treatment effect for macroinvertebrate metrics, May 1988 through April 1990. Circle = 6 replicate mesocosms; triangle = 3 replicate mesocosms; horizontal line = current EPA recommended criterion for biological significance.

replicates (Figure 4B). In most cases, macroinvertebrate taxa were identified to the genus level. Detection limits for the proportion of Chironominae/Orthocladiinae and proportion of collector-gatherers metric were both usually 10 to 20% with 6 replicates and 20 to 25% and 15 to 20%, respectively, with 3 replicates (Figures 4C and 4D). The proportion of Chironominae/Orthocladiinae metric represents the proportion of the numerically dominant taxon. Chironominae and Orthocladiinae are the two most common subfamilies of Chironomidae, which were impractical to identify to genus. The proportion of collector-gatherers metric is the numerical proportion of the most abundant functional feeding group, which ingests fine particles of detritus from sediment. The detection limits of the proportion of collector-gatherers metric were usually about the same as those for the proportion of Chironominae/Orthocladiinae metric, because the majority of the collector-gatherers were members of these two Chironomidae subfamilies.

Detection limits of other macroinvertebrate metrics were not within the suggested criteria for rejecting the inverted proportional means hypothesis on most dates. The proportion of predators metric had detection limits of 30 to 35% with 6 replicates and 40 to 50% with 3 replicates (Figure 4E). Predators were the second most abundant functional feeding group. We analyzed three density metrics, which proved to have similar detection limits. With 6 replicates, detection limits were usually 30 to 35% for total density, 35 to 40% for Chironominae/Orthocladiinae density, and 40 to 50% for *Caenis* density (Figures 4F through H). With 3 replicates, detection limits for the same metrics usually were about 50%, about 60%, and 60 to 70%, respectively.

Comparison of the 2 years of macroinvertebrate data may be especially meaningful. During the first year colonization and developmental changes were definitely taking place in the unmanaged mesocosms, whereas these ecological events may have stabilized somewhat during the second year. Therefore, results from the second year may better represent the variability that will be found in mature, reasonably natural mesocosms. Detection limits were generally similar for all metrics between the first and second year. Proportion of collector-gatherers had slightly higher detection limits during the second year, and proportion of predators had slightly lower detection limits. Detection limits of the two taxa richness metrics also decreased slightly during the second year. All other metrics had no noticeable changes between years.

DISCUSSION

Colonization

Current EPA guidelines for pesticide registration tests suggest that new pond mesocosms be allowed to "season" for 6 months if they are lined with sediments from an established pond or 12 months if they are lined with topsoil. The additional length of time for mesocosms lined with topsoil is to develop benthic biota.[1] Howick et al.[5] and Ferrington et al. (Chapter 14) have reported that lining mesocosms with sediment from an established pond can be successful. In this study, we found that new pond mesocosms lined with topsoil will not be completely ready for testing within 12 months by natural processes alone; however, such pond mesocosms can possibly be made ready for conducting pesticide registration tests within that time with some management. All taxonomic groups that the EPA guidelines[1] suggest should be present in a "natural assemblage" of pond biota arrived in the mesocosms by natural colonization mechanisms well before the end of the first year, except Gastropoda. Densities of two other common nonflying macroinvertebrates, Oligochaeta and Amphipoda, were lower than would be expected in shallow lentic environments. However, populations of these three groups can easily be "seeded" by selectively collecting them from an established pond or purchasing them from a commercial supplier.

Obtaining an even distribution of biota among all mesocosms is a greater problem for getting topsoil-lined mesocosms ready for pesticide registration tests within 12 months than is obtaining a collective community typical of shallow lentic environments. This study may represent a worst-case condition because there was no biological management. Plankton communities might be the easiest to make homogeneous among mesocosms. deNoyelles et al.[29] demonstrated the success of draining pond mesocosms and refilling them from a common reservoir shortly before beginning experiments. Mixing some sediments among mesocosms may also be helpful, because planktonic organisms pass through resting stages on the bottom.

The macroinvertebrate community was more homogeneous among mesocosms, probably because the dominant aquatic insects readily disperse by air as adults. There may be little, if any, need for biological management of macroinvertebrate communities in topsoil-lined mesocosms that are allowed to colonize naturally for 12 months. Although only a little over half of the taxa occurred in all mesocosms during the first year, the mean number of taxa per mesocosm (17) was uniform during the latter part of the first year (coefficient of variation about 20%). In addition, most of the taxa not found in all mesocosms were exceptionally rare. Taxa common to all mesocosms accounted for 95% of the total density of macroinvertebrates. Using some other sampling methods (e.g., dip net) in addition to artificial substrates may be more effective in collecting rare taxa, because standardized artificial substrates, such as the ones used in this study, have been shown to be selective.[30] Mixing some sediments may also be helpful for making the macroinvertebrate community more uniform among mesocosms, because the eggs of some insects occur on the bottom. In addition, mixing sediments may transport adults of other macroinvertebrates that reproduce in the water (Oligochaeta, Crustacea, Gastropoda). The first appearance of macrophytes has been reported to be an important factor in the colonization of new lentic environments;[10] therefore, seeding macrophytes in all mesocosms would probably promote homogeneity of the macroinvertebrate community.

Our analyses during the second year suggest that macroinvertebrate colonization of the mesocosms had largely stabilized by the end of the first year. The following evidence supports this conclusion: (1) only a few noninsect taxa occurred sporadically or in very low numbers; (2) mean number of taxa per mesocosm only changed from 17 to 18; (3) coefficient of variation for mean number of taxa decreased from 20 to 12%; (4) proportion of taxa found in all mesocosms only increased from 57 to 71%; (5) taxa found in all mesocosms accounted for 95% of total density in both years; and (6) the dominant individual taxon (Chironomidae) accounted for 85% of total density in both years.

Biometrics

Phytoplankton, zooplankton, and macroinvertebrates have been used to demonstrate primary or secondary adverse effects in environmental toxicology studies conducted in mesocosms (see Reference 4 for a review). For regulatory purposes, such as pesticide registration by EPA, it is essential that mesocosm tests be able to confidently detect changes in the biota caused by treatment. Inherent variability in test systems as natural as pond mesocosms make confident detection of differences difficult. Our results indicate that very few metrics will be useful with the inverted proportional means hypothesis currently recommended. Metrics involving number of taxa, such as overall taxa richness or a subset of pollution sensitive taxa, will often have detection limits lower than the 15% reduction criterion currently recommended for use by EPA. Metrics involving proportions of numerical abundance comprised by dominant taxa or functional groups also will often have detection limits lower than the 20% change criterion currently recommended by the EPA. Using other metrics considered in this study may lead to erroneous regulatory decisions, because high detection limits (from high variability among mesocosms) would cause the inverse proportional means hypothesis to be accepted regardless of treatment effects. Density metrics demonstrated the least ability to confidently detect differences between treatment and control mesocosms. For all metrics, using six instead of three replicate mesocosms will usually improve the ability to detect differences.

Even with six replicate mesocosms, the detection limits of the most precise metrics sometimes are higher than the criteria currently recommended to reject the inverted proportional means hypothesis. In addition, we believe that some of the less precise metrics contribute useful biological information that should be included in regulatory decisions. Based on the results of this study, criteria for acceptable percent changes in biota due to treatment effects should be higher, for example: (1) <20% for metrics involving number of taxa, (2) <40% for metrics involving proportions of numerical abundance comprised by dominant taxa or functional groups, and (3) <60% for density and biomass metrics. Direct measurement of zooplankton community productivity does not appear to be a usable metric, but community respiration could be used as a functional metric with a criterion for acceptable change at <40%. However, final criteria should only be established after comparing the results of a number of studies of different toxic substances. Although some of the recommended detection limits are lower than what we measured in this study, it should be kept in mind that the results of our studies may represent worst-case conditions that might be ameliorated by simple biological management of mesocosms.

It is important to consider ecological significance as well as statistical significance when making environmental regulatory decisions, but there are few empirical data on what constitutes a "significant" ecological change. Although the detection limits for proportions and abundances of biota in this study may seem high, much higher changes in these metrics have been reported from natural unperturbed ecosystems.[31] Ecosystems may be characterized by resistance and resilience,[32] and many of the organisms that were the subjects of this study may have rapid turnover rates because of short life histories and rapid growth. We suggest that the detection limits from this study would probably protect ecosystems adequately from adverse effects, because there may be considerable redundancy in ecosystem function and usually the transfer of energy between trophic levels is only about 10%.[33]

We conclude that the approach we have used for determining the level of change that can be detected statistically should be followed in any pond mesocosm study that is to be conducted according to an hypothesis-testing design. This analysis could also be performed after the fact to evaluate statistical significance in terms of the power of the experiment for detecting change. With some pretreatment data, this approach makes it possible to establish biological criteria that are realistic for a particular study and will also provide adequate environmental protection.

ACKNOWLEDGMENTS

Clayton Stunkard, University of Maryland and Statistical Policy Branch, Science, Economics and Statistics Division, U.S. Environmental Protection Agency, was kind enough to review this manuscript and offer valued comments on its presentation. We are grateful for the assistance of James L. Tramel, Jr. and W.B. Wilkinson, III for on-site administration and management of the experimental pond facility. We thank Frank deNoyelles, Jr., University of Kansas, and James H. Kennedy, University of North Texas, for advice and encouragement throughout the study. Persons who provided valuable technical assistance included: Stephen W. Hiner, Timothy J. Morgan, Michael O. West, T. Michael Williams, and Lourdes M. George. The project was supported by Virginia Polytechnic Institute and State University and Cooperative Agreements CR-814358-01-0 and CR-814358-02-1 with the Athens Environmental Research Laboratory of the U.S. Environmental Protection Agency. Although this research was funded in part by the EPA, it has not been subjected to the Agency's optional peer and policy review and, therefore, does not necessarily reflect the views of the Agency, and no official endorsement should be inferred.

REFERENCES

1. Touart, L.W. 1988. Aquatic mesocosm tests to support pesticide registrations. U.S. Environmental Protection Agency Hazard Evaluation Division. Washington, D.C.
2. Touart, L.W. and M.W. Slimak. 1989. Use of experimental ponds to assess the effects of a pesticide on the aquatic environment. In J.R. Voshell, Jr. ed., *Using Mesocosms to Assess the Aquatic Ecological Risk of Pesticides: Theory and Practice*. Miscellaneous Publication Number 75, Entomological Society of America, Lanham, MD. pp. 33–40.

3. Voshell, J.R., Jr. 1989. Introduction. In J.R. Voshell, Jr. ed., *Using Mesocosms to Assess the Aquatic Ecological Risk of Pesticides: Theory and Practice*. Miscellaneous Publication Number 75, Entomological Society of America, Lanham, MD. pp. 1–3.
4. Buikema, A.L. and J.R. Voshell, Jr. 1993. Toxicity studies using freshwater benthic invertebrates. In D.M. Rosenberg and V.H. Resh, eds., *Freshwater Biomonitoring and Benthic Macroinvertebrates*. Chapman and Hall, New York, NY, pp. 344–398.
5. Howick, G.L., J.M. Giddings, F. deNoyelles, L.C. Ferrington, Jr., W.D. Kettle, and D. Baker. 1990. Rapid establishment of test conditions and trophic level interactions in 0.04-ha earthen pond mesocosms. *Environmental Toxicology and Chemistry* 11:107–114.
6. Paterson, C.G. and C.H. Fernando. 1970. Benthic fauna colonization of a new reservoir with particular reference to the Chironomidae. *Journal of the Fisheries Research Board of Canada* 27:213–222.
7. Street, M. and G. Titmus. 1979. The colonization of experimental ponds by Chironomidae (Diptera). *Aquatic Insects* 4:233–244.
8. Danell, K. and K. Sjoberg. 1982. Successional patterns of plants, invertebrates, and ducks in a man-made lake. *Journal of Applied Ecology* 19:395–409.
9. Barnes, L.E. 1983. The colonization of ball-clay ponds by macroinvertebrates and macrophytes. *Freshwater Biology* 13:561–578.
10. Voshell, J.R. Jr. and G.M. Simmons, 1984. Colonization and succession of benthic macroinvertebrates in a new reservoir. *Hydrobiologia* 112:27–39.
11. Stunkard, C.L. 1989. Statistical Hypotheses at EPA. Report submitted to Statistical Policy Branch, Office of Policy, Planning, and Evaluation, U.S. Environmental Protection Agency. Washington, D.C.
12. Geisy, J.P. 1990. SETAC: Part of the solution or part of the problem? *Environmental Toxicology and Chemistry* 9:1327–1330.
13. Boyle, T.P. 1983. Role and application of semicontrolled ecosystem research in the assessment of environmental contaminants. Aquatic Toxicology and Hazard Assessment: Sixth Symposium, ASTM STP 802, W.E. Bishop, R.D. Cardwell, and B.B. Heidolph, eds., American Society for Testing and Materials, Philadelphia, 1983, pp. 406–413.
14. Elliott, J.M. 1977. Some methods for the statistical analysis of samples of benthic macroinvertebrates. 2nd ed. Scientific Publication No. 25. Freshwater Biological Association, Cumbria, England.
15. Green, R.H. 1979. *Sampling Design and Statistical Methods for Environmental Biologists*. John Wiley & Sons, New York, NY.
16. Sokal, R.R. and F.J. Rohlf. 1981. *Biometry*. W.H. Freeman and Company, San Francisco, CA.
17. Resh, V.H. and E.P. McElravy. 1993. Contemporary quantitative approaches to biomonitoring using benthic macroinvertebrates. In D.M. Rosenberg and V.H. Resh, eds., *Freshwater Biomonitoring and Benthic Macroinvertebrates*. Chapman and Hall, New York, NY, pp. 159–194.
18. Virginia Cooperative Extension Service. 1984. A handbook of agronomy. Virginia Polytechnic Institute and State University, Blacksburg, VA.
19. Layton, R.J. and J.R. Voshell, Jr. 1993. Colonization of new experimental ponds by benthic macroinvertebrates. *Environmental Entomology,* 20:110–117.
20. Jenkins, D.G. 1990. Structure and function of zooplankton colonization in twelve new experimental ponds. Ph.D. dissertation. Virginia Polytechnic Institute and State University, Blacksburg, VA.

21. American Public Health Association, American Water Works Association, and Water Pollution Control Federation. 1985. Standard methods for the examination of water and waste water. 16th edition. American Public Health Association. Washington, D.C.

22. Ganf, G.G. 1974. Phytoplankton biomass and distribution in a shallow eutrophic lake (Lake George, Uganda). *Oecologia* 16:9–29.

23. Rosenzweig, M.S. 1990. Phytoplankton colonization and seasonal succession in new experimental ponds. M.S. thesis. Virginia Polytechnic Institute and State University, Blacksburg, VA.

24. Owens, T.G. and F.D. King. 1975. The measurement of respiratory electron-transport-system activity in marine zooplankton. *Marine Biology* 30:27–36.

25. Brigham, A.R., W.W. Brigham, and A. Gnilka. 1982. Aquatic insects and oligochaetes of North and South Carolina. Midwest Aquatic Enterprises, Mahomet, IL.

26. Merritt, R.W. and K.W. Cummins. 1984. *An Introduction to the Aquatic Insects of North America.* Kendall/Hunt Publishing Company, Dubuque, IA.

27. Layton, R.J. 1989. Macroinvertebrate colonization and production in new experimental ponds. Ph.D. dissertation. Virginia Polytechnic Institute and State University, Blacksburg, VA.

28. Plafkin, J.L., M.T. Barbour, K.D. Porter, S.K. Gross, and R.M. Hughes. 1989. Rapid bioassessment protocols for use in streams and rivers: benthic macroinvertebrates and fish. EPA/444/4-89-001. U.S. Environmental Protection Agency, Office of Water, Washington, D.C.

29. deNoyelles, F., Jr., W.D. Kettle, C.H. Fromm, M.F. Moffett, and S.L. Dewey. 1989. Use of experimental ponds to assess the effects of a pesticide on the aquatic environment. In J.R. Voshell, Jr. ed., *Using Mesocosms to Assess the Aquatic Ecological Risk of Pesticides: Theory and Practice.* Miscellaneous Publication Number 75, Entomological Society of America, Lanham, MD. pp. 1–3.

30. Rosenberg, D.M. and V.H. Resh. 1982. The use of artificial substrates in the study of freshwater benthic macroinvertebrates. In J. Cairns, Jr. ed., *Artificial Substrates.* Ann Arbor Science Publishers Inc. Ann Arbor, MI. pp. 175–235.

31. Resh, V.H. and J.K. Jackson. 1993. Rapid assessment approaches to biomonitoring using benthic macroinvertebrates. In D.M. Rosenberg and V.H. Resh, eds., *Freshwater Biomonitoring and Benthic Macroinvertebrates.* Chapman and Hall, New York, NY, pp. 195–233.

32. Odum, E.P. 1983. *Basic Ecology.* Saunders College Publishing, Philadelphia, PA.

33. Smith, R.L. 1980. *Ecology and Field Biology.* Harper and Row, Publishers, New York, NY.

CHAPTER 11

Applying Concentration-Response Theory to Aquatic Enclosure Studies

Dean G. Thompson, Stephen B. Holmes, Douglas G. Pitt, Keith R. Solomon, and Kerrie L. Wainio-Keizer

Abstract: Exploratory analysis of field-derived data highlighted a number of important statistical and experimental design considerations for aquatic mesocosm studies. Abundance data for numerically dominant zooplankton taxa were characterized by distinct heterogeneity of variance (coefficient of variation range 10 to 173%) proportional to the mean and related to both time of observation (0 to 77 days postapplication) and taxonomic level of investigation. Log-transformation did not obviate violations in parametric statistical assumptions, necessitating the use of procedures robust to or not assuming variance homogeneity. Using appropriate RM-ANOVA techniques and multiple comparison tests, an overall LOEC value of 1.0 ppm was estimated for effects on numerical abundance. Log-transformation served to linearize the inherently curvelinear concentration-response relationship. Threshold and temporal effects on the concentration-response relationship resulted in significant lack of fit for the linear model in many cases. Statistically significant, adequately fitting models were observed in 21 of 63 cases, with 16 of these meeting the minimum criteria ($F_{regression}/F_{(0.05; 1,12)} > 4$) for predictive value of the model. A nonlinear regression technique using an exponential decline model wherein both a and b parameters were considered as linear functions of time provided good fit statistics for 5/8 taxa investigated. Examination of residuals provided *a posteriori* evidence that the nonlinear model adequately addressed the three-dimensional nature of the problem. EC_{50} values estimated using linear and nonlinear regression procedures differed markedly, ranging from 0.004 to 0.045 and 0.06 to 1.26, respectively. Potential improvements to experimental design and data analysis, including disproportionate replication and three-dimensional methods of data analysis, were identified in relation to regulatory assessment of aquatic mesocosm impact data.

0-87371-592-6/94/$0.00 + $.50
© 1994 by CRC Press, Inc.

INTRODUCTION

Any toxicological response is a function of both the concentration and duration of exposure,[1] a concept which is fundamental to the experimental design and statistical analysis of all toxicity investigations. Laboratory toxicity tests, conducted under highly controlled conditions and using well-defined protocols with preselected test concentrations, may be analyzed effectively by linear regression techniques following appropriate (probit) transformation of the data. Under these conditions, relatively precise toxicological endpoints (EC_{50} and LOEC) are derived and currently serve as the cornerstone of environmental risk assessment.

Recently, scientists have recognized that laboratory toxicity tests may be an insufficient basis for protecting the structure and function of higher levels of biological organization.[2,3] As a result, development of protocols for aquatic toxicity testing in the field has increased dramatically. Field-level toxicity testing has generally evolved from observational studies in natural ponds,[4-6] through split-pond studies, to current protocols employing multiple ponds[7-9] or *in situ* enclosures. The introduction of the mesocosm concept[10] was critical in the development of highly replicated and experimentally designed studies which are now relatively standard practice in aquatic impact assessment.[11-16]

Mesocosm studies have generally been conducted using balanced designs involving three replications of four test concentrations (including untreated controls).[17] The protocol is commonly referred to as the ANOVA design or approach since results are subjected to standard ANOVA hypothesis testing in an effort to detect differences between treatment and control means. Recently, research interest has focused on the potential benefits of designing mesocosm experiments with a series of test concentrations with unequal or no replication, a protocol commonly referred to as the regression (REG) approach.

The REG protocol is regarded favorably owing to potential economic advantages, statistical flexibility associated with replication requirements, the interpolative capability inherent in the regression model, and the fundamental relation of regression to the concentration-response relationship. Results from such experimental designs may be analyzed by linear techniques following appropriate transformation or nonlinear techniques which directly reflect the intrinsic curve linearity of concentration-response relationships.[1]

Substantial research efforts continue in an attempt to define the most suitable protocol for simulated field studies of aquatic impact. Recently, a compromise between these two basic experimental protocols (the "hybrid" design) has been suggested as an alternative approach.[18] Studies addressing the influence of innate characteristics of field-derived data on statistical inference methods are generally lacking. Further, the effectiveness of hybrid experimental designs has not been verified in actual field experiments. Graney et al.[19] recently reviewed the advantages and disadvantages of ANOVA and REG approaches using computer simulated data and identified the need to design a study which could be effectively analyzed by ANOVA and REG techniques.

In this paper, data derived from a field study on the impact of herbicides to freshwater plankton communities are used to examine advantages and disadvantages of repeated measures analysis of variance (RM-ANOVA), linear regression (L-REG), and nonlinear regression (NL-REG) methods of statistical analysis. In addition, the general acceptability of the "hybrid" experimental design for aquatic impact studies is evaluated. Results are used to identify the most appropriate statistical techniques for analyzing the data set *in toto* and to suggest improvements in aquatic mesocosm experimental design and analysis.

MATERIALS AND METHODS

Site Selection

The experimental lake is typical of small, boreal, lentic ecosystems which might be inadvertently oversprayed during aerial applications of herbicides in forest vegetation management. The study was conducted under environmental conditions typical of northern Ontario, to which resulting risk assessments might ultimately be applied. Enclosures were deployed in a shallow bay of Greenwater Lake, Laveyendre Township, Ontario, where the bottom slope was minimal and sediments were suitable for sealing the enclosures.

Experimental Design

Enclosures constructed of impervious polyethylene sidewalls suspended from wood/styrofoam floats and anchored into natural bottom sediments served as individual experimental units for this study (Figure 1). Enclosures were deployed in a completely randomized, balanced experimental design, involving two treatments (hexazinone and metsulfuron methyl) with four levels (nominal concentrations) of each treatment. Each treatment level including untreated control was replicated three times. The 27 enclosures were arranged side-by-side in a 9 × 3 matrix (Figure 1). Deployment of the enclosures involved attaching sidewalls to the floats and iron frame and then lowering the frame into the bottom sediments. In this way, relatively undisturbed columns of water inclusive of natural phytoplankton and zooplankton communities were isolated from the open lake. Following deployment of all enclosures, each experimental unit was inspected by scuba divers to ensure complete submersion of the bottom frames into the sediments. During inspection, the side wall of one enclosure (labeled 32, Figure 1) was observed to be resting on a large sunken log, prohibiting proper sealing. Removal of the log and resealing of the enclosure caused considerable resuspension of bottom sediments within the enclosure. The resultant nutrient enrichment of the water column stimulated plankton growth substantially beyond that of all other enclosures, ultimately necessitating removal of corresponding data (enclosure 32: treatment = hexazinone; level = 0.01 mg/l) as outliers in the statistical analysis.

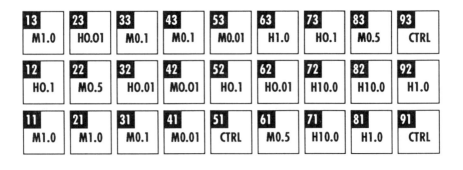

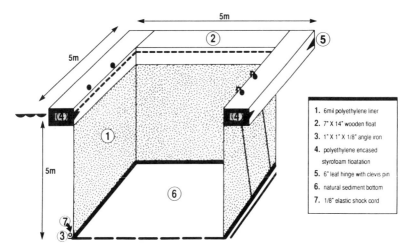

Figure 1.   Experimental layout and design detail for *in situ* aquatic enclosures.

Herbicides and Treatment Levels

Hexazinone (VELPAR* L 240 g/l), a member of the triazine class, and metsulfuron methyl (ESCORT* 60% DF), a member of the sulfonylurea class of herbicides, were used in this study. Hexazinone has recently received aerial registration for use in Canadian forest vegetation management and thus has the potential to enter natural bodies of water through drift or accidental overspray. Metsulfuron methyl, an experimental herbicide in Canadian forest vegetation management, has similar potential. In addition, both compounds are characterized by high water solubility and low sorption coefficients which suggest a high potential to move off treated sites via leaching or surface runoff. The use pattern and potential mobility dictated experimental treatment levels covering the full range of expected environmental concentrations (EEC) ranging from drift to accidental direct overspray scenarios.

* Registered trademark of E.I. DuPont de Nemours, Inc., Wilmington, DE.

Treatment levels for this experiment were established based upon limited published aquatic toxicity data and calculated worst-case expected environmental concentrations (EEC) (Figure 2). The worst case EEC, calculated as the concentration of the chemical resulting from the direct overspray at the maximum label rate (hexazinone @ 4.0; metsulfuron methyl @ 0.12 kg/ha ai) into a water body 0.5 m in depth, were 0.8 mg/l and 0.024 mg/l, respectively. Only published aquatic toxicity data for freshwater algae and zooplankton were considered since the experiment focused on direct and indirect impacts on these two communities. Nominal test concentrations for hexazinone (0.0, 0.01, 0.1, 1, 10 mg/l) and metsulfuron methyl (0, 0.01, 0.1, 0.5, 1.0 mg/l) were chosen to span the range of approximately 0.1 to 10 times the worst-case EEC and include the best available estimates of LOEC and EC_{50} for algae (Figure 2). The highest concentrations were employed to ensure that the majority of the species in the natural assemblage of phytoplankton and periphyton would respond significantly to at least one test concentration, thereby allowing interpolation of risk for the entire range of exposure scenarios, inclusive of overspray, drift, and runoff events.

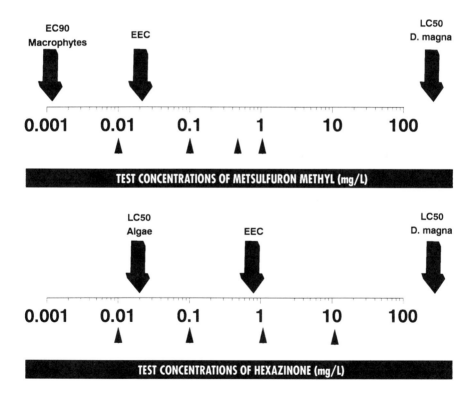

Figure 2. Graphical illustration of test concentrations in relation to selected toxicological endpoints for metsulfuron methyl and hexazinone herbicides.

General Assessment Parameters

Overall, toxicant impact was assessed primarily in terms of direct effects on numerical abundance of phytoplankton and community structure and secondary effects on zooplankton abundance and community structure. Indirect effects on system function in terms of water chemistry (dissolved oxygen, pH, major ions) and nutrient cycling were also investigated.

Data Subset for Exploratory Statistical Analysis

Abundance data for the four numerically dominant zooplankton taxa (*Keratella cochlearis* [Taxon 10]; *Ploesoma* spp. [Taxon 19]; *Bosmina* spp. [Taxon 32]; *Holopedium* spp. [Taxon 35]), as well as their respective higher taxonomic groups (total Rotifera [Taxon 29]; total Cladocera [Taxon 39]; total Copepoda [Taxon 79]; and total Zooplankton [Taxon 89]) comprised the data used in this analysis. The intent was to explore the general nature of these data and, from the exploratory analysis, choose the most suitable statistical approach for analysis and presentation of the data *in toto*. Zooplankton populations were sampled using an integrating tube sampler similar to that described by Solomon.[20] The sampler was fitted with a 30-micron NYTEX plankton net and bucket, allowing filtration of the single sample water volume of 30.8 l collected in the detachable bucket. For each enclosure, a total of 5 subsamples, taken to a constant depth of 2.5 m in a predetermined V-pattern,[21] were pooled to form a single composite field sample (5 × 30.8 = 154 l filtered water). Each enclosure was sampled 13 discrete times relative to the day of chemical application (i.e., days — 7, 0, 1, 2, 3, 7, 10, 14, 21, 28, 42, 56, 77). Hexazinone concentrations in the enclosures were verified by analyzing a subsample of depth-integrated (2 m) water samples for hexazinone residue using a validated capillary GLC technique employing thermoionic specific detection (TSD) of hexazinone. Details of the residue sampling methodology and the analytical technique are reported elsewhere.[22] Since observed mean concentrations were very similar to nominal concentrations at each of the treatment levels (see results), initial exploratory analyses were conducted using nominal concentrations as the independent variable.

STATISTICAL ANALYSIS

Initial Data Exploration

Basic descriptive statistics (means, standard deviations [STD] and coefficients of variation [CV]) were calculated to allow examination of inherent variability in the data. Variance homogeneity was evaluated using Levene's test ($p = 0.01$) for homogeneity of variances.[23] Diagnostic plots of means vs. variances and residuals were used to check for violations in standard parametric statistical assumptions (independence, normality, homogeneity of variance, additivity).[24] Log transformation of the data was employed in an attempt to correct for distinct heterogeneity of variance or other violations of parametric statistical assumptions.[24,25] The transformation used was log abundance = $\log_{10}$ (abundance + 0.006), where abundance is zooplankton counts (#/l) and where the scaling factor 0.006 is the lowest possible nonzero count added to the data to incorporate zero values in the original dataset without "swamping" the relative variability in the data.[26]

Repeated Measures ANOVA

Since preliminary examination of the data indicated distinct violations of parametric statistical assumptions, the data were log transformed prior to applying repeated measures analysis of variance (RM-ANOVA) using a mixed models univariate approach. The null hypothesis of no-effect of concentration $\times$ time interactions was tested ($p = 0.05$) using the RM-ANOVA F statistic associated with the interaction term.[27] Levene's test ($p = 0.01$) was used to test for equality of variance within the factors (time, hexazinone concentration, time *hexazinone concentration interaction).[23] Since Levene's test, applied to log-transformed data, was significant ($p < 0.01$) for both the time and hexazinone concentration factors in every case, Brown-Forsythe ANOVA ($p = 0.05$) was used to test the statistical significance of the concentration effect within a given day. The Brown-Forsythe ANOVA differs from standard ANOVA procedures in that it does not assume homogeneity of variance.[28] Finally, in cases where both RM-ANOVA and Brown-Forsythe ANOVA tests were significant, treatment means statistically different from controls within days were elucidated using Welch's approximate t-test. Welch's test, which assumes unequal variance, was conducted as described by Day and Quinn,[29] using Satterthwaite's adjusted degrees of freedom and constrained to a total of 40 comparisons. The procedure, as conducted, holds the experimentwise error rate (EER) less than alpha = 0.05, by using a Bonferroni adjustment of the comparison-wise significance level (i.e., alpha per number of comparisons = (0.05/40 = 0.001)). The approach was employed to control the high probability of a type I error which might otherwise result from such a large number of comparisons. Results of Welch's approximate t-test were compared to a similarly constrained Bonferroni multiple comparison procedure assuming homogeneity of variance. Lowest observable effect concentrations (LOEC) were considered to be those test concentrations yielding a statistically significant (Welch t-test) reduction in zooplankton abundance on two consecutive sampling days.

Linear Regression Analysis

Preliminary investigation of the concentration-response relationship was conducted using scatter plots of untransformed data. Scatter plots displayed patterns of exponential decline and thresholds reflective of inherent nonlinearity in theoretical concentration-response relationships.[1] Since initial exploration of the residuals about linear models clearly identified violations of normality, homoscedasticity, and additivity assumptions, log-transformations of both the dependent and independent data were employed in an attempt to linearize the relationship and confer homoscedasticity.[30]

Plankton are known to be characterized by "patchy-distribution".[31,32] As a result, substantial differences in initial population levels among replicate enclosures were anticipated. Therefore, the linear regression analysis used the initial abundance of a given taxon in each experimental unit as a covariate in the full model:

$$(Y + 0.006) = a + b_0(\log X + 0.001) + b_1(\log N_0) + b_2(\log N_0 x \log X) + \epsilon$$

where Y = abundance of zooplankton or phytoplankton taxa (organisms/l); a = intercept; $b_{0,1,2}$ = partial regression coefficients for toxicant concentration, covariate, and interaction terms, respectively; X = toxicant concentration (mg/l); N_0 = abundance of zooplankton or phytoplankton taxa in a given experimental unit on day 0; and ϵ = error.

A backward stepwise elimination process[30] was used to fit the model, with nonsignificant parameters eliminated until only parameters with significant ($p < 0.05$) partial regression coefficients remained.

Replication in the experimental design allowed for residual sums of squares to be broken down into pure error and lack of fit components to be used in testing for lack of fit.[24,30] In this case, a value of $F_{LOF} < F_{(0.05;3,9)}$ provided evidence to conclude adequate fit of the linear model. Further evaluation of adequately fitting linear models was conducted by examination of frequency distributions and plots of normal probability, residuals vs. predicted Y, and standardized residuals. Models were considered to have predictive value based on a minimally acceptable criterion of the $F_{regression}/F_{(0.05; 1,12)}$ ratio > 4.[30] Only predictive models were used to derive EC_{50} value estimates by inverse regression analysis.[25]

Nonlinear Regression Analysis

Based on initial examination of data scatterplots, an exponential decline model with a and b parameters as linear functions of day was used for nonlinear regression analysis. As for the L-REG procedure, initial abundance for each taxa was incorporated as a covariate in the full model:

$$Z = (a_0 + a_1 * t)e^{((b_0 + b_1*t)*X)} + c N_0$$

where Z = zooplankton taxa abundance (organisms per l); a and b = estimable parameters as linear functions of day; c = partial regression coefficient for the covariate; N_0 = covariate = average abundance of a taxa on day 0 (all enclosures); t = time (days postapplication); X = hexazinone concentration (mg/l).

Inclusion of zero within the 95% confidence intervals (CIs) identified parameters not significantly different from zero. Based on this criterion and using a reverse stepwise approach, the simplest model with all independent parameters significant was sought. Only models resulting in the same point of convergence (global minimum sums of squares) with test runs using several different starting parameters were entertained. In these cases, models were accepted if parameters were significant and not highly correlated ($r < 0.70$), if residuals plotted against predicted values complied with standard regression assumptions[30] and if the model adequately reflected observed trends in the data. Using accepted models, EC_{50} values for selected days of interest were calculated by inverse regression analysis with a specific input value for the time variable and $N_0/2$ used as the estimate of 50% of initial abundance for a given taxa.

RESULTS AND DISCUSSION

General Data Characteristics

Observed mean (n = 3) hexazinone concentrations on day 0 ($\pm$ standard deviations) were 0.032 ± 0.008; 0.102 ± 0.011; 1.1064 ± 0.078; 15.817 ± 1.534 mg/l, respectively, for the four treatment levels. The observed values were consistent with nominal concentrations (0.01, 0.1, 1.0, and 10 mg/l) with the exception of the highest treatment level. Since observed concentrations for the high test concentration averaged 11.276 mg/l on days 1 through 3, we suspect that higher than nominal concentrations for day 0 were an artifact of incomplete vertical mixing. Coefficients of variation ($<10\%$) for the three highest mean concentrations indicate relatively precise application. As expected, replication and quantitation at the lowest treatment level was less precise (CV = 26.7%). The precision and accuracy of hexazinone treatment applications, as confirmed by GLC-TSD quantitation of aqueous concentrations on day 0, supported the use of nominal concentrations for initial exploratory data analysis.

Based on average abundances during the period of observation, the zooplankton community in untreated control enclosures was clearly dominated by *Keratella cochlearis* (av 18.6 per l), with other dominant taxa being *Holopedium* spp. (av 7.7 per l), *Ploesoma* spp. (av 6.5 per l), and *Bosmina* spp. (av 6.0 per l). With the exception of *Ploesoma* spp., populations were relatively stable until approximately day 42 (12-09-89). After this time, a precipitous drop in numbers was generally observed, coincident with declining water temperatures and primary productivity, as typical for northern temperate lake systems. No simple explanation for the dramatic decline in abundance of *Ploesoma* spp. during the first week postapplication could be found; however, dramatic fluctuation in zooplankton populations in response to natural environmental stimuli is not uncommon. CVs for the three replicate controls were calculated for each day of observation within a given taxon. The range in CVs (8.1 to 173.2%) indicates the high variability among replicate untreated enclosures and is similar to values reported by Shaw et al.[33] Comparison of CVs within taxa over the period of observation suggested that variation in abundance estimates for the four dominant individual taxa were relatively similar (range of av CV = 83.9 to 97.5%) and substantially greater than those of higher taxonomic groups (range of av CV = 34.4 to 46.7%). An exception, Total Rotifera was characterized by average variation (av CV = 74.5%) approaching that of the individual taxa. The total zooplankton community (av 69.4 per l) was dominated by Rotifera (av 31.5 per l), with Copepoda (av 24.8 per l) and Cladocera (av 19.0 per l) represented in smaller numbers.

Examination of residuals using common diagnostic plots (Figure 3) revealed general trends of variance heterogeneity, confirmed by Levene's tests, which were typically highly significant ($p < 0.001$). Plots for taxon 89 — day 42 are presented as examples of typical trends observed in diagnostic analysis. For the nontransformed data a clear "S-pattern" trend was observed in normal probability plots (Figure 3a). Variance heterogeneity was clearly evident (Figure 3b notes 5 points superimposed at low predicted value) with variance inversely proportional to predicted values. Finally, the frequency distribution was somewhat abnormal. In our judgement, the variance heterogeneity and distribution of residuals observed for the raw data, clearly conflicted with the fundamental parametric statistical assumptions of normality and homoscedasticity. As recommended,[24,25,30] data were transformed in an attempt to normalize the error distribution and impart homoscedasticity, such that more powerful parametric statistical procedures could be used to test null hypotheses. Examination of similar diagnostic plots for log-transformed data showed improvement in residual characteristics particularly in the plot of predicted vs. residuals (compare Figure 3b and Figure 3e). Substantial reduction of the variance-mean correlation observed in the raw data (r = 0.88) resulted from log-transformation (r = 0.47). Levene's test for homogeneity of variance was generally less significant for log-transformed data but indicated that log-transformation did not confer complete homoscedasticity to all of the data. Similar effects observed for the majority of the data indicated that violations of normality and homoscedasticity assumptions were substantially alleviated, but not eliminated, by log-transformation. Thus, further exploratory statistical analysis was conducted using the log-transformed data.

RM-ANOVA

The RM-ANOVA F-statistic for time*hexazinone concentration interaction was significant ($p < 0.05$) for all taxa investigated. Levene's test ($p < 0.05$) confirmed that significant heterogeneity of variance remained in association with both factors (time and hexazinone concentration), even following log-transformation of the data. At the maximal values of either time or concentration factors, species numbers and abundance were reduced to very low levels, resulting in simplified zooplankton population structures relative to either within day controls or time zero. We postulate that significant variance heterogeneity will be typical of mesocosm experiments conducted with lengthy periods of observation or employing a wide range of test concentrations.

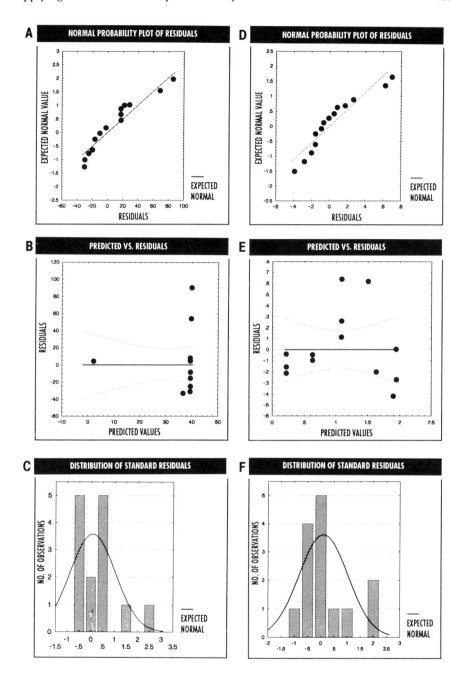

Figure 3.   Sample diagnostic plots for linear regression of untransformed (a,b,c) and log-transformed (d,e,f) data for Total Zooplankton.

Given significant variance heterogeneity even after log-transformation, the Browne-Forsythe ANOVA procedure, which does not assume equal variances, was used to further investigate the concentration-response relationship. In all cases, the F-statistic for the hexazinone concentration factor was highly significant ($p < 0.001$), indicating that averaged over all days significant treatment level effects had occurred. The statistical significance ($p = 0.001$) associated with the time factor in the RM-ANOVA analysis indicated that nontreatment related effects on the zooplankton population also occurred. Presumably, these reflect typical late-season declines in zooplankton populations in response to environmental cues (cooler water temperatures, lower food availability, etc.); however, these are not further considered here. Given the relationship between test concentrations and laboratory-based acute toxicity data (Figure 2), we suggest that impacts on zooplankton observed in this study are likely to be the result of secondary effects associated with reduced phytoplankton, oxygen depletion, or a combination of these and other environmental stress factors, rather than direct toxic effects of hexazinone alone.

Treatment means elucidated as significantly different from controls within a given day (as determined by Welch's approximate t-test; alpha = 0.05) are shown graphically in Figure 4. Although the data were clearly characterized by variance heterogeneity, results of the Bonferroni multiple comparison procedure assuming constant variance were equivalent to that of the Welch approximate t-test in 84% of the 340 comparisons tested. In the remaining 16% of comparisons, the Bonferroni procedure claimed significance where the Welch test did not. The degree of correlation in results between the Welch t-test and Bonferroni multiple comparison with pooled variance suggests that, in this case, the latter procedure was relatively robust to variance heterogeneity. Shaw et al.[33] noted that 20 to 50% change in biological parameters have been suggested as ecologically significant. Applying these criteria to this study, both the Bonferroni procedure and Welch approximate t-test retained sufficient power to detect differences of this magnitude in cases involving higher taxonomic groups and higher treatment levels which were characterized by relatively lower variability. However, the high inherent variability associated with comparisons involving lower treatment levels and/or individual taxa generally precluded differences below the 50% level from being detected as statistically significant. The power of these tests may be increased by altering the alpha level (i.e., to 0.10). However, due consideration must be given the increased probability of a type I error which is concomitant with this approach.

Threshold effects clearly evident in the data, as exemplified in plots for *Keratella cochlearis* (taxon 10) days 21 to 42 (Figure 4a) and *Bosmina* spp. (taxon 32) on day 56 and 77 (Figure 4c), were also evident in similar plots for phytoplankton (not presented here). These effects appear to be attributable to differential rates of impact and recovery, with populations exposed to the two lowest test concentrations (0.01 and 0.1 mg/l) requiring relatively longer periods of exposure to elicit the same reduction in abundance as compared to populations exposed to higher (1.0 and 10.0 mg/l) test concentrations. In addition, the zooplankton exposed to the two lowest concentration exhibited recovery, while those exposed to higher treatment levels did not.

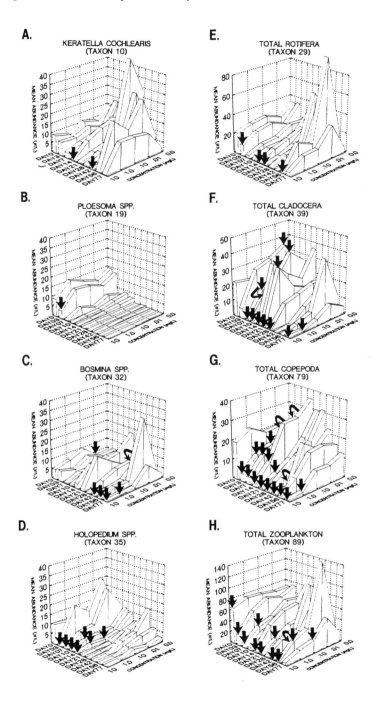

Figure 4. Three-dimensional block plots of mean abundance for dominant zooplankton taxa in relation to time and hexazinone concentration. Straight arrow denotes treatment means significantly different from controls within days as determined by Welch approximate *t*-test (alpha = 0.05). Curved arrow indicates significantly different treatment mean obscured by other data.

The data plots also provided evidence of stimulatory effects (mean abundance in hexazinone treatments substantially and/or significantly greater than controls), exemplified in graphics for *Holopedium* spp. (taxon 35), days 3 and 7 (Figure 4d) and total Cladocera (taxon 39), days 3 and 7 (Figure 4f). Although stimulatory effects were infrequently observed, particular association with the lowest test concentration tested (0.01 mg/l) and correlation of the phenomena to similar effects in phytoplankton data (not reported) lead us to conclude that they were real effects as opposed to artifacts.

Statistical differences in mean numerical abundance were associated primarily with the 1 and 10 mg/l test concentrations. Although spurious statistical differences in mean abundance were observed at lower treatment levels, we suggest that these are of questionable ecological significance given their duration, the natural variation in zooplankton population abundance, and the substantial fluctuations of zooplankton populations in response to natural stimuli. Therefore, considering statistically significant (via ANOVA) reductions in abundance on more than two consecutive days of observation as representative of an ecologically important effect, the LOEC for hexazinone impact on zooplankton may be approximated as 1 mg/l.

The RM-ANOVA technique is significantly constrained since it does not incorporate a functional model of the concentration-response relationship and thus lacks interpolative ability. The tiered approach of mixed effects, main effects, and multiple comparison testing is time consuming and cumbersome. Further, the advantage of testing specific null hypotheses of differences among means may be seriously impaired by the lack of power associated with comparison where variation is inherently high.

Linear Regression

As a preliminary step in the linear regression analysis, untransformed abundance data (#/l) were regressed against the independent variable of hexazinone concentration (mg/l). Fit statistics were generally poor and analysis of the residuals indicated violations of the basic assumptions of normality, homoscedasticity, and additivity. These results, together with the sigmoidal nature intrinsic to theoretical concentration-response relationships, led us to entertain log-transformation as a means of conferring linearity, normality, and homoscedasticity to the data. Applying reverse stepwise elimination to each of 12 days of observation within the eight taxa under investigation (96 cases) resulted in only 3 cases with the interaction term (b_2) significant, 17 cases in which the partial regression coefficient (b_1) for the covariate was significant, and 76 cases in which only the concentration term (b_0) was significant. Significance of the covariate term was generally (13/17 cases) restricted to the first week of observation. These results suggested that differences in initial zooplankton abundance were generally unimportant in defining the observed concentration-response during the period of primary interest and thus only the basic model [$\log(Y + 0.006) = a + b_0(\log X + 0.001)$] was further evaluated.

Fit statistics for regression of log transformed data using the basic linear model are provided in Table 1. In 63 of 95 cases, the basic linear model applied to log-transformed data was considered statistically significant ($p < 0.05$). None of the regressions were significant on day 0. Significant regressions were detected within the first week postapplication for a number of taxa, indicating a rapid response to hexazinone treatments. With the exception of *Bosmina* spp. (taxon 32), all taxa showed statistically significant concentration-response relationships by day 14, continuing generally through day 42. Regressions for *Bosmina* spp. became consistently significant beginning on day 28, with only the regression for day 77 being both statistically significant and exhibiting adequate fit.

The degree of replication in hybrid experimental designs (19) bestows significant power to the lack of fit test for regression models. Of the 63 statistically significant regressions, significant lack of fit (LOF F > $F_{(0.05,3,9)}$ = 5.08) was detected in 21 cases. The high proportion of statistically significant regressions with inadequate fit indicated that linearization by log-transformation often resulted in an unacceptable model. In this regard we concur with Neter et al.[34] who state that "clearly repeat observations are most valuable whenever we are not certain of the nature of the regression function. If at all possible, provision should be made for some replications." Although the theoretical nature of the concentration-response relationship is well understood, knowledge of how the theoretical relationship may be affected by various factors in field-toxicity testing is largely unknown.

Inadequate fit may be caused by significant leverage points (outliers), as well as general systematic departures from linearity (30). Plots of standardized residuals (residuals exceeding ± 2 STD) detected outliers in regressions for *Ploesoma* (taxon 19) and Total Cladocera (taxon 39). In both cases, the outlier was a sample estimate from the untreated control. Removal of the outlier and reanalysis resulted in significant improvement in the regression fit and increase in the magnitude of the slope (Table 1). Inherent nonlinearity observed in a number of the concentration-response relationships and reflected in values of the coefficient of determination (r^2) resulted largely from stimulatory and threshold effects as discussed previously. Threshold effects, caused primarily by differential magnitude of impact and rates of recovery within the two lowest test concentrations (0.01 and 0.1 mg/l) as compared to the two higher concentrations (1.0 and 10 mg/l), were clearly elucidated in the RM-ANOVA analysis. It is important to emphasize that such threshold levels are dynamic with respect to time, since more rapid recovery occurs in minimally affected populations. We postulate that threshold effects are likely to be typical of aquatic mesocosm impact studies, due to the lengthy periods of observation and a wide range of test concentrations employed. A wide range of test concentrations is advisable, since there are no adequate methods of determining *a priori* where the linear portion of the concentration-response curve will occur and because expected environmental concentrations in lentic aquatic systems may also vary widely depending upon the mechanism of input (i.e., accidental overspray, drift, runoff, etc.).[35]

Table 1. Fit Statistics for Linear Regression of Log-Transformed Data

Day	r^2	Model F	P	SEE	a	b_0	$SE(b_0)$	F/F_0	LOF
				Keratella cochlearis (taxon 10)					
0	0.07	0.93	0.354	0.210	0.932	−0.038	0.039	0	0.468
1	0.64	21.71	0.172	0.001	0.860	−0.150	0.032	3	0.669
2	0.63	20.54	0.001	0.197	0.865	−0.166	0.037	3	0.686
3	0.23	3.37	0.094	0.239	0.850	−0.083	0.045	1	26.933
7	0.54	14.12	0.003	0.225	0.646	−0.158	0.042	2	0.158
10	0.67	24.37	0.000	0.314	0.413	−0.289	0.059	4	1.660
14	0.62	19.95	0.001	0.417	0.456	−0.343	0.078	3	4.165
21	0.53	13.53	0.003	0.473	0.207	−0.325	0.088	2	0.876
28	0.49	11.42	0.006	0.546	0.341	−0.344	0.102	2	1.193
42	0.56	15.44	0.002	0.583	−0.218	−0.427	0.109	2	0.935
56	0.16	2.35	0.151	0.800	0.462	−0.229	0.150	0	6.638
77	0.04	0.45	0.516	0.723	0.448	−0.090	0.135	0	12.631
				Ploesoma spp. (taxon 19)					
0	0.04	0.55	0.472	0.196	1.050	−0.027	0.037	0	1.284
1	0.11	1.48	0.247	0.132	1.087	−0.030	0.025	0	0.431
2	0.04	0.52	0.484	0.170	0.935	−0.023	0.032	0	2.434
3	0.31	5.00	0.047	0.213	0.850	−0.090	0.040	1	7.450
7	0.21	3.10	0.104	0.478	0.008	−0.157	0.089	0	7.413
10	0.48	11.16	0.006	0.357	−0.241	−0.222	0.067	2	0.252
14	0.63	20.56	0.001	0.452	−0.755	−0.382	0.084	3	1.025
14[a]	0.87	48.57	0.0001	0.333	−0.749	−0.473	0.068	7	
28	0.45	9.72	0.009	0.430	−1.294	−0.250	0.080	1	0.306
42	0.17	2.41	0.146	0.716	−1.648	−0.208	0.134	0	3.380
56	0.03	0.40	0.541	0.504	−1.955	−0.059	0.094	0	0.106
77	0.11	1.57	0.235	0.519	−1.636	0.121	0.097	0	0.269
				Bosmina spp. (taxon 32)					
0	0.10	1.31	0.274	0.233	0.787	0.050	0.043	0	2.214
1	0.21	3.23	0.097	0.320	1.038	0.011	0.060	0	2.070
2	0.12	1.63	0.226	0.401	0.805	0.096	0.075	0	4.498
3	0.02	0.26	0.622	0.287	0.640	−0.028	0.054	0	0.872
7	0.30	5.05	0.044	0.455	0.131	−0.191	0.085	1	2.884
10	0.18	2.72	0.125	0.399	0.095	−0.122	0.074	0	1.973
14	0.24	3.86	0.073	0.553	0.522	−0.202	0.103	1	1.254
21	0.23	3.55	0.084	0.673	0.291	−0.236	0.126	1	13.427
28	0.41	8.21	0.014	0.609	−0.406	−0.326	0.114	1	8.166
42	0.48	11.19	0.006	0.528	−0.383	−0.329	0.099	2	15.475
56	0.48	10.92	0.006	0.733	−0.302	−0.451	0.137	2	7.790
				Holopedium spp. (taxon 35)					
0	0.19	2.78	0.121	0.417	0.959	0.130	0.078	0	1.641
1	0.00	0.00	0.976	0.323				0	1.718
2	0.32	5.57	0.036	0.955	0.113	−0.420	0.178	1	12.964
3	0.41	7.53	0.019	0.946	0.018	−0.492	0.179	1	0.425
7	0.55	14.86	0.002	0.848	−0.220	−0.609	0.158	2	29.743
10	0.57	16.15	0.002	0.748	−0.480	−0.561	0.140	2	9.494
14	0.41	8.18	0.014	1.004	−0.811	−0.535	0.187	1	3.597
21	0.44	9.33	0.010	0.900	−1.153	−0.516	0.169	1	0.907
28	0.38	7.36	0.019	1.062	−1.268	−1.537	0.198	1	0.488
42	0.53	13.42	0.003	0.960	−1.320	−0.656	0.179	2	0.466
56	0.16	2.33	0.153	1.289	−1.466	−0.367	0.240	0	1.060
77	0.19	2.90	0.115	1.025	−1.352	−0.325	0.191	0	1.939
				Total Rotifera (taxon 29)					
0	0.07	0.91	0.359	0.153	1.379	−0.027	0.029	0	0.323
1	0.49	11.33	0.006	0.127	1.346	−0.080	0.024	2	0.665
2	0.60	18.06	0.001	0.144	1.248	−0.114	0.027	3	0.427
3	0.40	7.30	0.020	0.194	1.184	−0.099	0.037	1	3.158
7	0.60	17.74	0.001	0.207	0.804	−0.163	0.039	3	1.430
10	0.68	26.05	0.000	0.287	0.562	−0.273	0.054	4	1.519

Table 1. Fit Statistics for Linear Regression of Log-Transformed Data (*continued*)

Day	r^2	Model F	P	SEE	a	b_0	SE(b_0)	F/F$_0$	LOF
				Total Rotifera (taxon 29)					
14	0.68	25.42	0.000	0.368	0.571	−0.346	0.069	4	3.915
21	0.75	36.39	0.000	0.354	0.373	−0.399	0.066	6	0.746
28	0.53	13.74	0.003	0.492	0.527	−0.340	0.092	2	1.809
42	0.67	24.53	0.000	0.496	0.092	−0.458	0.093	4	0.535
56	0.31	0.54	0.038	0.570	0.799	−2.481	0.106	0	3.362
77	0.01	0.12	0.735	0.542	0.822	−0.035	0.101	0	4.109
				Total Cladocera (taxon 39)					
0	0.17	2.38	0.149	0.175	1.341	0.050	0.033	0	1.304
1	0.05	0.66	0.433	0.162	1.511	0.025	0.030	0	0.610
2	0.05	0.65	0.437	0.188	1.306	−0.028	0.035	0	2.560
3	0.25	3.71	0.080	0.345	1.131	−0.126	0.065	1	−0.610
7	0.60	18.33	0.001	0.397	0.762	−0.317	0.074	3	77.359
10	0.54	14.16	0.003	0.326	0.753	−0.229	0.061	2	27.799
14	0.45	9.78	0.009	0.570	0.782	−0.332	0.106	1	44.459
21	0.35	6.50	0.025	0.680	0.608	−0.323	0.127	1	61.018
28	0.54	13.83	0.003	0.605	0.367	−0.419	0.113	2	25.333
42	0.73	32.78	0.000	0.448	−0.002	−0.478	0.084	5	4.677
42[a]	0.89	42.24	0.000	0.2947	0.005	−0.577	0.060	7	
56	0.60	18.16	0.001	0.633	−0.604	−0.503	0.118	3	11.108
77	0.70	27.93	0.000	0.518	−0.556	−0.511	0.097	4	9.174
				Total Copepoda (taxon 79)					
0	0.01	0.10	0.760	0.078	1.311	0.005	0.015	0	0.710
1	0.5	12.69	0.004	0.108	1.222	−0.072	0.020	2	0.370
2	0.79	44.24	0.000	0.147	0.959	−0.183	0.028	7	1.521
3	0.63	18.81	0.001	0.214	0.871	−0.176	0.041	3	−1.595
7	0.60	17.68	0.001	0.204	0.915	−0.160	0.038	3	5.776
10	0.77	41.14	0.000	0.194	0.812	−0.232	0.036	6	6.209
14	0.72	30.58	0.000	0.300	0.806	−0.310	0.056	5	9.082
21	0.82	55.94	0.000	0.249	0.527	−0.347	0.046	9	3.707
28	0.74	35.05	0.000	0.333	0.500	−0.368	0.062	5	2.331
42	0.76	38.07	0.000	0.383	0.272	−0.441	0.072	6	14.309
56	0.59	17.07	0.001	0.374	0.567	−0.289	0.070	3	2.679
77	0.77	39.15	0.000	0.310	0.165	−0.362	0.058	6	2.622
				Total Zooplankton (taxon 89)					
0	0.00	0.00	0.948	0.081	1.846	0.001	0.015	0	0.000
1	0.27	4.44	0.057	0.112	1.841	−0.044	0.021	1	0.193
2	0.63	20.70	0.001	0.117	1.679	−0.099	0.022	3	0.511
3	0.51	11.66	0.006	0.196	1.559	−0.127	0.037	2	3.473
7	0.69	27.01	0.000	0.198	1.286	−0.192	0.037	4	8.143
10	0.68	25.64	0.000	0.237	1.148	−0.224	0.044	4	2.850
14	0.64	21.26	0.001	0.360	1.199	−0.309	0.067	3	12.797
21	0.72	30.80	0.000	0.336	0.932	−0.348	0.063	5	4.759
28	0.67	24.46	0.000	0.389	0.985	−0.359	0.073	4	33.660
42	0.81	52.70	0.000	0.328	0.633	−0.443	0.061	8	3.623
56	0.49	11.63	0.005	0.436	1.080	−0.278	0.081	2	7.463
77	0.35	6.44	0.026	0.375	0.972	−0.178	0.070	1	5.636

[a] Regression statistics with outlier removed.

Note: Basic linear model is log(Y + 0.006) = a + b_0(logX + 0.001) + ϵ where Y = abundance (#/l); a = intercept; b_0 = partial regression coefficient for toxicant concentration; X = hexazinone concentration (mg/l); and ϵ = error. Model and error degrees of freedom were 1 and 12, respectively, in all cases. Ratio of F/F$_0$ > 4 indicates predictive value for the model [$F_{(0.05;1,12)}$ = 6.55]. Significant lack of fit (LOF) is indicated by value less than tabular $F_{(0.05;3,9 = 5.08)}$ for lack of fit and pure error components of residual SS. P = probability of a greater F statistic, SEE = standard error of the estimate.

Log-transformation provides a linearized "best-fit" model of an inherently curvilinear concentration-response relationship. As such, the linearized model is subject to error resulting from the influence of threshold effects and particularly where the thresholds change with time. In this situation, the slope of the regression is strongly affected by the test-concentrations employed. In our specific case, data for the 10 mg/l concentration clearly influenced the slope, as evidenced by the 15 to 50% difference in slope estimates, which result depending upon inclusion or exclusion of data for this treatment level (Table 2). Under such conditions, extrapolation of predicted response beyond the range of concentrations tested is clearly inadvisable. However, and perhaps more importantly, the insidious consequences of thresholds and inadequately fitting models on slope estimates must be carefully considered in the assessment of risk. Such consequences are exemplified by the fact that EC_{50} estimates derived from adequate linear models (Table 4) ranged from 0.004 to 0.045 mg/l, a result clearly inconsistent with the RM-ANOVA and nonlinear analysis. Further, since the estimated slope varies with time and since duration of impact is ecologically important, it would be generally inappropriate to assess the potential risk to aquatic systems on the basis of any single within-day regression regardless of how significant and adequate the fit. In this regard, simple evaluation of trends in slope estimates and regression significance of the linearized model provide substantial knowledge with respect to the influence of time on the concentration-response relationship and may serve as an effective method of screening data for maximal concentration-response.

Table 2. Comparison of Regression Statistics With n = 3 Data Points For 10 mg/l Treatment Level Included and Excluded

Day	Taxon	r^2	Model F	P	SEE	a	b_0	$SE(b_0)$
			10 mg/l Test Concentration Included					
10	10	0.67	24.37	0.000	0.314	0.413	−0.289	0.059
14[a]	19	0.87	48.57	0.0001	0.333	−0.749	−0.473	0.068
77	32	0.69	27.29	0.000	0.505	−0.807	−0.492	0.094
42	35	0.53	13.42	0.003	0.960	−1.320	−0.656	0.179
21	29	0.75	36.39	0.000	0.354	0.371	−0.399	0.067
42[a]	39	0.89	42.24	0.000	0.295	0.001	−0.577	0.060
21	79	0.82	55.94	0.000	0.249	0.527	−0.347	0.046
42	89	0.81	52.70	0.000	0.328	0.633	−0.443	0.061
			10 mg/l Test Concentration Excluded					
10	10	0.37	5.28	0.05	0.286	0.645	−0.172	0.075
14[a]	19	0.66	15.74	0.004	0.358	−0.636	−0.411	0.104
77	32	0.46	7.54	0.023	0.543	−0.604	−0.389	0.142
42	35	0.39	5.73	0.04	1.097	−1.381	−0.686	0.287
21	29	0.54	10.48	0.010	0.363	0.555	−0.307	0.095
42[a]	39	0.84	42.51	0.002	0.230	0.266	−0.433	0.066
21	79	0.73	24.66	0.007	0.178	0.759	−0.231	0.046
42	89	0.65	16.61	0.003	0.358	0.756	−0.381	0.094

[a] Regressions with outliers removed.

Note: P = probability of a greater F statistic, SEE = standard error of the estimate; a = intercept, b_0 = slope estimate and $SE(b_0)$ = standard error of the slope estimate.

Notwithstanding the aforementioned inadequacies of this approach, statistically significant linear models of log-transformed data adequately fitting the data were observed in 16 cases. These cases were most prevalent in higher taxonomic groups and concomitant with higher coefficients of determination (r^2). Such models were considered to have predictive value based on the minimal criterion of $F_{obs}/F_{(0.05, 1, 12)}$ > 4, as described by Draper and Smith[30] and were primarily associated with days 10 through 42, approximating the period of significant impact as determined by RM-ANOVA analysis. Examples of regression situations generally meeting criteria of adequate fit and predictive value are provided in Figure 5. For higher taxonomic groups, adequately fitting models of predictive value were characterized by highly significant ($p < 0.001$) negative slopes with values ranging from -0.35 to -0.48, and r^2 values exceeding 0.70. Regressions for individual taxa frequently exhibited significant concentration-related response ($p < 0.05$) with adequately fitting models (Table 1). However, predictive models were atypical for individual taxa, and r^2 values (range 0.53 to 0.89) were generally low in comparison to the higher levels of taxonomic classification. These results reflect the substantially greater variation which characterized the lower taxonomic levels and may be indicative of inadequacies in the experimental procedure *vis à vis* taxa low in abundance.

Nonlinear Regression

Following estimation of the initial starting values, the model was fit to data for total Copepoda (taxon 79) by day. Parameter estimates were then plotted against day to observe change in parameters over time. Results showed a general decreasing tendency for both a and b parameters with increasing time. Both trends were approximately linear; however, the plot for parameter b suggested a possible exponential decline over time. For simplicity, a and b parameters were considered as linear functions of day, thus assimilating time as a third dimension in the nonlinear model. The full nonlinear model was applied to total Copepoda (taxon 79) as a test case. Results indicated both the b_0 parameter and c parameter were insignificant. Further evaluation of various reduced models, suggested that the full model with the b_0 eliminated [Model 1: $Z = (a_0 + a_1*t)e(b_1*t*X) + cN_0$ as the most appropriate general model.

Model 1 was applied to data sets for each individual taxa and group. For all cases, several sets of starting parameters resulted in the same point of convergence, suggesting that the residual sums of squares was a global minimum. As observed for the linear regression approach, the covariate term was generally insignificant (Table 3); however, it was retained in the analysis, since elimination of the covariate term resulted in illogical negative values for the EC_{50} estimates and response surfaces inconsistent with observed values. Nonlinear regression using model 1 provided acceptable results in 5 of the 8 test cases. Of these 5 cases, best fit (greater than 75% of the total sums of squares accounted for by the model) was associated with the higher taxonomic groups. Predicted response surfaces, focusing on the concentration range of primary interest (0 to 1 mg/l), are presented graphically in

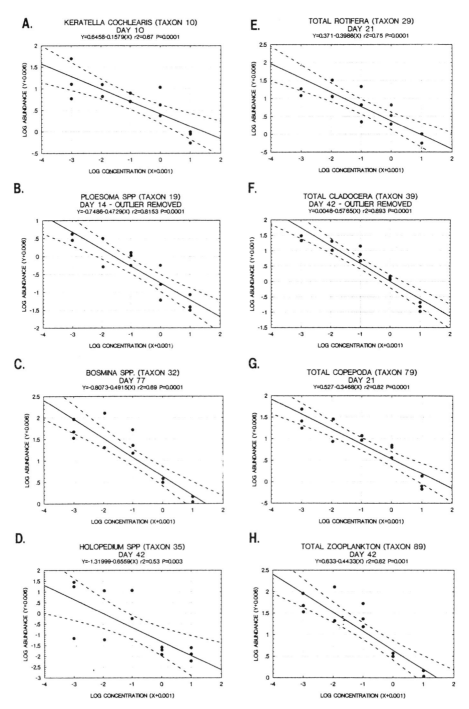

Figure 5. Selected linear regressions of log-transformed data showing impact of hexazinone on abundance of dominant zooplankton taxa.

Table 3. Fit Statistics For Two Nonlinear Regression Models

Taxa	Parameter Estimates (lower 95% CL; upper 95% CL)
Total Copepoda (Taxon 79)	
RegSS/Total SS	0.85
MSE	48.5
Min SS	7863.81
a_0	20.19 (17.42; 22.98)
a_1	−0.0679 (−0.13; −0.004)
b_1	−0.079 (−0.119; −0.0396)
c	0.09 (−0.082; −0.21)
N_0	20.652
Total Zooplankton (Taxon 89)	
RegSS/Total SS	0.76
MSE	914.0
Min SS	149000.0
a_0	63.29 (50.77; 75.80)
a_1	−0.32 (−0.59; −0.049)
b_1	−0.075 (−0.013; −0.022)
c	0.116 (−0.03; 0.26) NS
N_0	70.99
Total Cladocera (Taxon 39)	
RegSS/Total SS	0.8
MSE	100.16
Min SS	16236.92
a_0	24.609 (20.798; 28.42)
a_1	−0.253 (−0.335; −0.171)
b_1	−0.0196 (−0.034; −0.006)
c	0.107 (−0.04; 0.255) NS
N_0	21.21
Holopedium spp. (Taxon 35)	
RegSS/Total SS	0.65
MSE	34.64
Min SS	5646.61
a_0	10.55 (8.670; 12.439)
a_1	−0.141 (−0.199; −0.083)
b_1	−0.183 (−0.354; −0.121)
c	0.198 (0.082; 0.315)
N_0	9.517

Note: NS denotes parameters not significantly different from zero. N_0 is the average initial abundance at time 0 for 14 enclosures (omitting encl 32), a, b, and c are estimable partial regression coefficients for the nonlinear model 1: $Z = (a_0 + a_1{}^*t)e^{((b_0 + b_1{}^*t)^*X)} + c \times N_0$.

Figure 6. In general, the response surfaces adequately reflected the major trends observed in the data:

1. Abundance for all taxa and groups in untreated controls declines with time.
2. The magnitude of response is proportional to both increasing test concentration and time, with maximal effects at test concentrations (>1 mg/l) from which no recovery is observed.
3. Substantial reductions in abundance occur for all taxa exposed to concentrations exceeding approximately 0.2 mg/l.
4. Effects are readily apparent within the first week of observation, particularly in relation to the highest test concentrations (>1 mg/l).

In three cases (Total Rotifera [taxon 29], *Bosmina* spp. [taxon 32] and *Keratella cochlearis* [taxon 10]), models were considered unacceptable, being characterized by insignificance and collinearity among a variety of parameters, with only 42 to 52% of the total variation being accounted for by the model. Thus, the basic exponential decline model could not be adequately fit to all of the taxa, with poor model behavior correlated to taxa characterized by high variability. The three-dimensional plots (Figure 6) provide a facile method for interpolating the magnitude of response to be expected at any chosen point within the range of time and concentrations tested. Simple shading gradients highlight various regions of the response surface indicative of minimal, moderate, and substantial observed impact (black, gray, white, respectively). Unfortunately, although statistical confidence

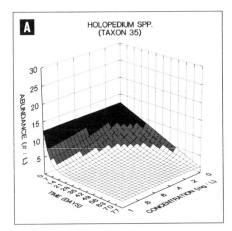
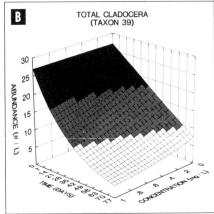
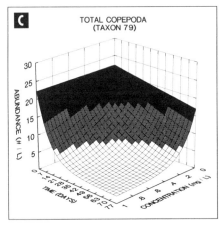
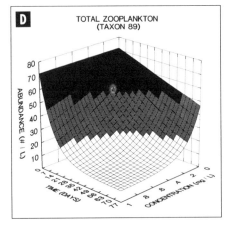

Figure 6.   Three-dimensional surface response plots generated by nonlinear regression model 1. Gradient shading represents increasing (black, gray, white) levels of impact at various levels of time and concentration variables.

Table 4. Comparison of EC$_{50}$ Values Estimated by Linear and Nonlinear Regression Models

Taxon	Day	Linear Model EC$_{50}$ Values	Nonlinear Model EC$_{50}$ Values
79	21	0.016	0.049
89	42	0.017	0.240
39	42	0.045	1.260
35	42	0.004	0.060

associated with interpolative estimates may be calculated mathematically, there is no simple graphical means for conveying precision about these estimates. A possible approach to resolving this problem may lie in fitting a curve to some chosen value (say lower 2 × 10%) of the residuals generated by the model. Similar techniques have been applied to two-dimensional concentration-response problems.[36] In effect, this approach would result in a "90 percentile" three-dimensional response surface. By interpolating from this graphic, investigators could theoretically be confident that 9 times out of 10 in similar repeated experiments, the reduction in abundance for a given time and concentration would not exceed the predicted value. The "90th percentile response surface" approach incorporates both the three-dimensional nature inherent to the problem and a statistical inference capability in a single graphic. Such capability may be particularly important in aquatic impact studies where expected concentrations vary with exposure scenarios and where the ecological significance of a given magnitude of impact is poorly understood, subjective, and dependent upon time to subsequent recovery. In this study the three-dimensional nonlinear model is less effective than possible, since the experimental design was not optimized to this particular approach. As a result, the lack of test concentrations between 1.0 and 10.0 mg/l significantly constrain interpretive potential in the model. Further investigation is required in terms of the practical utility and theoretical validity of the three-dimensional 90th percentile response-surface approach, and we suggest it here to stimulate the interest of other researchers in developing this or similar alternative approaches. Optimally, any such approach should take into account the nonconstant variance which may be considered typical of such data.

In cases where the model was deemed adequate, EC$_{50}$ values were derived from nonlinear models with specific inputs for the independent variable Y (time postapplication). For comparative purposes, these values are presented together with comparable values generated by the linear regression analysis of log transformed data (Table 4). EC$_{50}$ values estimated using the nonlinear technique (range 0.06 to 1.26) were within an order of magnitude of the NOEC estimated by RM-ANOVA. Comparable values estimated from the linear regression technique ranged from 0.004 to 0.045 mg/l and were generally inconsistent with both the RM-ANOVA results and graphical interpolation of the raw data. These results suggest that the nonlinear model provides substantially better approximation of the true concentration response and as a result a more realistic statistical endpoint estimate.

CONCLUSIONS

A comprehensive exploratory analysis of field-derived impact data highlighted a number of important considerations in relation to aquatic mesocosm experimental design and statistical analysis. Distinct heterogeneity of variance, inversely proportional to the mean and time of sampling postapplication, was the dominant characteristic of data sets for dominant zooplankton taxa and groups investigated. Variability was substantially greater for lower taxonomic levels of classification as compared to higher taxonomic groups. Such data characteristics, considered to be typical of field-derived aquatic impact experiments, have important implications on methods of statistical analysis.

In particular, transformation of the data did not completely resolve violations in standard parametric assumptions. Thus, the use of specific RM-ANOVA/multiple comparison procedures (assuming or robust to heterogeneity of variance) was most appropriate for testing hypotheses relating to treatment means. This approach clearly identified significant impact of hexazinone concentration on zooplankton abundance, particularly at the 1.0 and 10 mg/l treatment levels. Considering the data *in toto* an approximate LOEC was established as 1.0 mg/l, for hexazinone impact on zooplankton abundance. The RM-ANOVA approach was severely constrained by both inherent lack of interpolative ability and by the impingement of high and heterogeneous variation in the data on the power of statistical comparisons.

Stimulatory responses, differential species sensitivity, trophic interactions, nontreatment related stress factors, and differential recovery rates for various species may all contribute to threshold effects in observed concentration-response relationships in the field. These effects are particularly troublesome in statistical analysis and interpretation of data using linear regression of transformed data. Since mesocosm studies are conducted specifically to study natural populations, under relatively uncontrolled conditions with a wide range of exposure concentrations to simulate various exposure scenarios, these factors may always tend to influence such data. In this study, log-transformation of the data was somewhat effective in terms of reducing violations in regression assumptions and in linearizing the inherently curvilinear concentration-response relationship. Linear regression of log-transformed data resulted in models meeting minimal criteria for adequate fit and prediction in some cases. However, the threshold effects evident in the data appeared to result in EC_{50} estimates inconsistent with RM-ANOVA analysis and careful examination of the scatter-plots. As it is unlikely that any single transformation will confer adequate linearity to such concentration-response relationships over the observation period, alternative approaches are required. The use of range-finding tests, commonly employed in laboratory toxicity testing protocols, is logistically infeasible in simulated field studies and conflicts with the general purpose of employing a wide range of test concentrations predicated by expected exposure scenarios and differential sensitivity of the numerous species involved. Liber et al.[37] have approached the problem by using an unreplicated design employing a large number of treatment levels and selecting the linear-portion of the response curve *a posteriori*. This approach excludes valuable and costly data from the analysis. In addition, the lack of replication precludes unbiased estimation of error variance and the ability to assess the adequacy of fit, which is critical to satisfactory risk assessment.

Inherent nonlinearity in observed concentration-response relationships, inadequacy of transformation in linearizing the relation, and inadvisability of other "artificial" linearizing techniques led us to consider nonlinear regression techniques as generally more appropriate for analysis of field-derived concentration-response data. The three-dimensional nonlinear approach seems most appropriate since it more closely reflects the theoretical curvilinear nature of concentration-response relationships and effectively assimilates the dynamic influence of time. Although not well-suited to balanced "hybrid" experimental designs, the nonlinear approach obviates the need for data-transformation. Finding a simple, adequate functional form which relates directly to the chemistry-biology dynamics of concentration-response in freshwater-lentic systems may not be easy, but may, with further development, provide a technique for interpolative prediction with a stated level of statistical confidence. Modification of the experimental design to increase the number of test concentrations and reduce replication would be required to optimize the nonlinear, response-surface approach to aquatic mesocosm data studies.

The overall limiting factor to effective statistical analysis of data in this study was the extreme heterogeneity of variance, with variation being inversely proportional to means. These factors, postulated to be common in mesocosm impact data suggested further potential improvements in terms of optimal partitioning of resources between replication and number of concentrations tested. Since differences among replicates allow estimation of the random variation at any test concentration and also an effective method for assessing lack of fit in applied regression techniques, some degree of replication in the experimental design is well advised. However, disproportionate replication in favor of the low test concentrations should offset the effects of inherently high variance for these levels of the independent variable. Applying disproportionate replication and increasing the number of test concentrations would be a cost-effective method of offsetting the variance heterogeneity at the low concentrations and increasing confidence in prediction.

In conclusion, we draw attention to the fact that none of the statistical approaches investigated in this paper are completely satisfactory or universally applicable. Continued research comparing various statistical procedures as applied to field-generated data may ultimately provide defensible guidelines for suggested experimental design and methods of statistical analysis of mesocosm impact data. However, in keeping such efforts in perspective and qualifying the heavy statistical emphasis in this paper we quote Hurlbert[38] who noted that "biologists can often extract useful information from data that high-church statisticians would regard as hopeless, biology having advanced considerably before statistics emerged as its auxiliary discipline".

ACKNOWLEDGMENTS

The authors gratefully acknowledge the assistance of L. MacDonald, D. Thomas, S. Gazzola, D. Kreutzweiser, L. Sicoli, B. Staznik, and T. Buscarini in deployment and field sampling of the *in situ* enclosures. The authors also wish to express sincere appreciation to G. Stephenson (Ecological Services for Planning, Guelph, Ont.) for advice and training in enumeration and identification of zooplankton. Critical review provided by Dr. R. Regal (University of Minnesota), D. Kreutzweiser (FPMI), R. Fleming (FPMI), K. Day (Canadian Centre for Inland Waters), and two anonymous reviewers resulted in considerable improvements to this paper and are gratefully acknowledged. This research was jointly funded by the Forest Pest Management Institute and DuPont Canada Inc.

REFERENCES

1. Rand, G.M. and S.R. Petrocelli. 1985. Fundamentals of Aquatic Toxicology. Hemisphere Publishing Corp., Washington, D.C. 666 pp.
2. Cairns, J. Jr. 1984. Multispecies toxicity testing. *Bull. Ecolog. Soc. Am.* 65:301–304.
3. Schindler, D.W. 1987. Detecting ecosystem responses to anthropogenic stress. *Can. J. Fish. Aquat. Sci.* 44 (Suppl. 1):6–25.
4. Swingle, H.S. 1947. Alabama Polytechnical Institute, Agricultural Experiment Station. Bulletin No. 264, pp. 1–34.
5. Komarovsky, B. 1953. *Bull. Res. Counc. Isr.* 2:379–410.
6. Hepher, B. 1962. *Limnol. Oceanogr.* 7:131–136.
7. O'Brien, W.J. and F. deNoyelles, Jr. 1974. *Hydrobiologia* 44:91–104.
8. Crossland, N.O. 1982. Aquatic toxicology of cypermethrin II. Fate and biological effects in pond experiments. *Aquat. Toxicol.* 2:205–222.
9. deNoyelles, F., Jr. and W.D. Kettle. 1985. Experimental ponds for evaluating bioassay predictions. In: T.P. Boyle, ed., Validation and predictability of laboratory methods for assessing fate and effects of contaminants in aquatic ecosystems. ASTM STP 865. American Society for Testing and Materials, Philadelphia, pp. 91–103.
10. Odum, P. 1984. The mesocosm. *Bioscience* 34:558–562.
11. Lund, J.W.G. 1972. Preliminary observations on the use of large experimental tubes in lakes. *Verh. Internat. Verein. Limnol.* 18:71–77.
12. Beers, J.R., M.R. Reeve, and G.D. Grice. 1977. Controlled ecosystem pollution experiment. Effects of mercury on enclosed water columns. IV. Zooplankton population dynamics and production. *Mar. Sci. Comm.* 3:355–394.
13. Menzel, D.W. 1977. Summary of experimental results: controlled ecosystem pollution experiment. *Bull. Mar. Sci.* 27:142–145.
14. Solomon, K.R., K.E. Smith, G. Guest, J.Y. Yoo, and N.K. Kaushik. 1980. The use of limnocorrals in studying the effects of pesticides in the aquatic ecosystem. *Can. Tech. Rep. Fish. Aquat. Sci.* 975:1–9.
15. Kaushik, N.K., G.L. Stephenson, K.R. Solomon, and K.E. Day. 1985. Impact of permethrin on zooplankton communities in limnocorrals. *Can. J. Fish. Aquat. Sci.* 42:77–85.
16. Herman, D., N.K. Kaushik, and K.R. Solomon. 1986. Impact of atrazine on periphyton in freshwater enclosures and some ecological consequences. *Can. J. Fish. Aquat. Sci.* 43:1917–1925.

17. Touart, L.W. 1988. Aquatic mesocosm tests to support pesticide registration. Hazard evaluation division. U.S. EPA, EPA 540/09-88-035. Washington, D.C.

18. Touart, L.W. Regulatory endpoints and the experimental design of aquatic mesocosm tests, this volume, ch. 4.

19. Graney, R.L., J.P. Giesy, and D.M. DiToro. 1989. Mesocosm experimental design strategies: Advantages and disadvantages in ecological risk assessment. *Misc. Publ. Ent. Soc. Am.* 5:74–88.

20. Solomon, K.R., K. Smith, and G.L. Stephenson. 1982. Depth integrating samplers for use in limnocorrals. *Hydrobiologia* 94:71–75.

21. Stephenson, G.L., N.K. Kaushik, K.R. Solomon, and K.E. Day. 1986. Impact of permethrin on zooplankton communities in limnocorrals. *J. Fish. Aquat. Sci.* 42:77–85.

22. Thompson, D.G., L. MacDonald, and B. Staznik. 1992. Persistence of hexazinone and metsulfuron methyl in a boreal forest lake. *J. Agric. Food Chem.* (submitted).

23. Levene, H. 1960. Robust tests for equality of variances. In: Il Olkin (Ed). *Contributions to Probability and Statistics*. Stanford University Press. Stanford, California. pp. 278–292.

24. Sokal, R.R. and F.J. Rohlf. 1981. *Biometry* (2nd). W.H. Freemand and Company, New York. 859 pp.

25. Zar, J.H. 1974. *Biostatistical Analysis*. Prentice Hall Inc. Englewood Cliffs, NJ. 620 pp.

26. Regal, R. 1991. Personal communication. Department of Mathematics and Statistics, University of Minnesota, 3 pp.

27. Green, R.H. 1979. *Sampling Design and Statistical Methods for Environmental Biologists*. John Wiley & Sons Inc., New York, 257 pp.

28. Dixon, W.J., M.B. Brown, L. Engelman, and R.I. Jennrich. 1990. *BMDP Statistical Software Manual*. Volume I. University of California Press, Berkeley, 629 pp.

29. Day, R.W. and G.P. Quinn. 1989. Comparisons of treatments after and analysis of variance in ecology. *Ecol. Monogr.* 59:433–463.

30. Draper, N.R. and H. Smith. 1981. *Applied Regression Analysis* (2nd ed.). John Wiley & Sons Inc. New York, 709 pp.

31. Bloesch, J., P. Brossard, H. Bruhrer, H.R. Burgi and U. Uehlinger. 1988. Can results from limnocorral experiments be transferred to in situ conditions. *Hydrobiologia* 159:297–308.

32. Hamilton, P.B., G.S. Jackson, N.K. Kaushik, K.R. Solomon, and G.L. Stephenson. 1988. The impact of two applications of atrazine on the plankton communities of in-situ enclosures. *Aquat. Toxicol.* 13:123–40.

33. Shaw, J.L., M. Moore, J.H. Kennedy and I.R. Hill. Design and statistical analysis of field aquatic mesocosm studies. In R.L. Graney, J.H. Kennedy, and J.H. Rodgers, Eds., this volume, ch. 9.

34. Neter, J., W. Wasserman, and M.H. Kutner. 1985. *Applied Linear Statistical Models*. Richard D. Irwin, Homewood, IL, 1127 pp.

35. Thompson, D.G. 1992. Fate and impact of forest management herbicides in freshwater lentic ecosystems — A literature analysis. In preparation.

36. Pitt, D.G., R.A. Fleming, D.G. Thompson, and E.G. Kettela. Glyphosate efficacy on eastern Canadian forest weeds — Part II: Deposit-response relationships and crop tolerance. *Can. J. For. Res.* 22:1160–1171.

37. Liber, K., K.R. Solomon, N.K. Kaushik, and J.H. Carey. (1992). Experimental designs for aquatic mesocosm studies: a comparison of the ANOVA and REGRESSION design for assessing the impact of tetrachlorophenol on zooplankton populations in limnocorrals. *Environ. Toxicol. Chem.* 11:61–77.

38. Hurlbert, S.H. 1975. Secondary effects of pesticides on aquatic ecosystems. *Residue Reviews* 58:81–148.

CHAPTER 12

Optimal Design of Aquatic Field Studies

Ronald R. Regal and Stephen J. Lozano

Abstract: In designing field studies the experimenter must decide how many replicates to run at which concentrations. There is much current debate over the relative merits of different designs including "regression" and "ANOVA" designs. Optimal design techniques provide a framework for addressing choices in the design of field studies, largely because the first step in determining the optimal design is deciding the exact goal for the experiment. If one can specify the goals in a quantified form and specify a model governing the responses, optimal design techniques can be applied to find the best design or set of designs for addressing the question from the available resources. Uncertainty in specifying the model governing the responses can be quantified with a prior distribution of possible models. Numerical optimization techniques can then be used to optimize the objective under the assumed model. Optimal design techniques can be applied to assigning available units to given concentration levels or to the more difficult question of determining which concentration levels optimize the goals of the experiment.

INTRODUCTION

Under FIFRA (Federal Insecticide, Fungicide, and Rodenticide Act) the hazard of pesticides to an aquatic ecosystem may require a field test. Touart and Slimak[1] suggest that aquatic mesocosms (experimental ponds and *in situ* enclosures) provide the best alternative for conducting field tests. At present there is considerable debate over the merits of different experimental designs. Should one use a single treatment concentration along with control units? Should one use a few concentrations with replication of each treatment group? Should the replication be the same in each group or is it more important to replicate some treatments more than others? Should more doses be used with less replication, possibly even single replication at each dose? Much of the discussion of mesocosm design breaks into different opinions on the merits of "ANOVA" and "regression" designs. In short an "ANOVA" design uses a few concentrations with replications at each concentration and employs ANOVA multiple comparison techniques in the analysis. A "regression" design chooses more concentrations with little or no replication at each concentration and uses regression techniques to model the response. For the EPA the goal of field testing is to negate the presumption of hazard as indicated by laboratory toxicity tests.[1] Their perspective is to use ANOVA techniques to test for effects at concentrations around the expected environmental concentration (EEC) for *one* pattern of use for a pesticide. Graney et al.[2] assess the relative advantages of "regression" and "ANOVA" designs and conclude that regression designs are preferable for their goals. From the perspective of Graney et al.[2] testing around a particular EEC for a particular use pattern is not practical. They want an estimate of the effect over a range of concentrations from a no effect concentration to a concentration that completely disrupts the ecosystem. Their statistical interest then focuses on estimating a dose-response relationship for the effects. The differences between Touart and Slimak[1] and Graney et al.[2] in preferences for "ANOVA" or "regression" designs stem from their different objectives.

In this chapter we propose optimal design methodology as a framework for quantifying objectives and choosing a design which optimizes that objective. The possible designs encompass both "regression" and "ANOVA" designs, and optimal design techniques provide a method for determining what design optimizes the researcher's goals. As stated by Graney et al.[2] (p. 75), "In short, a researcher wants to assign resources in an optimum way to be able to interpret results and extrapolate to other conditions or situations." EPA perspective does not emphasize extrapolation as much as Graney et al.[2] do; however, from either perspective there is need to optimize the information from these costly, time-consuming studies. Our discussion here focuses on how results known in the field of optimal design can be brought to bear on the question of designing field studies. We begin by looking at the question of deciding how many replicates to run at each concentration if the concentrations and the total number of units have already been determined. We then review methods for deciding the optimum concentrations.

CHOICE OF REPLICATE ALLOCATION FOR GIVEN CONCENTRATIONS

We begin with an example of a specific field study, the 1990 azinphos-methyl study by the Center for Lake Superior and Environmental Studies of the University of Wisconsin-Superior and the Environmental Protection Agency in Duluth. The

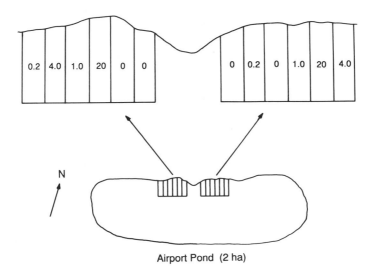

Figure 1. Azinphos-methyl concentrations (μg/l) for enclosures of 1990 littoral study at Airport Pond.

study site included 12 enclosures available in 2 blocks of 6 enclosures. The scientists decided to use four treatment concentrations, 0.2, 1.0, 4.0, and 20.0 μg/l, along with control enclosures. The physical layout of the design is shown in Figure 1. The enclosures were built to encompass approximately 25 m³ of water. Due to shoreline irregularities, the volumes of the enclosures ranged from 21.6 m³ to 27.2 m³. In this study, the range of concentrations was well within the range of concentrations that might arise from drift and runoff under normal application practices. However, the chance of reaching the highest concentration (20 μg/l) is less than for the lower concentrations. When determining the range of concentrations for the field study, lab LC_{50} values were used as guidelines. However, due to compensating mechanisms of field exposures, there was no assurance that a relatively high concentration would be ecologically significant in the field.

First we address statistical considerations in deciding how many enclosures to assign to each concentration, given that the concentrations and the total number of units to run have already been fixed. To find the optimal allocation to treatment groups for a given number of units, we need to define our goal and a model for the responses. One possible goal is to compare the results at each treatment concentration to control. Let τ_0, τ_1, τ_2, τ_3, and τ_4 be the effects of the control and four treatment concentrations respectively. Let $\hat{\tau}_i$ be the resulting estimator of τ_i under some constraint such as $\Sigma \tau_i = 0$. The variance of the comparison of a treatment concentration to control is $var(\hat{\tau}_i - \hat{\tau}_0)$. Hedayat et al.[3] consider the following objective functions for measuring how precisely all treatments can be compared to control.

$$minimize \sum_{i=1}^{4} var(\hat{\tau}_i - \hat{\tau}_0) \tag{1.1}$$

or

$$minimize \ max \ var(\hat{\tau}_i - \hat{\tau}_0) \tag{1.2}$$

The first criterion minimizes the total variance of treatment to control contrasts. The second criterion is a mini-max criterion which keeps as small as possible the variance for the least precise comparison to control. The objective functions (1.1) and (1.2) correspond, respectively, to A-optimality and MV-optimality. Both criteria put equal weight on all comparisons to control. These objective functions are not the only possibilities or the best choices for any particular experiment. We might, for example, want to put less weight on the comparison for the highest concentration which is less likely to occur in normal application practices. The methods described below can be adapted to different objective functions for particular experimental goals.

We now proceed to show a simple derivation of the optimal completely randomized design for objective function (1.1) when there are no blocks and variances are the same for all treatments. In this case criterion (1.1) is

$$minimize \ \sum_{i=1}^{4} \left(\frac{\sigma_0^2}{n_0} + \frac{\sigma_i^2}{n_i} \right)$$

which can be written

$$minimize \ \frac{4\sigma_0^2}{n_0} + \frac{\sigma_1^2}{n_1} + \frac{\sigma_2^2}{n_2} + \frac{\sigma_3^2}{n_3} + \frac{\sigma_4^2}{n_4} \tag{1.3}$$

If we could split the n's continuously, the function would be minimized by taking the partial derivatives with respect to the n_i and setting the derivatives to zero. Letting N be the total sample size,

$$n_0 = N - (n_1 + n_2 + n_3 + n_4)$$

Substituting for n_0 into (1.3) and setting the partial derivative with respect to n_i equal to zero gives

$$\frac{4\sigma_0^2}{[N - (n_1 + n_2 + n_3 + n_4)]^2} - \frac{\sigma_i^2}{n_i^2} = 0$$

Solving for n_0/n_i gives

$$\frac{n_0}{n_i} = \frac{2\sigma_0}{\sigma_1}$$

If we assume a model for responses with equal variances, then

$$n_0 = 2n_i$$

The control group is replicated twice as often as all other treatments. The control treatment is replicated more often because the control is used in every comparison.

For testing the hypotheses of no differences from control, the rule of $n_0 = \sqrt{v}n_i$ where v is the number of treatments is given by Dunnett[4] and Williams[5] for their multiple comparison tests of significance for each treatment vs. control. The derivation just given is for a design with no blocks. For a randomized block design, the optimal design still has $n_0/n_i = \sqrt{v}$ if that ratio fits exactly into blocks with the given number of experimental units. To show this for the design of Figure 1, we can use results in Hedayat et al.[3] for what they call one-way elimination of heterogeneity. Their results are for a more general situation where the number of experimental units in a block is not necessarily large enough to accommodate a complete replicate of the optimal design. An augmented balanced incomplete block design, denoted ABIB(v, b, k $-$ t; t), is a balanced incomplete block design in the v noncontrol treatments in b blocks of size k $-$ t with an extra t control unit in each block. According to the Family 2 result on page 465 of Hedayat et al.,[3] an ABIB design is A-optimal if

$$(k - t - 1)^2 + 1 \le t^2v \le (k - t)^2 \tag{1.4}$$

For a design with single units of the v noncontrol treatments in each block augmented by multiple control units, we claim the optimal number of control units in each block is

$$t = \sqrt{v} \tag{1.5}$$

The number of noncontrol units in each block is

$$k - t = v \tag{1.6}$$

Substituting (1.5) and (1.6) into (1.4) gives

$$(v - 1)^2 + 1 \le v^2 \le v^2$$

which is true for any $v \ge 1$. Hence a design with $\sqrt{v}$ control unit per block and $n_i = 1$, $i \ge 1$, unit for each noncontrol treatment is A-optimal. As indicated on page 468 of Hedayat et al.,[3] these designs are also MV-optimal.

In the azinphos-methyl study there were two blocks of six units each. The optimal design for either the total variance criterion or the criterion of minimizing the largest variance has two control enclosures and a single treatment concentration in each block. The design used in 1990 is shown in Figure 1. The positions of the units were assigned randomly within a block. The assumed model for the responses does not have to be used in the analysis of the data. Even if after collecting the data we find that the assumption of unequal variances is not appropriate, all is not lost. The data can still be analyzed accordingly. The design may no longer be optimal, however, when the variances are not equal. In the next section we show how to modify the derivations if unequal variances are anticipated at the design stage.

NONCONSTANT VARIANCES

If one accepts the goals as quantified in either objective function (1.1) or (1.2) and the assumption of equal variances, the design implemented in Figure 1 is optimal. The scientific questions must be addressed in the goal, and best judgment enters the choice of the assumed model for the responses. If either the goal or the assumed model governing the responses changes, the optimal design also changes. What exactly is entailed in specifying a model governing the responses depends on what parameters are needed to optimize the goal. For example in criterion (1.3) only the relative variances at the given conditions are required. In the next section we consider choosing the concentrations in order to estimate some parameter such as an LC_{50}. In this case we need to specify not only variances at particular concentrations but rather a broader model for the possible magnitudes and variabilities of responses over the whole range of concentrations we might choose to use.

In the case of the previous section where we optimize criteria such as (1.3), the assumption of equal variances will not always be appropriate. In lab and field experiments with control and treatments concentrations, it is many times observed that the control responses are more repeatable (have smaller variances) than treatment responses having partial effects. Organisms respond more variably to treatment stress than to control situations. The treatment conditions are harder to repeat and regulate than control conditions. On the other hand, results from extreme concentrations may have little absolute variation when the effect is very large. Suppose as an illustration the standard deviation in the treatments are $\sigma_0 = 0.7$, $\sigma_1 = \sigma_2 = \sigma_3 = 1.0$, and $\sigma_4 = 0.1$. The optimal allocation minimizing (1.3) would be

$$n_0 = 1.4n_1 = 1.4n_2 = 1.4n_3 = 14n_4 \qquad (1.7)$$

if that allocation fits the available sample size exactly. Consider a design with 12 enclosures and no blocks. A design with blocks could be handled, but we illustrate our point here using designs with no blocks. A design with 12 units cannot be found which has exactly the ratios of the sample sizes in (1.4). We need at least one enclosure for each of the four treatments plus control. Assigning five of the units to these treatments leaves seven units to assign. This leaves $_{(7+5-1)}C_4 = {}_{11}C_4 = 330$ ways to assign the remaining 7 enclosures to 1 of the 5 treatments. This is a manageable number of possibilities for computing the objective function, say (1.1), for each possible design and listing potential designs. An algorithm for running through all possible allocations is

$$\text{for } w_0 = 1 \text{ to } 8$$
$$n_0 = w_0$$
$$\text{for } w_1 = w_0 + 1 \text{ to } 9$$
$$n_1 = w_1 - w_0$$
$$\text{for } w_2 = w_1 + 1 \text{ to } 10$$
$$n_2 = w_2 - w_1$$
$$\text{for } w_3 = w_2 + 1 \text{ to } 11$$
$$n_3 = w_3 - w_2$$
$$n_4 = 12 - w_3$$

The algorithm moves the inner walls of the "walls and balls" representation of Feller.[6] To visualize the algorithm, we have to assign 1 of the 12 units to each of the 5 treatment/control conditions. This leaves 7 units to assign above and beyond the minimum of one in each group. A particular choice can be represented as seven balls and four walls separating the seven balls into five groups, corresponding to control and treatments one through four. For example, the configuration

$$0|00\|000|0$$

would represent one extra control unit (corresponding to having one ball partitioned off by the first wall), two extra treatment 1 units (corresponding to the two balls between the first and second walls), no extra treatment 2 units, three extra treatment 3 units and one extra treatment 4 unit. In the representation above, there are eleven positions, and four must be chosen for the wall positions. Hence as indicated above there are $_{11}C_4 = 330$ possible assignments. Call the positions of the walls w_0, w_1, w_2, w_3. The algorithm then runs through all possible placements of the walls and assigns the corresponding total number of units for each treatment.

Using $\sigma_0 = 0.7$, $\sigma_1 = \sigma_2 = \sigma_3 = 1.0$, and $\sigma_4 = 0.1$ gives a minimum value of the objective (1.1) of 1.83. The designs with a value for (1.1) within 10% of the optimal 1.83 (those less than about 2.0) are given in Table 1. The designs which attain the optimum have one unit at the highest concentration and three units elsewhere except one of the middle concentrations is assigned to only two units. Other designs are nearly as good. An advantage to checking all of the possibilities rather than using an algorithm to find the one and only optimal design is that competitive alternatives can be identified permitting one to see how much difference it makes if one chooses a suboptimal design. A design can be chosen from the competitive designs taking into account considerations not explicitly included in the objective function. Table 2 shows the designs close to optimum for minimizing the maximum variance for any comparison (1.2). The designs do not have the same ordering as in Table 1; the optimum design under one criterion is not the same as the optimal design under another criterion. These two particular criteria are similar, however, and the competitive designs are similar.

A special case of unequal variances would be the case of constant coefficients of variation. A logarithmic transformation would make the variances constant, but

Table 1. Designs to Minimize the Sum of Variances of Differences From Control

n_0	n_1	n_2	n_3	n_4	Variance Sum
3	3	3	2	1	1.830
3	3	2	3	1	1.830
3	2	3	3	1	1.830
4	3	2	2	1	1.833
4	2	3	2	1	1.833
4	2	2	3	1	1.833
5	2	2	2	1	1.902
3	4	2	2	1	1.913
3	2	4	2	1	1.913
3	2	2	4	1	1.913
2	3	3	3	1	1.990
3	3	2	2	2	1.992
3	2	3	2	2	1.992
3	2	2	3	2	1.992
4	2	2	2	2	1.995

Note: $\sigma_0 = 0.7$, $\sigma_1 = \sigma_2 = \sigma_3 = 1.0$, and $\sigma_4 = 0.1$.

we do not believe the objective is to minimize the sum of the variances in the log scale. We are concerned about our precision in estimating the ratio of each treatment mean to the control mean. If the ratio is small, it doesn't matter whether the ratio is 0.1 or 0.05; in either case the treatment causes a major effect. But it matters considerably whether the ratio is 1.0 (no effect) or 0.50 (a 50% decrease). However, on the log scale 0.1 and 0.05 are as far apart as 1.0 and 0.5. We are interested in the variability in the ratios in the original response scale. Using the delta method,[7] the variance of the ratio $\overline{Y}_i/\overline{Y}_0$ is approximately

$$var(\overline{Y}_i/\overline{Y}_0) \approx (1/\mu_0)^2 var(\overline{Y}_i) + (\mu_i/\mu_0^2)^2 var(\overline{Y}_0)$$

With constant coefficient of variation, $\sigma_i = c\mu_i$. Substituting this expression for the standard deviations and $var(\overline{Y}_i) = \sigma_i^2/n$ gives

$$var(\overline{Y}_i/\overline{Y}_0) \approx c^2(\mu_i/\mu_0)^2(1/n_i + 1/n_0)$$

The sum of variance objective analogous to (1.1) then becomes

$$\min \frac{\sum_{i=1}^{4} (\mu_i/\mu_0)^2}{n_0} + \frac{(\mu_1/\mu_0)^2}{n_1} + \frac{(\mu_2/\mu_0)^2}{n_2} + \frac{(\mu_3/\mu_0)^2}{n_3} + \frac{(\mu_4/\mu_0)^2}{n_4}$$

As an illustration, consider a model with $\mu_0 = 1.0$, $\mu_1 = 0.9$, $\mu_2 = 0.8$, $\mu_3 = 0.7$, and $\mu_4 = 0.2$. Table 3 shows the designs with objective values within 10% of the optimum. The optimum design has only one replicate at the highest concentration, four units at the control concentration, three units at the lowest noncontrol concentration, and two units at each of the middle concentrations. Again other

Table 2. Designs to Minimize the Largest Variance of Differences From Control

n_0	n_1	n_2	n_3	n_4	Maximum Variance
2	3	3	3	1	0.578
5	2	2	2	1	0.598
4	3	2	2	1	0.622
4	2	3	2	1	0.622
4	2	2	3	1	0.622
4	2	2	2	2	0.622
3	3	3	2	1	0.663
3	3	2	3	1	0.663
3	2	3	3	1	0.663
3	3	2	2	2	0.663
3	2	3	2	2	0.663
3	2	2	3	2	0.663
3	2	2	2	3	0.663
3	4	2	2	1	0.663
3	2	4	2	1	0.663
3	2	2	4	1	0.663

Note: $\sigma_0 = 0.7$, $\sigma_1 = \sigma_2 = \sigma_3 = 1.0$, and $\sigma_4 = 0.1$.

designs are almost as good in regards to this criterion. Table 4 gives some designs for minimizing the largest variance under these same equal coefficient of variation conditions. Again the designs are not exactly the same as in Table 3. As a special case if we are only comparing one treatment concentration to control, the optimal design would assign the same number of units to both groups.

The methods above allocate a given number of units to the treatment levels. In some situations the total number of units is dictated by cost or other external considerations. Ideally one would be able to also control the total number of units in order to assure sufficient power to detect important alternatives or sufficient precision for estimating effects. The methods above could be adapted to optimize the appropriate criterion for a fixed total number of units, and then progressively increase the total number of units until the observed optimum is deemed good enough.

Table 3. Designs to Minimize the Sum of Variances of Ratios to Control With Constant Coefficient of Variation

n_0	n_1	n_2	n_3	n_4	Variance Sum
4	3	2	2	1	1.37
4	2	3	2	1	1.40
5	2	2	2	1	1.41
4	2	2	3	1	1.42
3	3	3	2	1	1.43
3	3	2	3	1	1.45
3	4	2	2	1	1.47
3	2	3	3	1	1.48
4	2	2	2	2	1.49

Note: $\mu_0 = 1.0$, $\mu_1 = 0.9$, $\mu_2 = 0.8$, $\mu_3 = 0.7$, and $\mu_4 = 0.2$.

Both criteria (1.1) and (1.2) treat the comparisons as equally important. In some situations one of the comparisons may be more important to the scientific or regulatory questions. For example we may be most interested in comparing the effective environmental concentration (EEC) to control. In that case the objective function could be changed to weight that comparison more heavily than other comparisons.

In practice the relative variances within the groups are never known. One can, however, put a probability distribution on possible variance values. This distribution expresses one's beliefs about potential variance structures. Since the objective function (1.1) is linear in the variances, the expected value of the objective over the possible values of the variances is found by replacing the variances in (1.1) with our expected variances, regardless of the uncertainty in our prior belief about the variances. The relative variances in the treatment groups will also change for different endpoints. A total variance can also be taken over all endpoints. Again since the objective (1.1) is linear in the variances, the total variance of the contrasts will be minimized by using the average expected variance over all endpoints for each concentration.

CONSIDERATIONS IN CHOOSING THE NUMBER OF DOSES TO TEST

The previous section described a strategy for deciding how many units to allocate to each dose once the doses and total number of units have been chosen. The problem of deciding in addition what doses to use is more difficult. Using results known from optimal design theory, we discuss some considerations in this choice.

If the goal is to decide whether one particular predetermined concentration, say the EEC, has an effect or to estimate the size of the effect relative to control, all units should be assigned to only control or the EEC. Using the results above with $k = 1$ treatment, the sample sizes should be balanced if $\sigma_1^2 = \sigma_2^2$ and we are estimating the difference between effects. The sample sizes should also be balanced if we are estimating the ratio to control and the coefficient of variation is the same

Table 4. Designs to Minimize the Largest Variance of Ratios to Control With Constant Coefficient of Variation

n_0	n_1	n_2	n_3	n_4	Maximum Variance
4	3	2	2	1	0.480
3	4	2	2	1	0.533
3	3	2	3	1	0.540
3	3	3	2	1	0.540
3	3	2	2	2	0.540
5	2	2	2	1	0.567
6	2	2	1	1	0.572
5	2	2	1	2	0.588
5	2	3	1	1	0.588
5	3	2	1	1	0.588

Note: $\mu_0 = 1.0$, $\mu_1 = 0.9$, $\mu_2 = 0.8$, $\mu_3 = 0.7$, and $\mu_4 = 0.2$.

in both groups. In other situations the optimal assignment may not have equal allocation to each treatment group. If for example the EEC units are expected to be more variable than the control units and we are estimating differences, more units should be assigned to the more variable EEC group.

If, rather than comparing a single concentration to control, the goal is to obtain the best approximation of the response curve as in Graney et al.,[2] then the units should be assigned to a range of concentrations. Chaloner and Larntz[8] studied the estimation of the LC_{50} and other percentiles assuming a logistic response for percentage survival type data. The logistic survival rate at a concentration x would be modeled by

$$1/(1 + \exp(-\beta(x - LC_{50})))$$

In their formulation the experimenter specifies a prior distribution of the LC_{50} and the sharpness of the logistic response, β. The LC_{50} designates what concentration would kill (on the average) 50% of the organisms. The sharpness of the response, β, measures how quickly the survival rate decreases with increasing concentration. As a benchmark on the effect of β on logistic survival, an increase in concentration of $|3/\beta|$ above the LC_{50} causes the expected survival rate to decrease from 50% at the LC_{50} to 5% at $LC_{50} + |3/\beta|$. Equivalently, a concentration increase of $|6/\beta|$ corresponds to survival decreasing from 95% to 5%. A large absolute value of β means that the concentration does not have to change very much to cause a large change in survival. A small absolute β indicates a more gentle decrease in survival as the concentration increases. The experimenter specifies a range of values for the possible LC_{50} and the sharpness, β, of the response. The results of Chaloner and Larntz show that if a logistic response holds, the optimal number of doses to test increases as the uncertainty in the LC_{50} increases or the possible sharpness of the response increases. If the effect is potentially sharp and one is not certain of the range of doses where the effect occurs, then a number of doses must be used to increase the chance of catching where the effect occurs. At an extreme if the effect is sudden so that in essence only all or nothing effects occur, then doing replicates at the same dose adds nothing to one's information, and a finer grid of points gives one optimal narrowing in on the dose where the effect occurs. If the effect is gradual, one can (if the assumed logistic response model holds) interpolate or even extrapolate successfully given data with between no kill and total kill.

In field studies of ecosystems there is often large variability in the potential position of the LC_{50} or other percentiles compared to the sharpness of the response curve. As discussed, for example in Graney et al.,[2] for a single organism there is uncertainty in the dose causing a given effect. Laboratory results are available in planning the study, but the field results will not necessarily mimic the lab results. This uncertainty is the basis for the need to do field testing in the first place. Field results can differ from lab results for several reasons. For one, the test compound may affect some other part of the ecosystem causing a secondary effect on the organism of interest. For example a fish itself may not be affected directly by a

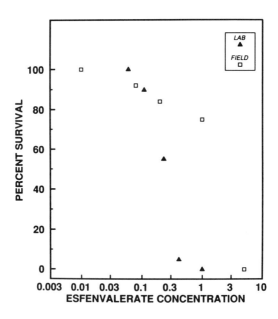

Figure 2.   Fathead minnow survival in lab and field studies.

given dose, but the dose may damage an invertebrate food species. Graney et al.[2] describe other mechanisms which contribute to field effects being less than predicted from lab tests. The range of possible LC_{50} values is extended further if there is interest in estimating response curves for multiple species. The response curves of two species of fish will not necessarily be the same. Some of these points are illustrated in Figures 2, 3, and 4 for laboratory and field results for bluegill, fathead minnows, and *Hyallela* exposed to esfenvalerate. The lab and field results are quite different for fatheads and *Hyallela*. In the lab 0% of the fatheads survived 1.0 μg/l, and only 5% survived 0.41 μg/l. In the field by comparison, 75% of the fatheads survived 1.0 μg/l. The responses for the three species are also quite different. None of the bluegills survived a field concentration of 1.0 μg/l, but 75% of the fatheads survived this concentration. Most of the fatheads and bluegills survived concentrations less than 0.2 μg/l, but in the field the *Hyallela* abundance was only 30% of control at 0.01 μg/l and no *Hyallela* were present at 0.08 μg/l. *Hyallela* is not of primary interest as an endpoint, but since *Hyallela* is an important food source for some fish in many ecosystems, the potential effect of esfenvalerate on *Hyallela* at very low concentrations would be important information in modeling the responses of fish populations to esfenvalerate when *Hyallela* is a major food component. If the goal is to model an entire ecosystem, a range of doses will be needed to model a range of possible response curves.

Thus for a field test, multiple doses would be required to find optimal estimates of the response curves even if a family of models for the response curve were known to be logistic, a relatively simple two parameter family. Generally, however, the possible models for field responses will be uncertain and a large number of parameters would be required to describe the range of possible models. With more parameters in the model, the number of support points in the optimal design will increase even more. In polynomial regression, for example, optimal designs for estimating the regression parameters (following particular criteria) have as many points as the degree of the polynomial.[9] For example if the model is really linear, two points as far apart as possible will be optimal. However, this design leaves no room for checking the model and makes no allowance for the possibility of curvature in the true form of the response. In field studies the exact parametric form of any response is not known exactly, and different responses will follow different models. In many instances one would use a simplified form of response curve to approximate the responses: for example, a spline smoothing curve, a piecewise linear fit, or some nonlinear family such as the logistic. The objective function to minimize would need to consider not only the precision of the estimates but also the probable biases or inaccuracies from using an approximation to the true form. For the approximate interpolation to be accurate as possible, one would again prefer more support points in the design than if the simple model were known to be in fact the truth.

Hence to obtain estimates of the response curves from a field study, especially a study interested in multiple endpoints with a range of possible response models, multiple doses will be needed. The optimum doses can be found if one can define a quantifiable objective and a range of possible models. The quantifiable objective

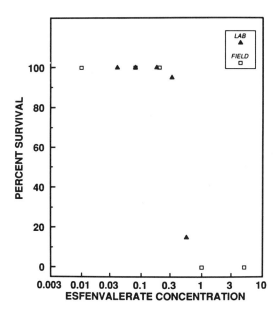

Figure 3.   Bluegill survival in lab and field studies.

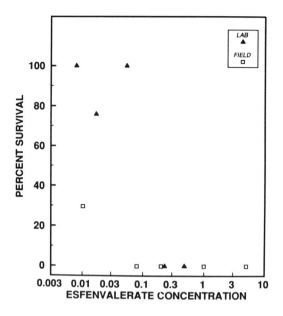

Figure 4.   *Hyallela* survival in lab and field studies.

would include the way in which the data are analyzed and the goals for which they are analyzed. Having defined the goal and a model for the data, optimization techniques such as those in Chaloner and Larntz[8] can be used to find the optimal design.

CONCLUSION

What we put forth here is a description of how optimal design methodology can be applied to the design of field studies. The results give no one design which is optimal for all questions and situations. As stated by Graney et al. (p. 77),[2] "A prerequisite to the design of any scientific experiment is a clear statement of the objectives or hypothesis to be tested." To decide the optimal design, the experimenter must first specify the goals of the study. Then optimal design techniques can be applied to find the best design or set of designs for answering the questions under the expected range of experimental conditions. The statistical methods and design need to be tailored to fit the goals, not vice versa. To use optimal design techniques, the goals must be stated in a quantifiable form. Then suppositions must be made about the model governing the generation of the data. Specifying the experimental conditions includes specifying what models may control the generation of the data. Are the variances in the groups equal? Does the response curve follow a logistic form? How quick or steep is the response? Uncertainty in answering those questions can be quantified in a prior probability distribution over the possible models. Numerical optimization techniques can then be applied to find a design which optimizes the goal or, better yet, to specify some possible designs which do a reasonable job. The model which is optimal will depend both on the scientific goals of the experiment and on the assumptions about what models the data will follow.

ACKNOWLEDGMENT

The authors thank two anonymous reviewers for their thorough and most helpful comments. This work was supported in part by EPA Cooperative Agreement No. CR-816758.

REFERENCES

1. Touart, L.W. and M.W. Slimak. 1989. Mesocosm approach for assessing the ecological risk of pesticides. In J.R. Voshell, Jr., ed., *Using Mesocosms to Assess the Aquatic Ecological Risk of Pesticides: Theory and Practice*. Miscellaneous Publications, Entomology Society of America, Lanham, MD.
2. Graney, R.L., J.P. Giesy, Jr., and D. DiToro. 1989. Mesocosm experimental designs: Advantages and disadvantages in ecological risk assessment. In J.R. Voshell, Jr., ed., *Using Mesocosms to Assess the Aquatic Ecological Risk of Pesticides: Theory and Practice*. Miscellaneous Publications, Entomology Society of America, Lanham, MD.
3. Hedayat, A.S., M. Jacroux, and D. Majumdar. 1988. Optimal designs for comparing test treatments with controls. *Statistical Science* 3:462–491.
4. Dunnett, C.W. 1955. A multiple comparisons procedure for comparing several treatments with a control. *Journal of the American Statistical Association* 50:1096–1121.
5. Williams, D.A. 1972. The comparison of several dose levels with a zero dose control. *Biometrics* 28:519–531.
6. Feller, W. 1968. *An Introduction to Probability Theory and Its Applications*. John Wiley & Sons, New York, p. 38.
7. Rice, John A. (1987) *Mathematical Statistics and Data Analysis,* Wadsworth, Monterey, CA.
8. Chaloner, K. and K. Larntz. 1989. Optimal Bayesian design applied to logistic regression experiments. *Journal of Statistical Planning and Inference* 21:191–208.
9. Chaloner, K. 1984. Optimal Bayesian design for linear models. *The Annals of Statistics* 12:283–300.

CHAPTER 13

Design of Mesocosm Studies, and Statistical Analysis of Data: Summary and Discussion

John H. Rodgers, Jr.

At this symposium, a series of six papers on mesocosm study design as well as statistical analyses were presented. These papers examined the role of mesocosms in pesticide risk assessment as well as the interface between regulation and science in pesticide registration. An assumption in risk assessment is that we are not going to live in a zero risk environment and therefore risk assessments for pesticides are necessary. The purposes of mesocosm studies have been identified as (1) to negate a presumption of unreasonable risk, and (2) to define the intensity and duration of adverse effects of pesticides in aquatic systems.

A variety of approaches were presented and have been historically utilized to design mesocosm studies for pesticide risk assessments. Each of these studies has been based upon a specific premise. That premise centers around "What is the question?". It is impossible to develop an optimal design for allocation of resources if the objective of the study is not clearly identified and stated. Thus the pivotal issue fostering much discussion and controversy has arisen from pursuit and identification of the illusive question.

A further challenge arises from the pesticide registration charge of negating a presumption of adverse or unacceptable risk. In order to develop information for a risk assessment that will stand the test of time, both type I errors and type II errors must be minimized in mesocosm studies. From the regulatory perspective, one would like to be assured that if effects exist in the study then they are readily identifiable. Further, it would seem important from a regulatory perspective to avoid concluding that there were no effects when in fact ecologically significant effects occurred at field application rates. From the registrant's perspective, one would not like to find apparent effects in these studies when in reality they do not exist. It is apparent that registrants would not like to conclude that there are effects in mesocosm studies when the effects will not be manifested in field exposures during actual use of the compound. This dilemma is further complicated by our ability to discern ecologically relevant effects once they have been statistically identified. Thus the primary role of statistics in design of mesocosm studies has been frequently suspended for practical or logistic reasons.

Based on the information presented in this series of six papers, it was apparent that there is little probability that mesocosm studies as currently designed with ANOVA approaches, *t*-test approaches, or regression approaches will find effects that may be manifested in the field. This conclusion is based on the result of reported variances and is also the expected result of inadequate replication, in most cases with three to four replicates at a treatment level. More than 100% change in many parameters would be needed in order to identify an effect as statistically significant. As we increase variance, the power is concomitantly decreased. Minimal replication (i.e., n = 3 − 4) almost assures that no differences will be found. A common axiom of statistics is that as you increase the number of replicates, the probability of finding effects increases dramatically. Therefore, the basic design as presented by Touart and Slimak (1989) has emphasized the probability of concluding there are no observable statistically significant effects when there actually are effects. This seems to be contrary to the purposes of the USEPA. Perhaps the guidance document can be treated as merely guidance, and more scientifically defensible and optimal experimental designs will produce more accurate data, conclusions, and regulatory decisions.

As currently utilized in evaluation of mesocosm experimental results, statistical assumptions are widely violated. For example, the assumption in Student's *t*-tests is (1) that the data are normally distributed, (2) that there are homogeneous variances, and (3) that the variables are continuous and not discrete. Further, utilization of nonparametric tests may also lead to accepting the notion of no effects when actually effects did occur. Frequently, nonparametric tests require a sample size of six or more replicate mesocosms to reject a false null hypothesis.

Thus, we have considerable disagreement and discussion regarding appropriate statistics. However, it appears that these issues can be at least partially resolved if concerns regarding the appropriate question or objectives are initially addressed. If the ecological parameters of regulatory concern are clearly stated, then an experiment can be designed to address those issues. In this manner, the study will not focus on irrelevant parameters and considerable effort will not be expended to gather superfluous information. Regulatory endpoints of concern may be focused on EC_{10}, EC_0, LOEC, or NOEC. Probablistic approaches will be most promising in dealing with regulatory issues that arise from these endpoints, as well as recognition that these endpoints are not based in single concentration parameters but probabilistic distributions of exposures under field conditions.

A clear message from the six papers was that our ability to detect differences is hampered by the lack of adequate replication and inadequate identification and focus on sensitive endpoints. These issues may be addressed by revised EPA protocols that permit flexibility in experimental designs following a clear definition of objectives. The current somewhat rigid mesocosm protocol of the EPA may hamper the development of information for accurate risk assessments for pesticides rather than enhancing that development.

REFERENCES

Touart, L.W. and Slimak, M.W. 1989. Mesocosm approach for assessing the ecological risk of pesticides. In J.R. Voshell, Jr., ed., *Using Mesocosms to Assess the Aquatic Ecological Risk of Pesticides: Theory and Practice*. Miscellaneous Publications, Entomology Society of America, Lanham, MD.

SECTION III

Management and Treatment of Mesocosms

INTRODUCTION

Since mesocosms are artificially constructed systems, their representativeness relative to "natural" ecosystems is a concern. Many of the case studies present in Section III of this book address the procedures used to establish mesocosm test systems. Colonization and equilibration are discussed in general terms; although, specific information is not provided on the representativeness of the test system. The first chapter in this section addresses this point for the macroinvertebrate fauna, specifically chironomids in a lentic system. Procedures for establishing the macroinvertebrate fauna in mesocosms are described and, after one year colonization, insect emergence from this test system is compared to historical emergence patterns for the geographical area.

Procedures for determining the appropriate dosing rates for mesocosm studies are not well developed. Exposure of aquatic ecosystems can occur via both aerial spray drift and surface runoff and accurate quantitative prediction for either route is extremely difficult. The second chapter in this section addresses the difficulty associated with establishing dose levels which are realistic and useful in making regulatory decisions. A case study for a pyrethroid insecticide is presented and suggestions for mesocosm loading procedures are provided.

The remaining two chapters in this section are essentially case studies providing information on the effects of specific toxicants on different types of mesocosms. These chapters were placed in this section because they address the importance of exposure in understanding the effects observed in aquatic field studies. Although all of the case studies discussed in Sections IV and V address the measurement of test concentrations, they, for the most part, focus on the specific effects which occurred in the studies. It is important to realize that the chemical's fate (i.e., persistence, bioavailability, etc.) is an important factor in determining the type and extent of the effects observed. The third chapter in this section discusses the importance of bioavailability in the effects observed and the fourth chapter addresses how a chemicals photolytic rate can directly influence length of exposure and thus the effects observed. Although both of these chapters contain other information, these have been separated into this section to help emphasize the importance of exposure in mesocosm test systems.

CHAPTER 14

Sediment Transfers and Representativeness of Mesocosm Test Fauna

Leonard C. Ferrington, Jr., Mary Anne Blackwood, Christopher A. Wright,
Tracey M. Anderson, and David S. Goldhammer

Abstract: Transfers of sediments from old farm ponds to newly constructed mesocosm test ponds in autumn facilitate rapid establishment of taxonomically rich communities of benthic organisms for use in studies of pesticide effects the following summer. It has been presumed, but not critically demonstrated, that the sediment transfers conserve patterns of benthic organization and thus provide a test fauna highly representative of the undisturbed source ponds. Comparisons of emergence composition and phenologies of Chironomidae (the dominant taxon in our mesocosm tests) from mesocosm ponds at the Nelson Environmental Studies Area with historical patterns of Chironomidae emergence in Kansas confirm high degrees of representativeness among Chironomidae emerging from mesocosm ponds. Species richness of Chironomidae emerging from mesocosm ponds within one year after seeding with transferred sediments is higher than recorded for ponds where aerial colonization by ovipositing females is the primary source of Chironomidae. Species richness, emergence composition, chronology, and phenology compare favorably with results of studies of Chironomidae emergence dynamics from small ponds with mature sediments.

INTRODUCTION

The United States Environmental Protection Agency has issued guidelines for aquatic field testing of pesticides for registration which require use of replicate experimental ponds, commonly referred to as mesocosms (Touart, 1988). Touart and Slimak (1989) discuss the mesocosm approach for assessing the ecological risk of pesticides and include a recommended protocol and required features of a mesocosm. Among other characteristics, mesocosms should contain representative proportions of the complete natural assemblage of organisms of the system that is being modeled (Touart and Slimak, 1989; Buikema and Voshell, 1993), including aquatic invertebrates. It seems logical to assume that newly constructed mesocosms may take one to several years to develop communities that are close to equilibrium conditions if dispersal is the primary mechanism for invertebrate colonization. Extended periods for colonization and development of the test fauna can add several years to a given pesticide registration test, thus tying up mesocosm test facilities and limiting the number of tests that can be conducted. Procedures that shorten the period of development of the representative test faunas are therefore preferred.

Over the past 3 years we have utilized transfers of sediments from old farm ponds to seed our mesocosms with invertebrate communities. Transfers have been made in fall, with monitoring initiated during winter or early the following spring. Howick et al. (1992) have demonstrated that sediment transfers produce taxonomically diverse communities in our mesocosms, with mature populations of emergent aquatic insects appearing early the following spring. In this chapter we present the results of our emergent insect monitoring studies and critically examine the species composition of emerging taxa. Chironomidae comprise the dominant component of aquatic insect emergence from our mesocosm test ponds (Howick et al., 1992); therefore, in this paper we focus on patterns of species richness, composition, phenology, and relative abundances of Chironomidae in order to address the question of representativeness of the test fauna residing in the mesocosms during the period from 6 to 12 months after sediment transfers.

METHODS AND MATERIALS

Figure 1 shows the arrangement of the mesocosm test ponds at the Nelson Environmental Studies Area (NESA) of the University of Kansas. In October 1987 sediments were transferred from a 20 + year-old farm pond (F1) to each of 14 refurbished mesocosm test ponds (M1). These mesocosm test ponds were built in 1977 and 1979 and were previously used in various research projects. All ponds were refurbished prior to sediment addition by use of a backhoe to scrape the sides and bottom of accumulated sediments down to the original subsoils. Sediments from the old farm pond were removed by crane and placed in trucks for transport to the mesocosm ponds. To reduce variation in the quality of sediments delivered to mesocosm ponds (due to differences in source locations within the farm pond), one batch of sediment was delivered to each mesocosm pond before adding subsequent batches. Four batches of sediment were delivered to each pond in this sequential manner. After the last batch was delivered, the sediments were spread across the bottoms of the mesocosm ponds and formed a layer of 10 to 15 cm of

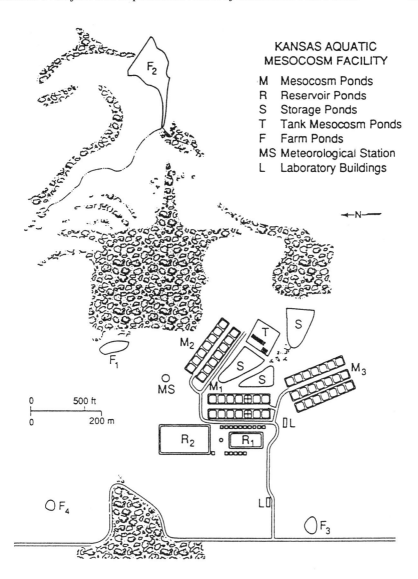

Figure 1. Map of the Kansas Aquatic Mesocosm Facility at the John H. Nelson Environmental Studies Area of the University of Kansas.

imported substrates. Water from a single reservoir (R1) was used to fill all ponds, and the ponds were chemically and physically monitored throughout the winter.

Sampling for emergent insects was initiated on 14 April, approximately 6 months after sediment transfers and coincident with increasing water temperatures. Samples were collected with a floating emergence trap that covered 0.56 square meters, and used plates coated with adhesive to retain emerged insects. Traps were collected and replaced every 7 days from 14 April through 19 October. Two trap samples were taken from each pond, corresponding to a shallow (less than 2.0 m) and deep zone (approximately 2.0 m). Insects were removed from the adhesive coated plates with solvent, then transferred into 80% ethanol for preservation until identified. Since some aquatic insects crawl up an emergent plant or other emergent physical structure before transforming to the winged adult phase (e.g., Odonata), small wire traps were placed in shallow water areas to document the emergence of these taxa from the mesocosms.

All male Chironomidae were identified to species or morphotype and counted for every other 7 day period. Selected specimens of each species were slide mounted for inspection and to provide vouchers for all identifications. Identifications of males were verified by consultation with other recognized Chironomidae taxonomists, and vouchers have been integrated into the permanent museum collections of the Water Quality and Freshwater Ecology Program of the Kansas Biological Survey. Females were identified to species and counted when recognizable; otherwise, they were counted at the tribe or subfamily level. Patterns of species richness were derived from data on males; however, relative abundance patterns for tribes or subfamilies were derived by summing the data for males and females.

The data for this chapter were generated as part of a pesticide reregistration study. Twelve of the original fourteen ponds were used in the study, two of which were held as controls. In order to focus on the question of representativeness, it is necessary to restrict our data analysis and discussion to patterns and trends over all 12 ponds up to the date of first application of pesticide, and for the duration of the project for the control ponds. Our last emergence sample event before pesticide addition ended 26 June 1988, resulting in six sets of emergence data for all ponds, plus eight additional sets of emergence data for the control ponds. The patterns of emergence composition, phenology, and numerically dominant taxa were compared to existing literature (when available) or unpublished historical data from files of the Kansas Biological Survey in order to evaluate the representativeness of the patterns derived in this study.

Coefficient of Community (CC) values, which measure the frequency of taxa shared among pairs of ponds (Whittaker, 1975), were sequentially calculated for all 12 ponds taken two at a time using the cumulative species composition up to the date of first application of pesticide. This resulted in 66 CC values. One additional CC value was calculated for the two control ponds, based upon the cumulative species list for the entire study.

RESULTS

Species Richness and Emergence Composition

The taxonomic composition of aquatic insects emerging from all mesocosm test ponds in 1988 is given in Table 1. Patterns of Chironomidae taxonomic composition for the study and control ponds, summarized both individually and collectively, are shown in Figure 2. Forty Chironomidae taxa emerged from the control ponds from

Table 1. List of Aquatic Insects Collected as Adults From Mesocosm Test Ponds in 1988

Coleoptera
  Dineutus sp.
  Hydroporus sp.
  Peltodytes sp.
  Tropisternus sp.
Diptera (Exclusive of Chironomidae)
  Ceratopogonidae
  Culicidae
  Ephydridae
  Muscidae
Ephemeroptera
  Caenis sp.
  Callibaetis sp.
Hemiptera
  Belostoma sp.
  Buenoa sp.
  Mesovelia sp.
  Neoplea striola
  Sigara sp.
Hymenoptera — Parasitic wasps, not identified
Odonata
  Anax sp.*
  Enallagma civile
  Enallagma aspersum
  Enallagma signatum
  Eurethymis sp.*
  Ischnura verticalis
  Ischnura sp.*
  Lestes sp.*
  Libellula sp.*
  Pantala sp.*
  Tetragoneuria sp.*
Trichoptera
  Hydroptilidae
  Leptoceridae
Chironomidae
  Chironominae
    Chironomini
      Apedilum elachistus
      Chironomus sp. 1
      Chironomus sp. 2
      Cladopelma collator
      Cryptochironomus spp.
      Cryptotendipes darbyi
      Cryptotendipes emorsus
      Demeijeria brachialis
      Dicrotendipes modestus
      Dicrotendipes nervosus gr. sp.
      Endochironomus nigricans
      Endochironomus subtendens

Chironomidae (cont'd)
  Chironominae
    Chironomini
      Glyptotendipes sp.
      Nilothauma sp.
      Parachironomus chaetaolus
      Parachironomus monochromus
      Parachironomus varus
      Parachironomus n. sp.
      Paralauterborniella nigrohalteralis
      Paratendipes albimanus
      Polypedilum cf. *floridense*
      Polypedilum illinoense
      Polypedilum ophioides
      Polypedilum simulans
      Polypedilum sp. nr. *simulans*
      Polypedilum trigonus
      Xenochironomus xenolabis
      Zavreliella varipennis
    Tanytarsini
      Cladotanytarsus sp.
      Constempellina n. sp.
      Tanytarsus sp. 1
      Tanytarsus sp. 2
    Pseudochironomini
      Pseudochironomus pseudoviridis
      Pseudochironomus rex
  Orthocladiinae
    Cricotopus sp. gr. 1
    Cricotopus sp. gr. 2
    Corynoneura spp.
    Hydrobaenus spp.
    Nanocladius spp.
    Parakiefferiella coronata
    Psectrocladius vernalis
    Psectrocladius spinifer
    Psectrocladius simulans
  Tanypodinae
    Coelotanypodini
      Clinotanypus pinguis
    Macropelopiini
      Psectrotanypus dyari
    Pentaneurini
      Ablabesmyia spp.
      Labrundinia spp.
      Paramerina smithae
      Telopelopia okoboji
    Procladiini
      Procladius bellus
      Procladius sublettei
    Tanypodini
      Tanypus neopunctipennis

Note: Taxa listed with asterisk after them were collected as exuviae from wire emergence traps. All other taxa were collected in floating emergence traps.

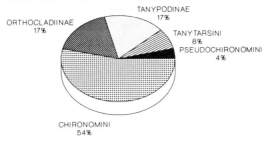

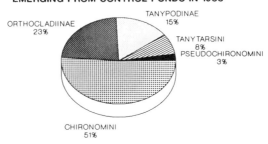

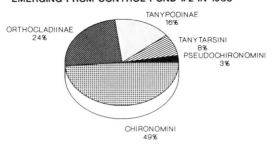

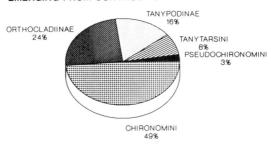

Figure 2.   Taxonomic composition of Chironomidae emerging from mesocosm ponds during 1988.

April through October 1988. The cumulative number of Chironomidae taxa emerging from all 12 ponds during the same time was 52; thus, the communities in control ponds represented 76.9% of total Chironomidae taxonomic richness. All taxa that were common, here defined as comprising 1.5% or more of total emergence when all 12 ponds were considered, were present in the controls. The species richness values for control ponds were equivalent, though not identical, for the study, with 37 taxa detected from each pond. The CC value for control ponds was 0.919, indicating a high degree of similarity in Chironomidae emergence composition.

Chironomini was the most species-rich tribe with 28, 18, and 18 species represented in the study and individual control ponds, respectively. Collectively, a total of 20 species of Chironomini were present in control ponds. Orthocladiinae and Tanypodinae were second most species rich with nine species each. All Orthocladiinae species were collected from each control pond. Only six of the nine Tanypodinae were collected from control ponds, with all six present in each control pond. Tanytarsini and Pseudochironomini were represented in the study by four and two species, respectively. The same three species of Tanytarsini were collected from each control pond. Only one of the Pseudochironomini species was present in the control ponds, and it was collected only infrequently in both ponds.

The total number of species detected in emergence traps for all ponds up to the date of first application of pesticide is given in Table 2. Species richness values up to the date of first application ranged from 20 to 30 taxa. The number of Chironomini emerging per pond during the preapplication period was most variable, with 5 to 12 taxa represented per pond. Tanytarsini was least variable, with 3 species detected in 11 ponds and 2 species detected in the remaining pond. Orthocladiinae and Tanypodinae were intermediate in variability, ranging from six to nine and five to seven taxa emerging per pond, respectively, during the preapplication period.

Table 2. Number of Chironomidae Taxa Collected in Emergence Traps, Summarized by Subfamily or Tribe For All 12 Mesocosm Test Ponds Through 26 June 1988

Pond #	Chironomini	Tanytarsini	Orthocladiinae	Tanypodinae	Pseudochironomini	Sum
1	7	3	8	5	0	23
2	8	3	9	6	1	27
3	12	3	7	6	1	29
4	7	3	7	5	1	23
5	10	3	9	5	1	28
6	11	3	9	5	0	28
7	5	3	7	5	1	21
8	5	2	6	7	0	20
9	6	3	7	6	0	22
10	11	3	9	6	1	30
11	9	3	7	6	1	26
12	5	3	7	5	1	21

Table 3. Coefficient of Community Values Calculated From Cumulative Emergence Data as of 26 June 1988

Pond #	2	3	4	5	6	7	8	9	10	11	12
1	0.800	0.731	0.793	0.784	0.824	0.773	0.698	0.844	0.792	0.776	0.818
2	—	0.786	0.760	0.727	0.836	0.750	0.638	0.776	0.807	0.830	0.750
3	—	—	0.885	0.731	0.807	0.720	0.612	0.667	0.746	0.800	0.760
4	—	—	—	0.745	0.824	0.773	0.698	0.711	0.792	0.776	0.864
5	—	—	—	—	0.714	0.735	0.708	0.720	0.725	0.741	0.776
6	—	—	—	—	—	0.694	0.667	0.760	0.793	0.741	0.776
7	—	—	—	—	—	—	0.732	0.791	0.784	0.894	0.810
8	—	—	—	—	—	—	—	—	0.769	0.792	0.791
9	—	—	—	—	—	—	—	—	0.769	0.792	0.791
10	—	—	—	—	—	—	—	—	—	0.786	0.784
11	—	—	—	—	—	—	—	—	—	—	0.766

Pseudochironomini were represented by a single species in 8 ponds during the same period.

Coefficient of Community values calculated from cumulative emergence as of 26 June based upon pairwise comparisons for all 12 ponds are given in Table 3. The CC values ranged from 0.612 to 0.894. The percent of the CC values between 0.681 and 0.840 was 85% and 61% ranged between 0.721 and 0.800. The distribution of CC values was unimodal and appeared to be normally distributed. When considered over the duration of the study, the Coefficient of Community value for the control ponds was 0.919.

Chronology of Emergence

The temporal pattern of first appearance in emergence trap samples is shown in Figure 3 for all 12 ponds to 26 June 1988 and in Figure 4 for control ponds through the study duration. The general shapes of the cumulative emergence curves are similar for each pond up to 26 June, despite differences in the total number of taxa detected per pond. If species that were rare (individually comprising less than 1.5% of total emergence) in the emergence samples are not considered, the variation in cumulative emergence curves is even further reduced. This indicates that there was very little temporal variation among ponds in the dates of initiation of emergence by the numerically dominant taxa.

Similar curves are also seen for the control ponds over the entire study period (Figure 4). Emergence was detected from both control ponds on the date of first sampling, and continued through the end of the project. By 15 June 50% of the taxa had initiated emergence, and more than 80% were detected by 13 July. The cumulative emergence curves flatten out between 13 July and 27 July, and only isolated specimens of rare species influence the curves during the remaining sample periods.

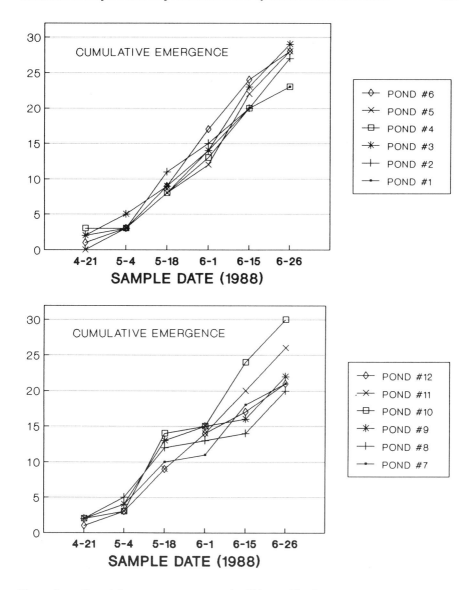

Figure 3.   Cumulative emergence curves for Chironomidae in mesocosm ponds up to the date of first application of pesticide in 1988.

Phenology Patterns

The seasonal patterns of phenology for Orthocladiinae, Tanytarsini, Tanypodinae, Chironomini, and Pseudochironomini are shown for the control ponds in Figures 5 and 6. In each figure the upper graph shows the number of taxa emerging for each taxonomic category by sample date. The lower graph shows the percentage of total emergence. Highest species richnesses were observed in June for control pond #2 and in June, July, and September for control pond #11. Both ponds showed a decline in emergence species richnesses near mid-August; however, the declines differed by one sample period. In both ponds the number of species emerging in April through early May and September through mid-October was similar, and less than throughout the summer.

Abundance Patterns

The emergence of numerically dominant taxa during spring, late spring/early summer, late summer, and fall is summarized in Table 4. During each time period, the five numerically dominant taxa constituted 80% or more of the emergence and consisted of species that are commonly encountered across eastern Kansas during the respective time periods. The dominant taxa during late spring and early summer in the control ponds were also the dominant taxa during the same periods in the remaining ten ponds used as treatment ponds later in midsummer.

DISCUSSION

In order to discuss Chironomidae faunal representativeness, it is first necessary to consider ways in which two or more faunas can vary. Taxonomic composition, species richness, and relative abundances of individual taxa are commonly used

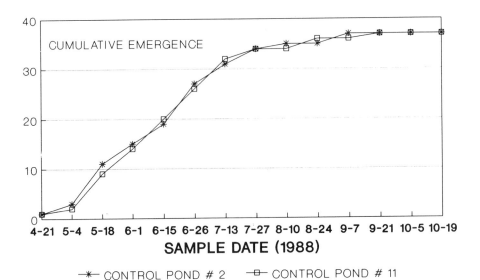

Figure 4.   Cumulative emergence curves for Chironomidae in mesocosm control ponds during April through October 1988.

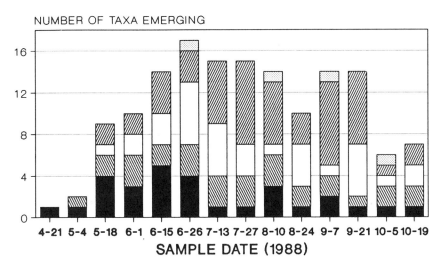

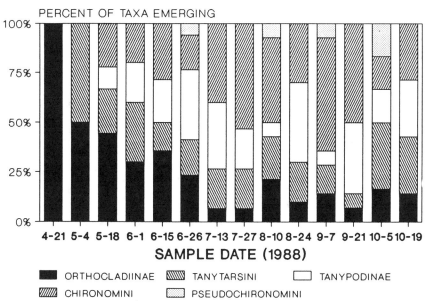

Figure 5. Phenological patterns of Chironomidae emergence from control pond #2 in 1988.

criteria for comparing faunas. In addition, two faunas can vary temporally, both in terms of the chronology of emergence and the patterns of phenology. Excessive divergence in one or more of these attributes may be considered as affecting representativeness. Consequently, in the following paragraphs we will critically evaluate the patterns of Chironomidae emergence from all these perspectives, in order to address the question of representativeness of the Chironomidae communities that developed in the NESA mesocosm ponds after transfer of sediments from a farm

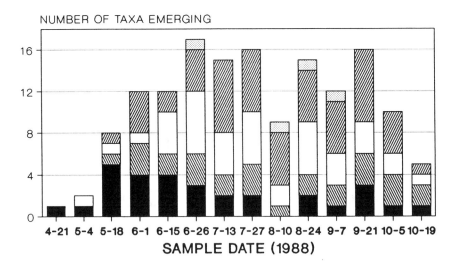

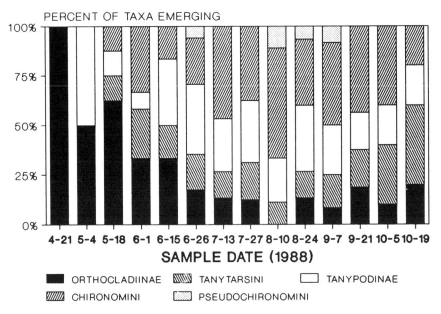

Figure 6.   Phenological patterns of Chironomidae emergence from control pond #11 in 1988.

pond. In addition, we will contrast our results with three recent studies conducted in different types of lentic systems where aerial dispersal was the principal mechanism of Chironomidae colonization.

Although there are numerous studies of Chironomidae emergence from lentic systems, relatively few are similar enough in design or analysis to allow detailed comparisons. We have extensively reviewed the studies and selected a subset which we feel provide for the most meaningful contrasts and comparisons. Studies by Street and Titmus (1979), Voshell and Simmons (1984), and Layton and Voshell (1991) allow for meaningful contrasts of our emergence patterns with patterns that derive from studies in which dispersal is the primary means of Chironomidae colonization of new lentic habitats. Studies by Carrillo (1974) and Dendy (1971) have several parallels to ours in design and data analysis, and although sediments were not transferred in these studies, we feel they also provide for meaningful comparisons with mature small ponds that have undergone periodic draining and exposure of benthic sediments.

Species Richness and Emergence Composition

Street and Titmus (1979) observed 22 and 28 taxa emerging from 6 newly constructed experimental ponds in the first and second years after construction, respectively. They compared emergence from these ponds with a nearby more mature pond that presumably served as a source of aerial colonization and concluded that within 2 years the Chironomidae of the experimental ponds came to resemble that of a much more mature system. By comparison, our sediment transfers produced

Table 4. Numerically Dominant Taxa in Emergence Trap Samples For Spring, Late Spring/Early Summer, Late Summer, and Early Fall

Spring: 14 April through 18 May 1988
    Hydrobaenus spp. (21%)
    Tanytarsus sp. 2 (21%)
    Procladius sublettei (15%)
    Psectrocladius vernalis (12%)
    Cricotopus sp. 1 (11%)

Late Spring/Early Summer: 26 May through 13 July 1988
    Cladotanytarsus spp. (60%)
    Tanytarsus sp. 2 (13%)
    Procladius sublettei (4%)
    Cricotopus sp. 1 (4%)
    Paramerina smithae (3%)

Late Summer: 20 July through 7 September 1988
    Cladotanytarsus spp. (68%)
    Polypedilum cf. *floridense* (4%)
    Procladius bellus (4%)
    Tanytarsus sp. 2 (4%)
    Labrundinia spp. (3%)

Fall: 14 September through 19 October 1988
    Cladotanytarsus spp. (46%)
    Procladius bellus (22%)
    Polypedilum cf. *floridense* (10%)
    Endochironomus nigricans (5%)
    Tanytarsus sp. 2 (3%)

Note: During each time period the five numerically dominant taxa constituted 80% or more of the emergence, and consisted of species that are commonly encountered across eastern Kansas during the respective time periods.

38% higher species richness during the first year after transfer than was achieved by the end of this two year study. Titmus (1979) studied emergence of Chironomidae from the more mature pond for 2 consecutive years and found 33 and 35 species emerging per year. Only 21 of the species were collected during both years, and the cumulative number of species emerging over the duration of the study was 47. These species richness values more closely approximate the numbers of taxa found during the NESA study and confirm that our sediment transfers result in species richness patterns that are comparable to those of mature ponds and higher than those attained after 2 years of aerial colonization of new ponds in England.

Voshell and Simmons (1984) demonstrated that the Chironomini genera *Dicrotendipes* and *Endochironomus* were among the first Chironomidae to colonize newly constructed Lake Anna in Virginia. *Ablabesmyia, Procladius, Chironomus, Cryptochironomus, Glyptotendipes,* and unidentified Tanytarsini became more abundant in subsequent years of the study. No Orthocladiinae or Pseudochironomini were detected throughout their study, which continued through 3 years after reservoir filling. Thus it appears that a few Chironomini and, to a lesser extent, Tanytarsini species predominate in larger reservoirs when aerial dispersal of adults is the primary mechanism of recruitment into the system, with minimal occurrence of Orthocladiinae during early stages of benthic chironomid community development.

In contrast to chironomid colonization of large reservoirs, Layton and Voshell (1991) found that an unidentifiable Orthocladiinae and a species of *Nanocladius* were among the early taxa to colonize 12 newly constructed experimental ponds that are similar in design to the ponds at NESA. Although these two Orthocladiinae were among the earliest colonizers, only *Nanocladius* was numerically abundant throughout the study. Other numerically dominant chironomid taxa were *Chironomus, Larsia, Tanytarsus,* and *Procladius,* with *Chironomus* accounting for more than half of the total organisms collected. Moreover, the average number of species for all 12 ponds did not exceed 15 taxa during the 14-month study.

Carrillo (1974) studied emergence dynamics of lentic Chironomidae over a 2-year period in a small artificial fish pond located on the edge of the Pymatuning Laboratory of Ecology (University of Pittsburgh) in northwest Pennsylvania. The pond is strikingly similar to our mesocosm ponds in terms of habitat complexity and maximum depth, although somewhat larger. More importantly, the pond is drained every other year in autumn for management purposes, and was drained for 4 weeks in November between the first and second year of the study. Although sediments were exposed and physically disturbed, they were not removed; thus, the benthic insects were subjected to similar, but not identical, conditions as occurred in our study when sediments were exposed for a period after transfer.

Species richnesses during the 2 years of Carrillo's study were 35 in the first year and 42 in the second year, roughly equivalent to the richnesses of 37 found in our control ponds. The patterns of taxonomic composition were also similar to our control ponds. Chironomini was most species rich in both years, with 19 and 22 taxa, comprising 54.3% and 52.4% of total emerging taxa, respectively, during each year. Orthocladiinae and Tanypodinae were intermediate in species richness, with five and seven species emerging during the first year, and five and 10 species during year two. Tanytarsini was represented by four species each year, the same number as present in our control ponds. No Pseudochironomini emerged during the first year of the study, but one species was detected during the second year (however, it was recorded as Chironomini).

Dendy (1971) presented emergence composition and phenology patterns of Chironomidae, summarized over a 13-year period, from experimental ponds at the Fisheries Research Unit of the Auburn University Agricultural Experiment Station near Auburn, AL. Data were presented for 12 small ponds referred to as "F" ponds and two larger ponds termed "S" ponds. The F ponds were 0.1 ha, varied in depth from 0.6 m to 1.4 m and were drained and refilled periodically, but without removal of sediments. The S ponds were considerably larger and managed differently; therefore, we restrict our comparisons to emergence from F ponds.

Collected in emergence traps from all ponds were 50 species of Chironomidae over the 13 years that were summarized, with 45 species reported from the F ponds. As in our study, Chironomini were most species rich with 26 species. In contrast, however, Tanypodinae and Tanytarsini were intermediate in species richness, both being represented by eight species. Orthocladiinae and Diamesinae were represented by three and one species, respectively. No Pseudochironomini were collected.

The differences in species richnesses of Tanytarsini and Orthocladiinae between the Auburn ponds and the NESA ponds are related to differences in taxonomic resolution and geographic location of the sites. In our study it was possible to discriminate three species groups of *Tanytarsus,* although additional species occurred in the ponds. By contrast, seven species of *Tanytarsus* were identified in Auburn. The Auburn site is more southerly than our site and experiences a warmer average temperature regime. This presumably results in lower average dissolved oxygen concentrations at the sediment/water interface, which would account for the reduction in species richness of Orthocladiinae.

No Coefficient of Community values were reported by either Street and Titmus or Dendy, and they cannot be calculated from data provided in the publications. We have, however, calculated CC values to compare emergence from year one to year two for both the Carrillo and Titmus studies. The values are 0.883 and 0.894, respectively, closely approximating our value of 0.919 for control ponds.

From these studies it can be seen that the species richness of emerging chironomids in the NESA control ponds was neither unusually low nor high relative to studies in comparable ponds. The relative numbers of taxa by subfamily or tribe also closely corresponded to the patterns seen in the other studies or can readily be explained as due to taxonomic or geographic differences. The patterns differed considerably from others observed when aerial dispersal is the principal mechanism of colonization. The CC values measuring the similarity in emergence composition from year to year in the Carrillo and Titmus studies show that there was roughly the same variation among our ponds during one year as can be expected to occur within one pond over consecutive years.

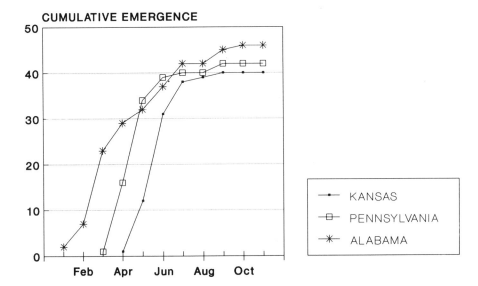

Figure 7.   Comparisons of cumulative emergence patterns of Chironomidae from ponds in Kansas, Pennsylvania, and Alabama.

Chronology of Emergence

Carrillo provides detailed chronology data for the second year of his study, which can be readily compared to our study (Figure 7). The temporal pattern of first appearance is similar in shape to the curves for NESA control ponds, although shifted to the left, indicating that taxa are beginning to emerge earlier in the season. By the end of the first week of May, 50% of taxa show up in emergence trap samples and 80% by the first week of June. This represents an approximate 5-week difference when compared to our curves and may be related to differences in mean daily water temperatures or cumulative degree days among the 2 study sites, since the chronologies determined for NESA control ponds closely correspond to the patterns seen in unpublished historical data in the Kansas Biological Survey files. As with our study, the shape of the emergence curve through midsummer and into fall in Pennsylvania was flat, and only slightly influenced by the appearance of rare taxa.

The temporal pattern of first appearance can also be reconstructed from the data provided by Dendy (Figure 7), though it should be cautioned that the resulting pattern represents an average pattern that derives from 13 years of emergence trapping. This pattern can be considered a more realistic pattern of emergence chronology for the Auburn facility than one derived from fewer years of study, and as such provides a better estimate of chronology than our study or those of Carrillo and Titmus.

The general shape of the curve for Dendy's data is similar to our curve but is also shifted to the left, even more so than in the study by Carrillo (Figure 7). In

Auburn 50% of taxa show up in emergence by the end of March and more than 80% by June. Emergence occurred sporadically throughout winter, usually after periods of milder weather, and the ponds were typically covered only briefly with ice. Consequently, the first appearances of taxa were spread out over a longer period of time, resulting in a lower slope of the curve during spring than seen for the studies in Kansas and Pennsylvania. The shape of the curve through midsummer and into the fall was flat, as in the other two studies, indicating that only a few taxa were collected for the first time during these seasons.

Phenology Patterns

The seasonal pattern of phenology for Orthocladiinae, Tanytarsini, Tanypodinae, Chironomini, and Pseudochironomini in NESA control ponds is commonly observed in historical data for eastern Kansas. When compared to Carrillo's results, some striking inconsistencies are observed. The most significant differences are seen in early spring emergence, which was dominated by *Procladius* cf. *culiciformis* (a Tanypodinae species which is probably *P. sublettei*) and two species of *Chironomus*. By contrast, species of *Hydrobaenus* and *Orthocladius* (both Orthocladiinae) and *Tanytarsus* sp. 2 dominated emergence in NESA control ponds during early spring. *Procladius sublettei* was also present in spring samples, but became most abundant after the peak emergence of *Hydrobaenus*. As a consequence of these differences, emergence in April and May was dominated by Tanypodinae and Chironomini in Carrillo's study, and by Orthocladiinae and Tanytarsini in NESA control ponds. This is most likely due to geographic differences in faunal composition rather than to sediment transfers, since *Hydrobaenus* is not common in streams and ponds in northeastern Pennsylvania (Ferrington, personal observations) but is a numerically dominant genus that is characteristic of late winter and early spring Chironomidae emergence in both lentic and lotic habitats in eastern Kansas.

The initial portion of the seasonal phenology patterns for the F ponds in Auburn was similar to the pattern observed in Kansas, with Orthocladiinae predominating during colder months early in the year. By contrast, however, Orthocladiinae emergence started earlier in the calendar year and either did not occur during summer months or was only represented by a single species.

The patterns of seasonal phenology among the three studies were quite similar through summer months. A common trend of all three studies is that Orthocladiinae comprised a smaller percent of taxa emerging during June through September, and Chironomini comprised a greater percentage. This pattern is also well supported in historical data for eastern Kansas.

The reduction in emergence species richness in NESA control ponds during mid-August is also well supported by historical data for eastern Kansas. This is commonly observed in both lentic and lotic systems, and is often accompanied by reduction in emergence densities (Howick et al., 1992). Carrillo observed reductions in species richness of emerging Chironomidae during August and early September, though not as great as in our control ponds. He also found reductions in emergence rates during July, approximately 4 weeks earlier than would be expected in Kansas. As was the case with the pattern of emergence chronology, this may represent differing thermal regimes of the two study sites.

By contrast, Dendy's data indicate reductions in species richnesses during May and August. This pattern most likely indicates that several of the species are multivoltine at the Alabama site, with one generation overwintering and emerging during March to April, followed by two summer generations, whose adults emerge in July and again in September to October.

Abundance Patterns

The numerically dominant taxa of each of the four time periods given in Table 4 are all commonly encountered during the respective time period across eastern Kansas, except for *Polypedilum* cf. *floridense. Procladius sublettei, Psectrocladius vernalis, Procladius bellus, Endochironomus nigricans,* and *Cladotanytarsus* spp. are widespread and common across North America. *Hydrobaenus* spp., *Tanytarsus* sp. 2, *Cricotopus* sp. 1, *Paramerina smithae,* and *Labrundinia* spp. are common across Kansas and throughout much of the midwest. *Polypedilum* cf. *floridense* is most likely an undescribed species, and its distribution is not known.

It is difficult to make precise species-by-species comparisons of abundances with other studies because of differing levels of identification and rapid changes in nomenclature that are common with Chironomidae. This type of fine scale comparison would require that voucher materials for all studies be located (if they still exist) and specimens studied to confirm the species involved in each study. However, one can make meaningful comparisons at the generic level, where a high degree of overlap can be demonstrated between our control ponds and the ponds studied by Carrillo and Dendy.

All genera recorded by Carrillo except for *Einfeldia, Guttipelopia, Acricotopus,* and *Limnophyes* were also present in our study. *Guttipelopia* and *Acricotopus* have not been historically collected in Kansas, and although they may be present in the state, it is clear that they are not common components of Chironomidae faunas in small local farm ponds. *Limnophyes* larvae are usually semiterrestrial, or live in wave zones or saturated soil along the shore line, and would not have been collected in our emergence traps, which were always placed away from the margins of the ponds. *Einfeldia* has been collected in light traps and other small ponds at NESA, but usually not in large numbers. The lack of this genus in our emergence traps could be related to its rarity in the ponds, or alternatively could indicate that larvae do not effectively survive the sediment transfer.

A similar high degree of overlap is also seen among our results and those of Dendy. Only *Einfeldia, Limnophyes, Goeldichironomus,* and *Diamesa* were not detected in our study. *Goeldichironomus* and *Diamesa* are both known historically from Kansas, but neither have been collected in emergence traps from experimental ponds at NESA. Ferrington and Crisp (1989) recorded *Goeldichironomus holoprasinus* from organically enriched streams, where it was restricted to sites that were most heavily influenced by discharges from waste-water treatment plants. Other historical data indicate similar patterns of occurrence for this genus in eastern Kansas. Four species of *Diamesa* are historically recorded from Kansas, but all appear to be restricted to lotic habitats with good to excellent water quality.

CONCLUDING REMARKS

In this paper we have attempted to determine if sediment transfers from mature farm ponds to mesocosm test ponds during fall produce representative benthic test faunas within 6 months to a year. We have analyzed our data with regard to historical data for eastern Kansas and other studies that closely resemble our study in terms of design and analysis but where aerial dispersal is the source of colonizing species. For discussion purposes we have focused on Chironomidae since they constitute a large proportion of total benthic species richness in our ponds and are also likely to comprise a large proportion in most small lentic systems that mesocosms are designed to mimic. We have stated that the patterns of taxonomic composition, emergence composition, chronology, phenology, and relative abundances of dominant taxa are similar to what can be expected based upon historical data.

We have compared our data with studies by Carrillo and Dendy, in which small ponds were drained but sediments were not removed during the period of study. We have shown that our emergence patterns are more similar to theirs than to the patterns that derive from studies where aerial dispersal was the primary source of Chironomidae during the first year after construction of ponds. We must therefore conclude that transfer of sediments during fall is an efficient way to seed newly constructed mesocosms, and that the physical rigors of the sediment transfer and subsequent periods of sediment exposure before ponds are filled are no more disruptive to the life cycles of the resident Chironomidae fauna in our ponds than periodic draining is to the faunas of other small ponds at other sites.

ACKNOWLEDGMENTS

We would like to acknowledge Springborn Laboratories, Inc., for cooperation and funding for portions of the research reported in this publication. Gregory L. Howick, Jeffrey M. Giddings, W.D. Kettle, and Frank deNoyelles were instrumental in various aspects of field sampling, coordination and constructive encouragement throughout the study. Kirk Larson, Paul Grant, and Debra Baker assisted in field collection and sample sorting. J. Reese Voshell kindly provided copies of submitted manuscripts and prepublication page proofs for our review. Robert L. Graney provided excellent comments during a poster session at the 1990 Annual SETAC Meeting. Judy McPherson assisted in typing and revising drafts of the manuscript and figures. This research was conducted through the Water Quality and Freshwater Ecology Program of the Kansas Biological Survey at the University of Kansas.

REFERENCES

Buikema, A.L. and J.R. Voshell. 1993. Toxicity studies using freshwater benthic inverte-brates. *In* Rosenberg, D.M. and V.H. Resh (eds.) *Freshwater Biomonitoring and Benthic Macroinvertebrates*. Chapman and Hall, New York, pp. 344–378.

Carrillo, R.J. 1974. Emergence dynamics of a lentic Chironomidae (Diptera) community in northeastern Pennsylvania. Ph.D. Dissertation, University of Pittsburgh, 307 pp.

Dendy, J.S. 1971. Phenology of midges in experimental ponds. *Can. Entomol.* 103:376–380.

Dewey, S.L. 1986. Effects of the herbicide Atrazine on aquatic insect community structure and emergence. *Ecology* 67:148–162.

Ferrington, L.C., Jr. and N.H. Crisp. 1989. Water chemistry characteristics of receiving streams and the occurrence of *Chironomus riparius* and other Chironomidae in Kansas. *Acta. Biol. Debr. Oecol. Hung.* 3:115–126.

Goldhammer, D.S., C.A. Wright, M.A. Blackwood, and L.C. Ferrington, Jr. 1993. Com-position and phenology of Chironomidae at the John H. Nelson Environmental Study Area, eastern Kansas, *Netherlands Journal of Aquatic Ecology*. In press.

Howick, G.L., J.M. Giddings, F. deNoyelles, L.C. Ferrington, Jr., W.D. Kettle and D. Baker. 1992. Rapid establishment of test conditions and trophic level interactions in 0.04-ha earthern ponds. *Env. Tox. Chem.,* 11:107–114.

Huggins, D.G. 1990. Ecotoxic effects of Atrazine on macroinvertebrates and its impact on ecosystem structure. Ph.D. Dissertation, Department of Civil Engineering, University of Kansas, Lawrence, 380 pp.

Layton, R.J. and J.R. Voshell. 1991. Colonization of new experimental ponds by benthic macroinvertebrates. *Environ. Entomol.* 20(1):110–117.

Street, M. and G. Titmus. 1979. The colonization of experimental ponds by Chironomidae (Diptera). *Aquatic Insects* 4:233–244.

Titmus, G. 1979. The emergence of midges (Diptera: Chironomidae) from a wet gravel-pit. *Freshwater Biology* 9:165–179.

Touart, L.W. 1988. Hazard evaluation division, technical guidance document: aquatic meso-cosm tests to support pesticide registrations. EPA-540/09-88-035. U.S. Environmental Protection Agency, Washington, D.C.

Touart, L.W. and M.W. Slimak. 1989. Mesocosm approach for assessing the ecological risk of pesticides. *Misc. Publ. Entomol. Soc. Am.* 75:33–40.

Voshell, J.R. and G.M. Simmons, Jr. 1984. Colonization and succession of benthic mac-roinvertebrates in a new reservoir. *Hydrobiologia* 112:27–39.

Whittaker, R.H. 1975. *Communities and Ecosystems*. Second Edition. Macmillan, New York, 385 pp.

QUESTIONS ASKED DURING POSTER SESSION

A question repeatedly asked by attendees of this conference was how repre-sentative are the test faunas on which pesticide effects studies are being made, and how do you evaluate representativeness?

ANSWER: Although in concept it is easy to accept the premise that the faunas of mesocosms should contain representative proportions of the complete natural assemblage of organisms of the system that is being modeled, as prescribed by Touart and Slimak (1989), in practice it is difficult to quantitatively determine if the test fauna is sufficiently representative of the natural system of choice. Part of the difficulty, of course, lies in the definition of the term representative, which can be variously interpreted, either loosely as meaning structurally similar, or more rigidly as inferring that all species should be represented in frequencies and/or abundances that are close to those empirically estimated from studies of a variety of naturally occurring systems in the geographic region of the mesocosm test facility. To a great extent this problem of definition can be resolved by more explicit criteria statements, and by development of more precise standards for evaluating if the test fauna is "acceptably representative".

The second difficulty associated with representativeness deals with the variability that can be found among an array of naturally occurring systems that form the basis of the type of system being modeled. Small lentic habitats, impoundments, or other naturally occurring standing waters which are physically mimicked by mesocosms may not be highly variable in faunal composition, but can differ widely in the relative abundances of aquatic macroinvertebrate species occurring within them. In order to evaluate the amount of variability characteristic of natural small lentic systems, numerous faunal studies would have to be made, and procedures would have to be developed to formulate a generalized definition that incorporates the natural variability of the faunas characteristic of the model system.

For some faunal elements such as fish, mayflies, and caddisflies, there may often be a large enough set of historical data available to form an acceptable, empirically based concept of taxonomic composition and relative abundance. However, for other groups such as the aquatic Diptera, which constitute a large percentage of total species richness in our mesocosms, there most likely would not be much historical data available. The lack of historical data is especially of concern, since the aquatic Diptera will often form a large percentage of total taxa and total specimens in most naturally occurring small lentic systems, as in this study. Without a historical data set, it becomes difficult to develop a concept of expected taxonomic composition and relative abundances against which representativeness questions can be gauged.

In previous studies performed at NESA, it has been repeatedly shown that the Chironomidae dominate emergence in experimental ponds that have been filled for two to several years (Dewey 1986, Goldhammer et al. submitted, Huggins 1990). This has enabled us to begin to differentiate between newly formed aggregates of Chironomidae resulting from dispersal and early colonization and more mature community structures that derive from successive years postcolonization. In addition, the staff of the Kansas Biological Survey (KBS) has conducted extensive surveys of aquatic macroinvertebrates, including Chironomidae, from other lentic habitats across Kansas during the past 12 + years. Consequently, it is possible to begin to develop an even more generalized concept of expected taxonomic composition and relative abundance for Chironomidae in small lentic systems. Although the KBS surveys were made using a variety of sampling gear and varied widely in effort or intensity, we feel that they form a justifiable basis for preliminary evaluation of the representativeness of the Chironomidae component of the mesocosm test fauna that developed in the control ponds after sediment transfer in our 1987 to 1988 study. We also feel that the evaluation of the representativeness of our fauna which we provide here will serve as a useful model for making future decisions regarding questions of representativeness and for further refinement of our concept of species richness and relative abundance criteria.

Question asked by Greg Howick: Can you make more precise comparisons of your data with other studies of small farm ponds near your study site?

ANSWER: There are no other studies of which I am aware for small farm ponds in eastern Kansas that have been sampled in a relatively similar manner. Therefore the answer to your question is no; more precise comparisons cannot be made. More importantly, though, I think your question points to the need for intensive regional studies of unmanipulated farm ponds, which can then serve as the standard for developing regionalized concepts of "faunal representativeness", and, perhaps as important, can give us more information regarding the natural variability in faunal composition among a random array of pond types. We simply need more background studies to better define our concepts of expected faunal compositions.

Question asked by Reece Voshell: At your latitude in the midwest I would expect that you have at least some winter emergence by Chironomidae, especially following periods of warm weather. How can you be sure that the early patterns of emergence presented in this poster result from larvae that were transferred with the sediments as opposed to oviposition and subsequent development by winter-emerging species?

ANSWER: It is true that there is often significant midwinter emergence by Chironomidae in eastern Kansas, especially during mild winters or brief periods of very warm weather. However, the taxa that characteristically emerge from small lentic habitats during November through late February are *Hydrobaenus* and *Orthocladius*. Taxa emerging from lotic habitats include these same two genera, plus the subgenus *Euorthocladius* (of *Orthocladius*), *Oliveridia, Diamesa, Eukiefferiella, Sympotthastia, Potthastia,* and *Chaetocladius*. None of these latter taxa occur in our experimental ponds, nor for that matter in other small ponds in eastern Kansas, to the best of my recollection. It is my impression that both *Hydrobaenus* and *Orthocladius* are univoltine at our latitude, therefore if adults did oviposit in the ponds, the larvae would estivate through the summer months and only resume development as water temperatures declined the following fall, since we do not find larvae of these taxa in the ponds through warm water months. I do not think that there is any cohort splitting that could result in some late spring emergence of larvae from egg masses oviposited during the preceding winter months, but I cannot rule this out. The limited benthic sampling that we have done in winter months, however, confirms that *Hydrobaenus* and *Orthocladius* larvae survive sediment transfers, along with larvae of several other taxa that emerged in spring and early summer.

CHAPTER 15

Spray Drift and Runoff Simulations of Foliar-Applied Pyrethroids to Aquatic Mesocosms: Rates, Frequencies, and Methods

Ian R. Hill, Kim Z. Travis, and Paul Ekoniak

Abstract: Agricultural applications of the foliar-applied pyrethroid insecticides can result in small quantities of these products entering aquatic environments by spray drift deposition or by runoff. A number of mesocosm studies have been carried out with these products to help quantify the effects of this entry on aquatic organisms and ecosystems. However, there has been relatively little detailed scientific consideration of the amounts of these chemicals entering aquatic environments during normal agricultural usage, nor of how to translate field aquatic exposure data into mesocosm treatment rates and procedures. Therefore, doubt is inevitably cast upon environmental risk assessments made with the results from these aquatic biological studies. This chapter assesses spray drift deposition and runoff for foliar-applied pyrethroids, using data from published and unpublished sources. Although at present there is a paucity of relevant information, the available data are discussed in order to derive a best estimate of application numbers, frequency, and rates of pyrethroids for mesocosm investigations.

INTRODUCTION

Over the past decade a considerable number of aquatic field studies have been carried out with the foliar-applied pyrethroid insecticides. The investigations have

0-87371-592-6/94/$0.00 + $.50
© 1994 by CRC Press, Inc.

been both in natural bodies of water such as ponds, lakes, and streams, and in replicated experimental ponds and enclosures referred to as mesocosms.

In the studies using natural ecosystems, the insecticides were applied either directly to the water surface or to crops adjacent to the body of water, mostly using recommended crop application rates, frequency, and equipment. Many of these studies have been previously discussed (Hill, 1985, 1989).

Several mesocosm studies with the pyrethroids have also been carried out or are currently underway. The investigators have used a variety of systems, including:

- Small concrete ponds (Hill, 1985, 1989)
- Small earthen ponds (Rawn et al., 1982)
- Partitioned field ponds (Crossland, 1982)
- Lake limnocorrals (Solomon et al., 1980; Smith et al., 1981)
- Littoral enclosures in a natural pond (Lozano et al., 1992)
- Large earth-lined ponds (Hill et al., 1988; Fairchild et al., 1989; Hill, 1989; Webber et al., 1989)

Several of the recent mesocosm investigations in large earth-lined ponds have been designed to meet the regulatory requirements of the United States Environmental Protection Agency (EPA). Such studies, complying with the EPA guideline protocol (Touart, 1988), are logistically complex, labor intensive, and extremely expensive; each costing in the order of $2 to 5 million. The data generated in these mesocosm tests done during the past 5 years are being used in making environmental risk assessments of the products tested in relation to their recommended agricultural use-patterns. In order to make such judgments it is critical that all aspects of the experimental program reflect or can be extrapolated to the product's normal field usage.

Over the past years there has been considerable discussion of mesocosm techniques, involving industry, academia, and regulators (EPA, 1986; NACA, 1986). However, most of this has been devoted to biological, physicochemical, and residue analytical procedures. A disproportionately small amount of time has been spent addressing the pesticide application issues and especially the treatment rates; at least in part because the practice is ahead of the scientific knowledge. Consequently, there is much disagreement on the doses that should be applied to mesocosms, and how these relate to the range of field aquatic Estimated Entry Rates (EER; the amount of pesticide entering an aquatic environment, for the total body of water) and Estimated Environmental Concentrations (EEC; the concentration of the pesticide in the body of water) that may result from agricultural use. This has caused inconsistencies in the application rates at which studies with comparable compounds have been carried out. Until the EERs and EECs can be defined, agreed upon, and understood in the context of the spectrum of typical field aquatic exposures, great difficulty will be experienced in extrapolating mesocosm data to natural field aquatic environments in order to perform realistic risk assessments.

It has been realized for some time that the available databases for both spray drift deposition and runoff are inadequate for reliable estimation of EERs or EECs, and therefore substantial international activity is in progress to address this. Representatives of EPA, government agencies, industry, and academia constitute the "Aquatic Effects Dialogue Group" (AEDG), organized under the auspices of The Conservation Foundation (World Wildlife Fund). The AEDG discussions are primarily concerned with the development of field aquatic testing procedures and include the refinement of procedures by which runoff EERs may be more accurately determined (AEDG, 1992). In addition much international work is under way to measure off-target spray drift deposition. In the U.S., a "Spray Drift Task Force" has been formed by 25 companies and, involving industrial and academic specialists, has embarked on a 5-year program of practical studies to produce a comprehensive spray drift database (SDTF, 1991; Drewes et al., 1990). Substantial amounts of work by industry and/or government institutions are also underway in other countries, for example, Canada and Germany.

This publication addresses one specific group of pesticides, the foliar-applied pyrethroid insecticides. The entry of these products into bodies of water, by spray drift deposition and runoff, during or following agricultural spraying of crops, is described together with the methods by which this might best be simulated in aquatic mesocosms.

In reviewing the literature, it soon became apparent that there was a paucity of relevant spray drift deposition and runoff data for the pyrethroids and for products with similar characteristics. Furthermore, it is evident that little additional or better information will become available in the next 2 years. Despite this, mesocosm studies of pyrethroids have and still are being required by government regulatory agencies. Therefore, we felt that it was desirable to use the data available (published and unpublished) in order to develop current best estimates of aquatic entry rates of these products and to propose mesocosm application scenarios. Our conclusions will most probably be modified or refined as the size of the exposure database increases.

PYRETHROID INSECTICIDES

The environmental fate and effects of the foliar-applied pyrethroid insecticides have been well described in the scientific literature. The most comprehensive reviews may be found within a series of papers on pyrethroid history, agronomy, mode of action, mammalian toxicology, and environmental behavior, edited by Leahey (1985).

Many photostable synthetic pyrethroid insecticides have now been developed for foliar applications to field crops. Some of the principal compounds are cyfluthrin, cypermethrin, deltamethrin, fenvalerate, lambda-cyhalothrin, permethrin, and tralomethrin. All of these products are marketed internationally. However, applications of these to any one crop-type are usually very comparable in all respects except active ingredient rate. For example, for control of pests on cotton in the USA, the majority of applications of pyrethroid products are made using:

- Emulsifiable concentrate (EC) formulations
- Ground spray-rigs or fixed-wing aircraft
- Water only as spray tank diluent
- Similar volumes per unit area of crop
- The same range of hydraulic nozzle types

Consequently the range of droplet sizes emitted during application, by any given spray equipment under comparable conditions, should be similar for each of these pyrethroid insecticides. The runoff characteristics of these compounds shortly after application are also very similar, since their physicochemical properties (strong adsorption to soil, very low water solubility, very low vapor pressure) fall within narrow ranges. Their rates of degradation in soil do, however, differ; therefore, with increasing time after application the amount of chemical available in the soil, for runoff, will be product dependent.

AQUATIC ENTRY ROUTES

During and following the agricultural application of pesticides there is the possibility that small proportions of these products may enter aquatic environments by several routes (Hill and Wright, 1978). The foliar-applied pyrethroids are all highly lipophilic and essentially nonvolatile. Consequently, for this group of compounds the only processes resulting in aquatic entry of amounts of potential environmental significance are spray drift deposition and soil surface runoff.

While a pesticide is being applied, a small proportion of droplets may be blown out of the crop target-area, and over adjacent land or water surfaces. This wind-blown pesticide in the air above the nontarget area is termed *spray drift*. Some of the spray drift particles may also impact on nontarget land or water surfaces, and this is referred to as *spray drift deposition*. These two terms, spray drift and spray drift deposition, are in no way equivalent. All data discussed in this paper are for spray drift deposition, which is the relevant quantity with regard to determining aquatic entry rates.

When the rate of precipitation (principally rainfall) exceeds the percolation rate through the soil profile, *runoff* will occur as the movement of water and soil particles across the land surface. Pesticide may be carried in the runoff, either in solution or adsorbed to particulate material. In practice, pyrethroid insecticides in runoff will almost exclusively be associated with the particulate phase. Runoff predominantly occurs on soils either of relatively low permeability or with subsurface drainage restrictions. However, following very heavy rainfall some runoff may occur on all but the most permeable of sandy soils.

SPRAY DRIFT DEPOSITION

The amounts of pyrethroid insecticides deposited on land or water adjacent to sprayed crops has been measured on relatively few occasions. The following discussion therefore considers data from a variety of sources:

- Published studies designed specifically to quantify downwind spray drift deposition, using mostly nonpyrethroid pesticides or "blank" formulations
- Studies of agricultural applications of pyrethroids, where off-target deposition was adequately quantified
- Mathematical models of spray drift deposition

Factors Influencing Spray Drift Deposition

A wide range of factors affect both spray drift and deposition of any drifting droplets. These have been fully described elsewhere (see, for example, Akesson and Yates, 1964, 1984) and are summarized here in Table 1. The majority of the parameters have the most influence on drift and deposition when spraying is done at high speeds and at the greatest heights above the crop canopy, as occurs with aircraft applications. Much less drift is apparent from ground boom sprayers.

In the following discussions of spray drift deposition of foliar-pyrethryoids, data are only taken from studies where the parameters are relevant to applications of these insecticides onto agricultural crops.

Literature Review

Selection Criteria

Within the scientific literature there are numerous studies of the spraying of pesticides onto bare soil, row crops, orchard trees, forest trees, etc. However, only a very small proportion of these quantify off-target spray drift deposition.

Spray drop size and spray equipment factors (see Table 1) can have a major influence upon the extent of spray drift deposition. This is illustrated in Figure 1, which shows a variation of three orders of magnitude in the spray drift deposition resulting from four different application procedures. The spray target can also substantially influence spray drift deposition rates, although mostly as a consequence of spraying height; the extremes of variation being evident with preemergence or

Table 1. Parameters That Can Influence the Dispersion and Deposition of Pesticide Spray Droplets On and Off the Target Area

Parameter	Influencing Factors
Spray drop sizes	Nozzle (atomizer) design Nozzle angle relative to airstream Liquid emission and airstream velocity Formulation constituents Spray tank diluent Volume and pesticide application rates
Spray equipment	Aircraft type, size and configuration Application height (e.g., ground vs. aerial)
Meteorology	Temperature and gradient from ground to above spray release height Wind velocity and gradient Relative humidity
Terrain and target	Land slopes and undulations "Crop" canopy structure and density and/or soil surface

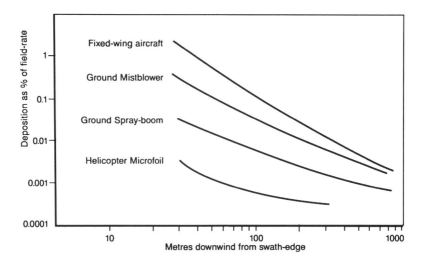

Figure 1.   Effect of different application procedures on spray drift deposition (representative data taken from the scientific literature; see, e.g., Akesson and Yates, 1964); fixed-wing and ground spray using conventional hydraulic nozzles.

early row crop applications *vs.* forestry pesticide treatments. Relatively few investigations have been done for each specific use pattern and formulation type, and thus only a small proportion of published data on spray drift deposition is appropriate to the agricultural uses of the pyrethroid insecticides. Therefore, a specific use pattern was selected for review here, and only literature studies of relevance to this were considered. In the U.S., among the highest rates of pyrethroid applications are the multiple sprays to cotton. Thus the following parameters were included in order to examine typical worst case pyrethroid spray drift deposition resulting from applications to cotton:

- Emulsifiable concentrate formulations, or similar; with or without a pesticide (pyrethroid or nonpyrethroid) active ingredient included
- Fixed-wing aircraft applications (ground booms are not discussed here, as they give substantially less spray drift deposition than aerial equipment; and mesocosm application rates are based on typical worst case aquatic entry rates)
- Water as the spray tank diluent
- Spray pressures and volume rates similar to those used for the pyrethroid insecticides
- Conventional hydraulic nozzle types recommended for spraying pyrethroid emulsifiable concentrates, or similar to these
- Applications to bare soil, pasture, or agricultural row crops
- Any meteorological condition; studies reported in the literature and described in this review had a maximum wind speed of 3.3 m/s

Within the above range of conditions, all available studies were reviewed. Thus, no selection for topographical factors was made. This was considered likely to give the widest possible spectrum of data, representative of that occurring during normal agricultural spraying of pyrethroid ECs, and would probably include typical worst case spray drift deposition.

Available Data

Using the criteria described above, a range of data relevant to pyrethroid use was abstracted from the scientific literature. The results are described in Table 2 and Figure 2 and show spray drift deposition only from fixed-wing aircraft spraying. Studies by Akesson and Yates (1964, 1984) were done without a pesticide, Frost and Ware (1970) used methoxychlor, those by Riley et al. (1989) were with the pyrethroid deltamethrin, and ICI (unpublished, 1982) used permethrin.

The downwind spray deposition data from the various studies (Figure 2) demonstrate about one order of magnitude variation. At the swath edge, deposition was as high as 50% of the nominal field rate, whereas at 100 m downwind, the deposit had declined to between 0.1 and 1%. An order of magnitude decline in deposition was apparent between 10 and 100 m downwind of the spray plane, and a similar amount between 100 and 1000 m.

Interpretation

Figure 3 shows how the literature data were typically generated. The spray application target was mostly grass pasture or bare soil, which often results in greater off-target drift than does a crop canopy. Drift deposition cards were placed downwind of the sprayed area. Applications were made either to a single swath (with multiple runs) or to several swaths upwind of the drift-deposit collection cards; and on all occasions the aircraft was flown directly above the target swaths. Consequently, the "zero-point" ("0 meters") from which spray drift deposition is quantified in Figure 2 relates to the downwind edge of the target swath, with the aircraft directly above the swath. If, in normal agricultural use of the pyrethroids, this same procedure of spraying directly above each target swath were to be followed when a wind is blowing, undesirable amounts of spray drift could be deposited on the adjacent nontarget area, as is demonstrated in Figure 4a.

However, in agricultural practice, the spray plane is not necessarily flown directly above the target swath. Instead, when a wind is blowing, the pilot "offsets" the aircraft upwind of the target swath (often termed "swath displacement" [Ware, 1983]). This is illustrated in Figure 4b and is done for several reasons:

- To obtain as even a pesticide coverage as is possible across the total target crop canopy, and thus provide efficient and cost-effective pest control
- To prevent or minimize contamination of crops in adjacent fields (as can be particularly important with aerial applications of herbicides)
- To reduce contamination of nearby noncropped areas and hence minimize the potential for nontarget effects on terrestrial or aquatic organisms

The necessity for this is illustrated in Figure 5, where the wind can be seen to displace the spray swath from the aircraft.

Table 2. Application and Meteorological Conditions for Literature Spray Drift Deposition Data Presented in Figure 2

Reference	Trial	Meteorology				Application Details						
		Temp (°C)	Gradient (°C)[a]	Wind Speed (m/sec)	Relative Humidity (%)	Fixed-Wing Aircraft	No. of Swath Widths	Area Sprayed (ha)	Pesticide Applied	Formulation	Volume (l/ha)	Nozzle Type
Akesson and Yates, 1964	15 June 1961	39	+4	2.2	43	Stearman	1[b]	—	None	10% oil in water[c]	80	D10-46
	16 June 1961	38	+5	3.3	34	Stearman	1[b]	—				
Akesson and Yates, 1984	Figure 5c	—	—	1.2	—	Stearman	1[b]	—	None	Water base[c]	—	D6-46
	Figure 6	—	—	—	—	—	1[b]	—				
Frost and Ware, 1970	August 1965 am	15	+4.4	0.9	—		9	5.6		EC	65	No. 8
	August 1965 pm	37	+0.3	1.5	—	Piper Pawnee	9	5.6	Methoxychlor			
	August 1967 am	28	-1.1	0.7	—		8	4.0				
	August 1967 pm	28	+1.7	<0.4	—		8	4.0				
Riley et al., 1989	Spray 1	13	Stable	3.2	76	Grumman Ag Cat	25	8	Deltamethrin	EC	28	D6-45
	Spray 3	20	Sl. stable	2.4	93		7	8				
ICI, 1982[d]		12	—	3.0	82	AT-301	1[b]	—	Permethrin	EC	27	#5-flood

a Positive values indicate increasing temperature with increasing height above ground.
b Carried out with multiple runs over a single swath width. In Figure 2 these data have been extrapolated to 10 swath widths.
c No additional details given; formulation may or may not be relevant to pyrethroid use.
d Unpublished data.

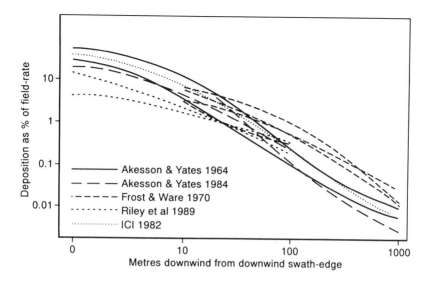

Figure 2. Spray drift deposition data from applications described in Table 2. The original data did not in all cases cover the range shown; values have been extrapolated wherever possible to extend the curve. See Table 2 for application and meteorological conditions.

The purpose of this chapter is to quantify the spray drift deposition entry of pyrethroids into aquatic environments under recommended agricultural use conditions. Figure 6 shows a typical aerial application adjacent to a pond when the wind is blowing towards the body of water. This illustrates the aircraft offsetting (as described above), but also shows the presence of an uncropped edge zone. The majority of fields next to an aquatic environment do have a strip of land next to the water that is not cultivated or cropped. This should be taken into account when interpreting the literature data in Figure 2. The assumptions used here for both these

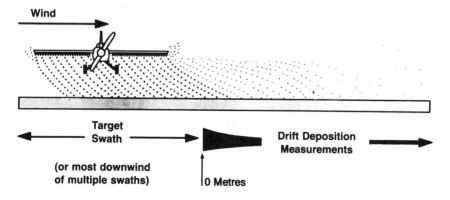

Figure 3. Position of aircraft and spray drift deposition measurements in relation to target swath, in literature studies of spray drift deposition.

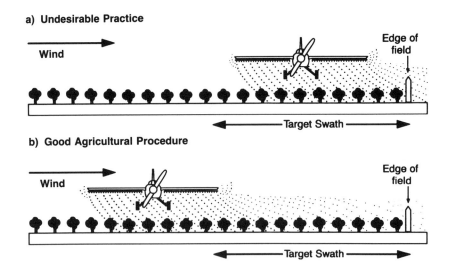

Figure 4.   Agricultural spraying of pesticides and aircraft "offsetting" under windy conditions. (a) Undesirable practice; aircraft directly over target swath, allowing substantial off-target spray drift deposition. (b) Good agricultural procedure; aircraft offset upwind giving good crop coverage and minimizing downwind spray drift deposition.

Figure 5.   Aerial crop spraying of a pyrethroid EC formulation, to show the effect of a crosswind on spray emission.

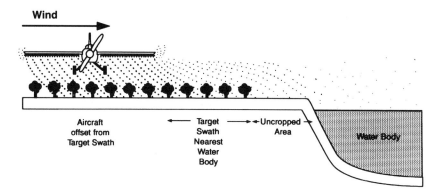

Figure 6.   Agricultural aerial application of a pesticide to a crop adjacent to and upwind of a body of water.

factors are presented in Table 3. A typical uncropped area next to a body of water is assumed to be 5 m wide. Spray plane offsets of 15 to 45 m at wind speeds of 1 to 4 m/s are proposed. These have been taken from the scientific literature (Akesson, 1984; Akesson et al., 1984; Taylor et al., 1984) and from discussions with spray plane pilots and with the California Agricultural Aircraft Association.

In the following discussion spray drift deposition is considered only in relation to a 100 m wide pond (equivalent to 1 ha [2.5 a]). This is used to represent a small field pond, which will be "worst-case" in respect of susceptibility to drift contamination. However, such small ponds on farmland rarely constitute significant biological environments, being commonly used for watering animals, crop irrigation, and other agronomic purposes. Unless they have permanent feeder streams, they are also susceptible to drying out during the summer.

Using the factors in Table 3, together with literature spray drift deposition data shown in Figure 2, the likely extent of deposition onto a body of water from agricultural use can be determined. This is presented in Table 4. In the "nearshore" littoral zone, spray drift deposition was in the range 0.8 to 7% of the nominal field rate. A further 100 m across the body of water and downwind, deposition fell to approximately 0.1 to 1.5%. Consequently, the average downwind deposition across a 100 m wide aquatic environment calculated from literature data varied between 0.3 and 2.7%.

Table 3. Factors Used to Convert Literature Data For Spray Drift Deposition to That Occurring in Normal Agricultural Use (See Figure 6)

	Distance Factors (m)		
Wind Speed (m/sec)	Uncropped Area	Spray Plane Offset Due to Wind	Total Downwind Distance From Edge of Aircraft Swath to Body of Water
1	5	15	20
2	5	30	35
4	5	45	50

Table 4. Spray Drift Deposition Onto a Body of Water

Reference	Trial	Aircraft Swath Edge to Pond Distance (m) From Table 3	Deposition (as % of Nominal Field Rate)[b] at Distances (m) From Edge of Water Body						
			Unconverted Average for 0–100 m[c]	Converted as in Table 3					Average For 0–100 m[d]
				0	10	20	50	100	
Akesson and Yates, 1964	15 June 1961	35	3.0	2.0	1.6	1.4	0.9	0.6	1.0
	16 June 1961	45	8.5	5.0	4.2	3.5	2.4	1.5	2.6
Akesson and Yates, 1984	Figure 5c	20	5.8	7.0	5.0	3.9	2.2	1.2	2.7
	Figure 6	—[e]	—[f]	—[f]	—[f]	—[f]	—[f]	—[f]	—[f]
Frost and Ware, 1970	August 1965 am	20	—[f]	3.4	2.4	1.9	1.1	0.6	1.3
	August 1965 pm	25	—[f]	3.9	3.0	2.3	1.3	0.6	1.6
	August 1967 am	15	—[f]	—[f]	1.3	0.9	0.5	0.3	0.7
	August 1967 pm	5	—[f]	—[f]	—[f]	1.4	1.0	0.6	1.0
Riley et al., 1989	Spray 1	45	1.1	0.9	0.7	0.6	0.5	0.3	0.5
	Spray 3	40	1.1	0.8	0.6	0.5	0.3	0.1	0.3
ICI, 1982[g]		40	5.8	3.6	2.8	2.3	1.4	0.7	1.6

a Each set of data corrected according to individual wind speed.
b For meteorological and application details — see Table 2.
c Assumes "0 meters" value in Figure 2 is at pond edge.
d Assumes body of water is 100 m wide.
e No windspeed specified.
f Insufficient data for calculation.
g Unpublished data.

Note: Determined from literature data (see Figure 2) converted with factors in Table 3.

Farmers commonly have substantially wider uncropped areas (buffer zones) next to ponds than the 5 m used in the scenario considered here. This will further reduce the amount of pesticide reaching the water body. In addition, in many instances, drift deposition may be reduced even more by the vegetation, shrubs, and trees growing on the margins of ponds, streams, and rivers. These will "trap" out a proportion of the drift before it reaches the body of water.

The following conclusions and observations may thus be drawn for pyrethroids from the spray drift deposition studies described in the scientific literature:

- Average downwind deposition onto a body of water 100 m wide is approximately 1% of the nominal field rate per application, with an overall range, due to meteorological and other conditions, of 0.3 to 2.7%. Downwind deposition may on a small proportion of occasions be slightly greater due to higher wind speeds than in the studies reviewed here or due to the spray plane pilot underestimating the offset required.
- Downwind deposition may be reduced by pond-edge vegetation, shrubs, and trees, and possibly by the crop canopy.
- Buffer zones wider than used in the above scenario will also reduce aquatic entry of pesticides.
- During a proportion of applications, bodies of water adjacent to a target crop will be upwind and receive no, or negligible, pyrethroid deposition.

Agricultural In-Use Determinations

The spray drift deposition observations from aerial applications of pyrethroid insecticides to two agricultural in-use studies are available to the authors, and these are described below.

In 1987, ICI aerially sprayed cypermethrin as an EC formulation onto 16 ha of cotton in a drainage basin surrounding a 3.4-ha pond of 100 m width. There was a 5 m wide uncropped zone around the pond. The land was undulating, with slopes of 4 to 12%, and the pond had several inlets (Figure 7). All of this made application difficult, and the spray plane had to be flown at a height of 3 to 9 m. This is considerably higher than the 1.5 to 2.5 m typically used in agricultural practice and increased the potential for spray drift deposition onto the pond. Ten applications of cypermethrin were made at weekly intervals, at 112 g ai/ha per occasion. The EC formulation was applied in water at 28 l/ha spray volume. The pilot sprayed the field using a series of premarked swaths (identical for all ten applications) and therefore did not carry out the conventional practice of upwind "swath offsetting", to reduce spray drift deposition onto the pond. The spray route and deposition card positions are shown in Figure 8. The spray was shut off where the target swath crossed an inlet of pond water, although this was difficult to achieve accurately.

The second study was carried out by the Ontario Ministry of Agriculture and Food, Ontario Ministry of Environment, and ICI Chipman (Frank et al., 1990). Two 6-ha potato fields were each aerially sprayed twice with permethrin EC, at 95 g ai/ha per occasion and 36 l/ha spray volume. The spray boom was approximately 2 m above the crop canopy, as is typical of normal agricultural practice. Deposition

Figure 7.   Alabama cypermethrin study: 16-ha cotton crop in a basin draining to a 3.4-ha farm pond. Crop aerially sprayed ten times (ICI, 1987; unpublished data).

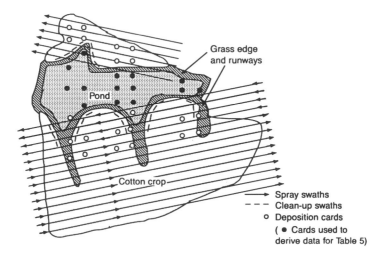

Figure 8.   Spray route and deposition card positions in Alabama cypermethrin study (ICI, 1987; unpublished data).

cards were laid in four opposing directions from the target area (Figure 9 shows one of the study sites) to determine maximum off-target residues.

The results of the above two studies are described in Table 5. At a downwind distance of 5 m (equivalent to near-shore littoral zone), deposition ranged from 1 to 10% of the nominal field application rate. However, the amount of pyrethroid spray drift deposition declined rapidly with distance downwind, to 0.1 to 1.6% by 50 m on all occasions. The average spray drift deposition of cypermethrin onto the 100 m wide pond was 1.9% of nominal field application rate. However, cypermethrin was sprayed onto the cotton crop from a height of 3 to 9 m and without any "swath offsetting", and thus gave a greater potential for drift than would occur under normal agricultural conditions. Extrapolating the permethrin data to a 100 m wide body of water, and allowing for an uncropped area 5 m wide, the average deposition onto the total water surface would have ranged from 0.4 to 0.6% of the nominal single field application rate.

Mathematical Models

The potential role of a spray drift and deposition mathematical model is to integrate experimental results and knowledge of the mechanisms involved. Such a model could then be used to assess the deposition of pyrethroid insecticides onto water bodies under a given set of conditions. The numerous factors which affect spray drift and deposition (Table 1) could then be varied in order to derive an overall picture of the problem, including "average" and "typical worst-case" scenarios.

Unfortunately, no such comprehensive modeling procedure has been fully developed. This, in part, results from a lack of sufficient deposition data to calibrate the model. The large number of application and environmental factors and the complexity of atmospheric conditions make model development very difficult. For

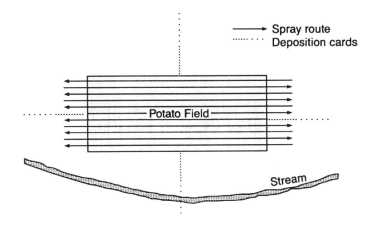

Figure 9. Spray route and deposition cards on a potato field aerially sprayed with permethrin (Frank et al., 1990).

Table 5. Spray Drift Deposition of Pyrethroids From In-Use Agricultural Studies

Study	Meteorology Temp (°C)	Gradient (°C)	Wind Speed (m/sec)	Wind Direction (°)[b]	Relative Humidity (%)	Deposition (% of Nominal Field Rate) at Distances — Downwind (m)[c] 5	50	100	0–100 av	0–100 av[e]	Upwind (m) 5	20
ICI, 1987; Cypermethrin[a]												
2 June	24	0	0.7	52	92	6.6	1.6	4.9[d]	3.9		Not applicable as pond sprayed on both sides (see Fig. 8); 100 m downwind was equivalent to 5 m upwind from opposite "shoreline".	
9 June	17	−0.3	0.2	18	93	6.6	0.8	0.7	1.9			
16 June	24	0	1.3	38	91	10	0.4	0.1	1.6			
23 June	28	+0.5	0.8	26	84	5.6	0.9	0.5	1.8			
30 June	21	−0.3	0.5	78	91	2.2	1.0	7.3	2.5			
7 July	20	−0.2	0.6	15	92	3.4	0.5	0.2	1.2			
14 July	22	0	0.6	66	90	7.8	1.2	1.1	2.3			
21 July	22	0	0.7	68	88	1.8	1.0	1.6	1.2			
28 July	22	+0.1	0.5	5	89	6	0.4	0.2	1.3			
4 August	25	+0.4	0.8	37	86	6	0.1	0.8	1.0			
Frank et al. 1990[f] Permethrin — Site 1	—[g]	—	0.9	40	—	5	0.1	<0.06	2	0.6	0.1	<0.06
— Site 2	—	—	2.9	40	—	5	0.1	<0.06	2	0.6	<0.1	<0.06
	—	—	<0.9	20	—	2	0.1	<0.06	1	0.4	1.5	<0.06
	—	—	<0.9	varied	—	1	0.2	<0.06	1	0.4	0.1	<0.06

a Ten weekly sprays each of 112 g ai/ha, commencing 2 June. Aerial applications at 130 mph, using Delevan Whirljet D-10 nozzles pointing vertically downwards, at 30 psi pressure. Wind was from south side of pond (see Figure 8) on all occasions; data above does not include the cards in the north shore inlets.
b Direction in relation to lines of deposition cards.
c ICI (1987) — distance from pond edge; Frank et al. (1990) — distance from crop edge.
d ICI (1987) — distance from pond edge.
e In cypermethrin study, residues at 100 m (far shore) contributed to by spraying swaths on "far shore" (see Figure 8).
e An allowance made for 5-m uncropped zone between target and nontarget areas.
f Two applications, each of 95 g ai/ha.
g Data not reported.

example, the chaotic nature and mathematical complexity of air turbulence restricts the applicability of many simple approaches to predicting the behavior of spray droplets. The difficulty is compounded by changes in droplet size due to solvent evaporation.

However, understanding of the processes involved has increased in recent years, and attempts to incorporate spray drift and deposition models into assessment procedures are beginning to appear. One example of this progress is the incorporation of aspects of the forest spraying FSCBG drift model (Dumbauld et al., 1980) into the Terrestrial Ecosystem Exposure Assessment Model (TEEAM; Dean et al., 1989). Thus, in TEEAM (which is still in the development stages) the input of pesticides to the terrestrial environment is determined by the spray drift and deposition portion of the model, rather than the more common procedure of assuming an application rate and perhaps an application efficiency factor (as, for example, with the PRZM model [Carsel et al., 1984]). Such modeling developments are not yet sufficiently advanced to be used for assessing pyrethroid spray drift deposition onto aquatic bodies and, therefore, we must currently rely entirely on data from relevant practical studies. However, the "Spray Drift Task Force" (described earlier) is, as part of its study program, examining the potential for developing the present models to provide environmentally realistic and predictive data.

Conclusions

Data for aerial applications of emulsifiable concentrate formulations, taken from the scientific literature (corrected for agricultural crop spraying techniques, as described above) and from in-use studies of pyrethroids, have been used to quantify the spectrum of spray drift deposition and the typical Maximum Estimated Entry Rate (MEER) as an average deposition onto a 100 m wide body of water adjacent to the target crop (e.g., cotton). The results suggest that entry ranges from 0 to 4% of the nominal field rate, with amounts of 2.0% and below being most common. However, these data are mostly generated under experimental conditions where the potential for drift deposition is substantially worse than occurs in normal agronomic practice. The typical pyrethroid MEER for a small farm pond, from agricultural applications, is therefore suggested here to be 2.0% of the field rate. Larger and more stable bodies of water will have lower MEERs. All the foliar-applied pyrethroids will have the same MEER value for the equivalent set of use conditions.

RUNOFF ENTRY

Pyrethroid runoff from sprayed fields into adjacent water bodies has not often been measured. It is therefore necessary to examine data from both direct and indirect sources:

- Published studies designed to specifically quantify the runoff of pyrethroids and that of compounds with similar physicochemical properties

- Agricultural applications of pyrethroids, where runoff into water bodies was adequately quantified
- Mathematical models of pesticide runoff

Factors Influencing Runoff Entry

Many factors affect the loss of pyrethroids from fields by runoff. Anything that affects the volume of runoff water or the erosion of soil by water will also influence the runoff of pyrethroids. These will combine with the factors affecting the availability of pyrethroids for runoff, such as crop interception and degradation, to determine the actual amount entering aquatic bodies.

The main factors are summarized in Table 6, with much fuller descriptions of runoff available elsewhere (e.g., Haan et al., 1982; Wauchope et al., 1977, 1978).

A general outline of the processes leading to the runoff of sediment (and therefore of pyrethroids) is shown in Figure 10. The water runoff processes in Figure 10 are represented as "transport" and these drive the soil and particle detachment and deposition processes. The whole runoff process has been divided into overland (or inter-rill) flow, rill flow, and channel (or concentrated) flow. While the physics of the mechanisms of these three components of runoff are similar, the balance of processes within them differs and so each requires separate consideration. For example, overland flow is often dominated by particle detachment, has low water depths, and occurs over short distances, whereas channel flow is dominated by transport, has greater water depths, and can be over larger distances. The concentration of pyrethroid in the sediment fractions present at each point in this system has also to be defined before an understanding of runoff can be translated into one of pyrethroid runoff.

Table 6. Factors Controlling the Runoff Entry of Pyrethroids Into Aquatic Bodies

Parameter	Influencing Factors
Weather	Rainfall amount and intensity
Soil conditions	Moisture content
	Infiltration rate
	Organic matter content
	Erodibility
Topography	Slope length and steepness
Cropping	Crop canopy cover
	Tillage practices
	Crop rotation
Conservation practices	Terracing
	Strip contouring
	Uncropped field boundaries
	Grassed channels
Pesticide properties	Degradation on leaves
	Wash-off from crop canopy
	Degradation on soil surface
	Adsorption to soil

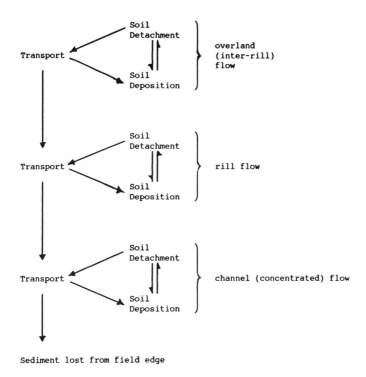

Figure 10.   Sediment runoff processes.

Literature Review

Selection Criteria

The scientific literature contains many studies of the runoff of pesticides from treated areas, but only a small proportion are relevant to the runoff of pyrethroids. Some studies take the form of monitoring surveys, where surface waters of various kinds have been analyzed for their pesticide content, but as it is generally not possible from these studies to quantify the runoff entry of pesticides from particular well-characterized source areas, they are not considered further. Other studies examined the runoff of pesticides from nonagricultural or forested areas. The processes and problems involved in these situations are mostly quite different to those that are important for runoff from agricultural fields; so only studies conducted in agricultural situations will be discussed here.

Although runoff water contains a high ratio of water to soil particles, pesticides which are very strongly adsorbed (such as pyrethroids and organochlorines) are still found mainly in the soil phase rather than in the water phase. The factors affecting the runoff of soil are not the same as those affecting the runoff of water. In practice this means that the literature on the runoff of the more water soluble and less strongly adsorbed pesticides, such as organophosphate insecticides and most herbicides, is not relevant to the consideration of pyrethroid runoff. Since the literature on pyrethroid runoff is small, studies on the runoff of other strongly adsorbed insecticides are also considered here. These include the organochlorines which are extremely persistent when compared with pyrethroids. The different persistence of the two groups has to be taken into account in the interpretation of the literature data.

In addition to the above restrictions on use of literature data, all runoff studies on products which were incorporated into the soil or which were applied as granules are also excluded from consideration here.

Available Data

Studies in the open literature which satisfied the above criteria are summarized in Table 7 for pyrethroid insecticides and Table 8 for organochlorine and other insecticides. There is a fairly small amount of data, and some of this is from in-furrow irrigation.

The length of study period represented by the data in Tables 7 and 8 was from the first pesticide application in a year to 50 days after the last application in the same year. The reason for excluding runoff events which occurred more than 50 days after the last application was to make the pyrethroid and organochlorine data more comparable. This had virtually no effect on the pyrethroid data, since pyrethroids are not persistent and runoff events more than 50 days after the last application generally contained little or no pyrethroid. An extreme case, where significant runoff after this cutoff might have been expected to occur, was the first set of data from the study of Carroll et al. (1981) on permethrin (Table 7). In this study there were only two small runoff events totalling 1.2 mm between the first application and 50 days after the last application; in the next 100 days there were 10 runoff events totalling 72 mm, yet these only resulted in the loss of 0.005% of the total permethrin applied. In contrast, the runoff of organochlorines has commonly been observed for several years after the last application, and at levels scarcely less than in the years of application (Epstein and Grant, 1968; Willis et al., 1976, 1983; McDowell et al., 1981). Removal of such long-term data is necessary in this review since it is not relevant to the consideration of the much less persistent pyrethroids.

Interpretation

The data in Tables 7 and 8 show a wide range of runoff residue amounts, displayed as a percentage of the total annual applied pesticide. This variability is particularly due to the nature of weather patterns with respect to the timing of pesticide applications. The consequence of this is that many pesticide applications will result in no pesticide runoff, some will lead to a measurable amount of runoff, and on rare occasions the runoff event may be substantial in size, far exceeding all other runoff events in that season. The data presented here include all these circumstances, hence their variability.

Table 7. Runoff From Spray Applications of Pyrethroid Insecticides

		Site			Application				Run-off				
Pesticide	Location	Area (ha)	Slope (%)	Soil Texture	Total (kg ai/ha)	No.	Rainfall (cm)	No. of Events	Total Water (cm)	Total Sediment (kg/ha)	Total Pesticide (g ai/ha)	% of Total Pesticide Applied	Reference
Fenvalerate	Imperial Valley, CA	28	0.1–0.2	Silty clay	0.57	3	30[b]	2	9	620	0.6	0.10	Spencer et al., 1985
		28	0.1 0.2	Silty clay	0.27	2	57[b]	4	9	1580	0.2	0.05	
		28	0.1 0.2	Silty clay	0.49	3	35[b]	4	7	390	1.0	0.21	
		28	0.1 0.2	Silty clay	1.23	7	47[b]	3[c]	9	300	1.1	0.09	
		28	0.1–0.2	—[a]	0.10	1	13[b]	1	1	90	N.D.	N.D.	
		14	0.1–0.2	—[a]	0.35	3	65[b]	3	15	100	0.5	0.14	
	LA	0.001	0.2	Silty loam	0.88	4	37	8	2	N.M.	0.2	0.02	Smith et al., 1983
		0.001	0.2	Silty loam	0.88	4	24	7	4	N.M.	5.3	0.60	
Permethrin	Imperial Valley, CA	28	0.1–0.2	Silty clay	0.71	5	44[b]	3	10	1310	1.1	0.15	Spencer et al., 1985
		28	0.1–0.2	Silty clay	0.51	3	37[b]	3	6	1160	N.D.	N.D.	
		28	0.1–0.2	Silty clay	1.22	7	57[b]	6	11	590	0.3	0.03	
		28	0.1–0.2	Silty clay	0.10	1	13[b]	1	3	170	N.D.	N.D.	
		28	0.1–0.2	—[a]	0.51	3	35[b]	3	5	240	N.D.	N.D.	
		14	0.1–0.2	—[a]	0.60	3	47[b]	2	11	30	0.5	0.09	
	LA	0.055	0.15	Silty clay	1.12	10	7	2[c]	1	N.M.	0.002	0.0002	Carroll et al., 1981
		0.055	0.15	Loam	1.12	10	46	10[c]	24	N.M.	2.6	0.23	

[a] Areas of silty clay, loam, and loamy sand.

[b] Furrow irrigation.

[c] Monitored events occurring more than 50 days after the final pesticide application were excluded (see text).

Note: N.D. = not detectable; N.M. = not measured.

Table 8. Runoff From Spray Applications of Nonpyrethroid Insecticides

Pesticide	Location	Site Area (ha)	Slope (%)	Soil Texture	Application Total (kg ai/ha)	No.	Rainfall (cm)	No. of Events	Total Volume (cm)	Run-off Total Sediment (kg/ha)	Total Pesticide (g ai/ha)	% of Total Pesticide Applied	Reference
DDT	Rocky Mount, NC	0.0017	4	Loamy sand	13.4	12	39	10	7		346	2.6	Sheets et al., 1972;
	Lewiston, NC	0.0017	2	Sandy loam	13.4	12	42	10[d]	9		395	2.9	Bradley et al., 1972
Endosulfan	Presque Isle, ME[a]	0.008	8	Gravelly loam	1.05	3	<45	8	3		3	0.3	Epstein and Grant, 1968
	Imperial Valley, CA	28	0.1–0.2	Silty clay	1.66	2	22[b]	3	3	90	3	0.2	Spencer et al., 1985
		53	0.1–0.2	Silty clay loam	5.95	5	23[b]	3	8		37	0.6	
Ethylan	Imperial Valley, CA	28	0.1–0.2	Silty clay	3.20	2	15[b]	3	2		0.30	0.008	
Methoxychlor	Coshocton, OH	0.0008	0.2	Silty loam	22.4	1		6[d]	0.4		0.09	0.0004	Edwards and Glass, 1971
Toxaphene	Rocky Mount, NC	0.0017	4	Loamy sand	27.0	12	39	10	7		84	0.3	Sheets et al., 1972;
		0.0017	4	Loamy sand	27.0[c]	12	19	6[d]	2		17	0.06	
	Lewiston, NC	0.0017	2	Sandy loam	27.0	12	42	10[d]	9		91	0.3	Bradley et al., 1972
		0.0017	2	Sandy loam	27.0[c]	12	28	9[d]	5		30	0.1	
	Clarksdale, MS	16	0.2	Silty clay	10.1[c]	6	45	7[d]	15	1600	28	0.3	Willis et al., 1976, 1983; McDowell et al., 1981
		16	0.2	Silty clay	9.0[c]	5		5[d]			14	0.2	

[a] Average of two plots with different crop rotation treatments.
[b] Furrow irrigation (ethylan study included 2 cm of natural rainfall).
[c] Some of the pesticide runoff was due to applications in the previous year.
[d] Monitored events occurring more than 50 days after the final pesticide application were excluded (see text).

Certain patterns emerge when the data are examined more closely. One is that the studies which used the smaller plots (as small as 8 m^2) tended to result in the larger amounts of pesticide runoff. This is as would be expected, but is difficult to confirm conclusively from the small dataset, and is confounded by the effects of slope on runoff. Losses by runoff from small plots (0.0008 to 0.055 ha) and larger sites (14 to 53 ha) ranged from 0.0004 to 2.9% of the total pesticide applied (Tables 7 and 8). When small plot studies are excluded from the data, the average annual pesticide loss, as a percentage of total applied, falls to a maximum of 0.6% for all compounds, and ranges from 0.03 to 0.21% for the pyrethroids alone.

Another pattern apparent from the data is that the nonpyrethroid insecticides generally result in more runoff than the pyrethroids. In most cases this is principally because their persistence results in there being more pesticide near the soil surface available for runoff than is the case with pyrethroids. To illustrate this, Table 9 shows the soil residues of DDT and pyrethroids which result from a cotton pest control program consisting of 10 to 12 pesticide applications. The data demonstrate that much more of the applied DDT is still present on the soil surface at the end of the spray season; whereas, at this same time, there remains only a small proportion of the total applied pyrethroid.

Agricultural In-Use Determinations

A field study of cypermethrin, with applications to 16 ha of cotton, has already been described (with respect to spray drift; ICI, 1987). This study was also used to determine the aquatic entry rate due to runoff of cypermethrin. The site was selected as being a worst case for runoff, it being an undulating field with slopes of 4 to 12% directly adjacent to a pond. It would not normally be considered suitable for the growing of row crops without extensive soil conservation practices.

For the whole period of the study, the volume of water entering the pond during runoff events was monitored, and for four of the events the runoff water was sampled and analyzed. This was done by taking water samples every 2 h from the 8 runoff channels just before they entered the pond. These samples were used to estimate the amount of cypermethrin entering the pond. Samples of pond water were also taken at intervals immediately following each runoff event, and used to obtain a secondary estimate of runoff entry.

Table 9. Soil Residues of DDT and Pyrethroids Immediately After a Series of 10–12 Applications to Cotton

Pesticide	Application		Soil Residue			
	Total (Kg ai/ha)	No.	Kg/ha	% of Total Applied	Days after Last Spray	Reference
DDT	13.4	12	2.7	20	11	Sheets et al., 1972;
	13.4	12	5.3	39	13	Bradley et al., 1972
Permethrin	1.12	10	0.058	5.1	4	Carroll et al., 1981
	1.12	10	0.023	2.0	9	
Cypermethrin	1.12	10	0.056	5.0	3	ICI, 1987[a]

[a] Unpublished data.

Table 10 shows the data for all runoff events during and shortly after the period over which cypermethrin applications were made. There were no runoff events after this until monitoring ceased at the end of October. Cypermethrin residues in the field soil were at a maximum after the second application (see footnote to Table 10) and despite eight subsequent applications, the residues never exceeded this level. The runoff percentages estimated from pond water samples were in good agreement with those from the runoff land channel samples, except for the event on the 21st of July. In the latter instance a cypermethrin application to the cotton crop was made on this date, only 9 h prior to runoff, and therefore the residues in the pond were from both spray drift and runoff entry.

The total annual cypermethrin runoff can be estimated from the data in Table 10. The two unsampled events were relatively small in terms of intensity and runoff volume, but if they are assumed to have produced cypermethrin runoff equal to the average of the other events, then the total annual runoff was about 0.08% of the total pesticide applied during the year (there were 10 applications). Individual events gave runoff ranging from 0.01 to approximately 0.3% of the single cypermethrin application rate.

Mathematical Models

The key to accurate prediction of pyrethroid runoff is the accurate prediction of soil particle runoff. The most widely used tool for estimating particle runoff from an area of land is the Universal Soil Loss Equation (USLE). This is an empirical equation which does not explicitly represent any of the processes of soil particle runoff. It is essentially an erosion equation based on data from small (22.1 m long) field plots, and so covers the overland (inter-rill) and perhaps the rill flow portions to the runoff process shown in Figure 10. The limitations of the USLE are

- It estimates long-term average sediment runoff (i.e., over many years), rather than loss for individual runoff events
- The erosion equation is based on small plot data, and thus the processes of channeling and deposition (which can dominate edge-of-field losses in typical agricultural situations) are included only to a very limited extent

Consequently the output from the USLE is termed "gross erosion" and is most applicable to the runoff from small plots rather than from whole fields.

The mathematical model SWRRB (Simulator for Water Resources in Rural Basins) is a model commonly used to assess the entry rate of pesticide into water bodies due to runoff (Computer Sciences Corporation, 1980). This model relies on a modified version of the USLE (called MUSLE) to predict sediment runoff and on the U.S. Soil Conservation Service (SCS) curve number technique to predict the water runoff. The modification to the USLE is intended to make the equation applicable to individual runoff events, rather than giving a long-term average value (Williams, 1975), and has been done by replacing the R factor (rainfall erosivity) with a factor depending on runoff volume and peak runoff rate. While SWRRB is

Table 10. Runoff Entry of Cypermethrin: In-Use Determination in Alabama (1987) and PRZM Model Predictions

| | | Alabama Field Study (see also Table 5)[a] | | | | Model Predictions | |
| | | | Pesticide Runoff | | | | |
Date 1987	Rain (cm)	Runoff Water Volume (m³)	Land[b] (g ai/ha)	Pond[c] (g ai/ha)	% of Single Application	Runoff Water Volume (m³)	Pesticide Runoff (g ai/ha)
13 Jun	3.7	498	0.01	<0.11	0.01	1129	0.91
17 Jun	3.0	1808	0.16	0.22	0.17	1334	1.8
7 Jul	2.8	765	—[d]	—[d]	—	882	—
21 Jul	3.0	230	0.08	1.6[e]	0.07	787	1.4
7 Aug	4.6	1981	0.34	0.26	0.27	3177	3.6
18 Aug	3.4	466	—[d]	—[d]	—	1685	—
Total runoff for 6 events[f]		5748 m³	~0.9 g ai/ha		~0.08% of total applied	8995 m³	~10 g ai/ha

[a] Applications were made over period 2 June to 4 August. Cypermethrin field soil residues (mg/kg dry wt, 0–10 cm depth; for 3 replicates and [mean]) were 12 June, 0.07–0.1 (0.08); 10 July, 0.05–0.1 (0.08); 7 August, 0.03–0.09 (0.06); 4 September, 0.02–0.05 (0.03); 2 October, 0.03–0.07 (0.04). Average deposition onto field cards at each interval was 42 to 81% (mean 58%) of the nominal applied rate. Crop canopy cover was 18 May, 1 to 5%; 8 June, 6 to 10%; 22 June, 11 to 21%; 6 July, 41 to 50%; 20 July, 71 to 80%; 3 August, 81 to 100%.

[b] Determined from runoff channel samples.

[c] Determined from pond water samples.

[d] Runoff event not sampled.

[e] Value not used in runoff calculations, as pond water residues were also contributed to by spray drift deposition from a cypermethrin application made earlier the same day.

[f] Where necessary, values estimated for unsampled runoff event.

Table 11. Prediction of Runoff Losses of Lambda-Cyhalothrin Using the SWRRB Model

Model Input		Parameter (per year)	Drainage Basin Tifton	Yazoo
Soil Kd	1000	Rainfall (cm)	130	133
Foliar half-life (days)	10	Runoff volume (cm)	33	60
Soil half-life (days)	30	No. of events: average	1.5	3.4
Enrichment ratio	1.5	maximum	2	7
Application rate (g ai/ha)	34	Pesticide runoff: average	0.3	0.8
Number of applications	12	(% annual applic.): maximum	1.4	2.2
Frequency of application (days)	7			

intended to be used for large drainage basins with sub-basins of varying properties, it is very often used with a single "sub-basin".

Table 11 gives details of an estimation of the potential for runoff entry of lambda-cyhalothrin, using SWRRB with the "inbuilt" climatic data set for the Tifton and Yazoo drainage basins. The maximum annual loss of the pyrethroid, by runoff, was predicted to be 1.4 to 2.2% of the total product applied during the year. Average annual loss was estimated at 0.3 to 0.8% of the total application. This is a higher level of runoff than is suggested by the literature data or ICI's Alabama cypermethrin study (both cypermethrin and lambda-cyhalothrin would be expected to give very similar rates of runoff). To further investigate this, the in-use data from the Alabama cypermethrin study was compared to model predictions for this site.

The PRZM (Carsel et al., 1984) model was used to carry out this comparison as its runoff component is very similar to SWRRB, but it is better documented and easier to use for sites not already within the model. The main difference between the two models with regard to runoff is that PRZM has a soil surface mixing zone of 5 cm, compared to 1 cm for SWRRB; thus cypermethrin runoff values given here for PRZM have been multiplied by 5 to allow for the fact that the cypermethrin will be present in the top 1 cm (cypermethrin Koc >100,000; very low soil mobility). The various input parameters for the model were all carefully selected, using the substantial quantity of site details that were recorded during the study. The PRZM control file (input parameters) and a summary of the input weather data are given in the Appendix. The results in Table 10 show that the total predicted runoff volume for the 6 events was only twice as high as the observed value, but the total predicted cypermethrin runoff for the 4 measured events was nearly 12× higher than observed. This corresponds to a large overprediction of the soil particle runoff. Some possible reasons for this difference between predictions from PRZM (and hence SWRRB) and the measured pyrethroid run-off are

- The USLE is designed to give long-term yearly average runoff, most of which occurs in the winter/spring before there is much crop cover. MUSLE, as implemented in PRZM, accounts for erosional differences between storms in terms of runoff volume and peak runoff rate, while the overall difference between fallow and cropping periods is dealt with by having each have a crop factor value. The effects of crop canopy development on the erosive potential of storms is not treated. Foliar applied pyrethroids are principally available for runoff in the summer and early fall. At this time the model may overestimate runoff, as the soil is intrinsically less vulnerable to runoff than it is in the winter, spring, or very early summer when there is little or no crop canopy cover.

- The ''small-plot'' orientation of the USLE fails when extended to whole fields, due to the very limited inclusion of channel and deposition processes (for example, delivery ratio). While these processes have been implicitly included in the development of the MUSLE (from the USLE), most parameters of the model are still exactly as they were in the USLE version, and inevitably reflect this origin. This will generally result in an unquantifiable overestimation of pyrethroid runoff.

- Soil particle runoff is much more difficult to predict than the total runoff volume. Therefore, the prediction of runoff for pesticides lost mainly in the soil phase (such as pyrethroids) is always likely to be less reliable than predictions for pesticides principally lost in the water phase.

- The models were originally intended to give relative estimates of the effects of different agronomic practices on soil runoff of pesticides, rather than to generate absolute values in individual situations.

Conclusions

Taking all the available information into consideration, it can be concluded that for foliar-applied pyrethroids the total annual field runoff from conventional applications to agricultural crops will generally be approximately 0.1% or less of the total applied during the season. This may vary slightly for different products, as a result of their different soil degradation rates, which affects the amount of chemical available for runoff. There is clearly a continuum of runoff vulnerability in the agronomic environment, with greatest loss from the most susceptible fields in years with the largest and most intensive rainfall events. The value of 0.1% or less, proposed above, is believed to represent runoff losses for the great majority of situations in which foliar-applied pyrethroids are likely to be used.

MESOCOSM APPLICATIONS

A wide variety of types of mesocosm have been constructed (see Hill, 1989; Solomon and Liber, 1988). For example:

- Limnocorrals, enclosures in open lake water. University of Guelph researchers have used 5 × 5-m units to study the biological effects of a range of pesticides, including permethrin

- Small, earth-lined or concrete ponds. These are generally no larger than 5 × 5 m and have been used for investigations of permethrin, cypermethrin, and lambda-cyhalothrin. Figure 11 is of such mesocosms used by ICI for studies with the latter two pyrethroids

- Littoral enclosures on the edge of a pond as used by researchers from the University of Wisconsin Superior and the U.S. Environmental Protection Agency at Duluth, MN. Among the pesticides studied by this group is the pyrethroid esfenvalerate

- Large, earth-lined ponds of approximately 0.05 to 0.1 ha. These are of a type and size recommended by the U.S. EPA for regulatory studies and have been used for studies of several pyrethroid insecticides, including lambda-cyhalothrin and esfenvalerate. Figure 12 shows the mesocosms used by ICI for lambda-cyhalothrin investigations

Application Methods

The synthetic pyrethroids can enter aquatic environments by both spray drift deposition and runoff. Therefore, methods have been devised to simulate both of these entry modes for mesocosms.

Drift droplets of pyrethroids depositing on natural bodies of water are likely to be mostly 150 μm or less in diameter. However, it is generally unwise to apply such small droplets to a mesocosm, as these will have a high tendency to drift onto and contaminate adjacent mesocosms. It is possible to protect small units from such cross contamination (Figure 13), but most investigators have also preferred to apply large droplet sizes. In ICI studies, nozzles have been of the "raindrop" type; for example, the Delavan D2.5 flood nozzle with 95% of droplets >200 μm diameter.

A variety of different spray drift application methods have been used, according to mesocosm size, facilities available, and investigator preferences. These have

Figure 11.   Set of eight 5 × 5-m concrete mesocosms used for studies with cypermethrin and lambda-cyhalothrin (by ICI).

Figure 12.   Set of 15 × 30-m earth-lined mesocosms used for studies with lambda-cyhalothrin (by ICI).

mostly involved boom and nozzle application close to the water surface. The booms
have been either of sufficient width to "sweep" the whole pond (Figure 14) or
have been narrower, such that several pond swaths are required. In the latter case
some workers have used booms attached to the back of a boat. Solomon et al.
(1980) used an alternative approach with permethrin and injected the pyrethroid
under the water surface.

   The simulation of runoff entry is more difficult than for spray drift, because of
the nature and volume of the material to be applied. In mesocosm studies of lambda-
cyhalothrin by ICI (Hill et al., 1988, 1991) a soil-water slurry was prepared. The
pyrethroid was mixed into the slurry, then left for approximately 24 h before applying
to the pond. The slurry was then sprayed evenly across the mesocosm surface (as
also was the spray drift simulation), using a boom and 1-cm diameter nozzles
(Figure 14). An alternative procedure has been used in studies carried out by Wildlife
International at Auburn, AL (Hutchinson, personal communication). Here, the
slurry was injected at a small number of points along the edge of the pond, in order
to more closely represent natural field entry. However, it is possible that runoff
applied evenly to the mesocosm may help minimize the variability between replicate
ponds, this being of potential value in enhancing the chances of statistically detecting

Figure 13.   Small mesocosms being sprayed with cypermethrin (by ICI). The shielded spray boom and the pond covers prevent cross contamination of adjacent units.

effects (Shaw et al., 1991). A number of other variations have been practiced; for example, in some studies the pyrethroid has been mixed into dry soil prior to application, and in others, the treated soil has been broadcast onto the water surface. Whatever the procedure, the method of premixing with soil should allow the pyrethroid to become strongly adsorbed to the particles, as would occur prior to runoff in the field.

There is no doubt that the overall procedure of application of spray drift and runoff to a mesocosm does not, nor cannot, absolutely represent field environmental processes. Some of the differences could result in the effects of the chemical on the mesocosm organisms being unrepresentative of the natural field impact, for example:

- All of the techniques for spray drift simulation place the chemical into the water column (using large droplets, subsurface injection, or mixing). However, in the field the small spray drift droplets may cause surface films or "near-surface" high concentrations of the pyrethroids. Thus, in mesocosms the concentrations to which organisms are exposed could differ from the field, as also could rates of degradation, dissipation, and adsorption.

- The mesocosm spray drift simulation is done evenly across the water surface, and runoff has been similarly applied by some investigators. In natural situations, neither spray drift deposition nor runoff entry are distributed evenly onto a body of water. A gradient exists for drift deposition entry, and point sources are common for runoff. Consequently, organisms in various areas or depths of a natural aquatic environment will be exposed to different concentrations of the pesticides, and should any effects occur near the shoreline, recolonization may occur from unaffected areas. In a mesocosm sprayed across the whole water surface using "high" volumes and large droplets, little differentiation will occur and all organisms will experience similar exposure. This mesocosm concentration may be lower than the exposure in some littoral areas of natural water bodies, but higher than areas away from the shoreline in these same habitats.

All of the above, and other features of the mesocosm, mean that great care has to be taken in extrapolating the data to determine field risk. The information and procedures necessary to do this are currently not available.

Application Rates and Frequencies

Foliar-applied pyrethroid insecticides are sprayed onto some crops (e.g., cotton) several times per season and a mesocosm study must adequately represent the spray drift and runoff that enters natural bodies of water as a result of this. In order to determine mesocosm application rates and frequencies, the agricultural field parameters to be represented must be defined. These should reflect the maximum number and minimum interval specified on the label, even where this far exceeds typical farming practice. In the following discussion the conditions described below (and which are appropriate for the majority of foliar-applied pyrethroids) are utilized:

- Application to cotton (generally the highest label rate)
- Maximum of 10 sprays per season, each at maximum application rate
- Minimum spray interval of 1 week
- Aerial application by conventional hydraulic nozzles (as this gives much more drift than ground sprays)
- Emulsifiable concentrate (EC) formulation

The typical MEERs to aquatic environments for spray drift and runoff under these conditions are derived from data provided earlier in this paper, and, for individual events, are proposed as:

- Spray drift typical MEER: 2% of the nominal field application rate (see Table 12).
- Runoff typical MEER: approximately 2% of the nominal field application rate. This value is calculated from an annual field runoff of 0.1% of the total annual pyrethroid application, as is described in Table 12. The actual rate may vary slightly for the different products, due to soil degradation rates affecting availability.

Figure 14.   Large earth-lined mesocosms sprayed with lambda-cyhalothrin (by ICI). (Top) Spraying large mesocosms with a boom mounted on the hydraulic arm of a tractor. (Bottom) Spray container and pressure cylinder remotely controlled from tractor cab.

Table 12. Proposed Sequence and Rates of Spray Drift and Runoff Applications of Foliar-Pyrethroids to Mesocosms, to Represent the Typical MEER

Entry Route	Typical Maximum Field Runoff		Typical Single Event MEER[a] (% of Field Rate)	Application to Mesocosm											
	Annual (% of Annual Application)	Single Event (% of Field Rate)		Scenario Number	Rate (% of Single Field Rate)[e]										
					Wk 1	2	3	4	5	6	7	8	9	10	
Spray drift	—	—	2[c]	1	1	1	1	1	1	1	1	1	1	1	
				2	2	1.5	1	0.5	0	2	1.5	1	0.5	0	
Runoff	0.1	0.2[b]	~2[d]	1	1	1	1	1	1	1	1	1	1	1	
				2	2	2		0.25		1.25		0.5			

[a] Typical maximum estimated entry rate, into aquatic environments.

[b] Calculated from the annual entry value; assuming 10 applications and 5 equal runoff events.

[c] Assumes a body of water 100 m wide, a 5-m buffer zone, and that an aircraft upwind offset is used to allow swath displacement.

[d] Calculated from the single event field runoff value, assuming a field drainage basin to water body ratio of 10:1. The actual value may vary slightly for different products.

[e] Average rate per event for each scenario is equal to 50% of the MEER.

Runoff entry into aquatic environments will occur several times per season, and with the foliar pyrethroids so also will spray drift entry; therefore, multiple applications of each must be made to the mesocosm. However, in the field the rates of entry from both routes will vary from very little (from zero in the case of spray drift) up to the MEER. Consequently if multiple applications to the mesocosm were all made at the typical MEER, this would grossly exaggerate the annual natural field exposure.

Table 12 demonstrates application scenarios for the "aerial cotton" conditions described above. A label scenario of 10 applications is used, and thus the same number of drift sprays are proposed. The number of runoff treatments suggested is 5, equivalent to the typical field occurrence and enabling them to be integrated into the study design in a manner that aids operational logistics. The rates of application for each route are shown as two scenarios: one with "constant" rates, the other "variable" and representing more closely the field. Each series of treatments is based on the average application being 50% of the typical single event MEER. There is little doubt that the "variable" proposal is scientifically more valid for representation of extreme conditions, with at least on some occasions the organisms being exposed to the higher field entry rates. However, most mesocosm studies involve highly complex programs of work, from which the data are provided to regulatory agencies. The authors thus favor the use of "constant" rates, for the following reasons:

- This allows comparison of the ecological effects from different products, by the regulatory agency. If a series of "variable" rates were used, such comparisons would only be possible if the same combination and sequence were agreed for each study, and this is unlikely to be readily achieved
- Logistically easier, with less chance of errors in a complex spraying program (in a 12 pond design with 3 treatments and a control and using the conditions described above, 150 large-scale applications will be made over a 10-week period)
- Will probably enable easier extrapolation to natural field conditions (when procedures are developed for this)

Runoff applications require the preparation of a pesticide-soil mixture. The annual amount of soil ("sediment yield") running off from agricultural fields varies very widely, from below 50 kg/ha to over two orders of magnitude above this. However, on the majority of occasions annual losses are less than 2000 kg/ha and values of 100 to 1000 kg/ha are typical of individual events. Based on these values ICI, in a lambda-cyhalothrin mesocosm study with 0.04 ha mesocosms, applied runoff as a slurry of 2500 l volume and 250-kg soil (equivalent to approximately 600-kg soil/ha field runoff entering into a pond one tenth the surface area of the field drainage basin) per event (Hill et al., 1993). Other workers have mostly added lower amounts of soil to mesocosms in runoff simulations.

A mesocosm investigation of the effects of a pyrethroid on aquatic organisms is normally done at more than one spray drift deposition and runoff rate, three rates typically being used. The rate described in Table 12, equivalent to the typical MEER, is commonly used as the midrate, with higher and lower rates included.

The highest of these three rates will be totally unrepresentative of any field condition, but in past studies has generally been required (by regulatory authorities) or included (by the investigators) to ensure that a "major effect" level is established and also to determine mesocosm "dose-responses" for sensitive organisms. Most recently, the EPA has specified studies using one exposure rate only, equivalent to the typical MEER. The authors of this chapter would prefer that the option always be available that the typical MEER be used to derive the highest rate tested. The inclusion of lower rates has considerable benefits, enabling determination of the spectrum of potential impact under a wide range of field conditions and the assessment of risk from different use-patterns and application rates.

The U.S. EPA has required that registration mesocosm studies of foliar-applied pyrethroids include treatments to stimulate both spray drift and runoff. This is reasonable in that some bodies of water in the major U.S. pyrethroid-use areas are susceptible to pesticide entry by both routes. However, in the design of any mesocosm study (different products, countries, and use-patterns), the pesticide application methods should be chosen to only reflect the significant exposure routes into the natural bodies of water being represented by the test.

Where both spray drift and runoff entry are possible, some natural aquatic environments will receive both. Consequently, in U.S. registration studies with foliar-applied pyrethroids, drift and runoff applications have been made to the same mesocosms at the same "rate" (for example, MEER). While this gives a representation of the worst case impact, as is required for regulatory hazard assessment, it does not allow development of a scientific understanding of the relative impact of the two entry routes. Applications of drift and runoff to separate mesocosms, together with "mixed" applications to other units, would enable a better extrapolation of mesocosm impact to the wide spectrum of natural aquatic environments. In the U.S., large ($\geq$400 m^2) mesocosms are required for regulatory studies, with several replicates per "rate". The scale of operations makes it logistically (and financially) difficult to do drift-only or runoff-only tests in addition to the regulatory requirements described above. In Europe, mesocosms have traditionally been smaller, most commonly $\leq$25 m^2. Internationally, mesocosm size and design (in relation to study objective) is currently receiving substantial attention by debate. It is to be hoped that the acceptability and use of small mesocosms will increase and that this will lead to more fundamental work being carried out, to augment the regulatory studies and develop a better fundamental understanding of aquatic impact in the wide range of natural field aquatic conditions.

CONCLUSIONS

The off-target deposition of droplets from agricultural pyrethroid sprays has been quantified in this chapter for fixed-wing aerial application (with conventional hydraulic nozzles) of emulsifiable concentrate formulations. The data from all investigations are in reasonable agreement, considering the spectrum of application and meteorological conditions under which they were carried out.

Runoff data have also been reviewed and, although from relatively few studies, do give a good indication of environmental entry rates.

All the conclusions made here should be taken as preliminary, in view of the small database of studies relevant to the synthetic pyrethroids. However, it is necessary to estimate values for typical MEERs for pyrethroid runoff and spray drift deposition, as registration agencies have and still are requiring mesocosm studies on these products. A substantial number of programs are underway in several countries to better quantify spray drift deposition of pesticides. These will examine pesticides with a wide range of properties and applied by various techniques. Over the next few years the spray drift deposition data from such studies will enable the proposals made in this chapter to be reexamined and refined. In a similar manner, procedures for field determinations and model predictions of runoff are presently receiving close scrutiny, such that exposure rates for mesocosms can in the future be determined with greater confidence.

Spray Drift

The downwind spray drift deposition data from the application scenario which have been considered show a variation of about one order of magnitude. During agricultural use, the range of concentrations of pyrethroid deposited onto a 100 m wide body of water downwind of the target area is likely to be within 0.3 to 2.5% of the field rate, and the typical MEER is considered to be 2%. The downwind littoral nearshore inevitably receives the greatest dose, with from 1 to 5% of the nominal field rate, while the farshore would receive only 0.1 to 1%. However, many bodies of water will not be downwind of spraying on all occasions, and therefore a proportion of applications will result in little or no spray drift deposition.

Pyrethroids are multiply applied to many crops; for example, up to about 10 applications per season on cotton in the USA. Therefore, spray drift deposition onto a pond of 100 m width will typically vary between 0 and 2.5%, with a seasonal average of approximately 1% or less of field rate. It is therefore proposed that, in mesocosm studies of the effects of pyrethroids on aquatic organisms, spray drift applications are made as follows:

- Multiple treatments, at the maximum number and minimum interval stated on the product label
- Surface sprays, evenly to whole pond, using "large droplet" nozzles
- 1% of single maximum field rate per application, to represent half of the typical MEER. This may be the highest rate tested
- Lower rates may be included to enable good extrapolation of data to natural aquatic environments (where wide variations in pesticide entry will occur) for risk assessment, and to allow interpretation to different use-patterns and rates

Runoff

The annual runoff of foliar pyrethroids will generally be less than 0.1% of the total annual application. Using a scenario of a field drainage basin to pond ratio of 10:1, it is proposed that in mesocosm studies of the effects of pyrethroids on aquatic organisms, runoff applications are made as follows:

- Five applications, simulating five runoff events (which is typical of field occurrence)
- Surface sprays, applied either evenly to the whole pond or at point-sources, although the latter may result in greater pond-to-pond variability and thus reduce the sensitivity of statistical analyses. A soil-water-pesticide mixture should be applied to ensure that the chemical is adsorbed to the soil particles, as it would be in field runoff
- Approximately 1% of single maximum field rate per application, to represent half of the typical MEER. This may be the highest rate tested
- Lower rates if necessary, to enable good extrapolation of data to the field in risk assessment, and to allow interpretation to a range of use-patterns and rates
- Where runoff and spray drift treatments are necessary, these may need to be applied to the same mesocosms to satisfy regulatory requirements. However, it may be scientifically valuable to also apply each to separate mesocosms

ACKNOWLEDGMENTS

The authors gratefully acknowledge valued discussions with many colleagues within and outside ICI, and whose experience in the scientific disciplines encompassing spray drift, runoff, and mesocosm techniques has contributed to our writing of this paper.

REFERENCES

AEDG. 1993. *Improving Aquatic Risk Assessments under FIFRA: Report of the Aquatic Effects Dialogue Group,* in press.

Akesson, N.B. and W.E. Yates. 1964. Problems relating to application of agricultural chemicals and resulting drift residues. *Ann. Rev. Entomol.* 9:285–318.

Akesson, N.B. and W.E. Yates. 1984. Physical parameters affecting aircraft spray application. In W.Y. Garner and J. Harvey, eds., *Chemical and Biological Controls in Forestry.* ACS Symposium Series 238, American Chemical Society, Seattle, Washington, pp. 95–115.

Akesson, N.B., W.E. Yates, and R.W. Brazelton. 1984. Guides for improving insecticide, herbicide and crop defoliant application efficiency. In *Professional Standards for Aerial Application of Pesticides in California, Volume II.* California Agricultural Aircraft Association Inc., Sacramento, California, pp. 69–77.

Akesson, N.B. 1984. Weather factors and plant protection. In *Professional Standards for Aerial Application of Pesticides in California, Volume I.* California Agricultural Aircraft Association Inc., Sacramento, California, pp. 145–155.

Bradley, J.R., T.J. Sheets and M.D. Jackson. 1972. DDT and toxaphene movement in and J.B. Graves. 1981. Permethrin concentration on cotton plants, persistence in soil, and loss in runoff. *J. Environ. Qual.* 10:497–500.

Carroll, B.R., G.H. Willis, and J.B. Graves. 1981. Permethrin concentration on cotton plants, persistence in soil, and loss in runoff. *J. Environ. Qual.* 10:497–500.

Carsel, R.F., C.N. Smith, L.A. Mulkey, J.D. Dean, and P. Jowide. 1984. *Users Manual for the Pesticide Root Zone Model (PRZM): Release 1.* Report No. EPA/600/3-84-109. U.S. Environmental Protection Agency, Athens, Georgia.

Crossland, N.O. 1982. Aquatic toxicology of cypermethrin. II. Fate and biological effects in pond experiments. *Aquat. Toxicol.* 2:205–222.

Computer Sciences Corporation. 1980. *Pesticide Run-off Simulator (SWRRB) — User's Manual.* Prepared for U.S. Environmental Protection Agency, Falls Church, Virginia.

Dean, J.D., K.A. Voos, R.W. Schanz, and B.P. Popenuck. 1989. *Terrestrial Ecosystems Exposure Assessment Model (TEEAM).* Report No. EPA/600/3-88-038. Prepared for U.S. Environmental Protection Agency, Athens, Georgia.

Drewes, H.R., J.W. Lauber, and J.D. Fish. 1990. The Spray Drift Task Force: Development of a drift study database for registration purposes. *Proceedings, Brighton Crop Protn. Conf. — Pests and Diseases.* Brighton, England, 1990, pp. 1053–1060.

Dumbauld, R.K., J.R. Bjorklund, and S.F. Saterlie. 1980. *Computer Models for Predicting Aircraft Spray Dispersion and Deposition Above and Within Forest Canopies: Users Manual for the FSCBG Computer Program.* Report No. 80-11. USDA Forest Service, Davis, California.

Edwards, W.M. and B.L. Glass. 1971. Methoxychlor and 2,4,5-T in lysimeter percolation and runoff water. *Bull. Environ. Contamin. Toxicol.* 6:81–84.

EPA. 1986. *U.S. Environmental Protection Agency, Aquatic Mesocosm Workshop.* George Mason University, Fairfax, Virginia, April 1986.

Epstein, E. and W.J. Grant. 1968. Chlorinated insecticides in runoff water as affected by crop rotation. *Soil Sci. Soc. Amer. Proc.* 32:423–426.

Fairchild, J.F. T.W. La Point, and J. Zajicek. 1989. Population, community and ecosystem-level responses of aquatic mesocosms to pulse-doses of a pyrethroid insecticide. *Symposium, Society of Environmental Toxicology and Chemistry.* Toronto, Canada, 1989, Abstract No. 416.

Frank, R., K. Johnson, H.E. Braun, C.G. Halliday, and J. Harvey. 1990. Monitoring air, soil, stream and fish for aerial drift of permethrin. Submitted to *Env. Montg. Assessment.*

Frost, K.R. and G.W. Ware. 1970. Pesticide drift from aerial and ground applications. *Agric. Eng.* Aug 1970: 460–464.

Haan, C.T., H.P. Johnson, and D.L. Brakensiek, ed. 1982. *Hydrologic Modelling of Small Watersheds.* ASAE Monograph no. 5; ASAE, St. Joseph, Michigan, USA, 533 pp.

Hill, I.R. 1985. Effects on non-target organisms in terrestrial and aquatic environments. In J.P. Leahey, ed., *The Pyrethroid Insecticides.* Taylor & Francis, London, pp. 151–262.

Hill, I.R. 1989. Aquatic organisms and pyrethroids. *Pestic. Sci.,* 27:429–465 (presented at the symposium of the American Association for the Advancement of Science, San Francisco, 1989; Session "The Pyrethroid Insecticides; A Scientific Advance for Human Welfare").

Hill, I.R., J.K. Runnalls, J.H. Kennedy, and P. Ekoniak. 1993. Lambda-cyhalothrin: a mesocosm study of its effects on aquatic organisms, this volume, Ch. 24.

Hill, I.R. and S.J.L. Wright, 1978. The behaviour and fate of pesticides in microbial environments. In I.R. Hill and S.J.L. Wright, eds., *Pesticide Microbiology. Microbiological Aspects of Pesticide Behaviour in the Environment.* Academic Press, London, pp. 79–136.

Leahey, J.P. ed. 1985. *The Pyrethroid Insecticides*. Taylor & Francis, London.

Lozano, S.J., S.L. O'Halloran, K.W. Sargent, and J.C. Brazner. 1992. Effects of esfenvalerate on aquatic organisms in littoral enclosures. *Environ. Toxicol. Chem.* 11:35–47.

McDowell, L.L., G.H. Willis, C.E. Murphree, L.M. Southwick, and S. Smith. 1981. Toxaphene and sediment yields in runoff from a Mississippi Delta watershed. *J. Environ. Qual.* 10:120–125.

NACA. 1986. *National Agricultural Chemicals Association, Aquatic Risk Assessment Workshop*. Washington, D.C., October 1986.

Rawn, G.P., G.R.B. Webster, and D.C.G. Muir. 1982. Fate of permethrin in model outdoor ponds. *J. Environ. Health* B17:463–486.

Riley, C.M., C.J. Wiesner, and W.R. Ernst. 1989. Off-target deposition and drift of aerially applied agricultural sprays. *Pestic. Sci.* 26:159–166.

SDTF. 1992. Task force seeks solid data on drift. *Agric. Aviation,* January 1992.

Shaw, S., M. Moore, J.H. Kennedy, and I.R. Hill. 1993. Design and statistical analysis of field aquatic mesocosm studies, this volume, Ch. 9.

Sheets, T.J., J.R. Bradley, and M.D. Jackson. 1972. *Contamination of Surface and Ground Water with Pesticides Applied to Cotton*. Water Resources Research Institute of the University of North Carolina, Report no. 60.

Smith, K., N.K. Kaushik, and K.R. Solomon. 1981. A comparison of the effects of three pesticides in a lake ecosystem using large volume (125 m³) *in situ* enclosures. *8th Annual Aquatic Toxicology Workshop*. Guelph, Canada.

Smith, S., T.E. Reagan, J.L. Flynn, and G.H. Willis. 1983, Azinphosmethyl and fenvalerate runoff loss from a sugarcane-insect IPM system. *J. Environ. Qual.* 12:534–537.

Solomon, K.R. and K. Liber. 1988. Fate of pesticides in aquatic mesocosm studies — an overview of methodology. *Proceedings, Brighton Crop Protn. Conf. — Pests and Diseases*. Brighton, England, 1988, pp. 139–147.

Solomon, K.R., K. Smith, G. Guest, J.Y. Yoo, and N.K. Kaushik. 1980. Use of limnocorrals in studying the effects of pesticides in the aquatic ecosystem. *Can. Tech. Rep. Fish Aquat. Sci.* 975:1–9.

Spencer, W.F., M.M. Cliath, J.W. Blair, and R.A. LeMert. 1985. *Transport of Pesticides from Irrigated Fields in Surface Runoff and Tile Drain Waters*. USDA-ARS, Conservation Research Report 31, 71 pp.

Taylor, J.E., C.W. Childs, L.A. Armijo, and J.J. Snyder. 1984. Flying techniques. In *Professional Standards for Aerial Application of Pesticides in California, Volume I*. California Agricultural Aircraft Association Inc., Sacramento, California. pp. 28–29.

Touart, L.W. 1988. *Aquatic Mesocosm Tests to Support Pesticide Registrations*. Hazard Evaluation Division Technical Guidance Document. Report No. EPA/540/09-88-035. U.S. Environmental Protection Agency, Washington, D.C.

Wauchope, R.D. 1988. The pesticide content of surface waters draining from agricultural fields — a review. *J. Environ. Qual.* 7:459–472.

Wauchope, R.D., K.E. Savage, and D.G. DeCoursey. 1977. Measurement of herbicides in run-off from agricultural areas, pp. 49–58. In *Research Methods in Weed Science*, 2nd ed. Southern Weed Sci. Soc.

Ware, G.W. 1983. *Reducing Pesticide Losses*. Manual, Cooperative Extension Service, Coll. Agric., Univ. Arizona.

Webber, E.C., D.R. Bayne, J.J. Dulka, and D.A. Wustner. 1989. Ecosystem-level of testing of esfenvalerate (Asana) in replicated aquatic mesocosms in east-central Alabama. *Symposium, Society of Environmental Toxicology and Chemistry.* Toronto, Canada, 1989, Abstract No. 418.

Williams, J.R. 1975. Sediment-yield prediction with Universal Equation using runoff energy factor. In *Present and Prospective Technology for Predicting Sediment Yields and Sources.* ARS-S-40. USDA-ARS, pp. 244–252.

Willis, G.H., L.L. McDowell, J.F. Parr, and C.E. Murphree. 1976. Pesticide concentrations and yields in runoff and sediment from a Mississippi Delta watershed. *Proceedings of the 3rd Federal Inter-agency Sedimentation Conference,* 1976. Sedimentation Committee, Water Resources Council, PB-245 100 3:53–64.

Willis, G.H., L.L. McDowell, C.E. Murphree, L.M. Southwick, and S. Smith. 1983. Pesticide concentrations and yields in runoff from silty soils in the lower Mississippi Valley. *J. Agric. Food Chem.* 31:1171–1177.

CHAPTER 16

Use of Mesocosm Data to Predict Effects in Aquatic Ecosystems: Limits to Interpretation

Thomas W. La Point and James F. Fairchild

Abstract: Aquatic mesocosm studies are being used to refute a presumption of risk derived from laboratory toxicity tests conducted under the Federal Insecticide, Fungicide, and Rodenticide Act (FIFRA). Mesocosm studies incorporate many biological, chemical, and physical characteristics of natural ecosystems. Hence, they serve as realistic surrogates of natural ecosystems and allow tests of pesticide effect at the population, community, and ecosystem level. We discuss two factors, ecosystem trophic status and organism life history, which influence the results derived from aquatic mesocosm studies. Trophic status influences the fate and effects of chemicals which strongly sorb or biologically degrade, yet may not be as important in the fate and effects of more water soluble chemicals. Life history traits of organisms and the intensity, frequency, and duration of the pesticide disturbance also determine the mesocosm response pattern.

INTRODUCTION

Experimental ecosystems (mesocosms) are presently being used to register pesticides for agricultural use under the Federal Insecticide, Fungicide, and Rodenticide Act (FIFRA). One purpose of the pond mesocosm test is to negate a presumption of hazard that has been shown in the lower-tiered testing.[1] Touart and Slimak[2] consider pond mesocosms to be valuable for determining the ecotoxicity of pesticides, as mesocosms simulate ponds, lakes, and riverine backwater habitats. However, successful extrapolation of response data collected from pond mesocosms to surface waters in agricultural regions will depend, at least in part, on physical and biological characteristics of the mesocosms. Trophic status, including nutrient concentration, suspended particulates, and dissolved and particulate organic matter has been shown to affect the bioavailability of chemicals in surface waters.[3,4] Morphoedaphic components of the environment (depth, slope, current, physical-chemical nature of the substrate) exert a strong influence on the nature of the biota in aquatic environments.[5-7] Also, response of the biota to a contaminant depends upon life history traits of individual species (body size, reproductive potential, and habitat and food requirements).[8-10]

Ongoing mesocosm studies across the U.S. are providing highly useful data and, indirectly, a "feedback loop" of information on what biological endpoints can be used to regulate pesticide registration. These mesocosm tests are producing fundamental information on how ecosystems are structured and on how they function. This information may be used to refine our understanding of how pesticide residues in the natural environment directly and indirectly affect nontarget aquatic species. In turn, the information on community responses may suggest a refinement in pesticide application methods or use patterns and ultimately optimize testing requirements for pesticide registration.

Ultimately, the usefulness of experimental ecosystems is dependent upon using them to refine predictions generated from laboratory studies on chemical fate and degradation and on the direct and indirect toxic effects to aquatic organisms.[11-13] Results from mesocosm testing can reveal subtle ecological information such as the numerous indirect responses to chemical perturbation, thereby integrating the multispecies interactions of predation, competition, symbioses, etc.[14] Further, their value extends to providing the capability of viewing an array of species with different life cycles responding with different sensitivities and recovery patterns.[15] Finally, when pond mesocosms are established in a concentration-response series, the sum total of indirect effects can help determine if the aquatic community responds in a threshold manner (cf. Reference 16) or if the response is log normally distributed across the range of pesticide concentrations.

In any event, such field studies provide descriptive information which may be used in risk-benefit analyses.[2] In field situations, direct and indirect effects must be observed in view of chemical, physical, and biotic factors which modify the effects predicted from laboratory testing. These include water quality, adsorption onto particulates and sediments, photolysis, hydrolysis, the fluctuating exposure regimen, and the nature of the aquatic ecosystem (number and kinds of species present, trophic status, etc.).

Ultimately, these factors may be important in establishing statistical and biological confidence limits for results derived from individual mesocosm tests. The purpose of this chapter is to review two decisive environmental factors which exert an influence on the bioavailability and effects of agricultural chemicals to nontarget aquatic organisms: trophic status and life history strategies.

METHODS AND MATERIALS

This chapter draws on the results of two studies performed at the National Fisheries Contaminant Research Center (NFCRC), a U.S. Fish & Wildlife Service facility located in Columbia, MO. During 1987 and 1988, two approaches to pesticide testing in the NFCRC experimental ponds were developed. A full description of the methods used in dosing, chemical, nutrient and water quality sampling, and biological sampling can be found in Fairchild et al.[17] Each year, 12 mesocosms were dosed with the synthetic pyrethroid, esfenvalerate, sold under the tradename Asana® by E.I. DuPont de Nemours Corporation. The mesocosms are clay lined and are fully drainable by virtue of a concrete drain structure. The surface area averages 600 m², depth varies from 0 to 2 m, and average mesocosm volume is 700 m³ (range 620 to 750 m³). Total organic carbon of the sediments averages 3.2%. Water quality of the mesocosms is typical of hard-water systems of the midwest.

Our 1987 study incorporated the simultaneous application of an insecticide and an herbicide. The concentrations used were chosen to represent a range covering "worst case" scenarios (such as direct overspraying); the concentrations chosen were meant to ensure that we measured a response to the pesticide in at least some of the mesocosms. Atrazine concentrations in the environment range from 0 to 2300 µg/l in runoff,[18,19] 0 to 88 µg/l in groundwater,[20] and from 0 to 70 µg/l in aquatic ecosystems.[21] We chose 50 µg/l as the nominal treatment concentration based on residues commonly found in runoff adjacent to intensively farmed land[18] and from published data indicating that 50 µg/l is the approximate threshold level for effects on algae and aquatic macrophytes in aquatic ecosystems.[22,23] Atrazine (2-chloro-4-ethylamino-6 isopropylamino-1,3,5 triazine) is a selective triazine herbicide. We empirically determined its half-life in the NFCRC mesocosms to be 41 d. In other systems, atrazine half-life has been estimated to be 2 months.[24,25]

The synthetic pyrethroid insecticide esfenvalerate (S)-cyano(3-phenoxyphenyl)methyl-(S)-4 chloro-alpha-(1-methylethyl) benzene acetate, 84% active technical formulation, was donated by the E.I. DuPont de Nemours Corporation for use in our studies. Esfenvalerate is an insecticide with a high toxicity (0.31 and 0.89 ppb for bluegill sunfish [*Lepomis macrochirus*] 96 h and *Daphnia magna* 48 h LC_{50}, respectively) and rapid aqueous dissipation[17] (Figure 1).

During 1987, we dosed two replicate ponds at each of six esfenvalerate concentrations (Table 1); esfenvalerate was dosed twice, six weeks apart. In addition, atrazine was added at one concentration, 50 µg/l, to six of the ponds. The 1988 study focused on the concentration-response to esfenvalerate alone. During this second year, we used four concentrations of three replicates each (Table 1). Esfenvalerate was dosed into the mesocosms 6 times, once every 2 weeks.

During the 1988 study, depth-integrated zooplankton samples were taken weekly using a PVC column sampler (1 m length, 5 cm diameter). In this study, two 6-l composite samples (each consisting of six 1-l subsamples) were taken each date. Zooplankton from each composite were concentrated by sieving through an 80-µm

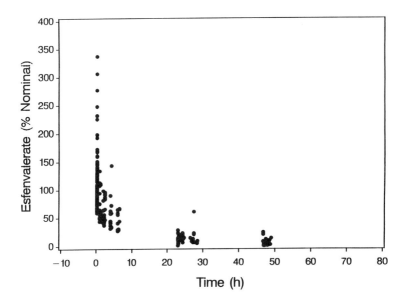

Figure 1.   Dissipation rate of esfenvalerate from the water column during 1988. Plotted are relative percentage of nominal concentrations from all dates and treatments. (From Fairchild et al., *Environ. Toxicol. Chem.* 11:115–129, 1992.)

Wisconsin plankton net, preserved in 1% (v/v) glycerin:ethanol, and counted under a microscope. All individuals were identified to phylum, order, or suborder. During 1988, zooplankton were sampled weekly using a battery-powered, metered pump. At each time, two replicate 114 l, depth-integrated composites (38 l from each of 0, 0.5, and 1.0 m depths) were taken, sieved through an 80-μm mesh Wisconsin plankton net, and preserved in 1% (v/v) glycerin:ethanol. Zooplankton data from each year are reported as numbers per unit volume.

Table 1. Dosing Intervals and Target Concentrations of Esfenvalerate and Atrazine Used in NFCRC Mesocosms During 1987 and 1988

		Experimental Design Studies	
1987	**Series**	**Esfenvalerate, μg/l**	**Atrazine, μg/l**
May 26	1	0, 0.25, 0.40, 0.67, 1.12, 1.71	0
	2	0, 0.25, 0.40, 0.67, 1.12, 1.71	50
July 8	1	0, 0.25, 0.40, 0.67, 1.12, 1.71	0
	2	0, 0.25, 0.40, 0.67, 1.12, 1.71	0
1988		**Esfenvalerate, μg/l**	
June 6		0, 0.25, 0.67, 1.71	
June 20		0, 0.25, 0.67, 1.71	
July 5		0, 0.25, 0.67, 1.71	
July 18		0, 0.25, 0.67, 1.71	
August 2		0, 0.25, 0.67, 1.71	
August 15		0, 0.25, 0.67, 1.71	

We calculated gross photosynthesis, community respiration, and net primary production from measures of dissolved oxygen taken at biweekly intervals in each mesocosm. Calculations for each mesocosm were based on measured dissolved oxygen and temperature at three depths (0, 0.5, and 1.0 m) and the diurnal oxygen method of McConnell.[26] Macrophyte shoot biomass and species composition were sampled monthly using a stove-pipe sampler.[27] Four samples, each encompassing 0.1 m^2 of the substrate, were taken within each mesocosm. Macrophytes were washed to remove sediments and then identified to genus. Subsequently, macrophytes were dried at 105°C for 24 h and weighed to estimate total biomass.

For the 1987 study, young-of-the-year (YOY) bluegill (mean length 4.18 cm, mean weight 1.39 g) were stocked into the mesocosms at a loading rate of 0.275 g/m^3. Upon test termination, stocked bluegill were removed, counted, and individual lengths and biomass measured.

In 1988 adult bluegill (*Lepomis macrochirus*) (mean length 11.7 cm, mean weight 27 g) were stocked into the mesocosms at a loading of 2 g/m^3, as recommended in the mesocosm protocol.[1] Upon study termination (early October, 5-month study duration), all adult bluegill were removed, counted, sexed, and individually measured for length (± 0.1 cm) and weight (± 0.1 g). Bluegill recruits were counted using hand-held counters and a subsample of 100 recruits from each mesocosm was used to measure larval length (± 0.1 cm) and weight (± 0.01 g).

Data analyses consisted of nonparametric analyses of variance of treatment means or linear regression of the biotic response vs. measured chemical concentrations. Significant differences among individual treatment means were determined using Fisher's Least Significant Differences (LSD) test.[28] All statistical analyses were performed using the Statistical Analysis System (SAS) on micro- or mainframe computing systems.

RESULTS AND DISCUSSION

Trophic Status and Habitat Structure

McCarthy and Bartell[4] provide an important modeling study on how ecosystem trophic status modifies the bioavailability of toxicants and decreases the subsequent dose to biota. Their model predicts that the association of a contaminant with dissolved organic material (DOM) or particulate organic material (POM) significantly lessens the bioavailability of a toxicant and, thus, the potential dose experienced by the organisms (Figure 2). Accrual of nutrients or allochthonous organic matter stimulates productivity and provides a ready source of particulate and dissolved sorbents in eutrophic systems. The dosage of the contaminant to which pelagic biota are exposed may be significantly reduced in eutrophic systems because of the interaction of the chemicals with sorbents. However, the degree of reduction in dose depends on certain biological and chemical characteristics. For example, it has been shown that the pelagic fauna mobilize chemicals by benthic feeding,[29]

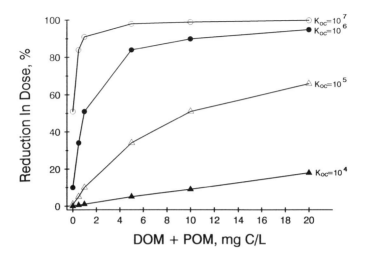

Figure 2.   Reductions in dose across fish gills as a function of dissolved and particulate organic material. The reduction in dose is greater for contaminants with higher hydrophobicity. From McCarthy, J. and Bartell, S., *Functional Testing of Aquatic Biota for Estimating Hazards of Chemicals,* ASTM STP 988, Philadelphia, PA, 3–16.

via food chain,[30] and via bioturbation of sediments, which can alter binding efficiencies.[31]

McCarthy and Bartell[4] conducted laboratory tests to confirm their model results. Their empirical results show that for periphyton, macrophytes, and detritivorous fish, the decrease in toxic effects in relation to increased POM was relatively minor, 10% or less. However, for phytoplankton, zooplankton, benthic invertebrates, and omnivorous fish, a greater reduction in toxic effects (15 to 30%) corresponded to increasing DOM in the system. Bacterial and zooplanktonic growth was dramatically enhanced as a function of available organic material and, presumably, a greater surface area for microbial activity.

Change in habitat structure would be an obvious consideration in successful extrapolation from mesocosms to natural aquatic environments. In our studies, there was an acute response of zooplankton to the addition of esfenvalerate in both herbicide treatments (with and without atrazine) (Figure 3). The "sensitivity" of the zooplankton community did not differ as a function of substitution of plant species stemming from atrazine. Functional redundancy of the macrophyte community precluded our measuring additive or synergistic effects of the atrazine and esfenvalerate combination on zooplankton or fish.

During 1987, *Naja* biomass increased in control mesocosms during the growing season. However, *Naja* declined in mesocosms treated with atrazine; the macrophyte community switched from one dominated by *Naja* to *Chara*. *Naja* continued to dominate in mesocosms lacking atrazine (Figure 4). However, neither total macrophyte biomass nor gross system productivity changed in response to the herbicide.

High primary production rates were reported to reduce the bioavailability of contaminants in an earlier field study conducted at this laboratory.[32] In that study, nitrogen and phosphorus were added into the experimental ponds to establish three levels of primary production: low nutrient phytoplankton (low primary production), high nutrient phytoplankton (medium primary production), and macrophyte dominated (high primary production). Pentachlorophenol (PCP) was added into these ponds in a single heavy dose (1 mg/l) and at multiple treatments of 0.2 and 0.4 mg/l applied at monthly intervals.

Higher levels of primary production corresponded to lower PCP residues and toxicity to fish. Yet, the decrease in PCP toxicity cannot be ascribed solely to an increase in primary production with a concomitant increase in dissolved and particulate organic matter. The results of Robinson-Wilson et al.[32] (Figure 5) indicate lower PCP concentrations in the high primary production ponds and that PCP concentrations were similar among the low and medium production ponds. Robinson-Wilson et al.[32] inferred that the larger surface area available within the macrophyte-dominated ponds contributed to an active epiphytic community which degraded the PCP. Fish survival was higher in the medium and high productivity mesocosms; however, this also presented an interaction between structure and function: fish survivorship was certainly a function of the decrease in PCP bioavailability, but may also have been a function of the better cover provided by macrophyte fronds.

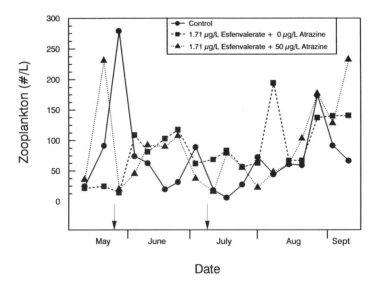

Figure 3.   Zooplankton density within the 1.71 μg/l esfenvalerate (high treatment) ponds with and without atrazine. The arrows indicate dates the mesocosms were dosed with esfenvalerate.

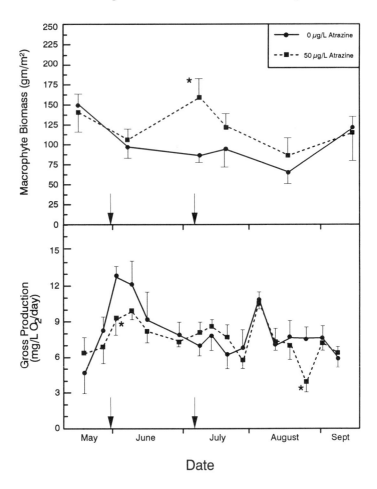

Figure 4.   Change in biomass of macrophytes in NFCRC mesocosms and gross primary production during 1987. The arrows indicate dates the mesocosms were dosed with esfenvalerate.

Life History Characteristics

The ability to extrapolate from mesocosm test results to natural ecosystems also depends upon life history characteristics of the resident organisms. During both studies of esfenvalerate, zooplankton abundance was a sensitive endpoint. Zooplankton abundance decreased at concentrations as low as 0.25 µg/l; however, rapid recovery rates of plankton populations led to small overall changes in abundance during 1987 (Figure 3). There are obvious differences in the sensitivity of different orders of zooplankton to esfenvalerate. Copepods appear to be very sensitive and the rotifers the least. Cladocera were declining rapidly in control ponds during June and July; the esfenvalerate accentuated the decline in treated ponds. Different rates of decline and growth in zooplankton as a result of permethrin exposure were observed by Kaushik et al.[33] They note zooplankton numerically respond to an insecticide in two ways, a direct reduction of Cladocera due to acute toxicity and an indirect numerical increase of Rotifers attributed to release from predatory or competitive pressures.

There is evidence that the recovery of zooplankton populations in mesocosms is a function of their relatively short generation times, the number of times they are exposed, and the persistence of the pesticide in the water column. In a study of carbaryl applied to experimental ponds, there was a minimal effect on cladocerans in ponds dosed one time at 0.1 mg/l carbaryl.[34] The time to 90% dissipation for carbaryl in the water column was 18 days. In ponds dosed twice, the cladocerans were able to recover within 10 to 15 days. However, in ponds dosed 10 times over a period of 20 days, each time at the 0.1 mg/l concentration, the cladocerans failed to recover.

In addition to the number of applications, the duration of exposure in mesocosms influences the biotic responses. For example, the persistence of atrazine in the water column may explain why Hamilton et al.[23] reported the striking similarity in response of the phytoplankton communities to atrazine between their study (in which atrazine was dosed twice over 35 days) and those of deNoyelles et al.,[35] in which atrazine was dosed once. The phytoplankton in their mesocosms may have been responding to essentially equivalent atrazine concentrations in both systems.

Differences in esfenvalerate dosing regimes between our 1987 and 1988 studies allow a comparison of the response of planktonic invertebrates to an application of

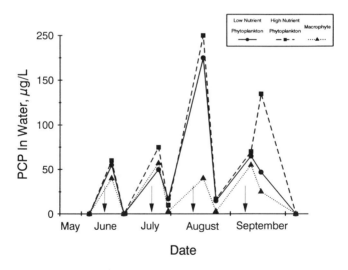

Figure 5. The decrease in pentachlorophenol (PCP) concentration in NFCRC mesocosms as a function of trophic status. The lowest PCP residues were measured in those mesocosms with the highest productivity. The arrows indicate dates the mesocosms were dosed with PCP. From Robinson-Wilson, E. et al., *Aquatic Toxicology and Hazard Assessment: Sixth Symposium,* ASTM STP 802, 239–251.

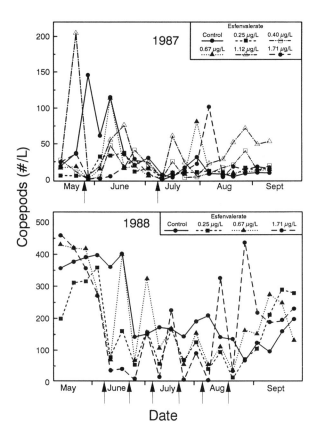

Figure 6.   Response of copepods in NFCRC mesocosms during 1987 and 1988.

a short-lived pesticide. Adding the pyrethroid twice to the mesocosms, 6 weeks apart, did not seriously perturb the ecosystems. Because esfenvalerate parent compound half-life in the water is 12 h, the biota would experience the applications as 2 independent acute events. The greater number of applications during 1988 led to greater differences among the copepods in the mesocosms (Figure 6). However, the ''sawtooth'' responses indicate the population growth rates were sufficient, under these conditions, to allow groups of zooplankton to recovery between dosing events.

The response of bluegill populations in our study supports the working hypothesis that large organisms, with longer life spans, integrate direct and indirect pesticide effects. The YOY produced per adult each year (Figure 7) was affected by esfenvalerate concentration. Whether by direct toxicity or by reduction in available food items, the YOY produced per female adult bluegill was reduced in the higher treatment ponds. During both years, there was a negative linear relationship between esfenvalerate concentration and the number of bluegill young produced (1987: r =

-0.45, $\alpha \leq 0.04$; 1988: $r = -0.75$, $\alpha \leq 0.005$). While ANOVA results show significant differences in YOY produced in 1988, larger within-pond variances among YOY produced in 1987 (and the fact there were only two replicate mesocosms per concentration) led to no significant differences among concentrations.

SUMMARY AND CONCLUSIONS

Pesticide residues in water bodies may be viewed as a chemical disturbance or perturbation to the ecosystem. As such, the direct effect of the chemical on the organisms needs to be studied relative to the potential for recovery and recolonization of the species. Southwood[36] states that an important aspect of the results of disturbance on communities is the rate at which the habitat becomes less suitable in relation to the requirements of the organisms. In effect, the number of dosing intervals and the concentration of the pesticide become points on the "disturbance"

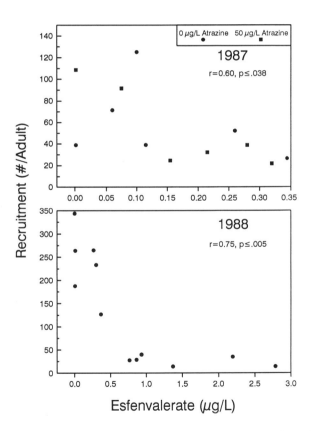

Figure 7.   Number of young-of-the-year (YOY) bluegill produced per adult in NFCRC mesocosms during 1987 and 1988.

and "adversity" axes described by Southwood.[35] Over sufficient time, the frequency, magnitude, and duration of disturbance have a strong influence on the biotic structure of communities. Further, trophic status influences the bioavailability of a pesticide; increased productivity can lead to greater quantities of dissolved and particulate material which bind to contaminants and effectively reduce the dose experienced by aquatic organisms in the water column. However, ecosystems with high productivity tend to be functionally redundant and effects on species may not be evidenced by changes in rates of primary production, respiration, or decomposition.

Life history characteristics, particularly body size and generation time, determine the rate of recovery of a population once it has been exposed to a pesticide. However, the long-term results of pesticide exposure on aquatic organisms also depends upon the availability of refugia, food-web relationships (including controls imposed on population size by its predators), and behavioral responses to avoid exposure, such as drift.[37] Exposure duration will be a function of the chemical nature of the pesticide and how often it gets into the ecosystem. Hence, the magnitude of effect (initial response and subsequent recovery) cannot be isolated from the persistence of the chemical in the water relative to the mean time between generations of exposed aquatic organisms. Extrapolating effects from mesocosms to plankton populations in ponds and lakes may be fairly accurate, particularly if the binding potential of the pesticide to organic carbon is known. However, for longer-lived aquatic populations, it will be difficult to extrapolate chronic effects of the pesticide, as measured in the laboratory, to natural systems.

If ecosystem-level tests are to successfully protect aquatic environments (from ponds and lakes to streams, rivers, and estuaries) the focus of mesocosm testing will need to be on how aquatic environments modify the bioavailability of the pesticide. Measured biotic responses in these systems need to match the appropriate scale of perturbation with life history characteristics of the species. Extrapolation of mesocosm data to natural ecosystems will require conservative estimates of hazard due to the limits on predictability in natural ecosystems. The effect of the pesticide on an ecosystem cannot be isolated from the life history characteristics of the resident organisms.

It is apparent from the results of our studies and many published (cf. References 4, 22, 34) that the measured biotic effects are a function of the duration and frequency of exposure, the size of the organisms exposed, and the trophic status and physical structure of the environment. The present knowledge base is adequate to predict direct toxicity of a pesticide to planktonic and benthic organisms. For instance, the direct toxicity of esfenvalerate to zooplankton and bluegill in the NFCRC ponds was demonstrated to be similar to that measured in pulsed exposures in laboratory tests (E. Little, personal communication). However, the indirect effects of pesticide contamination are much more difficult to predict and extrapolate. Unfortunately, we remain limited in our understanding of the quantitative nature of species interactions; our predictions of secondary effects are limited to qualitative changes in structure and function.

REFERENCES

1. Touart, L.W. 1988. Aquatic mesocosm tests to support pesticide registrations. EPA 540/09-28-0, U.S. Environmental Protection Agency, Washington, D.C.
2. Touart, L.W. and M.W. Slimak. 1989. Mesocosm approach for assessing the ecological risks of pesticides. In J.R. Voshell (ed.), *Using Mesocosms to Assess the Aquatic Ecological Risk of Pesticides: Theory and Practice,* Misc. Publ. Entomological Society of America, Nbr. 75. pp. 33–40.
3. Kaiser, K.L.E. 1980. Correlation and prediction of metal toxicity to aquatic biota. *Canadian Journal of Fisheries and Aquatic Sciences* 37:211–218.
4. McCarthy, J.F. and Bartell, S.M. 1988. How the trophic status of a community can alter the bioavailability and toxic effects of contaminants. In J. Cairns, Jr. and J.R. Pratt (eds.) *Functional Testing of Aquatic Biota for Estimating Hazards of Chemicals.* ASTM STP 988. American Society of Testing and Materials, Philadelphia, PA. pp. 3–16.
5. Minshall, G.W. 1984. Aquatic insect-substratum relationships. In V.H. Resh and D.M. Rosenberg (eds.), *The Ecology of Aquatic Insects.* Praeger Publishers, New York, NY. pp. 358–400.
6. Newbury, R.W. 1984. Hydrologic determinants of aquatic insect habitats. In V.H. Resh and D.M. Rosenberg (eds.), *The Ecology of Aquatic Insects.* Praeger Publishers, New York, NY. pp. 323–357.
7. Bloesch, J., P. Bossard, H. Buehrer, H.R. Buergi, and U. Uehlinger. 1988. Can results from limmocorral experiments be transferred to in situ conditions? *Hydrobiologia* 159:297–308.
8. Diner, M.P., E.P. Odum, and P.F. Hendrix. 1986. Comparison of the roles of Ostracods and Cladocerans in regulating community structure and metabolism in freshwater microcosms. *Hydrobiologia* 133:59–63.
9. Barnthouse, L.W., G.W. Suter II, A.E. Rosen, and J.J. Beauchamp. 1987. Estimating responses of fish populations to toxic contaminants. *Environmental Toxicology and Chemistry* 6:811–824.
10. Barnthouse, L.W., G.W. Suter II, and A.E. Rosen. 1990. Risks of toxic contaminants to exploited fish populations: Influence of life history, data uncertainty and exploitation intensity. *Environmental Toxicology and Chemistry* 9:297–311.
11. Kimball, K.D. and S.A. Levin. 1985. Limitations of laboratory bioassays: the need for ecosystem-level testing. *BioScience* 35(3):165–171.
12. Cairns, J., Jr. 1986. Overview. In: Cairns, J., Jr. (ed.) *Community Toxicity Testing.* ASTM STP 920. American Society for Testing and Materials, Philadelphia, PA. pp. 1–5.
13. La Point, T.W., J.F. Fairchild, E.E. Little, and S.E. Finger. 1989. Laboratory and field techniques in ecotoxicological research: Strengths and limitations. In A. Boudou and F. Ribeyre (eds.) *Aquatic Ecotoxicology: Fundamental Concepts and Methodologies,* Vol. II. CRC Press, Boca Raton, Florida. pp. 239–255.
14. Kelly, J.R. and M.A. Harwell. 1989. Indicators of ecosystem response and recovery. In: S.A. Levin, M.A. Harwell, J.R. Kelly, and K.D. Kimball (eds.) *Ecotoxicology: Problems and Approaches.* Springer-Verlag, New York. pp. 9–35.
15. Dewey, S.L. 1986. Effects of the herbicide atrazine on aquatic insect community structure and emergence. *Ecology* 67:148–162.

16. Graney, R.L., J.P. Giesy, Jr., and D. DiToro. 1989. Mesocosm experimental design strategies: Advantages and disadvantages in ecological risk assessment. In J.R. Voshell, Jr. (ed.) *Using Mesocosms to Assess the Aquatic Ecological Risk of Pesticides: Theory and Practice.* Misc. Publications of the Entomological Society of America 75:74–88.

17. Fairchild, J.F., T.W. La Point, J.L. Zajicek, M.N. Nelson, and F.J. Dwyer. 1992. Population, community, and ecosystem-level responses of aquatic mesocosms to pulsed doses of a pyrethroid insecticide. *Environmental Toxicology and Chemistry* 11:115–129.

18. Kadoum, A.M. and D.E. Mock. 1978. Herbicide and insecticide residues in tailwater pits: water and pit bottom soil from irrigated corn and sorghum fields. *Journal of Agricultural and Food Chemistry* 26:45–50.

19. Hall, J.K. 1974. Erosional losses of s-triazine herbicides. *Journal of Environmental Quality* 3:174–180.

20. Junk, G.A., R.F. Spalding, and J.J. Richards. 1980. Areal, vertical, and temporal differences in ground water chemistry. II. Organic constituents. *Journal of Environmental Quality* 9:479–482.

21. Richard, J.J., G.A. Junk, M.J. Avery, N.L. Nehring, J.S. Fritz, and H.J. Svec. 1975. Analysis of various Iowa waters for selected pesticides: Atrazine, DDE, and dieldrin — 1974. *Pesticide Monitoring Journal* 9:117–123.

22. Larsen, D.P., F. deNoyelles, F. Stay, and T. Shiroyama. 1986. Comparisons of single-species, microcosm, and experimental pond responses to atrazine exposure. *Environ. Toxicol. Chem.* 5:179–190.

23. Hamilton, P.B., G.S. Jackson, N.K. Kaushik, K.R. Solomon, and G.L. Stephenson. 1988. The impact of two applications of atrazine on the plankton communities of in situ enclosures. *Aquatic Toxicology* 13:123–140.

24. Jones, T.W., W.M. Kemp, J.C. Stevenson, and J.C. Means. 1982. Degradation of atrazine in estuarine water/sediment systems and soils. *Journal of Environmental Quality* 11:632–638.

25. deNoyelles, F., Jr., W.D. Kettle, C.H. Fromm, M.F. Moffett, and S.L. Dewey. 1989. Use of experimental ponds to assess the effects of a pesticide on the aquatic environment. In J.R. Voshell, Jr. (ed.) *Using Mesocosms to Assess the Aquatic Ecological Risk of Pesticides: Theory and Practice.* Misc. Publications of the Entomological Society of America 75:41–56.

26. McConnell, W.J. 1962. Productivity relation in carboy microcosms. *Limnol. Ocean.* 335–343.

27. Hynes, H.B.N. 1971. Benthos of Flowing Water. In J.A. Downing and F.H. Rigler (eds.). *A Manual on Methods for the Assessment of Secondary Productivity in Fresh Waters,* IBP Handbook 17. Blackwell Scientific Publications, Oxford, England, pp. 131–160.

28. Snedecor, G.W. and W.G. Cochran. 1967. *Statistical Methods,* 6th Ed. Iowa State University Press, Ames, Iowa. pp. 593.

29. Rubenstein, N., W.T. Gilliam, and N.R. Gregory. 1984. Dietary accumulation of PCBs from a contaminated sediment source by a demersal fish (*Leiostomus xanthurus*). *Aquatic Toxicology* 5:331–342.

30. Clark, K.E. and D. Mackay. 1991. Dietary uptake and biomagnification of four chlorinated hydrocarbons by guppies. *Environ. Toxicol. Chem.* 10:1205–1217.

31. Connell, D.W. 1990. Environmental routes leading to the bioaccumulation of lipophilic chemicals. Chapter 4 *in* D.W. Connell (ed.) *Bioaccumulation of Xenobiotic Compounds*. CRC Press, Inc., Boca Raton, Florida. 59–73.

32. Robinson-Wilson, E.F., T.P. Boyle, and J.D. Petty. 1983. Effects of increasing levels of primary production on pentachlorophenol residues in experimental pond ecosystems. In W.E. Bishop, R.D. Cardwell, and B.B. Heidolph (eds.) *Aquatic Toxicology and Hazard Assessment: Sixth Symposium*. ASTM STP 802. pp. 239–251.

33. Kaushik, N.K., G.L. Stephenson, K.R. Solomon, and K.E. Day. 1985. Impact of permethrin on zooplankton communities in limnocorrals. *Can. J. Fish. Aquat. Sci.* 42:77–85.

34. Hanazato, T. and M. Yasuno. 1990. Influence of persistence period of an insecticide on recovery patterns of a zooplankton community in experimental ponds. *Environmental Pollution* 67:109–122.

35. deNoyelles, F., Jr., W.D. Kettle, and D.E. Slinn. 1982. The response of planktonic communities in experimental ponds to atrazine, the most heavily used pesticide in the United States. *Ecology* 63:1285–1293.

36. Southwood, T.R.E. 1988. Tactics, strategies and templets. *Oikos* 52:3–18.

37. Cuffney, T.F., J.B. Wallace, and J.R. Webster. 1984. Pesticide manipulation of a headwater stream: Invertebrate responses and their significance for ecosystem-level processes. *Freshwater Invertebrate Biology* 3:153–171.

CHAPTER 17

Impact of 2,3,4,6-Tetrachlorophenol (DIATOX®) on Plankton Communities in Limnocorrals

Karsten Liber, Keith R. Solomon, Narinder K. Kaushik, and John H. Carey

Abstract: Two limnocorral (*in situ* enclosure) experiments were conducted to assess the effects of a commercial 2,3,4,6-tetrachlorophenol formulation (DIATOX®) on natural zooplankton communities. In Experiment 1, no significant reductions were noted in mean abundance of any major zooplankton taxa following a morning application with 0.75 mg a.i./l DIATOX®. Variability among zooplankton populations in replicate enclosures, especially controls, was high and was attributed to variable predation by uncontrolled populations of planktivorous fish (juvenile *Perca flavescens*). Lack of significant impact on zooplankton populations was attributed to rapid photodegradation of the surface applied chlorophenols (CPs), resulting in sublethal water column concentrations within 24 h of application. High variability among replicate enclosures made assessment of subtle impacts difficult.

In Experiment 2, the effects of 0.75 and 1.50 mg a.i./l DIATOX® were assessed following evening applications. Replication among zooplankton populations in control enclosures was significantly improved by removal of fish. The evening pesticide application delayed photodegradation and, at the 0.75 mg/l treatment, resulted in 1.6 times higher CP concentrations in the water 24 h after application than resulted from the same treatment in Experiment 1. Both Experiment 2 DIATOX® treatments resulted in significant reductions in all major zooplankton taxa. Rotifers were 1.5 times more sensitive than cladocerans and copepods, which responded similarly to the two treatments. All zooplankton populations had generally recovered by Days 14 to 28 and by Days 28 to 42 posttreatment at the low and high treatment concentration, respectively.

0-87371-592-6/94/$0.00 + $.50
© 1994 by CRC Press, Inc.

Direct effects on phytoplankton populations were minimal. There were no significant reductions in the abundances of major phyla (Cyanophyta, Chlorophyta, Chrysophyta), but 14 to 21% and 16 to 26% decreases were observed in dissolved oxygen concentrations at the low and high treatment, respectively. A 36-fold increase in total phytoplankton abundance was observed on Days 21 to 28 at the 1.50 mg/l DIATOX® treatment, largely accounted for by two filamentous Cyanophyta species (*Oscillatoria amphigranulata* and *O. lauterbornii*). This secondary effect most likely resulted from the reduction in total zooplankton abundance which reached a maximum of 93% on Day 7. Phytoplankton populations decreased to control levels by Day 42 as zooplankton populations recovered. Analysis of water chemistry data contributed no additional information on treatment effects.

INTRODUCTION

Single-species laboratory toxicity tests are a relatively simple way of evaluating the response of organisms to given levels of a toxicant. Their value is enhanced by the precision and repeatability of test results, and the use of standard methods for these tests provide a measure of relative toxicity among groups of organisms for a particular chemical, as well as a measure of relative toxicity among different chemicals. There are questions, however, about the accuracy with which results from these simple laboratory tests can predict the response(s) to similar chemical concentrations in complex, natural, aquatic systems. Most biologists agree that at each succeeding level of biological organization, new properties appear that would not have been evident even by the most intense and careful examination of lower levels of biological organization.[1] As a result, extrapolation from laboratory results to contaminant-related changes that may occur either directly or indirectly in the structure or function of an ecosystem are difficult to make with a high degree of certainty.[2]

Some of the more obvious problems associated with extrapolation from standardized laboratory toxicity tests result from short exposure times, unrealistic test conditions, limited number of test species, and the inability to assess recovery and indirect effects. Because natural populations are linked by a variety of ecological interactions, such as predation and competition, the response of any particular population to chemical exposure may depend on how the rest of the ecosystem is affected.[3] The drawbacks of standardized laboratory toxicity tests have thus driven the need for field testing at higher levels of biological organization and have led Cairns[4] to suggest that, when investigating contaminant impact on aquatic ecosystems, validation in field enclosures or natural systems should probably be mandatory. Similar reasoning resulted in the development of the limnocorral research program at the University of Guelph in 1979 to 1980. This program utilizes *in situ* enclosures (limnocorrals) to assess the aquatic fate and effects of pesticides, and it has shown limnocorrals to be both effective and reliable for such purposes.[5-10]

The pesticide investigated in this project was DIATOX®, a commercial 2,3,4,6-tetrachlorophenol (TeCP) formulation. In Canada prior to 1991, DIATOX® was used mainly by the lumber industry for short-term protection of freshly sawn wood from discoloration and deterioration by sapstain and mold fungi. Its use as an antisapstain product has now been discontinued in Canada. TeCP had been used for decades and was a common contaminant in many Canadian aquatic ecosystems. Concentrations in many Canadian surface waters were routinely in the range of 1 to 1000 ng/l,[11-16] with higher levels recorded near wood treatment facilities, pulp and paper mills, and some industrial discharges. In spite of its common usage, relatively little was known about the fate and toxicity of TeCP in natural environments. Its toxicity to aquatic organisms had been investigated in a few laboratory studies[17-22] and TeCP was generally considered to be slightly less toxic than pentachlorophenol (PCP).[11,17,20] The aquatic fate of TeCP had only been extrapolated from laboratory and field studies with PCP,[12,23-29] which suggested that TeCP should have a relatively short half-life in surface waters and that photolysis is the predominant mode of degradation. To our knowledge, no experimentation had been conducted with TeCP in a natural aquatic ecosystem to validate these assumptions.

Two enclosure experiments were thus undertaken to assess the impact of TeCP on a natural aquatic community representative of Canadian lentic ecosystems. Focus was placed on planktonic communities because of their central importance in most freshwater food webs. Experiment 1 tested the null hypothesis that single applications with 0.075 and 0.75 mg a.i./l commercial TeCP (DIATOX®) would have no effects on zooplankton abundance. Fish were uncontrolled and allowed to remain in all enclosures. Experiment 2 tested the null hypotheses that (1) removal of fish from all enclosures would not affect variability in zooplankton abundance among replicate enclosures, and (2) evening application would not delay degradation of chlorophenols and therefore would not result in different observations after single treatments with 0.75 and 1.50 mg a.i./l DIATOX® than those seen after the morning applications of Experiment 1 (i.e., treatments would have no impact on zooplankton abundance).

MATERIALS AND METHODS

Study Site

Both limnocorral experiments were conducted in Lake St. George, a 10.3-ha meso-eutrophic lake situated approximately 32 km north of Toronto, Ontario, Canada (43°57'30'' N, 79°25'45'' W). Lake St. George is a dimictic, clinograde, kettle lake comprised of two basins.[30] It has a maximum depth of 16 m, is ice-covered from approximately December to April, and the bottom is overlaid with an unconsolidated organic layer of variable depth. Lake St. George had been used for similar enclosure studies in the past and already contained a set of 12 reusable limnocorrals. Enclosures were of two similar types (both 5 × 5 m in area and located at a depth of 4.5 m), and details regarding their initial construction and installation have been described previously.[5,7,31,32] The mean and range of several limnological parameters measured within control enclosures were as follows: pH = 8.1 ± 0.3, total hardness

= 278 $\pm$ 10 mg/l as $CaCO_3$, dissolved oxygen concentration = 10.2 $\pm$ 3.6 mg/l, summer surface water temperature = 24 $\pm$ 4°C, dissolved organic carbon = 7.0 $\pm$ 0.7 mg/l, dissolved inorganic carbon = 34.7 $\pm$ 5.9 mg/l. These levels were not significantly different ($p > 0.05$) from those measured in adjacent, nonenclosed lake stations.

Prior to initiation of the study, the inside of each limnocorral was lined with a new UV-protected, 6-mil clear polyethylene plastic liner. This gave each enclosure a clean and identical inside surface and circumvented problems associated with biota growing on the original PVC walls. A slack of about 30 cm was maintained in the walls and liner to accommodate wave action and seasonal changes in water level. The installation of the liner was aided and checked by scuba diving to ensure that a proper seal in the sediment had been obtained.

Experimental Design

Both experiments were based on an analysis of variance (ANOVA) design and used the same set of enclosures. The row of 12 limnocorrals was divided into 3 blocks of 4 units and a randomized complete block design (RCBD) was used. The block design was chosen since one set of four enclosures was of a slightly different design[32] and to reduce potential problems associated with spatial heterogeneity within the lake. Each block received one replicate of each treatment and one untreated enclosure served as the control. Only three of the four enclosures in each block were used in either experiment. The fourth received a separate treatment and was not part of this study. The treatment concentrations for Experiment 1 were 0.075 and 0.75 mg a.i./l DIATOX®, and for Experiment 2, 0.75 and 1.50 mg a.i./l DIATOX®. Preliminary laboratory toxicity tests, conducted with mixed zooplankton collected from Lake St. George, had suggested that treatment concentrations should contain between 0.5 and 2.0 mg/l TeCP for significant reductions to be observed in the abundance of most dominant taxa. In addition, concentrations in this range would represent the upper level of what could be expected in freshwater environments adjacent to lumber treatment facilities and storage yards following heavy precipitation and runoff.[33] The concentrations used in Experiment 1 were therefore chosen to approximate a no observable effect concentration (NOEC) and an LC_{50} concentration for crustacean zooplankton, whereas both concentrations used in Experiment 2 were expected to cause significant changes (i.e., $\geq$50% mortality) in the aquatic communities. Enclosed zooplankton communities were relatively similar at the beginning of the two experiments.

The pesticide was applied as DIATOX® (Diachem Ltd., Richmond, B.C.), a commercial sodium salt formulation analytically shown to contain approximately 18.1% active ingredients (15.7% 2,3,4,6-tetrachlorophenol [TeCP] and 2.4% pentachlorophenol [PCP]). The remainder of the formulation was described as "unspecified buffers". Premeasured volumes of DIATOX® were mixed with 1 l of lake water and applied as a single surface treatment using a portable hand-pump sprayer with a single jet nozzle. Treatments, using the same DIATOX® stock solution, took place on the morning (10:25 to 11:45) of June 17 and evening (19:30 to 20:45) of August 13, 1987, for Experiments 1 and 2, respectively. Both were calm days and pesticide drift was negligible.

Aside from the time of pesticide application (morning vs. evening), there was one major difference between the two experiments. During Experiment 1, any fish trapped within the limnocorrals at the time the enclosure walls were lowered were allowed to remain there and no attempt was made to equalize numbers within the various enclosures. Previous limnocorral studies conducted in Lake St. George utilizing the same enclosures had shown that few planktivorous fish were, in fact, enclosed and that they did not present a problem. In these previous studies the enclosure walls were lowered earlier in the season, prior to the hatching of most larval fish (the most common fish species in the limnocorrals was juvenile yellow perch, *Perca flavescens*). For Experiment 2, a large nylon net (5 × 5 m, 2-mm mesh) was used to remove all fish from the enclosures. The net was attached to two 6-m long aluminum poles, and lead-weighted ribbon was sewn to its bottom edge. The net was carefully lowered along one wall to the lake bottom, slowly dragged across the enclosure, and fish scooped out. The netting procedure was repeated at least three times for each enclosure, or until fish were no longer caught. The netting took place 2 days after lowering the corral walls. Collection of the first pretreatment plankton samples followed 2 days later (Day − 6).

Sampling and Quantification Procedures

Chlorophenol Residues

Three types of samples were collected and analyzed for pesticide residues: water samples, sediment samples, and strips of plastic liner material (5 × 100 cm) suspended within the limnocorrals to monitor adsorption to the enclosure walls. Samples of all types were collected immediately prior to treatment and at selected posttreatment times from Day 1 (the surface applied pesticide was allowed to mix for ≥8 h before the first samples were taken) to the end of the two experiments. All samples were collected within a 3-h period on any given day.

Five 1-l water samples were collected with a depth integrating tube sampler[34] and pooled for each enclosure. A 500- to 1000-ml subsample for residue analysis was transferred to a glass jar, approximately 1 ml of concentrated HCl added, and the sample stored in dark, ice-packed coolers for transport back to the laboratory. Sediment samples were collected with a 5-cm diameter core sampler[35] and sectioned into two subsamples (0 to 5 cm and 5 to 15 cm depths) in the field. Two cores were taken from each corral at each sampling time and the two replicate sections pooled. Sediment samples were always collected after water samples to avoid contamination of the water by suspended material. All samples were brought back to the laboratory within 3 to 10 h and frozen at − 17°C for later analysis. One plastic strip (1000 cm²), suspended from a wooden float within each enclosure, was removed at each selected sampling time. Strips were gently shaken to remove excess water, carefully rolled up, and wrapped in aluminum foil. All strips were stored at $-17°C$ within 3 to 10 h of collection for later extraction and analysis. Analytical procedures, recovery efficiencies, and limits of detection and quantification are described in detail elsewhere.[36]

Zooplankton

All enclosures were sampled at regular, predetermined intervals during both the pre- and posttreatment periods. Zooplankton populations were sampled with a depth integrating tube sampler,[34] isolating a 60.5-l sample from the surface to a depth of 3.5 m. Five samples were collected from each enclosure in a "V" pattern,[37] filtered through a 30-μm Nytex® screen, and pooled for analysis. The pooled samples were rinsed into 250-ml glass jars with approximately 90 ml deionized water. They were preserved immediately by adding approximately 10 ml carbonated water, followed by approximately 100 ml of an 8% formalin (v/v)-6% sucrose (w/v) solution.[38,39] All plankton samples from the same day were collected within a 3-h period between 10:30 and 15:30.

Zooplankton were later identified and enumerated in the laboratory using a compound microscope equipped with a 3-ml counting chamber with a grid. After standardizing sample volumes (e.g., 50 to 1000 ml), aliquots of 1 to 3 ml, depending on the density of organisms, were transferred to the counting chamber and the total subsample counted. If necessary, additional aliquots were taken until 3 subsamples or 300 organisms were counted. Zooplankton data were divided taxonomically into macrozooplankton (Cladocera and Copepoda) and microzooplankton (Rotifera) for purposes of analysis. The Copepoda was further subdivided into adult and immature (nauplii and copepodites) individuals. All organisms were identified to species level when possible.[38,40,41]

Phytoplankton and Water Chemistry

Water samples for phytoplankton, chlorophyll *a*, and water chemistry analysis were collected simultaneously with water samples for chlorophenol residues. 100-ml subsamples were preserved immediately with 1 ml Lugol's iodine for later phytoplankton analysis. Chlorophyll *a* samples were stored in 500-ml amber glass bottles, while 1000-ml subsamples for determination of water chemistry were transferred to polypropylene plastic bottles. Chlorophyll *a* and water chemistry samples were immediately placed in ice-packed coolers for transport back to the laboratory. Preanalysis filtration required for determination of soluble reactive phosphorous (0.45-μm Sartorius filters), and for chlorophyll *a* and the particulate fractions of C and N (0.45-μm Whatman GFC glass-fiber filters), was carried out upon arrival at the laboratory approximately 3 to 10 h after sample collection. The processed and preserved samples were stored overnight in the dark at 4°C and transported in ice-packed coolers within 24 h to the Canada Center for Inland Waters, National Water Research Institute (Burlington, Ontario), where chemical analyses were performed. Analyses consisted of chlorophyll *a*, particulate organic carbon and nitrogen, dissolved organic and inorganic carbon, total Kjeldahl nitrogen, NO_2^-, NO_3^-, NH_4^+, total filtered and unfiltered phosphorous, and major ions (Ca^{2+}, Na^+, K^+, Cl^-).[42]

Phytoplankton samples were later processed separately. Subsamples (1.5, 2.5, or 5.0 ml) of the Lugol's preserved field samples were added to a 40-ml sedimentation chamber, topped up with distilled water, and allowed to settle for at least 24 h.[43] After settling, all samples were counted using an inverted compound microscope (Nikon DIAPHOT Type 108). Each algal species was enumerated only at the most appropriate magnification 150×, 300×, or 600×). At least 150 algal cells or 10 transects, whichever came first, were counted at each magnification for a total of 400 to 500 algal cells per sample. Samples were identified to genus level, and species when possible,[44-50] and counts expressed as cells per liter. Size measurements were also made for 10 cells of each species at each sampling time using an ocular micrometer. The resulting mean dimensions were then used to describe that species on that sampling day. Species were classified according to geometrical cell shapes selected to approximate the different shapes exhibited by freshwater algae (i.e., sphere, ellipse, rod, cone, etc.). The cellular volume, or biovolume, was then calculated in units of g/cm^3 as a measure of biomass.[51] The Shannon-Weiner index of species diversity was calculated and the relative percent abundance of major taxa in each corral was used to assess changes in community structure.

Water temperatures and dissolved oxygen (DO) concentrations were recorded both prior to sunrise (Day −6 to 14 only) and during midday in each corral at each sampling time with a dissolved oxygen-temperature meter (YSI Model 57, YSI, Yellow Springs, CO). Measurements were made at depths of 5, 50, 100, 200, and 300 cm. Average DO concentrations for the integrated water column were calculated using Equation 1, which takes the three measurements between 0 and 100 cm and incorporates them into one value representing this depth range. This value was then averaged with the 200-cm and 300-cm values to give an average DO concentration for the depth range 0 to 300 cm without introducing bias from the more numerous surface (0 to 100 cm) measurements.

$$\{\{([5 \text{ cm}] \times 0.1) + ([50 \text{ cm}] \times 0.4) + ([100 \text{ cm}] \times 0.5)\} \qquad (1)$$
$$+ [200 \text{ cm}] + [300 \text{ cm}]\} \div 3$$

(Numbers in [square brackets] are DO concentrations at that given depth.)

The relative transparency of the enclosed water was ascertained by Secchi readings prior to the collection of plankton samples. The pH of the integrated water sample was measured on site with a hand-held wand (Cole Palmer model 5985-75, Chicago, IL).

Visual Observations

Visual observations were also made of the enclosed systems and estimates of fish abundances recorded. Fish (juvenile yellow perch, *Perca flavescens*) usually schooled near the water surface until sufficiently disturbed by sampling procedures. As a result, a rough count could be made on each sample day prior to sampling. Reported estimates (see Table 2) were based on the maximum number observed in each corral during the pre- and posttreatment period of the study. Although these estimates varied slightly among sample days, and as a result may still be under-estimates, the relative abundances were usually the same.

Statistical Analyses

Significant differences among control and DIATOX® treated limnocorrals were determined using repeated measures ANOVAs[32] to standardize for pretreatment variability among enclosures. The repeated measures procedure used the difference between abundances on Days 0 and n $(n = 1$ to 43) for each enclosure as the response. By comparing posttreatment to pretreatment abundances within each enclosure, this type of analysis standardizes all enclosures with respect to Day 0. Zooplankton and phytoplankton data were subjected to a $\log_{10}$-transformation of the abundance per l data to correct for variance heterogeneity. Following transfor-mation, all underlying assumptions of the ANOVA model[52] were satisfied. Water chemistry data were analyzed as nontransformed values. All ANOVA evaluations were performed using a randomized complete block design (RCBD) and a signif-icance level of 95%. All analyses were carried out using PC SAS®, Release 6.03.[53]

The Shannon-Weiner diversity index (H') was calculated separately for the mac-rozooplankton, rotifer, and phytoplankton populations within each enclosure on each sample day using the formula:

$$H' = -\sum_{i=1}^{s} (p_i) \log_2(p_i) \qquad (2)$$

where s = number of taxa and p_i = the proportion of individuals of the total sample belonging to the ith taxon.

RESULTS

Experiment 1: Morning Treatment, Fish Present

Chlorophenol Residues

Mean TeCP and PCP concentrations of 320.5 µg/l and 52.4 µg/l, respectively, were detected in water samples 24 h after application with 0.75 mg/l DIATOX® (0.65 mg TeCP/l and 0.10 mg PCP/l). This translates to a mean percent reduction in nominal TeCP and PCP concentration during the first 24 h of 50.7% and 47.1%, respectively. After the first 24 h, the rate of dissipation of both compounds in the water column slowed progressively (Table 1). Both compounds displayed a similar dissipation rate with little variation among replicate enclosures. The estimated time for 50% reduction of nominal TeCP and PCP concentrations in the integrated water column was 1.0 and 1.1 days, respectively. Overall, >99% of the nominal concentrations disappeared from the water of all treated enclosures over the duration of the 43-day experiment. Adsorption of the chlorophenols to the sediments or enclosure walls was negligible. A more detailed description of persistence and fate can be found elsewhere.[36]

Impact of DIATOX® on Zooplankton Abundance

The mean macrozooplankton abundance in control corrals was not significantly different from the mean abundance in 0.75 mg/l DIATOX® treated corrals at any time during the study (Figure 1a). A repeated measures ANOVA revealed a significant difference in microzooplankton (rotifer) abundance between control and DIATOX® treated enclosures on Day 43 (Figure 1b). A breakdown of the macrozooplankton population into total Cladocera, total adult Copepoda, and total immature Copepoda similarly suggested that the mean abundances of DIATOX® treated populations were not significantly different from control populations. Similar analyses of the most dominant zooplankton species (*Daphnia galeata mendotae, Bosmina longirostris, Keratella cochlearis, Polyarthra* sp., mostly *P. euryptera* and *P. major*) also indicated that there were no significant differences between population abundances in treated and control corrals.

Subtle effects of the 0.75 mg/l DIATOX® treatment on zooplankton populations could possibly have been obscured by the high variation observed among replicate enclosures. High control variability was largely attributed to corral 7 which had considerably higher macrozooplankton and lower microzooplankton abundances than corrals 2 and 12 (Figure 2a,c). There was a negative correlation between macrozooplankton abundance and number of fish fry (*Perca flavescens*) observed in the control corrals (Figure 2 and Table 2). The three most abundant macrozooplankton species, *Bosmina longirostris, Ceriodaphnia lacustris,* and *Daphnia gal-*

Table 1. Mean (± SD) 2,3,4,6-Tetrachlorophenol (TeCP) and Pentachlorophenol (PCP) Concentrations in Integrated Water Samples From Three Replicate Limnocorrals After a Morning Treatment With 0.75 mg/l DIATOX® (Experiment 1)

Time (Days)	0.75 mg/l DIATOX®		
	TeCP (650)[a]	PCP (100)	Total (750)
1	320.5 ± 27.3	52.4 ± 3.9	372.9 ± 31.2
4	161.7 ± 18.2	25.1 ± 2.8	186.8 ± 21.0
8	96.4 ± 9.3	12.7 ± 1.3	109.1 ± 10.5
15	28.6 ± 5.3	1.8 ± 1.1	30.3 ± 6.4
29	7.0 ± 1.0	0.9 ± 0.2	7.9 ± 1.2
43	2.5 ± 0.5	0.1 ± 0.1	2.6 ± 0.5

[a] Values in brackets are nominal treatment concentrations in μg/l.

Note: All values are expressed as μg/l.

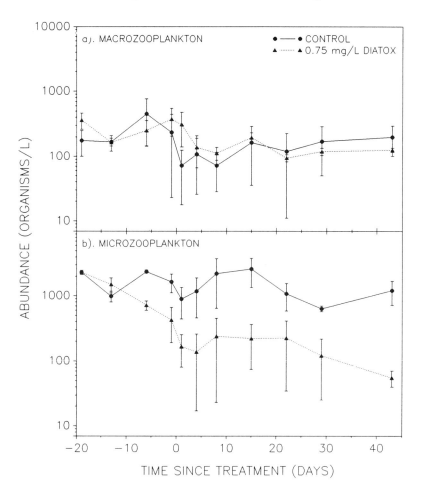

Figure 1.   Changes in (a) total macrozooplankton and (b) total microzooplankton (rotifer) abundance in control and 0.75 mg/l DIATOX® treated limnocorrals. Each point is the mean ± SE of three replicates. Treatment was applied on Day 0 (June 17, 1987).

eata mendotae, which accounted for >95% of the total adult macrozooplankton abundance in all instances but one, were present in considerably lower numbers in control corral 12 which had the greatest number of fish. *C. lacustris* and *D. g. mendotae* were nearly eliminated from corrals 2 and 12, but showed abundances of up to 200 organisms per l in corral 7. Differences in *B. longirostris* abundances between corrals 2 and 7 were not as great as for *C. lacustris* and *D. g. mendotae.* Total rotifer abundance showed a positive, although weaker, correlation with fish abundance (Figure 2 and Table 2). Rotifer numbers in corrals 2 and 12 remained relatively constant throughout the experiment, but corral 7 displayed a noticeable decrease between Days −1 and 29, largely due to decreases in the abundance of *Polyarthra* sp. and, to a lesser extent, *K. cochlearis.*

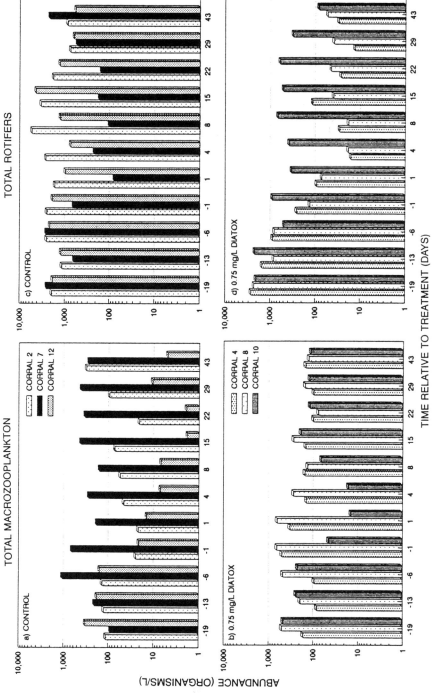

Figure 2. Abundance of total macrozooplankton and total rotifers in replicate control and 0.75 mg/l DIATOX® treated limnocorrals during Experiment 1. Treatment was applied on Day 0 (June 17, 1987).

Table 2. Estimated Number of Fish Fry (*Perca flavescens*, 1.5 to 3.0 cm) in Control and DIATOX® Treated Limnocorrals During the Pretreatment and Posttreatment Period of Experiment 1

Treatment	Corral No.	Maximum Number of Fish Observed	
		Pretreatment Period	**Posttreatment Period**
Control	2	30–40	30–40
	7	4	0
	12	50–60	50–60
0.75 mg/l	4	10–15	0
DIATOX®	8	10–15	0
	10	20–30	8–10

Note: All numbers are based on visual observations at time of sampling.

Variability among treated enclosures was generally less than among controls (Figure 2), presumably due to a treatment-related standardization of fish numbers. Dead perch fry were observed in all treated corrals from Days 1 to 7 posttreatment. All fish in corrals 4 and 8 were apparently eliminated by the DIATOX® treatment, but in corral 10 at least 8 to 10 fish (27 to 50%) survived. Corral 10, like the controls, displayed a negative correlation between number of fish fry and macro-zooplankton abundance between Days -1 and 8 (Figure 2b). Rotifer abundance in that same enclosure remained high for the duration of the experiment (Figure 2d). A slight posttreatment decrease was observed in rotifer numbers in corrals 4 and 8 (Figure 2d), although fish had been eliminated from both enclosures. This decrease was almost entirely due to a large decrease in *Polyarthra* sp., the most abundant rotifer taxon.

As a consequence of the inconclusive results observed after the 0.75 mg/l DIA-TOX® treatment, samples from the 0.075 mg/l treatment were not analyzed. Mean chlorophenol concentrations in these corrals on Day 1 posttreatment were 13.58 and 2.84 μg/l for TeCP and PCP, respectively — well below those predicted necessary for biological impact. Water chemistry data from Experiment 1 is not presented as high replicate variability obscured any treatment related trends.

Experiment 2: Evening Treatment, Fish Removed

Chlorophenol Residues

Mean total chlorophenol concentrations in integrated water samples collected after treatment with 0.75 and 1.50 mg/l DIATOX® are given in Table 3. The rate of chlorophenol dissipation during the first 24 h was reduced by the evening application and was significantly lower ($p < 0.05$) than the dissipation rate observed in Experiment 1. Relative to nominal, the mean TeCP concentration in 0.75 mg/l DIATOX® treated enclosures dropped by 20.4% during the first 24 h of Experiment 2, compared to 50.7% during the first 24 h of Experiment 1. Similarly, PCP levels only decreased by 8.0% over the 24-h period as compared to 47.1% at the same treatment concentration in Experiment 1 (Table 3). The estimated time for 50%

Table 3. Mean (±SD) 2,3,4,6-Tetrachlorophenol (TeCP) and Pentachlorophenol (PCP) Concentrations in Integrated Water Samples From Three Replicate Limnocorrals After Evening Treatments With 0.75 and 1.50 mg/l DIATOX® (Experiment 2)

Time (Days)	0.75 mg/l DIATOX®			1.50 mg/l DIATOX®		
	TeCP (650)ᵃ	PCP (100)	Total (750)	TeCP (1300)	PCP (200)	Total (1500)
0.4	598.2 ± 75.4	107.1 ± 18.4	705.3 ± 93.7	1129.1 ± 143.2	191.9 ± 26.0	1320.9 ± 167.1
1	516.4 ± 120.3	91.1 ± 18.8	607.5 ± 139.1	932.1 ± 158.4	160.8 ± 28.4	1092.8 ± 186.4
2	421.2 ± 62.8	70.6 ± 12.1	491.8 ± 74.7	944.1 ± 166.1	167.5 ± 34.4	1111.6 ± 200.0
4	280.5 ± 15.6	46.6 ± 1.9	327.1 ± 17.4	692.7 ± 45.2	121.9 ± 8.1	814.7 ± 53.0
7	168.8 ± 17.3	29.2 ± 4.4	198.0 ± 21.7	419.5 ± 28.1	61.3 ± 6.8	480.8 ± 34.7
14	59.7 ± 11.8	8.1 ± 2.0	67.9 ± 13.8	182.5 ± 30.0	28.2 ± 5.4	210.7 ± 35.4
21	28.1 ± 7.7	4.4 ± 1.4	32.5 ± 9.1	96.0 ± 16.0	14.0 ± 2.7	110.0 ± 18.8
28	17.7 ± 6.7	2.2 ± 1.3	19.9 ± 8.0	54.9 ± 8.5	8.3 ± 1.1	63.2 ± 9.6
42	4.5 ± 1.7	0.4 ± 0.3	4.8 ± 2.0	12.3 ± 3.2	1.7 ± 0.6	14.1 ± 3.7

ᵃ Values in brackets are nominal treatment concentrations in μg/l.

Note: All values are expressed as μg/l.

reduction of nominal TeCP and PCP concentrations in the water column ranged from 3.8 to 5.2 days and from 3.4 to 4.8 days, respectively. Observed concentrations at the end of the 63-day experiment were all <0.5% of the Day 1 (a.m.) concentration. Overall, dissipation patterns were similar to those observed in Experiment 1. A more detailed description of persistence and fate can be found elsewhere.[36]

Effect of Fish Removal on Zooplankton Variability among Limnocorrals

Removal of fish reduced variability among control enclosures (Figure 3). Total macrozooplankton abundances remained relatively constant over the duration of Experiment 2, with a mean coefficent of variation (CV) of 49% and a maximum difference between replicate corrals reaching a factor of 3 (Figure 3c). Variability in total macrozooplankton abundance among control corrals in Experiment 1 was significantly greater, reaching a maximum difference of 231 times (Figure 3a) and a CV of 157%. Total rotifer abundances showed a similar improvement in similarity, but exhibited a slight decrease over time. Maximum variability among replicate corrals in Experiment 2 was reduced to a factor of 2 to 3 (Figure 3d) with a CV of 56%, whereas maximum variability in Experiment 1 reached a factor of 53 (Figure 3b) with a CV of 123%.

Impact of DIATOX® on Zooplankton Abundance

Significant reductions in total macrozooplankton and total rotifer abundances were observed after treatment with both DIATOX® concentrations (Figure 4). A noticeable decrease in total macrozooplankton abundance was observed 1 d after treatment at the 0.75 mg/l concentration, but reductions were only significant on Days 4 and 7. Populations returned to control levels by Day 14 and remained there for the duration of the experiment (Figure 4a). The reduction in numbers in the 1.50 mg/l DIATOX® treated corrals was significant on Days 1 to 21, with control abundances approached by Day 28 posttreatment. Maximum reductions relative to control abundances (67.8% and 89.4% at the low and high treatment, respectively) were observed on Day 7 posttreatment. Rotifers were more sensitive to DIATOX® than macrozooplankton. Total rotifer abundances were significantly reduced on Days 1 to 28 at both treatment concentrations, with only a slightly greater reduction in mean numbers at the higher treatment (Figure 4b). Control numbers were again reached by Day 42 posttreatment. Maximum impact was observed on Day 7 for both treatments, with a percent reduction in abundance relative to controls of 99.1% and 99.7% for the low and high treatment, respectively.

Division of the macrozooplankton community into Cladocera (Figure 5a), adult Copepoda (Figure 5b), and immature Copepoda (Figure 5c) revealed similar responses by the various groups. All three groups exhibited significant reductions after either treatment, with a greater response at the 1.50 mg/l DIATOX® concentration. Maximum reductions relative to control abundances were 90.4%, 93.3%, and 92.9% for Cladocera, adult Copepoda, and immature Copepoda, respectively.

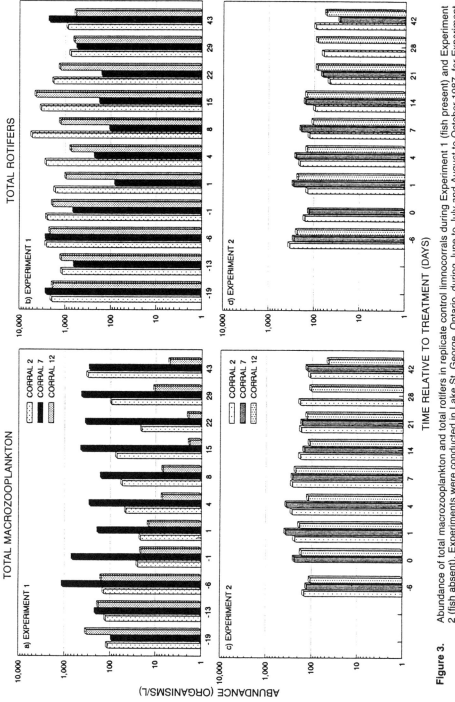

Figure 3. Abundance of total macrozooplankton and total rotifers in replicate control limnocorrals during Experiment 1 (fish present) and Experiment 2 (fish absent). Experiments were conducted in Lake St. George, Ontario, during June to July and August to October 1987, for Experiment 1 and 2, respectively.

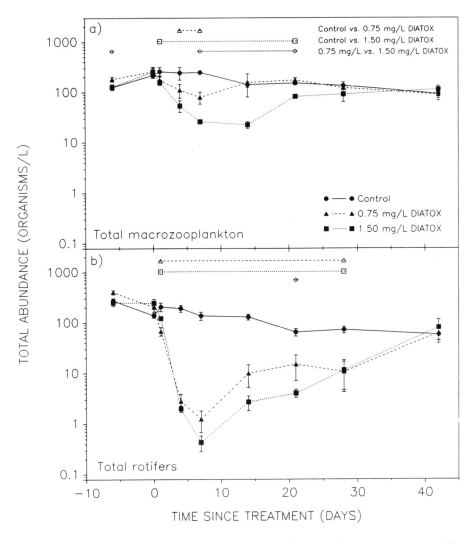

Figure 4.   Changes in (a) total macrozooplankton and (b) total rotifer abundance in control, 0.75 mg/l, and 1.50 mg/l DIATOX® treated limnocorrals. Each point is the mean ± SE of three replicates. Open symbols and horizontal lines at the top of each figure indicate days when treatments were significantly different ($p \leq 0.05$) as determined by a repeated measures ANOVA. Treatment took place on Day 0 (August 13, 1987).

All populations, except adult copepods at the 1.50 mg/l treatment, had recovered by Day 28 posttreatment. Similar analyses conducted with individual macrozooplankton species (Figure 6) revealed a significant decrease in the abundance of all major taxa, except *Mesocyclops edax* (Figure 6g). The seven species in Figure 6a to g accounted for >95% of the total macrozooplankton abundance in 87.3% of

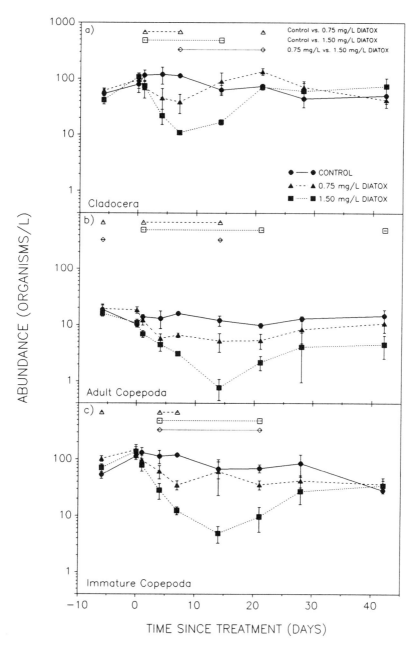

Figure 5.   Changes in (a) Cladocera, (b) adult Copepoda, and (c) immature Copepoda abundance in control, 0.75 mg/l, and 1.50 mg/l DIATOX® treated limnocorrals. Each point is the mean ± SE of three replicates. Open symbols and horizontal lines at the top of each figure indicate days when treatments were significantly different ($p \leq 0.05$) as determined by a repeated measures ANOVA. Treatment took place on Day 0 (August 13, 1987).

the samples analyzed and >90% in 96.2% of the samples. *Ceriodaphnia lacustris* (Figure 6b) was the most sensitive of the four main cladocerans, while *Skistodiaptomus oregonensis* (Figure 6e) and *Tropocyclops prasinus mexicanus* (Figure 6f) were the most sensitive copepods. The combined group of all other remaining macrozooplankton species (Figure 6h) was dominated by *Eubosmina coregoni* during the early period of the study, but the observed recovery at the 1.50 mg/l treatment

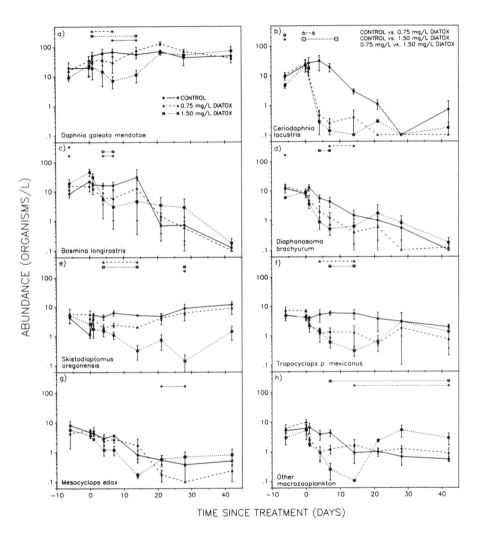

Figure 6.   Changes in abundance of dominant macrozooplankton taxa in control, 0.75 mg/l, and 1.50 mg/l DIATOX® treated limnocorrals. Each point is the mean ± SE of three replicates. Open symbols and horizontal lines at the top of each figure indicate days when treatments were significantly different ($p \leq 0.05$) as determined by a repeated measures ANOVA. Treatment took place on Day 0 (August 13, 1987).

on Days 21 to 42 was mainly due to a large increase in the abundance of *Scapho-leberis mucronata,* a species known to inhabit still, sheltered conditions.[54]

The enclosed rotifer communities were dominated by three taxa, *Keratella coch-learis, Polyarthra* sp., and *Conochilus unicornis*. Individuals from these taxa accounted for 93.5% of the total number of rotifers sampled over the duration of the experiment. Treatment-related reductions in the abundance of these species (Figure 7) revealed similar results to those obtained for the total Rotifera (Figure 4b). *K. cochlearis* populations were significantly impacted at both treatment concentrations (Figure 7a). *Polyarthra* sp. showed an obvious concentration-response relationship, with lower abundances relative to controls following the 1.50 mg/l treatment (Figure 7b). *C. unicornis* populations were significantly reduced by Days 1 to 4 and did not recover by Day 42 at either treatment concentration (Figure 7c). In fact, *C. unicornis* was eliminated from all but 4 of 36 posttreatment samples from Day 4 onwards. Other commonly encountered rotifer species, such as *Kellicotia longis-pina, Kellicotia bostoniensis, Chromagaster ovalis, Monostyla lunaris,* and *Lecane luna,* showed either no significant trends, or a similar posttreatment decrease at both treatment concentrations. These species, especially *C. ovalis,* were partially responsible for the observed recovery in total rotifer abundance (Figure 4b), but were too few in numbers at earlier times for a proper statistical evaluation of treatment impact.

Impact of DIATOX® on Zooplankton Community Structure

Community diversity, as measured by the Shannon-Weiner index, showed only minor posttreatment differences. Macrozooplankton diversity displayed slight posttreatment reductions at both DIATOX® treatments relative to controls, but mean values were only statistically lower on Day 14 (control = 1.689; 0.75 mg/l DIA-TOX® = 0.858; 1.50 mg/l DIATOX® = 0.749). Rotifer diversities showed no significant posttreatment reductions over the duration of the experiment.

The mean numbers of macrozooplankton taxa identified in the two treatments and the control were similar on all sample days, differing at most by 2 taxa (22%). The mean number of rotifer taxa identified was reduced by 13 to 50% in treated corrals, but there was little difference between the two treatments. The number of similar macrozooplankton taxa between any two treatments showed little change. The number of similar rotifer taxa showed a slight reduction between all treatment combinations from Days 4 to 28, reaching a low of 20% similarity between the control and the 1.50 mg/l DIATOX® treatment on Day 14. Calculations of the percent of the total zooplankton abundance accounted for by various numerically dominant taxa over the duration of the study revealed little additional information on DIATOX® impact on zooplankton community structure.

Impact of DIATOX® on Dissolved Oxygen, Chlorophyll a, and Secchi Disk Transparency

Midday mean dissolved oxygen (DO) concentrations showed a significant decrease on Days 4 to 7 and on Days 4 to 14 at the 0.75 mg/l and 1.50 mg/l DIATOX®

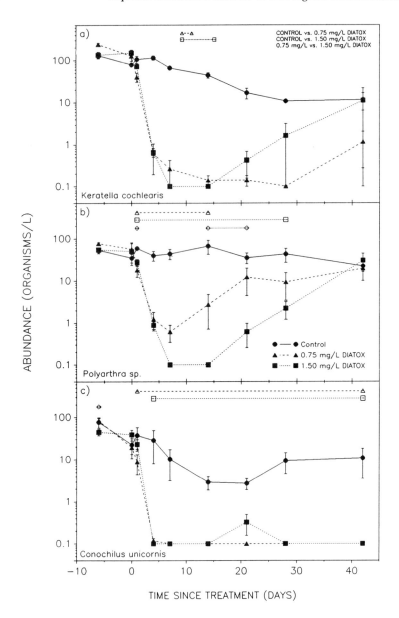

Figure 7. Changes in abundance of dominant rotifer taxa in control, 0.75 mg/l, and 1.50 mg/l DIATOX® treated limnocorrals. Each point is the mean ± SE of three replicates. Abundances of zero are shown as 0.1 organisms per l. Open symbols and horizontal lines at the top of each figure indicate days when treatments were significantly different ($p \leq 0.05$) as determined by a repeated measures ANOVA. Treatment took place on Day 0 (August 13, 1987).

treatments, respectively (Figure 8a). Mean DO levels remained lower in treated than in control corrals at most other posttreatment times, but variation among replicate enclosures rendered these levels nonsignificant ($p > 0.05$). Nighttime DO concentrations, in any treatment, were never significantly different from daytime concentrations, differing by ≤ 1.0 mg/l.

An initial posttreatment decrease in chlorophyll *a* concentration was observed after both DIATOX® treatments, but it was only significant at the 1.50 mg/l treatment on Day 7 (Figure 8b). This decrease was followed by a large increase on Days 14 and 28 at the 1.50 mg/l DIATOX® concentration and a smaller increase on Day 14 at the 0.75 mg/l DIATOX® concentration. Chlorophyll *a* levels in the 1.50 mg/l DIATOX® treated corrals again approached those of the controls by Day 42 posttreatment.

Relative transparency, as measured by Secchi depth, increased in all corrals, including controls, over the duration of the experiment (Figure 8c). In addition, both DIATOX® treatments resulted in a significant increase in transparency on Day 7. This was followed by a decrease in Secchi depth on Days 14 to 42 at the 1.50 mg/l DIATOX® concentration, corresponding to the increase in chlorophyll *a* concentration. The decrease in Secchi depth was only statistically significant on Days 14 and 21.

Impact of DIATOX® on Phytoplankton Abundance, Biovolume, Diversity, and Community Structure

There were no significant ($p > 0.05$) changes observed in total phytoplankton abundance during the first 7 d after treatment with 1.50 mg/l DIATOX®. An increase in total abundance was, however, observed on Days 14 to 28 with populations returning to control levels by Day 42 (Figure 9a). The observed increase resulted in a mean total abundance on Days 21 and 28 which was 36 times greater than in control corrals. This dramatic increase in abundance was not reflected in total biovolume (Figure 9b). Here, a slight decrease was recorded on Days 4 to 14, but this was not statistically significant ($p > 0.05$). Overall, total biovolume estimates were similar between treated and control enclosures over the duration of the experiment. Species diversity, as measured by the Shannon-Weiner index showed no immediate posttreatment response, but then decreased sharply on Day 14 and was significantly lower than the diversity in controls on Day 21 and 28 (Figure 9c). The diversity index rose again to control levels by Day 42. Phytoplankton samples from enclosures treated with 0.75 mg/l DIATOX® were not enumerated, since the 1.50 mg/l treatment had minimal direct effects on total phytoplankton abundance and biovolume, and since chlorophyll *a* concentrations in the 0.75 mg/l enclosures suggested no significant effects of this treatment.

The large increase observed in total phytoplankton abundance on Days 14 to 28 in 1.50 mg/l DIATOX® treated corrals was due to a large increase in the abundance of Cyanophyta (Figure 10a). Two filamentous species, *Oscillatoria amphigranulata* and *O. lauterbornii,* accounted for >99% of this increase. The small decrease

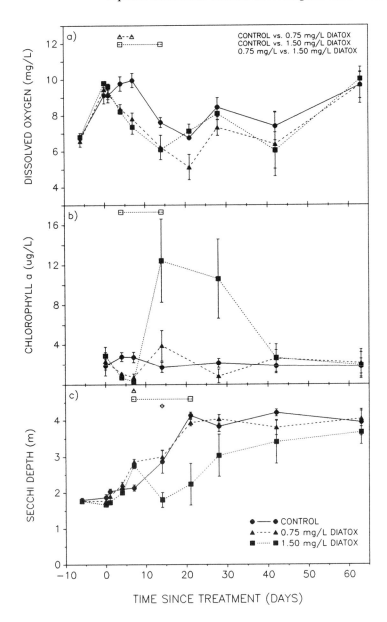

Figure 8. Changes in (a) dissolved oxygen concentration, (b) chlorophyll *a* concentration, and (c) Secchi depth in control, 0.75 mg/l, and 1.50 mg/l DIATOX® treated limnocorrals. Each point is the mean ± SE of three replicates. Open symbols and horizontal lines at the top of each figure indicate days when treatments were significantly different ($p \leqslant 0.05$) as determined by a repeated measures ANOVA. Treatment took place on Day 0 (August 13, 1987).

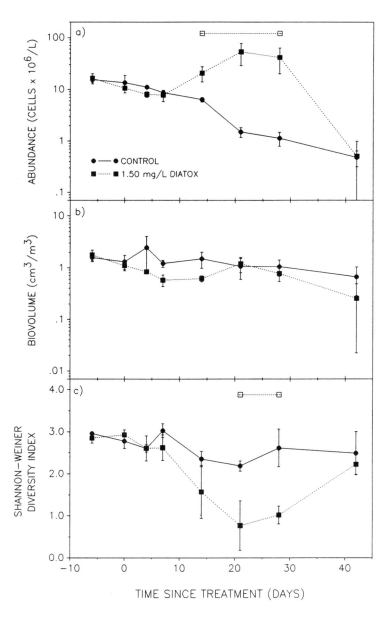

Figure 9.   Changes in (a) abundance, (b) biovolume, and (c) Shannon-Weiner diversity index of the total phytoplankton community in control and 1.50 mg/l DIATOX® treated limnocorrals. Each point is the mean ± SE of three replicates. Open symbols and horizontal lines at the top of each figure indicate days when treatments were significantly different ($p \leq 0.05$) as determined by a repeated measures ANOVA. Treatment took place on Day 0 (August 13, 1987).

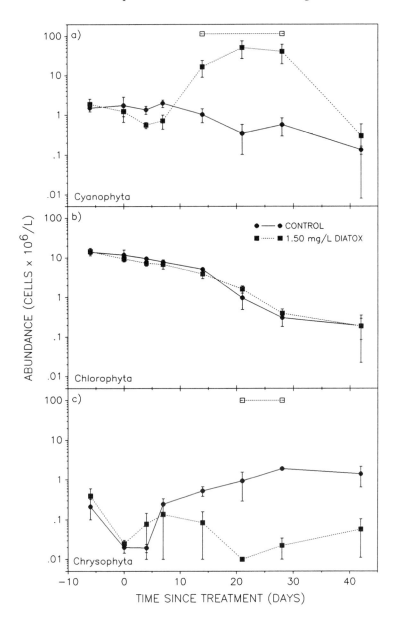

Figure 10. Changes in abundance of (a) Cyanophyta, (b) Chlorophyta, and (c) Chrysophyta in control and 1.50 mg/l DIATOX® treated limnocorrals. Each point is the mean ± SE of three replicates. Open symbols and horizontal lines at the top of each figure indicate days when treatments were significantly different ($p \leq 0.05$) as determined by a repeated measures ANOVA. Treatment took place on Day 0 (August 13, 1987).

observed in Cyanophyta abundance on Days 4 and 7 was due to reductions in the numbers of unicellular species such as *Chroococcus* sp. and *Synechocystis* sp., but this decrease was not statistically significant ($p > 0.05$). Chlorophyta species showed no change in response to the DIATOX® treatment (Figure 10b). Mean Chlorophyta abundances in treated corrals were nearly identical to those in control corrals at all times during the study, but numbers decreased steadily in all corrals over the duration of the experiment. There were no obvious effects on individual Chlorophyta species either, with *Dactylothece* sp., *Lagerheimia* sp., *Oocystis* sp., *Asterococcus* sp., and *Chlorella vulgaris* being the most abundant taxa. Chrysophyta abundance was reduced in treated enclosures from Days 14 to 42, but due to high within-treatment variability, this reduction was only significant on Days 21 and 28 (Figure 10c). The two main Chrysophyta species were *Dinobryon divergens* and *Mallomonas caudata*. *D. divergens* accounted for the largest percentage of the total Chrysophyta in control corrals, but was never detected in treated corrals from Days 14 to 28. Other phytoplankton groups contributed little to the overall percent abundance (Figure 11a). Dinophyceae and Euglenophyta species were rarely encountered in significant numbers and Bacillariophyceae (diatoms) reached a maximum of only 5.9% of the total phytoplankton abundance in control corrals on Day 21 (Figure 11a).

As a result of their large size, relative to the numerically dominant Cyanophyta and Chlorophyta species, Chrysophyta, Dinophyceae, and Bacillariophyceae species contributed significantly to the overall community biovolume (Figure 11b). Chrysophyta increased in importance in control enclosures from Day 7 onwards, but were never important in treated enclosures until Day 42. Dinophyceae accounted for 13 to 41% of the total biovolume in control corrals over the duration of the study, but were not encountered in treated corrals on Days 4 and 7. Their numbers remained low on Days 14 to 28, but regained significance on Day 42. Bacillariophyceae were also only significant in control corrals where they accounted for 16 to 29% of the total biovolume on Days 14 to 28 (Figure 11b).

Impact of DIATOX® on Water Chemistry

Few water chemistry variables were significantly affected by the DIATOX® treatments. Only particulate organic carbon (Figure 12a), particulate nitrogen (Figure 12b), and total unfiltered phosphorous (Figure 12c) showed obvious responses. Particulate organic carbon (POC) showed a significant posttreatment decrease at both DIATOX® treatment levels, with only a slightly greater response at the higher concentration. An increase in POC on Days 14 to 28 at the 1.50 mg/l DIATOX® concentration was apparent, but not significant ($p > 0.05$). Particulate nitrogen (PN) concentrations followed a similar pattern to POC, but showed greater variability. A significant reduction in PN was observed on Day 4 at the 0.75 mg/l treatment and from Days 4 to 7 at the 1.50 mg/l treatment. A large increase in PN followed on Days 14 to 42 at the high treatment concentration, but this was not statistically significant ($p > 0.05$). There was no detectable posttreatment decrease in total unfiltered phosphorous, but a large Day 28 increase was observed at the 1.50 mg/l DIATOX® treatment. High variability among the three replicate enclosures rendered this increase statistically insignificant ($p > 0.05$).

All other measured variables listed in Table 4 were not significantly affected. Most of these parameters (DIC, TKN, NH_4^+, NO_3^-, NO_2^-, TPF, SRP) exhibited

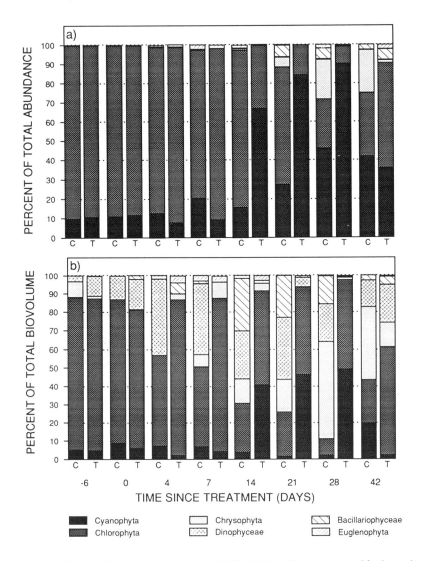

Figure 11. Percent of (a) total abundance and (b) total biovolume accounted for by various phytoplankton groups in control (C) and 1.50 mg/l DIATOX® treated (T) limnocorrals. Treatment took place immediately after collection of Day 0 samples (August 13, 1987).

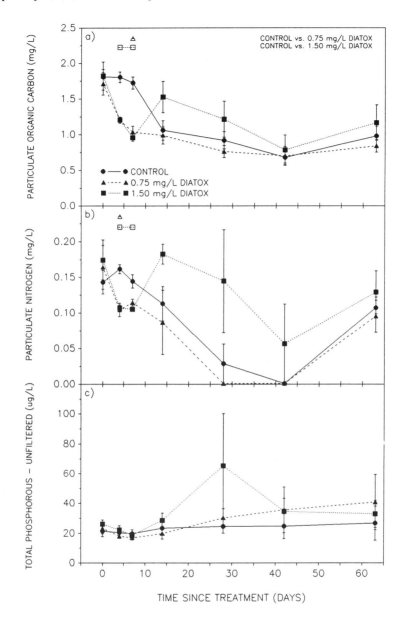

Figure 12. Changes in concentration of (a) particulate organic carbon, (b) particulate nitrogen, and (c) total unfiltered phosphorous in control and 1.50 mg/l DIATOX® treated limnocorrals. Each point is the mean ± SE of three replicates. Open symbols and horizontal lines at the top of each figure indicate days when treatments were significantly different ($p \leq 0.05$) as determined by a repeated measures ANOVA. Treatment took place on Day 0 (August 13, 1987).

either a gradual increase in all enclosures over the duration of the study, or an increase towards the end of the study (Days 42 to 63). In general, there was greater within-treatment variability associated with these increases. Other variables (DOC, K^+, Ca^{2+}) remained relatively constant over the duration of the study, whereas sulfate showed a steady decrease in all enclosures. There was no change in pH which remained at 8.1 ± 0.2 in all corrals during the experimental period.

DISCUSSION

Experiment 1: Morning Treatment, Fish Present

The high degree of variability observed in zooplankton abundance among replicate enclosures was attributed to the uneven distribution of planktivorous fish fry which were inadvertently enclosed within the limnocorrals. Since no effort was made at either removing or controlling fish in the enclosures, there was an uneven level of predation exerted on the enclosed zooplankton communities. Control corrals showed a strong negative correlation between total macrozooplankton abundance and number of fish, suggesting increased predation with increasing fish abundance. Of the major macrozooplankton species present, *C. reticulata* and *D. g. mendotae* appeared to be the preferred prey of the fish fry. The minimal predation pressure exerted by the low number of fish in corral 7 resulted in an increase in total macrozooplankton

Table 4. Mean and Range (in Brackets) of Various Water Chemistry Variables Monitored in Control and DIATOX Treated Limnocorrals Over the Duration of Experiment 2

Variable[a]	Mean (Range)			Common Trend
	Control	0.75 mg/l DIATOX®	1.50 mg/l DIATOX®	
Carbon (mg/l)				
DOC	7.3 (6.9–7.7)	7.4 (7.0–8.6)	7.4 (6.8–7.8)	Constant
DIC	33.8 (28.1–37.6)	34.5 (30.1–39.1)	34.3 (30.6–38.2)	Increase
Nitrogen (μg/l)				
TKN	684.4 (539–1583)	716.3 (537–1107)	677.9 (558–952)	Increase
NH_4^+	57.8 (5.0–190.0)	109.9 (5.0–356.0)	74.1 (5.0–311.0)	Increase
NO_3^-	9.5 (7.0–15.4)	10.2 (7.9–18.0)	9.8 (8.1–18.1)	Increase
NO_2^-	1.3 (0.7–3.0)	1.8 (0.7–5.0)	1.6 (0.6–5.0)	Increase
Phosphorus (μg/l)				
TPF	11.2 (6.7–38.9)	15.2 (3.6–57.9)	13.0 (7.3–36.2)	Increase
SRP	2.5 (0.4–25.0)	5.4 (0.6–45.0)	3.0 (0.7–17.3)	Increase
Major ions (mg/l)				
K^+	1.8 (1.8–1.9)	1.9 (1.8–2.0)	1.9 (1.7–2.0)	Constant
Ca^{2+}	52.5 (47.4–58.1)	51.9 (43.7–56.9)	52.9 (49.3–58.1)	Constant
SO_4^{2-}	16.1 (13.5–17.6)	16.1 (13.6–17.5)	16.0 (13.6–17.7)	Decrease

[a] DOC — dissolved organic carbon; DIC = dissolved inorganic carbon; TKN = total Kjeldahl nitrogen; NO_3^- = nitrate; NO_2^- = nitrite; TPF = total phosphorus-filtered; SRP = soluble reactive phosphorus; K^+ = potassium; Ca^{2+} = calcium; and SO_4^{2-} = sulfate.

Note: None of the listed parameters showed any statistically significant posttreatment response, but trends common to all corrals are given in the column to the far right.

abundance as would be predicted by trophic interaction theory.[55,56] This increase in macrozooplankton appeared to cause a corresponding decrease in total rotifer abundances. There could be several reasons for this decrease, including competition for food, space, and other resources, and damage to soft-bodied rotifer taxa (i.e., *Polyarthra*) by filter feeding *Daphnia*.[57] Similar relationships among fish, macro-zooplankton, and rotifers were observed in treated enclosures.

McQueen et al.[56] reviewed the literature on a variety of enclosure and small pond experiments which evaluated the impact of planktivores on zooplankton. They reported that in 10 of 11 studies, planktivorous fish selected against large zoo-plankton and that increased fish numbers resulted in reduced zooplankton biomass. In addition, Mills and Forney,[58] McQueen et al.,[59] and McQueen and Post[60] have all shown that juvenile perch biomasses of >20 to 30 kg/ha are sufficient to eliminate large-bodied cladocerans. Also, in a series of limnocorral (8 m diameter, 15 m deep, 750,000 l) experiments run during 1985 in Lake St. George, Ontario, McQueen and Post[60,61] showed that when the biomass of juvenile yellow perch reached >30 to 50 kg/ha, *Daphnia* (primarily large *D. g. mendotae*) biomass declined rapidly. If one assumes an average weight of 1 g per juvenile yellow perch, then a biomass of 20 to 30 kg/ha corresponds to 50 to 75 fish for a 5 × 5-m limnocorral. This compares favorably with our findings from Experiment 1, where corral 12 was estimated to have 50 to 60 fish (Table 2) and showed macrozooplankton densities reduced from 325 to 2 organisms per l over a 34-d period (Days − 19 to + 15) (Figure 2).

The 0.75 mg/l DIATOX® treatment was apparently more toxic to the perch fry than to the zooplankton. All fish disappeared in two of the treated enclosures and the third showed a 60 to 80% reduction. This finding was not unexpected as several researchers (see Jones[11]) have reported fish to be more sensitive to chlorophenols than zooplankton. The lack of effects of the 0.75-mg/l DIATOX® treatment on zooplankton populations suggests that chlorophenol levels during the first 48 to 96 h after treatment must have been below or near the no observable effect concentration (NOEC). This is supported by residue levels in integrated water samples (Table 1), which were below the estimated 48-h LC_{50} of 580 µg/l TeCP for Lake St. George *D. g. mendotae* derived under laboratory conditions (pH = 7.9 ± 0.2).[62]

The significantly lower than desired TeCP and PCP concentrations in water on Day 1 were attributed to rapid photodegradation as a result of the morning applied surface treatment. The significance of photodegradation in the dissipation of TeCP and PCP was substantiated by results from additional, concurrent experiments.[36] The lack of an obvious effect on zooplankton in this experiment can therefore be largely attributed to this rapid degradation of the pesticides within the first 24 h after treatment. Other contributing factors include: possible avoidance of high pes-ticide concentrations by some of the organisms within the first 24 h of treatment by moving deeper in the enclosures, high variability resulting from variable zoo-plankton predation by juvenile perch, and the relatively long pretreatment period (19 d) which enhanced divergence of the separate communities.

Experiment 2: Evening Treatment, Fish Absent

Impact of DIATOX® on Zooplankton Populations

The removal of fish from all enclosures enhanced similarity in zooplankton abundance among replicates. This was not surprising as the feeding of planktivorous fish can have a considerable impact on freshwater zooplankton community structure.[56,63] Although removal of fish from limnocorrals, or from mesocosms in general, is not always desirable, this data shows the degree of variability in zooplankton abundance (i.e., CVs up to 157%) which can result if planktivorous fish populations are not equal among all enclosures. A combination of the evening pesticide treatment and the removal of fish from all enclosures resulted in the detection of a significant impact on zooplankton populations at the 0.75 mg/l DIATOX® treatment, where none was seen in Experiment 1. As photolysis is the primary mechanism of chlorophenol degradation in surface waters,[12,23,24,64-66] the evening treatment was expected to delay and reduce TeCP and PCP degradation/transformation. This would expose organisms to higher chlorophenol concentrations than in Experiment 1. This was indeed the case as mean residue levels in the integrated water samples from Days 1 and 4 contained 516 μg/l TeCP + 91 μg/l PCP and 281 μg/l TeCP + 47 μg/l PCP, respectively (Table 3) — levels substantially higher than those on Days 1 and 4 in Experiment 1 (Table 1).

There were only minor differences in the response of the various macrozooplankton taxa to the DIATOX® treatments. Of the 7 most abundant species (Figure 6), 5 and 6 species were significantly affected by the 0.75 and 1.50 mg/l DIATOX® treatment, respectively. The response was always greater at the high treatment concentration and the percent reduction in abundance relative to controls was similar for most species. Interestingly, the most sensitive species was *Ceriodaphnia lacustris,* which supports the use of species of this genus in standard laboratory toxicity tests.

The greater sensitivity to DIATOX® exhibited by rotifers, as compared to macrozooplankton, was not expected. Previous limnocorral studies, also conducted in Lake St. George, Ontario, with the insecticides permethrin, fenvalerate, and methoxychlor, showed little or no toxic effects on rotifer populations. In fact, rotifers increased in abundance after treatment, in response to the release from competitive interactions with the treatment reduced macrozooplankton populations.[6,8,37] The reason for this difference in response by rotifers to these insecticides, as opposed to their response to chlorophenols (DIATOX®), is unknown, but could result from their different modes of action (i.e., neurotoxicants vs. metabolic inhibitors).

The three main rotifer taxa, *K. cochlearis, Polyarthra* sp., and *C. unicornis* (Figure 7), showed similar responses to the DIATOX® treatments. *Polyarthra* sp. were, however, the first to recover at the 0.75 mg/l treatment, returning to near control levels by Day 21. The recovery began between Days 7 and 14, when mean total chlorophenol concentrations were between 0.20 mg/l and 0.07 mg/l. Many other zooplankton populations also began to recover from the DIATOX® treatments when mean total chlorophenol levels were between 0.1 and 0.2 mg/l, suggesting that this range should be near the no-effect concentration. This is supported by data from laboratory toxicity tests conducted under similar pH and hardness, where NOECs of 0.1 and 0.25 mg/l TeCP were derived for *D. g. mendotae* in 48-h tests and for *D. magna* in a 21-day test, respectively.[62] Results by Crossland and Wolff,[23] who investigated the effects of PCP in outdoor experimental ponds, further support our field observations. They observed no significant direct impact ($p > 0.05$) on various macrozooplankton populations (Daphniidae, Diaptomidae, Cyclopidae, Nauplii) when PCP was repeatedly applied to the ponds, thereby maintaining an average concentration of 0.03 to 0.08 mg/l over a 43-day long experiment.

The similarity in response by the various zooplankton taxa to the DIATOX® treatments was also reflected in the lack of significant changes in community structure. In fact, the percent species composition in the two treatments was similar throughout the experiment. That most zooplankton taxa recovered to control levels by Day 42 posttreatment is a reflection of the short generation time and high reproductive rate of these organisms.

Assuming that the zooplankton species encountered in this study are representative of those in many natural lentic systems, and that most other zooplankton species would respond in a similar manner, DIATOX® concentrations ≥ 0.75 mg/l would be predicted to cause significant reductions in most natural zooplankton populations. Such concentrations are unlikely to occur over prolonged periods of time, but if the exposure was of a single dose (i.e., a spill or storm runoff) as reported here, then total population abundances should recover to control levels within approximately 40 days, with only minimal changes in community structure.

Impact of DIATOX® on Phytoplankton Populations

Although significant reductions (direct impact) in total phytoplankton abundance and biovolume, and in the abundance of Chlorophyta or Chrysophyta, were not observed, there was a noticeable reduction in unicellular Cyanophyta on Days 4 and 7 posttreatment. This decrease was not statistically significant ($p > 0.05$), but mean reductions relative to controls were 32 and 55% on Days 4 and 7, respectively. The significant reduction in chlorophyll *a* concentration on Day 7 at the 1.50 mg/l DIATOX® treatment similarly suggested that there was a negative impact on the phytoplankton community.

Other published data also suggest that blue-green algae may be more sensitive to chlorophenols than other common phytoplankton phyla. Palmer and Maloney[67] demonstrated that 2.0 mg/l NaPCP was toxic to blue-green algae, but only partially or nontoxic to green algae. Similarly, Brockway et al.[28] found that 1.0 mg/l PCP reduced the abundance of the less dominant blue-green algae genera *Anabaena* in laboratory microcosms, but did not affect dominant green algae and diatom genera.

Although we observed no statistically significant reductions in the abundance of dominant phytoplankton taxa, there was a significant posttreatment decline (16 to 26%) in dissolved oxygen (DO) concentrations after both DIATOX® treatments.

This suggested that the reduction (32 to 55%) in Cyanophyta abundance between Days 4 and 7 may have been sufficient to significantly reduce DO, or that DO, and thus possibly photosynthesis, was a more sensitive (less variable) indicator of chlorophenol stress than abundance. Similar reductions in photosynthesis have been reported by a number of other researchers[68-70] following PCP treatments of 0.1 to 0.5 mg PCP per l.

In our experiment, transparency in treated enclosures increased significantly on Day 7. This further suggested that an impact on phytoplankton abundance had occurred, since Secchi depth in lakes devoid of suspended solids is to a great extent a function of light absorption and scattering properties of algal cells.[71] Zooplankton populations were also severely diminished on Day 7, thus enhancing this increase in transparency. Changes in the particulate fractions of C, N, and P could similarly be explained by changes in the planktonic communities. In fact, examination of all water chemistry data revealed no additional meaningful information about the impact of DIATOX® on the test system, a conclusion similarly derived by Brockway et al.[28] in an experiment with PCP.

The most pronounced effect on the phytoplankton community was the increase in abundance of blue-green algae observed 14 to 28 d after the 1.50 mg/l DIATOX® treatment (Figure 10a). This delayed increase was a typical example of an indirect or secondary effect, but the exact cause is unclear. One possible explanation is that the reduction in total zooplankton abundance (93.1%), which resulted from the high DIATOX® treatment, reduced grazing pressure on the algal community. This may have allowed opportunistic phytoplankton species to flourish. This is supported by Haney,[72,73] who showed that *Daphnia* and other cladocerans were responsible for approximately 80% of the community grazing rate in a small lake and by data from other researchers[74,75] who have also observed phytoplankton "blooms", including one of filamentous cyanophytes, following severe reductions in zooplankton populations. Since the algal bloom in our study was not observed until Day 14 post-treatment, it is possible that phytoplankton growth/reproduction may have been initially suppressed by either the chlorophenols or their degradation products. Once residue concentrations dropped below effective levels, the most opportunistic taxon, in this case *Oscillatoria,* could rapidly multiply because of reduced competition for available resources and diminished zooplankton grazing pressure. That the increase was observed only in filamentous blue-green species may be indicative of the ungrazable or undigestible nature of these "poor food source" species to *Daphnia.*[60] In fact, cyanophytes have been reported to be relatively free from zooplankton grazing, appearing to inhibit zooplankton filtering rates either mechanically or chemically.[76] Furthermore, species of *Oscillatoria* are known to dominate many low turbulence, nutrient poor systems and display highest growth rates during circulation periods[77] (e.g., late season limnocorrals).

Similar observations of direct and indirect responses were reported by Schauerte et al.[26] who investigated the effects of 5 mg/l 2,4,6-TCP and 1 mg/l PCP on plankton population dynamics in separated compartments of an experimental pond. Although not conclusive, population densities of *Chroococcus limneticus* (Cyanophyta) decreased after treatment (primary effect), whereas *Trachelomonas hispida* and *Euglena acus* (Euglenophyta) increased in abundance after treatment. The increase in Euglenophyta (secondary effect) was attributed to the release from grazing pressure by *Daphnia* which were completely eliminated from treated compartments and did not recover during the 24-day experiment. Secondary effects have also been reported in other mesocosm studies with compounds such as atrazine[78] and coal-derived oil.[79]

CONCLUSION

When a toxicant enters an aquatic ecosystem, it affects not only populations of single species, but sets of interacting populations of different species. Since most species do not respond equally to all or any toxicant, differential effects should be expected. As a result of complex and often unknown interactions among these populations, it is often difficult to predict exactly what effects a toxicant will have. Therefore, when assessing the effects of a toxicant on aquatic ecosystems, one should not just focus on direct effects, but look further to possible secondary effects, resulting from the decline or elimination of sensitive populations. In fact, it would appear that most direct effects can be reasonably estimated from laboratory toxicity tests, whereas indirect effects are often more difficult to predict.

Limnocorrals have proven to be a valuable tool for the assessment of both direct and indirect effects of anthropogenic contaminants, as they enclose natural, interacting populations which are usually similar in both taxonomy and relative abundance to the populations in the surrounding lake. The simultaneous exposure of numerous species allows for the identification of sensitive taxa and observation of changes in community structure and function resulting from the decline or removal of such taxa. Limnocorrals also allow for the assessment of recovery of affected populations and associated shifts in community structure. In the work presented here, rotifers were identified as the most sensitive zooplankton taxon to DIATOX® and a bloom of blue-green algae resulted indirectly from the DIATOX® treatment most likely due to reductions in the abundance of zooplankton grazers. Recovery of most impacted populations took place by Day 42 posttreatment, but the resulting communities at the 1.50-mg/l DIATOX® treatment were substantially different from pretreatment and control populations.

ACKNOWLEDGMENTS

We gratefully acknowledge the assistance of John Warner with the collection of field data and the enumeration of zooplankton samples, and of Jim Virtue with the identification and enumeration of phytoplankton samples. We also thank Dr. D.R.S. Lean (CCIW) for providing the analysis of water chemistry samples. Financial support for this project was provided by Environment Canada through a DSS contract (02SEKW405-6-2060) and by the World Wildlife Toxicology Fund.

REFERENCES

1. Cairns, J. Jr. 1983. Are single species toxicity tests alone adequate for estimating environmental hazard? *Hydrobiologia* 100:47–57.
2. Boyle, T.P., S.E. Finger, R.L. Paulson, and C.F. Rabeni. 1985. Comparison of laboratory and field assessment of fluorene — Part II: Effects on the ecological structure and function of experimental pond ecosystems. In T.P. Boyle, ed., *Validation and Predictability of Laboratory Methods for Assessing the Fate and Effects of Contaminants in Aquatic Ecosystems,* STP 865. American Society for Testing and Materials, Philadelphia, PA, pp. 134–151.
3. Giddings, J.M. 1986. Protecting aquatic resources: An ecologist's perspective. In T.M. Poston and R. Purdy, eds., *Aquatic Toxicology and Environmental Fate,* Ninth Volume. STP 921. American Society for Testing and Materials, Philadelphia, PA, pp. 97–104.
4. Cairns, J. Jr. 1988. What constitutes field validation of predictions based on laboratory evidence? In W.J. Adams, G.A. Chapman and W.G. Landis, eds., *Aquatic Toxicology and Hazard Assessment,* Tenth Volume. STP 971. American Society for Testing and Materials, Philadelphia, PA. pp. 361–368.
5. Solomon, K.R., K. Smith, G. Guest, J.Y. Yoo, and N.K. Naushik. 1980. Use of limnocorrals in studying the effects of pesticides in the aquatic ecosystem. *Can. Tech. Rep. Fish. Aquat. Sci.* 975:1–9.
6. Kaushik, N.K., G.L. Stephenson, K.R. Solomon, and K.E. Day. 1985. Impact of permethrin on zooplankton communities in limnocorrals. *Can. J. Fish. Aquat. Sci.* 42:77–85.
7. Kaushik, N.K., K.R. Solomon, G.L. Stephenson, and K.E. Day. 1986. Use of limnocorrals in evaluating the effects of pesticides on zooplankton communities. In J. Cairns, Jr., ed., *Community Toxicity Testing.* STP 920. American Society for Testing and Materials, Philadelphia, PA. pp. 269–290.
8. Day, K.E., N.K. Kaushik, and K.R. Solomon. 1987. Impact of fenvalerate on enclosed freshwater planktonic communities and on *in situ* rates of filtration of zooplankton. *Can. J. Fish. Aquat. Sci.* 44:1714–1728.
9. Hamilton, P.B., G.S. Jackson, N.K. Kaushik, K.R. Solomon, and G.L. Stephenson. 1988. The impact of two applications of atrazine on the plankton communities of in situ enclosures. *Aquat. Toxicol.* 13:123–140.
10. Solomon, K.R., G.L. Stephenson, and N.K. Kaushik. 1989. Effects of methoxychlor on zooplankton in freshwater enclosures: Influence of enclosure size and number of applications. *Environ. Toxicol. Chem.* 8:659–669.
11. Jones, P.A. 1981. Chlorophenols and their impurities in the Canadian environment. Economic and Technical Review, March 1981. Report EPS 3-EC-81-2, Environmental Impact Control Directorate, Ottawa, Canada.
12. National Research Council of Canada. 1982. Chlorinated phenols: Criteria for environmental quality. National Research Council of Canada. NRCC No. 18578, Ottawa, Canada.
13. Fox, M.E. and S.R. Joshi. 1984. The fate of pentachlorophenol in the Bay of Quinte, Lake Ontario. *J. Great Lakes Res.* 10:190–196.
14. World Health Organization. 1987. Environmental Health Criteria 71: Pentachlorophenol. IPCS International Programme on Chemical Safety. World Health Organization, Geneva, Switzerland.

15. World Health Organization. 1989. Environmental Health Criteria 93: Chlorophenols other than pentachlorophenol. IPCS International Programme on Chemical Safety. World Health Organization, Geneva, Switzerland.

16. Carey, J.H., M.E. Fox, and J.H. Hart. 1988. Identity and distribution of chlorophenols in the north arm of the Fraser River estuary. *Water Poll. Research J. Canada* 23:31–44.

17. McLeese, D.W., V. Zitko, and M.R. Peterson. 1979. Structure-lethality relationships for phenols, anilines and other aromatic compounds in shrimp and clams. *Chemosphere* 2:53–57.

18. LeBlanc, G.A. 1980. Acute toxicity of priority pollutants to water flea *(Daphnia magna). Bull. Environ. Contam. Toxicol.* 24:684–691.

19. Buccafusco, R.J., S.J. Ells, and G.A. LeBlanc. 1981. Acute toxicity of priority pollutants to bluegill *(Lepomis macrochirus). Bull. Environ. Contam. Toxicol.* 26:446–452.

20. Hattula, M.L., V.-M. Wasenius, H. Reunanen, and A.U. Arstila. 1981. Acute toxicity of some chlorinated phenols, catechols and cresols to trout. *Bull. Environ. Contam. Toxicol.* 26:295–298.

21. Rao, K.R., F.R. Fox, P.J. Conklin, and A.C. Cantelmo. 1981. Comparative toxicology and pharmacology of chlorophenols: studies on the grass shrimp, *Paleomonetes pugio*. In F.J. Vernberg, A. Calabrese, F.P. Thurberg, and W.B. Vernberg, eds., *Biological Monitoring of Marine Pollutants*. Academic Press, New York, NY, pp. 37–72.

22. Ribo, J.M. and K.L.E. Kaiser. 1983. Effects of selected chemicals to photoluminescent bacteria and their correlations with acute and sublethal effects on other organisms. *Chemosphere* 12:1421–1442.

23. Crossland, N.O. and C.J.M. Wolff. 1985. Fate and biological effects of pentachlorophenol in outdoor ponds. *Environ. Toxicol. Chem.* 4:73–86.

24. Pignatello, J.J., M.M. Martinson, J.G. Steiert, R.E. Carlson, and R.L. Crawford. 1983. Biodegradation and photolysis of pentachlorophenol in artificial freshwater streams. *Appl. Environ. Microbiol.* 46:1024–1031.

25. Robinson-Wilson, E.F., T.P. Boyle, and J.D. Petty. 1983. Effects of increasing levels of primary production on pentachlorophenol residues in experimental pond ecosystems. In W.E. Bishop, R.D. Cardwell, and B.B. Heidolph, eds., *Aquatic Toxicology and Hazard Assessment:* Sixth symposium. STP 802. American Society for Testing and Materials, Philadelphia, PA. pp. 239–251.

26. Schauerte, W.S., J.P. Lay, W. Klein, and F. Korte. 1982. Influence of 2,4,6-trichlorophenol and pentachlorophenol on the biota of aquatic systems. *Chemosphere* 11:71–79.

27. Knowlton, M.F. and J.N. Huckins. 1983. Fate of radiolabeled sodium pentachlorophenate in littoral microcosms. *Bull. Environ. Contam. Toxicol.* 30:206–213.

28. Brockway, D.L., P.D. Smith, and F.E. Stancil. 1984. Fate and effects of pentachlorophenol in hard- and soft-water microcosms. *Chemosphere* 13:1363–1377.

29. Crosby, D.G. 1981. Environmental chemistry of pentachlorophenol. *Pure Appl. Chem.* 53:1051–1080.

30. Stirling, G. and D.J. McQueen. 1987. The cyclomorphic response of *Daphnia galeata mendotae:* polymorphism or phenotypic plasticity. *J. Plankton Res.* 9:1093–1112.

31. Solomon, K.R., J.Y. Yoo, D. Lean, N.K. Kaushik, K.E. Day, and G.L. Stephenson. 1985. Dissipation of permethrin in limnocorrals. *Can. J. Fish. Aquat. Sci.* 42:70–76.

32. Liber, K., N.K. Kaushik, K.R. Solomon, and J.H. Carey. 1992. Experimental designs for aquatic mesocosm studies: a comparison of the "ANOVA" and "regression" design for assessing the impact of tetrachlorophenol on zooplankton populations in limnocorrals. *Environ. Toxicol. Chem.* 11:61–77.

33. Krahn, P.K. and J.A. Shrimpton. 1988. Stormwater related chlorophenol releases from seven wood protection facilities in British Columbia. *Water Poll. Research J. Canada* 23:45–54.

34. Solomon, K.R., K. Smith, and G.L. Stephenson. 1982. Depth integrating samplers for use in limnocorrals. *Hydrobiologia* 94:71–75.

35. Solomon, K.R., C.S. Bowhey, K. Liber, and G.R. Stephenson. 1988. Persistence of Hexazinone (Velpar), Triclopyr (Garlon), and 2,4-D in a northern Ontario aquatic environment. *J. Agric. Food Chem.* 36:1314–1318.

36. Liber, K., K.R. Solomon, and J.H. Carey. Persistence and fate of 2,3,4,6-tetrachlorophenol and pentachlorophenol in limnocorrals. Submitted to *Environ. Toxicol. Chem.*

37. Stephenson, G.L., N.K. Kaushik, K.R. Solomon, and K. Day. 1986. Impact of methoxychlor on freshwater communities of plankton in limnocorrals. *Environ. Toxicol. Chem.* 5:587–603.

38. Pennak, R.W. 1978. *Fresh-water Invertebrates of the United States,* 2nd ed. John Wiley & Sons, New York, NY.

39. Haney, J.F. and D.J. Hall. 1973. Sugar-coated *Daphnia:* a preservation technique for Cladocera. *Limnol. Oceanogr.* 18:331–333.

40. Ward, H.B. and G.C. Whipple. 1959. *Fresh-water Biology,* 2nd ed. W.T. Edmondson, ed. John Wiley & Sons, Inc., New York, NY.

41. Grothe, D.W. and D.R. Grothe. 1977. An illustrated key to the planktonic rotifers of the Laurentian Great Lakes. United States Environmental Protection Agency, Region V, Central Regional Laboratory, Chicago, IL.

42. Environment Canada. 1979. *Analytical Methods Manual.* Environment Canada, Ottawa, Canada.

43. Lund, J.W.G., C. Kipling, and E.D. Le Cren. 1958. The inverted microscope method of estimating algal numbers and the statistical basis of estimations by counting. *Hydrobiologia* 11:143–170.

44. Prescott, G.W. 1962. *Algae of the Western Great Lakes Area.* Wm. C. Brown Publishers, Dubuque, IA.

45. Smith, G.M. 1950. *The Fresh-water Algae of the United States.* McGraw-Hill Book Company, Inc., New York, NY.

46. Findlay, D.L. and H.J. Kling. 1979. *A Species List and Pictorial Reference to the Phytoplankton of Central and Northern Canada,* Parts I and II. Fisheries and Marine Service, Manuscript Report No. 1503, Winnipeg, Canada.

47. Palmer, C.M. 1977. Algae and water pollution: An illustrated manual on the identification, significance and control of algae in water supplies and in polluted water. Office of research and development, United States Environmental Protection Agency, Cincinnati, OH.

48. Taft, C.E. and C.W. Taft. 1971. *The Algae of Western Lake Erie.* Bulletin of the Ohio Biological Survey New Series 4.

49. Patrick, R., C.W. Reimer, and S.-I. Yong. 1975. *The Diatoms of the United States,* Vol. 2, Part 1. Monographs of the Academy of Natural Sciences of Philadelphia, Philadelphia, PA.

50. Patrick, R. and C.W. Reimer. 1966. *The Diatoms of the United States,* Vol. 1. Monographs of the Academy of Natural Sciences of Philadelphia, Philadelphia, PA.
51. Janus, L.L. and H.C. Duthie. 1979. Phytoplankton composition and periodicity in a northeastern Quebec lake. *Hydrobiologia* 63:129–134.
52. Steel, R.D.G. and J.H. Torrie. 1980. *Principles and Procedures of Statistics: A Biometrical Approach.* Second Edition. McGraw-Hill Book Company, New York, NY.
53. SAS Institute. 1987. *SAS® STAT Guide for Personal Computers. Version 6.03.* Statistical Analysis Systems Inc., Cary, NC.
54. Fryer, G. 1985. Crustacean diversity in relation to the size of water bodies: some facts and problems. *Freshwater Biol.* 15:347–361.
55. Carpenter, S.R., J.F. Kitchell, and J.R. Hodgson. 1985. Cascading trophic interactions and lake productivity. *BioScience* 35:634–639.
56. McQueen, D.J., M.R.S. Johannes, J.R. Post, T.J. Stewart, and D.R.S. Lean. 1989. Bottom-up and top-down impacts on freshwater pelagic community structure. *Ecological Monogr.* 59:289–309.
57. Threlkeld, S.T. and E.M. Choinski. 1987. Rotifers, cladocerans and planktivorous fish: What are the major interactions? *Hydrobiologia* 147:239–243.
58. Mills, E.L. and J.L. Forney. 1983. Impact on *Daphnia pulex* of predation by young yellow perch in Oneida Lake, New York. *Trans. Am. Fish. Soc.* 112:154–161.
59. McQueen, D.J., J.R. Post, and E.L. Mills. 1986. Trophic relationships in freshwater pelagic ecosystems. *Can. J. Fish. Aquat. Sci.* 43:1571–1581.
60. McQueen, D.J. and J.R. Post. 1988. III. Lakes. 8. Mesocosms. Limnocorral studies of cascading trophic interactions. *Verh. Internat. Verein. Limnol.* 23:739–747.
61. McQueen, D.J. and J.R. Post. 1988. Cascading trophic interactions: Uncoupling at the zooplankton-phytoplankton link. *Hydrobiologia* 159:277–296.
62. Liber, K. 1990. Persistence and Biological Effects of a Commercial Tetrachlorophenol Formulation in Aquatic Ecosystems: Laboratory and Limnocorral Studies. Ph.D. Thesis. University of Guelph, Guelph, Ontario, Canada.
63. Wright, D.I. and W.J. O'Brien. 1984. Model analysis of the feeding ecology of a freshwater planktivorous fish. In D.G. Meyers and J.R. Strickler, eds., *Trophic Interactions within Aquatic Ecosystems,* AAAS selected symposium 85. Westview Press, Inc., Boulder, CO, pp. 243–267.
64. Hwang, H., R.E. Hodson, and R.F. Lee. 1987. Photolysis of phenol and chlorophenols in estuarine water. In R.G. Zika and W.J. Cooper, eds., *Photochemistry of Environmental Aquatic Systems.* American Chemical Society, Washington, D.C., pp. 27–43.
65. Wong, A.S. and D.G. Crosby. 1981. Photodecomposition of pentachlorophenol in water. *J. Agric. Food Chem.* 29:125–130.
66. Jones, P.A. 1984. Chlorophenols and their impurities in the Canadian environment: 1983 supplement. Economic and Technical Review, March 1984, Report EPS 3-EP-84-3, Environmental Protection Programs Directorate, Ottawa, Canada.
67. Palmer, C.M. and T.E. Maloney. 1955. Preliminary screening for potential algicides. *Ohio J. Sc.* 55:1–8.
68. Erickson, S.J. and C.E. Hawkins. 1980. Effects of halogenated organic compounds on photosynthesis in estuarine phytoplankton. *Bull. Environ. Contam. Toxicol.* 24:910–915.
69. Buikema, A.L., Jr., M.J. McGinniss, and J. Cairns, Jr. 1979. Phenolics in aquatic ecosystems: A selected review of recent literature. *Mar. Environ. Res.* 2:87–181.

70. Adema, D.M.M. and G.J. Vink. 1981. A comparative study of the toxicity of 1,1,2-trichloroethane, dieldrin, pentachlorophenol and 3,4 dichloroaniline for marine and fresh water organisms. *Chemosphere* 10:533–554.

71. Tilzer, M.M. 1988. Secchi disk-chlorophyll relationships in a lake with highly variable phytoplankton biomass. *Hydrobiologia* 162:163–171.

72. Haney, J.F. 1971. An *in situ* method for the measurement of zooplankton grazing rates. *Limnol. Oceanogr.* 16:970–977.

73. Haney, J.F. 1973. An *in situ* examination of the grazing activities of natural zooplankton communities. *Archiv. Hydrobiol.* 72:87–132.

74. Stephenson, R.R. and D.F. Kane. 1984. Persistence and effects of chemicals in small enclosures in ponds. *Arch. Environ. Contam. Toxicol.* 13:313–326.

75. Salki, A., M. Turner, K. Patalas, J. Rudd, and D. Findlay. 1985. The influence of fish-zooplankton-phytoplankton interactions on the results of selenium toxicity experiments within large enclosures. *Can. J. Fish. Aquat. Sci.* 42:1132–1143.

76. Richard, D.I., J.W. Small, Jr. and J.A. Osborne. 1985. Response of zooplankton to the reduction and elimination of submerged vegetation by grass carp and herbicide in four Florida lakes. *Hydrobiologia* 123:97–108.

77. Steinberg, C.E.W. and H.M. Hartmann. 1988. Planktonic bloom-forming Cyanobacteria and the eutrophication of lakes and rivers. *Freshwater Biol.* 20:279–287.

78. deNoyelles, F., W.D. Kettle, and D.E. Sinn. 1982. The response of plankton communities in experimental ponds to atrazine, the most heavily used pesticide in the United States. *Ecology* 63:1285–1293.

79. Giddings, J.M., P.J. Franco, R.M. Cushman, L.A. Hook, G.R. Southworth, and A.J. Stewart. 1984. Effects of chronic exposure to coal-derived oil on freshwater ecosystems: II. Experimental ponds. *Environ. Toxicol. Chem.* 3:465–488.

CHAPTER 18

Management and Treatment of Mesocosms: Summary and Discussion

Robert L. Graney

This section focused on two aspects of mesocosm testing: management of test systems (Chapter 14) and chemical exposure (Chapters 15 to 17). Although both of these aspects are critical to the conduct of mesocosm studies, little definitive information can be found in the literature which adequately addresses these issues. Case studies which have been published provide descriptive details on how the test systems were constructed and/or managed, but do not provide comparative discussions on how such practices may have influenced the test results. Similarly, there is considerable discussion about the importance of the exposure profile in determining the actual effects observed; however, there is little definitive information discussing the different approaches for dosing test systems and, once dosed, the influence of the compound's fate on the effects observed. The manuscripts provided in this section provided some insight into these areas.

There are numerous factors which must be considered when establishing an "artificial" aquatic ecosystem. Important factors include: (1) system construction, (2) source of inoculum, (3) time required for colonization, (4) control of system components such as macrophytes, (5) exclusion of undesirable components such as tadpoles, and (6) fish stocking and management. The first chapter in this section addressed one component of this, specifically the source of inoculum and its im-

0-87371-592-6/94/$0.00 + $.50
© 1994 by CRC Press, Inc.

portance in the development of the benthic fauna. The bottom line conclusion from this manuscript was that by "dredging" sediment from local ponds and transferring this sediment to mesocosms, the resulting benthic community is representative of the local fauna. The authors conclude that this aspect of mesocosm construction is acceptable for the "forced" colonization of artificial ecosystems. Also, that from an invertebrate perspective, ponds can be established which are representative of typical benthic fauna.

However, other aspects of mesocosm development are as yet unresolved. One component relates to macrophytes, which can be critical for the development and maintenance of an appropriate invertebrate fauna (deNoyelles and Dewey, Chapter 30). Macrophytes provide a physical substrate for colonization and a refugia for invertebrate prey species. Unfortunately, macrophytes can also cause problems in "artificial" test systems. Macrophyte development can vary among ponds, and due to their importance in the ecosystems, this can introduce unwanted variability in both structural and functional parameters. Excessive macrophyte growth can also cause sampling problems.

Another major concern in the development of mesocosm ecosystems is the establishment of a single species fish population, its influence on the test system, and its representativeness of a natural ecosystem. Most pond mesocosms are initially colonized without fish present. Invertebrate communities are established and develop without the influence of a top predator. In early spring, generally March to April depending on the geographical region, reproductively mature bluegill are introduced into the ponds. Reproduction begins almost immediately and, at the time of dosing, young bluegill are generally present in the system. The introduction of the "new" predator (young and adult bluegill) alters the structural and functional components of the ecosystem. Bluegill reproduction continues into the summer, and by test termination ($\sim$ October to November), young bluegill can number between 10 to 20,000 per pond (0.1 to 0.25-acre ponds). The impact of this density of bluegill on young invertebrate communities structure can be substantial. In addition, the growth of the juvenile fish may be constrained by the food supply and indirect effects may be more difficult to discern. Because of the continued, uncontrolled fish production through the summer, it is unlikely the system is in equilibrium or representative of a "stable" or natural aquatic community.

The importance of fish in the test system was further illustrated by Liber et al. (Chapter 17). In limnocorrals containing different densities of planktivorous fish, the variability in zooplankton numbers was much greater than in limnocorrals with no fish. Although this is not a surprising result, it does emphasize again the importance of "top down" control of fish and how they can influence the structure of the invertebrate community.

The Aquatic Effects Dialogue Group (AEDG) (Anonymous, 1992) spent a considerable amount of time discussing the use of more realistic fish populations and solutions to the fish "problem". Unfortunately, no data exists for making decisions on how to improve the system, and the AEDG concluded that prioritization of potential approaches is required. There is no doubt that more research is needed on the adaptation of some of the suggested improvements, which include the use of fish communities in mesocosm studies, addition of predators to control excessive fish production, use of enclosures, and the use of alternative species.

The bottom line question is "are the systems representative of typical aquatic ecosystems"? The manuscript by Ferrington et al., would seem to indicate that for the chironomed community, the answer is yes. Comparative evaluations of the other components of the benthic and plankton communities is necessary before definitive statements can be made for these specific components. Relative to fish, there is no doubt that the single-species population established is artificial and the excessive production of young bluegill has a direct impact on the system. It is unknown how this influences the ability to extrapolate to other types of ecosystems and whether such problems alter any regulatory decisions which may result from such studies.

Given the authors conclusion from Chapter 14 concerning the representativeness of the benthic fauna, can the next step be taken and conclude that the effects observed in the mesocosm test system would be representative of the effects observed if *similar exposure had occurred* in a real field situation? This may be an accurate extrapolation; however, the greatest uncertainty in such an extrapolation is associated with the phrase "if similar exposure had occurred". This leads into the subject matter of the remaining three chapters in this section, that is, the immense importance of "exposure" in simulated field studies. Relative to pond mesocosm studies, there are two fairly independent components to the exposure equation. First, there is the necessity to determine the concentrations which may be transported to aquatic habitats. For pesticides, this generally occurs via either overland surface runoff and/or aerial drift. Chapter 15 in this section addressed this component (Hill et al.). The second component is the actual exposure of aquatic organisms once the chemical is in the receiving systems. The physical/chemical properties of the compound and the compound's relative stability will determine the exposure profile within the system. Chapters 16 (La Point and Fairchild) and 17 (Liber et al.) address this component.

The chapter by Hill et al. provided an excellent overview of the critical components which must be considered when determining the dose levels for a mesocosm test. The factors are particularly relevant for a study designed to test an hypothesis concerning effects at a specific dose level (ANOVA approach; Graney et al., 1989; Touart, Chapter 4); however, they are not as critical for a study designed to develop dose-response information (Regression approach). The authors chose a particular class of pesticides, i.e., pyrethroids, and reviewed the procedures for estimating spray drift and runoff loading rates for a "typical" mesocosm study required by the EPA for regulatory compliance. Field studies were reviewed and typical loading rates and frequencies were recommended for pyrethroids. It is unnecessary to re-iterate the specific procedures they recommend for determining dosing rates; however, a restatement of some of their main points, perhaps from a somewhat different perspective, may be useful. Specifically:

1. The stochastic nature of both drift and runoff makes accurate predictions close to impossible. In the desire of regulatory authorities to be protective, this uncertainty can lead to an accumulation of worst case assumptions such that mesocosms are often tested under extreme exposure conditions. Such tests are difficult to interpret relative to "typical" conditions and often lessen the utility of mesocosm studies for risk managers attempting to evaluate the result of exposure mitigation measures or varied chemical use patterns (i.e., multiple crops or application procedures in many different geographical regions).

2. Factors such as buffer zones, good agricultural practices, and probabilities of occurrence must be considered when establishing the dose levels. Extensive efforts are being made by other government agencies (i.e., USDA) to reduce and/or control runoff and drift. These factors should be considered when determining the most appropriate dose levels.

3. The application procedure can influence the exposure profile within the test system. Since the results of the studies may be used by regulators in a comparative context, some form of "standardized" criteria should be established for treating the test systems. Treating across the pond surface with a soil slurry will certainly result in a different exposure profile than treating at a point source. Both approaches have positive and negative attributes which need to be clearly outlined and understood. For example, dosing with a soil slurry at a point source has been argued to be more realistic. However, if the appropriate soil particle sizes are not used when making the slurry for such dosings, exposure may be greatly underrepresented. The particle sizes which generally run off and which contain the greatest amount of pesticide load are smaller (i.e., clays), are not readily deposited and, due to turbulence in the receiving system, may be distributed throughout the pond. If top soil composed primarily of sand is used to make the soil slurry, depositing will occur extremely close to the point source and exposure in other parts of the system may be underrepresented. This may, for many products, be a minor factor; however, it emphasizes the importance and complexity associated with dosing the test system.

The uncertainty associated with the transport of chemicals to aquatic ecosystem is great. As discussed by Hill et al., this component of exposure must be addressed independent of the mesocosm study. Depending on the experimental design (Regression vs. ANOVA), this can occur either before or after the study is conducted (Graney et al., 1989). However, the fate of the compound within the pond system represents a critical attribute of the mesocosm study. Laboratory observed toxicity values do not represent the actual exposure under field conditions and may thus lead to inaccurate conclusions regarding the actual risk for a compound.

La Point and Fairchild (Chapter 16) discussed the importance of exposure relative to the trophic status of the system and the life history of the organisms. Understanding the influence of trophic status and other ecosystem properties on bioavailability is complex. Under eutrophic conditions the water column organic content (i.e., phytoplankton) is usually great. For hydrophobic compounds which bind to organic matter, bioavailability and thus toxicity may be reduced under these conditions. La Point and Fairchild review and discuss field data which tends to support this hypothesis. Other factors are also modified under euthrophic conditions such that bioavailability is not the only mechanism which may alter the exposure profile with ecosystem trophic status. For example, microbial concentrations may be greater in eutrophic waters such that degradation is greater. Conversely, under the same conditions, light penetration may be greatly reduced such that photolysis is minimized relative to that observed in oligotrophic conditions. In any case, the specific fate properties of the chemical will determine the influences of the trophic status on exposure, and thus effects, in aquatic systems.

Another example of the importance of chemical fate and the resultant exposure profile on the toxicity observed under field conditions is provided by Liber et al. (Chapter 17). Limnocorrals were treated with DIATOX®, a photolabile tetrachlorophenol, in the morning and the evening. Due to the immediate exposure to sunlight, the morning treatments resulted in a shorter exposure time (i.e., faster degradation). After the evening application, the water residues were more persistent and the effects observed were considerably greater. A factor as simple as the time of day the dosing was initiated had considerable impact on the effects observed. Given the size and complexity of some pond mesocosms, it often takes the entire day to dose all the test systems. If a photolabile compound is being investigated, this factor may add considerable variability to the results obtained.

Overall, these three chapters indicate the complexity and importance associated with dosing and exposure in mesocosm test systems. The effect observed can be influenced by the procedures used for dosing, the trophic status, the water quality of the systems, the life stage of the organisms, and the time of day the systems are dosed. Given the importance of all of these factors, it becomes difficult for a single study to provide predictive information regarding the likelihood of adverse effects occurring when an ecosystem is exposed to a specific chemical. It is impossible to know and understand all possible consequences associated with any action. Mesocosms are only surrogate ecosystems designed to provide additional information on the response of "natural" communities to a "typical" exposure profile which may occur. They cannot, and should not, be expected to address all possible factors or answer all possible questions. By understanding the processes which control the compound's fate (much of which is obtained in laboratory fate studies) and measuring the effects which occur under these defined exposure profiles, reasonable interpretations can be made regarding the potential effects which may occur in the field. Such interpretations can be greatly enhanced by understanding how the effects change with changing chemical concentrations.

REFERENCES

Anonymous. 1992. Improving Aquatic Risk Assessment Under FIFRA. Report of the Aquatic Effects Dialogue Group. 98 pages.

Graney, R.L., Gresy, J.P. and Toro, D. 1989. Mesocosm Experimental Design Strategies: Advantages and Disadvantages in Ecological Risk Assessment. In: *Using Mesocosms to Assess the Aquatic Ecological Risk of Pesticides: Theory and Practice*. Entomological Society of America. ISSN 0071-0717. pp. 74–88.

SECTION IV

Scaling

INTRODUCTION

The four chapters in this section focus on various aspects of scaling including physical, chemical, and biological or ecological. An experimental approach to evaluating the risk of pesticides in aquatic systems involves scaling in both time and space. For example, colonization rates of benthic invertebrates that may be important for evaluation of effects of insecticides can be accelerated by transplants of colonized sediments from uncontaminated ponds. Further, reciprocal transplants of sediments and water circulation among replicate mesocosms may minimize variance due to unequal colonization and permit pesticide effects to be statistically detected. In another example, effects observed in mesocosms must be scaled to other aquatic ecosystems in order to make defensible regulatory decisions. This is a fascinating area of ecological research that is unfortunately supported by relatively little data. These four chapters represent contributions to further understanding of some specific scaling concerns as applied to experimental evaluation of pesticide impacts for registration decisions.

Some of the chapters were concerned with physical scale or the minimum dimensions of aquatic ecosystems that could encompass the biological or ecological characteristics of concern that will be used to evaluate the effects of a pesticide.

Two chapters examine the fate and effects of a pyrethroid insecticide in pond mesocosms and outdoor microcosms. Chapter 21 reports that although exposure (aqueous half-life and sediment concentrations) was different in the two systems, parallel biological response patterns were observed for zooplankton, macroinvertebrate colonization and aquatic insect emergence. Fish reproduction in the pond mesocosms tended to decrease zooplankton densities and perhaps impacted other invertebrates. However, Morris et al. (Chapter 22) were able to delineate secondary effects on nonreproductive bluegill from prey depletion attributed to the pesticide in microcosms. This implies that relative sensitivity of scaled aquatic systems may be largely a function of the kind of experiments that are chosen, components, and conditions. Ultimately, careful professional judgment must be involved in order to extrapolate experimental results from any study to other sites and situations.

CHAPTER 19

Fate and Biological Effects of an Herbicide on Two Artificial Pond Ecosystems of Different Size

Fred Heimbach, Juergen Berndt, and Wolfgang Pflueger

Abstract: The fate and biological effects of Goltix®, a herbicide containing metamitron as the active ingredient, were studied in two artificial pond systems of different size. The smaller (outdoor microcosm) ponds (5-m³ stainless steel cylinders, 2 m diameter) contained a bottom layer of natural lake sediment. Benthic organisms, all trophic levels of planktonic organisms, and caged rainbow trout were examined. The larger (mesocosm) ponds (75 m² water surface, maximum water depth 1.75 m, average depth 1.0 m) had a sediment layer on the bottom and side slopes. In these ponds, aquatic plants were also examined. The herbicide was applied twice to each pond with a two-week interval: one pond of each type was treated with 0.5 kg AI/ha had a second one with 5 kg AI/ha. Over a period of 5 months, the effects on the ecosystem (phytoplankton, zooplankton, macrobenthos, aquatic plants, and fish), and the distribution and fate of this herbicide, were examined. The results show concordance between both pond systems, as has been shown in earlier studies with other pesticides. With Goltix®, almost no biological effects were observed, although the laboratory results on the phytoplankton species *Scenedesmus subspicatus* indicate a potential hazard to these organisms at the very high, exaggerated application rates chosen for the study. The missing effects on phytoplankton in these micro- and mesocosm studies are explained by the very short half-life of metamitron in water under natural conditions.

INTRODUCTION

The environmental fate and effects of pesticides in aquatic ecosystems have been investigated by various authors (e.g., References 1 through 7). Most of these studies have been performed to investigate the effects of insecticides on the aquatic fauna as laboratory tests indicate high intrinsic toxicities of some insecticides to these organisms; only a few results are available for herbicides (e.g., References 8 through 13).

The fate and biological effects of the herbicide Goltix® containing metamitron as AI were examined. The ecotoxicological laboratory data indicate a low potential toxicity to aquatic organisms. On the other hand, the results of the growth inhibition test of a green algae population (*Scenedesmus subspicatus*) indicate that effects on the aquatic flora cannot be excluded at high concentrations in aquatic ecosystems. The high application rate selected for use in this study was based on laboratory results to achieve biological effects in the pond systems. It does not reflect a realistic contamination of natural aquatic systems in agricultural practice. The pond systems used in this study, small scale model ecosystems in stainless steel basins of 5 m³ content (microcosm) and larger test ponds of 75 m² water surface (mesocosm) had to be compared for their use as aquatic ecosystems for a hazard assessment of pesticides. Previous studies on a synthetic pyrethroid insecticide in these ponds had already revealed their usefulness as model systems for insecticides.[14,15]

MATERIAL AND METHODS

Description of Small Ponds ("Microcosm")

The ponds consisted of three cylindrical stainless steel containers[15] which were connected with each other by wide locks (Figure 1). A 15-cm layer of natural sediment taken from a nearby lake was distributed on the bottom of the containers about 6 months before the application of Goltix®. The containers were filled with approximately 4.5 m³ of well water. Zooplankton and phytoplankton organisms introduced with the sediment developed large populations and uniformly colonized all three ponds. Small rainbow trout (*Oncorhynchus mykiss*, average body length 5.6 cm, weight 1.9 g) were placed in stainless steel wire mesh cages of 30 × 20 × 20 cm (mesh size 6 mm) and suspended in the water column six weeks before the first application. Six cages, each containing one fish, were placed at the side of the ponds at a water depth of about 60 cm. The trout relied on the zooplankton, which freely moved in and out of the cages, as a food source. This setup allows almost identical starting conditions in all ponds since the locks are open, and there is a free exchange of water and organisms. In earlier studies all three ponds were shown to be biologically and chemically identical after separation of the ponds by closing the locks; divergence of the systems occurred after about 10 weeks.[14,15]

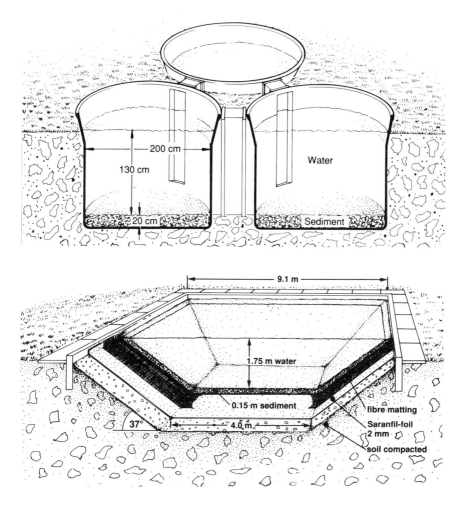

Figure 1. The stainless steel ponds (microcosm) (above) and the mesocosm ponds (below) filled with well water and natural lake sediment.

Description of Large Ponds ("Mesocosm")

The ponds consisted of four rectangular man-made ponds which were not connected with each other. The ponds were built as an upside down truncated pyramid. The water surface area was 75 m^2 (8.7 × 8.7 m), the bottom area 16 m^2. The maximum water depth was 1.75 m, the average depth 1.0 m. The ponds were lined with plastic sheets covered by a 15-cm layer of natural sediment taken from a nearby lake. To prevent sediment from sliding down the side slopes, a coconut fiber net was fixed on top of the plastic sheets. The sediment was distributed in the ponds and the ponds were filled with well water. Organisms introduced with the sediment, including several species of submergent and emergent aquatic macrophytes, developed large populations and uniformly colonized the ponds. The ponds were stocked with rainbow trout (*Oncorhyncus mykiss*, average body length 15.9 cm, weight 37 g), 14 free swimming fish per pond 6 weeks before the first application. To achieve identical starting conditions in all ponds as far as possible, the water of all ponds was mixed by pumping the water through pipes from 1 pond to the other for 5 weeks before the first application.

Test Compound and Dosing

The ponds were treated with a 70% water dispersible granule-formulation (WG) of Goltix®, a herbicide containing the AI metamitron (4-amino-3-methyl-6-phenyl-1,2,4-triazin-5(4H)-one).

For the studies in both pond systems, two applications of 0.5 and 5.0 kg metamitron/ha, respectively, were applied to the water surface of the ponds (corresponding to 0.7 and 7.0 kg Goltix® per ha, respectively) in a 14 day interval. This dosage corresponds to 10% and 100% of the normal agricultural application rate, and thus represents an exaggerated, unrealistic contamination. One microcosm pond was treated with the low dose, another one with the high dose, and the third one kept untreated as a control. For the mesocosm study two ponds served as control ponds while the low and high dosages were applied to the water surface of the third and fourth pond.

Goltix® was evenly applied to the water surface on the ponds of both sizes on May 22 and June 5, 1989 using usual agricultural spraying equipment.

Biological Sampling

Six samples each of phytoplankton, zooplankton, and benthic organisms were taken at regular intervals (Figures 3, 5, 6, 8) in all ponds. Water samples were collected with pumps as vertical, integrated cores and preserved for identification of organisms. A core substrate sampler was used to collect benthic invertebrates in microcosm ponds, while an Ekman grab substrate sampler was used in mesocosm ponds. The preserved zoo- and phytoplankton samples were evaluated individually on the sampling dates given in Figures 4 and 7; on all other sampling dates species were identified from a composite sample. Benthic organisms were identified from composite samples. The number of organisms per species was counted. Some of the less numerous organisms were only identified down to the genus level.

The occurrence of macrophytes was evaluated at regular intervals by subdividing the surface of the ponds into quadrants of 1 m $\times$ 1 m (Figure 9). The coverage of each quadrant was recorded from a portable walkway for each of the submergent and emergent species, and results were calculated as percent coverage of the water surface.

The trout in microcosm ponds were weighed and measured at regular intervals (Figure 2). The trout in mesocosm ponds were caught with nets for evaluation. As the macrophyte density increased during the study, the effectiveness of this method decreased. Additionally, 2 trout per pond were taken for residue analysis 1, 2, 3, and 4 months after the second application. Consequently, the results for the later period of the study given in Figure 2 refer to less organisms than at the beginning.

Chemical Analysis

The main physicochemical properties of the water such as pH, conductivity, and oxygen-concentration were determined at least at weekly intervals in all ponds using standard methods. Temperature was recorded continuously near the water surface and near bottom; in mesocosm ponds pH and oxygen-concentration were recorded continuously also. Other parameters such as nitrogen, phosphorus, and chlorophyll *a* were determined at the dates of biological sampling.

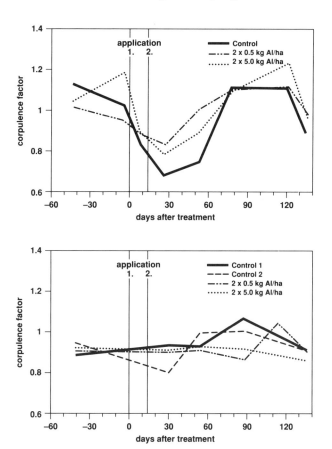

Figure 2. Corpulence factor of fish (*Oncorhyncus mykiss*) in microcosm ponds (above) and mesocosm ponds (below) sprayed with Goltix®. (Average of n = 6 caged trout in microcosms ponds, n = 14 free swimming trout in mesocosm ponds at the beginning; numbers decreasing for analysis samples.)

Samples for residue analysis from water, sediment, and fish were taken using the same biological sampling methods as given above. Samples of the water surface were taken with a fine mesh screen which was laid on the water surface for a few seconds. Macrophytes were collected at different places in each mesocosm pond by hand. The samples were frozen until analysis. Metamitron was extracted from the samples and determined by high pressure liquid chromatography.

RESULTS

Fish

Goltix® did not cause fish mortality (*Oncorhyncus mykiss*) and had no influence upon the development of the trout in either study (Figure 2). The corpulence factor of about 1.0 was constant in treated and control mesocosm ponds throughout the study period indicating that a normal amount of natural food was available for the trout (average body length at the end of the study 18.3 cm, average body weight 55 g). The corpulence factor was calculated from $k = 100 * w/l^3$, where w and l are the observed total weight and total length of a fish. The corpulence factor of trout in microcosm ponds decreased during the first 2 months of the study to 0.7 to 0.8 as a result of the limited food supply available to the caged fish. To avoid further stress or even mortality, the trout in microcosm ponds were fed small pellets of artificial commercial trout food 1 to 3 months after the first application. As a consequence, the corpulence factor increased during this period to normal values of 1.0 to 1.2 without differences between treated and control ponds (average body length at the end of the study 7.8 cm, average weight 4.7 g).

Zooplankton and Benthic Organisms

The application of Goltix® on the pond surface did not result in detectable effects upon the zooplankton and benthic organisms. The abundance of Cladocera, a major and sensitive group of zooplankton organisms, did not show any dose-related effect in either system (Figure 3). The fluctuations between sampling dates are within the same range in control and treated ponds. They correspond to natural variations of population densities of these organisms within the season.

A measure of similarity of the biocoenosis of different ponds, such as the Stander's index, can provide additional information on the comparison of treated and untreated ponds.[16-19] The variable in this permutation test is the ratio of "between-similarity" of all three or four ponds, respectively, and the "within-similarity" for each individual pond. A value of 1.0 means identity; markedly lower values indicate differences. The values for the zooplankton (Figure 4) show that initially all microcosm ponds can be considered to be identical without any significant divergence occurring further on. The setup of the four mesocosm ponds does not result initially in similarity values as high as in microcosm ponds. The Stander's index and the

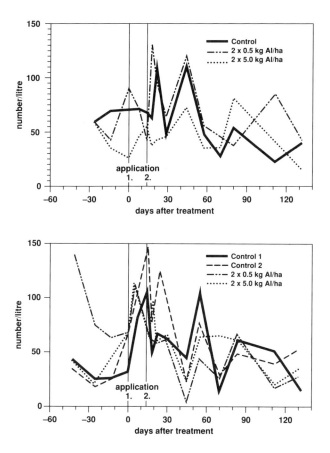

Figure 3. Number of Cladocera in microcosm ponds (above) and mesocosm ponds (below) sprayed with Goltix®.

critical limits for identity at $p = 0.05$ and 0.01 of mesocosm ponds (Figure 4) are markedly lower than those of microcosm ponds. Nevertheless, a relatively high similarity was recorded for about 2 months in mesocosm ponds as shown in Figure 4 for the high dosed and one control pond. Thereafter, significant divergence of the pond systems occurred, which cannot be interpreted as an effect of the application of Goltix® since the biocoenosis development diverged more and more between control ponds as well. As the mesocosm ponds were less similar even at the start of a study than microcosm ponds, differences between mesocosms occurring after some weeks have to be interpreted with care.

Phytoplankton

The total number of phytoplankton organisms in microcosm ponds was very similar in treated and untreated ponds before the first application and for a period of about 8 weeks thereafter. Then the higher dosed pond showed markedly higher cell numbers than the control and lower dosed pond (Figure 5), mainly caused by high numbers of Chlorophyta cells (see below). Differences in cell numbers of mesocosm ponds were higher than in microcosm ponds already before application. The cell numbers slightly increased after application in control and treated ponds for about 40 to 60 days without significant differences between treatments.

More than 30 phytoplankton species were identified in both systems (Figure 6), indicating diverse biocoenoses. The number of species fluctuated at different sampling dates from about 15 to 30 indicating no recognizable differences between treatments. (The Shannon-Weaver diversity index[20] was in the range of about 1.0 to 2.0.)

The similarity of the phytoplankton communities in microcosm ponds showed significant differences 56 days after the first application (Figure 7). This was mainly

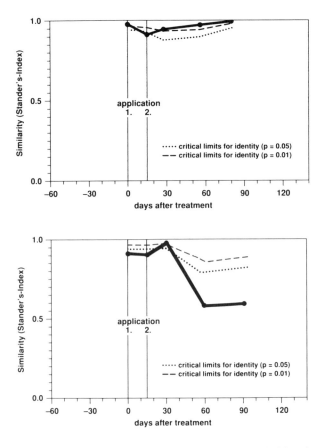

Figure 4. Similarity of zooplankton communities in microcosm ponds (above) and mesocosm ponds (below) sprayed with Goltix®. A value of 1.0 means identity; decreasing values indicate increasing differences.

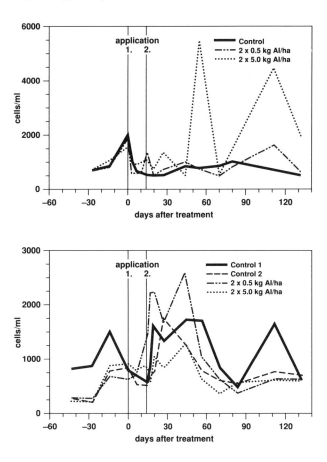

Figure 5.　Number of phytoplankton organisms in microcosm ponds (above) and mesocosm ponds (below) sprayed with Goltix®. The peaks in one microcosm pond are not considered as treatment related.

due to very high numbers of one Chlorophyta species (*Ankyra ancora*) in the high doses pond and some minor differences between abundances of several species of all genera. The spectrum of phytoplankton species was not influenced. This effect cannot be interpreted as treatment-related since earlier studies have shown that major divergences between microcosm ponds may naturally occur after about 8 to 10 weeks after separation.[14,15] In addition, the similarity index of mesocosm ponds was quite high even 8 weeks after application (Figure 7), indicating similar phytoplankton biocoenosis in all ponds.

Cryptophyta dominated the total cell numbers, followed by Diatomea, Cyanophyta, Chlorophyta, Conjugatophyceae, and Euglenophyta. As already stated for the total numbers of phytoplankton cells, also Cryptophyta (Figure 8) and the other phytoplankton groups did not show treatment-related differences between ponds in both systems.

Macrophytes

Eight species of emergent and submergent aquatic plants were derived from the sediment introduced to the mesocosm ponds: the submergent species *Ranunculus aquatilis, Zannichellia palustris, Callitriche* sp., *Potamogeton crispus, Potamogeton zizii,* and the emergent species *Alisma plantago-aquatica, Typha latifolia,* and *Sparganium erectum.* They did not spread uniformly within and between ponds. Nevertheless, the most abundant submergent species *R. aquatilis* covered 4 to 10 m² of the ponds and the most abundant emergent species *A. plantago-aquatica* 1 to 5 m². The results of the study showed no treatment-related effects on the abundance of these species (Figure 9) and all others. The reduction of *A. plantago-aquatica* in all ponds during the study was caused by a heavy aphid infestation which damaged the leaves of this species. During the first days after application,

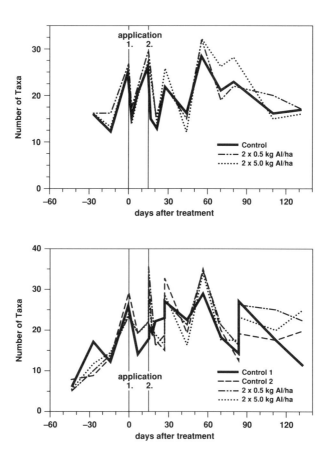

Figure 6.   Number of phytoplankton species in microcosm ponds (above) and mesocosm ponds (below) sprayed with Goltix®.

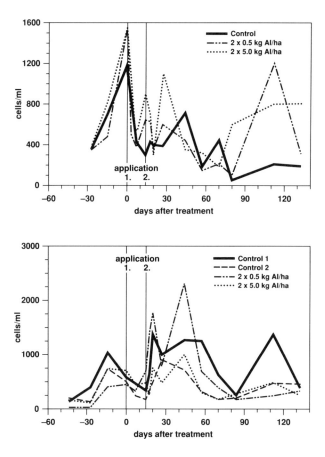

Figure 7. Number of Cryptophyta organisms in microcosm ponds (above) and mesocosm ponds (below) sprayed with Goltix®.

the leaves of some emergent plants in the higher dosed pond which had been directly sprayed with Goltix® showed some minor modifications in shape and color compared to untreated plants. This did not result in any recognizable further effects on the plants. After about 1 week, differences were no longer observed between treated and untreated plants. The application of Goltix® did not influence the further development of all species at both application rates.

Fate and Mass Balance of Goltix®

The solubility of metamitron, the AI of Goltix®, is high (1700 mg/l) and the adsorption low (KOC = 71 to 380 for 18 different soils), assuring a fast and uniform distribution of the AI in the water of the ponds. The hydrolysis stability of metamitron under laboratory conditions is low (half-life at pH 9: 10 days) and the light stability with a half-life of 30 min (direct photolysis) even lower. Consequently metamitron degrades rapidly under natural conditions in aquatic ecosystems to desaminometamitron, which itself undergoes photodegradation.

The fate of metamitron in the course of the experiments is shown in Figure 10 (percentage of total applied amount of metamitron; sum of first and second application after the second application). Only a small portion of the total amount applied was found at the water surface (<0.1%), where it declined rapidly. During the first days after application, the concentrations in the water increased and reached maximum values 1 to 7 days after application of 330 µg/l after the first application and 540 µg/l after the second in the high dosed mesocosm pond, 32 and 49 µg/l, respectively, in the low dosed one; 170 and 300 µg/l, respectively, in the high dosed microcosm, 16 and 28 µg/l, respectively, in the low dosed one; sum of metamitron and desaminometamitron each; the values declined rapidly. Metamitron

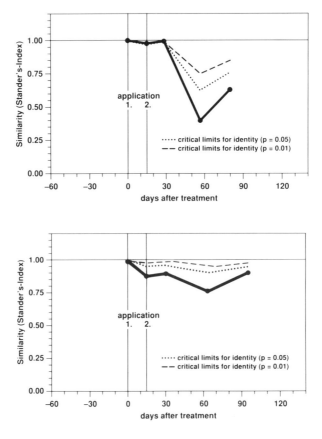

Figure 8.   Similarity of phytoplankton communities in microcosm ponds (above) and mesocosm ponds (below) sprayed with Goltix®. For further explanations, see Figure 4. The decrease after 7 weeks in the microcosm study is not attributed to treatments.

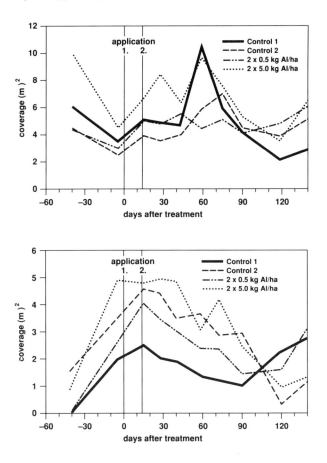

Figure 9.   Coverage of *Ranunculus aquatilis* (above) and *Alisma plantago-aquatica* (below) in mesocosm ponds sprayed with Goltix®.

was not detectable in the mesocosm pond water 14 days after each application. Desaminometamitron concentrations fell below or near to the detection limit of 1 µg/l after about 2 to 3 months. In microcosm ponds, metamitron concentrations were below the detection limit at all sampling dates, even shortly after application. The concentrations in these ponds refer to desaminometamitron only. At the end of the study after 4 months, the water in microcosm ponds still contained a small amount of desaminometamitron (7 to 9 µg/l).

Metamitron and desaminometamitron were not found in fish at all sampling dates 1, 2, 3, and 4 months after application in mesocosm ponds and after 4 months in microcosm ponds (detection limit 25 µg/kg). Maximum concentrations of 30 µg/ kg dry weight metamitron and 120 µg/kg desaminometamitron were analyzed in submergent macrophytes 11 days after the first application in the high dosed pond, 60 µg/kg and no desaminometamitron in the lower dosed one, respectively. Extrapolated to the biomass of macrophytes per pond, this results in less than 0.1%

of the total applied compound. The concentrations fell further during the course of the study. Two months after application, neither metamitron nor desaminometamitron was detected in macrophytes (detection limit 20 µg/kg).

In sediment, the concentrations of desaminometamitron increased during the first month after the first application to about 400 µg/kg dry weight sediment in the high dosed microcosm and about 30 µg/kg in the low dosed one. No metamitron was detected in any sediment sample from either system (detection limit 10 µg/l). The concentrations of the metabolite in mesocosm pond sediment increased during the first month to about 700 µg/kg in the high dosed pond and 85 µg/kg in the low dosed pond. The concentrations declined to about 150 µg/kg in the high dosed meso- and microcosm ponds and to 45 µg/kg in the low dosed ponds at the end of the study. The percentage of the total amount of the test compound applied and its main metabolite found in sediment at the end of the study was small (Figure 10).

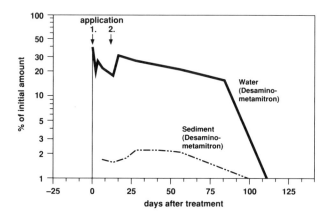

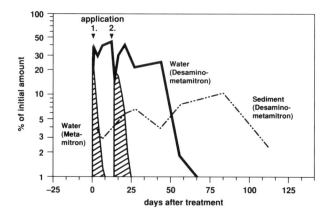

Figure 10.   Fate of Goltix® in microcosm ponds (above) and mesocosm ponds (below): distribution in different compartments. Two times 5 kg Al/ha had been sprayed onto the water surface.

Water Quality

Goltix® had no influence on the physicochemical water parameters. According to local climatic conditions, water temperature fluctuated between 15 to 23°C, identical in all ponds. The pH was similar in all ponds of both systems for about 2 months after application. Slightly higher pH values were recorded in one mesocosm control pond 2 to 3 months after application combined with slightly higher chlorophyll *a* concentrations. After 2 months, divergence occurred between individual ponds according to increasing biological differences. The pH values ranged from about 7.5 to 9.5.

Occasionally, changes in weather conditions caused a drastic reduction in dissolved oxygen concentrations in the ponds. As trout are sensitive to lack of oxygen, all ponds were then aerated with compressed air for several days. Oxygen concentrations fluctuated from about 7 to 13 mg/l and indicated no effects of Goltix®. Within the first few weeks after application, divergences of the main chemical parameters, as phosphorus, nitrogen, chlorophyll *a*, did not reach levels which might be interpreted as a direct or indirect effect of the application of Goltix®.

DISCUSSION

The application of 2 times 0.5 and 5.0 kg AI/ha, respectively, on the water surface of both pond systems did not affect trout and zooplankton. These results are not surprising as they can be extrapolated from laboratory studies. The acute toxicity of metamitron to rainbow trout is low. The lethal concentration causing 50% mortality (LC_{50}) after 96 h is 326 mg/l, following OECD test guideline No. 203.[21] The toxicity to water fleas (*Daphnia magna*) is low as well: the acute EC_{50} (48 h) is 102 mg/l (OECD guideline No. 202, Part I[21]) and the concentration without an effect on the number of offspring in a 21-day reproduction study is 32 mg/l (OECD guideline No. 202, Part II[21]). Compared to nominal and actual analyzed concentrations at both application rates, these laboratory results indicate a high safety margin for these organisms. The results of microcosm and mesocosm studies confirm this conclusion, showing no effect on zooplankton in either study. Similar conditions are reported from an atrazine study.[9,12]

To the contrary, the laboratory results on a phytoplankton species, *Scenedesmus subspicatus,* indicate a potential hazard to these organisms at concentrations that might be reached at the high application rates, which, however, are unrealistically high and do not reflect possible contaminations. The concentration causing a growth inhibition of 50% (EC_{50}) in a laboratory study is 0.9 mg/l (96 h) and the concentration without observed effects (NOEC) 0.3 mg/l (according to OECD guideline No. 201). In comparison to these results, the nominal concentrations of 0.5 mg/l (high dosage) in mesocosm ponds and 0.4 mg/l in microcosm ponds indicate a possible, though limited effect, on phytoplankton organisms. The analyzed concentrations confirmed the nominal ones very well. Nevertheless, no effects on phytoplankton organisms could be detected in either system, neither on the species number and diversity nor on single species, although the concentrations exceeded the NOEC of the laboratory study. This confirms the general observation that organisms in natural ecosystems frequently react less sensitively than in laboratory studies,[15,22] which are performed under high stress and do not include the fate and natural bioavailability of a test compound. The exposure is mostly reduced in the field compared to the exposure in laboratory studies. Some studies conducted showed that laboratory-derived single species data are comparable to those derived for natural algal communities.[23] The missing effects on phytoplankton in these micro- and mesocosm studies are explained by the very short half-life of metamitron in water (several hours under natural conditions). (The toxicity of the more persistent metabolite desaminometa-mitron to *Scenedesmus subspicatus* is low: the EC_{50} (96 h) is higher than 32 mg/l.) This might also contribute to the very weak and short-term symptoms of macrophytes in the mesocosm study. More stable herbicides may have more pronounced effects on phytoplankton or macrophytes.[9,10,12,13]

The distribution pattern of Goltix® was comparable in mesocosm and microcosm ponds. Metamitron degraded very quickly in both systems, but degraded faster in microcosm than in mesocosm ponds, while desaminometamitron seemed to degrade slightly slower in microcosm ponds. These differences may have been caused by slight differences in water quality such as pH and turbidity (light). Overall the fate of Goltix® was similar in both systems.

The biological effects of the application of Goltix® onto the water surface of the small scale microcosm ponds and the larger mesocosm ponds were identical: Goltix® did not influence the phytoplankton, zooplankton, benthic organisms, and rainbow trout in both systems. In addition, the growth and development of submergent and emergent aquatic plants of mesocosm ponds was not influenced. Leaves of emergent plants, which had been directly sprayed with the herbicide, showed only weak symptoms for several days without further effects on their development. Emergent macrophytes could not be investigated in microcosm ponds as these ponds do not offer shallow areas. The submergent species which had been introduced with the natural lake sediment have not been evaluated quantitatively in these ponds, as their abundance was low. A disadvantage of macrophytes is the increase of differences in phyto- and zooplankton communities between ponds.[14,15]

The suitability of these microcosm ponds as a model for the assessment of the potential hazards of herbicides for aquatic plants cannot sufficiently be discussed based on the results of this study. The database has to be enlarged in further research. Nevertheless, the results of these studies confirm the general conclusion that microcosm ponds met the required criteria for model ponds for the detection of possible effects of pesticides on aquatic ecosystems.[15] In comparison to mesocosm ponds, their use can result in considerable cost and time saving, while they provide almost the same or similar information.

ACKNOWLEDGMENT

The authors would like to thank Mrs. Margret Graef for her valuable technical assistance.

REFERENCES

1. Bloesch, J. 1988. Mesocosm studies. *Hydrobiologia* 159:221–222.
2. Crossland, N.O. 1988. Experimental design of pond studies. British Crop Protection Council, Ed., Environmental effects of pesticides, *Mono.* 40:231–236.
3. Crossland, N.O. and C.J.M. Wolff. 1988. Outdoor ponds: Their construction, management, and use in experimental ecotoxicity. In O. Hutzinger, Ed., *The Handbook of Environmental Chemistry,* Vol. 2, Part D — Reactions and Processes. Springer-Verlag, Berlin, pp. 51–67.
4. Giddings, J.M. and P.J. Franco. 1985. Calibration of laboratory bioassays with results from microcosms and ponds. In T.P. Boyle, Ed., *Validation and Predictability of Laboratory Methods for Assessing the Fate and Effects of Contaminants in Aquatic Ecosystems.* Special Technical Publication 865, American Society for Testing and Materials, Philadelphia, pp. 104–119.
5. deNoyelles, F. and W.D. Kettle. 1985. Experimental ponds for evaluating bioassay predictions. In T.P. Boyle, Ed., *Validation and Predictability of Laboratory Methods for Assessing the Fate and Effects of Contaminants in Aquatic Ecosystems.* Special Technical Publication 865, American Society for Testing and Materials, Philadelphia, pp. 91–103.
6. Solomon, K.R. and K. Liber. 1988. Fate of pesticides in aquatic mesocosm studies — an overview of methodology. Proceedings, Brighton Crop Protection Conference, Brighton, England, November 21–24, pp. 139–148.
7. Dortland, R.J. 1980. Toxicological evaluation of parathion and azinphosmethyl in freshwater model ecosystems. *Centre Agric. Publ. Docum.,* Wageningen, Netherlands. pp. 112.
8. Gunkel, G. 1984. Investigation of the ecotoxicological effect of a herbicide in an aquatic model ecosystem. II. Food chain significance and pesticide balance. *Arch. Hydrobiol. Suppl.* 69:130–168.
9. Hamilton, P.B., G.S. Jackson, N.K. Kaushik, K.R. Solomon and G.L. Stephenson. 1988. The impact of two applications of atrazine on the plankton communities of in situ enclosures. *Aquat. Toxicol.* 13:123–140.
10. Jenkins, D.G. and A.L. Buikema. 1990. Response of a winter plankton food web to simazine. *Environ. Toxicol. Chem.* 9:693–705.
11. Larsen, D.P., F. deNoyelles, F. Stay and T. Shiroyama. 1986. Comparisons of single-species, microcosm and experimental pond responses to atrazine exposure. *Environ. Toxicol. Chem.* 5:179–190.
12. deNoyelles, F., W.D. Kettle, C.H. Fromm, M.F. Moffett and S.L. Dewey. 1989. Use of experimental ponds to assess the effects of a pesticide on the aquatic environment. In J.R. Voshell, Ed., *Using Mesocosms to Assess the Aquatic Ecological Risk of Pesticides: Theory and Practice.* Entomological Society of America: Miscellaneous publications 75:1–88.

13. Stay, F.S., A. Kakto, C.M. Rohm, M.A. Fix and D.P. Larsen. 1989. The effects of atrazine on microcosms developed from four natural plankton communities. *Arch. Environ. Contam. Toxicol.* 18:866–875.

14. Pelzer, M. 1988. Zeigen parallel angesetzte Modelloekosysteme eine gleiche Dynamik? Master's Thesis, Technical University, Biology/Ecology, Aachen, Germany.

15. Heimbach, F., W. Pflueger and H.-T. Ratte. 1992. Use of small artificial ponds for assessment of hazards to aquatic ecosystems, *Environ. Toxicol. Chem.* 11:27–34.

16. Johnson, B.E. and D.F. Millie. 1982. The estimation and applicability of confidence intervals for Stander's-similarity index (SIMI) in algal assemblage comparisons. *Hydrobiologica* 89:3–8.

17. Smith, E.P. 1986. Randomized similarity analysis of multispecies laboratory and field studies. In A.H. El-Shaarawi and Kwiatkowski, R.E., Eds., *Statistical Aspects of Water Quality Monitoring.* Elsevier, New York, pp. 261–272.

18. Smith, E.P. and D. Mercante. 1989. Statistical concerns in the design and analysis of multispecies in microcosm and mesocosm experiments. *Toxicity Assessment: An International Journal* 4:129–147.

19. Tuchmann, M. and D.W. Blinn. 1979. Comparison of attached algal communities on natural and artificial substrate along a thermal gradient. *Br. Physiol. J.,* 14:243–254.

20. Washington, H.G. 1984. Diversity, biotic and similarity indices. *Water Research* 6:653–694.

21. Organization for Economic Cooperation and Development (OECD). Guidelines for testing of chemicals. No. 201: Alga, growth inhibition test (June 7, 1984). No. 202: Daphnia sp., acute immobilisation test and reproduction test (April 4, 1984). No. 203: Fish, acute toxicity test (April 4, 1984). Organization for Economic Cooperation and Development, Paris, France.

22. Hill, I.R. 1985. Effects on non-target organisms in terrestrial and aquatic environments. In J.P. Leahey, Ed., *The Pyrethroid Insecticides.* Taylor & Francis, London, pp. 151–262.

23. Lewis, M.A. 1990. Are laboratory-derived toxicity data for freshwater algae worth the effort? *Environ. Toxicol. Chem.* 9:1279–1284.

CHAPTER 20

Earthen Ponds vs. Fiberglass Tanks as Venues For Assessing the Impact of Pesticides on Aquatic Environments: A Parallel Study With Sulprofos

Gregory L. Howick, Frank deNoyelles, Jr., Jeffrey M. Giddings, and Robert L. Graney

Abstract: An aquatic, ecosystem level study with the insecticide sulprofos was conducted using twelve 470-m³ earthen ponds and twelve 11.0-m³ fiberglass tanks. For each test system, the initial benthic and limnetic communities came from the same sources. The ponds and tanks were stocked with adult and juvenile bluegill sunfish, respectively. Physical and chemical water quality parameters were similar in the ponds and tanks except that dissolved oxygen, hardness, and conductivity tended to be higher in the tanks while total suspended solids were higher in the ponds. Sulprofos concentrations in the water were usually slightly lower in the tanks. Sulprofos concentrations in the sediment were lower in the ponds. Phytoplankton chlorophyll *a* concentrations in the tanks and ponds were very similar both in terms of magnitude and variation. The responses to sulprofos of the zooplankton, benthic macroinvertebrates, and fish in the ponds and tanks were not always the same. The differences appeared to be related to differences in exposure to sulprofos or the differing abundance of bluegill. These results suggest that fiberglass tanks of this size could be cost-effective precursors to earthen ponds for pesticide registration tests.

0-87371-592-6/94/$0.00 + $.50

321

INTRODUCTION

For the final tier of testing for pesticide registration, the EPA has recommended the use of aquatic mesocosms with a minimum surface area of 0.04 ha and a minimum volume of 300 m^3.[1] The use of mesocosms to test ecosystem level responses to a pesticide represents a radical departure from the lower tiers of testing which use only single species and cannot assess the interactions among populations and communities. Mesocosm tests require large investments of time and money, and their outcomes cannot necessarily be predicted from the results of single-species tests.[2] Therefore, a smaller scale multispecies or ecosystem level test could be useful by providing preliminary data on the environmental fate and effects of a test compound. These data could serve as a basis to focus the experimental design of a mesocosm or more conclusively demonstrate whether or not a mesocosm test is necessary.

Various designs for laboratory scale (<1 m^3) multispecies tests have been proposed which range from highly controlled assemblages of introduced organisms (e.g., Reference 3) to systems based on natural sediments and water (e.g., Reference 4). Although these microcosms can provide some information on possible indirect effects of a pesticide, their small size tends to limit them to only representing one habitat type and makes the inclusion of fish difficult in any long term experimental design. In this chapter we present an experimental design for ecosystem-level tests which is based on 3.2-m diameter fiberglass tanks. By being roughly one to three orders of magnitude larger than laboratory microcosms, the tanks are large enough to support fish and have some habitat heterogeneity. The tanks are also roughly one order of magnitude smaller than pond mesocosms, and, therefore, less resource intensive. The usefulness of this design was assessed by comparing the performance of the tanks to pond mesocosms subject to the conditions of a pesticide registration study.

METHODS

Site

This study was conducted at the Kansas Aquatic Mesocosm Program's experimental pond facility at the John H. Nelson Environmental Study Area. The site is located 14 km north of Lawrence in northeastern Kansas and is operated by the University of Kansas. At the time this study was conducted, the mesocosm facility had 28 0.04-ha experimental ponds, two well water-fed source reservoirs (0.36 and 0.72 ha), three storage basins into which water from the experimental ponds is drained at the end of a study, and a watershed catchment pond to assure that potentially contaminated water could not leave the facility.

Pond and Tank Construction

Fourteen experimental ponds, 21.3 × 21.3 × 2.74 m deep with 2:1 sloping sides, were constructed in the summer through early fall of 1988. The ponds were excavated into high clay content soils and required no additional liners in order to retain water. The bottom half of the sloping sides of the ponds was covered with a 10- to 15-cm layer of sediment, thus forming a ring around the flat-bottomed deep portion of the ponds. The sediment was obtained from a nearby old farm pond which had been undisturbed for at least 20 years. The ponds were filled with water from the smaller of the two source reservoirs. This reservoir had been nearly drained then refilled with well water in November 1987. The plankton communities in this reservoir were monitored during 1988 and were found to have returned to a condition comparable to before draining by the time the new ponds were filled (unpublished data of J.M. Giddings, Springborn Laboratories, Inc.). When filled to a depth of 2.15 m, each pond contained 470 m^3 of water. The collection and spreading of the sediments and the addition of the water to the ponds took place during a 3-week period in late October through early November 1988. During this time, the sediments were exposed to air for approximately 10 days. The submerged macrophytes which subsequently grew in the ponds were restricted almost entirely to the area where the sediment was placed creating two distinct habitats: one open water and the other dominated by macrophytes. About 60 to 70% of each pond, as based on surface area, was relatively free of macrophytes. Construction of the mesocosms was completed in early April 1989 with the addition of 35, 10- to 15-cm bluegill (*Lepomis macrochirus*) to each pond.

Fourteen 3.2 m diameter, 1.52 m high fiberglass tanks were placed in two rows in a flat-bottomed storage basin adjacent to the pond mesocosms with the axis of the tank array oriented approximately north-south. The bottom of each tank was covered with a 10-cm layer of sediment from the aforementioned farm pond. The tanks were then filled to a depth of 1.37 m with water from the pond mesocosms resulting in a nominal operating volume of 11.0 m^3. The addition of the sediment and water took place during a 1-week period in late March 1989. The basin around the tanks was then filled with well water to a depth of 1.2 m to provide a thermal buffer. The macrophytes which naturally grew out of the sediment were periodically harvested to restrict their growth to approximately the northern one third of the tank. This was necessary to maintain proportions of open water and macrophyte dominated habitats similar to the ponds. About 65% of each tank, as based on surface area, was open water. Forty 3.0- to 4.5-cm bluegill were stocked in early April 1989.

Water levels in the ponds and tanks were maintained near normal operating levels by draining excess water into a storage basin or adding water to replace evaporative losses. Prior to treatment, water added came from the same reservoir used to initially fill the ponds. Well water was used during the treatment and posttreatment periods.

Treatment

The study was conducted with the formulated endproduct of sulprofos, Bolstar® 6L (EPA Reg. No. 3125-321, 731 g active ingredient [AI] per liter), an organophosphorus insecticide used primarily on cotton and soybeans. The treatment regime was designed to simulate the entry of the pesticide into aquatic environments as spray drift and surface runoff from treated fields. Spray drift was simulated by the application of aqueous solutions of Bolstar® to the surface of the ponds or tanks. Surface runoff was simulated by adding a clay-water mixture containing Bolstar®. The experimental design consisted of four treatment levels (including control) with three replicate ponds or tanks per treatment level. Two of the initial fourteen ponds and tanks were excluded from the study just before treatment began. One pond was eliminated because it had obviously higher turbidity than the others. A second pond was eliminated because a variety of water quality parameters were at or near the extremes of the ranges for the entire set of ponds. The tanks were more uniform and the choice was arbitrary. The two tanks removed from the experiment were those that occupied the same relative position in the tank array as the two eliminated mesocosm in the pond array.

The application rates were based on aerial-estimated environmental loadings rather than on target nominal concentrations. The spray drift application rates were based on 5, 2.5, and 1% of the maximum agricultural rate of 1680 g AI per ha per application. The resulting nominal concentrations in the ponds were 7.2, 3.6, and 1.4 μg/l. The loading rates for the simulated runoff events were 4.4, 2.2, and 0.88 g AI per pond for nominal concentrations of 9.4, 4.7, and 1.9 μg/l. The amounts of sulprofos added to the tanks were scaled in proportion to volume so that the nominal concentrations in the ponds and tanks would be the same. The frequency of spray drift applications was based on the maximum number of applications and the minimum time interval between applications as specified on the Bolstar® 6L label. The result was 4 applications at 5-d intervals beginning on 26 June 1989. The frequency of runoff simulations was 5 additions at 2-week intervals beginning on 29 June 1989. For both simulations, the ponds and tanks were treated on the same day and in ascending order of treatment level.

Sulprofos was applied to the ponds from the back of a small boat using a CO_2-powered backpack sprayer. The spray wand assembly resembled a "T" with 6 nozzles distributed along the top at 30 cm intervals. Each spray application consisted of 9 l of water and the measured amount of sulprofos. The wand was held over the water off the stern of the boat and the sulprofos mixture was sprayed onto the surface from a height of 2.5 to 5 cm as the boat was moved around the pond. Starting near the shore, movement was in ever tightening circles until the entire surface of the pond was covered and all of the sulprofos was discharged. Additional water was then added to the pressure tank as a rinse which was then sprayed onto the pond using the same procedure as for the original solution. The tanks were treated with simulated spray drift by adding the measured amount of Bolstar® 6L to 500 ml of water in a glass jar. After shaking, the solution was poured into the tank in approximately 50-ml portions while the tank was vigorously stirred with a canoe paddle. The jar was rinsed several times with water from the tank. The control ponds and tanks were not treated with simulated spray drift.

Sulprofos-treated clay slurries were added to the ponds using a recirculating mixing system. Water was pumped from 1 central point in each pond into a 200-l mixing chamber using a 3/4-hp pump. Water returned to the pond through three 3.8-cm PVC discharge pipes and was released just below the water surface about 1 m from shore at 3 widely spaced locations around the pond. The water recirculation continued for 6 h (a reasonable duration for a runoff event) for a total volume pumped of 57 m³. For each pond at each application event, 25 kg of dry clay was placed in a container and mixed with the measured amount of sulprofos which had been diluted in 100 ml of acetone. Control batches received only acetone. The clay/sulprofos mixture was added in approximately 1-kg portions to the mixing chamber at 7- to 8-min intervals during the first 3 to 4 h of the 6-h runoff simulation until all of the material had been dispensed. The swirling and turbulence caused by the discharge of water into the mixing chamber was sufficient to thoroughly disperse the clay and form a reasonably homogeneous slurry. Despite the point discharge configuration of the delivery system, each pond appeared to be uniformly turbid by the end of the recirculating period.

A scaled down and simplified method was used to simulate runoff into the tanks. For each tank at each treatment event, 0.62 kg of clay was mixed with the measured amount of Bolstar® 6L in a 4-l widemouthed plastic jar. Three liters of water from the tank was then added to the jar which was then sealed and shaken vigorously to create a slurry. The slurry was then slowly added to the tank in roughly 100-ml portions while the tank was stirred with a canoe paddle. Care was taken not to add any large chunks of undispersed clay into the tanks. Stirring of the tank continued until all of the slurry had been dispensed, the jar thoroughly rinsed, and the tank appeared uniformly turbid.

Monitoring

Many physical, chemical, and biological parameters were measured in both the ponds and tanks. Data and samples were collected from the ponds at varying intensities from late December 1988 through early October 1989. The tanks were sampled with less frequency relative to the ponds from early May through mid-September 1989. When data and samples were collected from the tanks, similar data and samples were also collected from the ponds within 5 days. Each pond contained two distinct habitats: a shallow zone dominated by submerged macrophytes and an open-water, central deep zone, which were usually sampled separately. Samples from the two areas in the tanks (macrophyte and no macrophytes) were a composite of equal numbers of subsamples collected from each area. In order to compare the ponds to the tanks, the results from the two zones in each pond were averaged with equal weight given to each zone. The data presented for the zooplankton and macroinvertebrate composition and density are from the one sample event collected less than 1 week before the beginning of the sulprofos treatment regime and from one sample event collected within 1 week after the completion of the treatment regime.

The same methods were used to collect data from the ponds and the tanks. Temperature, dissolved oxygen, and pH were measured during the late afternoon in the open water zones at 1 and 0.4 to 0.6 m for the ponds and tanks, respectively, using the appropriate meters. Water samples were collected using 3.8 cm diameter vertically integrating column samplers from which subsamples were removed for measurement of conductivity, hardness, alkalinity, total suspended solids, total organic carbon, dissolved organic carbon, and phytoplankton chlorophyll *a* and phaeophytin.[5] Sulprofos residues were measured in composite samples collected with 1.3 cm diameter, stainless steel, vertically integrating column samplers. A sufficient number of samples were collected from the high treatment level tanks and ponds after the first spray drift application to determine the half-life of sulprofos in the water column. Sediment samples for residue analysis were collected using 3, 2.5 cm diameter × 28 cm long stainless steel tubes attached to the end of a pole. The samples were analyzed using gas chromatography by Analytical Bio-Chemistry Laboratories, Inc., Columbia, MO. Zooplankters were collected after sundown with vertical hauls from botton to surface with a 20 cm diameter, 80-μm mesh Wisconsin net. Macroinvertebrates were collected on artificial benthic substrates (ABS) consisting of 14, 5.1 cm diameter × 5.1 cm high sinking plastic sewage treatment plant surface enhancers arranged in 2 layers of 7. The colonization period for the ABS was 4 weeks. Emergent insects were not collected from the tanks.

Fish were harvested from the tanks in mid-September 1989 by removing the remaining macrophytes, pumping the water out of the tanks, and removing the stranded fish by hand or with dip nets. Each fish collected from the tanks was measured for total length and weighed. Fish were similarly removed from the ponds in mid-November except that prior removal of macrophytes was not necessary. The adult fish from the ponds were individually measured for total length and weight. The juvenile fish from each pond were counted and weighed en masse.

RESULTS

The physical and chemical water quality parameters were not affected by sulprofos. No significant differences existed between the tanks and the ponds for temperature, pH, alkalinity, total organic carbon, and dissolved carbon (Table 1). Dissolved oxygen, hardness, and conductivity were significantly higher in the tanks. Total suspended solids (TSS) were lower in the tanks. The difference in TSS is not indicated as significant on Table 1 because a rain event dramatically increased the turbidity in the ponds and greatly inflated the variance around the mean difference. With that date excluded from the calculations, TSS is significantly lower in the tanks. The variance of all the water quality variables measured except total organic carbon tended to be less for the tanks (Table 1).

Table 1. Ranges, Mean CV (in Parentheses), and Differences Between Tanks and Ponds For Various Water Quality Parameters Over the Period From Early May Through Mid-September

Parameter	Tanks	Ponds	Difference Mean	SD	n
Temperature (°C)	16–31 (0.23)	16–29 (1.81)	0.7	1.50	9
Dissolved oxygen (mg/l)	11.0–16.6 (0.79)	8.2–13.3 (6.66)	2.8[a]	0.86	9
pH	8.7–10.1 (1.54)	8.5–9.7 (2.89)	0.4	0.56	9
Alkalinity (mg/l)	36–116 (11.4)	59–105 (16.3)	– 12.0	18.14	5
Hardness (mg/l)	82–153 (6.82)	69–126 (13.8)	24.8[b]	10.96	5
Conductivity (μmhos/cm)	221–348 (3.44)	188–271 (7.18)	62.6[b]	18.89	5
Total suspended solids (mg/l)	5.8–34.4 (32.4)	15.7–169.4 (60.5)	– 35.5	55.76	5
Total organic carbon (mg/l)	3.5–7.7 (21.5)	4.8–8.0 (11.5)	0.2	1.31	5
Dissolved organic carbon (mg/l)	3.1–7.2 (14.2)	3.3–7.3 (23.9)	0.4	2.17	5

[a] $p < 0.001$ based on a *t*-test for the hypothesis that the difference is equal to zero.
[b] $p < 0.01$.

Note: Each point in the range was the mean of the tanks or ponds for each sampling date.

Chlorophyll *a* concentrations, the only measure of phytoplankton standing crop made in both the tanks and the ponds, were quite similar in the tanks and ponds (Figure 1). With the exception of one date in early June, mean chlorophyll concentrations in the tanks and ponds followed the same trajectory through time. The variations in chlorophyll concentrations among the tanks and ponds on each date were remarkably similar (Figure 1).

The average concentrations of sulprofos in the water after each addition to the tanks or ponds were less than the calculated nominal concentrations (Table 2). Concentrations after spray drift applications were typically 10 to 20% below nominal in the tanks or the ponds. After the slurry additions, concentrations were usually 60 to 70% below nominal in the ponds and 80 to 90% below nominal in the tanks. Half-life of sulprofos was estimated to be 14.7 h in the ponds and 13.4 h in the

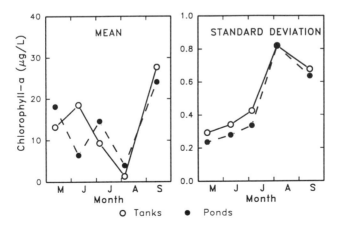

Figure 1. Mean and SD of chlorophyll *a* concentrations in the fiberglass tanks and ponds.

Table 2. Average Deviations From Nominal of Sulprofos Concentrations in Water in the Experimental Ponds and Fiberglass Tanks

Addition	Low[a]		Medium[a]		High	
Type	Ponds	Tanks	Ponds	Tanks	Ponds	Tanks
	Nominal concentrations (μg/l)					
Spray	1.4		3.6		7.2	
Slurry	1.9		4.7		9.4	
	Average percent deviation from nominal					
Spray	−30	18	−17	−10	−10	−22
Slurry	−70	−88	−61	−80	−61	−81

[a] Comparable samples were collected from both the tanks and ponds after the first and fourth spray application and after the first, third, and fifth slurry addition.

Note: Samples were collected within 1 h after the additions.

tanks (Table 3). Sediment concentrations of sulprofos within 2 h of the last slurry addition were highly variable, but, on average, were roughly 10 times greater in the tanks than in the ponds (Table 4).

The densities of the three major taxonomic groups of zooplankton differed considerably between the tanks and ponds just before treatment with sulprofos. In the tanks, the densities of cladocerans and copepods were generally 10 times greater and densities of rotifers were roughly 100 times greater than in the ponds (Figure 2). After the completion of the sulprofos treatments, the densities of cladocerans were significantly reduced relative to control at all treatment levels in both the tanks and ponds. Copepod densities were not affected in either the tanks or the ponds. For both cladocerans and copepods, the tenfold difference in densities between tanks and ponds observed before treatment was present after treatment but rotifer densities were now similar. In the tanks, rotifer densities were not affected by sulprofos. However, in the ponds, rotifer densities were significantly reduced at the high treatment level (Figure 3). Just prior to treatment, the number of zooplankton genera found in the tanks and ponds were slightly greater in the ponds for cladocerans, similar for copepods, and greater in the tanks for rotifers. After

Table 3. Mean Concentrations of Sulprofos (μg/l) in the Water in the High Treatment Level Tanks and Ponds After the First Slurry Addition

	Ponds	Tanks
Time after addition (h)		
0	6.57	7.43
8	5.00	3.73
24	2.30	2.37
48	0.71	0.54
Regression statistics		
Y-intercept (SE)	1.936 (0.057)	1.930 (0.199)
Slope (SE)	−0.047 (0.0016)	−0.0519 (0.0054)
r	0.999	0.989
Half-life (h)	14.7	13.4

Note: Regression statistics used to calculate the half-life of sulprofos were based on ln-transformed concentrations.

Table 4. Mean (SD) Concentrations (μg/kg) of Sulprofos in the Sediments of the Experimental Ponds and Fiberglass Tanks

	Control	Low	Medium	High
Ponds	ND[a]	ND	4.1 (4.2)	11 (8)
Tanks	ND	31 (32)	33 (28)	131 (117)

[a] Not detected <4.0 μg/kg.

Note: Data are from samples collected within 2 h after the last slurry addition.

treatment, cladoceran genera were more numerous in the tanks, copepod genera were slightly more abundant in the ponds, and the numbers of rotifer genera were similar to slightly lower in the tanks (Table 5).

The macroinvertebrate assemblages collected with the ABS were numerically dominated by chironomids. Just prior to the addition of the sulprofos, the densities of chironomids were quite similar in the tanks and ponds. Ephemeropterans were found in the ponds but not in the tanks whereas snails were fairly common in the tank samples but almost nonexistent in the pond samples (Figure 4). After treatment, chironomids densities were significantly reduced at the low, middle, and high treatment levels in the tanks but only at the high treatment level in the ponds. By this time, ephemeropteran densities had become quite similar in the tanks and ponds, and a significant reduction occurred at the high treatment level in both the tanks

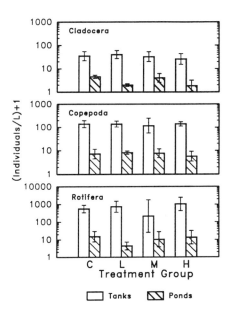

Figure 2. Densities of zooplankters (±1 SD) in the fiberglass tanks and ponds on 1 day within 1 week before treatment with sulprofos. The densities of copepodes include all life stages. C = control, L = low, M = middle, and H = high treatment levels; see Table 2 for concentrations.

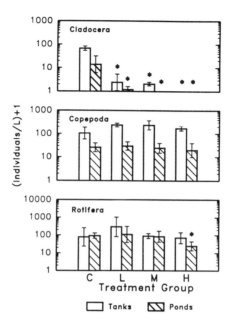

Figure 3.   Densities of zooplankters (±1 SD) in the fiberglass tanks and ponds on 1 day within 1 week after completion of the sulprofos treatment regime. * = significantly less than control ($p < 0.05$, Dunnett's one-tailed t-test).

and ponds. The densities of snails were also similar in the tanks and ponds and were not reduced by the sulprofos treatment (Figure 5). The taxonomic composition of the ABS collected macroinvertebrates was also dominated by chironomids. Both before and after treatment with sulprofos, more chironomid genera were found in samples from the ponds than from the tanks. However, the response to the sulprofos treatment was the same in the tanks and pond with noticeable reductions in generic richness occurring only at the high treatment level (Table 6).

Sulprofos did not have an effect on the adult bluegills in the ponds either in terms of survival, as measured by the number recovered, or in terms of growth, as measured by average individual mass or length (Figure 6). The number of juvenile bluegill recovered from the ponds varied considerably and was not significantly affected by sulprofos. However, the average mass of juvenile bluegill from the ponds was significantly reduced at the middle and high treatment levels with the average mass of a high treatment level fish being only about 50% of a control fish. The average length of juvenile fish in the ponds was also significantly reduced by sulprofos but only at the high treatment level (Figure 7). The response to sulprofos of the bluegill in the tanks resembled the response of the adult bluegill in the ponds in that the number recovered was not significantly affected. The average mass of the bluegill in the tanks also declined with increasing treatment level. However, the maximum reduction was only about 25% of the control and the difference was not statistically significant. As with the ponds, the average length of bluegill in the

Table 5. Number of Genera Within the Three Major Taxonomic Categories of Zooplankton Found in the Tanks and Ponds on One Day Within One Week Before, and on One Day Within One Week After, Treatment With Sulprofos

	Pretreatment		Posttreatment	
	Tanks	Ponds	Tanks	Ponds
Cladocera				
Control	3	4	6	3
Low	3	3	1	1
Middle	2	5	3	0
High	1	4	0	0
Copepoda				
Control	3	3	3	4
Low	3	3	3	4
Middle	3	3	3	5
High	3	3	4	4
Rotifera				
Control	8	4	9	9
Low	8	5	9	10
Middle	7	4	7	9
High	9	3	8	9

Note: Each number is the total found in three replicate tanks or ponds.

tanks was significantly reduced in the high treatment level (Figure 8) although the decrease in length relative to control for the juveniles from the ponds was 21% as compared to 10% for the tanks.

DISCUSSION

The fiberglass tank system was designed to be as similar to the experimental ponds as possible. The dimensions of the tanks provided the same surface area to volume ratio as found in the ponds, the water and sediments came from the same source, the tanks were surrounded by a large volume of water to provide similar temperature conditions, and the treatment with sulprofos occurred at the same times and at the same nominal concentrations. These efforts were successful in that many physical, chemical, and biological conditions in the tanks were very much like those in the ponds.

Perhaps of more interest, however, are the differences. Dissolved oxygen concentrations were consistently higher in the tanks than in the ponds. A typical explanation for this difference would be greater primary production in the tanks. This should then be accompanied by a relatively higher pH due to greater uptake of CO_2 and relatively lower alkalinity, hardness, and conductivity due the insolubility of calcium carbonate at high pHs.[6] The tanks and ponds did not follow this scenario since pH in the tanks was not significantly higher than in the ponds, and hardness and conductivity were lower in the ponds. However, these results are consistent with a lower rate of respiration in the tanks. This could be due to the differences in bottom area to volume ratio which is slightly lower in the tanks (0.73 m^{-1}) than in the ponds (0.82 m^{-1}). This difference in bottom area to volume ratio

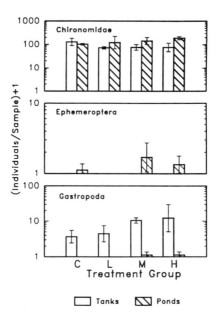

Figure 4.   Abundance of benthic macroinvertebrates (± 1 SD) in fiberglass tanks and ponds on 1 day within 1 week before treatment with sulprofos.

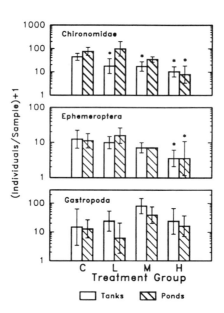

Figure 5.   Abundance of benthic macroinvertebrates (± 1 SD) in fiberglass tanks and ponds 1 day within 1 week after completion of the sulprofos treatment regime. * = significantly less than control ($p < 0.05$, Dunnett's one-tailed t-test).

Table 6. Number of Chironomid Genera Found in the Tanks and Ponds on One Day Within One Week Before, and on One Day Within One Week After, Treatment With Sulprofos

	Pretreatment		Posttreatment	
	Tanks	**Ponds**	**Tanks**	**Ponds**
Control	12	15	11	15
Low	9	15	12	19
Middle	11	17	11	13
High	9	18	6	8

Note: Each number is the total found in three replicate tanks or ponds.

is a result of the sloping sides of the ponds compared to the vertical sides of the tanks. Relative to the differences in TSS, the lack of a sloping shoreline by the tanks and their much smaller size prevented wind-generated water movements and runoff from rain events. Thus, resuspension, shoreline erosion, and new sediment input were greatly reduced or eliminated in the tanks resulting in lower TSS than in the ponds.

Sulprofos concentration were generally lower in the tanks, particularly after the slurry additions. Solomon et al.[7] noticed more rapid loss of methoxychlor from the smallest of three sizes of lake enclosures and attributed the difference to the greater wall surface area to volume ratio in the smaller limnocorrals. In our study, the difference in sulprofos concentrations after the slurry additions occurred within a few hours. This, in conjunction with the greater accumulation of sulprofos on the sediments of the tanks, suggests that the difference between the tanks and ponds was due to more rapid settling of the particle bound sulprofos in the tanks. This could have been caused by less thorough dispersion by the method used to dose the tanks, the shallower maximum depth of the tanks, and reduced wind-generated water movement.

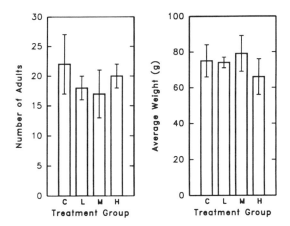

Figure 6. The number, average mass, and average length (±1 SD) of adult bluegill re-covered from the ponds.

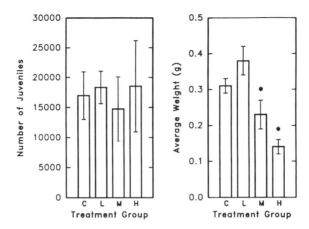

Figure 7.   The number, average mass, and average length (± 1 SD) of juvenile bluegill recovered from the ponds. * = significantly less than control ($p < 0.05$, Dunnett's one-tailed t-test).

The major difference in the zooplankton communities between the tanks and ponds was the persistently greater densities of crustacean zooplankton in the tanks. This difference was undoubtedly due to predation by the thousands of juvenile bluegill which were present in the ponds but not in the tanks. Despite the differences in densities, the responses of the zooplankton to sulprofos were similar enough in both experimental systems so that one would reach the same conclusions for the cladocerans and copepods.

The tanks and ponds yielded the same response to sulprofos for ephemeropterans and gastropods; however, the chironomids were impacted at lower nominal concentrations of sulprofos in tanks. The apparently greater sensitivity of the chironomids in the tanks is in agreement with the higher concentrations of sulprofos found in the tank sediments. This result underscores the importance of measuring the actual exposure to a test material faced by the organisms.

The fish from the ponds and tanks were not collected contemporaneously because one of the hoped for benefits of the tank protocol was a shorter, therefore, less resource-intensive study duration. Because of this and the different stocking regimes between the tanks and ponds, only the relative effects of the sulprofos on the fish can be compared. The earlier harvest date for the tanks means less recovery time for those fish from the effects of the sulprofos. In this regard, the relative impact of sulprofos could be expected to be greater in the tanks than in the ponds. However, the effects of sulprofos on the fish were observed to be greater in the ponds.

The responses of the bluegill to sulprofos may represent an interaction among juvenile bluegill recruitment, predation by bluegill, and the toxicity of sulprofos to zooplankters and macroinvertebrates. In the ponds, the recuitment of juvenile bluegill was largely uninhibited due to the lack of a predator. The result was the thousands of juvenile bluegill per pond we observed. Because of the small size of their mouths, the diet of these bluegill was limited to zooplankton[8] with the larger-bodied, slower moving forms being the preferred prey.[9] Consequently, the crustacean zooplankton densities in the ponds were low, and growth of the juvenile bluegill was limited by the availability of zooplankton. Because of this limitation, we would expect that the growth of the juvenile bluegill would be sensitive to any external impacts on the abundance of zooplankton. Therefore, it is reasonable to conclude that the decrease in average weight of juvenile bluegill at the two highest treatment levels in the ponds is an indirect effect due to the reduction in cladoceran abundance.

The situation in the tanks was much different. There, the juvenile bluegills did not reproduce, crustacean zooplankton remained relatively abundant, and growth of the fish was much greater than in the ponds. Also, as the bluegill in the tanks increased in size, their diet could expand to other food sources such as macroinvertebrates. Thus, when sulprofos reduced the abundance of cladocerans, the fish in the tanks had other sources of food available and the effect on their growth was much less. Therefore, it is possible that the unrealistic overabundance of juvenile bluegill in the ponds made those fish more sensitive to indirect effects of sulprofos.

Despite their physical differences, the tanks produced many dose-related results that were essentially the same as those found with the ponds particularly for the zooplankton and benthic macroinvertebrate communities. It was also our experience that the cost of the fiberglass tank test was less than 20% the cost of the pond mesocosm test. For these reasons, this fiberglass tank system could serve as a useful intermediate between single-species toxicity tests and pond mesocosm tests. However, the different fish communities influenced the abundance of zooplankton and the impact of sulprofos on the growth of juvenile bluegill. Further design refinement in the area of fish stocking, such as the use in the tanks of a fish that reproduces at a small size and the inclusion in both systems of a piscivorous fish would seem

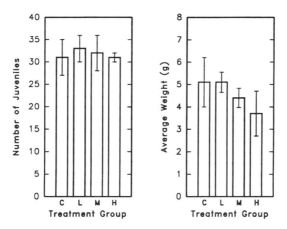

Figure 8.   The number, average mass, and average length (± 1 SD) of juvenile bluegill recovered from the tanks. * = significantly less than control ($p < 0.05$, Dunnett's one-tailed t-test).

desirable. The former would allow the assessment of the impacts of a pesticide on fish recruitment in the tanks and the latter would balance the predator-prey interactions in both test systems.

REFERENCES

1. Touart, L.W. 1988. Hazard evaluation division, technical guidance document: Aquatic mesocosm tests to support pesticide registrations. United States Environmental Protection Agency, Washington, DC.
2. deNoyelles, F., Jr. and W.D. Kettle. 1985. Experimental ponds for evaluating bioassay predictions. In T.P. Boyle, ed., *Validation and Predictability of Laboratory Methods for Assessing the Fate and Effects of Contaminants in Aquatic Ecosystems, ASTM STP 865.* American Society for Testing and Materials, Philadelphia, PA, pp. 91–103.
3. Taub, F.B. and P.L. Reed. 1982. Final report and protocol. Model ecosystems: Design, development, construction and testing. Department of Health and Human Services, Food and Drug Administration, Washington, DC.
4. Giddings, J.M. 1986. A microcosm procedure for determining safe levels of chemical exposure in shallow-water communities. In J. Cairns, Jr., ed., *Community Toxicity Testing,* ASTM STP 920. American Society for Testing and Materials, Philadelphia, PA, pp. 121–134.
5. American Public Health Association, American Water Works Association, and Water Pollution Control Federation. 1985. *Standard Methods for the Examination of Water and Wastewater,* 16th ed. American Public Health Association, Washington, DC.
6. Wetzel, R.G. 1975. *Limnology.* W.B. Saunders Co., Philadelphia, PA.
7. Solomon, K.R., G.L. Stephenson and N.K. Kaushik. 1989. Effects of Methoxychlor on zooplankton in freshwater enclosures: Influence of size and number of applications. *Environ. Toxicol. Chem.* 8:659–669.
8. Werner, E.E. 1974. The fish size, prey size, handling time relation in several sunfishes and implications. *J. Fisheries Res. Board Can.* 31:1531–1536.
9. O'Brien, W.J. 1979. The predator-prey interaction of planktivorous fish and zooplankton. *Am. Scientist* 76:572–581.

CHAPTER 21

Fate and Effects of Cyfluthrin (Pyrethroid Insecticide) in Pond Mesocosms and Concrete Microcosms

Philip C. Johnson, James H. Kennedy, R. Gregg Morris, Faithann E. Hambleton, and Robert L. Graney

Abstract: The chemical fate and biological impacts of cyfluthrin in aquatic ecosystems were investigated using mesocosms (634.7 m^3) and microcosms (1.9 m^3) during 1989. Ten spray drift and five soil runoff simulations were conducted. Pesticide loadings were scaled by system volume, with the same experimental design in ponds and microcosms. Aqueous cyfluthrin concentrations were similar among systems, but aqueous half-life and sediment concentrations were influenced by system scale. Biological effects (zooplankton, macroinvertebrate colonization, and aquatic insect emergence) showed parallel response patterns in both systems. Large cladocerans, mayflies, and Tanypodinae chironomid populations were reduced in the higher treatments, while oligochaetes, rotifers, and Chironominae chironomids were not reduced. Bluegill stocked in microcosms were sexually immature, while bluegill stocked in mesocosms were sexually mature resulting in large fish populations in mesocosms. Higher fish densities resulted in decreased zooplankton densities in mesocosms. Microcosms were less expensive to construct and were easier to sample. These data suggest that smaller scale systems may be cost effective alternatives for environmental effects testing.

0-87371-592-6/94/$0.00 + $.50
© 1994 by CRC Press, Inc.

INTRODUCTION

Single-species toxicity testing may not reflect toxicant impacts in complex eco-systems. Components of ecosystems function differently in isolation,[1] thus meso-cosms provide greater environmental realism. Using an experimental ecosystem such as a mesocosm, extrapolation is not from one level of biological organization to another (single species to ecosystem), but rather from a complex but somewhat artificial system to a complex natural system.[1,2] Cairns[3] makes no distinction be-tween microcosms and mesocosms because both consider higher levels of biological organization, resulting in greater environmental realism relative to laboratory bioas-says. Additionally, environmental fate and biological effects may be evaluated simultaneously in ecosystem testing.[3]

Currently, mesocosm testing is used as a supplement to laboratory toxicity eval-uations. Mesocosms represent the final level of a USEPA tiered testing hierarchy which includes chemical data, exposure modeling, and laboratory bioassays.[4] When it is determined using single-species tests that a chemical has an expected environ-mental concentration near the lowest effects levels, field testing is required to determine if risks to the environment exist.[5]

Large scale mesocosms (0.1-acre ponds) are commonly used for pesticide reg-istration studies. These experiments are expensive and logistically difficult. Use of microcosms might help reduce costs, enabling either testing of a larger number of potential toxicants or allowing greater replication (thus enhancing precision).

The question of physical scale in ecological testing has received considerable attention.[6-10] Some argue that whole-lake testing is necessary to achieve sufficient realism.[11] Unfortunately, lack of replication hampers the inferential capability of whole ecosystem manipulations.[4] Further testing at a range of physical sizes ranging from beaker scale to lakes will help address questions of scale.[4,12,13]

We compared the chemical fate and biological impacts of a pyrethroid insecticide (cyfluthrin) concurrently in outdoor 1.9-m^3 experimental tanks (microcosms) and 634.7 m^3 mesocosms from May to November 1989. This experimental design allowed evaluation of the validity of using microcosms as surrogates for required mesocosm tests. By testing at different scales we hoped to identify system param-eters which were either similar or dissimilar across test systems.

METHODS AND MATERIALS

Site Description and Test Initiation

Fourteen earthen ponds (mesocosms) were constructed in the cotton growing region of north Texas for use in the cyfluthrin pesticide reregistration study. The bottom of each pond was lined with clay and covered by approximately 15 cm of topsoil. Mesocosm dimensions were 30 m × 16 m at the surface (0.12 acre), maximum depth of 2 m with a 2:1 slope at each end. Cylindrical concrete micro-

Table 1. Physical Dimensions of Experimental Systems and Scaling Factors Between Mesocosms and Microcosms

	Microcosms	Mesocosms	Scaling[a]
Volume (m^3)	1.95	635	325
Surface area (m^2)	1.77	480	271
Wall + bottom area (m^2)	6.99	516	74
Surface/volume ratio	3.58	0.81	0.226

[a] Scaling = mesocosm/microcosm.

cosms (1.5 m × 1.3 m) were used as the smaller test systems. A layer of topsoil (ca. 10 cm) obtained from the same source as mesocosm soil was added to each tank. System metrics and scaling relationships among the two systems are summarized in Table 1.

Mesocosms were filled with water from an on-site maintenance pond, and microcosms were filled from an adjacent mesocosm. This allowed establishment of representative zooplankton and phytoplankton populations. Additional biological "inocula" (macroinvertebrates) were collected locally, ensuring the establishment of relatively diverse invertebrate communities. Natural colonization by insects occurred throughout the study.

Microcosms and mesocosms utilized separate water circulation systems (Figure 1; mesocosm recirculation shown), which served to mix water among all tanks/ponds during an equilibration period. Mixing was stopped and microcosms/mesocosms were isolated at test initiation. Evaporative losses for both systems were replaced with ground water from a deep water well.

Mesocosms were each stocked with 18 male and 18 female sexually mature bluegill sunfish, *Lepomis macrochirus* Rafinesque. Microcosms were each stocked with eight sexually immature bluegill sunfish. Although the carrying capacity of the microcosms was unknown, outdoor model ecosystems of moderately large size (>1000 l) are generally capable of supporting fish.[14]

Due to concerns regarding untested fish stocking levels, a single artificial refugia was placed in each microcosm. Refugia were cylinders of 0.25-inch mesh plastic netting (high density polyethylene) and were filled approximately halfway with plastic cylinders (5 cm OD × 5 cm high) manufactured as surface area enhancers for sewage treatment plants (Actifill "50" units; Norton Chemical Process Products, MA). Each refugia contained approximately 100 Actifill units. Refugia prevented invertebrates from being eliminated by fish predation. Smith[15] found that refugia allowed zooplankton populations to coexist with silver carp and channel catfish in 100-l tanks.

Pesticide Application

BAYTHROID®, the emulsified concentrate formulation of cyfluthrin, was used in this study. The pesticide application schedule in this study was determined from maximum loadings allowable under label directions. We conducted ten spray drift (SD) applications, each a week apart. Simulated pesticide runoff (RO) was modeled using 5 biweekly stormwater runoff events.

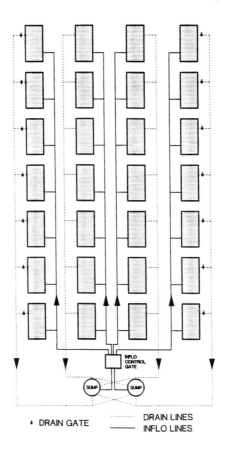

DRAIN LINES
* DRAIN GATE   INFLO LINES

Figure 1.   Water recirculation system for experimental mesocosms at the University of North
Texas Water Research Field Station. Water drained from each pond to a central
sump where water mixed and was then pumped back to the mesocosms. All units
were allowed to recirculate until treatment initiation, when ponds were isolated
from each other. A similar, scaled down recirculation system (not shown) was
constructed for concrete microcosms.

Treatments consisted of controls (DO; Dose 0) and four concentrations of cyfluthrin (D1 to D4). Replication in both mesocosms and microcosms was as follows:
D0 = 3, D1 = 2, D2 to D4 = 3. Microcosm and mesocosm pesticide applications
were performed concurrently and concentrations in microcosms were computed to
match mesocosm concentrations. Pesticide loadings were scaled down by volume,
as opposed to surface area. Pesticide loadings in microcosms would have been
slightly higher if concentrations were scaled by surface area (Table 1).

Treatments utilized varied SD·RO ratios, achieving four different loading rates
(Table 2). Spray drift D1 through D3 increased in concentration, while D4 spray
drift was identical to D3. Treatments D1 through D3 held runoff constant, while

D4 had a higher (5×) runoff value (Table 2). This allowed evaluation of the relative importance of spray drift (free pesticide) vs. runoff (sediment bound) input. Pesticide loadings corresponded to 1.0, 2.5, and 5.0% drift in the spray drift applications (percentage of field application rate), while runoff simulations were conducted at 0.3 and 1.5% runoff.

Microcosm drift applications were obtained by preparing a concentrated stock solution in water. Aliquots were pipetted into glass beakers and added to microcosms by pouring evenly across the water's surface. Mesocosm spray drift applications were more complex, utilizing a modified GAMACO® bridge spanner (Figure 2). Application stock solutions were added to 5-gallon stainless steel canisters which were pressurized using CO_2. Pressurized liquid was delivered to the ponds through 13 TK-552.5 flood jet R spray nozzles attached to the spanner, which were selected in order to minimize small droplet emission. The pesticide was introduced into mesocosms by driving the spanner at approximately 0.4 m/s while spraying 10 to 15 cm above the pond surface.

Runoff application soil slurries for mesocosms were prepared in cement mixers, where they were mixed for 1 h prior to application. Microcosm slurries were prepared in glass beakers and allowed to stand (covered with aluminum foil) for 1 h, simulating the mesocosm mixing period. Slurries were added to mesocosms using the bridge spanner, modified by the addition of Herd 1-92 K broadcast spreaders.

Table 2. Cyfluthrin (AI) Loading Rates and Residue Analysis Results

Treatment Level and Parameters	Spray Drift		Run Off	
	Microcosm	Mesocosm	Microcosm	Mesocosm
Dose 0				
Nominal concentration	0.0000	0.0000	0.0000	0.0000
Percent of nominal	—	—	—	—
Standard deviation	—	—	—	—
Dose 1				
Nominal concentration	0.0356	0.0356	0.2143	0.2143
Percent of nominal	81.92	75.76	44.51	44.75
Standard deviation	65.24	36.05	30.42	16.12
Dose 2				
Nominal concentration	0.0911	0.0911	0.2143	0.2143
Percent of nominal	90.16	63.44	32.76	56.53
Standard deviation	59.74	30.49	14.73	34.69
Dose 3				
Nominal concentration	0.1780	0.1780	0.2143	0.2143
Percent of nominal	86.86	71.89	50.89	49.95
Standard deviation	29.24	38.49	49.33	30.14
Dose 4				
Nominal concentration	0.1780	0.1780	1.0714	1.0714
Percent of nominal	92.39	82.01	39.56	42.40
Standard deviation	39.95	30.68	21.40	21.22

Note: Nominal concentrations are in ppb. Percent of nominal concentration, with standard deviation, obtained from water column samples collected one hour after treatment. Values represent a mean of surface and bottom measurements, averaged over the length of the application period.

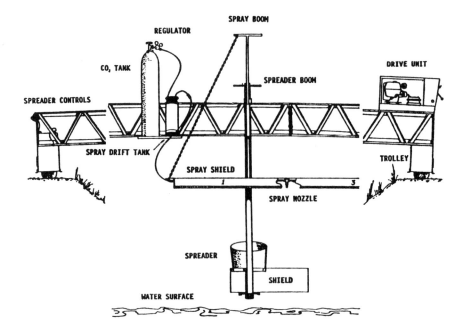

Figure 2. Modified bridge spanner used to apply cyfluthrin to mesocosms. Spray drift applications were accomplished using pressurized CO_2 delivery of stock solution to flood jet spray nozzles. Runoff applications used Herd broadcast spreaders to apply soil/water slurries to mesocosms.

Application was achieved by driving the spanner the length of the pond while broadcasting the soil/water mixture at approximately 38 to 40 cm above the surface. Microcosms were treated by pouring slurries evenly across the tank surface. Water column residues in both mesocosms and microcosms were collected 1 h after application for both RO and SD treatments.

Sample Collection and Analysis

Pyrethroid water-column residue samples were collected in Teflon bottles, transferred to 1-l volumetric flasks, and extracted with hexane. Water-column samples were collected 1 h after spray drift and runoff pesticide application (Tables 3 and 4). Water-column half-life residue samples were collected at 1, 8, 24, and 48 h after spray drift application.

Sediment samples were collected using a coring device which held a sleeve of butyrate tubing. Overlying water was drained from the cores in the field. Sediment cores were frozen and the top 2 cm were sectioned with a saw. Cyfluthrin was extracted from sediment samples with acetone. Acetonitrile was extracted from sediment samples with acetone. Acetonitrile was added to remove water from the acetone fraction, and this extract was evaporated to dryness by rotary evaporation. Hexane was added and samples were then sonicated.

Table 3. Application and Sampling Schedule for Microcosms

	Pretreatment			Treatment Period										Posttreatment								
Week #	−3	−2	−1	1	2	3	4	5	6	7	8	9	10	11	12	13	14	15	16	17	18	19
Microcosm application schedule				SD RO	SD	SD RO	SD	SD RO	SD	SD RO	SD	SD RO	SD									
Res. water				X		X	X	X	X	X	X	X		X		X	X	X				
Sed. cores			X																		X	
Water chem.		X		X		X		X	X	X				X	X	X	X	X	X	X		
Zooplankton	X			X		X		X		X		X		X		X		X		X		X
Emergence			X				X		X		X								X		X	
MAS samples		X	X	X	X	X	X	X	X	X	X	X	X	X	X	X	X	X	X	X	X	X
DO/temp/pH	X	X	X	X	X	X	X	X	X	X	X	X	X	X	X	X	X	X	X	X	X	X
Algal meas.			X	X		X		X		X		X		X		X		X		X		X

Note: SD = Spray drift pesticide application; RO = Soil runoff pesticide application; Res. water = residue samples from water column; Sed. cores = sediment cores for residue analysis; MAS samples = benthic insect colonization; Emergence = aquatic insect emergence (exuviae and/or emergence traps; and Algal meas. = phytoplankton and periphyton biomass and chlorophyll analysis.

Table 4. Application and Sampling Schedule for Mesocosms

	Pretreatment			Treatment Period										Posttreatment								
Week #	−3	−2	−1	1	2	3	4	5	6	7	8	9	10	11	12	13	14	15	16	17	18	19
Mesocosm application schedule				SD RO	SD	SD RO	SD	SD RO	SD	SD RO	SD	SD RO	SD									
Res. water			X	X	X	X	X	X	X	X	X	X	X	X		X		X				
Sed. cores			X	X	X	X	X	X	X	X	X	X	X	X		X				X	X	X
Water chem.		X				X	X	X	X	X	X	X				X	X	X	X	X		
Zooplankton		X	X	X		X		X		X		X		X	X	X	X	X	X	X		
Emergence	X		X		X		X		X		X		X		X		X		X		X	
MAS samples		X	X	X	X	X	X	X	X	X	X	X	X	X	X	X	X	X	X	X	X	X
DO/temp/pH	X	X	X	X	X	X	X	X	X	X	X	X	X	X	X	X	X	X	X	X	X	X
Algal meas.	X	X	X	X		X		X		X		X		X		X		X		X		X

Note: SD = Spray drift pesticide application; RO = Soil runoff pesticide application; Res. water = residue samples from water column; Sed. cores = sediment cores for residue analysis; MAS samples = benthic insect colonization; Emergence = aquatic insect emergence (exuviae and/or emergence traps; and Algal meas. = phytoplankton and periphyton biomass and chlorophyll analysis.

Cyfluthrin residues were analyzed using an HP-5890A gas chromatograph with an electron capture detector, utilizing cool on-column injections techniques.

Macroinvertebrate artificial substrate (MAS) samplers were used to estimate epibenthic macroinvertebrate population density. Colonization substrates were constructed using the Actifill surface enhancers described above. Actifill units (14 in mesocosm samplers, 8 in microcosm samplers) were bound together using plastic cable-ties and were weighted with pebble-filled 125-ml Nalgene bottles. Mosquito netting on the bottom of the samplers helped retain macroinvertebrates during retrieval. Mesocosms contained 6 littoral zone MAS samplers per pond, with a combined surface area of 1.615 m². Microcosms had two smaller MAS samplers per tank with a combined surface area of 0.308 m². Samplers were introduced into microcosms/mesocosms for 1 month, allowing colonization. The MAS samplers were then removed and scrubbed gently to dislodge organisms, which were retained on a 180-μm mesh sieve, and preserved in Kahle's solution.[16] Samplers were replaced to allow further colonization.

Benthic macroinvertebrates were collected from mesocosms only using an Ekman grab sampler. Samples were preserved in Kahle's solution[16] and stained with Rose Bengal.

Insect emergence from mesocosms was measured using pyramid-shaped floating emergence traps, similar to LeSage and Harrison.[17] Emergence from microcosms was quantified using two measures: adult emergence and final instar/pupal exuviae (cast exoskeletons). Emerging adult insects were collected with semisubmerged funnel traps (after Davies[18]). Exuviae were either skimmed from the water's surface using dip nets or collected above the waterline using forceps. Both exuviae and adult insects were preserved in Kahle's solution.

Zooplankton were collected with an integrated water column tube sampler (schedule R-4000 PVC tubing, 2'' diameter) in both systems. Using a 30-μm mesh net, 5 l per microcosm/mesocosm were concentrated and were preserved with Lugol's solution.

Phytoplankton were collected using the tube sampler described above, and a grab sample (250 ml) was preserved in Lugol's. Periphyton were allowed to colonize glass microscope slides for 2 weeks in floating periphytometers. Periphyton and planktonic chlorophyll analyses involved acetone extraction followed by spectrophotometric measurement.[19] Algal biomass was determined as ash-free dry weight by scraping slides into crucibles, obtaining dry weights, ignition in a muffle furnace, and reweighing.[19]

Water chemistry parameters such as total suspended solids, total phosphorus, nitrates, nitrites, ammonia, TOC, DOC, POC, and turbidity were analyzed using standard methods.[19] Dissolved oxygen, water temperature (surface and bottom), and surface pH were measured weekly using calibrated meters. Gross photosynthesis and community respiration were determined using the three point diel oxygen pulse method.[20]

Four critical time periods were selected for graphical representation of biological effects. These were (1) prior to pesticide application, week -1; (2) during the middle of the treatment period, week 5; (3) preceding the last pesticide application period, week 9; and (4) final week of experiment, post-application, week 19. Although data were collected more often (Tables 3 and 4), we felt that visual trends would be obscured by presenting all weeks, for all taxa, in both mesocosms and microcosms.

Statistical Analysis

To assess treatment effects, one-way analysis of variance procedures were computed for each week, for all water chemistry parameters, and for each taxon using SAS software. This was followed by Dunnett's multiple range tests (alpha = 0.05) to determine which treatments differed significantly from controls. Mesocosms and microcosms were evaluated separately.

Tabular comparisons of pesticide impacts in mesocosms and microcosms were conducted (Tables 5 and 6). Significant impacts, increases/decreases in treated tanks/ponds relative to controls, were determined during the 10-week application period (Dunnett's multiple range test; alpha = 0.05). Treatments were determined to differ from controls if Dunnett's values were significantly different for more than 1 week during the application period. One-time differences were judged to not be reflective of the general response pattern for the whole experiment.

We also quantified the cumulative number of significant differences for biological effects determined via Dunnett's multiple range test (alpha = 0.05) observed during the 19-week application and postapplication period (Figure 18). Unlike graphical representations for each taxon (Figures 5 to 17), all available data were used in these analyses (Tables 5 and 6, Figure 18).

RESULTS

Chemical Fate

Water column residues were collected near the surface of the water and near the bottom above the sediments 1 h after applying BAYTHROID®. Following spray drift treatments, surface values were generally higher than, or equal to, nominal values. Bottom values were generally lower than nominal concentrations. We calculated the mean concentrations across the experimental period, averaging surface and bottom concentrations. Microcosm spray drift values were consistently higher and closer to nominal targets than were mesocosm concentrations (Table 2). Water column residues following runoff treatments were similar in both mesocosms and microcosms (Table 2). Percent of nominal values following runoff treatments were consistently lower than after spray drift application.

Water column concentrations, following spray drift application, dissipated more rapidly in microcosms (22 h half-life) than in mesocosms (40 h half-life; Figure 3).

Sediment residues were generally higher in microcosms, with some exceptions (Figure 4).

Physical and Chemical Parameters

Water chemistry parameters showed no response to pesticide treatment in microcosms (Table 5). Some mesocosm parameters demonstrated significant differences between controls and treated ponds, but exposure-response relationships within treated mesocosms were rare. Alkalinity increased in both systems while hardness decreased in both. Microcosm alkalinity was consistently higher than mesocosm values (Table 5). Phosphate, nitrite, and ammonia in treated mesocosms were lower than controls, whereas microcosms showed no patterns related to pesticide input and no patterns through time. Total suspended solids and nitrates did not show

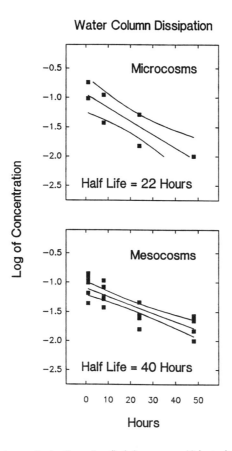

Figure 3.   Water column dissipation of cyfluthrin over a 48-h period in microcosms and mesocosms. Points are observed concentrations (ppb), log base 10, through time. Lines represent linear regression with 95% confidence intervals. Half-life values were computed from regressions.

Sediment Residue Values

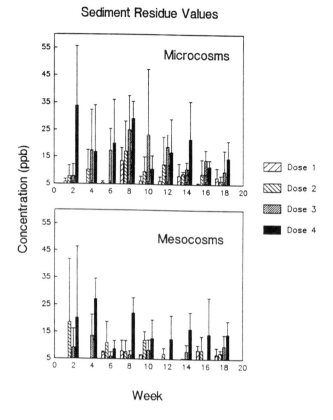

Figure 4.   Observed cyfluthrin concentrations in experimental systems obtained from anal-
              ysis of sediment cores. Vertical bars represent mean values (ppb); lines are one
              standard deviation. Each of four treatment levels are grouped contiguously during
              each sampling period.

treatment effects in either system. Organic carbon values (total, dissolved, and
particulate fractions) did not show trends with pesticide concentration and were
somewhat higher in microcosms relative to mesocosms (Table 5).

Mean turbidity values were affected in mesocosms, with controls significantly
higher than all treatments from week 3 to study termination (Table 5). High turbidities in control ponds might be linked with higher water quality values noted
above. Microcosms showed a more complex pattern, with controls higher than dose
4 at weeks 5, 7, and 11 through 19. Turbidity was consistently higher in mesocosms
relative to microcosms (Table 5). Higher turbidities were attributed to intense rains
in May and June. Mesocosms received soil inputs via runoff while microcosms did
not. This basic difference may have carried over into the development of macrophyte
communities in these systems, with more macrophytes in microcosms. Higher
turbidity may have reduced macrophyte development in mesocosms.

Mean dissolved oxygen, temperature, and pH showed no pesticide effects. Mean pH values were somewhat higher in microcosms compared to mesocosms (by ca. 0.5 pH units; Table 5).

Biological Effects

Fish

Fish responses in the mesocosms and microcosms are detailed in Chapter 22 of this volume.[21] Juvenile fish in microcosms did not reproduce. Adult bluegills in mesocosms produced large numbers of young. Fish biomass measured at the end of the experiment, scaled for system volume, was lower in mesocosms compared to microcosms (Table 6). Bluegill sunfish densities (fish per m³) at experiment termination were much higher in mesocosms. Mesocosms averaged 21.1 fish per m³ and 6.08 g/m³ of biomass for all ponds combined at study termination. Microcosms averaged 4.3 fish per m³ and 17.77 g/m³ of biomass for all tanks combined.[21]

Table 5. Summary of Pyrethroid Chemistry, Water Chemistry, and Phytoplankton Pigment/Biomass Analyses

Parameter	Microcosms Impact	Mesocosms Impact	Mesocosms Treat	Comparison of Concentrations
Residues				
Sediment	NA	NA	NA	Mic > Mes
Water half-life	NA	NA	NA	Mic < Mes
Percent target	NA	NA	NA	Mic ≥ Mes
Water chemistry				
Nitrogen	—	R	D1–4	Mic ≤ Mes
Phosphorus	—	R	D1–4	Mic < Mes
Alkalinity	—	—	—	Mic > Mes
Hardness	—	—	—	Mic ≥ Mes
TOC, DOC, POC	—	—	—	Mic ≤ Mes
TSS	—	—	—	Mic < Mes
Turbidity	—	R	D1–4	Mic < Mes
pH	—	—	—	Mic > Mes
Phytoplankton				
Chlorophyll *a*	—	—	—	Mic = Mes
Biomass	—	—	—	Mic = Mes

Note: Impact refers to significant increases (I) or reductions (R) in treated tanks/ponds relative to controls, using Dunnett's multiple range test (alpha = 0.05) observed during the ten week application period. Treatment (Treat) indicates the level(s) which were significantly impacted. Dashes (-) represent no significant differences. Concentrations in microcosms are compared with mesocosm values in the final column, indicating if one system had consistently greater values. Treatments were determined to differ from controls if Dunnett's values were significantly different for more than one week during the application period. One-time differences were judged to not be reflective of the general response pattern for the whole experiment (summarized here). NA, not applicable; I, increase; R, reduction; and —, no impact.

Table 6. Summary of Fish, Zooplankton, and Macroinvertebrates Population Responses

Parameter	Microcosms			Mesocosms			Scaled Population Densities
	Meth	Impact	Treat	Meth	Impact	Treat	
Fish							
Biomass (g/m³)	NA	—	—	NA	—	—	Mic > Mes
Density (#/m³)	NA	—	—	NA	—	—	Mic < Mes
Zooplankton (#/l)							
Rotifera	Tube	I	D2–4	Tube	—	—	Mic ≥ Mes
Cladocera	Tube	R	D4	Tube	*	*	Mic > Mes
Copepoda	Tube	R	D4	Tube	R	D1–4	Mic > Mes
Nauplii							
Macroinvertebrates (#/m²)							
Gastropoda	MAS	—	—	MAS	I	D1–4	Mic ≤ Mes
Oligochaeta							
Naididae	MAS	—	—	MAS	—	—	Mic < Mes
				Ekman	I	D1–4	
Amphipoda	MAS	R	D1–4	MAS	R	D1–4	Mic < Mes
Nematoda	MAS	*	*	MAS	—	—	Mic < Mes
				Ekman	I	D1,3	
Ephemeroptera							
Caenidae	MAS	R	D1–4	MAS	R	D1–4	Mic < Mes
	ET	*	*	ET	*	*	
	EXUV	—	—				
Baetidae	MAS	R	D1–4	MAS	R	D2–4	Mic > Mes
	ET	*	*	ET	*	*	
	EXUV	R	D1–4				
Trichoptera	MAS	R	D1–4	MAS	R	D1–4	Mic = Mes
	ET	*	*	ET	*	*	
	EXUV	*	*				
Diptera							
Chironomidae							
Chironominae	MAS	I	D1,D3	MAS	I	D1–4	Mic < Mes
	ET	—	—	ET	—	—	
	EXUV	—	—	Ekman	—	—	
Tanypodinae	MAS	R	D2–4	MAS	R	D4	Mic = Mes
	ET	R	D1–4	ET	R	D1–4	
	EXUV	R	D2–4	Ekman	I	D1	
Ceratopogonidae	MAS	*	*	MAS	I	D1–4	Mic < Mes
	ET	*	*	ET	I	D1–4	
	EXUV	*	*	Ekman	I	D1–2,4	
Chaoboridae	MAS	*	*	MAS	R	D2–4	Mic < Mes
	ET	*	*	ET	R	D1–4	
	EXUV	R	D1–4	Ekman	R	D1–4	
Odonata							
Libellulidae	MAS	—	—	MAS	—	—	Mic > Mes
	EXUV	—	—				
Coenagrionidae	MAS	—	—	MAS	—	—	Mic > Mes
	EXUV	—	—				
Coleoptera	MAS	R	D1–4	MAS	—	—	Mic ≤ Mes
Berosus	MAS	R	D1,3–4	MAS	R	D3–4	

Note: Method indicates the sampling method used. Impact refers to significant increases (I) or reductions (R) in treated tanks/ponds relative to controls, using Dunnett's multiple range test (alpha = 0.05). Dashes (—) indicate no significant differences. Treatment indicates the level(s) which were significantly impacted. Population densities in microcosms are compared with mesocosm values in the final column, indicating if one system had consistently greater values. Selection criteria outlined in Table 5 were used for inclusion in this table. NA, not applicable; and *, densities too low to evaluate.

Phytoplankton and Periphyton

Phytoplankton biomass levels (mg/l) were similar in mesocosms and microcosms. Treatment-related trends were rarely observed in either system. Phytoplankton chlorophyll *a* levels (mg/l) were generally higher in treated microcosms, compared with the controls. This trend was not seen in the mesocosms. Phytoplankton chlorophyll *a* levels were very similar in both systems and tracked each other well through time (Table 6). Surprisingly, periphyton chlorophyll *a* levels (mg/mm^2) were lower in microcosms. Microcosm periphyton biomass values (mg/mm^2) were either lower than mesocosms or were similar in concentration. Periphyton colonization was not influenced by treatment level.

Zooplankton

Total Cladocera were considerably more abundant in microcosms relative to mesocosms (Figure 5). Both pyrethroid-sensitive and insensitive taxa were represented in microcosm samples, but this distinction could not be made in mesocosms due to low numbers (Table 6). *Diaphanosoma brachyurum* (Figure 6) and *Chydorus sphaericus* (not shown) demonstrated population declines consistent with pesticide loadings. *Diaphanosoma brachyurum* were the dominant large-bodied Cladocera in microcosms. *Macrothrix rosea* did not show treatment related responses (Figure 6), while *Alona rustica* (not shown) were enhanced in the highest loadings relative to controls during week five, but were unaffected in any treatments at week nine.

Mature copepods were not abundant in either system. Nauplii were more abundant in microcosms compared to mesocosms (Figure 7). Nauplii in mesocosms exhibited treatment-related reductions at lower exposure levels than those seen in microcosms (Table 6).

Total rotifer populations were much higher in microcosms compared with mesocosms (Figure 8). Total rotifers were significantly enhanced in microcosm treatments D2 through D4 during the last application week (Table 6). A similar but nonsignificant trend was observed in mesocosms for treatments D3 to D4 during the same time period (Figure 8).

Similar rotifer taxa were found in mesocosms and microcosms, but the dominant species differed in these systems. *Brachionus* (*B. angularis* and *B. havanensis*) were most abundant in microcosms and were the driving force behind treatment enhancements during the last application. *Brachionus* were uncommon in mesocosms. Some taxa in mesocosms were rarely seen in microcosms. *Hexarthra mira*, for instance, was common during week five in mesocosms and *Keratella cochlearis* was abundant prior to treatment initiation, but both were rare in microcosms.

Other rotifers were common to both mesocosms and microcosms. *Filinia longiseta* showed similar enhancements with higher treatments in both systems (Figure 9). *Polyarthra remata* showed disparate trends in the two systems. Microcosms experienced enhancements with increasing treatment levels (Figure 10), but mesocosms showed either reductions (week five; "Mid") or slight enhancement (week nine; "Last").

Cladocera

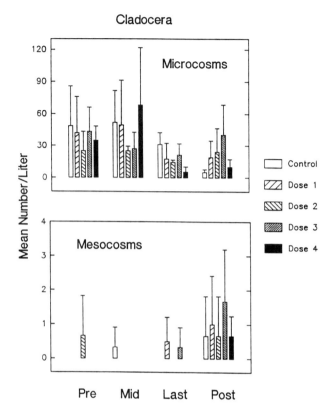

Figure 5.   Total Cladocera populations in experimental systems. Note difference in vertical scales between mesocosms and microcosms. Vertical bars represent mean values (number /l); lines are one standard deviation. Treatment levels are grouped together within four sampling periods. Pre = prior to pesticide application (week − 1), Mid = mid-treatment period (week 5), Last = last pesticide application period (week 9), Post = final week of experiment, post-application (week 19).

Macroinvertebrates

Macroinvertebrates identified as being sensitive or insensitive were similar in both systems. Artificial substrate colonization was effective in demonstrating pesticide impacts in some taxa. The mayfly *Callibaetis floridanus* was significantly reduced at all treatments (D1 to D4) in microcosms and reduced at levels D2 to D4 in mesocosms (Table 6). *Callibaetis* nymphs found in control microcosms were higher than those in mesocosms, and an exposure/response relationship was evident in the smaller systems (Figure 11). The mayfly *Caenis* experienced higher populations in mesocosms relative to microcosm controls (Figure 12). *Caenis* nymphs were significantly reduced at all treatments in both systems, as were trichopterans and amphipods (Table 6).

Chaoboridae MAS colonization was significantly decreased and Ceratopogonidae populations were significantly enhanced with treatment in mesocosms (Table 6). Parallel responses were not found in microcosms due to low numbers of these taxa in MAS samplers. Significant reductions of Tanypodinae chironomids were detected at lower levels in microcosms (D1 to D4) compared to mesocosms (D4 only; Table 6).

Other taxa colonizing substrates were identified as being insensitive. Gastropods (not shown) were not impacted by the pyrethroid. Oligochaetes (family Naididae) demonstrated very similar colonization patterns even though the absolute numbers differed greatly (Figure 13). Naididae showed a small increase in microcosms during the last treatment week. Substrate colonization may not be the most effective method for sampling Naididae, however, since Ekman grab samples from mesocosms revealed significant population increases in treated ponds not observed in MAS colonization (Table 6).

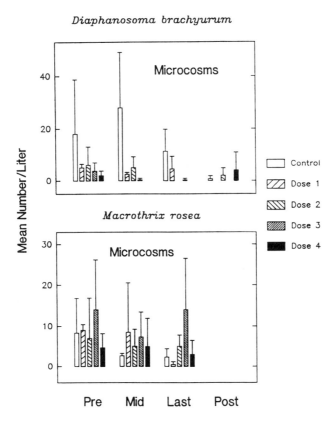

Figure 6.   Dominant cladocerans found in microcosms. A sensitive taxa (*Diaphanosoma brachyurum*) and insensitive taxa (*Macrothrix rosea*) are shown. Low sample sizes in mesocosms precluded this type of analysis.

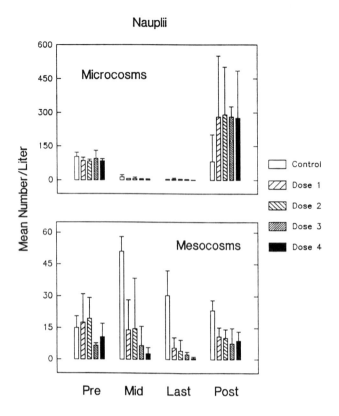

Figure 7. Nauplii (Copepoda) populations in experimental systems. Treatment/response relationship evident in mesocosms, not in microcosms. Lower numbers in mesocosms may reflect larval fish predation.

Large odonate predators were more abundant in microcosms compared with mesocosms (Table 6). Libellulidae (dragonfly larvae) numbers did not show consistent treatment-related patterns, with the possible exception of D4 in microcosms (Figure 14). Coenagrionidae (damselfly larvae) also lacked clear treatment-related patterns, being reduced in D4 microcosms and not affected in mesocosms (not shown).

Insect emergence appeared to be an excellent method of detecting pesticide/biotic relations. Both sensitive and insensitive taxa were identified, with fairly good agreement between mesocosms and microcosms. Exuviae collection in microcosms yielded larger sample sizes and larger species diversity compared with funnel traps in microcosms. Thus, all comparisons are of microcosm exuviae and mesocosm pyramid traps.

Chaoborus emergence (Figure 15) was among the most sensitive parameters measured, with similar patterns in both systems. Chironomids in the subfamily Tanypodinae were also identified to be sensitive to pesticide additions (Figure 16) while members of the subfamily Chironominae were not (Figure 17). One major difference in the emergence patterns among these systems was the greater number Ceratopogonidae (Diptera) emerging from mesocosms. Ceratopogonid emergence in mesocosms was significantly enhanced at all pesticide loadings (Table 6).

The total number of significnt differences detected over the course of the experiment increased with increasing treatment level (Figure 18). The total number of differences detected were generally higher in mesocosms, primarily due to MAS samples. In microcosms, exuviae collection identified more differences than did emergence traps. More differences in zooplankton densities were found in microcosms.

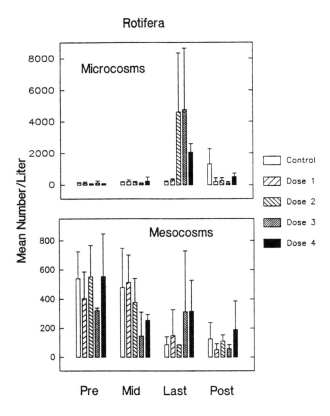

Figure 8. Total rotifer populations in experimental systems. Lower rotifer abundance in mesocosms may reflect larval fish predation. Vertical bars represent mean values (number/l); lines are one standard deviation. Total rotifers in mesocosms might be impacted by larval fish predation.

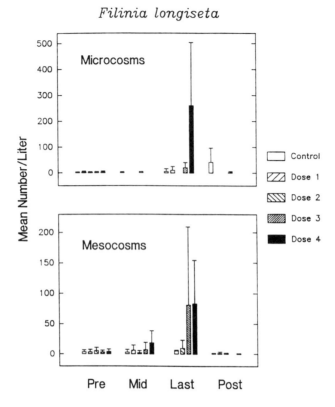

Figure 9.   *Filinia longiseta* (rotifer) populations in experimental systems. Vertical bars represent mean values (number /l); lines are one standard deviation. Enhancement at high pesticide levels is similar for both systems.

DISCUSSION

Chemical Fate

Most pyrethroids are very lipophilic, halogenated, and exhibit very low water solubilities,[22] and tend to adsorb to soils, sediments, and plant cuticles.[23,24] Cyfluthrin has a solubility of 1.3 to 2.6 ppb (Material Safety Data Sheet, BAYTHROID® technical grade compound).

Chemical fate characteristics of BAYTHROID® shared common patterns among the two systems but were apparently influenced by scaling relationships such as surface/volume ratios. High surface-to-volume ratios found in microcosms influence chemical fate and nutrient cycling.[25] Pyrethroids should sorb rapidly to the walls and bottom of microcosms, effectively reducing the amount of bioavailable pesticide relative to mesocosms. More macrophyte growth occurred in microcosms than in mesocosms, providing additional surface area for sorption. This was probably reflected in the shorter pesticide half-life in microcosms. The microcosm's half-life in this study may not be definitive, however, since fewer samples were collected compared to mesocosms, with resulting increases in 95% confidence intervals. It should also be noted that all microcosm values would fit within the observed range of mesocosm concentrations. Other studies indicate that pyrethroids show typical half-lives in water measured in hours or days.[26]

The method of pesticide application may have partially offset these potential sinks, with mesocosms receiving a spray application and microcosms treated with a larger droplet size. Since volatilization of sprayed pesticide may be an important loss mechanism in photostable pyrethroids,[23,27] microcosms actually achieved drift concentrations closer to target values (Table 2). Also, the pesticide was delivered to microcosms more rapidly since the application process was simpler. Finally, more complex application gear (bridge spanner with many feet of Teflon tubing,

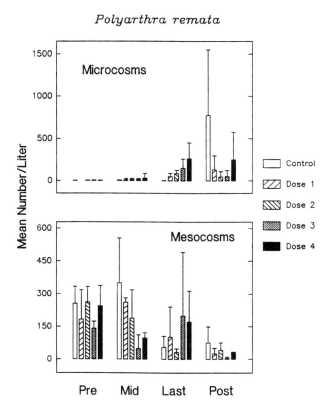

Figure 10. *Polyarthra remata* (rotifer) populations in experimental systems. Lack of correspondence between mesocosms and microcosms suggests that this taxa is influenced by secondary interactions more than direct pyrethroid impacts.

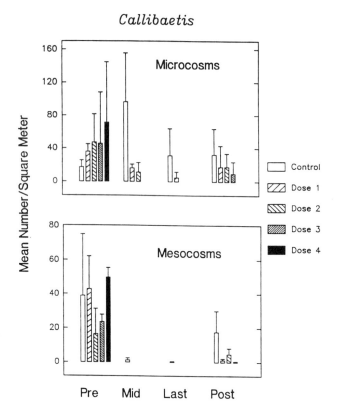

Figure 11.   *Callibaetis floridanus* (mayfly) populations colonizing artificial substrates. Vertical bars represent mean values (number/m²); lines are one standard deviation. This species is a mobile epiphytic/epibenthic group presumably vulnerable to fish predation.

spray nozzles, and stainless steel spray tank; Figure 2) provided more surface area for pre-application sorption.

Water column residues collected an hour after runoff applications were very similar in both systems, and probably reflect sedimentation of bound pyrethroid during this interval (Table 2). Since soil for runoff applications was obtained from a common stock, similar patterns in both systems were anticipated.

Pyrethroids show rapid but shallow penetration of sediments, with the majority of the pesticide within the upper 2 cm.[28,29] Values were somewhat higher in microcosms relative to mesocosms (Figure 4) and probably reflect scaling (surface area/volume) relations. Other factors, such as microbial activity, may influence pesticide concentrations in sediments; unfortunately, we did not measure these factors.

Biological Effects

Population level endpoints were emphasized in this study design and were correspondingly found to be more sensitive than functional parameters. Ecosystem function (nutrient cycling, etc.) may show few long-term effects of perturbation, since natural systems have feedback mechanisms that serve as buffers. Structural variables such as species composition are often affected more rapidly than functional variables in ecotoxicological studies.[14,30,31]

Fish

Fish do not usually exhibit acute mortality in pond systems when pyrethroids are applied at field rates.[23,32,33] We observed no pesticide-related mortality in our study.

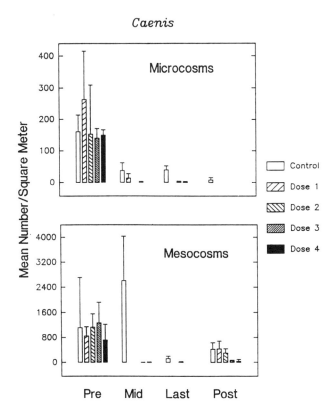

Figure 12. *Caenis* spp. (mayfly) populations colonizing artificial substrates. Vertical bars represent mean values (number/m²); lines are one standard deviation. Members of this genera are less exposed to fish predation.

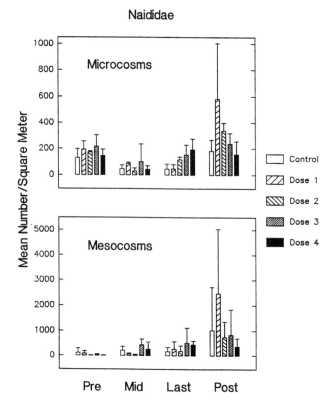

Figure 13.   Naididae (oligochaete) populations colonizing artificial substrates. These invertebrates are subject to predation by odonates and midges in the subfamily Tanypodinae.

Zooplankton

Studies using other pyrethroids have found cladocerans and chaoborids to be very sensitive, while copepods and ostracods were less affected, and rotifers often increased or were unaffected.[34-38] These results agree with effects observed in our study, particularly the microcosms (Table 6). Significant impacts on copepods were detected at lower levels in mesocosms compared with microcosms (Table 6).

Within the rotifers, we observed differing response patterns when comparing mesocosms with microcosms. *Filinia longiseta* showed general agreement, but *Polyarthra remata* showed variable responses. *Hexarthra mira, Keratella cochlearis,* and *Brachionus* spp. differed in their importance within the two systems. Increases in rotifer densities following reductions in large Cladocera are often attributed to interference and exploitative competition.[39]

Large reductions in copepod nauplii were observed in microcosms during the summer ("Mid" and "Last" periods; Figure 7). This trend was not observed in mesocosms. Summer minima in microcosms brought nauplii densities within the range observed in mesocosms. These population fluctuations were not linked to treatment levels, suggesting competitive or predatory impacts. Thus, secondary interactions showed more diversity compared with pesticide-induced effects.

Macroinvertebrates

Macroinvertebrates differ in sensitivity to pyrethroid exposure. Anderson[40] found decreasing sensitivity from amphipods > mayflies > stoneflies and caddisflies > snails. Mayflies, amphipods, water mites, and surface-dwelling groups (many hemipterans) are often strongly affected by pyrethroids.[32,33,35,38,41] Oligochaetes are often unaffected.[32,33] These trends agree with sensitivities found in our systems.

Midges commonly show differing susceptibility to pyrethroids, with the subfamily Tanypodinae generally more sensitive than the subfamily Chironominae. This difference is thought to reflect differences in exposure since many Chironominae

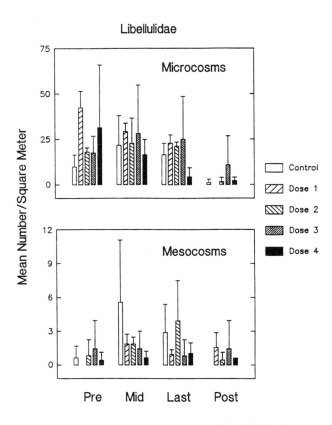

Figure 14.   Libellulidae (dragonfly) populations colonizing artificial substrates. Bluegill sunfish are known to select these odonates preferentially.[47]

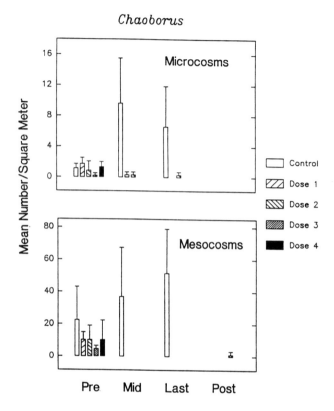

Figure 15. Emergence of *Chaoborus* (phantom midge) populations. Microcosm data are from exuviae collection; mesocosm data are from emergence traps. Vertical bars represent mean values (number/m^2); lines are one standard deviation. These dipterans are known to be very sensitive to pyrethroids.[37]

burrow within the sediment or construct feeding tubes.[38,42] These observations agree with our emergence and epibenthic colonization data in both systems.

More significant differences, both reductions and enhancements, were detected in mesocosms compared to microcosms, but similar trends were evident (Figure 18). This may reflect the generally larger sample sizes found in mesocosms, particularly for many macroinvertebrates. Zooplankton were as numerous, or more numerous, in microcosms (Table 6), and this is reflected in the number of differences identified. In general, a clear treatment-response relationship is seen for cumulative statistical differences over the experimental period in both systems.

Impacts of Fish

In addition to direct impacts of pesticide, secondary (indirect) impacts are observed in ecotoxicological manipulations due to reductions in predators and/or competitors. In these types of experimental systems, fish obviously have a large impact and determine the abundance of some taxa. Mesocosms were stocked with adult fish (mean length, at study initiation, of 13.197 cm ± 1.58 cm), which reproduced during the experimental period. An average of 12,961 (± 3872) juvenile fish, with a modal size class of 2.0 to 2.9 cm, were harvested from mesocosms at study termination. Microcosms were each stocked with 8 sexually immature bluegill sunfish (mean length, at study termination, of 6.87 cm ± 0.88 cm), which did not reproduce. These two systems thus experienced different predation pressures, with mesocosm bluegills ranging in size from small to very large.

Comparison of microcosms and mesocosms shows higher microcosm densities for cladocerans, copepods, rotifers, libellulid dragonflies, coenagrionid damselflies, and baetid mayflies (Table 6). Caenid mayflies, Chironominae midges, naidid oligochaetes, chaoborids, gastropods, and ceratopogonids were more abundant in mesocosms. Some taxa were equally represented in both systems, such as Tanypodinae midges and trichopterans (Table 6).

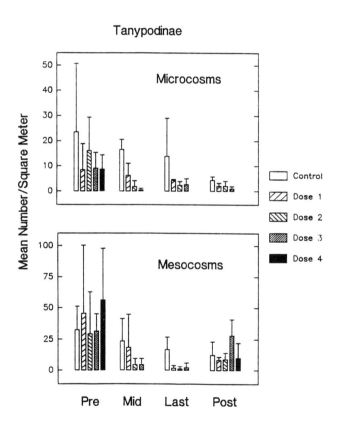

Figure 16. Emergence of Tanypodinae (Chironomidae) populations. Microcosm data are from exuviae collection; mesocosm data are from emergence traps. This is another pyrethroid sensitive taxa.

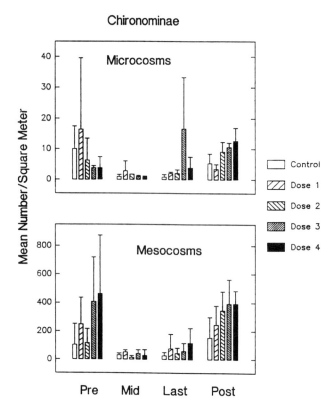

Figure 17. Emergence of Chironominae (Chironomidae) populations. Microcosm data are from exuviae collection; mesocosm data are from emergence traps. This group is less sensitive to pyrethroids.

Visual planktivores generally select prey based on prey size[43,44] and escape ability.[45,46] Thus, large and slow prey like *Daphnia* are subject to intense predation.[12,47,48] Larval bluegill sunfish feed on small rotifers such as *Polyarthra* and copepod nauplii initially, switching to other rotifers and cyclopoid copepods when fish reach approximately 7 mm in length.[49] Fish larger than 8 mm feed primarily on small cladocerans such as *Bosmina, Chydorus, Diaphanosoma brachyurum,* and *Alona.*[49] Selective predation of larval and small bluegill on rotifers and nauplii may explain observed reductions of these taxa within mesocosms relative to microcosms, since these small size classes were not present in microcosms.

Open water systems and planktivory have received more attention than fish impacts in littoral systems. Bluegill predation on benthic communities is also size selective, with bluegills considered to be "keystone predators".[50] Large or active prey are selected, increasing the densities of smaller, more sedentary prey.[50-52]

Many differences in macroinvertebrate population levels between mesocosms and microcosms can be explained when prey life-history and bluegill predation are considered. *Callibaetis* nymphs are "swimmers and climbers",[53] usually associated with aquatic macrophytes, suggesting increased exposure to fish predation.

Caenis mayfly immatures are classified as "sprawlers", inhabiting depositional areas and sediments and would be less vulnerable to fish predation.[53] Production of *Caenis simulans* (now *C. amica*) was enhanced in replicated ponds containing bluegills and lower in fishless treatments.[54] This parallels the higher *Caenis* population densities in mesocosms relative to microcosms.

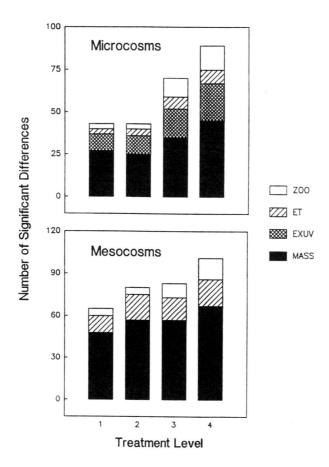

Figure 18. Number of significant differences detected by Dunnett's multiple range testing (alpha = 0.05) at each treatment level. Analyses were conducted for each taxa, for each sampling method (ET, EXUV, MASS, ZOO) during each week. All significant differences, both enhancements and reductions, were summed across the 19-week application and postapplication period. The number of differences for zooplankton are probably low, since zooplankton data were not available for weeks seven and fifteen. Weeks seven and fifteen were included for ET, EXUV, and MAS to provide the most complete data possible. Note that EXUV and ET emergence data are included for microcosms, while EXUV were not collected in mesocosms.

Odonates are usually reduced in enclosures or ponds containing bluegills.[47,54,55] This trend may reflect both direct predation of fish on odonates and competition between fish and surviving naiads for common prey items.[55] Size-selective bluegill predation results in high predation rates on large odonate naiads.[51] Odonates colonizing MAS samplers, primarily Libellulidae and Coenagrionidae, were more abundant in microcosms and reduced in mesocosms. Some taxa which did not readily colonize epibenthic substrates (Gomphidae, Aeshnidae) were not sampled effectively; thus, population impacts within these groups are currently unknown.

The reduced numbers of odonates in mesocosms relative to microcosms may reflect a combination of factors. First, a broader range of bluegill sizes were represented in mesocosms which might allow predation on a number of odonate size classes simultaneously. Microcosm bluegills were juvenile fish and may have experienced gape limitations when handling the largest odonate naiads. Alternatively, many Libellulidae and Coenagrionidae are associated with aquatic macrophytes, which were more abundant in microcosms.

At a general level, chironomid larvae are considered to be preferred bluegill prey.[56] Close examination of chironomid responses to fish predation reveal more complex relationships. Gilinsky[57] found that chironomids responded variably to fish predation depending upon ecological niche, habitat sampled, and season of the year. We observed somewhat higher Tanypodinae and much higher Chironominae populations in mesocosms relative to microcosms.

Macrophytes may provide a refuge for macroinvertebrates by reducing the searching efficiency of bluegill.[57,58] Macrophyte density has also been correlated with survival of *Daphnia* populations.[59] Even though macrophytes theoretically reduce search efficiency, macrophytes also provide habitat for epiphytic invertebrates. Bluegill utilization of epiphytic prey may be much greater than predation upon benthic organisms.[56] Thus, invertebrate production and fish growth may be maximized at some intermediate macrophyte density.[60] Higher densities of potential bluegill prey in our microcosms compared to mesocosms probably reflect higher macrophyte densities and the artificial refugia present, in addition to the obvious differences in fish loadings.

We utilized fishless microcosms to assist in partitioning fish effects from pesticide stress. We observed large impacts on epibenthic/epiphytic macroinvertebrates and emergent insects.[21] Impacts on emerging insects were more pronounced than impacts on macroinvertebrates colonizing substrates.[21]

Other Factors

Differences in community structure between the mesocosms and microcosms may also reflect colonization phenology. Odonate community composition, for instance, is influenced by which species first colonize a pond.[51,61] Also, the surface area of water bodies may influence colonization rates,[62] and the presence of suitable oviposition sites may help determine insect community composition.

Finally, microcosms lack a depth gradient (vertical walls are present). The lack of a sloped bank may influence emergence success for some groups of insects (such as some odonates, mayflies, and beetles). For example, we noted a number of drowned odonates in some extra microcosms not in the study design. These tanks lacked artificial refugia, and emerging odonates apparently had difficulty clinging to the smooth surface. Refugia, when present, were used extensively by emerging odonates. This suggests that some sort of emergence route linking the microcosm sediments to the surface may be useful and could be incorporated in future microcosm designs.

SUMMARY

Chemical fate was influenced by system size and application methodology. Characterization of sensitive and insensitive taxa was similar among the two systems. Responses of sensitive populations showed corresponding patterns but differences in absolute numbers among the two systems. These differences in absolute numbers and dominant taxa may be attributable to differences in bluegill predation pressures, habitat (macrophytes), and may also reflect taxa-specific colonization of these systems.

It has long been recognized that microcosms would be cost effective models for determining environmental fate of chemicals[63] and that microcosm size is an important variable.[64,65] This study should contribute to the growing information regarding scaling of ecological systems. We feel that microcosms hold considerable promise as a supplement to current methods used for evaluation of pesticide impacts in aquatic systems. While some differences were observed among the two systems, the same sensitive taxa were identified at similar exposure levels in microcosms and mesocosms. The relation of chemical fate to system size deserves greater evaluation in future studies. Smaller scale systems might eventually be used as an intermediate tier in the risk assessment process.

ACKNOWLEDGMENTS

The research for this chapter was conducted at the University of North Texas, Denton.

Any project of this magnitude involves the assistance of many people. We wish to thank all the employees of the University of North Texas Water Research Field Station who made this study possible, particularly L. J. Scott who conducted the residue chemistry and Roxanne Montadon for coordinating mesocosm fieldwork. Shaun Gibson prepared the figure of the bridge spanner. Dr. John H. Rodgers, Jr., University of Mississippi, was involved during experimental design and set up. We are particularly grateful to Mobay Corporation, Kansas City, U.S. for funding this research.

LITERATURE CITED

1. Cairns, J., Jr. 1988. Should regulatory criteria and standards be based on multispecies evidence? *Environ. Profess.* 10: 157–165.
2. Cairns, J., Jr. and E.P. Smith. 1989. Developing a statistical support system for environmental hazard evaluation. *Hydrobiologia* 184: 143–151.
3. Cairns, J., Jr. 1988. Putting the eco in ecotoxicology. *Reg. Toxicol. Pharmacol.* 8: 226–238.
4. La Point, T.W. and J.A. Perry. 1989. Use of experimental ecosystems in regulatory decision making. *Environ. Management* 13: 539–544.
5. Bascietto, J., D. Hinckley, J. Plafkin and M. Smilak. 1990. Ecotoxicity and ecological risk assessment: Regulatory applications at EPA. *Environ. Sci. Technol.* 24: 10–15.
6. Stephenson, G.L., P. Hamilton, N.K. Kaushik, J.B. Robinson and K.R. Solomon. 1984. Spatial distribution of plankton in enclosures of three sizes, *Can. J. Fish. Aquat. Sci.* 41: 1048–1054.
7. Kaushik, N.K., K.R. Solomon, G.L. Stephenson and K.E. Day. 1986. Use of limnocorrals in evaluating the effects of pesticides on zooplankton communities. J.C. Cairns, Jr., [ed.]. *Community Toxicity Testing.* ASTM Special Technical Publication 920. ASTM, Philadelphia, PA, pp. 269–290.
8. Frost, T.M., D.L. DeAngelis, S.M. Bartell, D.J. Hall and S.H. Hurlbert. 1988. Scale in the design and interpretation of aquatic community research. In: Carpenter, S.R., [ed.]. *Complex Interactions in Lake Communities.* Springer-Verlag, New York, pp. 229–258.
9. Solomon, K.R., G.L. Stephenson and N.K. Kaushik. 1989. Effects of methoxychlor on zooplankton in freshwater enclosures: influence of enclosure size and number of applications. *Environ. Toxicol. Chem.* 8: 659–669.
10. Menge, B.A. and A.M. Olson. 1990. Role of scale and environmental factors in regulation of community structure. *Trends in Ecol. and Evolution* 5: 52–57.
11. Carpenter, S.R. and J.F. Kitchell. 1988. Consumer control of lake productivity. *Bioscience* 38: 764–769.
12. Stein, R.A., S.T. Threlkeld, C.D. Sandgren, W.G. Sprules, L. Pearson, E.E. Werner, W.E. Neill and S.I. Dodson. 1987. Size structured interactions in lake communities. In: Carpenter, S.R., [ed.]. *Complex Interactions in Lake Communities.* Springer-Verlag, New York, pp. 161–179.
13. Schindler, D.W. 1990. Experimental perturbations of whole lakes as tests of hypotheses concerning ecosystem structure and function. *Oikos* 57: 25–41.
14. Giddings, J.M. 1980. Types of aquatic microcosms and their research applications. In: *Microcosms in Ecological Research.* J.P. Geisy, [ed.]. DOE Symposium Series 52, Technical Information Center, U.S. Dept. of Energy, pp. 248–266.
15. Smith, D.W. 1985. Biological control of excessive phytoplankton growth and the enhancement of aquacultural production. *Can. J. Fish. Aquat. Sci.* 42: 1940–1945.
16. Borror, D.J., C.A. Triplehorn, and N.F. Johnson. 1989. *An Introduction to the Study of Insects.* Saunders College Publishing, Philadelphia, pg. 761.
17. LeSage, L. and A.D. Harrison. 1979. Improved traps and techniques for the study of emerging aquatic insects. *Ent. News* 90: 65–78.
18. Davies, I.J. 1984. Sampling insect emergence. Chapter 6. In: J.A. Downing and F.H. Rigler. *A Manual on Methods for the Assessment of Secondary Productivity in Fresh Waters.* IBP Handbook 17. Blackwell Scientific Press, Oxford, pp. 161–227.

19. APHA. 1985. *Standard Methods for the Examination of Water and Wastewater.* American Public Health Association, Washington, D.C. 16th edition.

20. Lind, O.T. 1979. *Handbook of Common Methods in Limnology.* C.V. Mosby Company, St. Louis, MO, 147–149.

21. Morris, R.G., J.H. Kennedy, P.C. Johnson, and F.E. Hambleton. Pyrethroid insecticide effects on bluegill sunfish in microcosms and mesocosms and bluegill impact on microcosm fauna. Chapter 22. In: *Aquatic Mesocosm Studies in Ecological Risk Assessment.* CRC Press, Boca Raton, FL, 1993.

22. Coates, J.R. and S.P. Bradbury. 1989. Aquatic Toxicology of synthetic pyrethroid insecticides. *Environ. Toxicol. Chem.* 8: 359.

23. Rawn, G.P., G.R.B. Webster and D.C.G. Muir. 1982. Fate of permethrin in model outdoor ponds. *Environ. Sci. Health B* 17: 463–486.

24. Muir, D.C.G., G.P. Rawn and N.P. Grift. 1985. Fate of the pyrethroid insecticide deltamethrin in small ponds: A mass balance study. *J. Agric. Food. Chem.* 33: 603–609.

25. Dudzik, M., J. Harte, A. Jassby, E. Lapan, D. Levy and J. Rees. 1979. Some considerations in the design of aquatic microcosms for plankton studies. *Intern. J. Environ. Studies* 13: 125–130.

26. Smith, T.M. and G.W. Stratton. 1986. Effects of synthetic pyrethroid insecticides on nontarget organisms. *Res. Rev.* 97: 93–120.

27. Maguire, R.J., J.H. Carey, J.H. Hart, R.J. Tkacz and H. Lee. 1989. Persistence and fate of deltamethrin sprayed on a pond. *J. Agric. Food Chem.* 37: 1153–1159.

28. Sharom, M.S. and K.R. Solomon. 1981. Adsorption-desorption, degradation, and distribution of permethrin in aqueous systems. *J. Agric. Food Chem.* 29: 1122–1125.

29. Solomon, K.R., J.Y. Yoo, D. Lean, N.K. Kaushik, K.E. Day and G.L. Stephenson. 1985. Dissipation of permethrin in limnocorrals. *Can. J. Fish. Aquat. Sci.* 42: 70–76.

30. Odum, E.P. 1985. Trends expected in stressed ecosystems. *Bioscience* 35: 419–422.

31. Schindler, D.W. 1987. Detecting ecosystem responses to anthropogenic stress. *Can. J. Fish. Aquat. Sci.* 44 (Suppl. 1): 6–25.

32. Shires, S.W. and D. Bennett. 1985. Contamination and effects in freshwater ditches resulting from an aerial application of cypermethrin. *Ecotoxicol. Environ. Saf.* 9: 145–158.

33. Hill, I.R. 1985. Effects on non-target organisms in terrestrial and aquatic environments. In: J.P. Leahey [ed.]. *The Pyrethroid Insecticides.* Taylor & Francis. Philadelphia, pp. 151–262.

34. Kaushik, N.K., G.L. Stephenson, K.R. Solomon and K.E. Day. 1985. Impact of permethrin on zooplankton communities in limnocorrals. *Can. J. Fish. Aquat. Sci.* 42: 77–85.

35. Helson, B.V. and G.A. Surgeoner. 1986. Efficacy of cypermethrin for the control of mosquito larvae and pupae, and impact on non-target organisms, including fish. *J. Amer. Mosq. Control Assoc.* 2: 269–275.

36. Day, K.E., N.K. Kaushik and K.R. Solomon. 1987. Impact of fenvalerate on enclosed freshwater planktonic communities and on in situ rates of filtration of zooplankton. *Can. J. Fish. Aquat. Sci.* 44: 1714–1728.

37. Yasuno, M., T. Hanazato, T. Iwakuma, K. Takamura, R. Ueno and N. Takamura. 1988. Effects of permethrin on phytoplankton and zooplankton in an enclosure ecosystem in a pond. *Hydrobiologia* 159: 247–258.

38. Hill, I.R., S.T. Hadfield, J.H. Kennedy and P. Ekoniak. 1988. Assessment of the impact of PP321 on aquatic ecosystems using tenth-acre experimental ponds. Brighton Crop Protection Conference — Pests and Diseases, pp. 309–318.
39. MacIssac, H.J. and J.J. Gilbert. 1989. Competition between rotifers and cladocerans of different body sizes. *Oecologia* 81: 295–301.
40. Anderson, R.L. 1982. Toxicity of fenvalerate and permethrin to several nontarget aquatic invertebrates. *Environ. Entomol.* 11: 1251–1257.
41. Crossland, N.O. 1982. Aquatic toxicology of cypermethrin. II. Fate and biological effects in pond experiments. *Aquatic Toxicology* 2: 205–222.
42. Mulla, M.S., H.A. Darwazeh and G. Majori. 1975. Field efficacy of some promising mosquito larvicides and their effects on nontarget organisms. *Mosquito News* 35: 179–185.
43. Brooks, J.L. and S.I. Dodson. 1965. Predation, body size, and composition of plankton. *Science*, 150: 28–35.
44. Werner, E.E. and D.J. Hall. 1974. Optimal foraging and the size selection of prey by the bluegill sunfish (*Lepomis macrochirus*). *Ecology* 55: 1042–1052.
45. Confer, J.L. and P.I. Blades. 1975. Omnivorous zooplankton and planktivorous fish. *Limnol. Oceanogr.* 20: 571–579.
46. Drenner, R.W. and S.R. McComas. 1980. The roles of zooplankter escape ability and fish size selectivity in the selective feeding and impact of planktivorous fish. In: W.C. Kerfoot [ed.]. *Evolution and Ecology of Zooplankton Communities*. University Press of New England, Hanover, NH, 587–593.
47. Hambright, K.D., R.J. Trebatoski and R.W. Drenner. 1986. Experimental study of the impacts of bluegill (*Lepomis macrochirus*) and largemouth bass (*Micropterus salmoides*) on pond community structure. *Can. J. Fish. Aquat. Sci.* 43: 1171–1176.
48. Vanni, M.J. 1987. Indirect effect of predators on age-structured prey populations: Planktivorous fish and zooplankton. In: Kerfoot, W.C. and A. Sih [eds.]. *Predation: Direct and Indirect Impacts on Aquatic Communities*. University Press of New England, Hanover, NH, pp. 149–160.
49. Siefert, R.E. 1972. First food of larval yellow perch, white sucker, bluegill, emerald shiner and rainbow trout. *Trans. Amer. Fish. Soc.* 101: 219–225.
50. Butler, M.J., IV. 1989. Community responses to variable predation: Field studies with sunfish and freshwater invertebrates. *Ecolog. Monogr.* 59: 311–328.
51. Morin, P.J. 1984. Odonate guild composition: Experiments with colonization history and fish predation. *Ecology* 65: 1866–1873.
52. Mittelbach, G.G. 1988. Competition among refuging sunfishes and effects of fish density on littoral zone invertebrates. *Ecology* 69: 614–623.
53. Merritt, R.W. and K.W. Cummins. 1978. *An Introduction to the Aquatic Insects of North America*. Kendall/Hunt, Dubuque, Iowa, USA.
54. Hall, D.J., W.E. Cooper and E.E. Werner. 1970. An experimental approach to the production dynamics and structure of freshwater animal communities. *Limnol. Oeanogr.* 15: 839–928.
55. Morin, P.J. 1984. The impact of fish exclusion on the abundance and species composition of larval odonates: Results of short-term experiments in a North Carolina farm pond. *Ecology* 65: 53–60.
56. Schramm, H.L., Jr. and K.J. Jirka. 1989. Epiphytic macroinvertebrates as a food resource for bluegills in Florida lakes. *Trans. Am. Fish. Soc.* 118: 416–426.

57. Gilinsky, E. 1984. The role of fish predation and spatial heterogeneity in determining benthic community structure. *Ecology* 65: 455–468.

58. Loucks, O.L. 1985. Looking for surprise in managing stressed ecosystems. *Bioscience* 35: 428–432.

59. Wright, D. and J. Shapiro. 1990. Refuge availability: a key to understanding the summer disappearance of *Daphnia. Fresh. Biol.* 24: 43–62.

60. Crowder, L.B. and W.E. Cooper. 1982. Habitat structural complexity and the interaction between bluegills and their prey. *Ecology* 63: 1802–1813.

61. Benke, A.C. 1978. Interactions among coexisting predators — A field experiment with dragonfly larvae. *J. Anim. Ecol.* 47: 335–350.

62. Friday, L.E. 1987. The diversity of macroinvertebrate and macrophyte communities in ponds. *Freshwater Biology* 18: 87–104.

63. Draggan, S. 1976. The microcosm as a tool for estimation of environmental transport of toxic materials. *Int. J. Environ. Studies* 10: 65–70.

64. Giesy, J.P. and P.M. Allred. 1985. Replicability of aquatic multispecies test systems. In: Cairns, J., Jr. [ed.]. *Multispecies Toxicity Testing.* Permagon Press, New York, NY, pp. 187–247.

65. Neuhold, J.M. 1986. Toward a meaningful interaction between ecology and aquatic toxicology. In: Poston, T.M. and R. Purdy [eds.]. *Aquatic Toxicology and Environmental Fate: Ninth Volume, ASTM STP 921.* American Society for Testing and Materials, Philadelphia, pp. 11–21.

CHAPTER **22**

Pyrethroid Insecticide Effects on Bluegill Sunfish in Microcosms and Mesocosms and Bluegill Impact on Microcosm Fauna

R. Gregg Morris, James H. Kennedy, Philip C. Johnson, and Faithann E. Hambleton

Abstract: Concurrent aquatic microcosm and and mesocosm experiments evaluating the ecological impact and fate of the pyrethroid insecticide cyfluthrin were conducted at the University of North Texas Water Research Field Station. Sixteen 1.9 m³ microcosms and fourteen 634.7 m³ mesocosms were established using water, sediments, biological inoculum, and bluegill sunfish (*Lepomis machrochirus* Rafinesque) from the same sources. Cyfluthrin impacts were not observed on mesocosm bluegill, but a slight decrease in growth was observed in the microcosm bluegill.

Two different types of microcosm controls were established, with and without young-of-the year bluegill. The fishless controls were established to partition the predation impacts of bluegill from the insecticide effects in the microcosms. Bluegill did not strongly impact epibenthic colonizing insects in our system but fed on epiphytic and emerging insects through-out, i.e., *Callibaetis,* diptera pupae, and odonates. The exuviae collection proved to be a sensitive measure of emergence. Bluegill effects on zooplankton populations in the micro-cosms were typical in that larger organisms were preferentially selected. Rotifers were numerically dominant in tanks with fish as a result of reduction of larger competitors and invertebrate predators.

Microcosms and mesocosms exhibited very similar trends with regards to insecticide effects on invertebrate population dynamics. The mesocosms and microcosms supported similar zooplankton populations. Microcosms produced fewer total organisms but supported the same sensitive taxa and provided adequate supplies for bluegill growth to occur normally.

0-87371-592-6/94/$0.00 + $.50

INTRODUCTION

The impacts of pesticides on ecosystems, particularly aquatic systems, are among the driving forces behind pesticide registration under The Federal Insecticide, Fungicide, and Rodenticide Act (FIFRA). Pesticides have been released into the environment in tremendous quantities, e.g., approximately 4 billion pounds per year in 1981.[1]

Much of the data used for evaluating the environmental hazard posed by a particular chemical are based on single-species laboratory toxicity tests.[2] Components of ecosystems do not function the same in isolation,[3] and single-species tests may not protect all species which might be exposed to a given toxicant.[2] This dilemma is addressed using mesocosm research which incorporates ecological functions as well as multispecies testing capabilities. Mesocosm testing comprises the final tier for registration and reregistration of many compounds. The mesocosm protocol requires tenth acre (0.04 ha) ponds containing breeding populations of bluegill sunfish (*Lepomis machrochirus* Rafinesque) with well-established populations of invertebrates and macrophytes.[4]

Because many potentially environmentally hazardous compounds are produced annually, it would be useful to establish environmentally realistic, cost effective testing methods for ecosystem level effects. The use of large outdoor "microcosms" may permit the evaluation of ecosystem level effects at a fraction of the cost incurred when using larger systems.

Model ecosystems have been shown to be useful in examining and evaluating the hazards of xenobiotics in aquatic ecosystems.[5-11] Functional studies of aquatic ecosystems such as energy flow, nutrient and material cycling, and homeostasis have been facilitated through the use of microcosms.[11] Food chain interactions and other community and population activities have also been studied using microcosms.[11] A microcosm, as defined for this study, is a 1950-l water tank containing sediments, plankton and benthic invertebrate populations, aquatic plants, and fish.

Many fish species have been used in a wide variety of laboratory and field ecological studies.[6-10] Fish are regarded as integrators of ecological parameters affecting lower trophic levels. Bluegill are considered to be relatively sensitive to pesticides, are widely available, and much laboratory toxicity testing information and ecological data exist for this species.

Epiphytic and benthic macroinvertebrates and zooplankton are known to be major dietary constituents of bluegill.[12-14] Juvenile bluegill shift diet preference with age and in response to largemouth bass predation pressure. Relative visibility of different prey sizes is important in bluegill prey selection.[14] Bluegill have been described as "keystone" predators on benthic invertebrates.[15]

In this study we discuss the comparison of the impact of the pyrethroid insecticide, Baythroid®, on microcosm and mesocosm bluegill populations, and the impact of bluegill predation on invertebrate populations in the microcosms. Fishless controls were established in order to evaluate the impacts of bluegill predation on invertebrate populations.

Pyrethroids have a relatively short half-life in the environment, do not bioaccumulate, and are relatively nontoxic to man. The pyrethroid insecticides are highly effective neurotoxins for most insect pests and interefere with sodium-potassium neural transport.[16] This ultimately disrupts normal nerve function via the depolarization of the neural membrane.[17]

Pyrethroids may also interfere with osmoregularity processes in fish.[16] The selective toxicity of pyrethroids to vertebrates in order of decreasing sensitivity is fish ≫ amphibians > mammals > birds.[18] Synthetic pyrethroids are reported to be extremely toxic to aquatic insects and crustaceans while mollusks are unaffected.[19]

Baythroid® (cyfluthrin), a synthetic pyrethroid insecticide (molecular weight: 434.2, water solubility: 1.3 to 2.6 ppb) is produced and sold by the MOBAY Chemical Corporation.[20] This insecticide was introduced as a cotton insecticide in 1981 and received conditional registration which is now due for amendment. The reported Baythroid® 96 h LC_{50} for bluegill sunfish is 0.0015 ppm and for rainbow trout, 0.0006 ppm.[20]

METHODS

Sixteen 1.9-m³ concrete tanks (microcosms) and fourteen 0.12-acre (0.04 ha) earthen ponds (mesocosms) containing water, invertebrates, bluegill, and sediments from a common source were established at the University of North Texas Water Research Field Station in Denton, TX in 1988 to 1989. Water in the microcosms and mesocosms was circulated among the tanks and ponds in each system prior to the initial insecticide dosing in July 1989, at which time circulation was ended. Pesticide-screened well water was used to replace evaporative losses in both systems.

Young-of-the-year bluegill sunfish (YOY) were obtained from the Southwest Texas Hatchery in Terrell, TX. Fish were acclimated and handled according to Brauhn[22] before being placed into a holding tank on the microcosm circulation system. After 1 week of acclimation, group weights of 8 juvenile bluegill were obtained. Each group of YOY were then placed into 14 randomly selected microcosms; total number of fish stocked = 112. Three controls with fish (CWF = controls with fish) and eleven treated microcosms containing fish were also established. Two microcosms were not stocked with fish (NFC = no fish controls) to evaluate bluegill predation impacts on invertebrates. In each mesocosm 18 pairs of adult bluegill were placed and fry (1 to 2 cm) were observed within 1 month after stocking. All microcosm and mesocosm fish were removed, counted, length and weight determined 10 weeks after the final dosing period (end of sampling week 19, week of November 22, 1990). Juvenile bluegill from comparable size classes in the microcosms and mesocosms were used to compare growth.

PESTICIDE APPLICATION

Baythroid®, the emulsifiable concentrate formulation of cyfluthrin, was utilized in these studies, Pyrethroid exposure levels (Table 1) for individual microcosms and mesocosms were randomly assigned. Ten weekly spray drift (SD) and five bimonthly sediment slurry runoff applications (RO) were employed in both systems. Dosing was initiated the first week of July 1989 and concluded 10 weeks later, the second week of September (sampling week 10). Microcosm Baythroid® treatments were scaled down by volume from mesocosm treatments. Microcosm NFCs were treated exactly as all other tanks with the exception that no control runoff or spray drift simulations were added. Pyrethroid residue samples were analyzed using an HP-5890A gas chromatograph with an electron capture detector, using cool on-column injection methods.[21] Mesocosm and microcosm pyrethroid treatment methods are discussed in more detail in Johnson et al. (Chapter 21).

Five-minute observation periods were conducted on the microcosms and fifteen-minute observations were implemented on the mesocosms. Observations were obtained immediately before (pre) and 1 h after (post) insecticide dosing. Any bluegill activity such as gasping at the surface, feeding, or breeding activity was recorded.

SAMPLE COLLECTION AND ANALYSIS

Invertebrate sampling was initiated the week of June 5, 1989 (sampling week −3) and terminated sampling week 19. Microcosm insect emergence was monitored by two different methods, emergence traps and exuvia collection. Emergent adults were sampled by a semisubmerged cone sampler,[23] whereas insect exuviae[24] were removed from the entire surface of each tank via dip nets (mesh size = 110 × 120 μm) and preserved in Kahle's solution.[25] Mesocosm insect emergence was evaluated by floating pyramid emergence traps.

Invertebrate colonization was sampled by macro-invertebrate artificial substrates or "MAS" samplers (composed of Actifil "50" units; Norton Chemical Process Products, MA). MAS samplers were allowed to colonize for 1 month prior to sampling. All collected invertebrates were preserved in Kahle's solution.

Table 1. Target and Actual Water Column Concentrations of Baythroid® as Determined by Gas Chromatograph Analysis From Samples Obtained 1 h After Each Spray Drift (SD) and Runoff (RO) Treatment in Microcosms

Treatment	SD (ppb)		RO (ppb)	
	Target	Actual	Target	Actual
D0	0.0000	0.0000	0.0000	0.0000
D1	0.0356	0.0270	0.2143	0.0960
D2	0.0911	0.0570	0.2143	0.1210
D3	0.1780	0.1320	0.2143	0.1070
D4	0.1780	0.1450	1.0714	0.4540

Note: D0 = control, D1 through D4 = Dose 1 through Dose 4.

Cylindrical fish exclusion cages (refugia) were installed in all microcosms to provide refuge for macroinvertebrates and zooplankton. The refugia encompassed approximately 4.2% and 3.8% of the surface area and total volume of the microcosms, respectively, and contained colonization substrates. Refugia prevented the elimination of zooplankton populations in 1000-l tanks populated with silver carp and channel catfish.[26] Wright and Shapiro[27] also concluded that refuge availability plays a major role in *Daphnia* survival and population dynamics.

Zooplankton and phytoplankton were sampled by an integrated water column tube sampler (butyrate plastic tube) and were preserved in Lugol's solution.[28] Periphyton colonization was measured using glass microscope slides in floating periphytometers which were allowed to colonize for 2 weeks prior to sampling.[28] Periphyton and planktonic chlorophyll *a* were analyzed by acetone extraction followed by spectrophotometric measurement.[28] Algal biomass was determined as ashfree dry weight.[28]

Physical and chemical parameters measured in both studies included total organic carbon (TOC), particulate organic carbon (POC), total suspended solids (TSS), ammonia, nitrites, nitrates, pH, dissolved oxygen, temperature, total phosphorus, alkalinity, turbidity, and hardness following standard methods.[28] Gross photosynthesis and community respiration were determined using the three point diel oxygen pulse method.[28]

STATISTICAL ANALYSIS

NFCs were compared to CWF using the Student's paired *t*-test.[29] Bluegill growth among treatment levels, microcosms, and mesocosms were analyzed using the General Linear Model Analysis of Variance (ANOVA) and Dunnett's Multiple Range Test (MRT) for comparison of controls to treatments unless otherwise noted. Alpha levels for all statistical analyses were set at 0.05. Arrangement of microcosm treatments was two NFCs, three CWF, two Dose 1 (D1), three Dose 2 (D2), three Dose 3 (D3) tanks. Mesocosm treatments were three CWF, two Dose 1, three Dose 2, three Dose 3 ponds.

RESULTS

Chemical Fate

Microcosm water column residues, collected 1 h after each spray drift treatment, were consistently higher and closer to nominal targets than were mesocosm residues (Table 1). Percentage recovery for the two systems (microcosm and mesocosm) averaged over the whole experiment were spray drift (95 and 74) and runoff (42 and 48). Highest measured water column concentrations for the microcosm and mesocosms were 0.600 and 0.539 ppb, respectively, 1 h after runoff applications

Water Column Dissipation

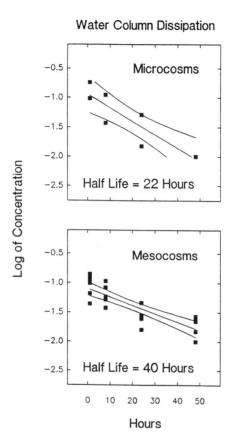

Figure 1. Baythroid® half-life in the microcosms and mesocosms. Error bars indicate 95% confidence intervals.

in both systems. Baythroid®'s half-life was 22 h in the microcosms and 40 h in the mesocosms (Figure 1). For further mesocosm and microcosm comparisons of fate and environmental effects of cyfluthrin, see Johnson et al.[21]

PHYSICAL AND CHEMICAL PARAMETERS

Microcosm and mesocosm water chemistry parameters exhibited no major responses to pesticide treatment.[21] There was little difference between microcosm NFCs and CWF water quality parameters. Mean NFC TOC levels were higher for 63% of the study duration, primarily from week 11 on, though only statistically different at week 11.

Mean ammonia levels were higher in the CWF at the start and end of sampling; however, no other trends were apparent and ammonia levels were never significantly different between NFC and CWF. The higher ammonia levels in CWF were not unexpected as ammonia is excreted by most teleost fish.[30] The elevated ammonia levels suggest that microcosm macrophytes, algae, and nitrification bacteria did not utilize all available ammonia. This may have been due to the flora having been acclimated to lower ammonia levels prior to fish addition or to lower initial flora population sizes. The higher ammonia levels at the end of the experiment are consistent with seasonal decreases in plant photosynthesis,[31] hence lower uptake of ammonia.

Mean pH values for CWF were slightly lower for most of the study, but no significant differences were noted among treatments. Mean dissolved oxygen levels (DO) were lower in CWF after sampling week 4, and significantly lower at sampling week 6. Both surface and bottom dissolved oxygen levels were significantly different at this point (NFC surface DO = 11.02, CWF surface DO = 8.8, p = 0.024).

Mean turbidity was usually higher in the CWF than in the tanks exposed to the pyrethroid from sampling weeks 2 until study termination. This difference was only significant at sampling weeks 5 (CWF > D4: df = 8, T = 4.886) and 7 (CWF > D2, D4: df = 8, T = 4.886). D4 turbidity levels were considerably lower than CWF after week 10, but were not significantly different due to high among tank variance. Neither gross photosynthesis, community respiration, periphyton colonization, or planktonic chlorophyll *a* revealed significant differences between NFC and CWF.

EFFECTS ON LEPOMIS

No pyrethroid attributable bluegill mortality was observed for the duration of the experiment at any dose level in the microcosms or mesocosms. Only one microcosm fish in D1 was unaccounted for at the termination of the experiment (microcosm YOY recovery: 111/112). Mean microcosm bluegill group weight gains ranged from 0.819 gm (D3) to 2.36 gm (CWF) (Figure 2). ANOVA analysis of the daily YOY weight gain per m^3 water yielded a significant difference between Dose 0 and Dose 3 (Tukey's MRT, alpha = 0.1, p = 0.06). Weight gain and total length were negatively correlated, weakly, with cyfluthrin dose level (Spearman Correlation, r = -0.312 for length, r = -0.302 for weight). Condition factors exhibited no significant differences for microcosm bluegill at any dose level.

In the mesocosms, no pyrethroid impacts could be determined for tagged adult bluegill, total bluegill numbers, or total biomass of fish. The mean number of juvenile bluegill recovered from the mesocosms were lower, though not significantly so, in the controls than in the treated ponds (Figure 3). Total numbers per m^3 of water in the mesocosms were considerably higher than microcosm loadings by the end of this study (Table 2). This disparity in population characteristics was not surprising as considerable reproduction occurred in the mesocosms. Mean weights of juvenile bluegill in the mesocosms were higher than those of the microcosm bluegill at all dose levels and for all size classes, except for the 9-cm size classes. Figures 4A and 4B demonstrate trends observed at all exposure levels with the exception that the upper size class was only higher in CWF.

Five-minute visual observations on each tank, pre- and postdosing, indicated no differences in microcosm fish activity among dose levels before or after dosing. No indications of stress were observed such as gasping, escape behavior, etc. at any time, before or after pyrethroid exposure. The primary activities observed were normal feeding and swimming behavior. Total frequency of bluegill sightings (pre + post) increased markedly after the onset of dosing in D4 by the second week. Afer the third week, total frequency of microcosm bluegill sightings were much higher in dose 4 tanks than any other microcosms, being significantly different from controls at weeks 5 and 9. Total fish sightings followed an exposure-response pattern consistently by sampling week 3 until the end of dosing (Figure 5). Fifteen-minute visual observations in mesocosms exhibited similar trends in bluegill activity.

Timed observations of bluegill in the microcosms revealed feeding on insects which fell into the microcosms. They actively hunted spiders on the sides of the tanks and fed on odonate eggs as they were deposited on the water surface. These observations are understandable when considering that feeding on terrestrial insects is an adaptive response observed in many fish when faced with low prey populations reduced by insecticide exposure.[32,33] Terrestrial insects have also been shown to be normal prey items for adult bluegill in Bull Shoals Lake and De-Gray Reservoir in Arkansas.[59,60]

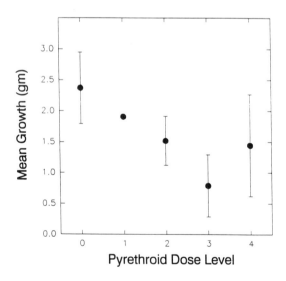

Figure 2.   Average growth (Final group wt. − initial group wt.) per dose level of cyfluthrin for microcosm bluegills. Error bars indicate ± one standard deviation.

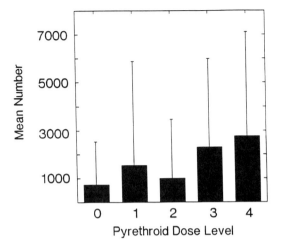

Figure 3. Average final weight of juvenile bluegills from mesocosms for each dose level of cyfluthrin.

BLUEGILL EFFECTS ON INVERTEBRATE SPECIES

Benthic Colonization

There was little difference in benthic colonization of MAS samplers between NFCs and CWF until week 5. At this time, mean total numbers began diverging and invertebrate density in CWF became significantly higher by week 15. There were no significant differences for taxa richness between treatments. *Callibaetis floridanus* and *Caenis* populations, the dominant mayflies in the microcosms, were never significantly different between treatments at any time.

Larval chironomid populations exhibited a general increase in both treatments from week 1 to week 19 at which time the mean numbers in NFCs were larger than in the CWF (375 vs. 200 per m²). High population variance in the CWF

Table 2. Final Mean Numbers and Biomass of Bluegills in the Mesocosms and Microcosms

	Dose 0	Dose 1	Dose 2	Dose 3	Dose 4
Mean Bluegill Count/Cubic Meter Water Volume					
Mesocosms	14.19	30.34	20.85	18.13	21.84
Microcosms	4.4	4.1	4.4	4.4	4.4
Mean Biomass/Cubic Meter Water Volume					
Mesocosms	4.76	7.7	5.52	5.99	6.43
Microcosms	20.33	18.07	15.61	16.86	18.51

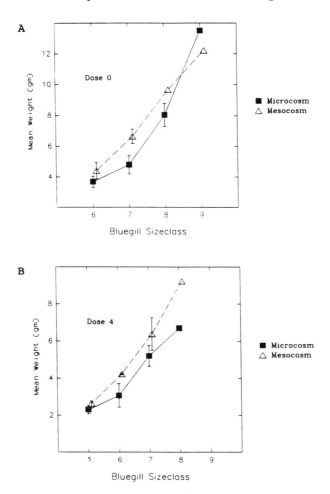

Figure 4.   (A) Average weights of comparable bluegill size classes from the mesocosms and microcosms for cyfluthrin Dose 0. (B) Average weights of comparable bluegill size classes from the mesocosms and microcosms for cyfluthrin Dose 4.

precluded any determination of statistical significance. The mean NFC chironomid population numbers were higher for 67% of the experiment. Chironomid subfamilies Tanypodinae and Chironominae did not exhibit major differences, though Chironominae numbers were higher in NFCs than CWF about 80% of the time. Tanypodinae did not exhibit this trend, and both populations gradually increased from sampling week 10 in both treatments.

Hydracarina exhibited low population numbers in both NFC and CWF until sampling week 11 at which time a large population increase began in the CWF. This increase remained statistically significant until the end of sampling and was observed, to a greater or lesser extent, in all microcosms containing fish. Amphipod populations were sparse and never differed among treatments. Amphipod populations steadily declined throughout the experiment until they were virtually nonexistent by week 15.

Benthic zygopteran populations, primarily Coenagrionidae, exhibited a steady numerical increase in NFCs and CWF beginning on week 5 (Figure 6A). NFC populations were significantly higher by week 15 while CWF populations decreased slightly at week 19. Anisopteran populations increased in both treatments from the onset of sampling. However, they increased much more rapidly in the NFCs until week 5. At this time both populations began decreasing and approached zero by the last sampling week in November.

Planorbid and physid gastropod populations both decreased by week 5 in each treatment, followed by an increase in number toward the end of sampling. Neither group exhibited any other trends or differences between treatments. Coleoptera, primarily *Berosus* larvae, exhibited higher mean populations in NFCs throughout the study. However, they were significantly different from CWF only at sampling week -1. These populations steadily declined in both treatments throughout the experiment.

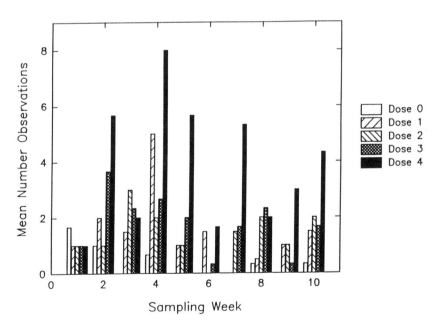

Figure 5. Total frequency of sightings (pre + post dosing observations) of bluegill in microcosms according to sampling week and cyfluthrin exposure.

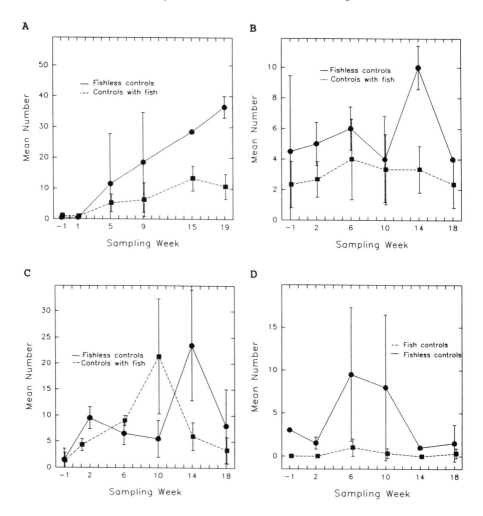

Figure 6. (A) MAS colonization, mean numbers of zygoptera sampled from the microcosms with and without fish. (B) Taxa richness for emergence trap samples. (C) Chironomidae emergent adults from emergence traps. (D) *Callibaetis floridanus* as sampled by emergence traps. Error bars indicate ± one standard deviation.

Emergence Traps

Total number of invertebrates collected in emergence traps from NFCs were usually greater than those in CWF. Average taxa richness was consistently higher in NFC and was significantly higher than CWF at sampling week 14 (Figure 6B). Total dipteran numbers (primarily Chironomidae) were higher in NFCs in 66% of the samples analyzed and significantly higher at weeks 2 and 14 (Figure 6C).

Subfamilies Tanypodinae and Chironominae and family Chaoboridae exhibited different trends in emergence. Tanypodinae emergence did not reflect predation by bluegill. However, Tanypodinae in CWF emerged in large numbers at week 10, in numbers significantly higher than that for NFCs. Tanypodinae in NFCs exhibited a bimodal emergence pattern with higher total numbers at weeks 2 and 14. Mean NFC numbers were approximately double those of CWF at these times. Chironominae emergence in the NFCs was significantly higher than that in CWF at week 14. Chironominae populations also exhibited different emergence trends between treatments (Figure 6D). Chaoborid emergence revealed higher, but not statistically significant, mean numbers from CWF for weeks 6 and 10. Emergence traps indicated low numbers emerging in either treatment.

Total Ephemeroptera emergence was significantly greater from NFCs for weeks − 1, 6, and 10, and mean numbers were consistently higher throughout the study. Caenid and baetid mayflies both exhibited greater emergence throughout the study from NFCs, revealing peak emergence periods at week 6. *Callibaetis floridanus* emergence from CWF was consistently lower throughout the study (Figure 6D). Trichoptera numbers were low throughout the study and exhibited no trends related to bluegill predation. The emergence traps revealed almost nonexistent odonate emergence; however, exuviae collection provided very different information.

Exuviae

Exuviae taxa richness was nearly double that collected from emergence traps for both treatments. Mean NFC taxa richness was always higher than CWF (Figure 7A). In general, the number of exuviae collected were higher than the number of emerging insects and significantly higher in NFCs compared to CWF for all weeks except week 2 (Figure 7B). Chironomidae pupal exuviae collection revealed a statistically significant emergence peak at week 14 in the NFCs. Emergence trap data confirmed the peak in emergence. Chironomid emergence from CWF remained relatively stable but low throughout this study (Figure 7C).

Chaoborid populations were well represented in exuviae and were significantly higher in NFCs throughout most of the study. Two emergence peaks were observed, weeks − 1 and 6 in NFCs, with a single emergence peak in CWF. Chaoboridae maintained low, stable emergence patterns in CWF. Maximum population numbers were at week − 1 (NFC = 175 vs. CWF = 10) and emergence decreased to just above 0 in both treatments by week 19. Chaoborous populations were revealed to be much higher than originally estimated by MAS samplers, or in emergence traps by exuviae collection.

Exuvial numbers for Zygoptera (Figure 7D), Anisoptera, and baetid and caenid mayflies exhibited similar responses. Numbers of exuvia collected from NFCs were much higher in NFCs than CWF. Each of these taxa showed exuvial production peaks at slightly different times. Exuvial production better described emergence for most groups than did our emergence traps.

Zooplankton

Total zooplankton numbers were slightly higher in the CWF. Average zooplankton taxa richness was generally higher in the CWF but decreased steadily until the end of the study (Figure 8A). Statistically significant differences between treatments were only exhibited at sampling week − 1. Total zooplankton revealed no discernible difference between treatments and followed similar trends over time. However, cladoceran populations were always higher in NFCs and they nearly doubled CWF numbers by week 19. Cladoceran populations in CWF had dropped to near 0 by this time (Figure 8B).

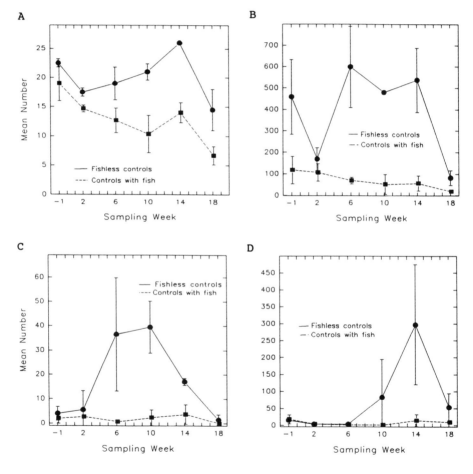

Figure 7.   (A) Taxa richness for insect exuviae per week. (B) Total numbers of exuviae collected per week. (C) Average number of Chironomidae pupae exuviae collected per week. (D) Average number of zygopteran exuviae collected per week. Error bars indicate ± one standard deviation.

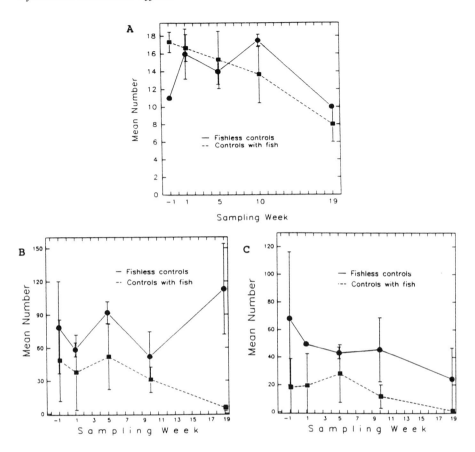

Figure 8. (A) Taxa richness for zooplankton. (B) Mean numbers of Cladoceran zooplankton per liter. (C) Mean numbers of *Diaphanosoma brachyurum* per liter. Error bars indicate ± one standard deviation.

Two dominant cladoceran taxa revealed different population responses to bluegill predation. *Macrothrix rosea* numbers were generally low in both CWF and NFCs. CWF generally contained higher, though no significantly so, populations throughout the study. *Diaphanosoma brachyurum* exhibited a population pattern exactly opposite to *Macrothrix*, i.e., they were always higher in NFC (Figure 8C). Both treatment populations followed similar patterns over time. *Chydorous sphaericus* numbers were generally lower in NFCs. This species exhibited two population peaks, at sampling weeks 5 and 10. At this times their numbers increased to levels greater than the population in CWF. The CWF exhibited low ($< = 20$ per l), but relatively stable *Chydorous* populations approaching 0 by week 19. *Alona rustica* populations exhibited a population peak at sampling week 5 (mean no. per l CWF = 9 and NFC = 5 per l) and decreased to near 0 by week 19 in both treatments. Total mean numbers were slightly higher for NFCs except at week −1.

Total numbers of rotifers were generally higher in CWF than in NFCs and increased dramatically in CWF at weeks 5 and 19. At week 19, mean total rotifer numbers were more than 10 times those of the NFC (Figure 9A). Orders Ploima and Floculariaceae were well represented in the microcosms and each group exhibited different trends both among and between genera. Order Ploima was represented by three species of Brachionus, *B. angularis*, *B. havanaensis*, and *B. quadridentata*, and *Monostyla*. Total numbers of *Brachionus* spp. in CWF were nearly double NFC for sampling weeks 1 and 19. No statistically significant differences were detected due to high variance among the tanks.

B. angularis populations revealed population peaks at weeks 1 and 19 in the CWF and were the dominant *Brachionus* species (max = av of 800 per l CWF). They were the driving factor behind the trends observed for total rotifers (Figure 9B). *B. havanaenis* populations were generally low but exhibited a population peak at week 10 in the NFCs. Other rotifers present in both treatments included *Monostyla bulla*, *Filinia longiseta*, and *Polyarthra remata*. Each of these populations varied markedly in population dynamics. All appeared to be higher in CWF at certain times. Notably, numbers were higher for weeks 1 and 19 for *F. longiseta* and week 19 for the other groups. The CWF generally revealed more striking population increases which were not reflected in NFCs.

DISCUSSION

Chemical Fate

The chemical fate of cyfluthrin observed in the microcosms and mesocosms exhibited common patterns.[21] However, the mesocosm cyfluthrin half-life was al-

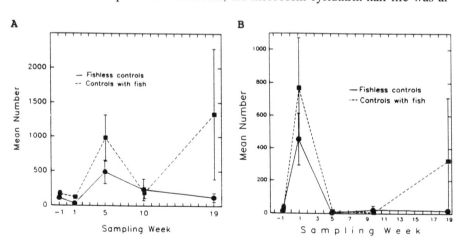

Figure 9.   (A) Mean numbers of rotifers per liter. (B) Mean numbers of *Brachionus angularis* per liter. Error bars indicate ± one standard deviation.

most double that of the microcosms. The difference in the half-life of cyfluthrin may be attributable to: (1) the edge effect with the microcosms having a greater surface to volume ratio, (2) the higher density of the macrophytes in the microcosms giving more surface area for the adherence of the pyrethroid, and (3) sampling bias from having one sample weighting the 48 h half-life measurement, thus strongly influencing the regression line. A larger number of replicates were used in the mesocosms to determine half-life at this time, and more replications from the microcosms would have yielded a tighter confidence interval and a better estimate of the half-life of cyfluthrin.

Bluegill

Fish have generally not exhibited mortality during field trials when pyrethroids have been applied at,[8,34-36] or greatly in excess of, field rates.[37] Shires[9] showed cypermethrin toxicity to rainbow trout and carp at high insecticide loadings in 1-m^3 outdoor enclosures in ponds. However, he concluded that fish mortality was unlikely to result from the use of this pesticide under normal agricultural practices.

The slight but significant decrease in bluegill growth observed in the pyrethroid-exposed microcosms did not appear to stem from direct pyrethroid toxicity. Instead, diminished growth apparently resulted from the insecticidal depletion of already limited prey populations, i.e., large cladocera, caenid, baetid mayflies, immature odonates, and emerging dipterans. The microcosm refugia and dense macrophyte populations prevented the bluegill from depleting certain sensitive species which were reduced by the fish in the mesocosm project,[21] which had lower macrophyte densities. This allowed the maintenance of adequate prey items in the untreated microcosms. The sensitive prey populations appeared to have been decreased below optimal levels for bluegill sustenance by the application of the pyrethroid.

Increased observations of bluegill activity in dosed microcosms may also indicate that more feeding time was necessary for bluegill to obtain adequate food due to cyfluthrin impacts on prey items. Alternatively, reductions in turbidity of treated microcosms may have increased bluegill visibility over controls. Fish activity did not appear to be caused from pesticide poisoning as no coughing, muscular spasms, stress bars, flight behavior, or other indicators of stress were apparent 1 h after dosing. Microcosm bluegill were observed feeding normally after dosing at all cyfluthrin dose levels throughout the experiment.

The decreased microcosm and mesocosm turbidity levels in dosed tanks, possibly due to pyrethroid particulate removal, may be the major contributing factor to increased frequency of bluegill sightings. The high octanol/water partitioning coefficient of cyfluthrin, 420,000,[21] and its tendency to adsorb in general may have been a factor in the increased numbers of observations by acting as a particulate flocculent. Day and Kaushik[38] observed that fenvalerate caused *Chlamydomonas* algae to stick on *Daphnia* to such an extent that the *Daphnia* were unable to move or feed. This effectively removed both organisms from the water column.

The lower water solubility and proclivity of cyfluthrin to adsorb to particles may also have been a factor in the failure to observe acute toxic impacts on *Lepomis*. Cyfluthrin may adhere to particulates tightly enough to be rapidly rendered non-bioavailable to fish. This binding apparently reduced free pesticide concentrations to sublethal levels for bluegill. Sublethal effects such as increased activity due to chronic pyrethroid irritation may have been a factor but separating increased activity due to irritation and increased frequency of sightings due to decreased turbidity was not possible under these experimental conditions.

Kingsbury's study using high levels of permethrin in Canadian streams[32] revealed that many fish, particularly brook trout, switched to an almost exclusively terrestrial diet when aquatic prey was depleted by the pesticide. This is consistent with our observations of microcosm bluegill feeding on spiders, etc. The ecological flexibility of bluegill has been demonstrated through habitat switching in the face of congeneric competition.[39] Niche shifts in maturing bluegill which resulted in ontogenetic diet shifts from littoral prey to zooplankton[41,42] have also been recorded. Adult bluegill have been reported to feed at the surface when mayflies and other aquatic insects are emerging.[43] Bluegill are also known to feed on terrestrial insects under normal conditions.[59,60] Dietary plasticity has also been seen in normally planktivorous fish which were observed feeding on adult chironomids during peak emergence periods.[43]

The ecological plasticity of bluegill sunfish and many other fish is notable. While preferring a particular size prey or prey item, many fish will take suboptimal prey items when faced with low food densities. This would increase the difficulty in isolating secondary effects of prey organism depletion by pesticides from fish predation in mesocosm studies. Using microcosms, we were able to delineate secondary effects on bluegill from prey depletion. These secondary effects might be less obvious in mesocosms due to variability in reproductive success among ponds and increased access to terrestrial prey sources. In this case study, the use of prereproductive YOY bluegill enabled greater control of stochastic factors resulting in clearer effects on fish.

Invertebrates

Aquatic community ecologists have long known of the effects of predators on the structure of prey assemblages,[44-46] and our results mirrored those of other studies[12,15,47-49] as far as bluegill impact on benthic macroinvertebrates and epiphytic macroinvertebrates are concerned. Our experiment also suggests that bluegill exert a strong impact on emergent insects as measured by emergence traps and exuviae production. Hall et al.[12] observed similar results in their 1970 studies investigating bluegill predation/nutrient enrichment studies. Iwakuma et al.[43] observed that the predominantly planktivorous whitefish (*Coregonus lavaretus maraenus*) fed intensively on freshly emerged adult chironomids (*T. akamusi*) to the exclusion of its normal cladoceran prey. Adult bluegill are regarded as being primarily planktivorous,[40] though many categorize bluegill as generalist predators[15,41,46,48-50,58,59] and insectivores.[42,58,59] It is evident that the diet of bluegill varies depending upon fish age,[40] body size,[42] habitat,[39,40,42,47-50] prey size, abundance and availability,[42,43,46-48] and the quantity and types of competitors[39] and predators[50] present. Thus numerous factors will influence bluegill predation impacts on invertebrate communities.

Bluegill, as "keystone" predators, were shown to define the species composition of certain odonate communities.[51] Vadas, however, concluded that since omnivory is often important in freshwater systems, fish may not always control the abundance of their prey.[52] Certain invertebrate populations appeared to be indirectly influenced by bluegill predation on invertebrate predators. Anisopterans and coenagrionids were reduced by bluegill in CWF. Oligochaete and hydracarina populations exhibited dramatic population increases as zygoptera, and to a lesser extent anisoptera, numbers decreased. These groups are listed as Zygoptera (*Pyrrhosoma*) prey by Lawton[53] and were probably prey items for the dominant Coenagrionidae (*Enallagma*) found in our microcosms. Benthic oligochaetes and hydracarina are not listed as prey items for bluegill though they may be taken.

The opportunistic predation of bluegill was easily observable in our microcosms. Species richness and total invertebrate numbers were not decreased in epibenthic colonization where prey were not readily accessible to fish. They also probably consumed epiphytic invertebrates and those clinging to the microcosm walls. Exuviae production and emergent adult numbers markedly decreased for most taxa in the CWF. This indicated that bluegill were feeding on these invertebrates as the immatures molted or attempted to emerge. Finally, they may be actually eating the empty exuviae.

Zooplankton

Bluegill predation on zooplankton revealed both direct and indirect effects. Cladocerans were reduced in CWF, and the larger taxa, *Diaphanosoma* sp., was the main group severely reduced. Rotifers exhibited increased populations in CWF. This relationship is consistent with observations that fish predation pressures on zooplankton populations result in smaller sizes of cladocerans and larger populations of smaller sized cladocerans and rotifers.[12,54]

CONCLUSION

Microcosms have been recognized as cost effective models for determining environmental fate of chemicals.[55] Microcosm size is also an important variable to consider.[56,57] Our study suggests that microcosms are environmentally realistic testing systems for evaluating certain ecological relationships and the impacts of pesticides in the environment. Realistic ecosystem level information was obtained from this study, similar to numerous other microcosm, mesocosm, pond, and lake studies. Certain sensitive taxa, i.e., large cladocerans, maintained larger population numbers in the microcosms which were severely reduced in the mesocosms by YOY bluegill produced in the mesocosms.[21] Additional benefits of using microcosms would include: (1) lower construction and manpower costs, (2) relative ease and efficiency with which the microcosms may be observed and sampled, (3) additional replicates may be easily added for increased statistical control, and (4) more control over experimental population numbers and sizes since reproduction and YOY recruitment was not included in the experimental design.

Bluegill in the microcosms revealed a response to Baythroid® that was not observed in the mesocosms, i.e., the decreased growth observed at upper cyfluthrin treatment levels. The lack of effects observed in the mesocosms were probably due to the nature of the population dynamics in fish which, while adding reproduction and juvenile growth as testing endpoints, adds numerous stochastic factors such as variable fecundity, reproductive success, adult predation on juveniles, etc.

Juvenile bluegill were shown to impact many taxa, both directly and indirectly. Bluegill affected microcosm invertebrate emergence and exuviae production more than benthic substrate colonization. The emergence phase in many aquatic insects' life cycles may cause this stage to be more vulnerable to fish predation. Fishless controls (NFCs) aided in the partitioning of these effects. If microcosms are used to evaluate the impacts of chemicals on the environment, the fish population must be carefully evaluated since available resources are frequently limiting. Fish loadings must be sufficiently low such that overcropping of prey items at all treatment levels does not obscure secondary effects related to the toxicant being evaluated.

ACKNOWLEDGMENTS

I would like to thank Dr. Robert Graney of MOBAY Corporation for his input and making funding possible, and Dr. John Rodgers, Jr. for his aid in forming the project proposal. I also would like to thank the employees at the Water Research Field Station for their inestimable aid.

REFERENCES

1. Weir, D. and M. Shapiro. 1981. *Circle of Poison*. Institute for Food and Development Policy, San Francisco. p. 6.
2. Larsen, D.P., F. DeNoyelles, Jr., F. Stay, and T. Shiroyama. 1986. Comparisons of single-species, microcosm and experimental pond responses to atrazine exposure. *Environ. Toxicol. Chem.* 5:179–190.
3. Cairns, J., Jr. 1988. Should regulatory criteria and standards be based on multispecies evidence? *Environ. Profess.* 10:157–165.
4. LaPoint, T.W. and J.A. Perry. 1989. Use of experimental ecosystems in regulatory decision making. *Environ. Management* 13:539–544.

5. Hurlbert, S.H., M.S. Mulla, and H.R. Wilson, 1972. Effects of an organophosphorus insecticide on the phytoplankton, zooplankton, and insect populations of freshwater ponds. *Eco. Monogr.* 42:269–299.

6. Boyle, T.P. 1980. Effects of the aquatic herbicide 2,4-D DMA on the ecology of experimental ponds. *Environ. Pollut. (Series A)* 21:35–49.

7. Rawn, G.P., G.R.B. Webster, and D.C.G. Muir. 1982. Fate of permethrin in model outdoor ponds. *Environ. Sci. Health* B17:463–486.

8. Hill, I.R., S.T. Hatfield, J.H. Kennedy, and P. Ekoniak. 1988. Assessment of the impact of PP321 on aquatic ecosystems using tenth-acre experimental ponds. *Brighton Crop Prot. Conf. — Pests and Diseases,* pp. 309–318.

9. Shires, S.W. 1983. The use of small enclosures to assess the toxic effects of cypermethrin on fish under field conditions. *Pest. Sci.* 14:475–480.

10. Crossland, N.O. 1984. Fate and effects of methyl parathion in outdoor ponds and laboratory aquaria. *Ecotox. Environ. Safety* 8:482–495.

11. Rodgers, J.H., K.L. Dickson, F.Y. Saleh, and C.A. Staples. 1983. Use of microcosms to study transport, transformation and fate of organics in aquatic systems. *Environ. Toxicol. Chem.* 2:155–167.

12. Hall, D.J., W.E. Cooper, and E.E. Werner. 1970. An experimental approach to production dynamics and structure of freshwater communities. *Limnol. Oceanog.* 15:839–928.

13. McLane, A.J. 1974. *Field Guide to Fresh Water Fishes of North America.* Holt, Rinehart and Winston, New York, New York, pp. 120–123.

14. Werner, E.E. and D.J. Hall. 1988. Ontogenetic habitat shifts in bluegill: the foraging rate-predation risk trade-off. *Ecology* 69(5):1352–1366.

15. Butler, M.J., IV. 1989. Community responses to variable predation: field studies with sunfish and freshwater invertebrates. *Ecolog. Monogr.* 59:311–328.

16. Dyer, S.D., J.R. Coats, S.P. Bradbury, G.J. Atchison, and J.M. Clark. 1989. Effects of water hardness and salinity on the acute toxicity and uptake of fenvalerate by bluegill (*Lepomis machrochirus*). *Bull. Environ. Contam. Toxicol.* 42:359–366.

17. C.J. Marshall and M.W. Brooks. 1989. Neurotoxicology of pyrethroids: single or multiple mechanisms of action? *Environ. Toxicol. Chem.* 8:361–372.

18. Haya, K. 1989. Toxicity of pyrethroid insecticides to fish. *Environ. Toxicol. Chem.* 8:381–391.

19. Solomon, K.R. K.M. Lloyd, J.R. Roberts, M.H. Akhtar, J.R. Coats, P.D. Kingsbury, H.-W. Leung, H.T.J. Mount, and L.O. Ruzo. 1986. Assoc. Comm. Sci. Crit. Environm. Qual. Pyrethroids: their effects on aquatic and terrestrial ecosystems. Nat. Res. Coun. Can. NRCC/CNRC, Ottawa, Canada. 146–183.

20. MOBAY Chemical Corporation. 1985. Material Safety Data Sheet Mobay Chemical Corporation, a Bayer USA Inc. Company, Kansas City, MO, 1–4.

21. Johnson, P.C., J.H. Kennedy, R.G. Morris, and F.E. Hambleton. 1993. Fate and effects of cyfluthrin (pyrethroid insecticide) in pond mesocosms and concrete microcosms, this volume, ch 21.

22. Brauhn, J.L., R.A. Schoettger, and L.H. Mueller. 1975. Acquisition and culture of research fish: rainbow trout, fathead minnows, channel catfish, and bluegills. Nat. Environ. Res. Center, Off. Res. Dev. U.S. E.P.A., Corvallis, Oregon, EPA-660/3-75-011.

23. Davies, I.J. 1984. Sampling insect emergence. In J.A. Downing and F.H. Rigler. *A Manual on Methods of the Assessment of Secondary Productivity in Fresh Waters,* IBP Handbook. Blackwell Scientific Press, Oxford.

24. Wilson, R.S. and P.L. Bright. 1973. The Use of Chironomid Pupal Exuviae for Characterizing Streams. *Freshwat. Biol.* 3:283–302.

25. Borror, D.J., C.A. Triplehorn, and N.F. Johnson. 1989. *An Introduction to the Study of Insects.* Sixth Edition. Saunders College Publishing, Philadelphia. pp. 761.

26. Smith, D.W. 1985. Biological control of excessive phytoplankton growth and the enhancement of aquacultural production. *Can. J. Fish. Aquat. Sci.* 42:1940–1945.

27. Wright, D. and J. Shapiro. 1990. Refuge availability: a key to understanding the summer disappearance of *Daphnia. Freshwater Biology* 24:43–62.

28. APHA. 1985. Standard methods for the examination of water and wastewater. American Public Health Association, Washington, D.C. 16th edition.

29. Statistical Analysis Systems, SAS. 1987. Institute, Inc., Cary, N.C.

30. Lehninger, A.L. 1982. *Principles of Biochemistry,* Worth Publishers, Inc., New York, New York. pp. 548.

31. Wetzel, R. G. *Limnology* Second Edition. Saunders College Publishing, Philadelphia, pp. 168.

32. Kingsbury, P.D. 1986. Aquatic toxicity of pyrethroids in Solomon, K.R., K.M. Lloyd, J.R. Roberts, M.H. Akhtar, J.R. Coats, P.D. Kingsbury, H.-W. Leung, H.T.J. Mount, and L.O. Ruzo. Assoc. Comm. Sci. Crit. Environm. Qual. *Pyrethroids: Their Effects on Aquatic and Terrestrial Ecosystems.* Nat. Res. Coun. Can. NRCC/CNRC, Ottawa, Canada.

33. Kreutzweiser, D.P. 1990. Response of a brook trout (*Salvelinus frontalis*) population to a reduction in stream benthos following an insecticide treatment. *Can. J. Fish Aquat. Sci.* 47:1387–1401.

34. Crossland, N.O. 1982. Aquatic toxicology of cypermethrin. II. Fate and biological effects in pond experiments. *Aquat. Toxicol.* 2:205–222.

35. Crossland, N.O., S.W. Shires, and D. Bennett. 1982. Aquatic toxicology of cypermethrin. III. Fate and biological effects of spray drift deposits in fresh water adjacent to agricultural land. *Aquat. Toxicol.* 2:253–270.

36. Shires, S.W. and D. Bennett. 1985. Contamination and effects in freshwater ditches resulting from an aerial application of cypermethrin. *Ecotoxicol. Environ. Saf.* 9:145–158.

37. Muir, D.C.G., G.P. Rawn, and N.P. Grift. 1985. Fate of the pyrethroid insecticide deltamethrin in small pools: a mass balance study. *J. Agric. Food Chem.* 33:603–609.

38. Day, K. and N.K. Kaushik. 1987. The adsorption of fenvalerate to laboratory glassware and the alga *Chlamydomonas reinhhardii,* and its effect on uptake of the pesticide by *Daphnia galatea mendotae. Aquat. Toxicol.* 10(2/3):131–142.

39. Werner, E.E. and D.J. Hall. 1979. Foraging efficiency and habitat switching in competing sunfishes. *Ecology* 60(2):256–264.

40. Osenberg, C.W., E.E. Werner, G.G. Mittelbach, and D.J. Hall. 1988. Growth patterns in bluegill (*Lepomis machrochirus*) and pumpkinseed (*L. gibbosus*) sunfish: environmental variation and the importance of ontogenetic niche shifts. *Can. J. Fish. Aquat. Sci.* 45:17–26.

41. Mittelbach, G.G. 1981. Foraging efficiency and body size: a study of optimal diet and habitat use by bluegills. *Ecology* 62(5):1370–1386.

42. Pflieger, W.L. 1975. *The Fishes of Missouri*. Missouri Department of Conservation. p. 270.
43. Iwakuma, T., H. Hayashi, I. Yasuda, T. Hanazato, and K. Takada. 1990. Impact of whitefish on an enclosure ecosystem in a shallow eutrophic lake: changes in nutrient concentrations, phytoplankton and zoobenthos. *Hydrobiol.* 200/201:141–152.
44. Carpenter, S.R. and J.F. Kitchell. 1988. Consumer control of lake productivity. *BioScience* 38(11):764–769.
45. Harrass, M.C. and F.B. Taub. 1985. Effects of small fish predation on microcosm community bioassays. Aquat. Toxicol. Haz. Assess.: Seventh Sympos. ASTM STP 854. R.D. Cardwell, R. Purdy, and R.C. Bahner, Eds. Am. Soc. Test. Mat. pp. 177–133.
46. Thorp, J.H. and E.A. Bergey. 1981. Field experiments on responses of a freshwater benthic macroinvertebrate community to vertebrate predators. *Ecology* 62(2):365–375.
47. Schramm, H.L. and K.J. Jirka. 1989. Epiphytic macroinvertebrates as a food resource for bluegills in Florida lakes. *Trans. Am. Fish. Soc.* 118:416–426.
48. Mittlebach, G.G. 1984. Predation and resource partitioning in two sunfishes (Centrarchidae). *Ecology* 65(2):499–513.
49. Ehlinger, T.J. 1990. Habitat choice and phenotype-limited feeding efficiency in bluegill: individual differences and trophic polymorphism. *Ecology* 71(3):886–896.
50. Werner, E.E., J.F. Gilliam, D.J. Hall, and G.G. Mittlebach. 1983. An experimental test of the effects of predation risk on habitat use in fish. *Ecology* 64:1540–1548.
51. Morin, P.J. 1984. Odonate guild composition: experiments with colonization history and fish predation. *Ecology* 65:53–60.
52. Vadas, R.L. 1990. The importance of omnivory and predation regulation of prey in freshwater fish assemblages of North America. *Envir. Biol. Fish.* 27:285–302.
53. Lawton, J.H. 1968. Feeding and food energy assimilation in larvae of the damselfly *Pyrrosoma nymphula* (Sulz.) Odonata: Zygoptera) *Ecology* 24:669–689.
54. Keerfoot, W.C. 1981. Long-term replacement cycles in cladoceran communities: a history of predation. *Ecology* 62(1):216–233.
55. Draggan, S. 1976. The microcosm as a tool for estimation of environmental transport of toxic materials. *Int. J. Environ. Studies* 10:65–70.
56. Giesy, J.P. and P.M. Allred. 1985. Replicability of aquatic multispecies test systems. In Cairns, J., Jr. [ed.] *Multispecies Toxicity Testing*. Pergamon Press, New York, NY, pp. 187–247.
57. Neuhold, J.M. 1986. Toward a meaningful interaction between ecology and aquatic toxicology. In Poston, T.M. and R. Purdy [eds.]. *Aquatic Toxicology and Environmental Fate: Ninth Volume*, ASTM STP 921. American Soc. Test. Mat., Philadelphia, pp. 11–21.
58. Applegate, R.L., J.W. Mullan, and D.I. Morais. 1966. Food and growth of six centrarchids from shoreline areas of Bull Shoals Reservoir. Proc. 19th Ann. Conf. Southeastern Assoc. Game and Fish Comm. pp. 469–482.
59. Bryant, H.E. and T.E. Moen. 1980. Food of bluegill and longear sunfish in DeGray Reservoir, Arkansas, 1976. *Proc. Ark. Acad. Sci.* 34:31–33.

CHAPTER 23

Scaling: Summary and Discussion

John H. Rodgers, Jr.

The preceding chapters pose interesting questions and present data related to scaling of parameters that may be important in assessing the ecological effects of pesticides in aquatic systems. These issues range from the predictability of mesocosm results from microcosms to appropriate strategies for scaling pesticide exposure from spray drift and runoff from rainfall events in these experimental studies.

The purpose of this chapter is to capture some of the discussion, comments, and questions that were initiated at the workshop in response to ideas presented in these and other chapters. This discussion frequently revolved around risk assessment of pesticides and the role of scaling in time and space. Many previous concerns regarding risk assessment of pesticides have involved translations of laboratory results to field considerations. This nonlinearity in transfer has provided part of the impetus for this workshop since results from lower tier laboratory studies have prompted a judgment of unacceptable risk and the need for a mesocosm study.

Numerous issues were explored relative to scaling in time and space. Scaling has its roots in basic stimulus-response biology. In the specific cases considered here, we are primarily concerned with exposure to a pesticide and the consequent direct and indirect effects from that exposure in aquatic ecosystems. Concerns with the reality of exposure in laboratory experiments have given rise to the notion that we can obtain more realistic exposures in field situations. These field situations would include both outdoor microcosms, mesocosms, and pond systems that are subject to environmental conditions that may modify exposure. In other words, the

0-87371-592-6/94/$0.00 + $.50
© 1994 by CRC Press, Inc.

field exposures incorporate more of the factors that may infuence exposure and subsequent responses of interest. Temporal scalings of field exposures have centered upon using realistic timing of applications, realistic numbers of applications or treatments, and analytical data to verify realistic field half lives. The obvious difference between field exposures and laboratory exposures arises from factors inherent in field situations that alter exposure to a particular pesticide or test chemical. One might expect that we could better mimic field responses in laboratory experiments if we more closely simulate field exposures and vice versa.

Complex and multispecies systems in the field have been proposed as a solution for nonaccurate translation of laboratory to field results. Field situations incorporate a variety of species and interactions ranging across all biological kingdoms and many levels of ecological organization. Thus, field exposures and properly designed field studies may incorporate both direct and indirect effects that can be observed and quantified. At this symposium, questions were raised regarding the ability of scaled systems to encompass important field responses.

WHAT DID WE LEARN FROM THE PRESENTATIONS?

Johnson et al. (Chapter 21) presented a comparison of microcosm and mesocosm responses to a pyrethoid. Although there was considerable differences in exposure, as indicated by the factor of two differences in half lives of the compound in these systems, there were insufficient differences to conclude that microcosm results were different from mesocosm. However, Morris et al. (Chapter 22) in the same study concluded that bluegill growth was affected in microcosms and no effects were observed in mesocosms. These effects were observed in microcosms despite the fact that there was considerable difference between exposures in microcosms and mesocosms. Heimbach et al. (Chapter 19) similarly found that small pond microcosms could produce results that did not differ greatly from mesocosm results. These conclusions were based upon exposure of these systems to an herbicide. However, the interpretation of these results may be somewhat limited due to the fact that little effects were observed from the herbicide. Howick et al. (Chapter 20) reported differences in responses of invertebrates and fish to an insecticide in outdoor mesocosms and microcosms. Based on information supplied by LaPoint and Fairchild (Chapter 16), this is not surprising. Rather, it would be expected to be an exception if the responses were precisely scaled. This would suggest that we understand all factors influencing exposure and scale these accurately as well as all of the biological/ecological responses. Hill et al. (Chapter 15) suggest that we should pay considerable attention to pesticide exposure in order to more readily interpret responses in experimental field studies. It is often useful to return to the first principles of aquatic ecotoxicology when designing experiments to negate presumptions of adverse ecological risk for a pesticide.

UNRESOLVED ISSUES AND RESEARCH NEEDS

Despite the excellent presentations at this symposium, numerous issues remain unresolved relative to scaling, fostering a variety of research needs. A primary unresolved scaling issue is to come to grips with delineation of the question of concern or objective in ecological risk assessment for pesticides in aquatic systems. In order to scale aquatic systems in time and space and design experiments to address negating a presumption of unacceptable risk due to utilization of a pesticide, one needs to clearly state the species, the processes, or the ecological interactions that are of concern and must be protected. One reason for the requisite clarity of purpose for these experiments is to ensure that the spatial and temporal scaling has included the ability to encompass the ecological effects of concern. If this is the case, we can only discover the indirect or unanticipated effects that we engineer or build into these systems. We can also ''discover'' artifacts of this tinkering.

Model aquatic systems such as microcosms or mesocosms have been used for many decades to evaluate the effects of pesticides and other materials on aquatic systems. Frequently these systems are used in an iterative process to study in detail some ecological behavior or activity that has been discerned based on field observations. Frequently, parameters that are measured in microcosms or mesocosms are experimental and relatively sensitive parameters. For example, it has been almost useless to evaluate aquatic plant biomass as wet weight or dry weight in response to herbicide activity even for herbicides registered for use on the aquatic vegetation. What we are expecting the plant to do is to die and decay before we can discern a change in biomass and conclude that we have a response to herbicide exposure. There are many more sensitive techniques and approaches for evaluating aquatic plant responses to herbicides. However, relatively few of these approaches have apparently been incorporated into studies for ecological risk assessment.

Other questions arise where fish are of primary concern in the ecological risk assessment of the pesticide. For example, mesocosm and microcosm studies frequently employ a fish such as bluegill as the sole species to represent all of the fish that may be exposed to a particular pesticide. If effects are not observed on the numbers of bluegill produced, then the conclusion is that there were no observable effects. However, given that an efficient predator on bluegills was included in a mesocosm study, the results may be entirely different. This would be anticipated especially if there were considerable differences between the response of bluegill and the predator to the pesticide. For example, if the predator were even temporarily less affected than the bluegill from the exposure to this pesticide, then the predator may efficiently harvest most of the affected fish. This would be manifested in effects on the prey as well as the predator in subsequent years. If the predator (bass) was considerably more affected than the prey (bluegill), then one might expect to see loss of predator and overgrowth of the bluebill in subsequent years. These kinds of higher order interactions or indirect effects are not captured in most of the current mesocosm designs for pesticide registration.

Based on presentations at this symposium, there is apparent interest in shrinking mesocosm studies spatially. Several cautions were offered at the symposium. Shrinking in size may eliminate some of the behaviors or responses that could be observed in larger scale studies. Shrinking in size may also enhance some of the behaviors and result in effects being detected in microcosms that may not be manifested in the field. However, this hypersensitivity should be tempered with the notion that the physical act of sampling of the small scale systems may produce a much larger impact that similar sampling would produce in larger scale systems, thereby further confounding results and inhibiting the ability to detect effects to pesticides. Clearly, scaling contains numerous unresolved issues and needs additional research. One proposal from this workshop was that an additional symposium or activity was needed to discuss scaling and utilization of microcosms for ecological risk assessment for pesticides.

SECTION V

Case Histories

INTRODUCTION

All chapters in this section discuss the results of ecosystem level research that were conducted in various types of surrogate aquatic test systems. An assessment of the ecological risk of pesticides is required under the United States Federal Insecticide, Fungicide, and Rodenticide Act (FIFRA). Prior to the U.S. EPA's adoption of the mesocosm technique as part of the ecological risk assessment of pesticides, aquatic ecosystem level impacts were evaluated in natural systems that were exposed to the test chemicals during the course of normal farm practices. While these types of studies provided realism in terms of environmental fate of the compound and exposure to the aquatic ecosystem, they were difficult to evaluate, in part, because of insufficient or no replication and a high degree of variability associated with the variables being measured. In the mid-1980s the U.S. EPA adopted the use of mesocosms (experimental ponds) as surrogate natural systems in which ecosystem level effects of pesticides could be measured and included in the ecological risk assessment process.[1] The first three chapters in this section, Hill et al. (Chapter 24), Giddings et al. (Chapter 25), and Mayasich et al. (Chapter 26), discuss the results of research that used mesocosms for assessing the ecological risk of pesticides. The chapter by Hill et al. discusses the results of the first mesocosm study, submitted to the U.S. EPA in 1987, for support of an agricultural chemical registration. This study was completed before the establishment of the U.S. EPA guidance document and was based in part on discussions from a workshop of academia, industry, and federal government scientists held at George Mason

University, Fairfax, VA, on 8 to 9 April 1986. This chapter was the most detailed of the group. Hill et al., in addition to their research results, presents a good chronology of field testing up to the acceptance of mesocosms as the highest tier test for pesticide registration in the United States. Mayasich et al. and Giddings et al. are additional case studies of research that used the mesocosm approach for assessing the ecological risk of pesticides. The Hill et al. and Mayasich et al. chapters evaluated the influence of a pyrethroid applied to the aquatic ecosystem. The Giddings team worked with an organophosphorus insecticide. Both of these chemicals are primarily used on cotton for the control of the boll weevil. Although there are differences, all of these mesocosm studies have very similar study designs.

The rest of the chapters in this section use a variety of surrogate aquatic systems to evaluate the effects of various xenobiotics on specific components of the aquatic ecosystems. In another type of lentic water system, Lucassen and Leeuwangh (Chapter 27) used unreplicated small ponds to measure the effects of the organo-phosphorus pesticide, Dursban, on the zooplankton community. Data collected in this study is evaluated using correspondence analysis, an ordination technique, which allows the investigators to examine ecological interrelationships.

The final chapter by Belanger et al. (Chapter 28) discusses the use of indoor experimental streams (lotic mesocosms) to evaluate the impact of a surfactant (surface active agent) on an algal periphyton community. Belanger and his associates include a number of community comparison indexes when analyzing their data.

CHAPTER 24

Lambda-Cyhalothrin: A Mesocosm Study of Its Effects on Aquatic Organisms

Ian R. Hill, Jill K. Runnalls, James H. Kennedy, and Paul Ekoniak

Abstract: The potential for effects of lambda-cyhalothrin on aquatic organisms under field conditions was studied using 16 mesocosms, each being a pond 15 × 30 m with a maximum depth of 2 m. Sets of four replicate mesocosms were treated with three rates of the pyrethroid insecticide and control. The product was applied to the total water surface as multiple applications of simulated spray drift (12×; weekly) and runoff (6×; biweekly), having commenced June 1986. The middle of the three treatment rates was designed to represent the typical maximum estimated entry rate (MEER) into natural aquatic environments, and each mesocosm was dosed for spray drift at 0.5% of the field rate and for runoff at 1.5% of the field rate. The high and low rates were 10× and 0.1× the middle rate, respectively.

A wide range of parameters was studied, pre-, during, and postapplication, over a 24-week period. Residues in water and hydrosoil were determined and physicochemical characteristics monitored. The biology of the various trophic levels was studied: microbes, algae, macrophytes, zooplankton, macroinvertebrates, and fish.

Residues of lambda-cyhalothrin in the water column declined rapidly after applications, with less than 20% of the applied dose remaining after 3 days. Physicochemical characteristics of the system, such as temperature, dissolved oxygen, pH, conductivity, turbidity, and alkalinity, were not altered by the test chemical. Neither were there significant effects at any of the treatment rates on hydrosoil microbes, phytoplankton, periphyton, filamentous algae, and macrophytes.

Zooplankton populations were only affected at the highest rate, with a reduction in copepod numbers. A substantial proportion of the macroinvertebrate families were also impacted by the high rate of lambda-cyhalothrin. Some changes in macroinvertebrate populations occurred at the middle rate, but there was a negligible effect at the low rate.

By the end of the study period, bluegill sunfish numbers had increased and biomass reduced at all three treatment rates, with no dose-related response. However, the data shows this was probably not due to lambda-cyhalothrin, but possibly the result of fish overcrowding and competition with tadpoles for food or habitat.

The mesocosm test is a severe test of exposure, and thus may substantially overestimate typical field risk. It is concluded that the study shows that lambda-cyhalothrin is unlikely to cause adverse effects on overall populations or productivity of aquatic ecosystems, although some minor or transient effects may occasionally occur.

INTRODUCTION

Aquatic toxicity testing of pesticides to generate data for inclusion in registration submissions to the U.S. Environmental Protection Agency (EPA) is carried out using a tiered hazard assessment scheme. Lower tier studies involve acute and chronic single-species tests done under controlled laboratory conditions, often with maintained concentrations of the test chemical in clean water. The toxicological endpoints of these studies (e.g., EC_{50}, LC_{50}, and NOEL), together with estimates of environmental exposure concentrations, are used by the EPA to determine the potential for any adverse effects following normal agronomic use. Defined relationships ("quotients") between the measured laboratory effects and the estimated field exposure are used to quantify the margin of environmental safety and, where necessary, trigger the need for higher tier field studies.

Many types of field procedures have been developed to study the fate and effects of pesticides in aquatic environments (Lundgren, 1985; Solomon and Liber, 1988; Hill, 1985, 1989). In the early 1980s, the EPA favored the use of farm ponds as the highest tier test. However, problems of representational sampling of the entire system and the absence of a control (and replication) presented substantial difficulties in interpretation of the data. In 1985, the EPA indicated that, in general, farm ponds were unacceptable and that future regulatory studies should be done in mesocosms (experimental ponds). Over the next 3 years, the EPA staff, in consultation with scientific experts from government agencies, industry, and academia, refined their ideas. A guideline protocol was subsequently published as the basis for the highest aquatic tier test for pesticides (Touart, 1988 [see also Touart and Slimak, 1989]).

Mesocosms can have substantial advantages over other test systems. They allow the use of a range of treatment concentrations, enable replication and statistical analysis, and also minimize biological and physicochemical variation between the test units. However, they are not without certain problems; for example, our ability to determine application rates that realistically represent natural aquatic entry of pesticides is currently far from satisfactory. Furthermore, mesocosms are not natural systems, and their size together with the pesticide application methodology may exaggerate natural impact, especially with multiple-application products (for example, pyrethroid insecticides). As yet, a method of extrapolating from mesocosm data to natural aquatic environments has not been developed.

A substantial number of mesocosm studies has now been completed by industry as part of the pesticide registration process, and most of these have closely followed the EPA guideline protocol. Typically three treatment rates and a control have been included, with three or four replicate mesocosms per group, although an alternative study design with more treatment rates and fewer replicates has also been used (Mayasich et al., 1993). The mesocosms are required to be at least 0.04 ha (0.1 acre) in size and up to 2 m deep. The systems are seeded and aged for periods of up to 1 year, to allow the development of a diverse population of phytoplankton, periphyton, zooplankton, and macroinvertebrates. Known numbers of fish are placed into each mesocosm. The entire site of mesocosms usually has a circulation system, used in the pretreatment period to minimize mesocosm-to-mesocosm variations in water quality and biota.

Nonregistration type mesocosm studies have also been done, by organizations such as the U.S. Fish and Wildlife Service, Columbia, MO (Fairchild et al., 1992), and the EPA Environmental Research Laboratory, Duluth, MN in collaboration with the University of Wisconsin-Superior. These have been designed in part to examine environmental impact of the products studied, but also to address issues of test methodology and the validity of the mesocosm approach.

During 1990, representatives of the EPA, government agencies, industry, and academia formed the "Aquatic Effects Dialogue Group" under the auspices of The Conservation Foundation (World Wildlife Fund), in order to review the procedures, processes, and options for higher tier field aquatic testing of pesticides.

The mesocosm study of the pyrethroid lambda-cyhalothrin (trade name, Karate™) presented in this paper was carried out by ICI at Goldsboro, NC in 1985 to 1986. It was the first study of its type to be carried out for regulatory submission to EPA and was required because laboratory ecotoxicological tests (in clean water using maintained concentrations of the test chemical) showed that this and other pyrethroids might have the potential for impact on natural aquatic environments, following agricultural use. The lambda-cyhalothrin study was done prior to publication of the EPA guideline by Touart (1988). The protocol was developed and agreed by the ICI and EPA following detailed and cooperative discussions of the mesocosm approach, and was the forerunner to the EPA mesocosm guideline protocol.

METHODS AND MATERIALS *

Mesocosm Description

During 1985 12 ponds, each 30 m × 30 m, were excavated at the ICI Americas Eastern Research Center, Goldsboro, NC (Plate 1a). The ponds were designed to

Plate 1a.   Mesocosm site. Aerial view after construction in 1985.

Plate 1b.   Mesocosm site. Division of ponds; concrete "logs" being lowered to secure divider into hydrosoil.

Plate 1c. Mesocosm site. Mesocosms after division, with on-site laboratory and "stock-pond" in rearground.

have a water depth ranging from 15 cm at the shallow end to 2m at the deepest point (Figure 1), maintained in the pretreatment period by an overflow pipe. Each pond was lined with a 15-cm layer of clay covered with a 10 cm depth of sandy loam soil (pH 6.1, organic matter 1.1%). A 0.9-m high bank surrounded each pond to prevent cross contamination between ponds and runoff entry from the surrounding land.

The 12 ponds were each divided longitudinally in early 1986, creating mesocosms of 15 m × 30 m (0.04 ha, 0.1 a) with a volume of 450 m³ (Figure 1, Plate 1c). Each mesocosm retained a shallow and a deep end. The division was done using a 1-mm thick reinforced terylene (Hypalon™) barrier, with an inverted T-joint sealed into the hydrosoil (Figure 2, Plate 1b). A sealed foam flotation tube was built into the divider, and sufficient flexibility was allowed for this to rise up to 0.6 m above the overflow level. In addition, a 40-cm vertical fin was provided to prevent "wave-splash" from mesocosm to mesocosm, and to minimize any cross contamination during pesticide application. At termination of the study, the dividers were all found to be undamaged and hydrosoil seals intact.

The mesocosms were set out in two rows and were interconnected by a water circulation system (Figure 3, Plate 1), to enable continuous mixing in the period before pesticide application. Overflow pipes from each mesocosm allowed passive flow to a pumping chamber, from where water was mixed and returned to all

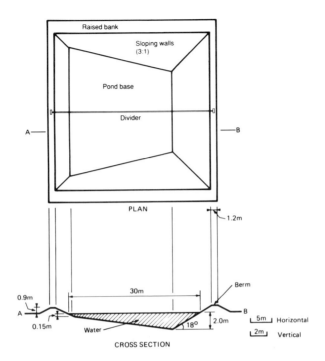

Figure 1.   Mesocosm design.

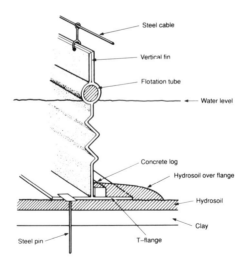

Figure 2.   Cross-section of mesocosm divider.

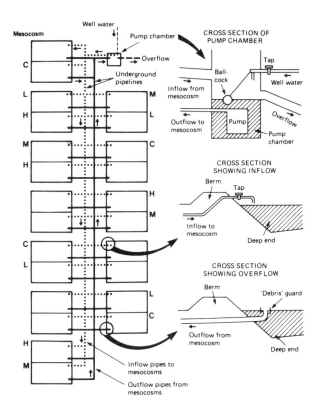

Figure 3. Mesocosm layout, water circulation system, and allocation to treatments (C = control mesocosms; L = low rate lambda-cyhalothrin; M = mid rate; H = high rate).

mesocosms. The chamber had an overflow, and also a well-water inlet, for automatic maintenance of mesocosm water levels.

Sixteen of the 0.04 ha mesocosms were used to evaluate the potential impact of lambda-cyhalothrin on aquatic ecosystems (the remaining eight pairs of mesocosms were allocated to other work).

Biological Establishment of Mesocosms

The mesocosms were initially filled with water in August 1985, from a large adjacent natural pond containing substantial and diverse populations of aquatic organisms. All water added to the mesocosms was passed through a 2.5-mm screen, to prevent fish and minimize the number of fish eggs being transferred. In addition, inoculations were made of macroinvertebrates collected from several local ponds.

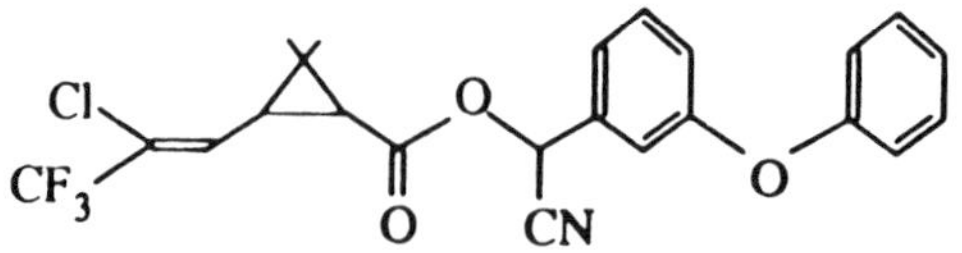

Figure 4. Structure of lambda-cyhalothrin.

The aquatic macrophyte *Ludwigia uruguayensis* was planted in 1985, in the shallow region and around the perimeter of each mesocosm (Plate 1c). The central area of each mesocosm was kept free of macrophytes to facilitate sampling procedures. Additionally, artificial "weed-beds" were placed in each mesocosm, along the edges of the sampling zones. These are included to provide refuge for macroinvertebrates, due to concerns that the bluegill (see below) would overpredate their food supply. The "weed-beds" were made of plastic strands (Fish-Hab™; Berkeley Products, Spirit Lake, IA).

The mesocosms were fertilized only in the period prior to pesticide treatment. Wheat shorts (Bartlett Milling Co, Statesville, NC; crude protein >15%, crude fat >3.5%, crude fiber <8%) were added on three occasions; 8, 4, and 1 week before pesticide application, evenly broadcast at 1.4 kg per mesocosm.

Sexually mature bluegill sunfish (*Lepomis macrochirus*) of 11 to 16 cm total length were stocked in the mesocosms in May 1986 2 weeks before the first application of lambda-cyhalothrin. A total of 25 fish, with a combined weight of approximately 1 kg, were placed in each mesocosm. During the initial 5 days, the fish were kept in holding cages for observation, and over this period any dead or abnormally behaving fish were replaced. The initial intention was to add 13 females and 12 males to each mesocosm, and the fish were sexed based on external secondary sexual characteristics, by the source hatchery.

Pesticide Treatments

The mesocosm study was carried out using lambda-cyhalothrin, a racemic mixture of the enantiomers (*S*)-α-cyano-3-phenoxybenzyl (1*R*)-*cis*-3-(Z-2-chloro-3,3,3-trifluoropropy-1-enyl)-2,2-dimethylcyclopropanecarboxylate and (*R*)-α-cyano-3-phenoxybenzyl (1*S*)-*cis*-3-(Z-2-chloro-3,3,3-trifluoroprop-1-enyl)-2,2-dimethylcyclopropanecarboxylate, designated enantiomer pair B (Figure 4). The possible enantiomer pairs are A to D and A′ to D′, with Z- and E-configurations, respectively. The A and B isomers have a *cis*-configuration about the cyclopropane ring and the C and D isomers a *trans*-configuration.

The mesocosms were treated with the commercial formulation Karate™, an emulsifiable concentrate (EC) formulation of lambda-cyhalothrin (13.8% active ingredient w/w, containing >98% enantiomer pair B).

The application program began in early June 1986. Three rates (high, mid, and low) of lambda-cyhalothrin were applied, with four replicates for each rate. Four mesocosms were left as controls. Mesocosms were allocated to three treatment

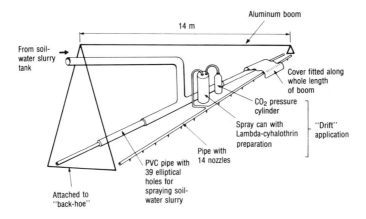

Figure 5.   Diagram of pesticide application boom system.

groups (Figure 3). A selection was initially made using a random numbers table, then adjusted to prevent high and mid rate mesocosms being next to controls, to minimize the potential for cross contamination (this required one control to be moved and two pairs of low and high rate mesocosms to be exchanged).

The mid rate treatments were calculated from the typical MEER into aquatic environments from aerial spraying of cotton at the maximum U.S. 1986 Karate™ (lambda-cyhalothrin) label rate (the present maximum label rate is one third higher). The low and high rates were $10 \times$ below and above the mid rate, respectively. The MEERs were calculated using the principles described by Hill et al. (1993), although the rates in the 1986 lambda-cyhalothrin study reported here are slightly different to those now proposed by Hill et al. (1993).

Applications to simulate both "spray drift" and "runoff" entry were made to each mesocosm by spraying evenly over the whole water surface using a traveling boom which spanned the 15 m mesocosm width. A 14-m boom was constructed and attached to the arm of a "back-hoe". The vehicle traveled along the mesocosm raised bank, using the articulated arm to set the nozzles 0.5 m above water level (Figure 5, Plate 2).

Spray Drift Simulation

Twelve spray drift entry applications were made to each treated mesocosm at weekly intervals, commencing in the week beginning 10 June 1986. The high, mid, and low rates were equivalent to 5%, 0.5%, and 0.05% of the maximum U.S. 1986 Karate™ label field application rate for cotton, as shown in Table 1.

The application system consisted of a CO_2-pressure cylinder (activated by a remote switch in the driver's cabin) linked to a 20-l reservoir containing the Karate™, and 14 Delevan D2.5 "flood" nozzles at 1 m spacing on the boom. Tests showed that even applications were made, with individual nozzles almost always spraying volumes within 10% of the mean for the boom. Applications were made at 0.4

Plate 2a. Application of lambda-cyhalothrin to mesocosms. Spray-drift simulation.

Plate 2b. Application of lambda-cyhalothrin to mesocosms. Run-off simulation, with slurry mixing tank in foreground (emergence traps can be seen on adjacent mesocosms).

Table 1. Lambda-Cyhalothrin Mesocosm Application Rates

Entry Route	Lambda-Cyhalothrin Application Rate		Nominal Rates per Application						Number and Frequency of Applications
	Rate	% of Max Field Rate[a]	Lambda-Cyhalothrin[b]		Water		Soil (Wet Wt)		
			g ai/ha	mg ai/ Mesocosm[c]	l/ha	l/ Mesocosm	kg/ha	kg/ Mesocosm	
Spray drift	High	5.0	1.7	72.0	285	12	—	—	12 apps; weekly intervals
	Mid	0.5	0.17	7.2	285	12	—	—	
	Low	0.05	0.017	0.72	285	12	—	—	
	Control	—[d]	—	—	—	—	—	—	
Runoff	High	15.0	5.0	210.0	60,000	2,500	6,000	250	6 apps; every two weeks
	Mid	1.5	0.5	21.0	60,000	2,500	6,000	250	
	Low	0.15	0.05	2.1	60,000	2,500	6,000	250	
	Control	—[e]	—	—	60,000	2,500	6,000	250	

a At time of study, maximum field rate was 34 g ai/ha (0.03 lb ai/a). The maximum rate has now been raised by one third.
b Applied as the EC formulation Karate™.
c From accurate dimensions; 29 × 14.5 m (0.0421 ha) at overflow.
d No application made.
e Application of soil only.

m/s boom speed and 20 psi system pressure. This procedure was designed to deliver the nominal 12 l of spray solution (Table 1) in one pass of the mesocosm (Plate 2a). Of the 144 sprays made, over 92% delivered volumes within 5% of nominal.

Runoff Simulation

Runoff tretments were applied as soil-water slurries, on six occasions at two weekly intervals to every mesocosm. The first runoff application was 3 days after the initial spray drift treatment. The high, mid, and low rates were equivalent to 15%, 1.5%, and 0.15% of the maximum U.S. label rate for cotton, as shown in Table 1. The control mesocosms were also sprayed with a soil-water slurry but not containing the insecticide. The sediment load of 250 kg wet weight applied to each mesocosm (in the slurry) was calculated from literature data. Sediment losses from fields in substantial runoff events are reported as commonly ranging from 100 to 1000 kg dry wt/ha. In the present study, the loading to each mesocosm was equivalent to approximately 600 kg dry wt/ha lost from a land area $10 \times$ the size of the recipient pond.

The slurry was prepared from 250 kg of a sandy loam soil and 2500 l of well water (see Hill et al., 1993). After 30 minutes mixing, the Karate™ was added and mixed for a further 15 min. The slurry was then left to stand for approximately 24 h before remixing and applying to the mesocosm (Plate 2b). Laboratory data suggested that the lambda-cyhalothrin would be relatively stable over this period. However, analyses were carried out to quantify any loss of parent chemical due to hydrolysis or microbial degradation.

The system was comprised of a 3800-l mixing tank with a circulation pump and separate pump to pressurize the spray boom. The boom had 39 nozzles at 34-cm centers, each being a hole of approximately 1.0 cm diameter. In order to distribute the soil-water slurry evenly across the mesocosm, the boom pipe diameter was progressively stepped down from the center to each end. Tests showed applications to be even across the mesocosms, with most individual nozzles dispensing volumes within 15% of the mean.

Applications were made at 0.4 m/s boom speed. The total volume of slurry was applied in three passes of the mesocosm.

Mesocosm-to-Mesocosm Cross Contamination

The possibility of small amounts of cross contamination during spraying is inevitable in studies with mesocosms in close proximity to each other. Therefore methods used were designed to prevent or minimize any such occurrence, as follows:

- Flood nozzles were used for drift applications and had a specification of 95% of droplets larger than 200 μm. Prior to the study, deposition was measured at 0.25 to 0.5 m from the end of the spray boom (using dyes) and found to be <0.5% of the nominal spray rate.
- A cover was fitted over the top and down the edges of the boom (Figure 5).

- The mesocosm raised banks were 0.9 m above the water surface at overflow level and the mesocosm divider had a vertical "fin" to 40 cm above the water.
- Deposition cards on mesocosms surrounding those being sprayed were analyzed on a number of occasions. Residues of lambda-cyhalothrin during the study were always below the limit of determination (0.02 to 0.08% of the rate onto the adjacent mesocosm being sprayed).
- All equipment systems were washed internally and externally between every mesocosm application.
- The distribution of mesocosms on the site was such that only low rate mesocosms were next to controls. This was used to minimize any impact on the latter, should cross contamination have occurred.

The methods used and the data obtained showed that negligible, if any, mesocosm cross contamination occurred, such that during the application of lambda-cyhalothrin to any one mesocosm, there would not have had been any effect on the biota in the adjacent mesocosms.

Mesocosm Sampling

In each mesocosm the majority of sampling techniques, biological observations, and direct physicochemical measurements were carried out within two defined zones (Figure 6). The shallow zone had an average depth of 0.5 m and the deep zone was up to 2 m deep. Subsamples were taken, for compositing, from 3 areas (I to III, Figure 6) of each zone. Sampling was mostly carried out from boats, but visual assessments were made from the shoreline.

The sampling schedule in the pretreatment year (September to November 1985) was for physicochemical and biological parameters, and consisted of 6 samplings at 2-week intervals from September to November. In the lambda-cyhalothrin treatment year (1986), sampling began in May, 3 weeks before the first application, and continued until November when the fish were harvested. Residues were determined, and physicochemical and biological variates measured at all time intervals. Most samples were collected at 2- to 3-week intervals, but visual mesocosm observations were made several times per week during the spraying period, and emergence trap samples were removed twice each week.

Lambda-Cyhalothrin Residue Analyses

Application Preparations

Samples for residue analysis were taken from all spray drift preparations immediately before and after spraying and from each runoff slurry immediately before application.

Spray drift samples were immediately shaken with hexane and stored at $< -18°C$ until analyzed. The hexane phase was later removed, dried with sodium sulfate, and an aliquot analyzed by gas-liquid chromatography (GLC) using electron capture

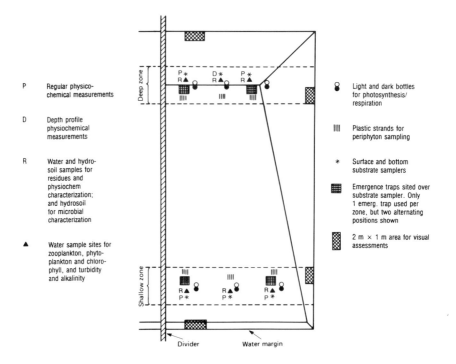

P — Regular physico-chemical measurements

D — Depth profile physiochemical measurements

R — Water and hydro-soil samples for residues and physiochem characterization; and hydrosoil for microbial characterization

▲ — Water sample sites for zooplankton, phyto-plankton and chloro-phyll, and turbidity and alkalinity

☺ — Light and dark bottles for photosynthesis/respiration

|||| — Plastic strands for periphyton sampling

* — Surface and bottom substrate samplers

▦ — Emergence traps sited over substrate sampler. Only 1 emerg. trap used per zone, but two alternating positions shown

▨ — 2 m × 1 m area for visual assessments

Deep zone

Shallow zone

Divider   Water margin

Figure 6.   Mesocosm sampling positions.

detection and external standardization. Analytical recoveries from laboratory "spiked" samples averaged 90%.

The runoff preparations were sampled from a range of depths in each mixing tank, composited, and stored at $< -18°C$. For analyses, the solid and aqueous phases were separated by centrifugation. The soil was extracted with acetonitrile and the extract recombined with the aqueous phase. The acetonitrile-water mixture was extracted with hexane and lambda-cyhalothrin residues determined by the procedures described above for mesocosm water. Analytical recoveries from laboratory "spiked" samples averaged 86%.

Mesocosm Water

Mesocosm water was sampled in each area of both zones; at mid-depth in the shallow zone and from both near the surface and close to the hydrosoil in the deep zone. Residues were collected using an adsorptive matrix technique, with the adsorbent material packed into a cartridge. The matrix consisted of two adsorbent layers, Sepralyte SAX trimethylaminopropyl and Sepralyte C8. Studies with ^{14}C-lambda-cyhalothrin showed that this system, used at experimentally optimized flow rates and volumes, gave radiochemical recoveries of $>85\%$, even in the presence of suspended particulate material or planktonic organisms (Hadfield et al., 1992).

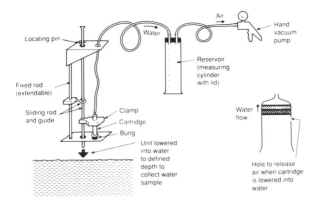

Figure 7. Sampling device for collection of mesocosm water for residue analysis.

Activated cartridges were held in a sampling device (Figure 7, Plate 3c) which enabled each cartridge to be opened at a defined depth. A prescribed volume of water was passed through the adsorbent, using a vacuum hand pump. Cartridges were stored at $< -18°C$ until analyzed.

Analysis of the lambda-cyhalothrin residue was carried out by fortifying the adsorbent with an internal standard followed by solvent elution, clean up, and capillary GLC using electron capture detection. The B enantiomer pair (lambda-cyhalothrin) was resolved from its epimer, the A enantiomer pair, and from any other isomers. The limit of determination (LOD) for each enantiomer pair was set at 1 ng/l mesocosm water. Mass spectrometry (MS) was used to confirm residues on a representative number of occasions.

Mesocosm Hydrosoil

Hydrosoil was collected from both zones by a coring technique using 5 cm diameter thin-walled plastic tubes (Figure 8, Plate 3). Grab or suction methods were not used as these are essentially inefficient, unrepresentative, and nonquantitative, and also can cause major disturbance of the mesocosm bottom.

The cores were cut, prior to analysis, into 2.5-cm or 5-cm deep fractions while frozen. Analysis of lambda-cyhalothrin was by acetonitrile reflux extraction, clean up, and GLC/MS as described for mesocosm water analysis. The LOD for each enantiomer pair was 0.2 μg/kg dry wt hydrosoil.

Physicochemical Determinations

At the start and end of the 1986 study period, the well water (used to top up mesocosm levels), mesocosm water, and hydrosoil were analyzed for a wide range of organochlorine and organophosphorus pesticides. None were found to be present

Plate 3a.   Hydrosoil coring and water sampling for residue analysis. Draining off water from deep zone water sampler.

above the limits of determination (10 to 100 ng/l in water and 0.05 to 1.0 mg/kg in hydrosoil).

In addition, at each sampling interval in 1985 and 1986, mesocosm water was measured *in situ* at 25 to 50 cm depth for the following parameters:

- pH; by glass electrode
- Temperature; by probe (also measured at 25 cm above hydrosoil in deep zone)
- Temperature range since previous sampling; by submerged "max-min" mercury thermometer
- Dissolved oxygen; by probe (also measured at 25 cm above hydrosoil in deep zone)
- Conductivity; by probe

Water samples taken as integrated depth cores (Plate 4a) were also returned to the laboratory for determination of:

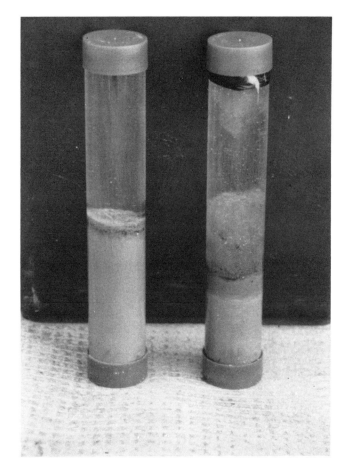

Plate 3b. Hydrosoil coring and water sampling for residue analysis. Hydrosoil cores in sampling core tubes.

- Turbidity; using a turbidimeter
- Alkalinity; by titration (Lind, 1979)

Biological Assessments

Microbial Populations in Hydrosoil

Hydrosoil samples were collected immediately prior to pesticide applications and at study termination, for determining viable plate counts of microorganisms as follows:

- Total aerobic and anaerobic bacteria; using casein-peptone-starch agar
- Total actinomycetes; on Jensen's casein-dextrose agar
- Total fungal propagules; on rose bengal chloramphenicol-peptone-dextrose agar

Plate 3c.	Hydrosoil coring and water sampling for residue analysis. Water sampling device with cartridge.

The plates were incubated inverted at 20°C, for 1 to 2 weeks. Anaerobic cultures were grown in 10% carbon dioxide in nitrogen.

Phytoplankton

Water samples were collected as vertical, integrated cores from the water surface to 10 cm above the hydrosoil, using a 5 cm diameter graduated transparent tube (Plate 4a). The sample was used for a range of techniques described below:

- Cell numbers and identification; samples preserved in Lugol's iodine were examined using a settling chamber and inverted microscope
- Cell volumes and biomass; also by microscopy on preserved samples, with volume calculations based on formulae that idealize cell shapes as single or combinations of geometric solids
- Chlorophyll a; the photosynthetic pigment and its degradate phaeophytin a were determined by 0.45-μm membrane filtration, ethanol-water extraction (78°C for 5 minutes), and HPLC analysis

Photosynthesis and respiration were also measured. Gross primary productivity was determined (in the treatment year [1986] only) *in situ* in the mesocosm using light and dark bottles (method based on Gaarder and Gran, 1927) held vertically (Vollenweider, 1974). Calculations were done according to Franson, 1985.

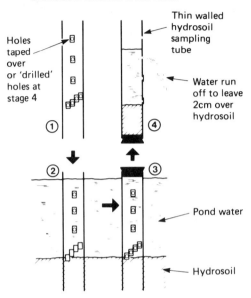

a   SHALLOW WATER SAMPLING

Holes taped over or 'drilled' holes at stage 4

Thin walled hydrosoil sampling tube

Water run off to leave 2cm over hydrosoil

Pond water

Hydrosoil

b   DEEP WATER SAMPLING

Perspex/ Plexiglass extension tube

Holes taped over

Thin walled tube held in 'recess' in extension tube, by a pin

Pond water

Hydrosoil

Thin walled hydrosoil sampling tube

Figure 8.   Hydrosoil sampling method. (a) Shallow zone sampler. (b) Deep zone sampler.

Plate 4a. Biological sampling procedures. Integrated depth water core sample for phytoplankton, zooplankton, and physicochemical determinations.

In addition, community metabolism was calculated from ''dusk-dawn-dusk'' *in-situ* dissolved oxygen data, collected using the method suggested by McConnell (1962) and described by Lind (1979).

Periphyton

Periphyton was defined as the ''plant'' communities growing on submerged surfaces. Samples were collected (in the treatment year [1986] only) by an integrated depth method using artificial substrates. Plastic strands of 0.15 cm diameter, with a known surface area were suspended from the water surface to just above the hydrosoil. The substrates were allowed to colonize for 7 to 14 days before collection. Periphyton growth was stripped from the strands and examined as described below:

Plate 4b. Biological sampling procedures. Seine netting for fish harvest.

- Cell numbers and identification; as for phytoplankton
- Cell volumes and biomass; as for phytoplankton
- Chlorophyll *a*; as for phytoplankton
- Total dry weight biomass and autotrophic index; samples were filtered (ash-free filters) and the filters dried at 105°C to a constant weight. The filters were then combusted at 500°C, rewetted, dried at 105°C, and weighed. Calculations were done as described by Franson (1985)

Filamentous Algae and Macrophytes

The distributions of the filamentous algae and macrophytes were determined as follows:

- Mapping of mesocosm coverage; at each sampling interval, and estimating precentage cover from these using a planimeter
- Visual identification of macrophytes
- Monthly identification of filamentous algae; using samples collected and preserved in Lugol's iodine
- Quantification of filamentous algae; into broad categories, from rare-dominant
- Macrophyte biomass; determined by air drying quadrat samples only at study termination

Plate 4c.   Biological sampling procedures. Bluegill nesting sites.

Zooplankton

Samples for analysis of zooplankton were taken from the integrated core of water described for phytoplankton assessment. Up to 4 l of mesocosm water was concentrated through a 60-μm mesh and preserved in Lugol's iodine. The following examinations were done:

- Cell numbers and identification; by a settling (concentration) technique and examination on an inverted microscope
- Size categories; taxa identified in the treatment year (1986) were placed in the two fish food size categories described by Bougis (1976); as microzooplankton of 20 to 200 μm length and macrozooplankton of 200 to 2000 μm length

Macroinvertebrates

Populations of macroinvertebrates were studied using artificial substrates at two depths, emergence traps and visual observations of the mesocosms. The emergence traps were used to help interpret population fluctuations of the mesocosm insects and as an aid to species identification. Information on some nektonic forms was often only available from the visual assessments. The methods of use are described below:

Plate 4d.   Biological sampling procedures. Bluegill harvested, with "centimeter group" sizing scale.

- Artificial substrates; these were assemblages of plastic cylinders manufactured for surface area enhancement in sewerage treatment systems (Norton Chemical Process Products, England). Units were prepared that floated at the water surface and others that rested on the hydrosoil (Figure 9). These were left to colonize for 2 weeks prior to removal. Organisms were washed and brushed from the substrates, and concentrated through a 0.3-mm mesh. Organisms were immediately counted and identified and separated into three categories: live, abnormal (showing abnormal behavior), and dead. These substrates have been found to be excellent refuges for a very wide range of macroinvertebrates and also offer the benefit of not requiring disassembly to remove the organisms.
- Emergence traps; floating box traps, 1 m × 1 m × 0.15 m high, were placed in each zone (Plates 1c and 2b) and assessed twice weekly. Insects were removed from the traps by a modified hand-held vacuum cleaner. The "catch" was preserved in 70% ethanol for later counting and identification.
- Visual assessments were made from two m² shoreline quadrats (Figure 6) and from whole mesocosm observations.

Bluegill Sunfish

During visual observations of the shoreline quadrats and the whole mesocosm, numbers and behavior of adult and young fish were recorded.

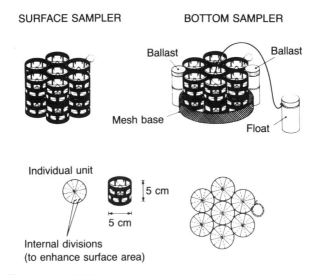

Figure 9.   Artificial substrates for sampling macroinvertebrates.

Fish were harvested in November 1986 by lowering mesocosm water levels and seine netting with a 0.6-cm mesh (Plate 4b). The mesocosm bottom was carefully examined and stranded fish collected. All fish collected (Plate 4d) were assigned to centimeter groups based on maximum length (1 = <1.5 cm; 2 = 1.51 to 2.5 cm; 3 = 2.51 to 3.5 cm, etc). Collective weights and numbers were determined for each size group.

Statistical Analyses

An analysis of variance (ANOVA) was carried out on the data. If the F-test from the analysis was significant, then the pooled estimate of error was used to calculate a two-sided Student's *t*-test. The latter was then used to compare the mean response in the control group with the mean response in each treated group. The null hypothesis under test was that "there is no difference between each treatment group and the control". The level of significance used was 5%.

RESULTS

Physicochemical and biological assessments carried out on the 12 ponds in the pretreatment year (before division, 1985) showed them all to be very similar, with a very small number of random statistically significant differences.

Diverse populations of plankton and macroinvertebrates developed in 1985. These are illustrated in Figures 10 and 11, as overall means for the 12 ponds. The major groups of macroinvertebrates were the Anisoptera, Zygoptera, Coleoptera (mainly Dytiscidae), Ephemeroptera (Baetidae and Caenidae), Hemiptera, and Diptera (mainly Chironomidae), although a number of other groups were present. The ponds also

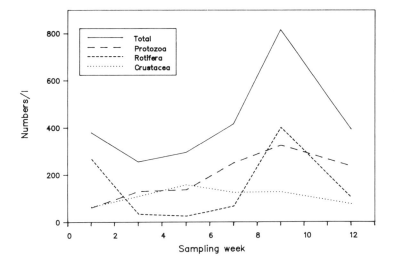

Figure 10. Zooplankton development in mesocosms in pretreatment year; mean of all 12 ponds (September to November 1985).

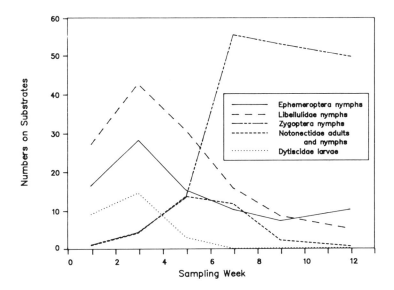

Figure 11. Macroinvertebrate on substrates in mesocosms in pretreatment year (1985); mean of all 12 ponds (September to November 1985).

had similar physicochemical characteristics; dissolved oxygen was never less than 7 mg/l and pH values ranged from 7 to 9 over the study period. At the final sampling conductivity was between 100 and 130 μS/cm.

Similar populations of organisms were present in all mesocosms in the weeks prior to applying lambda-cyhalothrin in 1986, although for some parameters statistically significant differences were observed in this period. The sections below describe the observations on residues, physicochemical characteristics, and biological assessments for the four treatment groups during the treatment year.

Residues of Lambda-Cyhalothrin

Spray Tanks

The mean amounts of lambda-cyhalothrin (isomer B) applied over the 12 weekly spray drift applications were very close to nominal, especially at the low rate (104 to 111% of nominal for the 4 replicate mesocosms; mean 106%) and mid rate (88 to 106%; mean 100%). Analytical data showed that the high rate dose was 79 to 81% of nominal.

Analysis of the runoff preparations immediately prior to application (after 24 h aging per mixing) showed lambda-cyhalothrin (isomer B) to be present at 70 to 80% of nominal at all rates. Examination of the slurry tank mixing system and the sampling procedures and analytical procedures showed that the sedimentation rate of the sediment particles resulted in the slurries being "undersampled". As >96% of the lambda-cyhalothrin was adsorbed to the particles, it is very probable that the slurry applications were in reality close to nominal concentrations.

Some epimerization occurred during the preparation of both the spray drift and runoff applications, as approximately 10% of the A isomer was found to be present on all occasions.

Mesocosm Water

The water residue concentrations of lambda-cyhalothrin were determined in detail at the high and mid rates and regularly in control mesocosms. Only a small proportion of the low rate samples were analyzed. All the low rate samples done were below the LOD, as would be expected, from calculation of nominal concentrations and also from interpolation from the high and mid rate results. The data are shown in Table 2 and Figure 12.

The residues in replicate mesocosms were generally very similar at each sampling interval. Furthermore the concentrations at the three sampled positions in each mesocosm were comparable. The amounts of lambda-cyhalothrin in the mid rate mesocosms, where above the LOD were almost exactly one tenth of those at the high rate, reflecting the application levels. Residues in the low rate and control mesocosms were below the LOD on all occasions analyzed. The concentrations in the water declined rapidly, following both spray drift and runoff applications, with half lives of approximately 1 day.

Lambda-cyhalothrin in the water was comprised of the isomer B (the active ingredient of the applied product) and its epimer A, formed in the mesocosm water. The ratio of B:A was approximately 1:1 at all times. All other isomers were below the LOD. Laboratory studies have shown epimerization of B to A in water above

Table 2. Residues of Lambda-Cyhalothrin (as Isomers A and B) in Mesocosm Water

Sampling Time[a]	Residues (ng/l)[b]													
	High Rate				Mid Rate				Low Rate		Control			
	1[c]	2	3	4	1	2	3	4	1	2	1	2	3	4
Preapplication	<2	<2	<2	<2	<2	<2	<2	<2	<2	<2	<2			
D1 + 1 day	<10	<10			<2	<3					<2	<2		
+ 2 days	<2	<2			<2	<2					<2			
R1 + 1 day	20	30			<2	<3					<2			
+ 3 days	3	4	d	d	<2	<2	<2	<2	<2	<2	<2		<2	
R2 + 3 days	5	4	4	13	<2	<2	<2	<2			<2	<2		<2
R3 + 3 days	d	d	36	6	<2	<2	<2	<2	<2	<2	d	<2	<2	<2
R4 + 1 day	90	98			10	10					<2			
+ 3 days	18	23	31	d	<2	<2	2	d	<2	d	<2		<2	
D8 + 1 day	27	33			2	4					<2			
+ 2 days	d	16			±	2					<2			
+ 4 days	5	4			<2	<2						<2		
+ 6 days	3	2			<2	<2						<2		
R5 + 3 days	13	12	d	21	<2	2	d	<2			<2	<2		<2
R6 + 3 days	18	34	31	27	4	2	2	3			<2		<2	<2
D12 + 2 wks	7	4	4	4	<2	<2	d	<2			<2	<2		
+ 4 wks	16	4	<2	<2		<2			<2	<2	<2		<2	<2
+ 6 wks	d	d	<2	<2							d	<2		
+ 9 wks	<2	<2	<2	<2							<2		<2	

[a] D1 to D12 = spray drift applications; at weekly intervals, commencing early June 1986.
R1 to R6 = runoff applications; at biweekly intervals, commencing 3 days after D1.
[b] Data are generally means of 3 analyses; shallow zone, and deep zone surface and bottom.
[c] Replicate microcosms.
[d] Sample lost, faulty, damaged, or contaminated.

pH 7, whereas photolysis results in formation of the *trans*-isomers. The *E*-isomers in laboratory studies were always small. Thus, the mesocosm residues were not surprising, as the water was pH >7. The absence of *trans*-isomers in the mesocosms was also to be expected, as photolytic processes are generally of little significance in natural waters, due to the extent of adsorption and scattering of sunlight near the water surface (Zepp and Cline, 1977).

Hydrosoil

The results from analyses of the hydrosoil in the high and mid rate mesocosms are shown in Table 3, together with data from control mesocosms. The low rate mesocosm hydrosoil samples were not analyzed, as interpolation from the high and low rate data suggested that these would all have been at or below the LOD.

The residues in hydrosoil would be expected to be quite variable across the mesocosm bottom, and with depth, due to

- The sloping mesocosm bottom possibly resulting in differential sedimentation of particles, and movement of sedimented soil particles down the slope
- Possible incorporation of surface residues by organisms such as oligochaetes (Karickhoff and Morris, 1985)
- Disturbance and movement of hydrosoil by fish, especially during nesting (see Plate 4c)

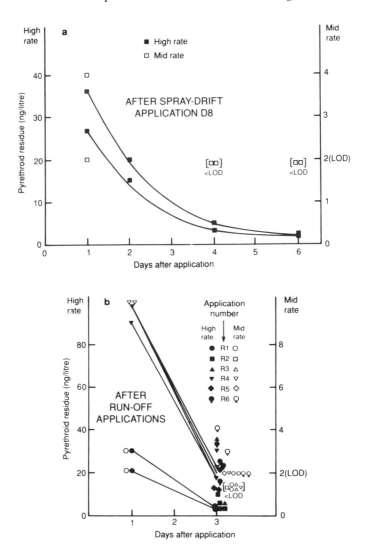

Figure 12.   Rate of disappearance of lambda-cyhalothrin residues (as isomers A and B) in water from high-rate and midrate mesocosms. (a) Following spray drift application D8. (b) Following runoff applications R1 and R4. (See Table 2 for application dates).

Despite this potential for variability, the hydrosoil residues were similar in replicate mesocosms and in the two zones in each mesocosm, demonstrating the evenness of the application procedures. Although the concentrations of lambda-cyhalothrin in the mid rate mesocosms were very close to the LOD, it was apparent that they were approximately one tenth of those in the high-rate mesocosms, reflecting the application rates. As with the water samples, epimerization was evident, with B:A ratios being generally 2:1.

Table 3. Residues of Lambda-Cyhalothrin (as Isomers A and B) in Hydrosoil

Sampling Time[a]	Residues (µg/kg dry wt)[b,c]											
	High rate				Mid Rate				Control			
	1[d]		2		1		2		1		2	
	S[e]	D	S	D	S	D	S	D	S	D	S	D
Preapplication	<0.2	<0.2	<0.2	<0.2	<0.4	<0.4	<0.4	<0.8	<0.4			<0.8
R1 + 3 days	6.4	3.6	8.9	2.1	<1.2	1.8	<1.2	<1.6	<0.4		<0.8	
R2 + 3 days	21	13	19	5.1	1.0	4.0	6.8	2.6	<0.5	<1.0		
R3 + 3 days	13	12	20	13	c	6.4	1.6	9.6	<0.4		<0.8	<0.8
R4 + 3 days	15	28	21	22	2.6	2.8	1.6	1.6	<0.4		<0.8	
R5 + 3 days	22	46	38	29	1.4	2.4	1.8	1.6	<0.4			
R6 + 3 days	27	53	34	23	2.2	c	3.0	4.2	<0.4			
D12 + 2 weeks	22	30	14	22	1.0	1.8	2.0	1.4	<0.4			
+ 4 weeks	14	36	14	59	<1.2	2.0	2.4	2.8	<0.4			
+ 6 weeks	22	48	13	34	3.0	2.5	4.0	3.8	<0.4		<0.8	
+ 9 weeks	27	25	44	18	<1.2	1.4	1.6	2.0	<0.4			

[a] R1 to R6 = runoff applications, every 2 weeks commencing early June. D12 = final spraydrift application.
[b] Data are total residues in depth fractions analyzed; calculated as all in the 0 to 2.5 cm layer.
[c] Sample lost or contaminated.
[d] Replicate mesocosms.
[e] S = shallow zone; D = deep zone.

Residues were principally in the top 2.5 cm of the hydrosoil core. At greater depths, amounts were below the LOD in the early stages of the study, but increased with time, to approximately 20% of the total hydrosoil residue in the postapplication period.

Residue Recoveries

Residue recoveries could only be obtained from the high rate mesocosms, as at the mid rate a substantial proportion of the water analyses were below the LOD. Mesocosm water samples taken 1 day after spray drift application numbers 1 and 8 showed <6% and 20%, respectively, of the nominally applied lambda-cyhalothrin to be in the water column. A day after runoff treatment numbers 1 and 4, mesocosm water samples contained 5% and 20%, respectively, of the nominal application rate. Losses from the water column were probably the result of deposition onto the mesocosm bottom and of adsorption to suspended particles (followed by deposition), macrophytes, and filamentous algae.

The total recovered lambda-cyhalothrin residues, from water and hydrosoil, are shown in Figure 13, together with the nominal applications. During the treatment phase, recoveries ranged from 23 to 47% (mean 32%) of applied, while values in the posttreatment period were 15 to 31% (mean 23%) of the total dosed to the mesocosms.

Discussion

Lambda-cyhalothrin residue analyses were carried out to achieve several objectives:

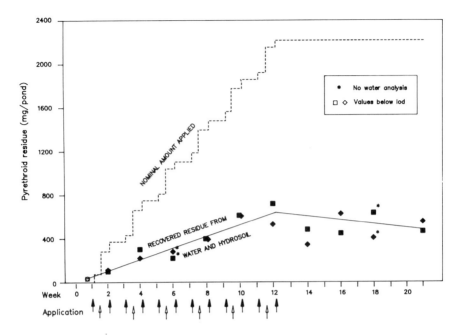

Figure 13. Total recovery of residues from lambda-cyhalothrin (isomers A and B) in two high rate mesocosms, in comparison to the cumulative nominal applications (assuming no losses or degradation).

- To validate the application procedures
- To show that the mesocosm exposure concentrations at all rates were in the same ratio as applied (high rate $10 \times$ and low rate $0.1 \times$ the mid rate)
- To demonstrate the reproducibility of exposure concentrations between replicates
- To verify that lambda-cyhalothrin in the test system was behaving as would be expected, based on its chemical and physicochemical properties

The measured lambda-cyhalothrin concentrations in the spray tanks, the calibration of the application equipment, observations of the spraying process, and the pyrethroid residues in the water and hydrosoil demonstrated that:

- The application methodology was efficient, reliable, and consistent. All applications were shown analytically to be at least 70 to 80% of nominal, although in reality most were probably even closer to nominal.
- The spray drift and runoff treatments were dosed evenly across the mesocosms.
- The concentrations of lambda-cyhalothrin in replicate mesocosms were comparable, and the water and hydrosoil residues were relatively evenly distributed within each mesocosm (measured in high- and midrate only).
- The residues in the mid rate mesocosms were one tenth of those in the high rate, accurately reflecting the treatment levels (low-rate mesocosm residues were below the analytical LOD).

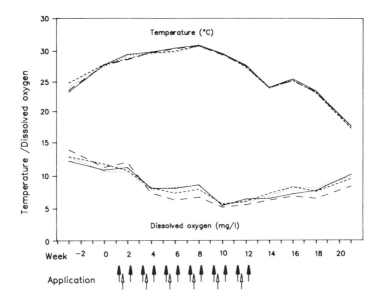

Figure 14. Midday measurements of dissolved oxygen, and temperature in the water column. Mean of shallow and deep zone 25 cm depth measurements.

Thus the first three objectives of the residue program, outlined above, were achieved, although the conclusions for the low rate can only be arrived at by interpolation. The final objective was also satisfied, in that the distribution and behavior (adsorption and isomerization) of the product in the mesocosm was according to predictions based upon laboratory studies. Furthermore, the residue distributions and recoveries were very similar to those reported for other pyrethroids in the scientific literature (Hill, 1985).

Physicochemical Assessments

There were no statistically significant differences in any physicochemical parameters resulting from the application of lambda-cyhalothrin at any rate. Dissolved oxygen (DO) and temperature midday measurements taken at 25 cm depth of water are shown in Figure 14. DO fell gradually from 12 to 14 mg/l before the application of lambda-cyhalothrin to 5 to 6 mg/l in midsummer, then increased to 8 to 10 mg/l by the study end. This principally reflected the water solubility of oxygen in relation to temperature, with water temperature reaching a maximum of 30°C in midsummer.

Dusk-dawn-midday-dusk measurements of DO and temperature showed little difference between any of the mesocosms. Both parameters were highest at dusk, following the pattern described by Lind (1979) and Boyd (1979). Such daily fluctuations of DO concentrations in the euphotic zone occur because during daylight hours oxygen is produced by photosynthesis faster than it is used in respiration, whereas at night, oxygen levels fall as a result of continued respiration in the absence of photosynthesis. Daily variation in DO was generally 2 to 5 mg/l, while temperature flucutated by 1 to 3°C. Least variation was seen in the deeper water, but differences were small and there was no thermocline.

Water pH varied between 7.0 and 9.5, conductivity was 170 to 270 μS/cm, and total alkalinity 65 to 90 mg $CaCO_3$ per l. Water turbidity was measured in nephelometric turbidity units (NTU). Turbidity was low pretreatment, at 1.5 NTUs, rising in general to approximately 4 NTUs during the pesticide application period and slightly higher posttreatment. A smaller number of mesocosms (not dose-related) had exceptionally high values at one sampling time only, in late June.

Biological Assessments

Microorganisms in Hydrosoil

There was little difference between viable numbers of microbes in the hydrosoil at any treatment rate, measured before and after the lambda-cyhalothrin applications. Aerobic bacteria were of the order of 1 to 4 $\times$ 10^7/g, actinomycetes 0.5 to 3 $\times$ 10^6/g, anaerobic bacteria 1 to 8 $\times$ 10^5/g, and fungi 3 to 22 $\times$ 10^3/g. The mesocosms all possessed a substantial, viable microflora, which was unaffected by lambda-cyhalothrin, as would be expected from laboratory studies of pyrethroids and microorganisms (Hill, 1985).

Phytoplankton

A range of parameters was measured to determine if lambda-cyhalothrin had any effects on phytoplankton populations. A very small number of statistically significant differences were apparent; however, these are considered to be random occurrences. Laboratory studies have shown algae to be very insensitive to the pyrethroid insecticides (Hill, 1985). The data suggested that there would be little likelihood of any direct effect of lambda-cyhalothrin on the phytoplankton in the mesocosms, even at the highest rate. This is substantiated by the results of the mesocosm study, which demonstrate that these algal populations were unaffected by the test chemical.

Cell Numbers, Biomass, and Identification Cell numbers and biomass are shown in Figure 15. Numbers in the control mesocosms were slightly greater than in any of the lambda-cyhalothrin treated mesocosms, although not significantly so. However, it is unlikely that there was an effect of the pyrethroid as:

- The same was apparent on one of the pretreatment sampling occasions
- There was no dose-rate relationship
- There were no differences in biomass

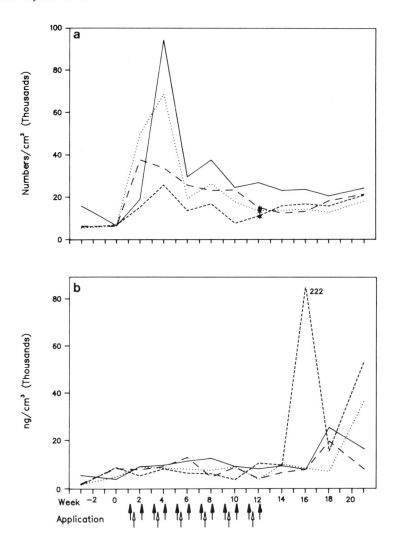

Figure 15. Total phytoplankton. (a) Cell numbers. (b) Biomass.

A very substantial, but statistically nonsignificant, increase in biomass on one date in the midrate mesocosms was due to the presence of *Spirogyra* sp. in one of the replicates. This filamentous member of the Chlorophyta constituted only 3% of the cell numbers but over 90% of the sample biomass in the mesocosm.

One hundred and seventy three taxa were identified. Figure 16 shows the relative numbers of the various phyla of algae. The Chlorophyta were generally the most abundant group. The Cyanophyta were also present in abundant numbers, but as a much smaller proportion of the biomass. These two groups at times accounted for

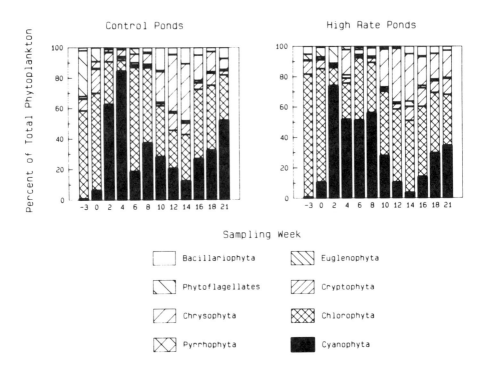

Figure 16.   Proportional distribution of phytoplankton numbers.

around 90% of the phytoplankton cell numbers. No one species retained dominance, but the most abundant cyanophytes were *Oscillatoria amphibia, Aphanizomenon flosaquae, Lyngbya limnetica,* and *Rhaphidiopsis curvata,* while the most commonly found chlorophytes were *Gloeocystis planktonica, Oocystis pusilla, Schroderia setigera, Scenedesmus quadricauda,* and unidentified chloroflagellates and coccoid cells.

The dominance of chlorophytes and cyanophytes, particularly in unfertilized ponds, has been noted by Boyd (1979) at Auburn University, AL, where these groups comprised 90% or more of the phytoplankton population in almost all ponds during the summer months.

In the present study, the Chrysophyta became abundant later in the study, with *Dinobryon bavaricum* and unidentified small coccoid cells making up the majority of this group.

Considerable variation in populations was evident between replicates sampled at any one time, although the trends described above remained clear. Such variation is common for phytoplankton, as supported by Boyd's (1979) statement that ''ponds which are located in the same vicinity, have similar water quality and receive identical treatments seldom have identical abundances of phytoplankton''.

The data indicated that lambda-cyhalothrin had no effect on the phytoplankton cell numbers and their biomass.

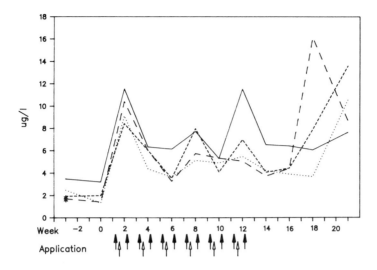

Figure 17.   Chlorophyll *a* concentrations.

Chlorophyll a *and Phaeophytin* a   Chlorophyll *a* levels in the mesocosms fluc-tuated throughout the study, but were mostly between 3 and 17 μg/l (Figure 17). This compares well with an average of 7 μg/l found by Boyd (1973) during the summer months in unfertilized ponds at Auburn University, Alabama, U.S. Similar levels were also reported by O'Brien and de Noyelles (1974) in 0.1-ha ponds at Cornell University, New York, U.S. Phaeophytin *a* was mostly present at one tenth or less of the chlorophyll concentrations.

The control mesocosms generally had slightly higher chlorophyll concentrations than the treated mesocosms, both before and during pesticide treatments. There were no statistically significant differences at any time during the study, and lambda-cyhalothrin was not considered to have had any effect on chlorophyll levels in the mesocosms.

Photosynthesis and Respiration   Measurements with light and dark bottles showed there to be no effects of the test chemical on photosynthesis, respiration, and gross productivity (Figure 18). Mean values for gross productivity, calculated from the dissolved oxygen changes in the light and dark bottles, fluctuated between 50 and 180 mg C fixed per m^3/h from June to September, and rose to over 200 mg C in October. This compares closely to values of 180 mg C fixed per m^3/h measured by Boyd (1973) in unfertilized ponds at Auburn University.

Community metabolism was calculated using dusk-dawn-dusk *in-situ* determi-nations of mesocosm water dissolved oxygen (DO). There were no major differences in community photosynthesis or respiration between any of the treatment groups, with values mostly between 2 and 10 mg DO per l. The ratio of photosynthesis to respiration, shown in Figure 19, was generally around 1.0, following the pattern described by Lind (1979).

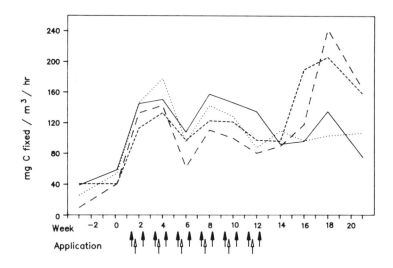

Figure 18.   Gross productivity from light and dark bottle measurements.

The values obtained by this method, designed by McConnell (1962) for small bodies of water with minimal surface agitation, provide only approximations of mesocosm ''activity'' without reference to individual constituents of the community.

Comparison of Phytoplankton Assessments   Results from determinations of numbers, biomass, chlorophyll, and gross productivity are shown for control mesocosms in Figure 20. As would be expected, numbers do not necessarily correlate

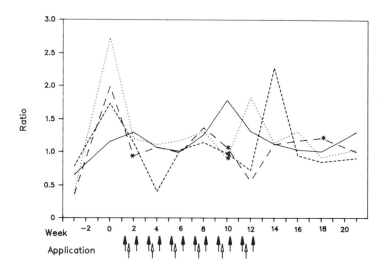

Figure 19.   Ratio of photosynthesis:respiration as determined by dusk-dawn-dusk dissolved oxygen measurements.

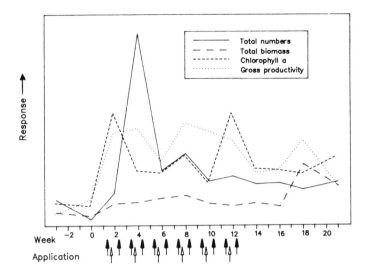

Figure 20.   Comparison of phytoplankton assessment techniques; in control mesocosms (arbitrary scales).

with biomass. For example, at week 4 a large numerical increase was not reflected in biomass, and this was due to a proliferation in some mesocosms of *Oscillatoria amphibia,* a cyanophyte of small cell volume (46 μm^3).

Biological activity is not necessarily a function of either numbers or biomass, as the phytoplankton population consists of many species which may respond differently to the prevailing conditions. Thus, neither chlorophyll nor gross productivity measurements reflected the pattern in total cell numbers or their biomass.

Periphyton

Several parameters were measured for periphyton. These were cell numbers, identification, chlorophyll, biomass by volume and by weighing, and autotrophic index. As with the phytoplankton, no effects of lambda-cyhalothrin were seen at any treatment rate.

Cell Numbers and Identification   Cell numbers generally ranged from 1000 to 3000/mm^2, with species of the filamentous chlorophytes *Mougeotia* and *Oedogonium* dominant (Figure 21). Cyanophyta and Bacillariophyta made up most of the remainder of the cells, with the commonest cyanophytes being *Calothrix* spp., *Lyngbya limnetica, Nostoc* sp., and *Phormidum sp.* and bacillariophytes *Achnanthes minutissima, Nitzshia palea* var. *debilis,* and *Synedra rumpens.*

Chlorophyll a *and phaeophytin* a decreased in the control and treatment groups, from approximately 3 to 10 mg/m^2 to 1 mg/m^2 at the end of the study period. This paralleled the trends seen in periphyton biomass. Phaeophytin levels were around 0.1 mg/m^2.

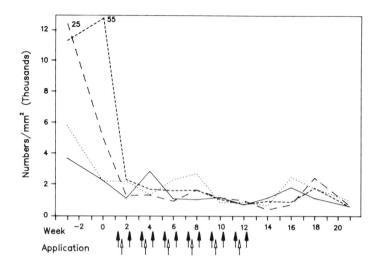

Figure 21.   Total periphyton cell numbers.

Biomass and autotrophic index values were unaffected by lambda-cyhalothrin at any of the three treatment rates. Both parameters followed the same general trend as chlorophyll *a* and algal numbers with the highest values in the spring. Mean total biomass by weight was around 8000 mg/m² pretreatment, declining to less than 1000 mg/m² at the end of October (week 21; Figure 22). Biomass by volume measurements showed a very similar trend to those obtained by weight, but prior to week 6 and after week 16 gave substantially higher values.

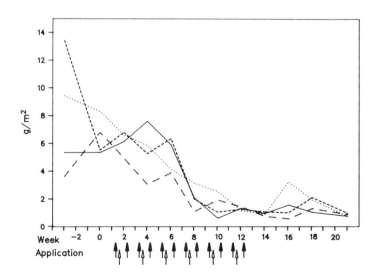

Figure 22.   Total biomass by weight of periphyton.

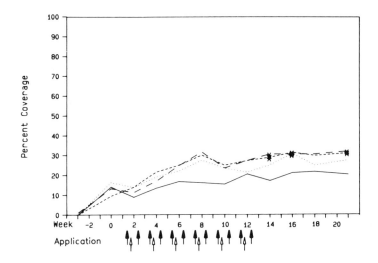

Figure 23.   Ludwigia mesocosm surface cover.

Filamentous Algae and Macrophytes

Both filamentous algae and macrophyte development (Plate 1c) were temporally and spatially similar in all mesocosms. Filamentous algae covered 40 to 50% of the mesocosm prior to the first application, but this progressively declined to 20% by week 4 and 10% by week 10.

Ludwigia uruguayensis was dominant throughout the study comprising 75 to 100% of the macrophyte cover and greater than 97% of the macrophyte biomass at the study end. The *Ludwigia* increased from 10% water surface cover in early June to around 20 to 30% by the end of July, and remained at this until study termination. The *Ludwigia* cover was generally one third greater at all treatment rates, than in control mesocosms (Figure 23), although it is very unlikely that this was an insecticide-related effect (see Hill, 1985). Two of the control mesocosms had a consistently lower macrophyte cover than the other two controls, often over 50% lower.

The total mesocosm coverage by filamentous algae and macrophytes was sub-stantially less than the sum of the two components, as after spraying started, filamentous algae principally remained within the shoreline *Ludwigia* beds.

Zooplankton

One hundred and forty eight taxa were identified during the study, representing the three major groups: Protozoa (60 taxa), Rotifera (64 taxa), and Crustacea (24 taxa). The protozoans were generally the most numerous group followed by the rotifers, then the crustaceans. This relationship is commonly found where these three groups are enumerated (Pace and Orcutt, 1981).

The total zooplankton numbers increased with time in all mesocosms, from around 500 cells per l in May to over 13,000 cells per l in October (Figure 24). Throughout most of the study there were no differences in overall numbers between any of the four test groups. However, on three occasions (weeks 10, 12, and 14), numbers in the control mesocosms showed a substantial increase compared to the treated mesocosms, with rotifers responsible. This was not believed to have been due to the test chemical, and is discussed further below. It is therefore likely that there was not an effect of lambda-cyhalothrin on the total zooplankton numbers.

Protozoans   Lambda-cyhalothrin had no effect on the overall protozoan population (60% of the zooplankton cell numbers) or its component taxa. Total cell numbers in all mesocosm groups were approximately 300 per l from the start of the study until week 12, then showed substantial increases (Figures 25a). Relative abundance of the component taxa was the same in all treatment groups and controls (Figure 25b). *Strombidium delicatissimum* formed up to 60% of protozoans in the early part of the study. The genus *Difflugia,* dominant from week 12, was composed of seven species of which *D. limnetica* was the most common.

Rotiferans   Rotifers constituted above one third of the overall zooplankton community. The total numbers increased from approximately 50 cells per l in the pretreatment period to 1500 by the end of the study (Figure 26a). Numbers were similar in all mesocosms until week 10, when a value of 50,000 cells per l in the deep zone of one of the four replicate control mesocosms distorted the control mean. Reexamination of the sample revealed the presence of a clump of filamentous algae containing a very high density of an unidentified rotifer. This data point was therefore considered unrepresentative. However, the control mesocosms did maintain substantially higher numbers of rotifers than the treated groups until week 14,

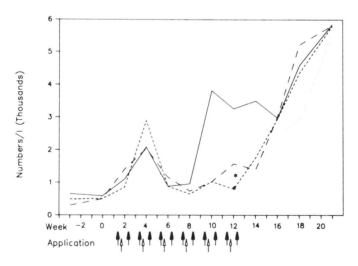

Figure 24.   Total numbers of zooplankton.

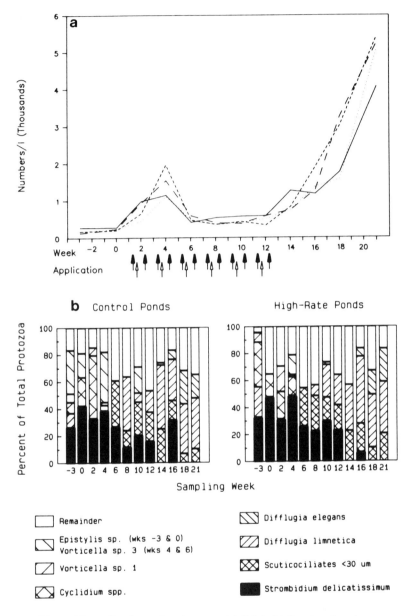

Figure 25. Numbers of protozoans. (a) Total. (b) Species abundance.

due mainly to *Polyarthra remata* at week 12 and *Keratella cochlearis* at week 14. During the latter part of the study, the numbers of rotifers in the mesocosms treated with lambda-cyhalothrin were often significantly lower than in the control group. However, this was not rate related, with the high-rate rotifers only significantly reduced on one occasion. It is therefore unlikely that the differences observed were

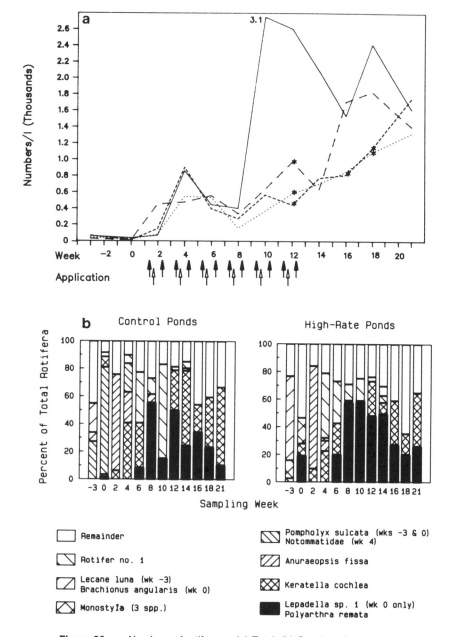

Figure 26.   Numbers of rotiferans. (a) Total. (b) Species abundance.

due to the test chemical, but more probably were the result of normal population interactions and variability.

Figure 26b shows the relative abundance of the constituent rotifer taxa in control and high-rate mesocosms (low- and midrates were very similar). Only three taxa

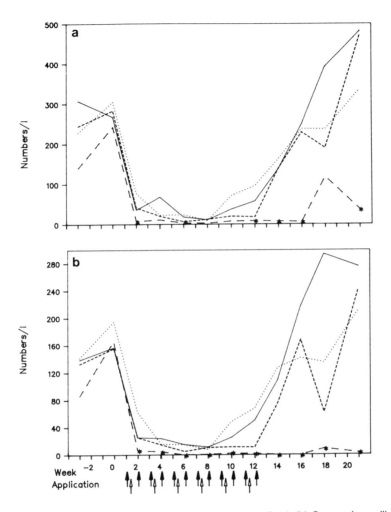

Figure 27. Numbers of zooplankton crustaceans. (a) Total. (b) Copepod nauplii.

were of major importance over substantial periods of the study; these were *Poly-arthra remata, Keratella cochlea,* and an unidentified species.

Crustaceans The planktonic crustaceans represented less than 6% of the overall zooplankton numbers in the mesocosms. Prior to application of lambda-cyhalothrin cell numbers were approximately 250 per l. However, these declined in all meso-cosm groups (treated and control) to around 20 cells per l immediately after the first sprays of the test chemical (Figure 27a). As this occurred in all mesocosms, it was probably due to one or both of the following:

- Introduction of and predation by the fish
- The application of the soil-water slurries and the effect of soil particles on filter feeding (McCabe and O'Brien, 1983; Martin and Klaine, 1987).

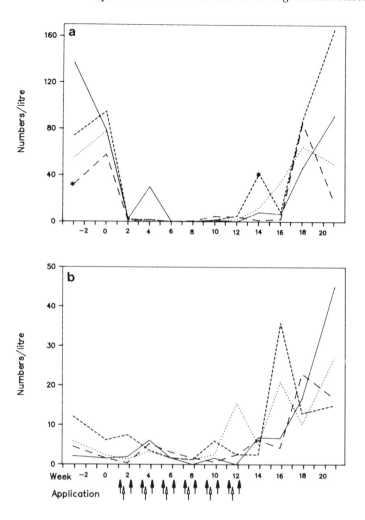

Figure 28. Numbers of zooplankton crustaceans. (a) Cladocera. (b) Ostracoda.

While the early decline in crustacean numbers was not related to lambda-cyhalothrin, there was a subsequent effect of the product on crustaceans in the high rate mesocosms only (Figure 27a).

The Copepoda generally made up between 60% and 80% of the crustaceans, and copepod nauplii were 90% of these, except in the high-rate mesocosms, where effects due to lambda-cyhalothrin occurred on nauplii (Figure 27b), copepodites, and adult copepods. Cladocerans and ostracods were unaffected at all treatment rates (Figures 28a, b).

Size Distribution The distribution of microzooplankton and macrozooplankton are shown in Table 4. Overall, the distributions showed no effects due to lambda-cyhalothrin. The effects noted on Crustacean copepods at the high rate were masked by numbers of unaffected organisms in other groups. Only the macrozooplankton showed the early decline in all mesocosm groups, described earlier as due to fish predation and/or sediment particle effects on filter feeders.

The distribution of the zooplankton size groups is important to young fish feeding on these organisms. This is further discussed later in this chapter.

Macroinvertebrates

Substrate sampler, emergence trap, and visual assessment data are discussed in this section. Unless stated otherwise, all values presented are for the total organisms collected in a mesocosm at the sampling occasion. As the same procedure was used each time, these values are relative to one another.

Some macroinvertebrates were affected by lambda-cyhalothrin, but generally in the high-rate mesocosms only. A few effects were noted at the midrate and none at the low rate. The organisms affected were those that have been found to be most sensitive to pyrethroids under laboratory conditions (Hill, 1985). These were groups with species that move over the surface of mesocosm substrates (herpobenthos), swim within the water profile (nekton), or inhabit the water surface (neuston). They included the Ephemeroptera (Baetidae and Caenidae), Zygoptera (Coenagrionidae), Hemiptera (Gerridae, Notonectidae, and Veliidae), Coleoptera (Haliplidae), Trichoptera (Leptoceridae), and Diptera (Chironomidae: Tanypodinae).

Mollusks were unaffected at all rates. The invertebrates living within the hydrosoil, i.e., Oligochaeta and Chironomidae: Chironominae, were also unaffected by any of the rates of lambda-cyhalothrin.

The total number of macroinvertebrates in all mesocosm groups is shown in Figure 29a. This showed substantial increases in total organisms throughout the

Table 4. Zooplankton Size Distribution

	Numbers of Organisms per l (Treatment Means)											
	Microzooplankton				Macrozooplankton				Macrozooplankton as Percent of Total Zooplankton			
Week No.	Control	Low Rate	Mid Rate	High Rate	Control	Low Rate	Mid Rate	High Rate	Control	Low Rate	Mid Rate	High Rate
−3	337	209	236	158	317	234	250	142	48	53	51	47
0	325	262	218	256	267	311	283	244	45	54	56	49
2	1,062	536	800	1,394	39	82	55	19	4	13	6	1
4	1,969	1,596	2,833	1,997	118	104	58	44	6	6	2	2
6	814	944	848	1,107	66	80	28	47	8	8	3	4
8	914	409	615	688	42	18	30	46	4	3	5	6
10	3,639	711	959	966	174	144	65	61	5	17	6	6
12	3,116	950	706	1,434	136	237	92	140	4	20	12	9
14	3,270	1,396	1,540	1,268	228	232	222	129	7	14	13	9
16	2,422	2,160	2,630	2,873	549	310	321	134	18	13	11	4
18	3,846	2,497	3,753	4,586	740	412	598	633	16	14	14	12
21	4,956	6,640	12,133	10,087	1,215	819	1,597	608	20	11	12	6

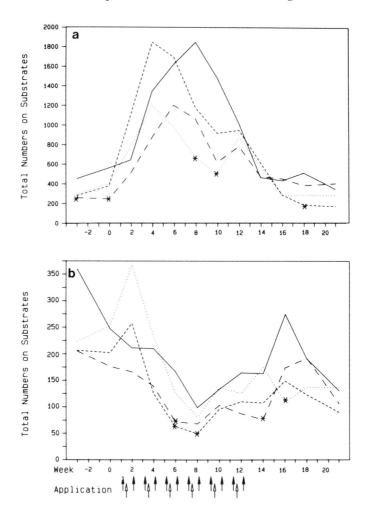

Figure 29.   Macroinvertebrates. (a) Total. (b) Fish-food.

summer months, and during the spraying period, followed by a decline in the late
summer and fall. Figure 29b shows only those organisms generally considered
important as fish food (Ewars and Boesel, 1935; Moffett and Hunt, 1943; Swingle,
1949; Seaburg and Moyle, 1964; Turner, 1955; Flemer and Woolcott, 1966; Etnier,
1971). Thus the taxa and life cycle stages included were as follows: Ephemeroptera
nymphs, Trichoptera larvae and pupae, Crustacea-Malacostraca, Diptera larvae,
and pupae. Neither total nor total fish-food macroinvertebrates showed differences
between treated and control mesocosms that could be ascribed to lambda-cyhalothrin. There were effects of lambda-cyhalothrin on some of the macroinvertebrate
fish-food groups, especially at the high rate (see following sections). However,
total numbers of fish-food organisms were dominated by dipterans, which were

less affected overall than any of the other fish-food groups, particularly the Chironominae on which the insecticide had no impact even at the high rate. Between weeks 8 and 12, total organism numbers were greatest in controls, and some statistically significant effects were apparent. However, these were principally due to mollusk numbers and were not rate related; furthermore, the differences were no greater than occurred in the pretreatment period when statistically significant differences were also evident. The concentration of lambda-cyhalothrin applied to the high rate was two orders of magnitude above the low rate, and laboratory data clearly shows that any effects on macroinvertebrates would follow a dose-related response over this range. The sections below individually examine the various taxa present in the mesocosm, as assessed by one or more of the sampling procedures.

Turbellaria   Low numbers of Planariidae were present on substrates and there were no statistical differences between treatments. Numbers increased to 12 per mesocosm (total on substrates) during June, decreasing thereafter. Very low numbers of Typhloplanidae were collected.

Mollusca   The mollusks in the mesocosms were all unaffected by lambda-cyhalothrin, as would be suggested by laboratory data showing the low sensitivity of these organisms to pyrethroids (Hill, 1985). Snails representing four families were present, the commonest being Physidae (*Physa* sp.) and Planorbidae (*Helisoma* sp.). These showed substantial but nonsignificant and nonrate-related differences between mesocosm groups. Numbers peaked in July and August at over 1000 per mesocosm substrates (Physidae, Figure 30a).

Low numbers of Lymnaeidae and Ancylidae were collected from substrates. In addition to the gastropods, very low numbers of the bivalve *Anodonta (Utterbackia) imbecilis* were found in all treatment groups when the mesocosms were drained for fish harvest. No relationship with lambda-cyhalothrin treatment was noted.

Oligochaeta   The oligochaete worms were present in substantial numbers and were unaffected by lambda-cyhalothrin (Figure 30b). Two major peaks of numbers were evident in all mesocosms during the treatment period, the first in early July and the second in late August, with 350 to 450 organisms per mesocosm substrates. In the high-rate mesocosms, the oligochaete numbers were lower than for all other treatment groups on the two sampling occasions during the first peak (although not significantly different from the controls). From week 8 to the end of the study, the oligochaete numbers were similar in all the mesocosm groups. It thus seems unlikely that the high rate was showing an effect of treatment.

Nearly all the oligochaetes collected on substrates were Naididae. *Dero* was the most common genus with *Chaetogaster* and *Stylaria* occasionally present. Whitly (1982) reported that naidid worm population increases are frequently seen during the summer. Very low numbers of Tubificidae were present, primarily *Limnodribus hoffmeisteri*.

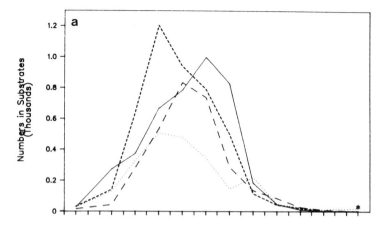

Figure 30a.   Macroinvertebrate numbers. Physidae on substrates.

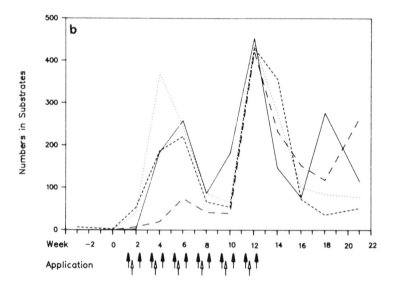

Figure 30b.   Macroinvertebrate numbers. Oligochaeta on substrates.

Hydracarina   Numbers of the Hydracarina water mites were low and variable. However, numbers were reduced rapidly by the high-rate lambda-cyhalothrin treatment, and to a lesser extent at the midrate. While values for these two rates were significantly reduced on only a few occasions, it was probable that the reductions were due to the pesticide. No effect was evident in the low-rate mesocosms.

Ephemeroptera   Two families of mayflies were present in the mesocosms, Baetidae and Caenidae, observed on substrate samplers and in emergence traps. Both of these are known to be highly sensitive to the pyrethroid insecticides (Hill, 1985).

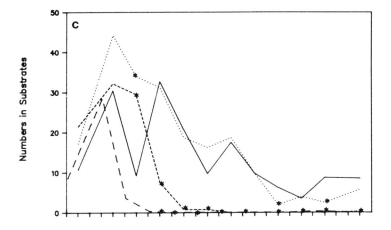

Figure 30c. Macroinvertebrate numbers. Caenidae nymphs on substrates.

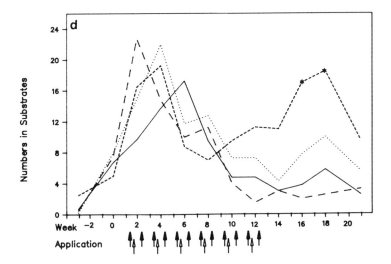

Figure 30d. Macroinvertebrate numbers. Libellulidae nymphs on substrates.

The Baetidae, represented by the single species *Callibaetis* nr. *Floridanus,* was the most numerous mayfly family. A reduction in baetid nymphs on substrate samplers was evident in the high-rate mesocosms at the sampling following the first application of lambda-cyhalothrin, but not significant at this time. The effect became more severe, and statistically significant, at later samplings with evidence also of effects at the midrate. A slight but significant reduction of baetids also occurred at the low rate, on two occasions only in the latter part of the treatment program, with no difference to the control posttreatment. No significant differences were seen in *Callibaetis* adults in emergence traps throughout the study, although after an early peak in June numbers were very low in all mesocosms.

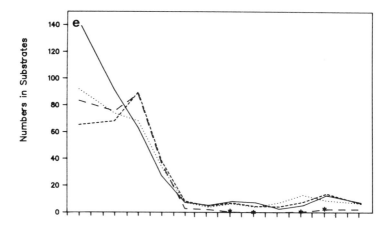

Figure 30e.   Macroinvertebrate numbers. Zygoptera nymphs on substrates.

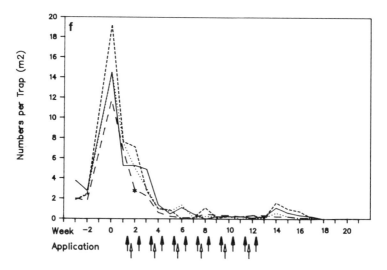

Figure 30f.   Macroinvertebrate numbers. Zygoptera adults in emergence traps.

The Caenidae were represented by a single genus, *Caenis*. While numbers in the control and the low rate mesocosms were very similar throughout the study, those in the high and mid rates were affected by lambda-cyhalothrin (Figure 30c). These organisms were virtually eliminated at the high rate during the treatment period, but were seen to be reappearing in this and the mid rate posttreatment, although still significantly lower than the controls.

Baetidae showed a rapid decline in all mesocosms (treatment and control) in mid-June, possibly due to a natural population cycle or alternatively as a result of predation or the effects of the particulate material in the soil:water slurries. During the pretreatment year (1985), *Callibaetis* showed no preference for either surface

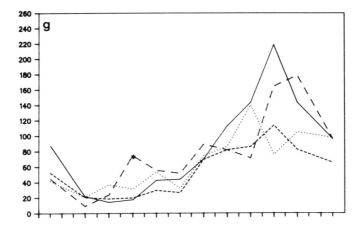

Figure 30g. Macroinvertebrate numbers. Chironomidae larvae on substrates.

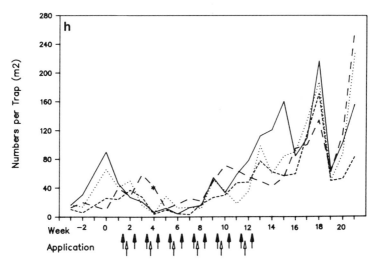

Figure 30h. Macroinvertebrate numbers. Chironominae adults in emergence traps.

or bottom substrates in the shallow zone and a distinct selection for the surface substrates in the deep zone. During the treatment year (1986), greater numbers of *Callibaetis* were collected from bottom substrates in the deep zone. The lower light levels at this depth may have afforded some protection from the introduced bluegill sunfish, a known visual predator (O'Brien, 1979).

Odonata The Anisoptera (dragonflies) were represented by the families Aeshnidae, Gomphidae, and Libellulidae (Figure 30d). Substrate sampler counts showed lambda-cyhalothrin to have had no effect on these organisms at any rate. However, on occasions, dead or abnormally behaving adult libellulids were noted along the shoreline or on the mesocosm surface after pesticide applications. Recently emerged

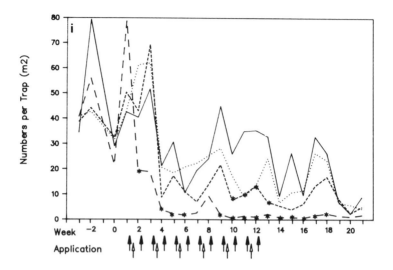

Figure 30i.   Macroinvertebrate numbers. Tanypodinae adults in emergence traps.

adults or those ovipositing at or immediately after spraying were especially vulnerable. Both *Erythemis simplicolis* and *Pantala* sp. were seen flying over mesocosms and also ovipositing during the application period.

The Zygoptera (damselflies) in the mesocosms all belonged to the family Coenagrionidae, with *Ischnura posita* most common. A major reduction in numbers occurred in all mesocosms through June, probably due to fish predation (Healy, 1984). Substrate sampler collections showed that there was no effect of lambda-cyhalothrin on this group at the mid and low rates (Figure 30e). At the high rate zygopteran numbers were depressed, but showed signs of posttreatment recovery. Emergence traps showed a flush of emergence in late May and early June with only low numbers of adults trapped thereafter, in any mesocosm (Figure 30f).

Hemiptera   Six families were found on substrate samplers: Belostomatidae, Corixidae, Gerridae, Naucoridae, Notonectidae, and Veliidae. The substrate samplers were not generally "attractive" to any of these organisms. The most efficient method for assessing hemipterans proved to be visual observation of the mesocosm quadrats and total surface, although data from this method inevitably has high variability.

Based on visual observations, Belostomatidae numbers were low and only seen between weeks 4 and 12, with no effect of lambda-cyhalothrin at the mid and low rate. The most severe effects were noted with the surface-living Gerridae and Veliidae, particularly during the 1 h posttreatment assessments. Overall, significant effects on these taxa occurred at the high and mid rates only.

The Nototectidae were the most abundant family, with mean visual quadrat counts of generally 10 to 100 per 2 min assessment time in May and June. A general decline occurred in all mesocosms through June, probably due to bluegill predation.

In the high rate mesocosms, this was more pronounced, probably due to the test chemical.

The Notonectidae spend a large proportion of their time just under the water surface, while the Gerridae and Veliidae are confined to living on the water surface (except during winged dispersal). These three families are thus especially at risk from the direct sprays of lambda-cyhalothrin to the total water surface of the mesocosms and the effects seen were not unexpected, particularly as they had no way of avoiding the chemical applied to their entire habitat.

Coleoptera The families Dytiscidae, Gyrinidae, Haliplidae, Hydrophilidae, and Elmidae were all present, but only regularly recorded by visual assessments. Gyrinid larvae were present on the substrate samplers, but in fairly low numbers.

Quadrat visual counts showed effects on the Haliplidae at the high rate only, and not on the other groups at any rate. Whole mesocosm observations showed a substantial number of adult aquatic Coleoptera to have been affected by the high rate applications, often 20 to 30 per mesocosm within 24 h of spraying. These observations also indicated a much smaller impact at the mid rate, with generally one to five or less affected adults being noted.

Trichoptera Two families were present on substrate samplers and in emergence traps. The most numerous organisms were *Oecetis inconspicua,* of the Leptoceridae. Hydroptilidae (e.g., *Oxyethira* sp.) were also present. Numbers on substrate samplers were small and variable. Although there were few statistically significant differences, the results clearly indicated a high-rate effect, and possibly some changes in the mid rate. By the end of the study, mid rate mesocosms were clearly the same as controls. Emergence traps showed peaks of hatching in May and August, with numbers at the latter data reflecting the substrate data.

Diptera The Chironomidae were the most abundant family in both the substrates and emergence traps, followed by the Ceratopogonidae. Chaoborids were also fairly common in the emergence traps, and other families present included the Culicidae, Tipulidae, Stratiomyidae, Syrphidae, and Tabanidae.

The overall numbers of Chironomidae from substrates were not significantly different for any of the mesocosm groups (Figure 30g). Numbers declined immediately after the introduction of the fish in late May, although there was a hatching peak at the same time. In early June, approximately 20 chironomid larvae were collected from the mesocosm substrates, with the numbers increasing progressively through the treatment period to around 100 to 150 in September. Emergence trap numbers of chironomid adults increased from around 60 per trap in May to nearer 200 at the end of October. At the subfamily level, however, while the sedentary Chironominae (usually confined to cases in the hydrosoil) were unaffected by lambda-cyhalothrin (Figure 30h), the free-living Tanypodinae were clearly affected at the high rate and probably the mid rate (Figure 30i).

Ceratopogonid larvae declined in numbers throughout the study period. In the high-rate mesocosms, there were significantly fewer larvae than in the controls, both before and during the pesticide applications. However, it is likely that lambda-cyhalothrin did suppress numbers at the high rate.

Chaoborid emergence trap counts, although variable, did indicate the likelihood of effects at the high rate. All the other dipteran groups not discussed above were too low in numbers and too variable to draw any conclusions.

Bluegill Sunfish

Throughout the study period the visual assessments were used to observe the bluegill sunfish. Adults were seen actively swimming and at all times behaving normally. The turbidity of the mesocosms prevented any observation of nesting, but young fish were regularly recorded. In the quantitative quadrat observations, immature fish varied from 0 to around 40 per 2-min period, with no apparent effect at any of the treatment rates. The young bluegill were always seen to behave normally, and at no time during the study were any dead fish found.

Fish were also harvested at the end of the study, in November 1986. Approximately 300,000 fish were physically collected from the 16 mesocosms, ranging from 14,000 to 22,000 per mesocosm (7 to 14 kg wet weight). All were bluegills except for a very small number of top minnows (*Gambusia sp.*: 14 in one mesocosm, 1 in another). The fish looked healthy, and visual examination found only one specimen with an abnormality. Tadpoles were harvested at the same time, with numbers varying enormously, from 0 to 12,000 per mesocosm, the latter weighing 37 kg (over 4 times the weight of fish in this mesocosm) (Table 5).

The fish condition at harvest was calculated from the total weight and numbers of fish in each size group, in comparison to values published by Swingle (1949). The adult fish in all mesocosm groups had condition factors generally between 0.94 and 1.0, close to Swingle's optimum value for small farm ponds. Similarly, there was no difference between condition factors for the young fish (10 cm or less) size groups in any of the treatments or control. However, the young fish values ranged generally from 0.75 to 0.85, somewhat below the optimum, and in this insistence probably indicative of overcrowding in control and treatment mesocosms.

The data for the fish and tadpole harvest are presented in Figures 31a and b and Table 5, and are described below.

Adult Fish Numbers and Biomass   Although only 25 fish were stocked in each mesocosm, the numbers of harvested adult fish (greater than 10 cm length) varied between 24 and 47 per mesocosm. The average number in each of the mesocosm groups was 29 to 31. It is probable that the increased number of adults was the result of a small number of bluegill larvae or eggs entered the mesocosm in the pretreatment year at the time of stocking. To achieve adult size in the period available, their growth would have had to have been somewhat faster than the typical rate proposed by Carlander (1977). However, the abundance of food and lack of competition could have made this possible. Examination of the internal organs of adult fish at harvest showed there to be an average ratio of females to males of approximately 2:1 in all mesocosm groups.

Table 5. Fish and Tadpole Harvest Data

| Parameter[a] | Mean of Four Replicate Mesocosms | | | | Standard Error of Mean | Control vs. Treated Groups | | | | | | Statistical Comparison of Low, Mid, and High Groups |
| | Control | Low | Mid | High | | % Difference from Control[b] | | | Statistical Analysis[c] | | | |
						Low	Mid	High	Low	Mid	High	
Number												
Young fish	16,113	18,781	19,338	18,918	1796	+17	+20	+17	ns	ns	ns	ns
Adult fish	29[d]	31	30	29	3.5	+6	+3	0	ns	ns	ns	ns
Adult female fish	21[d]	18	20	18	2.5	−1	+8	−4	ns	ns	ns	ns
Young/adult fish	556[d]	611	639	652	109	+10	+15	+17	ns	ns	ns	ns
Young/adult female	718[d]	1,015	955	1,051	191	+19	+12	+24	ns	ns	ns	ns
Fish size groups					Size group data are presented in Figure 31							
Tadpoles	2,258	3,970	3,656	3,030	2,076	+75	+62	+34	ns	ns	sl	ns
Biomass (Kg)												
Young fish	9.76	6.05	6.16	7.03	0.93	−38	−37	−28	sl	sl	sl	ns
Adult fish	2.47	2.78	2.54	2.47	0.17	+12	+3	0	ns	ns	ns	ns
Fish size groups					Size group data are presented in Figure 31							
Tadpoles	8.88	13.94	16.79	16.66	7.45	+57	+89	+88	ns	ns	ns	ns
Young fish & tadpoles	18.64	19.99	22.95	23.69	7.21	+7	+23	+27	ns	ns	ns	ns
All fish & tadpoles	21.11	22.77	25.49	26.17	7.17	+8	+20	+23	ns	ns	ns	ns

[a] Young = ≤10 cm size groups; adult = >10 cm size groups.
[b] + values = greater than control; − values = less than control.
[c] ns = not significantly different; sl = significantly lower; ($p = 5\%$).
[d] Values from only three of mesocosms, as data not recorded for fourth.

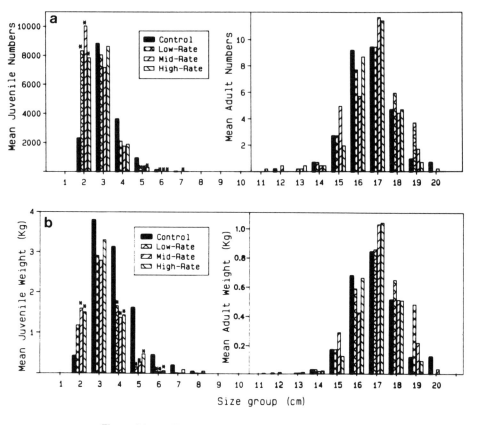

Figure 31.   Bluegill sunfish. (a) Numbers. (b) Biomass.

There were no significant differences betweens the control and treated groups in either numbers or biomass of total, female or male adult fish.

Total Young Fish and Tadpole Numbers and Biomass   Extremely large numbers of young fish were harvested from all mesocosms, and none of the lambda-cyhalothrin treatments were significantly different from the control. However, 17 to 20% more fish were present in the 3 treatment groups compared to the control. The biomass of young fish in the treated mesocosm groups was 28 to 38% less than in the control, this difference being statistically significant. The reduced biomass in the lambda-cyhalothrin mesocosms could well have resulted from the greater numbers of young and the subsequent increase in competition for food or habitat reducing fish growth rates.

There were also substantially more tadpoles in the treated mesocosm groups, giving increases of 34 to 75% by number and 57 to 89% by weight, compared to controls. These differences between treatment and control groups were not significant, due to the extremely high variability in the replicate mesocosms. In presenting this data, it is recognized that no account can be taken of tadpoles that have left the mesocosms during the year by metamorphosing into adults.

Tadpoles during their development are initially herbivorous, feeding on algae. As they start to metamorphose, the mouth develops horny plates and feeding changes to a more carnivorous diet. The majority of the tadpoles in the mesocosms were a large species and included the bullfrog (*Rana catesbeina*). Their average weights ranged from 4 to 24g, equivalent to bluegill sunfish of 6 to 11 cm in length. It is therefore likely that the tadpoles competed with bluegills for food and/or habitat, and more so in the treated mesocosms where tadpole numbers were highest. It would thus appear even more likely that the reduced fish biomass in the treated mesocosms resulted from increased competition.

In terms of total biomass of young fish plus tadpoles, the treated mesocosms were 7 to 27% higher than the controls, but this was not statistically significant.

Young Fish Size Groups Division of the young fish into size groups demonstrates how the overall differences in numbers and biomass between control and treated groups was expressed (Figure 31a and b). In all three lambda-cyhalothrin treated mesocosm groups, the 4- to 6-cm fish groups contained significantly fewer fish numbers and a lower biomass than the controls. However, in the treated mesocosm groups, the 2-cm group contained significantly more fish and a greater biomass than the equivalent control mesocosm group.

Interpretation of Fish Data If the test chemical was the cause of the reduced growth rate of young fish, the mechanism must have been either direct or indirect (or a combination of these). There appears to be good reasons to strongly doubt whether either of these processes could have occurred.

The potential for direct effects can be considered by comparing the concentration of lambda-cyhalothrin in the mesocosms with laboratory toxicity data for fish. The data are shown in Table 6B. The laboratory no-observed-effect-concentration (NOEC) for fish reproduction and survival is 30 ng/l and for growth is $\geqslant 130$ ng/l (ZENECA, unpublished data). Therefore reproduction and survival are more sensitive parameters, to lambda-cyhalothrin, than is growth of the young fish. As there was no reduction in numbers of young fish in the treated mesocosms, it therefore seems unlikely that the test chemical could have influenced the fish growth rates.

Furthermore, the range of concentrations of lambda-cyhalothrin in the mesocosm water was such that it is improbable that a direct effect could have occurred on any fish parameter at all three of the treatment rates. Table 6a demonstrates that at the high rate the estimated mean total concentration of the pyrethroid in the water was approximately 50 ng/l. However, adsorption has been shown to reduce bioavailability by up to two orders of magnitude (Hill, 1985). Therefore even at the highest rate in the mesocosm study, the biological activity of lambda-cyhalothrin was less than the laboratory fish life cycle NOEC. In the low-rate mesocosms, the bioavailable concentration was over two orders of magnitude below the overall laboratory NOEC and over three orders below that for fish growth. Even the nominal application dose in the low-rate mesocosms was one order of magnitude below the overall NOEC. This reinforces the proposal that the pyrethroid would not have had any direct effect on fish at all three treatment rates in the mesocosm study.

Table 6. Comparison of (A) Concentrations of Lambda-Cyhalothrin in Mesocosms With (B) Laboratory Fish Toxicity Data

| | | Concentration of Lambda-Cyhalothrin in Water (ng/l) | | | |
| | | Nominal at Application[a] | | Estimated Mean over Application Period[b] | |
(A)	Mesocosm Lambda-Cyhalothrin Rate	Spray Drift	Runoff	Total	Bioavailable
	High	160	470	~50	≥10
	Mid	16	47	~5	≥1
	Low	1.6	4.7	~0.5	≥0.1

| | | Response (Measured ng/l) | |
(B)	Fish Laboratory Test[c] and Parameter	LC_{50}	NOEC[d]
	Bluegill acute, 96 h	200	30
	Fathead minnow full life cycle		
	Egg production		60
	Egg hatching		60
	28- and 56-day young survival		30
	150 to 300 day survival		≥130
	28, 56, and 300 day growth		≥130

[a] If all of applied dose evenly mixed in water.
[b] In solution or adsorbed.
[c] In clean water under maintained concentrations of test chemical.
[d] No observed adverse effect concentration.

The possibility of an indirect effect of lambda-cyhalothrin on mesocosm fish can also be examined, in relation to available food sources. There were very different levels of effect on zooplankton and macroinvertebrate populations at the three treatment rates, as described ealier in this paper and summarized in Table 7. At the high rate a substantial impact was observed on the many of the invertebrate groups. At the midrate, the effect was much reduced, and at the low rate was negligible. With regard to the diversity of zooplankton food supply for the larval fish, there was no difference between control and the mid or low rates for either the total population or the component taxa. Consequently, fish in the low rate and in the control mesocosms had a virtually identical food supply, whereas at the high rate the distribution of food organisms was substantially changed. As there was no significant difference in the fish populations between any of the three mesocosm treatment rates, it is unlikely that an indirect (food supply) effect of lambda-cy-halothrin occurred.

Only bluegill sunfish were used to stock the mesocosms, and no predator fish such as bass were included. Thus all the mesocosms became overcrowded with young fish, and some also with tadpoles. The reason for the greater number of young fish in the three treatment groups (with no difference between these) is unclear. The numbers of the young fish harvested result from a balance between birth and survival, and there is little doubt that, in the test mesocosms, predation by adult bluegill on young fish will have had a major influence on the survival rate. In this regard it is interesting to note that the three treatment groups had somewhat more macrophyte cover than the control group (Figure 23), and it can

Table 7. Summary of the Effects of Lambda-Cyhalothrin on Mesocosms: Physicochemical and Biological Parameters

Parameter	Effect, in Comparison to Control Mesocosms[a]		
	Low Rate	Medium Rate	High Rate
Physicochemical	−	−	−
Microbial (hydrosoil)	−	−	−
Phytoplankton & Periphyton			
Cell numbers, volume, biomass	−	−	−
Taxonomic groups	−	−	−
Activity	−	−	−
Filamentous algae	−	−	−
Macrophytes	(+)	(+)	(+)
Zooplankton			
Protozoa	−	−	−
Rotifera	−	−	−
Crustacea	−	−	+ +(r)
Macroinvertebrates			
Turbellaria	−	−	−
Mollusca	−	−	−
Oligochaeta	−	−	−
Hydracarina	−	+(−)	+ +(−)
Ephemeroptera — Baetidae	+(R)	+ +(r)	+ +(r)
Caenidae	−	+ +(r)	+ +(r)
Odonata — Anisoptera	−	−	−
Zygoptera	−	−	+r
Hemiptera — Belostomatidae	−	−	+ +(−)
Gerridae	−	+ +(r)	+ +(nr)
Notonectidae	−	−	+ +(−)
Veliidae	−	+ +(−)	+ +(−)
Coleoptera — Hydrophilidae	−	−	+(−)
Haliplidae	−	−	+ +(−)
Trichoptera — Leptoceridae	−	+(R)	+ +(nr)
Diptera — Ceratopogonidae	−	−	+(R)
Chironominae	−	−	−
Tanypodinae	−	+(R)	+ +(r)
Fish (*Lepomis macrochirus*)			
Activity	−	−	−
Numbers	(+)	(+)	(+)
Weight	(+)	(+)	(+)

[a] −, No effect; +, minor effect; + +, major effect; (+), different to control, with no dose-response, and unlikely to have been caused by lambda-cyhalothrin; (R), full recovery by study end; (r), partial recovery; (nr), no recovery; (−), not possible to judge recovery.

be speculated that this additional refuge resulted in less predation and greater numbers of young. However, the study data and the inherent variability does not enable interpretation of the complex interactions between fish and tadpoles and their competition for food or habitat, or of the problems resulting from predation or overcrowding. While the differences between treated and control mesocosms cannot unequivocally be attributed to specific nonchemical factors, it is possible to conclude that it is highly unlikely that they were due to either direct or indirect effects of the test chemical.

DISCUSSION AND CONCLUSIONS

The lambda-cyhalothrin mesocosm study described in this paper was the first of several pesticide studies done for regulatory submission to the U.S. EPA. Conclusions, regarding lambda-cyhalothrin, are drawn below, from the study data, but the opportunity is also taken to discuss methodological aspects of this type of investigation.

Methodology

Higher tier aquatic studies (for example, single-species laboratory life cycle and multispecies field tests) with pesticides are required by the EPA on a case-by-case basis, to determine whether or not a product may cause adverse effects in natural aquatic environments when lower tier single-species laboratory toxicity tests do not allow this judgment to be made. A wide variety of multispecies aquatic testing procedures have been developed for determination of pesticide fate and/or effects, ranging from laboratory model ecosystems and microcosms to field mesocosms of many types and sizes (Lundgren, 1985; Solomon and Liber, 1988; Hill, 1985, 1989). The laboratory-based tests are of relatively low cost and enable a large number of experimental variables to be investigated. However, at present, these systems do not adequately represent the complexity of interactions occurring in field communities and it is not possible to confidently extrapolate from laboratory to field. Field mesocosms, with sets of replicate units, offer a potentially more acceptable alternative, in that all trophic levels can be incuded, together with diverse populations.

There are a wide variety of mesocosm designs already developed. While both "enclosed" (e.g., pond enclosures) and "flowing" (e.g., artificial streams) systems are feasible, the latter are least well developed and understood. In regard to the interpretation of mesocosm data relative to natural environments, it seems possible that extrapolation from an "enclosed" system to both static and flowing bodies of water will be somewhat easier than from a "flowing" design. Furthermore, "enclosed" systems would seem to offer a more severe test of ecosystem effects by pesticides, in that organisms in them will generally be exposed to these "pulse-entry" chemicals for longer periods of time. Therefore, further discussion here will relate only to mesocosms based on "enclosed" systems such as ponds and lakes. Solomon and Liber (1988) and Hill (1989) have reviewed the use of, and data from, purpose-built ponds and isolated enclosures within large natural bodies of water. The mesocosm design currently preferred by the EPA uses large (≥ 0.04 ha, 400 m^2) soil- or hydrosoil-lined units. However, the methodology is still in its infancy, with many factors requiring to be modified or investigated further. These include:

- Size of mesocosm. Pesticide effects have been studied in mesocosms varying in size from <5 to 800 m^2, although until recently without side-by-side comparisons of different size units. Studies reviewed by Hill (1985, 1989) do suggest that mesocosms of 25 m^2 may be just as sensitive to pesticides as the much larger units.

Recent studies in even smaller systems are presented in this publication and should help to define the sizes of mesocosms suitable for regulatory studies of the aquatic impact of pesticides.

- Numbers of replicates. This is infuenced by a variety of factors, including operational logistics, cost, statistical procedures, and biological variability. Some of these factors are addressed in this publication, by Shaw et al. (1993). Current studies have, in general, used three to four replicates.
- Study endpoints. In mesocosm studies for regulatory use, the EPA have to date required an extensive range of chemical, physicochemical, and biological variates to be measured at regular intervals. Present experience suggests that a more cost-effective process would be to use laboratory data to identify toxicological effects of potential concern, and to design the mesocosm sampling program towards the indicated structural and functional parameters.
- Biological composition. That the mesocosm system should incude several trophic levels, with diverse populations of phytoplankton, zooplankton and macroinvertebrates, is unlikely to be disputed for studies of the type described in this chapter. However, opinion on fish populations is divided. The present investigation demonstrated that inclusion of a single species (bluegill sunfish) of mature fish can lead to overcrowding by the progeny, with consequent effects on food organisms and the young fish themselves. The author's view is that, at present, regulatory mesocosm studies involving fish reproduction cannot be recommended due to their complexity, the unnatural interactions that result, and the difficulties in interpreting the data. A number of alternative approaches are possible:

  - Do not include fish, and judge impact from laboratory fish data and mesocosm effects on lower trophic levels
  - Use much fewer mature fish, although this may well increase variability
  - Use a species of fish that reproduces less prolifically than the bluegill
  - Include a predator fish, such as largemouth bass (*Micropterus salmoides*), in addition to the bluegill. This is possibly the best potential option, but insufficient data has been published to determine either the optimum species ratio or the resulting variability

- Treatment rates. The rates of application chosen to simulate natural field spray drift entry into aquatic ecosystems are presently selected from a relatively small database with insufficient information on any one product type and use pattern. Runoff entry concentrations are mostly determined using mathematical models such as SWRRB (Computer Sciences Corporation, 1990), which has not yet been sufficiently calibrated to accurately reflect the natural field process. In the majority of studies for the EPA, three rates have been applied. The middle rate has been determined from the typical MEER, thus by definition approaches the worst case. A higher and a lower rate are then selected. However, an alternative option is that the typical MEER is used to define the highest of the three mesocosm treatment rates (Hill et al., 1993).
- Interpretation of results. For pesticides applied several times per year (such as lambda-cyhalothrin), multiple applications are made to the mesocosms, with each at the "maximum" rate selected (i.e., each midrate application relating to the typical MEER). Furthermore, the mesocosms are relatively small bodies of water

treated over the whole surface at the same rate, whereas the natural field environments they represent are exposed to a gradient of concentrations across the water body. Consequently, effects seen in mesocosms are likely to be much more severe than those that might occur in natural aquatic environments following agricultural use. Methods of extrapolating from the experimental mesocosms to the field have yet to be considered and developed.

- Recovery and reinvasion by organisms in field environments inevitably follow natural and 'man-made' perturbations. This is presently difficult to simulate in mesocosms.

The above issues have been under debate or investigation. Mesocosm design, biological endpoints, statistical interpretation, and runoff entry rates were considered in the U.S. by the "Aquatic Effects Dialogue Group" (see Introduction), while in Europe workshops were held to address similar issues. In the U.S. the "Spray-Drift Task Force" formed in 1990 by 23 companies, and involving industrial, government, and academic specialists, has embarked on a 5-year program of practical studies to produce a comprehensive spray drift database. Substantial programs of work by industry and/or government institutions are also underway in other countries, for example, Canada and Germany.

While the authors support the concept of mesocosm studies, they also believe that it is possible to use much smaller systems than those reported here, to predict the field behavior of pesticides. In Europe, mesocosms are generally less than 100 m², and commonly no larger than 25 m². Increasing experience may enable us to better understand the effects of system size and to provide confidence in data derived from simpler (and less costly) studies. This will be easiest within "families" of chemically related pesticides, where the individual products exhibit similar spectra of aquatic effects.

Lambda-Cyhalothrin

The mesocosms used for the lambda-cyhalothrin study showed a high degree of similarity prior to treatment, with regard to both physicochemical and most biological parameters. Diverse populations developed, with organisms typical of lentic water communities.

All the pyrethroid applications were close to the nominal rate and evenly applied to the whole water surface. Residues in the water declined rapidly after both spray drift and runoff applications, with less than 20% remaining after 3 days. This is similar to data reported for other pyrethroids (Hill, 1985). The hydrosoil residues increased during the application period, with an indication of a decline following this, although the study was too short to determine a half-life.

The different components of the mesocosm showed widely differing responses to the test chemical (summarized in Table 7) reflecting the relative sensitivities previously observed in laboratory single-species tests (Hill, 1985). Several of the parameters measured were unaffected at any of the three treatment rates. These included all physicochemical properties, hydrosoil microbes, phytoplankton, periphyton, and filamentous algae.

The overall zooplankton numbers were not significantly affected, even at the high rate where copepod numbers were substantially reduced. There were no effects seen on the Protozoa, Rotifera, or the crustacean Cladocera and Ostracod at any rate of lambda-cyhalothrin, and copepods were unaffected at the mid and low rates. The crustacean populations did, however, undergo a decline in treated and control mesocosms in early June, soon after the treatments started. Two mechanisms possibly responsible were fish predation (fish had only recently been introduced) and inhibition of filter-feeding by the soil-water slurries used to simulate runoff.

While a few macroinvertebrate taxa were not affected at any rate, many others were affected severely at the high rate. A few effects were still apparent at the mid rate and virtually none at the low rate. The groups that were most sensitive lived in the water column (e.g., Ephemeroptera, Zygoptera, Chironomidae: Tanypodinae) or on its surface (e.g., Gerridae). Those macroinvertebrates living within the hydrosoil were unaffected at all rates (i.e., Oligochaeta, Chironomidae: Chironominae). Many of the macroinvertebrates in treated and control mesocosms were also reduced in numbers soon after the fish were introduced.

In all three lambda-cyhalothrin treatment groups there were substantially more young fish, with significantly less biomass, than in the controls. However, there was no difference between the three treated groups. Consideration of this absence of a dose-response, together with the major differences in fish-food organisms in the mesocosms at the three rates and laboratory fish full life cycle NOELs, suggests that the differences seen in the fish populations were not due to the test chemical. No unequivocal explanation of the changes can be made, but possible reasons exist. The increase in numbers of young fish in the treated mesocosms could have been due to the somewhat larger beds of macrophytes present in these mesocosms providing extra refuge from adult fish predation. The reduced biomass may have resulted from increased competition for food or habitat by the additional numbers of young fish and/or by the tadpole populations which were also highest in the treated mesocosms.

The overall data generated and the excessive severity of the mesocosm test system in comparison with the predicted entry of lambda-cyhalothrin into natural bodies of water allows us to conclude that it is unlikely that agricultural use of lambda-cyhalothrin will cause adverse effects on overall populations or productivity in environmental aquatic ecosystems. Some minor effects may occasionally be observed, but these will be transient.

ACKNOWLEDGMENTS

The authors are grateful to the many people who have contributed to this study, especially to P.D. Askew, E. Bolygo, W. Cody, J.F.H. Cole, G. Cullen, D. Farmer, E. Farrelly, P.D. Francis, S.T. Hadfield, M.J. Hamer, V. Kennedy, M. Moore, and J.L. Neal.

REFERENCES

Bougis, P. 1976. *Marine Plankton Ecology.* North-Holland Publishing Co., Amsterdam, Oxford.

Boyd, C.E. 1979. *Water Quality in Warmwater Fish Ponds.* Auburn University Agricultural Experiment Station, 359 pp.

Carlander, K.D. 1977. *Handbook of Freshwater Fishery Biology, Volume 2.* The Iowa State University Press, Ames, Iowa, 431 pp.

Computer Sciences Corporation. 1980. *Pesticide Run-Off Simulator (SWRRB)* — *Users Manual.* U.S. Environmental Protection Agency, Falls Church, Virginia.

Etnier, D.A. 1971. Food of three species of sunfishes (*Lepomis,* Centrarchidae) and their hybrids in three Minnesota lakes. *Trans. Am. Fish Soc.* 100: 124–128.

Ewars, L.A. and M.W. Boesel. 1935. The food of some Buckeye Lake fishes. *Trans. Am. Fish Soc.* 65: 57–70.

Fairchild, J.F., T.W. La Point, J.L. Zajicek, M.K. Nelson, F.J. Dwyer and P.A. Lovely. 1992. Population, community and ecosystem-level responses of aquatic mesocosms to pulsed doses of a pyrethroid insecticide. *Environ. Tox. Chem.* 11: 115–129.

Flemer, D.A. and W.S. Woolcott. 1966. Food habits and distribution of the fishes of Tuckahoe Creek, Virginia, with special emphasis on the bluegill, *Lepomis macrochirus* Rafinesque. *Chesapeake Sci.* 7: 75–89.

Franson, M.A.H. 1985. *Standard Methods for the Examination of Water and Wastewater,* 16th ed. American Public Health Association, Washington, D.C.

Gaarder, T. and M.M. Gran. 1927. Investigations of the production of plankton in Oslo Fjord. *Rapp. Process-Verbaux, Reunions, Cons. Perma. Int. Explor. Mer.* 42.

Hadfield, S.T., J.K. Sadler, E. Bolygo and I.R. Hill. 1992. Development and validation of residue methods for the determination of the pyrethroids lambda-cyhalothrin and cypermethrin in natural waters. *Pestic. Sci.* 34: 207–213.

Healy, M. 1984. Fish predation on aquatic insects. In V.H. Resh and D.M. Rosenburg, eds. *The Ecology of Aquatic Insects.* Praeger, New York, pp. 255–288.

Hill, I.R. 1985. Effects of nontarget organisms in terrestrial and aquatic environments. In J.P. Leahey, ed., *The Pyrethroid Insecticides,* Taylor and Francis, London. pp. 151–262.

Hill, I.R. 1989. Aquatic organisms and pyrethroids. *Pestic. Sci.* 27: 429–465.

Hill, I.R., K.Z. Travis and P. Ekoniak. 1993. Spray-drift and run-off applications of foliar-applied pyrethroids to aquatic mesocosms: rates, frequencies and methods, this volume, Ch. 15.

Karickhoff, S.W. and K.R. Morris. 1985. Impact of Tubificid Oligochaetes on Pollutant Transport in Bottom Sediments. *Environ. Sci. Technol.* 19: 51–56.

Lind, O.T. 1979. *Handbook of Common Methods in Limnology.* The C V Mosby Co., St. Louis, Missouri.

Lozano, S.J., L. Heinis, S. O'Halloran, M. Knuth and R. Seifert. 1991. Comparison of the distribution and effects of three pesticides in littoral enclosures, submitted.

Lundgren, A. 1985. Model ecosystems as a tool in freshwater and marine research. *Archives fur Hydrobiol. Suppl.* 70: 157–196.

Martin, J.R. and S.J. Klaine. 1987. The Effects of Suspended Solids on the Partial Life-cycle Toxicity of Atrazine to Daphnia pulex. *SETAC 8th Annual Meeting; Environmental Risk: Recognition, Assessment and Management.* Pensacola, Florida.

Mayasich, J.M., J.H. Kennedy and J.S. O'Grodnick. 1993. Evaluation of the effects of tralomethrin on aquatic ecosystems, this volume, Ch. 26.

McCabe, G.D. and W.J. O'Brien. 1983. The effect of suspended silt on feeding and reproduction of *Daphnia pulex. The American Midland Naturalist* 101: 324–337.

McConnell, W.J. 1962. Productivity relations in carboy microcosms. *Limnol. Oceanogr.* 7: 335–343.

Moffett, J.W. and B.P. Hunt. 1943. Winter feeding habits of bluegills, *Lepomis macrochirus* Rafinesque, and yellow perch, *Perca flavescens* (Mitchell), in Cedar Lake, Washtenaw County, Michigan. *Trans. Am. Fish Soc.* 73: 231–242.

O'Brien, W.J. and F. de Noyelles Jr. 1974. Relationship between nutrient concentration, phytoplankton density and zooplankton density in nutrient enriched experimental ponds. *Hydrobiologia* 44: 105–125.

O'Brien, N.J. 1979. The predator-prey interactions of fish and zooplankton. *Amer. Sci.* 67: 572–581.

Pace, M.L. and J.D. Orcutt Jr. 1981. The relative importance of protozoans, rotifers and crustaceans in a freshwater zooplankton community. *Limnol. Oceanogr.* 26: 822–830.

Seaburg, K.G. and J.B. Moyle. 1964. Feeding habits, digestive rates and growth of some Minnesota warmwater fishes. *Trans. Am. Fish Soc.* 101: 219–225.

Shaw, J.L. M. Moore, J.H. Kennedy, and I.R. Hill. 1993. Design and statistical analysis of field aquatic mesocosm studies, this volume, Ch. 9.

Solomon, K.R. and K. Liber. 1988. Fate of pesticides in aquatic mesocosm studies — an overview of methodology. *Brighton Crop Protn. Conf., Pests and Diseases,* 1988: 139–147.

Swingle, H.S. 1949. Experiments with combinations of largemouth black bass, bluegills and minnows in ponds. *Trans. Am. Fish Soc.* 76: 46–62.

Touart, L.W. 1988. Aquatic mesocosm tests to support pesticide registrations. Hazard Evaluation Division Technical Guidance Document, EPA-540/09-88-035. U.S. Environmental Protection Agency, Washington, D.C.

Touart, L.W. and M.W. Slimak. 1989. Mesocosm approach for assessing ecological risk of pesticides. In J. Reese Voshell, ed. *Using Mesocosms to Assess the Aquatic Ecological Risk of Pesticides: Theory and Practice,* Ento. Soc. Amer., Misc. Publicn. MPPEAL, No. 75, pp. 33–40.

Turner, W.R. 1955. Food habits of the bluegill *Lepomis macrochirus* (Rafinesque), in eighteen Kentucky Farm ponds during April and May. *Trans. Kentucky Acad. Sci.* 16: 98–101.7.

Vollenweider, R.A. 1974. *A Manual of Methods for Measuring Primary Productivity* in *Aquatic Environments,* 2nd ed. IBP Handbook No. 12. Blackwell Scientific Publications, London.

Whitly, L.S. 1982. Aquatic Oligochaeta. In A.R. Brigham and W.V. Brigham, eds. *Aquatic Insects and Oligochaetes of North and South Carolina.* Midwest Aquatic Enterprises, Manomet, Illinois, pp. 2.1–2.29.

Zepp, R.G. and D.M. Cline. 1977. Rates of Direct Photolysis in Aquatic Environment. *Environ. Sci. Technol.* 11: 359–366.

CHAPTER 25

The Fate and Effects of Guthion (Azinphos Methyl) in Mesocosms

Jeffrey M. Giddings, Ronald C. Biever, Raymond L. Helm, Gregory L. Howick, and Frank J. deNoyelles, Jr.

Abstract: We measured the ecological effects of an organophosphorus insecticide, Guthion, in 12 tenth-acre pond mesocosms. The ponds received Guthion at five exposure levels selected to bracket the toxicity range for aquatic biota. Treatments were repeated eight times at weekly intervals in a manner simulating agricultural runoff. Guthion residues disappeared from the water nearly completely within 1 week of each application. We interpreted Guthion effects by examining exposure-response relationships to identify a response threshold (the EC_0, determined by linear regression) for each taxon. Overall, fish and certain groups of insects were the most sensitive components of the ecosystem to Guthion. Other insects, as well as cladocerans and rotifers, were less sensitive, but higher Guthion exposure levels reduced their abundance. Copepods, snails, and plants were unaffected, and some of these groups increased in response to the loss of fish and invertebrate predators and competitors. Mesocosm results generally agreed with laboratory toxicity data.

INTRODUCTION

Guthion (azinphos methyl) is an organophosphorus insecticide used to control many pests on field crops, vegetables, fruits, nuts, and ornamental and shade trees. In laboratory toxicity tests, Guthion is acutely toxic to fish and freshwater invertebrates at concentrations from 2.7 to 40 µg/l (Table 1). Chronic effects occur at

0-87371-592-6/94/$0.00 + $.50

Table 1. Toxicity of Guthion (Azinphos Methyl) to Aquatic and Marine Organisms, as Measured in Laboratory Toxicity Tests

Acute Toxicity	Ref.
Mysid shrimp, LC_{50} = 0.22 to 0.23 µg/l	6,7
Sheepshead minnow, LC_{50} = 1.9 to 2.7 µg/l	8,9
Rainbow trout, LC_{50} = 3.0 µg/l	10
Daphnia magna, LC_{50} = 1.1 to 4.4 µg/l	11,12
Stoneflies, LC_{50} = 8.5 to 22 µg/l	13
Bluegill sunfish, LC_{50} = 40 µg/l	14
Green algae, EC_{50} = 3610 µg/l	15
Eastern oyster, EC_{50} = 4700 µg/l	16

Chronic Toxicity	Ref.
Daphnia magna LOEC = 0.40 µg/l	17
Fathead minnow LOEC = 0.51 µg/l	18
Rainbow trout LOEC = 0.98 to 1.1 µg/l	19,20

0.4 to 1 µg/l. Laboratory results, however, may not accurately indicate the potential for effects of pesticides under field conditions. The Environmental Protection Agency (EPA) required a mesocosm study with Guthion, to help assess the environmental safety of the product under the Federal Insecticide, Fungicide, and Rodenticide Act (FIFRA). Springborn Laboratories, with assistance from the University of Kansas, conducted the mesocosm study in 1987 to 1988 at the Kansas Aquatic Mesocosm Facility near Lawrence, KS.

The study generally followed the EPA's mesocosm guidelines.[1] We established 12 ponds containing water, sediment, and a biological community including fish. After allowing the ponds to develop for 8 months, we treated them with Guthion and monitored their response. We compared treated ponds with untreated controls to evaluate the ecological effects of Guthion exposure.

Because Guthion is used for a variety of crops under a wide range of conditions, expected environmental concentrations (EECs) vary widely. We applied Guthion at five different rates, with two replicate mesocosms at each treatment level plus two untreated controls. These treatment regimes created a gradient of exposure levels representing different agricultural situations. We expected the highest treatment level to produce acute effects, and the lowest level to have no ecological impact. Five experimental treatment levels were sufficient to define the exposure-response relationships for important ecosystem components and for the ecosystem as a whole. This would allow prediction of potential effects over a range of EECs.

This paper describes the ecological responses measured over the gradient of pesticide exposure levels. We used a combination of approaches to analyze and interpret the mesocosms results: hypothesis testing (Williams' Test), linear regression, and professional judgment. We also compared the mesocosm results with laboratory data on acute and chronic toxicity of Guthion to single species.

METHODS

Mesocosm Construction

The mesocosms were located at the Nelson Environmental Study Area outside Lawrence, KS. The ponds were square, approximately 20 m on a side and 400 m² (0.1 acres) in area, with a volume about 400 m³ when filled to an operating depth of 2 m. The sides of the ponds had a 2:1 slope. The pond bottoms were level and were constructed of packed clay. Approximately 30 cm of natural sediment accumulated in each pond between 1977 to 1979, when the ponds were constructed, and November 1987.

We renovated the ponds for this study in November 1987. Several inches of accumulated sediment were removed and replaced with sediment from a nearby farm pond. The source pond had not received pesticide runoff or drift for at least 10 years, and the sediment contained no pesticides at detectable levels. After the new sediment was in place, the mesocosms were filled with aged pond water (also pesticide-free) from a reservoir on the site. Added to each pond in late May 1988 were 50 adult bluegill sunfish (*Lepomis macrochirus*), with an average weight of 59 g. The sex of the bluegill was not determined.

During the late spring and early summer, dense macrophyte beds (primarily *Potamogeton* spp., *Chara* sp., and *Najas* sp.) grew throughout the mesocosms. The growth of *Potamogeton* was greatest in the areas where the farm pond sediment had been spread (the deep central portion of each pond), while *Chara* and *Najas* were prevalent in the shallow areas.

Ecological Monitoring

Chemical and biological monitoring was conducted from November 1987 through October 1988. Measurements included water quality (temperature, conductivity, dissolved oxygen, pH, hardness, alkalinity, suspended solids, total Kjeldahl nitrogen, and total phosphorus); phytoplankton pigments and taxonomic abundance; periphyton pigments, dry weight and taxonomic abundance; macrophyte biomass and distribution; zooplankton abundance; benthic macroinvertebrates; emergent insects; and numbers, lengths, and weights of surviving bluegill sunfish. Only zooplankton, benthic macroinvertebrates, emergent insects, and fish are discussed in this paper. Methods used for these groups were as follows:

Zooplankton

Zooplankton samples were taken every 4 weeks from November 1987 until April 1988, and at weekly intervals thereafter. Vertical (bottom-to-surface) hauls were collected after dark using a 20 cm diameter, 80-μm mesh plankton net. Usually, two tows were collected at each of two sampling zones (one shallow, one deep) in each mesocosm. Tow depths were recorded to allow calculation of the water volume sampled. Samples were preserved with 4% w/v sucrose-formalin. Zooplankton were identified (usually to species) and counted using a zooplankton counting wheel. During the weekly sampling phase, samples taken every other week were examined and the others were saved for possible future analysis.

Benthic Macroinvertebrates

Benthic samples were collected from two locations in each pond (one shallow and one deep) using a standard 230-cm^2 Ekman dredge. To reduce the disturbance to the benthic community, only half of the ponds were sampled at each monthly sampling event. Three samples taken at each location were combined in a plastic tub. The samples were rinsed through a 250-μm sieve using tap water. Material retained in the sieve was preserved with 70% ethanol. Organisms in the samples were sorted from debris and counted and identified (usually to genus) under a dissecting or compound microscope.

Emergent Insects

Beginning in April 1988, emergent adult insects were collected using semisubmerged inverted funnel traps described by Howick and Ferrington.[2] Two traps were placed in each pond, one in the shallow area and one in the deep area. Each trap consisted of a clear plastic underwater pyramid (75 × 75 cm base) rising to a floating 10-cm diameter PVC cylinder emergence area covered with a plate coated with adhesive (Bird Tanglefoot). Pupae ascending beneath the traps were funneled into the floating cylinder and, as emergent adults, became trapped on the adhesive when they attempted to disperse from the water surface. The adhesive-coated plates with captured insects were removed and replaced with fresh plates weekly. The collected plates were placed in individual storage containers and returned to the field laboratory for processing. Adhesive and insects outside the emergent insect zone (the area over the PVC cylinder) were scraped from the plates and discarded. The plates were then soaked in paint thinner for approximately 12 h to dissolve the adhesive. The insects were removed from the paint thinner and placed in glass vials with 95% ethanol.

Because odonates and some other insects normally emerge by crawling out of the water on a solid surface (such as a macrophyte shoot), they are rarely found in samples from the emergence traps. To estimate the relative emergent rates for these insects, vertical screens consisting of 1 cm × 1 cm mesh hardware cloth were placed in shallow water near the north and south shore of each mesocosm. The screens had a surface area of 0.75 m^2 and extended from the sediment to approximately 30 cm above the water surface. Insect exuviae were removed from the screens weekly and preserved in 70% ethanol for identification and enumeration.

Emergent insects were counted and identified (usually to species) under a dissecting microscope.

Fish

Four baited minnow traps were placed in each pond (one in each corner) for 24-h periods from three to seven times per week beginning on July 12, 1988. The fish collected in the traps were gently transferred to a shallow tray containing water and a ruler. The fish and ruler were photographed for later determination of individual lengths, and the fish were immediately returned to the ponds from which they were collected. The total length (head to tip of tail) of each fish was determined from the photographs using a caliper. The ruler in the photograph was used to measure the distance between caliper ends. Fish lengths were recorded to the nearest 0.1 cm.

At the termination of the study, the mesocosms were drained and all remaining fish were collected. Adult fish were weighed, measured, and dissected to determine their sex. The combined weight of all juveniles was determined for each pond, and a random sample of 100 juvenile fish from each pond were individually weighed and measured.

Treatment with Guthion

The method and frequency of Guthion treatment were selected to represent agricultural use on cotton. The Guthion label for cotton allows up to six applications at weekly intervals during the middle of the growing season (July and August). Each application on the crop presents a risk of spray drift deposition into adjacent water bodies. Additional input to water bodies may come as surface runoff during and after rainstorms. Computer simulation of field runoff showed that Guthion might be transported into aquatic systems six or eight times in a typical year (Peter Coody, Mobay Corporation, personal communication). The combination of spray drift and surface runoff could therefore result in irregular pulses of Guthion into a pond.

To represent this exposure pattern in a generalized way, we added Guthion to the mesocosms eight times at weekly intervals between July 19 and September 6, 1988. Because the computer model predicted that most Guthion input would occur as surface runoff rather than spray drift, we simplified the treatment regime by simulating only runoff, not drift.

The system used to deliver the pesticide to the water is shown in Figure 1. Water was pumped from one point in each pond into a mixing tank. A stock solution of Guthion 35 WP (a wettable powder formulation containing 29% active ingredient) in pond water was pumped from a reservoir into the mixing tank where it mixed with the recirculating pond water. The mixture flowed back to the pond at 3 points, entering just below the surface about 1 m from shore. The water recirculation continued for 6 h, with Guthion added to the mixing tank continuously during the

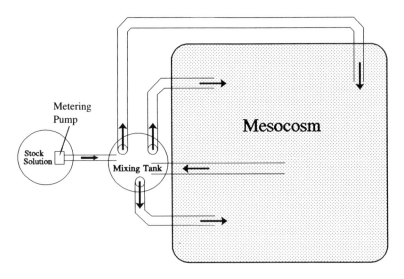

Figure 1. Schematic drawing of system used to deliver Guthion to mesocosms. See text for explanation.

first 4 h of this interval. The total volume of water recirculated during an application event was approximately 57 m³, about 15% of the volume of the pond.

A stock solution of Guthion 35 WP was prepared for each treated mesocosm by adding a measured mass of formulation to a reservoir containing 200 l of pond water. A submersed recirculation pump in the stock solution reservoir kept the solution homogeneous. A second pump transferred the stock solution from the reservoir into the mixing tank at a rate of approximately 50 l/h, while pond water circulated through the mixing tank at a rate of approximately 10,000 l/h. By this procedure, each mesocosm (including the controls) received the same hydrologic treatment, approximating the flow of water into and out of a pond during a typical runoff event. Guthion loading rate — the mass of Guthion added to each stock solution reservoir — was the only variable factor.

We selected the five Guthion treatment levels based on laboratory toxicity test results, as described in the Introduction, and results of a preliminary microcosm study. The intended maximum Guthion concentrations in the water column (the concentrations expected immediately after each application) were 0.056, 0.28, 1.3, 6.7, and 34 µg/l. Two ponds were assigned to each treatment level. Two ponds served as controls; they were treated exactly as the other ponds, but no Guthion was added to the stock solution reservoir.

Azinphos Methyl Analysis

Water and sediment were sampled periodically and analyzed for azinphos methyl and its oxygen analog degradation product. Depth-integrated water column samples were collected from each of the four quarters of the pond and composited. They were extracted by chloroform partition on the day of collection, and the extracts were evaporated to dryness under nitrogen. The dried extracts were shipped overnight to the analytical laboratory, where they were redissolved and analyzed by High Pressure Liquid Chromatography (HPLC).

Sediment samples (5 cm deep) from each quarter of the pond were collected with a stainless steel coring device, composited, frozen, and shipped to the analytical laboratory. Azinphos methyl and its oxygen analog were extracted from the sediment by refluxing with 50% methanol:50% dichloromethane. The extract was partitioned with dichloromethane, dried through sodium sulfate, and evaporated to dryness. The residues were dissolved in toluene, submitted to silica gel column chromatography, and the azinphos methyl residues were collected. The solvent was evaporated to dryness, and the residues were dissolved in acetonitrile:water and analyzed by HPLC.

The water and sediment samples were analyzed against external standards prepared to bracket the expected azinphos methyl concentrations. Concentrations of azinphos methyl and its oxygen analog were determined by linear regression analysis. Quality control samples (water and sediment from control ponds, fortified in the field with known quantities of azinphos methyl and its oxygen analog and extracted concurrently with the pond samples) were used to determine the accuracy of the procedure. With one exception (water samples on Day 38 after application), recovery from quality control samples was within acceptable limits (70 to 120%). Reported concentrations were not corrected for quality control recovery. Detection limits for both azinphos methyl and the oxygen analog were 0.02 to 0.04 μg/l in water and 25 μg/kg in sediment.

Statistical Analysis

The study was designed to allow statistical comparisons between control mesocosms and each of the five Guthion treatment groups by Williams' Test.[3,4] Taxonomic data (abundance of zooplankton, benthic macroinvertebrates, and emergent insects) were natural log transformed before analysis; fish survival, length, and weight data were not transformed. Williams' t statistic was compared with tabulated critical values of $p = 0.05$. The TOXSTAT program (David Gulley, University of Wyoming) was used for the computations.

Linear regression was used to analyze exposure-response relationships for many of the measured parameters. Taxonomic data and average Guthion concentrations were natural log transformed before analysis. For each parameter, the regression was based on the portion of the exposure gradient where the exposure-response relationship was the strongest. In most cases this range included five ponds. The regression equation was then used to estimate the EC_0 — the concentration at which the parameter value was equal to the average for the controls.

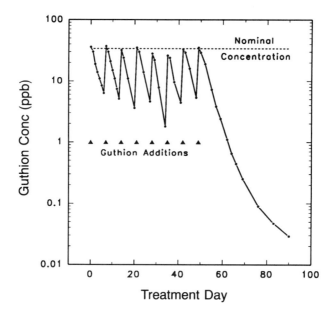

Figure 2.   Measured azinphos methyl concentrations (μg/l) in mesocosm water during and after Guthion treatment. Data are for one pond from the highest treatment level.

RESULTS

Azinphos Methyl Concentrations

Azinphos methyl concentrations in water were highest immediately after each treatment, then declined quickly. The disappearance half-life ranged from 1 to 2 days. Concentrations declined by 90% or more (often to below detection limits) before each subsequent application. The resulting exposure regime was a saw-toothed pattern with eight peaks corresponding to the eight applications (Figure 2). The average of the peak concentrations in each mesocosm ranged from 68 to 98% of the predicted concentrations (Table 2).

Though the peak concentrations were similar between replicates at each exposure level, there were differences in the persistence of Guthion after each treatment. For example, peak concentrations in the two replicates at the highest treatment level averaged 33 μg/l and 26 μg/l, respectively. These mesocosms had average concentrations of 4.8 μg/l and 1.8 μg/l, respectively (15 and 7% of peak concentrations), 6 days after each treatment. The difference reflected small differences in disappearance half lives: for the two ponds, half lives averaged 2.1 days and 1.4 days. The differences in Guthion disappearance rates were related to the pH of the pond water. (Guthion hydrolyzes more rapidly under alkaline conditions than under neutral or acidic conditions. More information on Guthion residues in the mesocosms will be provided in a subsequent article.)

Table 2. Azinphos Methyl Concentrations (µg/l) in Water of Mesocosms Treated With Guthion

Azinphos Methyl Concentration (µg/l)		
Nominal[a]	Mean Peak[b]	55-Day Average[c]
0.056	0.054	0.023
0.056	0.055	0.027
0.28	0.19	0.078
0.28	0.24	0.13
1.3	0.89	0.29
1.3	1.0	0.45
6.7	4.2	1.2
6.7	6.2	3.1
34	26	10
34	33	16

[a] Expected concentration immediately after each Guthion addition, based on quantity of Guthion added and estimated volume of ponds.
[b] Mean of 8 peak concentrations measured within 4 h after each Guthion addition.
[c] Average concentration during the 55-day treatment period, calculated by integrating the area under the concentration vs. time curve for each pond.

The significance of the between-replicate differences in Guthion persistence was evident when the biological data were examined (see below). Between-replicate differences in many of the measured parameters, especially fish survival and growth, corresponded to differences in Guthion persistence. To assist in interpreting the biological results, we calculated the average Guthion concentration in each pond over the 8-week treatment period. This was done by integrating the area under each concentration vs. time curve for Days 0 through 55, and dividing the area by 55. The original experimental design — five loading rates in duplicate — actually resulted in about nine different average concentrations ranging from 0.023 µg/l to 16 µg/l (Table 2).

No measurable concentrations of azinphos methyl were found in the sediment. The oxygen analog degradation product was not detected in the water or the sediment.

Fish

Bluegill began reproducing in the mesocosms before the first Guthion application. A few juveniles were large enough (approximately 2 cm in length) to be caught in the minnow traps in July. Larger numbers of juveniles were trapped in August, September, and October.

No bluegill sunfish were trapped from any of the 3 highest dosed mesocosms (peak concentrations ≥6.2 µg/l, average concentrations ≥3.1 µg/l). Bluegill were caught in all of the other mesocosms. During August, the 3- to 4-cm size class was the most frequently trapped size class in the controls and lower dosed mesocosms (≤0.13 µg/l), and the 2- to 3-cm size class was the most frequently trapped class in the higher dosed mesocosms (0.29 to 1.2 µg/l). Juveniles larger than 4 cm were trapped only in mesocosms with average concentrations ≤0.13 µg/l. In September and October, the 3- to 4-cm size class was the most frequently trapped size class in all of the mesocosms ⩽1.2 μg/l, and juveniles greater than 4 cm were trapped in almost all of the mesocosms ⩽1.2 μg/l.

Dead adult and juvenile bluegill were observed within 1 day of the first Guthion application in the 2 highest dosed mesocosms (peak concentrations 26 and 33 μg/l), and within 2 days in the 2 mesocosms at the next highest dose (peak concentrations 4.2 and 6.2 μg/l). Dead adults began to appear in the mesocosm 2 days after the second Guthion application, with a peak concentration of 1.0 μg/l. By the end of the treatment period, 36 to 40 dead adults had been collected from the 3 mesocosms with the highest average Guthion concentrations (⩾3.1 μg/l), 8 from the mesocosm with an average 1.2 μg/l, 7 from the mesocosm with an average 0.45 μg/l, and 1 from one of the control mesocosms.

When the ponds were drained in November, adult bluegill sunfish were recovered from all mesocosms except the 3 with average Guthion concentrations ⩾3.1 μg/l (Table 3). Only 1 adult was found in the 1.2-μg/l mesocosm, and 23 adults were found in the 0.45-μg/l mesocosm. The other mesocosms contained 36 to 44 adults at the termination of the study. Not all of the originally stocked fish were accounted for; some fish may have become entangled in the macrophyte beds before dying.

The total biomass of adult bluegill at study termination ranged from 2800 to 3400 g in the mesocosms with average concentrations ⩽0.29 μg/l (Table 3). Total adult biomass in the 0.45-μg/l mesocosm was only 1600 g. In the mesocosms with Guthion concentrations ⩽0.13 μg/l, average individual adult weights ranged from 75 to 84 g and average lengths ranged from 15.6 to 16.2 cm. Significantly lower adult lengths and weights occurred in the 0.29- and 0.45-μg/l mesocosms. The single surviving adult in the 1.2 μg/l mesocosm was a large fish, weighing 107 g and measuring 17.5 cm.

No juvenile bluegill were found at the termination of the study in the mesocosms with ⩾3.1 μg/l (Table 3). In the 1.2-μg/l mesocosm, more than 4000 juveniles were harvested, with a total biomass of nearly 900 g, even though only one surviving adult was found in this pond. In the 0.45-μg/l mesocosm, where adult survival was reduced by approximately half, juvenile abundance and biomass were in the same range as the controls. Juvenile abundance and total biomass were highest in the 0.29-μg/l mesocosm. Average weights and lengths of juveniles showed no concentration-related trends, ranging from 0.13 to 0.32 g and from 2.0 to 2.9 cm, respectively, in all ponds where juvenile fish were recovered.

Because spawning began before Guthion treatment, it is possible (not probable) that the juveniles we recovered at the end of the study were all spawned before the first Guthion exposure. If exposure had begun before spawning, effects on juvenile production might have been more severe.

Thus, Guthion treatments that resulted in average concentrations ⩾3.1 μg/l completely eliminated bluegill sunfish from the mesocosms. The treatment resulting in 1.2 μg/l allowed only 1 adult to survive, but the production of juvenile bluegill was nearly half that of the lower dosed ponds. The 0.45-μg/l treatment contained only half of the original number of adults, but juvenile production was similar to controls. The 0.45- and 0.29-μg/l mesocosms contained smaller adults than controls.

Table 3. Numbers, Lengths, and Weights of Bluegill Sunfish Collected From Mesocosms at the Termination of the Study

Nominal Concentration (µg/l)	55-d Average Concentration (µg/l)	Number of Adults	Adult Biomass (g)	Average Adult Weight (g)	Average Adult Length (cm)	Number of Juveniles	Juvenile Biomass (g)	Average Juvenile Weight (g)	Average Juvenile Length (cm)	Total Biomass (g)
Control	0	44	3,377	77	15.6	7,869	1,017	0.13	2.15	4,394
Control	0	39	2,909	75	15.9	9,213	1,917	0.21	2.32	4,826
0.056	0.023	36	2,795	78	15.6	7,050	2,275	0.32	2.92	5,070
0.056	0.027	38	2,890	76	15.6	6,135	1,363	0.22	2.33	4,253
0.28	0.078	37	3,059	83	16.2	11,932	1,989	0.17	2.22	5,048
0.28	0.13	39	3,270	84	16.0	7,902	1,270	0.16	2.14	4,540
1.3	0.29	42	3,051	73[a]	15.2[a]	14,307	2,375	0.17	2.24	5,426
1.3	0.45	23	1,567	68[a]	15.1[a]	9,589	1,297	0.14	2.02	2,864
6.7	1.2	1[a]	107[a]	107	17.5	4,248[a]	882	0.21	2.36	989[a]
6.7	3.1	0[a]	0[a]			0[a]	0			0[a]
34	10	0[a]	0[a]			0[a]	0[a]			0[a]
34	16	0[a]	0[a]			0[a]	0[a]			0[a]

[a] Average for treatment group was significantly less than average for controls, according to Williams' Test ($p \leq 0.5$). (Average length and weight data from the two highest treatment groups were excluded from the analysis.)

Table 4. Summary of Regression Analysis of Mesocosm Data

Dependent Variable	Exposure Range (μg/l)	Model	r^2	n	EC_0 (μg/l)
Adult bluegill number	0.29–3.1	y = 13.36–17.37x	0.84	4	0.20
Adult bluegill biomass	0.29–3.1	y = 957.4–1239x	0.83	4	0.17
Juvenile bluegill number	0.29–3.1	y = 5985–5819x	0.97	4	0.64
Juvenile bluegill biomass	0.29–3.1	y = 977.3–892.4x	0.91	4	0.58
Total bluegill biomass	0.29–3.1	y = 1935–2132x	0.89	4	0.29
Benthic Dipterans	0.45–16	y = 4.677–0.707x	0.92	5	1.26
Emergent Dipterans	0.45–16	y = 6.667–0.557x	0.86	5	1.43
Benthic Chironomidae	0.45–16	y = 4.116–0.741x	0.83	5	1.23
Emergent Chironomidae	0.45–16	y = 5.879–0.695x	0.71	5	0.61
Benthic Chironomini	0.45–16	y = 2.903–0.941x	0.88	5	0.91
Emergent Chironomini	0.29–16	y = 3.990–0.525x	0.74	6	0.26
Benthic Pseudochironomini	0.29–16	y = 1.392–0.994x	0.76	6	0.20
Emergent Tanytarsini	0.45–16	y = 4.842–1.045x	0.76	5	0.42
Benthic Ceratopogonidae	0.45–16	y = 3.131–1.036x	0.75	5	0.98
Emergent Ceratopogonidae	1.2–16	y = 6.552–0.780x	0.85	4	7.74
Benthic Ephemeropterans	0.29–10	y = 1.780–1.425x	0.97	5	0.59
Emergent Ephemeropterans	0.45–16	y = 2.879–1.437x	0.80	5	1.19
Emergent Odonates	0.45–16	y = 4.305–0.743x	0.82	5	16.65

Note: For each regression, the independent variable was the 55-day average Guthion concentration (natural log-transformed). Taxonomic data (all parameters except fish) were natural log-transformed before analysis. For each parameter analyzed, the exposure range where the exposure-response relationship was the strongest was included in the analysis. EC_0 values were calculated using the model; y was set equal to the control mean, and the equation was solved for x.

Williams' Test indicated significant reductions in adult and juvenile numbers, adult biomass, and total bluegill biomass at the two highest treatment levels (average concentrations ≥ 1.2 μg/l). Juvenile biomass was significantly reduced only at the highest treatment level. The average length and weight of adults were significantly reduced at the third highest treatment level (the 0.29-μg/l and 0.45-μg/l mesocosms).

Data on adult bluegill number and biomass, juvenile number and biomass, and total bluegill biomass were subjected to regression analysis. For each variable, a strong exposure-response relationship was apparent in the exposure range between 0.29 μg/l and 3.1 μg/l. Regressions of each variable against the natural logarithm of the 55-day average Guthion concentration resulted in r^2 values ranging from 0.831 to 0.973 (Table 4). The calculated EC_0 values were 0.20 μg/l and 0.17 μg/l for adult bluegill number and biomass, respectively; 0.29 μg/l for total bluegill biomass; and 0.64 μg/l and 0.58 μg/l for juvenile number and biomass, respectively.

Zooplankton

A wide variety of zooplankton were observed in the ponds (Table 5). Abundances varied seasonally, with peaks of most taxa occurring in late spring or early summer. During the Guthion treatment period, the control mesocosms were dominated by rotifers and copepods; cladocerans were relatively scarce.

Table 5. Zooplankton Taxa Identified in Samples Taken During Guthion Mesocosm Study

Copepods	*Alonella exisa*
Diaptomidae	Macrothricidae
Diaptomus pallidus	*Macrothrix* sp.
Cyclopidae	*Ilyocryptus spinifer*
Cyclops bicuspidatus	**Rotifers**
Tropocyclops prasinus	*Filinia* sp.
Mesocyclops edax	*Polyarthra* sp.
Cladocera	*Keratella* sp.
Daphnidae	*Brachionus* sp.
Daphnia ambigua	*Trichotria* sp.
Daphnia pulex	*Platyias* sp.
Simocephalus vetulus	*Asplanchna* sp.
Simocephalus serrulatus	*Monostyla* sp.
Ceriodaphnia reticulata	*Lecane* sp.
Scapholeberis kingi	*Trichocerca* sp.
Sididae	*Cephalodella* sp.
Diaphanosoma brachurum	**Other**
Chydoridae	Ostracoda
Chydorus sphaericus	Amphipoda
Alona sp.	*Hyalella azteca*
Leydigia acanthocercoides	Diptera
Kurzia latissima	*Chaoborus* spp.

The most obvious influence of Guthion on the zooplankton community was a significant increase in copepod abundance in the three highest treatment levels (Table 6). This was probably an indirect effect related to the decline of the bluegill sunfish, which are planktivorous as juveniles. Reductions in the abundance of competitors (rotifers and cladocerans) may also have contributed to the expansion of copepod populations.

Cladocerans were most abundant in the spring (Table 7). They remained present at low densities during the treatment and posttreatment periods in the mesocosms with average Guthion concentrations ≤ 1.3 µg/l. In the two mesocosms with the highest Guthion concentrations, however, no cladocerans were observed after Guthion treatment until the last sample event in October, and in the 3.1-µg/l mesocosm, cladocerans were not observed from the middle of the treatment period until October. These data suggest that cladocerans were affected by exposure to Guthion concentrations at least as low as 3.1 µg/l. However, the reductions were statistically significant only once — in the highest treatment level at the beginning of the treatment period. On the last sample event of the study, cladoceran abundances were significantly greater than contols in the three highest treatment groups.

Rotifer densities were highest at intermediate exposure levels (Table 8). There was a steady decrease in rotifer density with increasing concentration above 0.45 µg/l, and no rotifers were found at the highest exposure level (16 µg/l) at the end of the treatment period. As with the cladocerans, the impact of the higher Guthion exposures on rotifers was difficult to assess. There were no statistically significant differences among treatment levels.

Because there were no consistent exposure-response trends in zooplankton abundance, the data were not analyzed by linear regression and EC_0 values were not calculated.

Table 6. Density of Copepods (Individuals per l, Average For Deep and Shallow Sampling Zones) in Mesocosms Before, During, and After Treatment With Guthion

Concentration		Sample Date (1988)													
Nominal	Average	4/4	5/2	5/16	5/30	6/13	6/27	7/11	7/25	8/8	8/22	9/5	9/19	10/3	10/17
Control	0	16	129	83	44	11	15	7	3	2	1	1	1	5	7
Control	0	42	110	51	12	8	20	7	10	8	3	3	6	10	14
0.056	0.023	75	71	61	46	31	19	20	5	12	7	7	6	5	18
0.056	0.027	53	73	24	23	21	3	14	6	4	2	3	6	8	10
0.28	0.078	72	112	55	23	12	21	8	5	7	3	3	6	3	2
0.28	0.13	67	86	104	17	9	13	1	2	7	1	1	1	4	8
1.3	0.29	19	81	26	23	16	18	15	2	7	2	5[a]	10	19	9
1.3	0.45	102	53	75	32	22	12	4	4	10	5	16[a]	11	6	6
6.7	1.2	85	38	52	10	28	17	12	6	17[a]	4	10[a]	16[a]	3	18
6.7	3.1	42	114	62	36	6	4	8	30	40[a]	13	27[a]	12[a]	4	18
34	10	86	48	33	30	13	11	8	51	45[a]	16[a]	30[a]	37[a]	55	40[a]
34	16	36	80	38	39	27	6	10	5	56[a]	48[a]	11[a]	12[a]	7	32[a]

[a] Average for treatment group was significantly greater than average for control mesocosms, according to Williams' Test ($p \leq 0.05$).

Note: The first Guthion treatment was on July 18 and the last was on September 6.

Table 7. Density of Cladocerans (Individuals per l, Average For Deep and Shallow Sampling Zones) in Mesocosms Before, During, and After Treatment With Guthion

| Concentration | | Sample Date (1988) | | | | | | | | | | | | | |
Nominal	Average	4/4	5/2	5/16	5/30	6/13	6/27	7/11	7/25	8/8	8/22	9/5	9/19	10/3	10/17
Control	0	5	8	5	4	27	6	2	1	0	0	0	0	1	0
Control	0	3	17	7	4	74	1	2	4	1	1	0	5	1	0
0.056	0.023	11	7	7	16	18	0	1	7	8	2	1	2	1	0
0.056	0.027	4	4	9	5	41	2	1	6	1	1	0	0	0	0
0.28	0.078	9	12	8	28	34	0	3	5	1	0	0	1	0	0
0.28	0.13	7	8	6	7	39	2	0	0	1	0	0	1	0	1
1.3	0.29	5	4	4	26	34	2	2	2	7	3	9	9	4	3[b]
1.3	0.45	23	30	8	15	61	2	1	1	1	0	0	2	0	1[b]
6.7	1.2	16	30	7	12	4	1	3	7	1	1	1	7	1	4[b]
6.7	3.1	6	5	4	8	47	0	1	1	1	0	0	0	0	5[b]
34	10	8	12	6	28	35	2	3	0[a]	0	0	0	0	0	3[b]
34	16	9	5	5	6	174	1	1	0[a]	0	0	0	0	0	26[b]

[a] Average for treatment group was significantly less than average for control mesocosms, according to Williams' Test ($p \leq 0.05$).
[b] Average for treatment group was significantly greater than average for control mesocosms, according to Williams' Test ($p \leq 0.05$).

Note: The first Guthion treatment was on July 18 and the last was on September 6.

Table 8. Density of Rotifers (Individuals per l, Average For Deep and Shallow Sampling Zones) in Mesocosms Before, During, and After Treatment With Guthion

Concentration							Sample Date (1988)									
Nominal	Average	4/4	5/2	5/16	5/30	6/13	6/27	7/11	7/25	8/8	8/22	9/5	9/19	10/3	10/17
Control	0	0	0	2	15	232	31	347	40	17	3	2	4	5	14
Control	0	2	1	0	0	11	487	97	59	24	7	28	6	19	31
0.056	0.023	1	0	0	0	3	360	246	121	46	22	54	21	18	75
0.056	0.027	2	0	0	1	12	109	371	40	23	15	4	10	8	28
0.28	0.078	14	0	0	0	20	654	48	19	12	3	4	5	12	30
0.28	0.13	1	0	3	18	29	161	169	6	9	11	6	9	5	8
1.3	0.29	0	1	1	14	138	30	344	78	15	5	2	12	77	71
1.3	0.45	13	0	0	0	2	74	111	68	12	16	88	70	21	31
6.7	1.2	3	0	0	0	8	1196	36	119	38	5	50	11	18	63
6.7	3.1	1	1	0	80	220	227	176	119	13	30	16	2	4	162
34	10	10	0	0	0	6	167	154	51	61	5	9	10	30	15
34	16	2	0	0	3	52	84	240	43	4	2	0	1	1	13

Note: The first Guthion treatment was on July 18 and the last was on September 6.

Table 9. Macroinvertebrate Taxa Identified in Benthic Samples Taken During Guthion Mesocosm Study

Diptera	*Callibaetis* sp.	**Trichoptera**	Hydrophilidae
Chironomidae	Caenidae	Hydroptilidae	*Berosus* sp.
Chironomini	*Caenis* sp.	*Oxyethira* sp.	**Hemiptera**
Pseudochironomini	Ephemeridae	Leptoceridae	Corixidae
Tanytarsini	*Hexagenia* sp.	*Oecetis* sp.	Pleidae
Corynoneurini	**Odonata**	Phrygoneidae	*Neoplia striodus*
Metrioenemini	Aeshnidae	*Agrypnia* sp.	**Lepidoptera**
Orthocladiini	*Anax* sp.	Polycentropidae	*Parponyx* sp.
Coelotanypodini	Coenogrionidae	*Polycentropus* sp.	*Synclita* sp.
Pentaneurini	*Enallagma* sp.	**Coleoptera**	**Megaloptera**
Procladiini	*Ishnura* sp.	Chrysomelidae	*Sialis* sp.
Ceratopogonidae	Cordulidae	*Donacia* sp.	**Other**
Chaoboridae	*Epicordula* sp.	Dytiscidae	Oligochaeta
Chaoborus sp.	*Tetragoneuria* sp.	*Agabus* sp.	Hirudinea
Muscidae	Gomphidae	*Hydroporus* sp.	Acarina
Sciaridae	*Gomphus* sp.	*Ilybius* sp.	Amphipoda
Tabanidae	Libellulidae	Elmidae	*Hyalella azteca*
Chrysops sp.	*Celithemis* sp.	*Dubirophia* sp.	Copepoda
Tabannus sp.	*Libellula* sp.	Gyrinidae	Ostracoda
Tipulidae	*Pachydiplex* sp.	*Dineatus* sp.	Mollusca
Ephemeroptera	*Sympetrum* sp.	Halipidae	Gastropoda
Baetidae	*Tramea* sp.	*Halipidus* sp.	Pelecypoda

Macroinvertebrates

Macroinvertebrate taxa identified in the benthic samples are listed in Table 9, and those in the emergent insect samples are listed in Table 10. Numbers of macroinvertebrates collected in the benthic grabs and the insect emergence traps were quite variable in each pond over time, and among ponds on any given sample event. However, trends were clarified by summing the numbers of benthic invertebrates and emergent insects collected from each pond over the entire treatment and posttreatment interval.

The results are plotted against exposure levels in Figures 3 through 9. Each plot indicates treatment levels that were significantly different from controls, according to Williams' Test. The regression line (see Table 4) for each parameter is also shown.

The pond insect communities were dominated by dipterans. Benthic samples from ponds exposed to average Guthion concentrations $\leqslant 1.2$ µg/l contained 59 to 212 dipterans per pond (Figure 3). At exposure levels of 3.1 µg/l and higher, only 16 to 29 dipterans were collected per pond. In the emergent insect samples, the reduction in dipterans occurred at the next higher exposure level: ponds exposed to 3.1 µg/l or less had emergence rates ranging from 400 to 1000 individuals over the treatment and posttreatment period, while ponds exposed to 10 and 16 µg/l had emergence rates of 247 and 118 individuals per pond, respectively.

Williams' Test indicated significant reductions in dipteran densities at the highest treatment level (ponds with average Guthion concentrations of 10 and 16 µg/l). Regressions of dipteran density (log-transformed) against average Guthion concentration (also log-transformed) over the range between 0.45 µg/l and 16 µg/l yielded

Table 10. Taxa Identified in Emergent Insect Samples Taken During Guthion Mesocosm Study

Diptera	*Hydrobaenus* spp.
Ceratopogonidae	*Namocladius* sp.
Chaoboridae	*Parakiefferiella* cf. *coronata*
Chaoborus spp.	*Psectrocladius* cf. *vernalis*
Chironomidae	*Psectrocladius* sp.
Apedilum elachistus	*Clintanypus pinguis*
Chironomus spp.	*Psectrotanypus dyari*
Cladopelam collator	*Ablabesmyia* spp.
Cryptochironomus sp.	*Labrundinia* sp.
Cryptotendipes darbyi	*Larsia* sp.
Cryptotendipes emersus	*Paramerina smithae*
Demeijerea brachialis	*Telopelopia okoboji*
Dicrotendipes modestus	*Procladius bellus*
Dicrotendipes nervosus	*Procladius subletti*
Endochironomus nigricans	*Procladius* sp.
Endochironomus subtendens	*Tanypus neopunctipennis*
Glyptotendipes sp.	Culicidae
Nilothauma sp.	Ephydridae
Parachironomus chaetaolus	**Ephemeroptera**
Parachironomus monochromus	Baetidae
Parachironomus varus	*Callibaetis* sp.
Parachironomus sp.	Caenidae
Paralauterborniella nigrohalterai	*Caenis* sp.
Paratendipes albimanus	**Odonata**
Polypediulum cf. *floridense*	Coenagrionidae
Polypediulum illinoense	*Enallagma aspersum*
Polypediulum ophioides	*Enallagma civile*
Polypediulum simulans	*Enallagma signotum*
Polypediulum cf. *simulans*	*Ishnura verticalis*
Polypediulum sordans	**Hymenoptera**
Polypediulum trigonus	**Hemiptera**
Xenochironomus xenolabis	Mesovellidae
Zavreliella varipennis	*Mesovelia* sp.
Pseudochironomus cf. *Pseudoviridis*	Pleidae
Pseudochironomus rex	*Neoplea* sp.
Pseudochironomus cf. *richardsoni*	Notonectidae
Cladotanytarsus sp.	*Buenoa* sp.
Constempellina n. sp.	Corixidae
Tanytarsus spp.	*Sigara* sp.
Corynoneura sp.	**Trichoptera**
Cricotopus spp.	Hydrotilidae

r^2 values of 0.928 and 0.869 for benthic and emergent dipterans, respectively. The corresponding EC_0 values were 1.26 µg/l and 1.43 µg/l, respectively.

The abundances of certain dipteran families and tribes were affected at lower exposure levels than total dipteran abundance. The most abundant dipterans were midges of the family Chironomidae. Benthic and emergent Chironomidae were most abundant in the 0.45 µg/l pond, and declined at exposure levels ≥1.2 µg/l (Figure 4). Although the trends were apparent graphically, only the two highest dosed ponds (10 and 16 µg/l) were significantly different from controls according to Williams' Test. EC_0 values for benthic and emergent Chironomidae were 1.23 µg/l and 0.61 µg/l, respectively (r^2 values 0.839 and 0.713, respectively, over the exposure range from 0.45 µg/l to 16 µg/l).

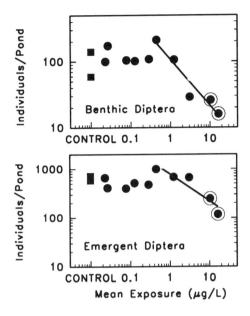

Figure 3. Numbers of benthic dipterans and emergent dipterans collected from mesocosms during and after treatment with Guthion. Values shown represent the total numbers of individuals collected in Ekman grabs or in floating emergence traps in each pond throughout the treatment and posttreatment periods. Data are averages for samples collected from shallow and deep zones of each pond. Squares represent data from control ponds; circles represent data from ponds treated with Guthion. Circles surrounded by rings indicate significant treatment effects ($p \leq 0.05$) according to Williams' Test. The lines through the data points represent the regression equations shown in Table 4.

The tribe Chironomini (a subtaxon of the Chironomidae) was also reduced at exposure levels ≥ 1.2 µg/l (Figure 5). Williams' Test corroborated this conclusion for the emergent Chironomini. The reductions in benthic Chironomini were significant only at the 10 and 16 µg/l exposure levels. EC_0 values were 0.91 µg/l for benthic Chironomini and 0.26 µg/l for emergent Chironomini (r^2 values 0.884 and 0.745, respectively).

Emergent insects of the tribe Tanytarsini were most abundant at 0.45 µg/l, with significantly reduced abundance at ≥ 1.3 µg/l (Figure 6). The EC_0 was 0.42 µg/l ($r^2 = 0.768$).

Tanytarsini were not common in the benthic samples, but another tribe, Pseudochironomini, was found consistently in the benthic samples. Abundances of benthic Pseudochironomini were lowest in the ponds exposed to ≥ 3.1 µg/l (Figure 6), and reductions were significant at ≥ 1.2 µg/l. The EC_0 was 0.20 µg/l ($r^2 = 0.761$).

The other dipteran family commonly found in benthic and emergent samples from the mesocosms was the Ceratopogonidae. Benthic Ceratopogonidae abundance was low in 1 control mesocosm, but ranged from 27 to 69 individuals per pond in

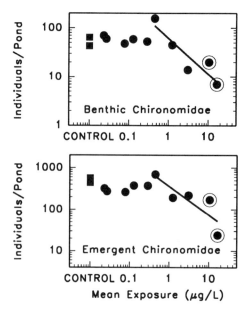

Figure 4. Numbers of benthic Chironomidae and emergent Chironomidae collected from mesocosms during and after treatment with Guthion. See Figure 3 for explanation of data and symbols.

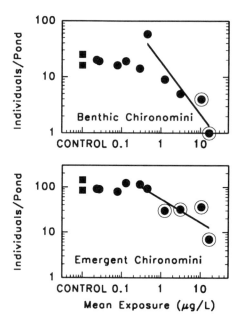

Figure 5. Numbers of benthic Chironomini and emergent Chironomini collected from mesocosms during and after treatment with Guthion. See Figure 3 for explanation of data and symbols.

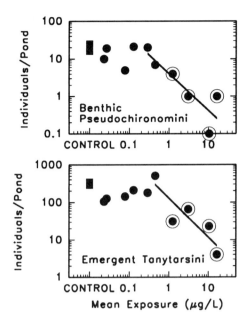

Figure 6. Numbers of benthic Pseudochironomini and emergent Tanytarsini collected from mesocosms during and after treatment with Guthion. See Figure 3 for explanation of data and symbols.

the other mesocosms at exposure levels ≤ 1.2 µg/l (Figure 7). Individuals of this family were less abundant at ≥ 3.1 µg/l. In the emergence samples, Ceratopogonidae were most abundant in the ponds exposed to 1.2 and 3.1 µg/l, and least abundant in ponds exposed to 10 and 16 µg/l (Figure 7). None of these differences was significant according to Williams' Test. The EC_0 values for benthic and emergent Ceratopogonidae were 0.98 µg/l and 7.74 µg/l, respectively ($r^2 = 0.756$ and 0.857, respectively).

Two other orders of insects — ephemeropterans (mayflies) and odonates (dragonflies and damselfilies) — were found in the ponds consistently. Ephemeropterans occured in low numbers in both benthic and emergent samples. Data for the benthic samples suggest that ephemeropterans were reduced at exposure levels of 1.2 and 3.1 µg/l, and were eliminated at 10 and 16 µg/l (Figure 8). In emergent samples no reduction in Ephemeroptera was apparent at 3.1 µg/l or less, but only 1 individual was found at 10 µg/l and none at 16 µg/l (Figure 8). According to Williams' Test, abundances of benthic and emergent ephemeropterans were significantly reduced in the 10-µg/l and 16-µg/l mesocosms. The EC_0 values were 0.59 µg/l for benthic ephemeropterans and 1.19 µg/l for emergent ephemeropterans ($r^2 = 0.970$ and 0.803, respectively).

Odonates occurred in emergent samples, but were rarely found in benthic samples. A loose inverse relationship between exposure levels and emergent odonate numbers was apparent at exposures ≤ 0.45 µg/l (Figure 9). Odonate numbers declined with

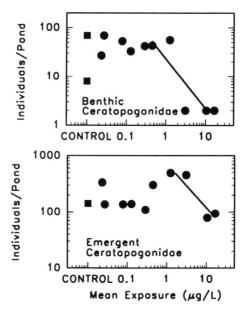

Figure 7.   Numbers of benthic Ceratopogonidae and emergent Ceratopogonidae collected from mesocosms during and after treatment with Guthion. See Figure 3 for explanation of data and symbols.

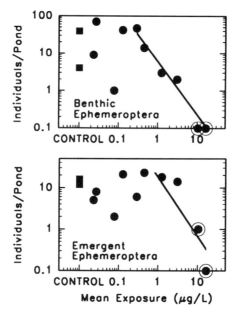

Figure 8.   Numbers of benthic ephemeropterans and emergent ephemeropterans collected from mesocosms during and after treatment with Guthion. See Figure 3 for explanation of data and symbols.

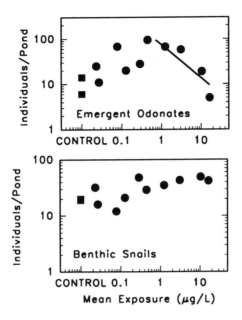

Figure 9. Numbers of emergent odonates and benthic snails collected from mesocosms during and after treatment with Guthion. Values shown for emergent Odonates represent the total number of exuvia collected from emergence screens in each pond throughout the treatment and posttreatment periods. See Figure 3 for explanation of benthic snail data. Symbols as in Figure 3.

increasing exposure above this level, but did not become significantly different from controls according to Williams' Test. The EC_0 was 16.7 $\mu g/l$ ($r^2 = 0.827$).

Benthic samples from ponds exposed to ≤ 0.13 $\mu g/l$ contained 12 to 32 snails per pond (Figure 9). The abundance of snails in ponds exposed to ≥ 0.29 $\mu g/l$ ranged from 29 to 49 per pond, significantly greater than controls.

DISCUSSION

Because of the variability among replicates, the power of Williams' Test to detect treatment effects was low for many measured parameters. For most measurements, the minimum significant difference was equivalent to a reduction of 30% or more relative to controls. Often, differences appeared treatment-related based on visual examination of the exposure-response trends but were not identified as such by Williams' Test. (Dunnett's Test and several nonparametric statistical tests were also tried, but they were even less sensitive than Williams' Test.)

Regression analysis was successful in describing the exposure-response relationship for most parameters over a defined exposure range. The EC_0 values were generally consistent with visual interpretation of the data. Each estimated EC_0 indicated the response threshold for a given parameter — i.e., the maximum exposure level that caused no effect on that parameter. Because of the steep slopes of the exposure-response relationships, relatively large effects were possible at exposures not far in excess of the EC_0. A fivefold increase in exposure concentration resulted in approximately 75 to 80% reductions in bluegill survival and biomass and in the density of most insect taxa.

Hypothesis tests such as Williams' Test address the question: is an observed difference between treated ponds and controls statistically significant? Because of natural variability, this question is often difficult to answer unequivocally unless treatment effects are very large. Moreover, statistical significance does not necessarily indicate ecological significance.

Using the regression approach, the definition of statistical and ecological significance is not as much an issue. Exposure levels above the EC_0 can be considered hazardous because they approach levels that cause unquestionably adverse effects. It is reasonable to infer that any exposure above the EC_0 for a parameter constitutes an ecological risk for that parameter because there is little or no margin for safety before effects begin to occur.

The usefulness of the regression approach depends on well-defined exposure-response relationships. This is best achieved when studies include closely spaced exposure levels around the range where the relationship is steepest. To establish the exposure-response trend, it is important to include some exposure levels high enough to cause severe effects. Replication within treatment levels is not necessary, but replication of controls is highly desirable because the EC_0 is defined in terms of average control behavior.

Statistics are not the only means of determining ecological effects. For some parameters in this study (such as cladoceran, rotifer, and odonate abundances), differences occurred that were suggestive of treatment effects but that were detected by neither hypothesis testing nor regression analysis. In these cases, conservative interpretation would support the presumption that effects occurred even though they could not be demonstrated statistically.

The overall pattern of ecosystem response is depicted in Table 11. No EC_0 values or significant reductions occurred at ≤ 0.13 µg/l. At 0.29 µg/l, EC_0 values were exceeded for adult bluegill growth, and for two major insect taxa, while abundances of snails and copepods increased. At 1.2 µg/l, effects were evident for adult and juvenile bluegill and for most insect taxa. Thus, exposure levels that caused unmistakable impact throughout the ecosystem were less than ten times higher than the level that caused no observable effects.

Mesocosm results are compared with laboratory results in Figure 10. The lowest observable effect concentration (LOEC) for Guthion in the mesocosms was 0.29 µg/l (average concentration over 55 days). The lowest LOEC in chronic laboratory toxicity tests with freshwater organisms was 0.40 µg/l (*Daphnia magna*). Severe effects occurred in the pond with an average concentration of 1.2 µg/l (average concentration peak of 4.2 µg/l). This was near the lower end of the acute toxicity range for freshwater organisms (1.1 to 4.4 µg/l for *D. magna*, 3.0 µg/l for rainbow trout).

Table 11. Summary of the Ecological Effects of Guthion in Mesocosms

	Guthion Concentration (µg/l)					
Nominal Loading Rate	1.3	1.3	6.7	6.7	34	34
Average Concentration Maximum	0.89	1.0	4.2	6.2	26	33
55-Day Mean Exposure	0.29	0.45	1.2	3.1	10	16
Adult Bluegill number, biomass	↓	↓	↓	↓	↓	↓
Juvenile Bluegill number, biomass			↓	↓	↓	↓
Cladocerans				?	↓	↓
Rotifers						?
Copepods	↑	↑	↑	↑	↑	↑
Dipterans (Benthic)					↓	↓
Dipterans (Emergent)					↓	↓
Chironomidae (Benthic)					↓	↓
Chironomidae (Emergent)					↓	↓
Chironomini (Benthic)					↓	↓
Chironomini (Emergent)			↓	↓	↓	↓
Pseudochironomini (Benthic)			↓	↓	↓	↓
Tanytarsini (Emergent)			↓	↓	↓	↓
Ceratopogonidae (Benthic)						
Ceratopogonidae (Emergent)						
Ephemeropterans (Benthic)					↓	↓
Ephemeropterans (Emergent)					↓	↓
Odonates (Emergent)					?	?
Snails (Benthic)	↑	↑	↑	↑	↑	↑

Note: Each column represents the response of one pond. ↓ indicates a significant negative effect, ↑ indicates a significant positive effect, according to Williams' Test ($p = 0.05$). Shading indicates concentration greater than the EC_0 calculated by linear regression. ? indicates presumed negative effect, based on conservative interpretation of data.

There is no "best" way to define exposure in these pulsed situations. The experimental design was based on Guthion loading rates (i.e., the mass of Guthion entering each pond). We intended to interpret effects based on maximum concentrations in water, but found that effects were even more closely related to average concentrations. The exposure regimes could also have been described in terms of the frequency of exceedence of a specified concentration, the duration of exceedence of a specified concentration, or other calculations. There is yet no theoretical basis for predicting which aspect of exposure is most closely related to ecological impact. Furthermore, these pulsed exposure regimes, which are typical of actual-use situations, are very different from those in laboratory toxicity tests. Comparing pulsed EEC or MEC (measured environmental concentration) with laboratory-derived LC_{50} or NOEC could lead to erroneous assessments of environmental hazard. Efforts are needed to design laboratory tests with exposure regimes that are relevant to nature.

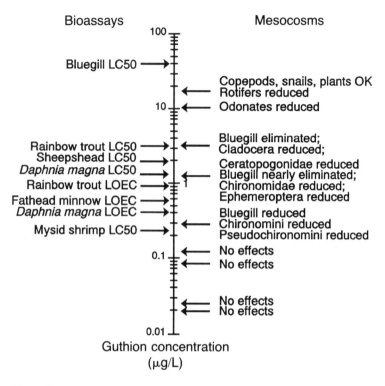

Figure 10.   Comparison of bioassay results (see Table 1) with observed responses of mesocosms to Guthion exposure.

ACKNOWLEDGMENTS

This study was sponsored by the Agricultural Chemicals Division of Mobay Corporation, Stilwell, KS. Dr. Robert Graney, Mobay Corporation, was the program monitor. Timothy Kendall (Springborn Laboratories, Inc.) analyzed the water and sediment for azinphos methyl and its oxygen analog. Dr. Leonard Ferrington (University of Kansas), with assistance from his graduate students, counted and identified the benthic invertebrates and emergent insects. Dr. Dean Kettle (University of Kansas) measured and identified the macrophytes. The authors thank the many other individuals, too numerous to list here, who participated in this study.

REFERENCES

1. U.S. Environmental Protection Agency. 1988. *Hazard Evaluation Division Technical Guidance Document. Aquatic Mesocosm Tests to Support Pesticide Registrations.* EPA-540/09-88-035. Office of Pesticides Programs, U.S. Environmental Protection Agency, Washington, DC. March 1988.

2. American Public Health Association, American Water Works Association, and Water Pollution Control Federation. 1985. *Standard Methods for the Examination of Water and Wastewater. 16th Edition.*

3. Williams, D.A. 1971. A test for differences between treatment means when several dose levels are compared with a zero dose control. *Biometrics* 27:103–117.

4. Williams, D.A. 1972. The comparison of several dose levels with a zero dose control. *Biometics* 28:519–531.

5. Howick, G.L., and L.C. Ferrington. 1988. An economical insect emergence trap for use in experimental ponds. *Abstracts, Ninth Annual Meeting, Society of Environmental Toxicology & Chemistry,* Arlington, VA, November 1988.

6. Springborn Life Sciences, Inc. 1988. Acute toxicity of technical grade azinphos-methyl (trade name Guthion) to mysid shrimp (*Mysidopsis bahia*) under flow-through conditions. SLS Report #87-9-2513. Wareham, MA.

7. Boeri, R.L. 1989. Flow-through, acute toxicity of Guthion 2L to mysids, *Mysidopsis bahia.* Enseco Inc., Marblehead, MA.

8. Boeri, R.L. 1989. Flow-through, acute toxicity of Guthion 2L to sheepshead minnows, *Cyprinodon variegatus.* Enseco Inc., Marblehead, MA.

9. Springborn Life Sciences, Inc. 1988. Acute toxicity of technical grade azinphos-methyl (trade name Guthion) to sheepshead minnow (*Cyprinodon variegatus*) under flow-through conditions. SLS Report #87-5-2504. Wareham, MA.

10. Carlisle, J.C. 1984. Acute toxicity of azinphos-methyl (Guthion) technical to rainbow trout. Mobay Chemical Corporation, Stilwell, KS.

11. Lamb, D.W. 1980. Azinphos-methyl (Guthion) technical acute toxicity to *Daphnia magna.* Mobay Chemical Corporation, Stilwell, KS.

12. Springborn Life Sciences, Inc. 1987. Acute toxicity of Guthion 50 WP to daphnids, *Daphnia magna,* under flow-through conditions. SLS Report #87-8-2466. Wareham, MA.

13. Jensen, L.D., and A.R. Gaufin. 1964. Effects of ten organic insecticides on two species of stonefly naiads. *Trans. Am. Fisheries Soc.,* 93:27–34.

14. Mobay Chemical Corporation. 1978. Acute toxicity of Guthion 2S to bluegill and rainbow trout. Mobay Report No. 66046. Kansas City, MO.

15. Heimbach, F. 1985. Growth inhibition of green algae (*Scenedesmus subspicatus*) by azinphos-methyl (technical), Bayer AG, Leverkusen, Germany.

16. Springborn Life Sciences, Inc. 1987. Acute toxicity of technical azinphos-methyl (trade name Guthion) to eastern oysters (*Crassostrea virginica*) under flow-through conditions. SLS Report #87-11-2564. Wareham, MA.

17. Analytical Bio-Chemistry Laboratories, Inc. 1984. Chronic toxicity of [^{14}C]Guthion to *Daphnia magna* under flow-through test conditions. ABC Report #31802. Columbia, MO.

18. Adelman, I.R., L.L. Smith, Jr., and G.D. Siesennop. 1976. Chronic toxicity of Guthion to the fathead minnow (*Pimephales promelas* Rafinesque). *Bull. Env. Contam. Toxicol.,* 15:726–733.

19. Springborn Life Sciences, Inc. 1988. The toxicity of technical grade azinphos-methyl (trade name Guthion) to rainbow trout (*Salmo gairdneri*) embryos and larvae. SLS report #87-11-2561. Wareham, MA.

20. Carlisle, J.C. 1985. Toxicity of azinphos-methyl (Guthion technical) to rainbow trout early life stages. Mobay Chemical Corporation, Stilwell, KS.

CHAPTER 26

Evaluation of the Ecological and Biological Effects of Tralomethrin Utilizing an Experimental Pond System

Joseph M. Mayasich, James H. Kennedy, and Joseph S. O'Grodnick

Abstract: Fourteen mesocosms were constructed during the summer and fall of 1987 at the Univeristy of North Texas. In June of 1988, a treatment program with the synthetic pyrethroid tralomethrin was initiated. The program featured ten weekly spray applications to the mesocosm surfaces to simulate spray drift and five biweekly applications as a fortified soil slurry to simulate runoff. Including the control, six levels of tralomethrin were established by the treatment regime. Triplicate mesocosms were monitored for the control and a dose representing the estimated environmental concentrations of tralomethrin for loading via drift (68.5 ng/l) and runoff (218.9 ng/l). Duplicate mesocosms were monitored at the other four dose levels. Measured levels of tralomethrin in water samples closely paralleled nominal levels. Tralomethrin residues rapidly declined after applications and no residues were detected in the water 2 weeks after the final application. Physicochemical characteristics of the water were not affected by tralomethrin. Mollusks, fish, phytoplankton, macrophytes, and most larval invertebrates inhabiting the hydrosoil were unaffected by tralomethrin. Tralomethrin did significantly reduce abundances of certain macroinvertebrates and zooplankton, relative to the control. The effect of tralomethrin on Ephemeroptera larvae differed considerably between Caenidae and Baetidae. The abundances of Caenidae larvae in treated mesocosms were frequently detected as significantly less than in the control. In contrast, the abundances of Baetidae larvae in treated and control mesocosms were never detected as significantly different. For zooplankton, the densities of Copepods and nauplii in treated mesocosms were

```
Common Name:    Tralomethrin
```

```
Chemical Name: (1R,3S)3[1(1'RS)(1',2',2',2',-tetrabromoethyl)] -2,
               2-dimethylcyclopropanecarboxylic acid (S)-alpha-cyano-3-
               phenoxybenzyl ester
```

```
CAS No.:   66841-25-6
```

```
Trade Names:    Scout^R Insecticide (3.75% [w/w] Tralomethrin)
                Scout X-Tra^TM Insecticide (10.30% [w/w] Tralomethrin)
```

Structure:

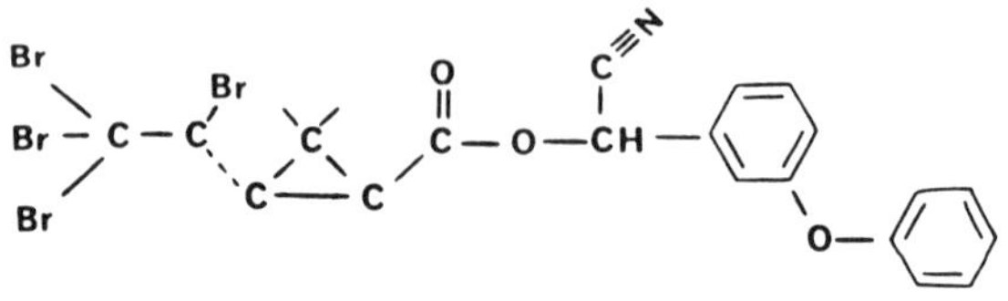

Figure 1.   Test substance information.

each detected significantly less than in the control more frequently than when density comparison were made utilizing Cladocera data. The study design and endpoints measured were more appropriate for assessing structural rather than functional properties of an aquatic ecosystem. The results indicate that tralomethrin can alter the structure of an aquatic ecosystem. The alteration in structure appears to be ephemeral in character and could be more thoroughly investigated in future studies which are narrower in scope and longer in duration.

INTRODUCTION

Tralomethrin is a synthetic pyrethroid discovered and manufactured by Roussel-Uclaf in France (Figure 1). Tralomethrin is the active ingredient of Scout® insecticide which is marketed in the U.S. by Hoechst-Roussel Agri-Vet Company. The Federal Insecticide, Fungicide, and Rodenticide Act, as amended (FIFRA, P.L. 92-516), specifies that for any insecticide the United States Environmental Protection Agency (EPA) must determine that the product will not cause unreasonable adverse effects on the environment. The law further states that the EPA must specify what data are necessary to make this determination. For Scout® insecticide, the EPA specified that a tier IV aquatic field test[1,2] be conducted to assess the effects of tralomethrin on aquatic ecosystems.

The purpose of the tier IV field test is to determine if a hazard, identified by comparison of laboratory toxicity results with a compound-specific estimated environmental concentration (EEC) value, is negated under field conditions. The procedure for comparing the laboratory toxicity results to the compound-specific

EEC is well defined.[3] The test systems used in aquatic field studies have undergone a transition from natural ponds to simulated aquatic ecosystems, i.e., mesocosms.[4-9] Irrespective of the test system, any ecosystem-level investigation has inherent advantages and disadvantages.[10] The ability to examine a range of treatment levels using relatively standardized experimental units, with replication, weighed heavily in the decision to use mesocosms as the test systems to evaluate the effects of tralomethrin on aquatic ecosystems.

It is important to monitor and evaluate both structural and functional parameters when investigating any ecosystem, terrestrial or aquatic. Aquatic ecosystems are structurally complex entities comprised of multiple interacting trophic levels. The detection of treatment effects on aquatic ecosystem function may be precluded by the functional redundancy afforded to the ecosystem by its structural complexity.[11] Structural complexity and functional redundancy can also work to an investigator's advantage by easing requirements to rigorously sample and monitor all parameters of each trophic level of an aquatic ecosystem.

The objectives of the study were to evaluate the effects of tralomethrin on the structure and, to some extent, the function of a simulated aquatic ecosystem by collecting biological, water quality and tralomethrin residue data. This manuscript will focus on representative study findings which are related to the objectives.

MATERIALS AND METHODS

Study Site

A site was selected in northern Texas near the University of North Texas at Denton. The site had previously produced a single crop (Sudan hay) annually over the 10 years prior to selection. No crop protection products were applied to the site during this 10-year period. Fourteen mesocosms, each approximately 0.05 ha., were established on the site (Figure 2). Each mesocosm was 30 m by 16 m, with a maximum water depth of 2 m, and sloped (3:1) end walls. The mesocosms were lined with clay followed by 15 cm of topsoil. All mesocosms were filled with water from a common maintenance pond. The water was selectively screened (0.6-cm mesh) to prevent passage of fish or fish eggs and yet allow for zooplankton and phytoplankton colonization. The water was circulated among the mesocosms, using a pumping system, for 6 months. During this period of time, it is estimated that the total volume of water in a given mesocosm (635,000 l) was completely replaced 4 times.

Colonization

In addition to the organisms introduced while the mesocosms were being filled, the mesocosms were seeded as equally as possible with biota and sediment from

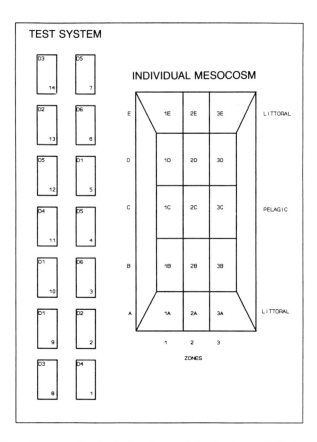

Figure 2.   Diagram of entire test system and details of an individual mesocosm.

local well-established natural ponds. Sweep nets were used to collect samples of insects and macroinvertebrates. Care was taken to ensure that juvenile fish, fish eggs, and ichthyoplankton were not introduced during the seeding effort. Sprigs of the aquatic macrophyte, *Potamogeton nodosus,* were transplanted in the littoral zones of each mesocosm. A total of 18 female and 17 male mature bluegill sunfish (*Lepomis macrochirus* Rafinesque) at a combined weight of approximately 1.5 kg were added to each mesocosm. Each group of 35 fish were comprised of individuals equivalent in size (maximum range per group ≤5 cm). The fish were obtained from a local hatchery. Upon completion of all seeding and stocking efforts, the mesocosms were then allowed to equilibrate for approximately 1 month prior to tralomethrin application.

Sampling

Each mesocosm was divided into three zones (oriented lengthwise) and five regions (oriented widthwise), two littoral and three pelagic (Figure 2). For each sampling period, a location was randomly selected within the pelagic and littoral regions of each zone. Samples for water quality and residue analysis were composited by mesocosm. For most biological analyses, one littoral and one pelagic composite sample was collected per mesocosm.

Sampling was organized into four phases: preliminary, pretreatment, treatment, and posttreatment (Table 1). Measurements of dissolved oxygen and temperature (YSI Model 54APB) and pH (Orion Model SA250) were made *in situ*. The oxygen measurements were used to monitor ecosystem metabolism in each mesocosm via the three-point diel oxygen method.[12] Water for residue analysis was collected in 500-ml Teflon bottles attached to a pole at 3 lengths so as to obtain representative top, middle, and bottom samples when submerged. For other water quality determinations (alkalinity, hardness, total phosphate and suspended solids, nitrogen and organic carbon) and some biological analysis (zoo- and phytoplankton, and photosynthetic pigments), samples were collected with a section of clear PVC pipe (schedule R-4000) lowered vertically through the water column to 10 cm above the sediment. Aliquots (2 to 5 l) were concentrated through a plankton net (80-μm mesh) and then preserved in Lugol's solution (1 to 2%) for subsequent identification and enumeration of zooplankton. Photosynthetic pigments were determined spectrophotometrically.[13] Macroinvertebrates were sampled with submerged artificial substrate units constructed from plastic cylinders (5 cm O.D. by 5 cm high) and with floating emergence traps similar to those described by Lesage and Harrison.[14] Fish were harvested by seining (3-mm mesh bag seine) and manually searching each drained (3-mm mesh screened pump) mesocosm. Hydrosoil cores were collected in detachable butyl plastic tubes fitted to the end of schedule R-4000 PVC pipe.

Mesocosm Dosing

The actual test substance was an emulsifiable concentrate formulation of tralomethrin containing 35.94 g of tralomethrin (active ingredient) per liter of formulation. The test substance was applied so as to simulate two routes of exposure: spray drift and runoff.

The number of applications and the nominal levels which define the spray drift and runoff treatments are given in Table 2. The spray drift and runoff levels which define dose level 5 represent the EEC for tralomethrin. The nominal levels represent percentages of the maximum labeled application rate (22.4 g a.i./ha). The total number (10) and schedule (7-day interval) for the spray drift applications were based on the product label. For runoff applications, the total number (5), schedule (14-day interval), and percentages of the application rate were based on estimates from the Simulator for Water Resources in Rural Basins (SWRRB) model.[15]

All applications, spray drift and runoff, were made with a modified Gamaco® highway bridge spanner. Spray drift applications were made by evenly spraying appropriately diluted (water) quantities of test substance across the surface of a

Table 1. Field Sampling Activities and Schedule

Sampling Activity	Preliminary			Pretreatment				Treatment					
Date (Mon–Sun)	Mar 21–27	Apr 18–24	May 16–22	May 23–29	May 30–Jun 5	Jun 6–12	Jun 13–19	Jun 20–26	Jun 27–Jul 5	Jul 4–10	Jul 11–17	Jul 18–24	Jul 25–31
Week #	−13	−9	−5	−4	−3	−2	−1	1	2	3	4	5	6
Runoff app. (RO)								RO		RO		RO	
Spray drift app. (SD)								SD	SD	SD	SD	SD	SD
Biological (B)	B	B	B		B		B		B		B		B
Water quality (W)	W	W	W		W		W		W		W		W
DO, temp, and pH (W1)	W1	W1	W1		W1		W1	W1	W1	W1	W1	W1	W1
Emergence traps (ET)						ET		ET		ET		ET	
Zooplankton (Z)	Z	Z	Z		Z	Z	Z	Z	Z	Z	Z	Z	Z
Nonquant. visuals (V)			V	V	V	V	V	V	V	V	V	V	V
Soil/sediment character (SC)							SC						
Water residue (RW)								RW	RW	RW	RW	RW	RW
Soil/sediment residue (RS)							RS		RS		RS		RS
Fish residue (RF)				RF								RF	
Sediment traps (ST)								ST		ST		ST	
Meteorological data (M)				M	M	M	M	M	M	M	M	M	M
				Daily	Daily	Daily	Daily	Daily	Daily	Daily	Daily	Daily	Daily
Fish harvest (F)													

Sampling Activity	Treatment (Continued)					Posttreatment										
Date (Mon–Sun)	Aug 1–7	Aug 8–14	Aug 15–21	Aug 22–28	Aug 29–Sep 4	Sept 5–11	Sept 12–18	Sept 19–25	Sept 26–Oct 2	Oct 3–9	Oct 10–16	Oct 17–23	Oct 24–30	Oct 31–Nov 6	Nov 7–13	Nov 14–20
Week #	7	8	9	10	11	12	13	14	15	16	17	18	19	20	21	22
Runoff app. (RO)	RO		RO													
Spray drift app. (SD)	SD	SD	SD	SD												
Biological (B)		B		B		B		B		B	B	B	B	B		
Water quality (W)		W		W		W		W		W	W	W	W	W		
DO, temp, and pH (W1)	W1	W1	W1	W1	W1	W1	W1	W1	W1	W1	W1	W1	W1	W1		
Emergence traps (ET)	ET		ET		ET		ET		ET		ET	ET	ET			
Zooplankton (Z)	Z	Z	Z	Z	Z	Z	Z	Z	Z	Z	Z	Z	Z	Z		
Nonquant. visuals (V)	V	V	V	V	V	V	V	V	V	V	V	V	V			
Soil/sediment character (SC)														SC		
Water residue (RW)	RW	RW	RW	RW	RW	RW		RW			RW	RW		RW		
Soil/sediment residue (RS)	RS	RS		RS	RS	RS		RS		RS	RS	RS		RS		
Fish residue (RF)					RF									RF	RF	RF
Sediment traps (ST)	ST		ST													
Meteorological data (M)	M	M	M	M	M	M	M	M	M	M	M	M	M	M		
	Daily	Daily	Daily	Daily	Daily	Daily	Daily	Daily	Daily	Daily	Daily	Daily	Daily	Daily		
Fish harvest (F)														F	F	F

Frequency notes (Nonquant. visuals): pretreatment/treatment = Daily; posttreatment (weeks 12–19) = 1×/WK.

Table 2. Tralomethrin Spray Drift and Runoff Treatment Rates

	Dose					
	1 (control)	2	3	4	5	6
Spray Drift						
% of Max. Label Rate	0	0.2	0.7	2.0	5.0	20.0
Mass (mg)[a]	0	2.0	6.8	20.3	50.7	202.9
Nominal Concn. (ppt)[b]	0	2.7	9.2	27.4	68.5	274.1
Measured Concn. (ppt)[c]	0	1.8	5.2	18.4	35.6	134.3
% of Nominal Concn.	—	67	56	67	52	49
Runoff						
% of Max. Labvel Rate	0	0.8	2.0	4.0	8.0	32.0
Mass (mg)[a]	0	16.2	40.6	81.2	162.0	648.0
Nominal Concn. (ppt)[b]	0	21.9	54.8	109.7	218.9	875.1
Measured Concn. (ppt)[c]	0	12.0	26.8	43.9	107.3	280.0
% of Nominal Concn.	—	55	49	40	49	32
Nominal Total Mass (mg)	0	18.2	47.4	101.5	212.7	850.9
Allocated to mesocosm #[d]	5,9,10	2,13	8,14	1,11	4,7,12	3,6

[a] Values represent the amount of tralomethrin, as active ingredient, added per mesocosm per application.
[b] Values represent the concentration of tralomethrin in water immediately following application assuming no degradation or accumulation over time and complete homogeneity within the water column.
[c] Values are means, n = 10 for drift and n = 5 for runoff.
[d] Refer to Figure 2 for relative proximities of mesocosms.

Note: Dose 5 represents the EEC.

mesocosm. Runoff applications were evenly applied to the surface of a mesocosm as a homogeneous soil/water slurry appropriately fortified with test substance. The amount of soil added to each mesocosm per runoff application was 117.9 kg, corrected for moisture. Control mesocosms were sham treated for runoff (unfortified soil/water slurry) but not for spray drift (formulation blank).

Experimental Design and Statistical Analysis

The study was conducted as a completely randomized design[16] which featured six levels of tralomethrin exposure, including the control. Treatments were randomly allocated to duplicate mesocosms, except at the control and dose level 5 treatments, which were allocated to triplicate mesocosms (Table 2). Additional replication was used for these two treatments because their comparison represented the effect of tralomethrin at the EEC. Data were subjected to one-way analysis of variance (ANOVA) followed by treatment mean separation by Tukey's honestly significant difference (HSD), if the ANOVA was significant. Some data required transformation in order to meet the assumptions of ANOVA. Data were analyzed within sampling periods not across (over time) so as to reduce the confounding influence of temporal effects.

Analytical Chemistry Methods

One-liter composite water samples were extracted with hexane or hexanes. After liquid-liquid extraction, the organic layer was allowed to separate. For sediment, the top 1 cm layer of cores were extracted with hexane and methylene chloride by vortexing and sonication. The extracts were concentrated by rotary evaporation and then analyzed by gas chromatography. The instrument was a Hewlett Packard 5890 equipped with an electron capture detector. A J&W DB1 capillary column (30 m × 0.32 mm, 0.25-micron film thickness) and an uncoated megabore precolumn (5 m × 0.53 mm) were used. The carrier gas was hydrogen (15 psi at 15 ml/min) and the make-up gas was a mixture of argon and methane (90:10). The temperature program (°C) was 75 (hold 1 min), 40 per min to 120 (no hold), 30 per min to 210 (no hold), and 8 per min to 290 (hold 10 min). Cool on-column injection was used and volumes ranged from 0.5 to 5.0 µl. Tralomethrin is converted to *cis*-delta-methrin, a facile degradate, when introduced to the column environment required for chromatographic separation. The analytical method also did not resolve *trans*-deltamethrin, another degradate, from tralomethrin. Therefore, all residues are reported as the sum of three analytes (tralomethrin, *cis*-, and *trans*-deltamethrin). The analytical method had a limit of detection of 1 ppt in water and 1 ppb in sediment.

RESULTS AND DISCUSSION

The measured levels of tralomethrin in the water column, applied as simulated spray drift and runoff, were approximately 50% of the nominal level for each treatment (Table 2). These data verify, analytically, that an exposure gradient was established and maintained. The recoveries for concurrent lab fortifications averaged 136% of the nominal fortified level (10 ppt) for water and 125% of the nominal fortified level (8 ppb) for sediment.

The physicochemical characteristics of the water, phytoplankton and periphyton chlorophyll *a* and biomass, metabolism (community photosynthesis, respiration, and their ratio), and densities of aquatic macrophytes and mollusks did not respond in a dose-dependent manner. Statistical analyses of these data did not identify significant treatment effects. Data on physicochemical charcteristics, photosynthetic pigments, phytoplankton biomass, and ecosystem metabolism were representative of seasonal or life history related variability expected within aquatic ecosystems.[12,17] The tolerance of mollusks to pyrethroids is documented in the literature.[18,19]

Densities of aquatic invertebrates did respond in a dose-dependent manner, and statistical analyses of these data did identify significant treatment effects. Given the insecticidal properties of pyrethroids, these results are not surprising and are in

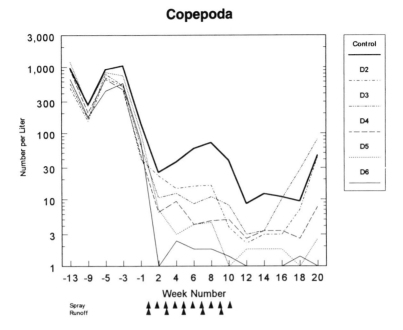

Figure 3.   Mean densities of Copepoda (no./l) recorded over time.

agreement with the literature.[18,20-24] Approximately 100 zooplankton taxa were identified.[25] The effect of tralomethrin on crustaceans was very evident and therefore will be focused on as representative of the responses of zooplankton to tralomethrin. The densities of three groups of Crustacea (Copepoda, Nauplii, and Cladocera) were declining in all mesocosms at the end of the pretreatment phase and then remained low throughout the treatment and posttreatment phases (Figures 3, 4, and 5). The universal decrease in density at the end of the pretreatment phase was probably due to cropping by young bluegill. Despite these low densities, significant differences between control and treatment means were detected, especially for mature and immature (nauplii) copepod densities (Table 3). It is important to note that Copepoda and Cladocera did not respond equivalently to the exposure gradient. This indicates that differential responses to tralomethrin are evident between orders (Cladocera vs. Copepoda) and within a class (Crustacea). In contrast, the similarity of the responses of Copepoda and nauplii indicates that zooplankton sensitivity to tralomethrin is not necessarily dependent on stage of development. It should be noted that the responses of the mature and immature copepods to the exposure gradient are variable in magnitude, and somewhat transitory.

For immature macroinvertebrates, MAS sampler data on the responses of larvae representing three families (Baetidae, Caenidae, and Heptageniidae) of the order Ephemeroptera will be focused on. As with zooplankton, a differential response was evident. However, the differential response among larval mayflies was evident

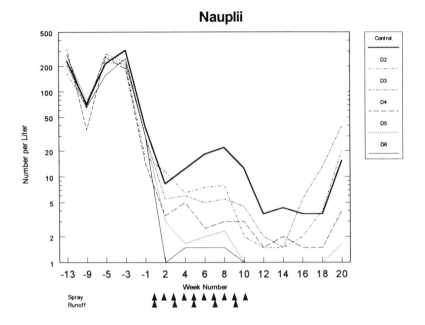

Figure 4.   Mean densities of nauplii (no./l) recorded over time.

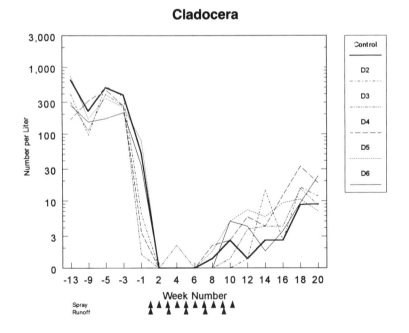

Figure 5.   Mean densities of Cladocera (no./l) recorded over time.

Table 3. Summary of Statistical Analysis of Crustacea Densities (no./l)

Taxa	Comparison Ctrl. vs. Dose[a]	Treatment 2	4	6	8	10	Posttreatment 12	14	16	18	20
Copepoda	2			*							
	3			*							
	4			*	*						
	5	*b	*	*	*	*		*			*
	6	*	*	*	*	*		*			*
Nauplii	2			*							
	3			*							
	4			*	*						
	5		*	*	*	*					*
	6	*	*	*	*	*					*
Cladocera					No Significant Differences						

a Dose range represented as 2 (lowest) to 6 (highest).
b Mean for this dose significantly less than control mean (Tukey's HSD, $p < 0.20$).

between families within an order and not between orders within a class. The abundances of Caenidae (*Caenis* sp.) (Figure 6) remained higher than that of Heptageniidae (Figure 7) or Baetidae (Figure 8) in the controls throughout the treatment and posttreatment phases. This was reflected in the statistical analysis which indicated that larval caenids were more sensitive to tralomethrin than larvae of the

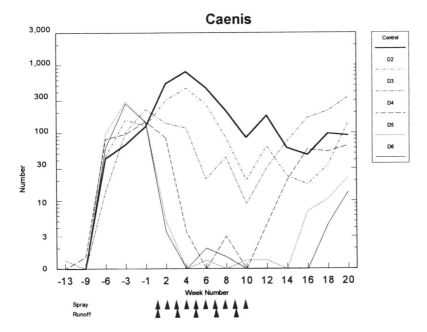

Figure 6. Mean numbers of Caenidae larvae (*Caenis* sp.) collected in MAS samplers over time.

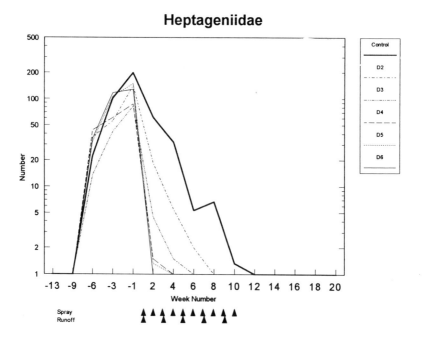

Figure 7. Mean numbers of Heptageniidae larvae collected in MAS samplers over time.

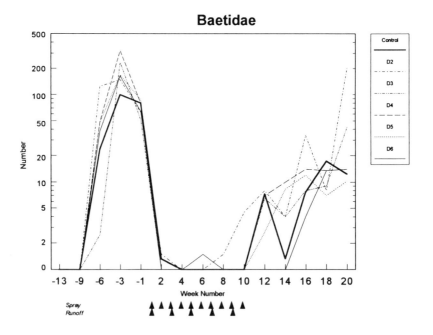

Figure 8. Mean numbers of Baetidae larvae collected in MAS samplers over time.

Table 4. Summary of Statistical Analysis of Ephemeroptera Larvae Abundances in MAS Samples

Taxa	Comparison Ctrl. vs. Dose[a]	Weeks Postapplication One									
		Treatment					Posttreatment				
		2	4	6	8	10	12	14	16	18	20
Caenidae	2					*	*				
	3		*			*	*				
	4		*	*	*	*	*				
	5	*b	*	*	*	*	*	*	*		
	6	*	*	*	*	*	*	*	*	*	
Heptageniidae	2	*	*								
	3	*	*	*							
	4	*	*	*							
	5	*	*	*							
	6	*	*	*							
Baetidae	No Significant Differences										

[a] Dose range represented as 2 (lowest) to 6 (highest).
[b] Mean for this dose significantly less than control mean (Tukey's HSD, $p < 0.20$).

other two mayfly families (Table 4). The response of caenid larvae to tralomethrin is one of the most dramatic recorded for any organism monitored in the entire study.

Adult midges, family Chironomidae, were commonly found in the emergence trap samples. At this level of taxonomic resolution, these invertebrates did not appear responsive to the exposure gradient (Figure 9). However, following enumeration at the subfamily level, statistical analysis indicated that Tanypodinae were more sensitive to tralomethrin than Orthocladinae, while Chironominae were not significantly affected at all (Table 5). The larval habitat preferences of these subfamilies probably contributed to the differences in response to tralomethrin by influencing the level of exposure. Tanypodinae are free-living in the water column, Orthocladinae are somewhat sedentary building loose cases around themselves, while Chironominae are predominantly sedentary and confined to the cases they construct.[26]

It is well documented that pyrethroids are highly toxic to fish, according to laboratory assay results.[27-30] In contrast to the effects on invertebrates, the effects of tralomethrin on fish in this field study do not agree well with the literature. The lack of agreement between the laboratory assay results and those of this field study may be due to differences in the bioavailability of tralomethrin in the test systems. Approximately 207,000 bluegills were harvested from the mesocosms at the end of the study. Treating the mesocosms with tralomethrin did not significantly reduce average numbers or weights of fish, relative to the control, and no dose-related trends were evident upon segementation of these data into size categories (Figures 10 and 11). There were no observations of fish mortality throughout the study. Even though tralomethrin did, in some instances, reduce numbers of invertebrates, this effect did not appear to manifest itself adversely within the bluegill populations.

Although brief, the preceding discussion of representative results do identify the principal study finding to be that tralomethrin significantly affects aquatic inver-

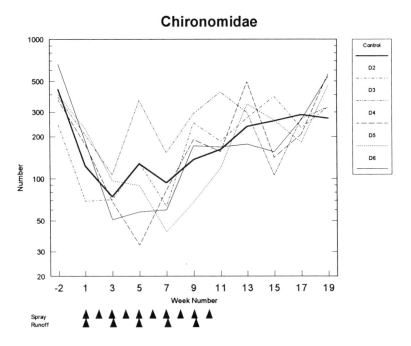

Figure 9. Mean numbers of Chironomidae collected in emergence traps over time.

tebrates and, therefore, the structure of aquatic ecosystems. To present this finding in greater detail, the effects of tralomethrin on all aquatic invertebrates are summarized for the entire study (Figure 12). This summary is presented for various levels of alpha because the EPA has recently indicated a preference for using alpha levels greater than the traditional 0.05 for hypothesis testing (Stunkard, C.L., 1990. Tests of Proportional Means for Mesocosm Studies. Symposium Presentation, Simulated Field Testing in Aquatic Ecological Risk Assessment. 11th annual SETAC meeting). It should be emphasized that the level of alpha defines the probability of committing a type I error, i.e., rejection of a true null hypothesis, and that levels of alpha and statistical significance are inversely related.[31] In the context of the

Table 5. Summary of the Statistical Analysis of Chironomidae Abundances in Emergence Traps

		Weeks Postapplication One									
		Treatment					Posttreatment				
Taxa	Comparison Ctrl. vs. Dose[a]	1	3	5	7	9	11	13	15	17	19
Tanypodinae	5						*				
	6			*b		*	*				
Orthocladinae	6			*		*					
Chironominae		No Significant Differences									

[a] Highest dose represented by 6.
[b] Mean for this dose significantly less than control mean (Tukey's HSD, $p < 0.20$).

FISH AVERAGE NUMBER

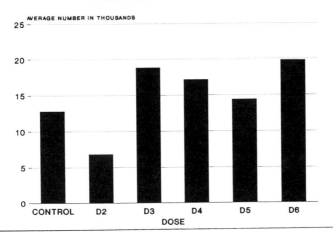

Size♦	Control	D2	D3	D4	D5	D6
(cm)			Estimated Average Number			
1 - 1.9	5263	860	2985	3192	3245	8061
2 - 2.9	5449	3763	11912	11030	8876	10403
3 - 3.9	1680	1826	3266	2470	1984	1028
4 - 4.9	251	301	559	352	180	225
5 - 5.9	72	40*	53	34*	29*	45*
6 - 6.9	17*	11*	15*	23*	14*	14*
7 - 7.9	4*	6*	3*	12*	4*	11*
8 - 8.9	2*	0	2*	3*	1*	4*
9 - 9.9	0	1*	0	0	0	2*
10 - 10.9	0	0	0	1*	1*	0
11 - 11.9	2*	1*	4*	3*	2*	2*
12 - 12.9	2*	4*	3*	3*	4*	3*
13 - 13.9	8*	9*	7*	3*	4*	6*
14 - 14.9	10*	13*	12*	13*	14*	13*
15 - 15.9	7*	5*	6*	10*	7*	8*
16 - 16.9	3*	1*	2*	1*	3*	2*
17 - 17.9	0	1*	1*	0	0	1*
	12771	6840	18827	17148	14368	19825

* Actual Count

♦ No fish less than 1.0 cm were collected.

Figure 10.   Mean numbers of bluegills harvested at study termination.

FISH AVERAGE WEIGHT

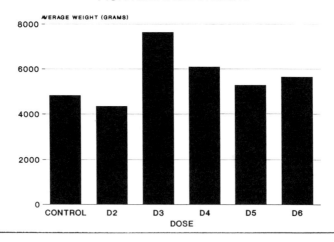

Size♦	Control	D2	D3	D4	D5	D6
(cm)			(grams)			
1 - 1.9	439.03	88.57	314.26	300.65	319.00	819.18
2 - 2.9	1157.99	862.72	2627.01	2156.77	1724.34	1995.30
3 - 3.9	926.32	1102.20	2042.44	1351.50	1088.64	579.38
4 - 4.9	306.48	366.19	674.48	405.13	208.56	277.95
5 - 5.9	162.77	88.43	121.44	77.85	66.21	103.95
6 - 6.9	65.49	42.00	60.91	99.65	54.39	54.31
7 - 7.9	25.60	35.36	20.17	76.09	21.12	65.73
8 - 8.9	21.65	0.00	15.94	28.10	8.45	30.66
9 - 9.9	4.58	12.81	0.00	0.00	4.00	20.92
10 - 10.9	0.00	0.00	0.00	9.38	11.85	0.00
11 - 11.9	60.58	10.48	99.87	61.16	41.40	48.02
12 - 12.9	74.59	133.45	96.96	93.10	132.70	74.71
13 - 13.9	355.95	399.18	317.72	126.37	150.87	241.09
14 - 14.9	522.04	727.07	617.58	650.88	707.36	676.51
15 - 15.9	467.43	329.03	406.89	590.58	463.81	468.04
16 - 16.9	231.67	59.53	174.03	65.78	227.98	107.76
17 - 17.9	0.00	84.72	48.75	0.00	34.57	74.78
	4822.17	4341.74	7638.43	6092.99	5265.25	5638.28

♦ No fish less than 1.0 cm were collected.

Figure 11. Mean weights of bluegills harvested at study termination.

current study, the chance of falsely concluding that tralomethrin produces a significant effect increases with the level of alpha. Aside from the influence of varying the criteria for hypothesis testing, data in Figure 12 indicate that immature macroinvertebrates (MAS) were more affected by tralomethrin than mature macroinvertebrates (ET) or zooplankton.

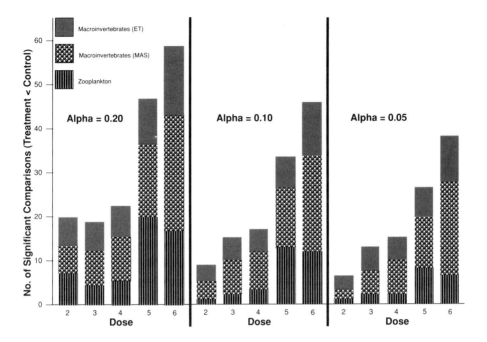

Figure 12.   Summary of statistically significant comparisons indicating treatment mean <
control mean for all invertebrates over the entire study. Means compared using
Tukey's HSD at alpha = 0.20, 0.10, and 0.05.

Future aquatic field studies with pyrethroids would benefit from being narrower
in focus, somewhat smaller in overall scale, and from replication over time, e.g.,
two seasons. Regarding focus, efforts and resources should be primarily allocated
to evaluating effects on invertebrates and discerning how the associated structural
changes to the ecosystem relate to its function and overall stability. deNoyelles and
associates have recently completed essential basic research which relates ecosystem
structure and function to stability via a modeling approach (see Chapters 31 and
32). The next step should be a strong applied research effort to develop practical
field techniques and methodologies which generate data more appropriate for assessing ecosystem function and stability. Regarding scale, the advantages and disadvantages of microcosms have been recently discussed (1991 SETAC/Resolve
Workshop on Aquatic Microcosms for Ecological Assessment of Pesticides). Microcosms could afford investigators with greater power to detect significant effects
because they may be less inherently variable than mesocosms and because they
may be more feasible, economically and logistically, to replicate within experimental
designs. Considerable basic research is still needed to optimize critiera, such as
physical size and biological composition, before microcosms can be fully accepted
as experimental units in ecotoxicological studies conducted for regulatory purposes.
Concerning duration, regardless of the discipline, field studies are traditionally
conducted over a period of at least 2 years to evaluate the influence of seasonal

effects. This fundamental principle has been ignored in many of the aquatic field studies designed to test current ecotoxicological hypotheses.

ACKNOWLEDGMENT

We thank the staff at the University of North Texas Water Research Field Station for their assistance in the conduct of the study, and A. Lewis and N. Herrera of Hoechst-Roussel Agri-Vet Co. for their assistance in preparation of the manuscript.

REFERENCES

1. Ecological Effects Branch. 1982. Pesticide Assessment Guidelines, Subdivision E, Hazard Evaluation: Wildlife and Aquatic Organisms. EPA 540/09-82-024. U.S. Environmental Protection Agency, Office of Pesticide and Toxic Substances, Washington, DC.
2. Touart, L.W. 1988. Aquatic mesocosm tests to support pesticide registrations. EPA 540/09-88-035. Hazard Evaluation Division Technical Guidance Document. U.S. Environmental Protection Agency, Office of Pesticide Programs, Washington, DC.
3. Urban, D.J. and N.J. Cook. 1986. Hazard Evaluation Division, Standard Evaluation Procedure, Ecological Risk Assessment. EPA 540/09-85-001. U.S. Environmental Protection Agency, Office of Pesticide Programs, Washington, DC.
4. Crossland, N.O. 1982. Aquatic toxicology of cypermethrin. II. Fate and biological effects in pond experiments. *Aquat. Toxicol.* 2:205–222.
5. Crossland, N.O. and J.M. Hillaby. 1985. Fate and effects of 3,4-dichloroaniline in the laboratory and in outdoor ponds: II. Chronic toxicity to *Daphnia* spp. and other invertebrates. *Environ. Toxicol. Chem.* 4:489–499.
6. Larsen, D.P., F. deNoyelles, F. Stay and T. Shiroyama. 1986. Comparisons of single-species, microcosm and experimental pond responses to atrazine exposure. *Environ. Toxicol. Chem.* 5:179–190.
7. Odum, E.P. 1984. The mesocosm. *Bioscience* 34:558–562.
8. Solomon, K.R. and K. Liber. 1988. Fate of pesticides in aquatic mesocosm studies — An overview of methodology. *Proceedings,* Brighton Crop Protection Conference, Brighton, UK, November 21–24, pp. 139–148.
9. Stay, F.S., A. Katko, C.M. Rohn, M.A. Fix and D.P. Larson. 1989. The effects of atrazine on mesocosms developed from four natural plankton communities. *Arch. Environ. Contam. Toxicol.* 18:866–875.
10. Perry, J.A. and N.H. Troelstrup, Jr. 1988. Whole ecosystem manipulation: A productive avenue for test system research? *Environ. Toxicol. Chem.* 7:941–951.
11. Cairns, J., Jr. and J.R. Pratt. 1985. Multispecies toxicity testing using indigenous organisms — a new, cost-effective approach to ecosystem protection. *TAPPI Proceedings,* 1985 Environmental Conference, Atlanta, GA, pp. 149–159.
12. Lind, O.T. 1979. *Handbook of Common Methods in Limnology.* C.V. Mosby, St. Louis, MO.

13. American Public Health Association (APHA). 1985. *Standard Methods for the Examination of Water and Wastewater.* 16th Edition.
14. Lesage, L. and A.D. Harrison. 1979. Improved traps and techniques for the study of emerging aquatic insects. *Entomol. News* 90:65–78.
15. Computer Sciences Corporation. 1980. *Pesticide Run-off Simulator (SWRRB) — Users Manual.* U.S. Environmental Protection Agency, Falls Church, VA.
16. Cochran, W.G. and G.M. Cox. 1957. *Experimental Designs.* John Wiley & Sons, New York, NY.
17. Boyd, C.E. 1979. *Water Quality in Warmwater Fish Ponds.* Auburn University Agricultural Experiment Station, Auburn, AL.
18. Anderson, R.L. 1982. Toxicity of fenvalerate and permethrin to several nontarget aquatic invertebrates. *Environ. Entomol.* 9:436–439.
19. Spehar, R.L., D.K. Tanner and B.R. Nordling. 1983. Toxicity of the synthetic pyrethroids, permethrin and AC 222,705 and their accumulation in early life stages of fathead minnows and snails. *Aquat. Toxicol.* 3:171–182.
20. Anderson, R.L. 1989. Toxicity of synthetic pyrethroids to freshwater invertebrates. *Environ. Toxicol. Chem.* 8:403–410.
21. Coates, R.L., D.M., Symonik, S.P. Bradbury, S.D. Dyer, L.K. Timson and G.J. Atchison. 1989. Toxicology of synthetic pyrethroids in aquatic organisms: An overview. *Environ. Toxicol. Chem.* 8:671–679.
22. Day, K.E. 1989. Acute, chronic and sublethal effects of synthetic pyrethroids on freshwater zooplankton. *Environ. Toxicol. Chem.* 8:411–416.
23. Hill, I.R. 1989. Aquatic organisms and pyrethroids. *Pestic. Sci.* 27:429–465.
24. Mokry, L.E. and K.D. Hoagland. 1990. Acute toxicities of five synthetic pyrethroid insecticides to *Daphnia magna* and *Ceriodaphnia dubia. Environ. Toxicol. Chem.* 9:1045–1051.
25. Pennak, R.W. 1978. *Freshwater Invertebrates of the United States,* 2nd ed. John Wiley & Sons, New York, NY.
26. Oliver, D.R. 1971. Life history of the Chironomidae. *Ann. Rev. Entomol.* 16:211–230.
27. Bradbury, S.P. and J.R. Coats. 1989. Toxicokinetics and toxicodynamics of pyrethroid insecticides in fish. *Environ. Toxicol. Chem.* 8:373–380.
28. Bradbury, S.P., D.M. Symonik, J.R. Coats and G.J. Atchison. 1987. Toxicity of fenvalerate and its constituent isomers to the fathead minnow (*Pimephales promelas*) and bluegill *(Lepomis macrochirus). Bull. Environ. Contam. Toxicol.* 38:727–735.
29. Haya, K. 1989. Toxicity of pyrethroid insecticides to fish. *Environ. Toxicol. Chem.* 8:381–391.
30. Smith, T.M. and G.W. Stratton. 1986. Effects of synthetic pyrethroid insecticides on nontarget organisms. *Residue Reviews* 97:93–120.
31. Sokal, R.R. and F.J. Rohlf. 1981. *Biometry: The Principles and Practice of Statistics in Biological Research,* 2nd ed. W.H. Freeman and Company, New York, NY.
32. Dewey, S.L. and F. deNoyelles, Jr. 1993. On the use of ecosystem stability measurements in ecological effects testing, this volume, Ch. 31.
33. Johnson, M.L., D.G. Huggins and F. deNoyelles, Jr. 1993. Structural equation modeling and ecosystem analysis, this volume Ch. 32.

CHAPTER 27

Response of Zooplankton to Dursban® 4E Insecticide in a Pond Experiment

Willem G.H. Lucassen and Peter Leeuwangh

Abstract: Dursban® 4E, an organophosphorus insecticide (A.I. chlorpyrifos) was applied to two experimental ponds (initial concentrations 1.7 and 16.5 µg/l), while a third pond served as a control. This resulted in an immediate and complete loss of the cladocerans. Concentration-dependent adverse effects occurred on copepod nauplii. From 5 weeks after application, nauplii densities exceeded those of the control, presumably the result of the absence of *Daphnia pulex* and *Daphnia longispina*. There were no observable effects on copepodids plus adult copepods; however, the low level of identification may have obscured toxic effects. Chlorpyrifos had no effect on ostracods. As a secondary effect, the number of rotifers increased substantially within 7 to 14 days after application. This coincided with the disappearance of *Daphnia*. The rotifer *Keratella* sp. became the dominant zooplankton species. Within 2 months after application, the carnivorous rotifer *Asplanchna* sp. presumably reduced the herbivorous rotifer populations and copepod nauplii.

Ordination by correspondence analysis proved effective in visualizing the effects of Dursban® 4E and time on the structure of the zooplankton community. The result of this analysis concurs with data from the literature of several field tests with Dursban®. A Monte Carlo permutation test suggested a significant ($p = 0.01$) reaction of the species to the insecticide treatment.

0-87371-592-6/94/$0.00 + $.50
© 1994 by CRC Press, Inc.

INTRODUCTION

The agricultural use of pesticides may lead to contamination of freshwater eco-systems by drift, leaching, or runoff.[1-4] This sometimes caused mortality of fish and other susceptible aquatic organisms. An increasing awareness by the public and by governmental agencies to the side effects of pesticides currently results in a more stringent registration and regulations for the use of pesticides.

The potential environmental impact of a pesticide for aquatic life is assessed by integrating information about the environmental releases, the fate of the pesticide, and its biological effects. Current procedures in the Netherlands for quantitatively estimating fate and effects are based essentially on physicochemical properties of the pesticide, laboratory data on sorption to sediment, transformation rate and products, bioaccumulation, and on concentration-effect relationships as studied in single-species laboratory toxicity tests: the primary effect. The impact of a pesticide on an aquatic ecosystem, however, is more than its primary toxicological effect on susceptible species. Also included are secondary effects that follow and result from these primary effects. Secondary effects often result from disturbances in ecological predator-prey-competitor relationships.[5]

Although primary toxicity can be studied in the laboratory, extrapolation to the complexity of ecosystems can only be superficial because new physical, chemical, and biological interactions are present in ecosystems that are not incorporated in the laboratory tests.[6] Assessing the *full impact* of a pesticide requires the recording of structural and functional changes for some time following exposure in test systems with ecological realism.

As part of this study we examined primary and secondary effects of Dursban® 4E (A.I. chlorpyrifos) on zooplankton. The responses of zooplankton in 3 artificial ponds were studied weekly for 14 weeks after spraying, and then once a month in springtime, 8 and 9 months after spraying. The results are compared with responses of natural zooplankton communities to formulations of Dursban® that have been reported previously.[7-14] Responses of *Daphnia pulex* in bioassays with water from the high treatment pond are related to laboratory toxicity data.[15,16]

Correspondence analysis and a similarity index are used to describe the effects of chlorpyrifos on the seasonal successions of the zooplankton community structure.

MATERIALS AND METHODS

Insecticide

The substance used in this study is Dursban® 4E, an emulsifiable formulation of 48% A.I. (w/v) chlorpyrifos [$O,O,$-diethyl O-(3,5,6-trichloro-2-pyridyl) phos-phorothioate], supplied by DowElanco Europe. For physicochemical properties see Marshall and Roberts.[17]

Ponds and Application Rates

Three identical ponds located in Wageningen (the Netherlands) were constructed in the summer of 1987. All ponds were shallow, ditch-like excavations measuring 3 × 8 m, lined with a waterproof nontoxic PVC sheet. Water depths varied between 0.5 and 0.7 m and volumes from 8.6 to 10.2 m³. The ponds were filled with 0.25-m natural sediment taken from the Wezelse Plas, an unpolluted lake with moderate nutrient concentrations near Nijmegen.

The sediment was characterized as sandy loam with 2.8% organic matter (loss on ignition) and 1.9% elementary carbon. A biocoenosis developed, originating from the sediment, during the course of 1 year. The dominant aquatic plants in 1988 were *Elodea nuttallii* and *Chara* sp., covering about 40% and 10% of the sediment surface, respectively. The aquatic plants were cut back to a height of 0.1 m in the pond that was to receive the highest dose at 7 and 2 weeks before pesticide application. In the control and low treatment ponds, the aquatic plants were cut back to a height of 0.1 m only 2 weeks before pesticide application. Macrophytes were removed completely from all ponds in a 1-m transversal zone at 7 and 2 weeks before pesticide application.

Dursban® 4E was applied to the surface of 2 ponds on August 30, 1988 (week 35) using a hand-held spraying boom. One pond was treated with 103.3 g A.I./ha; the second one with 11.3 g A.I./ha. Concentrations of chlorpyrifos measured in the water column 1 day after spraying were 16.5 and 1.7 µg/l, respectively. At that time the pesticide was distributed homogeneously through the water column. The third pond was used as a control.

Physicochemical Characteristics of Pond Water

Conductivity, pH, oxygen, and minimum and maximum temperatures were measured weekly between 12.00 and 13.00 h from 6 weeks prior to pesticide application to 15 weeks after application.

Temperature was measured in one pond only, and was found to represent all three.

Water samples for chlorpyrifos residue analysis were taken from the center of each pond 1 day prior to application and at 0.5, 1, 2, 4, 8, 16, 36, and 64 days after application. Four samples (200 to 300 ml) per date per pond were collected from different sites ranging from the pond surface to a few cm above the sediment using a vertically integrating tube sampler. The vertical mixing rate of chlorpyrifos in the water column after spraying was measured for 4 days in both treated ponds. To this end, horizontal laminar water samples were taken via sucking tubes at 10, 25, and 40 cm below the surface. Total organic carbon was measured in all ponds at 0.5, 1, 2, 4, 8, and 16 days after application in unfiltered water samples.

Insecticide Extraction and Analysis

Water samples (200 to 300 ml) were shaken for 1 h with n-hexane (75 ml). If necessary the extracts were concentrated by evaporating 25 ml to near dryness after addition of 50 μl butylstearate solution in n-hexane (1 mg/ml) as "keeper". The residue was taken up into 1.5 ml n-hexane. Chlorpyrifos was determined by splitless injection of 3 μl into a gas chromatograph. GLC operating parameters for column: wide-bore WCOT fused silica capillary, coated with CP Sil 5CB, length 25 m, temperature 240°C, nitrogen flow 10 ml/min; injection block: temperature 250°C; detector: NPD, temperature 280°C, hydrogen flow 3.5 ml/min; auxiliary gases: nitrogen 20 ml/min and air 90 ml/min. Retention time of chlorpyrifos was approximately 4 min. The detection limit was 2.5 pg.

Zooplankton Sampling, Identification, and Statistical Analysis

Samples of zooplankton were taken 1 week, 1 day before application of the insecticide (on August 30, 1988), 1 and 2 days after application, and then every week for 14 weeks after application. In 1989, 8 and 9 months after spraying, 1 sample was taken each month to investigate eventual zooplankton recovery.

Zooplankton was collected using an acrylate sampling tube, volume 1.5 l, which isolated a vertical water column from the surface to a few cm above the sediment. This was conducted twice in the middle, and on either side of the pond in a zone devoid of macrophytes. The 4 samples were combined and a subsample (3 to 4 l) was taken. Zooplankton was preserved by addition of 5 to 10 ml of a J. KJ solution (2 g of iodine and 3 g of potassium iodine per l) and concentrated by sedimentation. The zooplankton concentrate was preserved after 1 day in 4% formalin.

Zooplankton counts were made at 25 × magnification. Subsample volumes were chosen so that at least 25% of the rotifers and nauplii were counted. The entire sample was counted for cladocerans, copepods, and ostracods. Cladocerans and rotifers were identified to genus or species level.[18-20] Copepods were divided into nauplii and copepodids plus adults and were not identified. Ostracods were not identified.

Estimates of the biomass of the organisms were made measuring their dimensions, assuming a specific weight of 1 kg/l. Estimated error >25%.[21]

To visualize the effects of treatments and time on the species composition, correspondence analysis was performed using all samples.[22] This resulted in an ordination biplot, consisting of a sample plot and a species plot. The sample plot exhibits differences between samples: samples (indicated by dots in Figure 4b to d) lying far apart resemble less in species composition than those that lie close together. The species plot reveals occurrence of species: closely located species tend to occur in the same samples. This ordination assumes a unimodal response model; the species plot therefore shows the optima of the species. In superimposing the species and sample plots, the abundance of a given species is high in samples that lie near its optimum and is low in samples that lie far away. The plots only give an approximation of reality since ordination cannot be fully realized in two dimensions. The quality of this approximation is given by the eigenvalues of the first two axes. If the eigenvalues are large with respect to those of the higher order axes, the diagram represents the major differences between samples in species composition.

Canonical correspondence analysis[22] was performed to explain the species responses by ordination axes that are constrained to linear combinations of the exposure to Dursban® 4E. A Monte Carlo permutation test[22] was used to investigate the statistical significance of the effects of the insecticide.

Similarity indices between control and treated ponds were computed to describe whether the treated ponds resemble the control at different times. The similarity ratio according to Ball[22] compares samples on the basis of quantitative data. The index is rather sensitive to the total sample, dominant species, and species richness.

Bioassays

Four-day bioassays were carried out with *Daphnia pulex* in the laboratory[23] to relate toxicity studied in acute "protocol" toxicity tests to toxicity in pond water. Potential recovery (criterium: reproduction) was also determined. Filtered water (pore size 0.5 mm) from the control pond and high treatment pond was collected for 4 successive days, starting on 28, 42, and 69 days after application. Immobility and reproduction were measured in a semistatic test with pond water that was renewed daily. Duplicate tests were run, each with 10 *D. pulex* that were collected from the control pond. During the tests water samples were analyzed daily for chlorpyrifos.

RESULTS

Physicochemical Characteristics of Pond Water

Chlorpyrifos was unevenly distributed over the water column during the first 10 h. In that period the highest concentrations measured in the high and low treatment pond were 26 and 2.3 μg/l respectively; the lowest concentrations measured were 18 and 1.9 μg/l. From day 1 on, chlorpyrifos was mixed completely. The residues in water declined rapidly at first (partitioning) and then more gradually from 16.5 to 0.047 μg/l after 80 days in the high treatment pond, and from 1.7 to 0.005 μg/ l in the low treatment pond, respectively (Figure 1). The residue in the high treatment pond after 7 months was 0.015 μg/l. Traces of chlorpyrifos were measured from day 2 until 16 in the control pond: maximum concentration 0.015 μg/l; min. concentration 0.004 μg/l.

During the first 16 days after application of the insecticide, mean total organic carbon (S.D.) in water of the low and high treatment pond was 21.2 (3.6) and 19.5 (1.4) mg C/l respectively; TOC in the control pond was 26 (3.0) mg C/l. In this

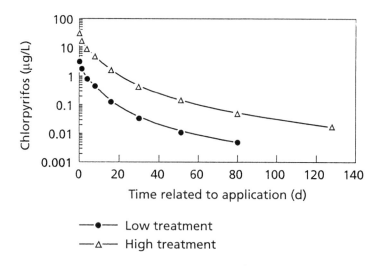

Figure 1.   Concentration of chlorpyrifos in water of two ponds after application of Dursban®
4E.

period less than 8% of chlorpyrifos was bound to organic molecules in water of
both treated ponds (separation of bound/nonbound chlorpyrifos by dialysis; S. Van
Gool, personal communication). Therefore bioavailability of chlorpyrifos in laboratory toxicity test and pond were thought to be almost similar.

It is assumed that there were no acute or secondary effects on oxygen, conductivity, or pH (Figure 2) since the differences per parameter between the three ponds
were similar before and after application of the pesticide. No explanation can be
given for the somewhat higher conductivity observed in the control pond.

Cladocerans

Of the cladocerans, *Chydorus* sp., *Daphnia pulex, Daphnia longispina,* and
Grabtolebris testudinaria were most abundant. The total number was highest in the
control pond. One day after application the cladocerans had disappeared completely
from both treated ponds (Figure 3A). The last time *D. pulex* was seen in the control
pond was 3 months after application. *D. longispina, Chydorus* sp., and *Grabtolebris
testudinaria* were found in this pond after 8 months. *Chydorus* sp. was found again
in the low treatment pond after 8 months, as well as low numbers of *Simocephalus*
sp., a species not seen before. The total number of cladocerans in control and low
treatment ponds were similar after 9 months. The first cladoceran (*D. longispina*)
reappeared in the high treatment pond in low numbers after 9 months.

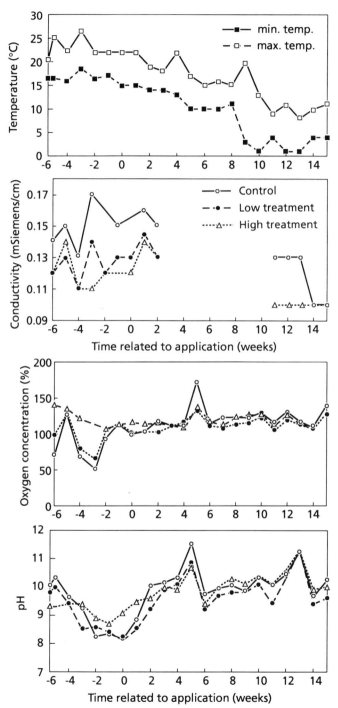

Figure 2. Temperature, oxygen concentration, conductivity, and pH as measured in pond water before and after application of Dursban® 4E.

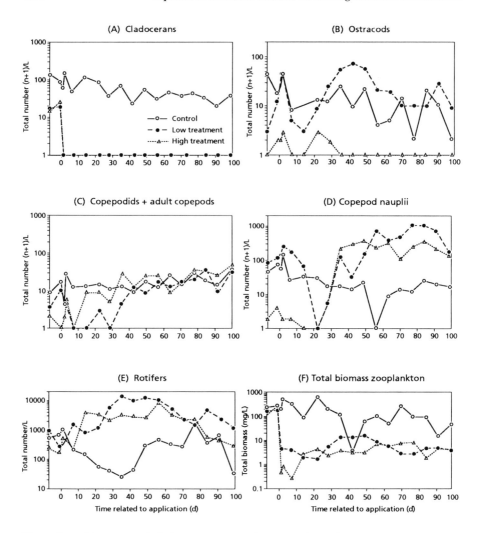

Figure 3.   Responses of cladocerans, ostracods, copepods, rotifers, and total biomass of zooplankton to the application of Dursban® 4E. Exposure concentrations can be read from Figure 1.

Ostracods and Copepods

Before application, ostracods and copepods were almost absent in the pond that was to receive the high treatment (Figures 3B, 3C, and 3D). Ostracods were found in very low numbers in the high treatment pond until day 28; after that they disappeared and were not found even after 9 months. The number of ostracods in the low treatment pond was similar to that of the control (Figure 3B).

Copepods (mainly nauplii and some copepodids) were seen in larger numbers from day 35 after application. The absolute number of nauplii in both treated ponds remained at 10-fold the control during the following 2 months. After 8 months the numbers of nauplii and copepods were similar in all ponds.

Rotifers

Before application the rotifers *Synchaeta* spp. were dominant in the ponds that were to receive Dursban® 4E. *Synchaeta* spp. and *Polyarthra* sp. were found in the control pond in equally high numbers. After application, dominant species in the control pond were *Synchaeta* spp., *Keratella* sp., and *Trichocerca* sp.

The number of rotifers increased in both treated ponds shortly after application (Figure 3E). The *Keratella* sp. rotifer became the dominant zooplankton species. The numbers of rotifers in the treated ponds were 50- to 100-fold the control after 1 month. In this period, codominant species in the high treatment pond were *Synchaeta* spp., *Polyarthra* sp., *Asplanchna* sp., *Brachionus* spp., *Anureopsis* sp., *Filinia* sp., and *Trichocerca* sp. The codominant species in the low treatment pond were *Synchaeta* spp.

Asplanchna sp. was seen only in ponds sprayed with Dursban® 4E. After 8 and 9 months the total number of rotifers in the control pond exceeded that in the treated ponds (2- to 15-fold). The main species in all ponds was *Keratella* sp.; *Asplanchna* sp. was absent.

Effect on Zooplankton Biomass

A drop in the total biomass of zooplankton occurred immediately after application due to the disappearance of *Daphnia*. Zooplankton biomass was about 10% of the control during 3 months (Figure 3F).

Correspondence Analysis

Ordination of the data with correspondence analysis resulted in biplot diagrams presented in Figure 4. The sample scores (Figure 4b to d) show that the species composition in the samples of the low treatment pond before application of Dursban® 4E resembled that of the control pond, since the samples (indicated by the dots A to B) almost have the same position in Figures 4b and 4c. The position of the samples A to B from the high treatment pond (Figure 4d), and therefore their species composition, was different.

The effect of the Dursban® 4E application on species composition is shown in the sample scores. Species composition of the samples of both treated ponds moves away from the control (Figure 4b to d) and back again after 2 to 3 months. The divergence from the control is at a maximum for the highest treated pond. The difference between control and treated ponds is at a minimum after the winter period (samples indicated by dots D to E).

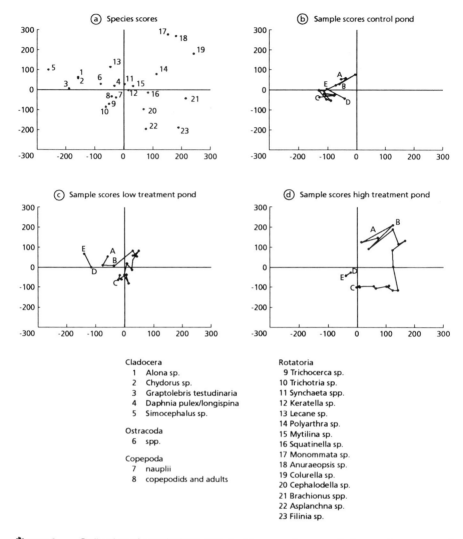

Figure 4.   Ordination of zooplankton data by correspondence analysis: species score and sample scores. Species in species score plot are indicated by numbers. Individual samples in sample scores indicated by dots. Time relative to application in sample scores marked by A through E. A: 1 week before application; B: 1 day after; C: 99 days after; D: 240 days after; E: 268 days after application. Eigenvalue of X-axis: 0.25; eigenvalue of Y-axis: 0.21.

The species to the left of the species scores (Figure 4a) represent those species that are characteristic for the control pond (particularly cladocerans). Species to the right (rotifers) are typical after chlorpyrifos application.

The eigenvalues of the X and Y axes of the samples plot are 0.25 and 0.21, respectively. The percentage variance of the species data represented by the first and second axes are 20% and 16%, respectively.

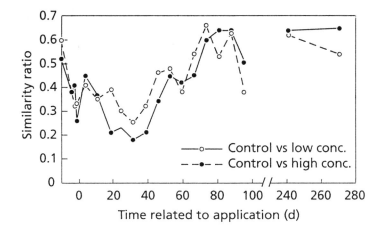

Figure 5.   Similarity ratio according to Ball.[22] Comparison of samples on the basis of quantitative data of 23 groups within the zooplankton.

Canonical correspondence analysis (CCA) indicated that species composition and the insecticide treatment are strongly correlated (>0.9 and >0.8, respectively) for the first and second CCA axis. A Monte Carlo permutation test resulted in a p value of 0.01 for the first canonical axis, suggesting a significant reaction of the species to the insecticide treatment.

Similarity Ratio

The similarity ratio of zooplankton in the treated ponds 2 to 3 months after application of the insecticide is the same as it was when Dursban® 4E was applied (Figure 5).

Bioassay with *Daphnia pulex*

Table 1 shows immobility of *Daphnia pulex* in 1- and 2-day bioassays with pond water. Reproduction was observed from day 69 (chlorpyrifos concentration in pond water 0.065 μg/l).

DISCUSSION

Cladocerans were the first species to disappear after application of Dursban® 4E (Figure 3A). The high susceptibility and therefore the direct toxic action (primary effect) can be demonstrated by comparing laboratory data on acute toxicity for *Daphnia pulex* with exposure concentrations and effects in the pond study. Van der Hoeven and Oldersma studied acute laboratory toxicity of Dursban® 4E for *D.*

Table 1. Immobility of *Daphnia pulex* Exposed to Chlorpyrifos in Pond Water

Effect	Days of Exposure	Chlorpyrifos μg/l	Expected EC Value (95% conf. limit)
30% imm.	1	0.45	0.31 (0.22–0.42)
60% imm.	1	0.45	0.47 (0.36–0.61)
20% imm.	2	0.21	0.12 (0.07–0.22)

Note: Days of exposure and measured chlorpyrifos concentration are indicated. EC values (μg/l) that would give the same effect within the specified time were calculated from a protocol laboratory toxicity test (Reference 16; Van der Hoeven, IVM-TNO, personal communication).

pulex:[16] immobility (EC_{50}) was induced at 0.42 and 0.21 μg/l when exposed for 24 or 48 h. LC_{50} (24 h) was 2.6 μg/l. Since hardly any difference was observed between the number of immobile and killed *Daphnia* after 48 h, LC_{50} (48 h) was the same as EC_{50} (48 h). Adjuvants (Dursban® 4Blank) were nontoxic (immobility) at the highest concentration tested (91 μg/l).

Residues in pond water (Figure 1) during the first 2 days decreased from 26 to 13 μg/l (high treatment pond) and 2.3 to 1.1 μg/l (low treatment pond). Therefore, the absence of *D. pulex* in pond water samples on the first day can be explained by immobility and/or mortality. This conclusion is confirmed by bioassays with pond water (Table 1). Very good agreement was observed between toxicity in the protocol toxicity test and in pond water, assuming that the bioavailability of chlorpyrifos for *Daphnia* was high (see Results).

When chlorpyrifos concentration in pond water had dropped to 0.065 μg/l, *D. pulex* was again able to reproduce (Table 1). This observation is in agreement with the NOEC for Dursban® 4E (17 days; immobility, mortality, reproduction; test conducted according to OECD guideline no. 202) of 0.1 μg/l.[15] Conditions for reproduction of *D. pulex* in the low treatment pond were met before day 30 when chlorpyrifos concentration had dropped to 0.03 μg/l. However, no recovery of *D. pulex* within 3 months after application of the insecticide was observed. Although potential recovery was possible from the beginning of October, the fact that it did not occur might have been normal considering the life cycle of cladocerans (reproduction is low and egg development is slow in autumn).

Considering the fact that cladocerans in the high treatment pond were observed in low numbers, even after 9 months, restoration might have slowed down due to the absence of any survivors. This concurs with Ali and Mulla[11] who observed that the chlorpyrifos impact was much more severe and longer lasting when lakes were treated entirely instead of partially.

Recovery of cladocerans in the low treatment pond (*Chydorus* sp., *Simocephalus* sp.) must have begun between 3 and 8 months after application. The first cladoceran (*D. longispina*) was found in the high treatment pond after 9 months. Unfortunately there is no information on the susceptibility of *Chydorus* sp., *Simocephalus* sp., or *D. longispina*. However, high susceptibility of other cladoceran species for chlorpyrifos has been noted by several authors (Table 2).

Table 2. Chlorpyrifos Toxicity to Zooplankton and Recovery as Observed in Field Studies

Species	Conc.[a,b,c] (µg/l)	Effect After (% Reduction)	Population Increased After (Weeks)	Observation (Weeks)	Ref.
Cladocera					
Moina micrura	<0.8[c]	Week 1 (>95)	2	3	7
Moina micrura	<4[a]	Week 1 (100)	3	3	7
Moina micrura	7.2[b]	Week 1 (99+)	*	2	8
Daphnia pulex	7.4[b]	Week 1 (85–98)	3	3	11
Daphnia galeata	7.4[b]	Week 1 (85–98)	3	3	11
Bosmina longirostris	7.4[b]	Week 1 (reduction)	3	3	11
not specified	0.2[a]	Week 1 (>80)	*	5	13,14
not specified	1.4–4.7[a]	Week 1 (>99)	*	7	12
Copepods					
Acanthocyclops sp.	0.2[a]	Week 1(>80)	—	5	13,14
Cyclops vernalis	<0.8[c]	Week 1 (52–92)	2	3	7
Cyclops vernalis (copepodids)	7.2[b]	Week 1 (60–98)	2	2	8
Cyclops sp.	7.4[b]	No effect		3	11
Diaptomus pallidus	10[c]	No effect		3	7
Diaptomus sp.	7.4[b]	No effect		3	11
not specified	0.2–6[a]	Week 1–3 (0–80)	5	5	13,14
Copepod nauplii	1.4–4.7[a]	Week 1 (delay)	3	7	12
Ostracods					
Cyprinotus sp.	3[b]	Week 1 (80)	2	14	11
Cyprinotus sp.	7.4[b]	Week 1–4 (60–90)	7	8	11
Rotifers					
Asplanchna brightwelli	72[b]	No effect	1,2	2	8
Asplanchna brightwelli	97[a]	No effect	2	3	7
5 herbivorous sp.	72[b]	No effect	1	2	8
not specified	1.4–6[a]	Week 3–5 (some)	*	5	13,14

[a] Measured.
[b] Nominal.
[c] Estimated on basis of measured concentrations of other applications.

Note: Concentration 24 h after application; * = no increase observed, — = not indicated. Of References 7 and 8, only results after first application are presented.

Chlorpyrifos seems to have had no or only a slight direct toxic effect on ostracods in both treated ponds (Figure 3B). A more definite interpretation is made impossible by the low numbers in the high treatment pond. The failure of ostracods to develop in this pond remains unclear. Food seems not to be a limiting factor since rotifers thrive. Physicochemical characteristics of water in all ponds varied in a similar way during the experiment (Figure 2) and therefore can give no explanation. There were no indications of a secondary effect.

Interpretation of the possible effect of Dursban® 4E on copepodids plus adults is complicated by the low and fluctuating numbers in both treated ponds shortly before and after application (Figure 3D). Since no concentration-effect relation was observed, there is no evidence that the insecticide was toxic. The low level of identification of the copepods may obscure the existence of interspecific differences within the subclass. Such differences were demonstrated for *Acanthocyclops* sp. and *Diaptomus* spp. in other pond studies (Table 2).

Although the copepod nauplii were not enumerated to genus level, Figure 3D shows that their numbers in both treated ponds were reduced considerably from day 14. This finding concurs with earlier observations.[8,12] Reduction follows the low numbers of adult copepods in both treated ponds. From day 35 on, nauplii densities exceeded those in the control pond 10- to 50-fold. The increase in nauplii might be associated with the disappearance of cladocerans due to a decreased competition for algae as a food resource.

Initially rotifers were the most abundant group within the zooplankton. Their numbers decreased in the control pond from day 1 to 35 (Figure 3 E). In contrast, rotifers in both treated ponds increased within 7 to 14 days of application. None of the genera present in the treated ponds decreased in number, indicating a very low rotifer susceptibility for Dursban® 4E.

There is evidence that rotifers became abundant as a result of the disappearance of *Daphnia*. Gilbert[24] suggests that *Daphnia* (>1.2 mm) can suppress rotifers both through competition for algae as a shared limited food resource and through mechanical interference when rotifers are swept into the branchial chambers of a feeding *Daphnia*. *Daphnia* interference can impose very high mortality rates on susceptible rotifer species even when *Daphnia* populations are low.[25] Increase of rotifers following a decrease of cladocerans was also observed after application of other insecticides.[26,27] Although rotifers are prey for certain copepods,[28] there is no indication that this played a role in the treated ponds since copepod numbers in control and treated ponds were the same.

In the low treatment pond, the herbivorous rotifer genera *Synchaeta, Keratella,* and finally the carnivorous *Asplanchna* increased from day 7, 14, and 56, respectively. Increase started in the high treatment pond with *Anuraeopsis* (day 1 to 14), followed by *Keratella, Polyarthra,* and *Brachionus* on day 14. *Filinia* sp. was observed in larger numbers from day 35. *Asplanchna* was seen first on day 42. All species mentioned (with the exception of *Asplanchna*) are relatively small. Those species were more likely to be damaged by mechanical interference by passing into the branchial chamber of *Daphnia* than the bigger ones. So following the disappearance of *Daphnia*, these small species were most likely to be at an advantage.[24]

Reduction of rotifers and copepod nauplii in both treated ponds, presumably by *Asplanchna,* followed within 3 to 6 weeks after the first observation of this carnivorous rotifer (day 42). These results confirm earlier observations (Table 2). *Chaoborus crystallinus,* of which the early-instar larvae predate on rotifers, was present in very low numbers in the control and low treatment pond, and absent in the high treatment pond, respectively (R. van Wijngaarden, personal communication). Therefore, this species is thought not to have contributed to the disappearance of the rotifers.

The drop in zooplankton biomass resulting from the disappearance of cladocerans was made up partially by the rotifers (Figure 3F). Biomass remained lower than in the control pond during the first 3 months; this may have had negative effects on the development and growth of carnivorous macroinvertebrates.

Scaling of taxonomical groups of zooplankton to their susceptibility for Dursban® 4E roughly ranges from cladocerans, copepods, ostracods to rotifers (experimental results and Table 2). Cladocerans were found to be most susceptible. In a pond study by Hurlbert et al.,[7] *Cyclops vernalis* was found to be almost as susceptible as the cladoceran *Moina micrura.* Due to interspecific differences, susceptibility to chlorpyrifos within the subclass of copepods is similar to that of the ostracods: some adult copepods being as or less susceptible than the ostracods mentioned. Rotifers were found to be very tolerant and opportunistic organisms.

Ordination of the data by correspondence analysis presents an indication of susceptibility of the species following chemical stress. One must realize, however, that the plot (Figure 4a), among other variations, exhibits both primary toxic effects and secondary effects. Most susceptible species (cladocerans) are found to the left of the graph, while the more tolerant species (rotifers) are on the right. Copepods and ostracods lay in between. This result concurs with data from literature of several field tests (Table 2). The sample plots (Figure 4b to d) demonstrate the changes in species composition and abundance following application of Dursban® 4E. It also suggests that the effect is transient; ultimately, both treatment ponds have much in common with the control pond. A similar result is seen by comparing the samples on the basis of the similarity ratio (Figure 5).

At the start of the experiment, the sample scores indicate that the high treatment pond deviates from the control pond. This could be the result of removing macrophytes from the high treatment pond 7 weeks before application. The low similarity ratio observed before application of the pesticide in both treated ponds cannot be explained by the macrophyte removal, since that was done only in the high treatment pond.

Although the use of correspondence analysis to ordinate the data presented here must be considered as tentative (low abundance of some organisms; no replicates), it is thought that statistical methods that visualize long-term processes at community structure level might be helpful in describing the threat of toxic chemicals on ecosystems.

ACKNOWLEDGMENT

The research was supported by the Directorate for Agricultural Research, and by the Dutch Ministry of Education and Science's "Program Committee for Toxicological Research (STO)". We thank Steven Crum for the chemical analysis of water samples, René van Wijngaarden for the biotests, Miss Saskia van Gool for bioavailability studies, Jan Oude Voshaar for statistical support, and DowElanco Europe for their interest in the research program and for providing us with Dursban® 4E and its adjuvants.

REFERENCES

1. Miles, J.R., E.F. Bolton and C.R. Harris. 1976. Insecticide and nutrient transport in water, related to agricultural land use of a stream basin in Ontario, Canada. *Arch. Environ. Contam. Toxicol.* 5:119–128.
2. Wanchope, R.D. 1987. The pesticide content of surface water drainage from agricultural fields, *J. Environm. Quality* 7:459–472.
3. Thomas, K. and B.C. Nicholson. 1989. Pesticide losses in runoff from a horticultural catchment in South Australia and their relevance to stream and reservoir water quality. *Environmental Technology Letters* 10:117–129.
4. Wan, M.T. 1989. Levels of selected pesticides in farm ditches leading to rivers in the lower mainland of British Columbia. *J. of Environm. Sci. & Health B* 24:183–203.
5. Hurlbert, S.H. 1975. Secondary effects of pesticides on aquatic ecosystems. *Residue Reviews* 57:81–148.
6. Cairns, J. Jr. and B.R. Niederlehner. 1987. Problems associated with selecting the most sensitive species for toxicity testing. *Hydrobiologia* 153:87–94.
7. Hurlbert, S.H., M.S. Mulla, J.O. Keith, W.E. Westlak and M.E. Dusch. 1970. Biological effects and persistence of Dursban® in freshwater ponds. *J. Econ. Entomo* 63:43–52.
8. Hurlbert, S.H., M.S. Mulla and H.R. Willson. 1972. Effects of an organophosphorus insecticide on the phytoplankton, zooplankton, and insect population of freshwater ponds. *Ecological Monographs* 42:269–299.
9. Nelson, J.H. and E.S. Evans. 1973. Field evaluations of the larvicidal effectiveness. Effects on non-target species and environmental residues of a slow-release polymer formulation of chlorpyrifos. U.S. Army Environmental Hygiene Agency, Aberdeen Proving Ground, Maryland. Rep. No. 44-002-73/75, March–October 1973. Data mentioned in Reference 17.
10. Cooney, J.C. and E. Pickard. 1974. Field tests with Abate and Dursban® insecticides for control of flood water mosquitoes in the Tennessee Valley Region. *Mosq. News* 34:12–22. Data mentioned in Reference 17.
11. Ali, A. and M.S. Mulla. 1978. Effects of chironomid larvicides and diflubenzuron on non-target invertebrates in residential recreational lakes. *Environmental Entomology* 7:21–27.

12. Hughes, D.N., M.G. Boyer, M.H. Papst, C.D. Fowle, G.A.V. Rees and P. Baulu. 1980. Persistence of three organophosphorus insecticides in artificial ponds and some biological implications. *Arch. Environ. Contam. & Toxicol.* 9:269–279.
13. Brazner, J.C., S.J. Lozano, M.L. Knuth, L.J. Heinis, D.A. Jensen, K.W. Sargent, S.L. O'Halloran, S.L. Bertelsen, D.K. Tanner and E.R. Kline. 1987. The effects of Chlorpyrifos on a Natural Aquatic System: A Research Design for Littoral Enclosure Studies and Preliminary Data Report. U.S. E.P.A., Environmental Research Laboratory — Duluth, Pesticide Branch, Duluth and University of Wisconsin-Superior, Center for Lake Superior Environmental Studies, Superior, pp. 139.
14. Brazner, J.C. and E.R. Kline. 1990. Effects of chlorpyrifos on the diet and growth of larval fathead minnows, Pimephales promelas, in littoral enclosures. *J. Fish. Aquat. Sci.* 47:1157–1165.
15. Hoeven, N. van der. 1989. The chronic toxicity of the insecticide Dursban® 4E (A.I. chlorpyrifos) and its formulation products, Dursban® 4 Blank to *Daphnia pulex*. Report no. R 89/014. Netherlands Organization for Applied Scientific Research, pp. 24.
16. Hoeven, N. Van der and H. Oldersma. 1989. The acute toxicity of chlorpyrifos, Dursban® 4E, an insecticide containing chlorpyrifos and its formulation products, Dursban® 4 Blank to *Daphnia pulex*. Report no. R 89/013. Netherlands Organization for Applied Scientific Research, pp. 13.
17. Marshall, W.K. and J.R. Roberts. 1978. Ecotoxicology of Chlorpyrifos. National Research Council of Canada, *NRCC No. 16709*. Ottawa, pp. 37.
18. Streble, H. and D. Krauter. 1982. Das Leben im Wassertropfen. Mikroflora und Mikrofauna des Süsswassers. Franck'sche Verlagshandlung. W. Keller and Co, Stuttgart, pp. 355.
19. Leentvaar, P. 1978. De Nederlandse kieuwpootkreeften en watervlooien (Branchiopoda-Crustacea). *K.N.N.V.-Mededeling No. 127*. Hoogwoud, pp. 32.
20. Scourfield, D.J. and J.P. Harding. 1966. A key to British freshwater Cladocera with notes to their ecology. Freshwater Biological Association. Scientific publication, 3rd edition, pp. 55.
21. Geelen, J.F.M. 1969. Vergelijkend planktononderzoek in twee Hatertse vennen. Ph.D. thesis. University of Nijmegen, The Netherlands.
22. Ter Braak, C.J.F. 1987. In: Jongman, R.H.G., C.J.F. ter Braak, and O.F.R. van Tongeren (Eds.). Data analysis in community and landscape ecology. Pudoc Wageningen. *ISBN 90-220-0908-4*, pp. 292.
23. EPA 1975. Methods for acute toxicity tests with fish, macroinvertebrates and amphibians. *Ecological Research Series EPA*. 660/3-75-009.
24. Gilbert, J.J. 1988. Suppression of rotifer populations by *Daphnia*: a review of the evidence, the mechanisms, and the effects on zooplankton community structure. *Limnol. Oceanogr.* 33:1286–1303.
25. Burns, C.W. and J.J. Gilbert. 1986. Effects of daphnid size and density on interference between *Daphnia* and *Keratella cochlearis*. *Limnol. Oceanogr.* 31:859–866.
26. Kaushik, N.K., G.L. Stephenson, K.R. Solomon and K.E. Day. 1985. Impact of permethrin on zooplankton communities in limnocorrals. *Can. J. Fish. Aquat. Sci.* 42:77–85.
27. Papst, M.H. and M.G. Boyer, 1980. Effects of two organophosphorus insecticides on the chlorophyll a and pheopigment concentrations of standing ponds. *Hydrobiologia* 69:245–250..
28. Gulatti, R.D. 1978. The ecology of common planktonic crustacea of the freshwaters in the Netherlands. *Hydrobiologia* 59:101–112.

CHAPTER 28

Algal Periphyton Structure and Function in Response to Consumer Chemicals in Stream Mesocosms

Scott E. Belanger, James B. Barnum, Daniel M. Woltering, John W. Bowling, Roy M. Ventullo, Scott D. Schermerhorn, and Rex L. Lowe

Abstract: An experimental stream facility (ESF) has been designed to evaluate the fate and effects of consumer product chemicals as a component of municipal wastewater treatment plant (WWTP) effluent discharged to streams and rivers. The ESF is equipped with provisions for eight, 11-m experimental stream channels, computer-assisted flow control, and water quality monitoring. The stream channels were used in preliminary investigations in fall 1989 and spring 1990 to evaluate the responses of indigenous biota to the mono-alkyl quaternary cationic surfactant, lauryl trimethyl ammonium chloride (C_{12} TMAC). In this chapter, periphytic algal responses to C_{12} TMAC exposure for 8 weeks at concentrations ranging from 0 to 1250 μg/l are presented. Because consumer product chemicals are introduced into the aquatic environment as a portion of WWTP effluent, all experimental streams received 10% secondary WWTP effluent with 90% parent river water (Lower East Fork of the Little Miami River, Ohio). Structural and functional endpoints at the population and community level were employed. Endpoints varied widely in sensitivity (reflecting enhancement or impairment depending on exposure concentration) and across seasons. In general, structure and function were impaired at 250 to 1250 μg/l, concentrations well above those that would be found at 100% industry-wide use of the surfactant. Advantages of the stream mesocosm approach are discussed in the context of risk assessments for consumer product chemicals.

INTRODUCTION

Environmental risk assessment strategies have become increasingly more complex in recent years. The case for generating information using progressively realistic approaches to assess risk to aquatic systems up to the ecosystem level was summarized in the early 1980s.[1] Today, one of these strategies involves the use of experimental ecosystems. Outdoor pond mesocosms are being employed in tier IV assessments of pesticides under the Federal Insecticide, Fungicide, and Rodenticide Act,[2] and stream mesocosms have been used to evaluate the environmental safety of a diverse array of chemicals introduced to surface water.

Household cleaning products are manufactured in large quantities and are disposed into the environment either as solid waste or via "down-the-drain" following consumer use. Thus, quantities of chemical ingredients reach municipal wastewater treatment plants (WWTP). Introduction to surface waters can occur if biodegradation and/or removal in treatment facilities is less than 100%. Ultimately, the detergent industry must be concerned with the safety of chemicals that enter surface water as a portion of WWTP final effluent. In response to this concern, the Procter & Gamble Company has developed an indoor experimental stream facility (ESF) to evaluate the environmental safety of consumer product chemicals under semicontrolled, yet environmentally realistic, conditions. This chapter will review the current operation and level of ecotoxicological understanding of this system and outline current approaches being used to evaluate periphytic algal communities of the system.

Experimental evaluations at the ESF began in 1989. The goals of these preliminary investigations were to:

1. Condition and evaluate computer-assisted operations of the facility, including flow and chemical monitoring
2. Refine experimental methods for periphytic and macroinvertebrate communities and fish population testing
3. Evaluate a model surfactant (surface active agent) under conditions defined for anticipated future experiments at the ESF. The surfactant chosen was lauryl trimethyl ammonium chloride (C_{12} TMAC), a compound with a well-described safety profile.[3]

TMAC is a monoalkyl ammonium compound. Previous studies have shown single-species algal toxicity tests to be more sensitive to this class of chemicals than invertebrate or fish tests.[3] Phytoplankton enclosure and in-stream dosing studies with TMAC have suggested single-species tests were overly conservative in predicting toxicity profiles.[4,5] The focus of this chapter will be to describe functional and structural responses of native algal periphyton in stream mesocosms exposed to TMAC and to provide comparisons with historical laboratory and field environmental data. Time course and dose-dependent changes at the population and community level were evaluated in two 11-week experiments, one in the fall of 1989 and one in the spring of 1990.

DESCRIPTION OF THE FACILITY AND SITE

Lower East Fork and Wastewater Treatment Plant Siting

The Procter & Gamble Company's ESF is located on the Lower East Fork of the Little Miami River (LEFR) in Clermont County, near Milford, OH. The ESF consists of a 325-m² building with provisions for eight 11-m experimental stream channels. A 51-m² trailer provides office and laboratory space for the facility. The facility also includes a river water intake on the bank of the LEFR and three sewage effluent pumping stations located within the Lower East Fork Regional WWTP adjacent to the ESF.

Water for experimental purposes at the ESF is supplied by the LEFR, a fifth order tributary of the Little Miami River that flows through southwestern Ohio. The LEFR was designated an Exceptional Warmwater Habitat in 1989 by the Ohio Environmental Protection Agency. The LEFR is 132 km long, has a drainage area of 1297 km², and descends some 191 m from its origin in New Vienna, OH to its convergence with the Little Miami River near Milford, OH, an average gradient of 1.4 m/km. Most of the LEFR watershed immediately adjacent to the stream consists of heavily wooded slopes and riparian zones. Seven population centers are located in the valleys along the river and its five major tributaries. The mainstem of the LEFR is impounded by dams at three locations upstream of the ESF. Two of these dams are low head dams creating small impoundments to provide municipal drinking water supplies, while the third dam forms the William H. Harsha Reservoir (East Fork Lake) that was constructed by the U. S. Army Corps of Engineers in 1978. The reservoir was constructed for multiple uses: flood control, recreation, water quality control, and municipal water supplies and has a seasonal pool surface area of 8.7 km². The EFL represents the largest single impact on the LEFR and is located 24 km upstream of the ESF. Designed low flow in the LEFR, controlled by the upstream dam, is 30 CFS. In 1990, flow ranged from 43 to 8490 CFS with a mean flow of 956 CFS based upon U.S. Geological Survey discharge data at Perrintown, OH (approximately 2 km upstream of the ESF intake).

Experimental Stream Facility

River water from the LEFR is supplied continuously to the ESF at a rate of 688 l/min (Figure 1). The water is pumped from a wet well intake structure on the bank of the LEFR upstream of the WWTP using submersible pumps (see Appendix B for sources, model numbers, and exact functions of all equipment mentioned in this section. Additional details are available upon request). The water travels some 488 m in a 0.15-m diameter polyvinyl chloride pipe to the stream building where it is placed into a 1740-l fiberglass head tank from which it is distributed by gravity to the research streams at the desired flow rate.

Sewage effluents utilized in experiments at the ESF are provided by the LEFR WWTP located adjacent to the ESF and operated by the Clermont County, OH, Sewer District. The WWTP serves approximately 42,000 residents and processes

Experimental Stream Facility Flow Chart

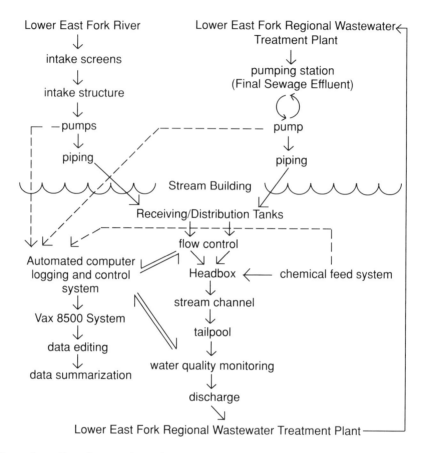

Figure 1.   Flow diagram of experimental stream facility water inputs and outputs, and critical sensor and operations positions.

an average of 2.1×10^7 l/day. Treatment capacity for the WWTP is 2.6×10^7 l/day. The Lower East Fork plant provides advanced secondary treatment of municipal wastewater. Treatment is accomplished using fine screens as part of primary treatment and rotating biological contactors followed by clarification, rapids and filtration, and chlorine disinfection.

Sewage effluents are supplied continuously to the ESF at a rate of 265 l/min by three pumping stations located within the WWTP. Each pumping station supplies a different type of treated municipal effluent: primary (grit removal only), secondary (grit removal plus biological treatment and clarification), and final (grit removal with biological treatment, clarifications, and sand filtration). Once pumped to the ESF, each sewage effluent type is placed into a 1113-l head tank from which it can be distributed to the stream channels at the desired dilution rate. Holding time within a tank is kept at a minimum in an attempt to provide the freshest sewage effluent possible for distribution to the experimental stream channels. Experiments at the ESF to date have only utilized final sewage effluent.

The river water and sewage effluent delivery rates to each experimental stream channel are controlled and monitored continuously. Water quality parameters (temperature, specific conductance, pH, and dissolved oxygen) are monitored continuously and automatically logged into an electronic database for each stream channel as well as the river water and sewage effluent. Test chemical addition is controlled by metering pumps and monitored for failure by a computerized alarm system. Artificial lighting is controlled to track the natural photo period and the lighting intensity is monitored continuously and logged to the electronic database. Effluents from all experimental systems within the stream building are combined and their flow continuously monitored as it is returned to the WWTP for treatment.

Control of river water and sewage effluent delivery to the headbox of each stream is accomplished by a computer interfaced automated system that includes: magnetic flowmeters, actuator operated control valves (combination of a control and diaphragm valve), a process controller (μMAC 5000), a personal computer (IBM PC-AT), and a specialized data logging and control software package (CAMM Version 3.2). Magnetic flowmeters and the actuator operated control valves installed on each river water and sewage effluent supply line in concert with the μMAC 5000 process controller regulate the rate of river water (151 l/min) and sewage effluent (15 l/min) delivered to the headbox of each stream. The process controller continuously (15-s intervals) acquires and processes information from the magnetic flowmeters and compares captured data against a preselected set point (flow value) and appropriate signals are sent to the actuator causing it to open or close the control valve to maintain a constant flow rate. Flow rate values are automatically logged into a daily database file at 5-min intervals by the IBM PC-AT computer and the CAMM software package. All experiments conducted at the ESF, to date, have utilized river water and sewage effluent delivery rates which have yielded an instream sewage concentration of 10% final effluent.

Water quality monitoring within the ESF is accomplished with instrumentation clusters located at the tailpool of each stream channel and at both the river water and sewage effluent head tanks. Each instrumentation cluster continuously monitors temperature, specific conductance, pH, and dissolved oxygen. The instruments utilized to monitor the various water quality parameters are a platinum resistance temperature sensor coupled with a temperature transmitter; an electrolytes conductivity sensor and transmitter; a combination reference/pH electrode coupled with a pION transmitter; and a membrane type dissolved oxygen sensor coupled to a dissolved oxygen analyzer. Readings from all instruments are monitored continuously using the computer interfaced automated system incorporating the μMAC 5000, IBM PC-AT, and CAMM data logging and control software package that provides flow control. However, for water quality monitoring, the μMAC 5000 only functions as a data acquisition system and provides no control functions. The μMAC 5000 acquires sensor reading updates at 30-s intervals from all instruments and the IBM PC-AT computer and CAMM software package automatically log readings into a daily database file at 5-mm intervals. To ensure that the water quality instrumentation readings logged are accurate, sensors are cleaned daily, and weekly calibrations are performed on all instruments. Weekly calibration procedures include electronic calibrations, measurements made using reference standards, and/or measurements made using independent methods or instruments where appropriate. The automated data acquisition system also monitors other environmental parameters and equipment failure alarms for the ESF. The internal temperature, humidity, and lighting intensity in the ESF are monitored as well. The equipment failure alarms monitored are for river water pumps and flow controllers, sewage pumps and flow controllers, test chemical feed pumps, power outages, and water quality instrument failure.

Overhead experimental lighting in the stream building is provided by thirty 1000-watt metal halide arc lamps. This lighting system provides a photo spectrum similar to natural sunlight and is arranged to emit even light intensities across the interior of the stream building. Photosynthetically active radiation monitored at the surface of the stream channels using LI-COR Quantum Sensors is 68 μmol/s/m². Photo period (sunrise/sunset) lighting control is accomplished using a Suntracker that automatically adjusts the photo period for seasonal changes in the number of daylight hours based on time, date, and latitude of the location. A sunrise and sunset twilight period is simulated by staging each of 6 lighting circuits over a 2.1 h interval.

A chemical feed system delivers test chemical to the headbox of each stream channel. Each chemical feed system consists of a 454-l stainless steel stock tank, a variable speed mixer, a diaphragm metering pump, a pressure switch, and high-density polyethylene tubing leading to an injection port. Test chemical is pumped from the stainless steel stock tank by a diaphragm metering pump, through a pressure switch and tubing, into a static on-line mixer, and ultimately to the headbox (142 l) for each stream. Each headbox functions as a mixing chamnber for the river water, sewage effluent, and test chemical prior to entering the stream channel. Dye studies and chemical analyses of TMAC indicate that complete mixing of components occurs within the headboxes. The pressure switch is interfaced with the automated data accusation system and provides an alarm signal upon failure of the metering pump.

Four experimental stream channels were utilized in both the fall 1989 and spring 1990 experiments. The stream channels were constructed of plywood and lined with a clear transparent 10-mil polyvinyl film material that was siliconed and stapled into place. Each stream had an upstream section devoted to algal and microbial studies, a downstream section devoted to macroinvertebrate studies, and a tailpool utilized for water quality monitoring and single-species exposure studies. The algal study area had a slope of 1% and was 4.3 m long by 0.29 m wide. The depth of water in the algal segment was between 1.5 to 2.5 cm. Current velocity ranged from 64.3 cm/s at the head to 44.9 cm/s at the lower end of this segment. The macroinvertebrate study area had a slope of 5% and was 4.3 m long by 0.53 m wide. Water depth was 1.5 to 3.0 cm in the invertebrate segment. The current velocity in this segment of the stream ranged from 55.1 cm/s to 58.0 cm/s depending on the turbulence. The tailpool of each stream was 1.2 m long, 0.61 m wide, and

0.9 m deep (total volume of 568 l). Discharge from the stream channel was through a weir located on the downstream end of the tailpool. All discharges were routed back to the WWTP for treatment.

MATERIALS AND METHODS

General Experimental Design

Four stream channels were used in each study. In both experiments, final WWTP effluent was added proportionately by volume to LEFR water at 1:9 v/v. In the fall study, TMAC was metered by the chemical feed system to each stream in order to achieve target concentrations of 50, 250, and 1250 µg/l. One stream was not treated and served as a control. Streams were not replicated and multiple samples for various response indices were taken from each stream. These concentrations spanned the range of likely no effect to highly effectual concentrations of the test chemical.[3] In the spring study, two streams were dosed at 50 and 250 µg/l and one stream served as a reference control. The fourth stream was used to address experimental stream gradient and sedimentation questions which will not be addressed here. Streams were allowed to be colonized for approximately 2 to 3 weeks prior to addition of the chemical. Depending on the parameter (biological vs. chemical), samples were collected weekly to biweekly for 8 weeks following the commencement of dosing (details for various components will be given below). New unglazed clay tile substrates (surface area of 18,206 mm^2 in the fall and 9103 mm^2 in the spring) were placed in the streams prior to the acclimation period. In this manner, time-course effects and concentration-dependent responses were followed for the duration of the study for both biological and chemical phenomena.

Algal Periphyton Community Structure

Tile substrates were evaluated at preselected intervals for taxonomic composition (Table 1). Tiles were removed from each stream and scraped clean using a razor blade. Samples were preserved in fall 1989 with formalin (final concentration of 5%) and in spring 1990 with buffered glutaraldehyde (final concentration of 3%). Selection of samples on each date was determined using a stratified random design. Tiles were equally represented from upper, middle, and lower reaches of the periphyton evaluation section and from left, middle, and right hand sides of the streambed. In addition to experimental stream periphyton samples in the fall study, algal taxonomic composition of the LEFR was evaluated to gauge the potential species pool drawn to the ESF streams. However, specific LEFR algal sampling methods and LEFR-ESF comparison results will be detailed in a separate manuscript (Belanger and Barnum, unpublished data). In brief, ESF-LEFR algal communities were structurally and functionally similar. Tile substrates (18,206 mm^2) were anchored to submerged stainless steel plates secured to the streambed in the LEFR approximately 0.2 km below the Clermont County Wastewater Treatment Plant final effluent

Table 1. Synoptic Overview of TMAC Experiments Conducted at the P&G Experimental Stream Facility in 1989 and 1990

Date of Experiment	Sampling Days	Replicates	Biology/Chemistry Sampling
		Experiment 1	
Aug. 21, 1989	Acclimation period	—	Biology: none Chemistry/sensors: (light, flow, D.O., temperature conductivity, pH)
Sept. 12 to Nov. 9, 1989	0, 14, 28, 42, 56	3	Biology: algal taxonomy, relative abundance, biovolume, diversity, community similarity
	0, 14, 28, 42, 56	2	Biology: chlorophyll *a*, primary productivity; weekly for all streams
		1	Chemistry: TMAC, background water quality
	All days	5-min intervals	Chemistry/sensors: (light, flow, D.O., temperature, conductivity, pH)
		Experiment 2	
Mar. 27, 1990	Acclimation period	—	Biology: none Chemistry/sensors: (light flow, D.O., temperature, conductivity, pH)
Apr. 16 to Jun. 11, 1990	0, 7, 14, 28, 42, 56	5	Biology: algal taxonomy, relative abundance, biovolume, diversity, community similarity
	−4, 8, 16, 24, 38, 52	3	Biology: chlorophyll *a*, primary productivity
	Weekly for all streams	1	Chemistry: TMAC, background water quality
	All days	5-min intervals	Chemistry/sensors: (light, flow, D.O., temperature, conductivity, pH)

outfall. The samplers were located midstream in a run dominated by gravel substrate a 0.2 to 0.5 m in depth (river depth was a function of river discharge). The calculated effluent dilution based on conductivity measurements at this site was between 5 to 25% effluent depending on runoff conditions. Collected periphyton samples were evaluated to determine diatom and soft-bodied algal taxa. Samples were initially scanned at $400\times$ and $1000\times$ using a Palmer-Maloney nannoplankton counting chamber (0.1 ml) to determine the general health of the sample. Algal units were rated as live vs. dead if chloroplasts were robust. Total algal abundances and counts of diatoms (determined to the lowest practical taxon) were made using the Palmer-Maloney chamber. Samples were analyzed by counting random fields or transects depending on the density of the sample. The first 500 algal units were enumerated and recorded. If a taxon comprised 20% or more of the first 100 units in the sample, the taxon was deleted from the remaining 400 units to be counted and identified. Diatom samples were cleaned of organic debris by nitric acid digestion techniques.

Species identifications of diatoms were confirmed by permanent diatom slides and applied to counting chamber data. Taxonomic literature included Hustedt;[6] Patrick and Reimer;[7,8] Prescott;[9] Bourelly;[10-12] Drouet and Dailey;[13] Germain;[14] and specialized monographs for specific families or genera of algae. Algal biovolumes for identified taxa were calculated using cell geometry measurements and applied to mensuration formulas.[15] Raw data was used to discern the presence of dominant taxa in each sample. Dominance was defined *a priori* as any taxon that comprised 5% or greater of the cell counts or biovolume in any given sample. Thereafter, it would be evaluated as a dominant taxon.

Various descriptors of population and communities were derived from the above analyses: total algal cell density, total algal biovolume (as a correlate of biomass), taxa richness, percent living algae, relative cell abundance by taxon, and relative biovolume abundance. In addition, Shannon-Weaver diversity for cell counts and biovolume data were calculated to integrate the concepts of species richness and evenness of distribution for sampled communities.[16] Community similarity was evaluated using biovolume relative abundance data to derive the cosine measure, Stander's SIMI.[17,18]

Chlorophyll Biomass and Primary Productivity

Tiles for chlorophyll *a* and primary productivity measurements were collected separately from the tiles sampled separately for community structure. Duplicate samples were prepared in the fall study and triplicate samples were prepared in the spring. The methods are briefly summarized below.

Tiles were removed from the streams and incubated between 11:00 a.m. and 3:00 p.m. in recirculating plexiglass chambers containing 400 ml stream water. Radiotracer (5 μCi NaH^{14}CO$_3$) was added to each chamber, the chamber was sealed, and water recirculated by peristaltic pumps under the light and temperature conditions in the streams. Periphyton photosynthate production (primary production) and chlorophyll *a* were measured simultaneously using the DMSO extraction methods of Palumbo et al.[19] Radioactivity measurements were converted to ^{14}C-photosynthate calculated on a per cm^2 basis after correction for inorganic carbon. Expressions of primary productivity were ultimately as μg C/cm^2/day. Chlorophyll *a* specific primary production was also calculated (mg C/mg chlorophyll *a* per day).

Chlorophyll *a* was determined by adding 90% acetone to a subsample of DMSO extract to a volume of 40 ml. Sample were centrifuged at 5000× for 10 min in the dark. Samples were read in a spectrophotometer at 750 nm and 664 nm using a 5-cm cuvette. Phaeophytin *a* was determined following acidification with HCl at the same absorbances used above. Chlorophyll *a* content of the sample was determined using the formula in *Standard Methods for the Examination of Water and Wastewater*[20] and was expressed on an aerial basis.

Water Quality

Water quality was monitored in two ways. In-stream sensor data (temperature, pH, conductivity, and dissolved oxygen) were integrated daily over the life of the experiment for the purpose of this study. In addition, samples were taken weekly in the tail of the stream for TMAC, pH, total alkalinity, chloride, ammonia-N, total-Kjeldahl-N, Nitrate-N, total-P, ortho-P, sulfate, total organic carbon, dissolved

Table 2. Water Quality Characteristics (Ranges of Mean Values Monitored on a Weekly Basis) of Stream Channels Dosed With C_{12} TMAC in the Fall 1989 and Spring 1990 Experiments

Chemistry Parameter	Fall 1989	Spring 1990
pH (SU)	7.3–7.3	7.4–7.4
Total Alkalinity (mg/l)	104 109	107–107
Chloride (mg/l)	115 119	17.6–18.6
Ammonia-N (mg/l)	0.7–0.9	0.9–1.1
Total Kjeldahl-N (mg/l)	1.1–1.3	1.6–1.8
Nitrate-N (mg/l)	1.4–1.7	2.0–2.1
Ortho-P (mg/l)	0.4–0.4	0.3–0.3
Total-P (mg/l)	0.4–0.4	0.4–0.4
Sulfate (mg/l)	43.0–44.5	34.2–35.2
TOC (mg/l)	6.9–7.3	11.9–13.7
DOC (mg/l)	4.4–4.9	5.8–6.5
Calcium (mg/l)	41.4–42.1	36.7–41.0
Magnesium (mg/l)	10.1–10.3	10.4–10.5
Potassium (mg/l)	4.5–4.6	4.3–4.3
Sodium (mg/l)	17.5–17.6	11.1–11.2

Note: Sample sizes were 8 in the fall and 9 in the spring.

organic carbon, calcium, magnesium, potassium, and sodium (Table 2). With the exception of TMAC, which was analyzed by analytical personnel at Procter & Gamble, all other samples were evaluated using U.S. EPA-approved methods[21] at International Technology Corporation (Cincinnati, OH). TMAC concentrations in the water column were determined by a chemical specific method (Table 3). In brief, upon collection, samples were fixed with formalin and Neodol (Shell Oil Co.) to reduce adsorption and microbial loss, followed by alkaline adjustment to pH 14 with NaOH. Samples were then passed through an anion exchange resin (Bio-Rad XAD 1 × 4×, hydroxide form). Column effluent was transferred to a separatory funnel and neutralized with HCl. Linear alkyl benzene sulfonate was added to ion pair with TMAC and then extracted by chloroform (3 times, 150 ml). An HPLC (high pressure liquid chromatography) method for separation and quantification was used for the ultimate analysis. TMAC was quantitated using a Waters Association, Model 820 Chromatographic system fitted with a Wescan Conductivity Detector.[22] An external standardization method with side-by-side reference spiked control samples was employed. Minimum detection limits were 50 ng on column.

Table 3. Theoretical and Actual Concentrations of TMAC Measured in the Final Pool of ESF Streams During the Fall 1989 and Spring 1990 Experiments

Theoretical Concentration (μg/l)	Fall 1989	Spring 1990
0 (control stream)	12.9 ± 12.2	0.1 ± 0.1
50	49.7 ± 11.3	43.2 ± 16.9
250	234.8 ± 56.9	185.1 ± 67.2
1250	1151 ± 124	—
LEFR	—	0.7 ± 1.1
100% effluent	—	0.4 ± 0.1

Note: Means ± 1 S.D. are indicated.

Statistics and Data Management

Due to the large and complex nature of the data set, raw data were entered into spreadsheets (LOTUS 1-2-3 Version 2.01, LOTUS Development Corporation, Cambridge, MA, U.S.). Statistical analyses were performed by uploading spreadsheets (ASCII format) and executing analyses on the Procter & Gamble VAX 8500 System (Digital Equipment, Inc.). Normality was assumed following tests of homogeneity of variance, and all data were analyzed parametrically. Because the vast majority of analyses involved determining exposure-response characteristics of stream communities, analyses of variance were performed. These included evaluations of density, biovolume, diversity, chlorophyll *a*, and primary productivity for community-level data and relative cell and biovolume abundance data for dominant populations. If significant differences were indicated (ANOVA F), a least significant differences procedure was used to detect which treatments were significantly different. All evaluations used a critical *p* value of 0.05 and were two-tailed because it could not be predicted *a priori* if TMAC would enhance or inhibit certain community and population level parameters. The software used was a customized package developed for Procter & Gamble on the VAX 8500 System.

Similarity evaluations involved the use of randomization or permutation tests which are effective in evaluating the degree of similarity within vs. between experimental treatments. A focused evaluation of permutation tests of biological data is given in Smith et al.[23] including the use of Stander's SIMI, although the procedure is easily generalized to include any correlation-type matrix.

RESULTS

Periphyton Community Structure

The ESF periphyton community was dominated by diatoms with over 186 diatom taxa during the two studies (Appendix A). Slightly greater than 200 total taxa (diatom and soft algae) were found in experimental streams or on tiles in the Lower East Fork and evaluated concurrently with the fall 1989 study. The periphyton was especially rich in keeled Nitzchiaceae and biraphid Naviculaceae.

Total biovolume, cell abundance, and chlorophyll *a* were used as indicators of periphytic standing crop biomass during both studies. In the fall 1989 investigation, algal biovolume was initially very low ($<10^6$ μm^3/mm^2, approximately 300 cells per mm^2, approximately 1 μg Chl *a* per cm^2, Figures 2, 3). All three estimators of biomass steadily increased during the initial 4 to 6 weeks without strong evidence of effects from TMAC exposure. By the 56th day of exposure, control biovolumes increased significantly and all TMAC treatments remained near levels previously established. Effects were clearly dose dependent and significant (Table 4). During the spring investigation, algal communities appeared to be more variable and patchy than in the fall and were never clearly affected by TMAC in a dose-dependent

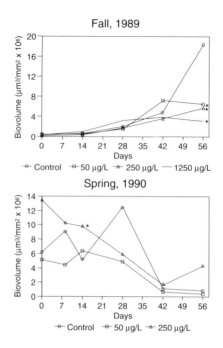

Figure 2. Total biovolume abundance for the fall 1989 (A) and spring 1990 (B) experiments. Sample sizes were 3 and 5, respectively. Groups that were significantly different from the control are indicated by an asterisk.

fashion (Table 5, Figure 2). Biovolumes and abundances peaked early in the spring study and declined thereafter.

During the course of the study, several minor and one major storm event occurred in southwestern Ohio resulting in increased discharge of the LEFR (Figure 4) with subsequent changes in total suspended solids, temperature, and water chemistry. While these factors probably exerted an effect on the algae, all ESF stream channels experienced the same changes in water quality. The use of a control channel allows for comparisons of treatments to a reference condition. The nature of the effects will be evaluated during long-term operation of the ESF. By the conclusion of the experiment, periphyton in the 50 μg/l exposed streams were lower in biovolume than in control streams, but not cell density. The community in the 250 μg/l stream was greater in biovolume, but lower in total cell density. Chlorophyll biomass was elevated on day 8 at 250 μg/l and on day 16 at 50 μg/l. During the following time periods, chlorophyll biomass was not significantly different across treatments.

Taxon richness was generally enhanced by TMAC exposure. In the fall study, the community at 1250 μg/l had significantly greater numbers of taxa present (mean of 64 per sample) compared to the control (45 taxa) (Table 4). In the spring, the 50 and 250 μg/l exposed communities also had significantly greater numbers of taxa than the control by the end of the experiment (25, 30, and 36 for 0, 50, and 250 μg/l, respectively).

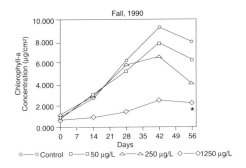

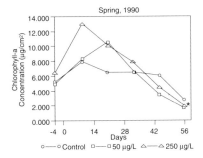

Figure 3.   Chlorophyll *a* biomass measured in periphyton for the fall 1989 (A) and spring 1990 (B) experiments. Sample sizes were 2 and 3, respectively. Groups that were significantly different from the control are indicated by an asterisk.

In the spring study, the percentage of cells with viable chloroplasts (i.e., were "living" cells) was evaluated. Generally, greater than 80% of all cells observed at any time were living, although the percentage tended to decline as communities matured (Table 5). By the end of the experiment, 0, 50, and 250 µg/l exposed

Table 4. Summary of Significant Differences for Total Algal Biovolume, Abundance, and Taxon Richness For the Fall 1989 TMAC Experiment

	Duration of Exposure (Days)				
Treatment (µg/l)	0	14	28	42	56
Total Abundance					
0					
50					(−)
250					(−)
1250					(−)
Taxon Richness					
0					
50					
250					
1250				(+)	(+)

Note: Comparisons are made with the effluent control at alpha = 0.05. Significantly greater response is indicated with a (+) and a significantly lower response is indicated with a (−).

Table 5. Summary of Significant Differences For Total Algal Biovolume, Abundance, Taxon Richness, and Percent Living Cells During the Spring 1990 TMAC Experiment

	Duration of Exposure (Days)					
Treatment (μg/l)	0	8	14	28	42	56
Total Abundance						
50						(−)
250			(+)			(+)
Taxon Richness						
50					(−)	(+)
250						(+)
Percent Living Cells						
50						
250		(+)	(+)			(−)

Note: Comparisons are made with the effluent control (0 μg/l) at alpha = 0.05. Significantly greater response is indicated with a (+) and a significantly lower response is indicated with a (−).

communities had 93.5, 89.5, and 80.6% living cells present, respectively. The latter treatment was significantly lower.

Shannon-Weiner diversity (H'), using biovolume relative abundance data in the fall, reflected the increased taxonomic richness observed with increasing TMAC concentration. Through the 8-week experiment, diversity was positively affected (Figure 5) and significant differences were consistently present at 1250 μg/l and often at 250 μg/l. During the spring study, when the highest test concentration was 250 μg/l, diversity was generally not affected. H' increased during the study and by the conclusion of experiment, communities varied little from each other with respect to diversity.

Community similarity analysis (SIMI) was used to simultaneously compare community structure across treatments. Initial community structure was highly similar (0.965 to nearly 1.000) across all treatments (Figure 6). By 2 weeks of exposure, periphyton SIMI was affected in a dose-dependent fashion in the fall. The 1250 μg/l community diverged quickly and significantly such that by 4 weeks, percent similarity was 0.02 compared to the control. In the second half of the experiment this treatment converged with the control and by the conclusion of the experiment control-treatment similarities ranged from 0.84 to 0.95. These findings were in contrast to the responses of the spring assemblages where the periphytic communities remained highly similar through the first 28 days (SIMI range from 0.958 to near 1.000). In the second month of exposure, treatments gradually departed from the control and differences were significant for contol to 50 and 50 to 250 μg/l comparisons.

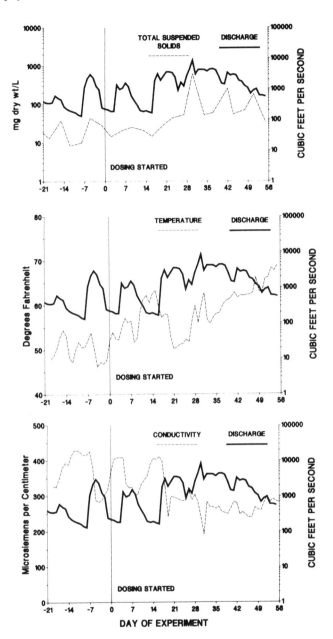

Figure 4.   Representative monitoring data collected for the spring 1990 experiments. Lower East Fork of the Little Miami River (LEFR) discharge, total suspended solids, temperature, and conductivity for the study period are presented. Discharge information was gathered at the U.S. Geological Survey Gauging station at Perrintown, OH upstream of the ESF river water intake. Suspended solids were recorded at the river water intake structure for the ESF. Temperature and conductivity information was collected by sensor monitors at the river water head tank in the ESF.

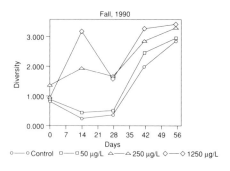

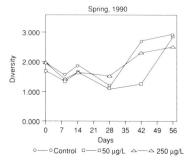

Figure 5.   Shannon-Weiner diversity based upon biovolume relative abundance for all taxa during the fall 1989 (A) and spring 1990 (B) experiments. Groups that were significantly different from the control are indicated by an asterisk.

Dominant Taxa

Fifteen taxa were identified as dominant in the fall. Twenty taxa were dominants in the spring. Of the 15 fall dominants, 9 were consistently present at some density that allowed for meaningful evaluations. *Cyclotella pseudostelligera, Nitzschia palea, Navicula symmetrica,* and *Melosira varians* were most often inversely affected by TMAC exposure (i.e., increased in abundance with increasing TMAC concentration, Figure 7). Total biovolume of these taxa, at 1250 μg/l and after 2 or more weeks of exposure, was approximately 4.8 × 10⁵ μm³/mm² or 15.2% of the community biovolume. *C. pseudostelligera* responded positively to 1250 μg/l for the initial 6 weeks of exposure after which it was significantly reduced compared to the control treatment. This was similar to *Navicula symmetrica* which increased rapidly at the high exposure concentration and later reduced while control densities continued to increase. *Nitzschia palea* and *M. varians* were consistently higher in density in TMAC exposed streams throughout the study.

Five additional taxa (*Navicula graciloids, Navicula salinarum, Nitzschia kutzingiana, Nitzschia frustulum* v. *perminuta*, and *Cocconeis placentula* v. *euglypta*) were depressed in abundance with increasing TMAC concentration (Figure 8). Further, these five taxa tracked closely with total biovolume curves (Figure 2). All 5 taxa were significantly reduced at 250 and 1250 μg/l, except *N. Kutzingiana* which

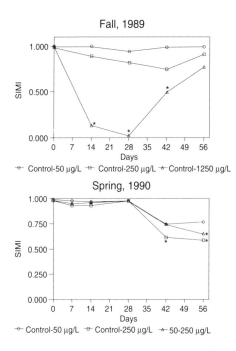

Figure 6. Periphyton community similarity using SIMI and biovolume relative abundance data for the fall 1989 (A) and spring 1990 experiments (B). Groups that are significantly different from the control are indicated by an asterisk.

was reduced only at 1250 µg/l, by 56 days. *C. placentula* v. *euglypta, N. graciloides,* and *N. frustulum* v. *perminuta* were also reduced at 50 µg/l. In total, the control abundances of these 5 taxa accounted for 9.7×10^6 µm³/mm² of the community biovolume (52.7%) by the final day of the experiment.

In the spring study, 23 taxa were dominant at some point in the 8-week exposure period. These taxa comprised 4 groups: early colonizers that were not affected by TMAC exposure and declined in abundance from day 0 (*Navicula cryptocephala, N. radiosa* v. *tenella, N. salinarum* v. *intermedia, S. ovata,* and *Schizothrix calcicola*); taxa that reached maximum abundances between the 7th and 28th day of exposure (*Diatoma vulgare, Fragilaria vaucheria, Gomphonema olivaceum, N. gregaria, N. luzonensis, N. viridula,* and *Stephanodiscus invisitatus*); taxa that increased in abundance by TMAC exposure (*Achnanthes minutissima, Nitzschia acicularis,* and *N. palea*); and taxa which decreased in abundance with TMAC exposure (*A. lanceolata, Amphora veneta, G. angustatum, M. varians, N. frustulum* v. *perminuta,* and *Rhoicosphenia curvata*) (Figures 9 and 10). By the conclusion of the study, dominant taxa accounted for over 90% of the biovolume in all periphytic communities. Taxa that increased in abundance with TMAC exposure accounted for 9% of the community biovolume at 250 µg/l vs. 3.6% in the control by day 56. Further, those that were reduced in abundance with TMAC exposure comprised 45% of the control community biovolume vs. 12% at 250 µg/l.

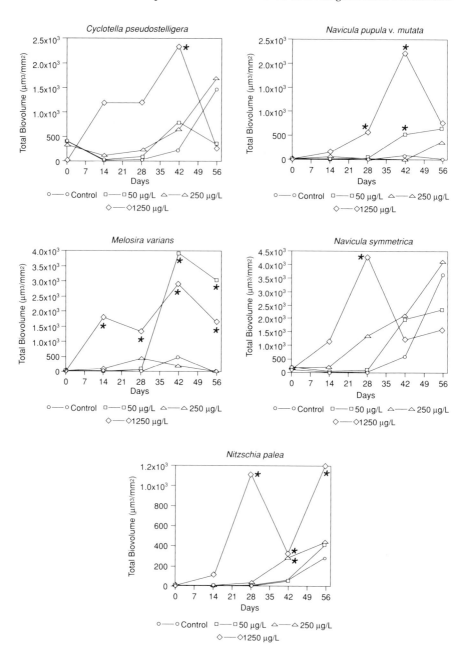

Figure 7.   Dominant taxa that increased in biovolume abundance with increasing TMAC exposure concentrations during the fall 1989 study. Groups that increased significantly relative to the control population are indicated by an asterisk.

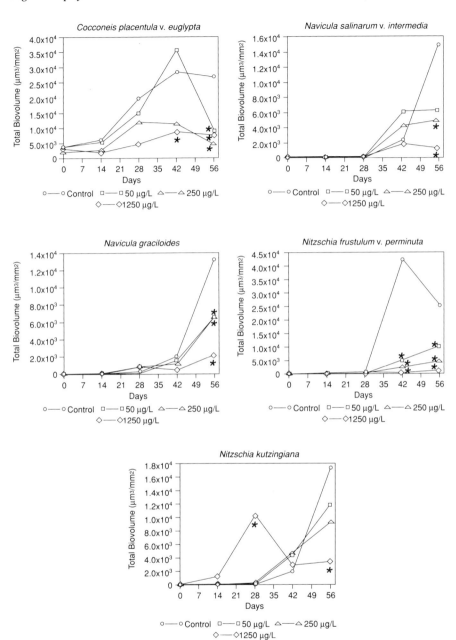

Figure 8. Dominant taxa that were reduced in biovolume abundance with increasing TMAC exposure concentrations during the fall 1989 experiment. Groups that were significantly lower relative to the control are indicated by an asterisk.

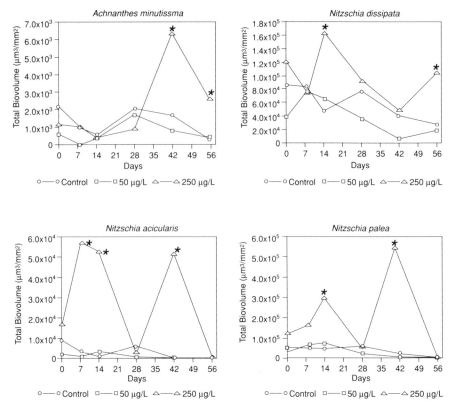

Figure 9. Dominant taxa that increased in biovolume abundance with increasing TMAC exposure concentration during the spring 1990 study. Groups that were significantly greater than the control population are indicated by an asterisk.

Community Primary Productivity

In the fall study, primary productivity tended to increase with time across all treatments except for the 1250 µg/l exposed stream (Figure 11). Reductions were significant at 1250 µg/l and dose-dependent by the end of the experiment. In the spring experiment, control primary productivity did not increase to levels observed in the previous study. TMAC exposure positively affected primary productivity by the sixth week of exposure, but were reversed in the final sampling. The 50 µg/l exposed communities were not significantly affected at any time.

DISCUSSION

Experimental stream systems have been used for many years for the investigation of effects of toxic material exposure on ecosystem structure and function. The value of complex, simulated ecosystems in reducing uncertainties in laboratory to field

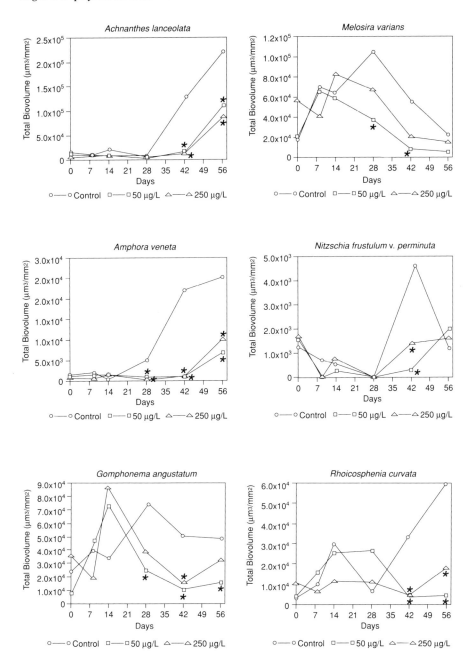

Figure 10.   Dominant taxa that were reduced in biovolume abundance with increasing TMAC exposure concentration during the spring 1990 study. Groups that were significantly lower than the control population are indicated by an asterisk.

Fall, 1989

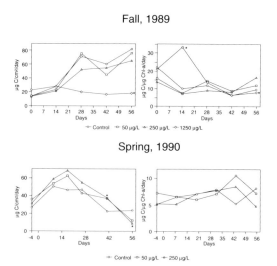

Spring, 1990

Figure 11. Periphyton primary productivity and chlorophyll *a* specific primary productivity during the fall 1989 and spring 1990 experiments. Groups that were significantly different from the control are indicated by an asterisk.

extrapolations of effect and validating biotic response to stressors has been presented.[1,24] Such systems allow for simultaneous exposure and response to chemical addition of numerous species and trophic levels. Effects at the species or population level can be direct (e.g., mortality, individual growth) and indirect (e.g., predation, competition) and offer ecological insights into environmental behavior of chemicals beyond the standard laboratory-based approaches. The ESF was developed to aid the understanding of consumer product chemical behavior (i.e., fate, exposure, and effects) in stream systems using a unique design that incorporates a combination of river water with a low percentage of secondarily treated wastewater treatment plant effluent. This matrix is most appropriate for "down the drain" ingredients used in consumer products. Existing programs include algal (this study, also Reference 25) and heterotrophic periphyton, macroinvertebrate communities[26] and selected macroinvertebrate population studies,[27] fish population studies, and classical ecotoxicological acute and chronic evaluations. Thus, a comprehensive view of structural and functional integrity of streams potentially exposed to consumer chemicals can be determined. ESF operations (e.g., light, stream flow, effluent dilution, physicochemical sensors, and chemical monitoring programs) allow tracking of key abiotic environmental variables that may affect biotic responses.

The mesocosm approach allows the integration of structural and functional testing simultaneously at all levels of biological organization. Thus, a holistic view of the maintenance of ecosystem integrity can be ascertained. In this study, some structural endpoints (e.g., total community biovolume in the fall) indicated that TMAC impacted periphytic communities, whereas a functional endpoint did not statistically indicate impairment. This is in part a reflection of reduced sampling effort for functional endpoints in this study, because they are inherently time consuming and must be performed on living samples. Increased sampling would probably result in a close correlation between structural (biovolume) and functional (primary productivity) impairment. Cairns[28] presented three scenarios for the interplay between structure and function: (1) structural changes could be induced that were not accompanied by functional change; (2) functional shifts could be found that were reflected by structural attributes; or (3) structural and functional damage co-occur because both are intimately associated. In the specific example of TMAC, certain structural attributes of communities were impacted negatively (biovolume), others positively (increased diversity), while productivity was reduced, although not significantly. The third alternative of Cairns[28] appears most plausible here if not constrained by small sample sizes.

Importantly, algal communities grew approximately tenfold or greater in the presence of C_{12} TMAC, even at the highest exposure (1250 μg/l), relative to day 0. Percent living cells and chlorophyll *a* biomass were not affected at the lowest concentration (50 μg/l) as were primary productivity and chlorophyll *a* specific primary productivity. The weight of the evidence indicates that effects at the community and population level are most likely to be observed at concentrations above 250 μg/l.

Responses of the ESF algal communities to TMAC were consistent with results of previous studies. Lewis et al.[4] evaluated periphytic responses of C_{12} TMAC dosed in flow through periphyton chambers in the Little Miami River, Ohio and phytoplankton enclosures in Acton Lake, Ohio for 21 days. Community similarity was the most sensitive index of effect and no-observed-effect concentrations were 0.96 and 0.03 mg/l (measured concentrations) for the two systems, respectively. Photosynthetic activity was similarly affected. Single-species toxicity tests determined that 96-h EC_{50}s for population growth were 0.06 to 0.20 mg/l nominal concentration depending on the test species.[5] These findings are relatively consistent with the most sensitive taxa in ESF TMAC studies. Community biovolume and cell abundance were reduced at 49 μg/l (measured concentration) in the fall and were not dose dependently affected in the spring at 185 μg/l.

The stream mesocosm approach is useful in identifying seasonal differences in biotic responses to chemical exposure associated with abiotic factors (e.g., temperature, types and amount of suspended particulates, nutrients, etc.). Seasonal differences in response to TMAC would be expected as a result of community structure alone, although an *a priori* prediction of the most sensitive seasonal community would not be possible. Of the 22 dominant taxa in the 2 studies, only 7 are common to both seasons, and of these only 2 (*Melosira varians* and *Nitzschia frustulum* v. *perminuta*) had significant exposure-responses. Differences between seasons in response to exposure at the population and community level are likely to be a combined function of biotic (species composition, abundances, changes in the LEFR source pool of species) and abiotic (temperature, solids, nutrients, WWTP effluent/efficiency at the time of testing) that should be factored into a comprehensive risk assessment program for future chemicals to be tested at the facility. Over time, a sufficient database may develop to allow prediction of the critical period(s) of the year to conduct testing.

In the ESF, TMAC tended to increase taxonomic richness and diversity. These findings were not previously noted by the Lewis studies.[4,5] The LEFR has an extremely rich algal flora. The position of the ESF in the LEFR drainage system allows for contributions from different ecological systems to the artificial streams. Immediately upstream of the ESF are several small first through fourth order streams (e.g., Shaylor Run, Reference 29) that probably uniquely contribute to the source pool of benthic species. Further, a flood control structure (William H. Harsha Reservoir) may be the source for many planktonic species in the ESF. The ESF was strongly dominated by *Navicula* and *Nitzschia* species at 1250 μg/l. At 1250 μg/l in the fall of 1989, suspended particulates interacted with TMAC resulting in enhanced deposition of solids on tiles and in crevices. Such a clear increase in suspended solid deposition on tiles at lower concentrations, relative to the control condition, was not observed. The loose, unaggregated substrate at 1250 μg/l certainly favored keeled taxa (e.g., *Nitzschia* and *Surirella*) which was reflected in the increased diversity for this treatment. Community similarity provided a convenient measure to compare community structure between treatments. It is important to note that initial community structure across the streams was very high, and often in excess of 0.99 (where 1 indicates complete similarity). SIMI was chosen for this evaluation based upon the widespread use of the index in algal community ecology, a supportable statistical framework, and ease of interpretation.[18,23,30] In the 8-week TMAC studies, SIMI was transiently affected in the fall 1989 study; however, in the spring, exposed communities diverged over time from the control and 185 μg/l was significantly different after 42 days of exposure. Differences can be attributed to increased richness and diversity induced by TMAC exposure.

Typically, risk assessments have included algal single-species testing in pure culture with a focus on a few selected taxa (e.g., *Selenastrum capricornutum*, *Navicula pelliculosa*, and *Microcystis aeruginosa*). However, most ecotoxicologists feel uncomfortable in interpreting resulting point estimates of effect (e.g., EC_{50}s from algal single species tests). This is due to questions regarding the relevance of the test species to the receiving system that the ecotoxicologist will be attempting to protect, the high densities of algae tested in closed exposure vessels, dissipation of the chemical over time or a changing concentration of chemical per cell as densities change dramatically during the test, unusually high macronutrient levels in the test media, and fixed (not necessarily realistic) molar ratios of physiological salts.[31]

Experimental ecosystem studies, such as those at the ESF, can provide simultaneous community and population level evaluations in flow through conditions. It is often hypothesized that protection at the community level will be provided by testing at the population level with key species (highly sensitive and/or keystone taxa). In other words, interactions between populations are not always considered significant. If this hypothesis is true, taxa exposed in the community setting that respond by increased or reduced abundance may prove viable alternatives to standard test species (i.e., they would eliminate or reduce many of the questions surrounding relevancy of the single species approach). The cosmopolitan nature of many algae make such findings more broadly applicable. It would be informative to determine the population response of species affected in the streams (for example, *Navicula graciloides,* a diatom affected at 50 to 1250 µg/l) in isolation and pure culture. If responses in pure culture differ substantially from that in streams, one may investigate further to determine if this is due to: (1) differences inherent to exposure design (static vs. flow through), (2) differences due to long-term exposure of a population that may allow for adaptation, or (3) population interactions that are unaccounted for by using single-species isolates (e.g., selective grazing pressure or species associations such as epiphytic diatoms on filamentous green algae).

Exposure to consumer product chemicals in receiving systems usually occurs in a complex matrix that includes WWTP effluent. ESF streams received 10% WWTP effluent and the dominant periphytic taxa were oligosaprobic-mesosaprobic species.[32,33] By definition these taxa are indicative of high nutrient, but not organically putrid, waters. Only *Nitzschia palea,* a taxon that peaked at 1250 µg/l in the fall study, was polysaprobic, or indicative of organically putrid conditions. However, in spite of the nutrient-enriched waters, even the lowest diversity stream had a mean number of taxa of 30 over all periods and a diversity value of 2.5 or greater.

Structural integrity of the algal community was maintained at 50 to 250 µg/l, and functional integrity as measured by primary productivity was found at similar levels. Primary productivity measurements were highly reproducible in the control streams during both studies resulting in good statistical sensitivity for the endpoint even with very small sample sizes (2 to 3) compared to structural evaluations (3 to 5 replicates).[34] If sample sizes for primary productivity were increased, this measure would exceed the sensitivity of most other endpoints. Caution should be exercised in using any one measure alone to understand the maintenance of community and population integrity. As an example, during the spring study, the 250 µg/l treatment had significantly greater primary productivity than controls on day 42, but significantly less by day 56. The reason for this is unclear. Biovolume of control and 250 µg/l treatments were stable during this period, and such a large shift in community function would not have been suggested from structural data.

In summary, the ESF offers many unique advantages for the investigation of environmental effects of consumer product chemicals. Key advantages include environmental control (flow, light, chemical exposure) and the use of a variable, but realistic exposure matrix (tracked by physicochemical factors, such as stream and effluent temperatures, suspended particulates, conductivity, etc.). For the study presented here, periphyton generally increased in biomass over time, but the 1250 µg/l treatment lagged behind in chlorophyll and productivity. In the spring, periphyton abundance peaked near the beginning of the study period and then generally declined; TMAC at 250 µg/l and less had little or no effect. In both studies, taxonomic richness increased with exposure. Algal periphytic community and population responses were interpretable and relatively consistent with information already known for this well-studied chemical.[3] The algal periphyton are but one component of the ecosystem established in the streams. Further characterization of the heterotrophic microflora, microfauna, and macrofauna in streams exposed to TMAC is underway which will be integrated with the findings presented here with the tacit goal of improved understanding of aquatic ecosystem, community, and population responses to consumer product chemicals.

APPENDIX A.

Species list of taxa at the Experimental Stream Facility (ESF) and Lower East Fork of the Little Miami River (LEFR) in Fall 1989 and Spring 1990 Experiments. Taxa that were dominant (cell density or biovolume) are indicated by a bold ''x''.

Taxon	Fall ESF	Fall LEFR	Spring ESF
Bacillariophyta			
Achnanthes biporoma Hohn & Hellerman	x	x	
Achnanthes clevei Grun.	x	x	x
Achnanthes detha Hohn & Hellerm.			x
Achnanthes exigua Grun.			x
Achnanthes lanceolata Breb. ex Kutz.	x	x	**x**
Achnanthes lanceolata v. *dubia* Grun.	x	x	
Achnanthes linearis fo. *curta* H.L. Sm.	x	x	
Achnanthes microcephala (Kutz.) Grun.	x	x	
Achnanthes minutissima Kutz.	x	x	**x**
Achnanthes minutissima v. *saprophila* Koba. & Maya.	x	x	
Achnanthes pinnata Hust.	x		
Achnanthes sp.	x		
Amphipleura pellucida (Kutz.) Kutz.	x		x
Amphora birugula Hohn	x		
Amphora ovalis (Kutz.) Kutz.	x	x	x
Amphora ovalis v. *pediculus* (Kutz.) V.H. ex DeT.	x	x	
Amphora perpusilla (Grun.) Grun.	x	**x**	
Amphora submontana Hust.	x	x	x
Amphora veneta Kutz	x		x
Asterionella formosa Hassal.			x
Biddulphia laevis Ehr.		**x**	
Bacillaria paradoxa Gmelin	x	x	x
Caloneis bacillum (Grun.) Cl.	x	x	x
Caloneis hyalina Hust.			x
Caloneis lewisii Patr.		x	
Caloneis ventricosa v. *trunculata* (Grun.) Meist.		x	
Cocconeis diminuta Pant.	x		x
Cocconeis pediculus Ehr.	x	x	x
Cocconeis placentula v. *euglypta* (Ehr.) V.H.	**x**	**x**	x
Cocconeis placentula v. *lineata* (Ehr.) Cl.	**x**	**x**	x
Cocconeis placentula Ehr.			x
Cyclotella aliquantula Hohn & Hellerman	x		
Cyclotella atomus Hust.	x	x	x
Cyclotella comensis Grun.			x
Cyclotella meneghiniana Kutz.	x	x	x
Cyclotella ocellata Pant.			x
Cyclotella pseudostelligera Hust.	x	**x**	x
Cyclotella stelligera Cl. & Grun.	x	x	x
Cyclotella striata (Kutz.) Grun.	x		
Cymbella affinis Kutz.	x	x	x
Cymbella cistula (Ehr.) Kirchn.			x
Cymbella microcephala Grun.	x		
Cymbella minuta Hilse ex Rabh.	x		x
Cymbella prostrata (Berk.) Cl.	x		x
Cymbella prostrata v. *auerswaldii* (Rabh.) Patr.	x	x	
Cymbella sinuata Greg.	x	x	x
Cymbella sinuata fo. *antiqua* (Grun.) Reim.	x	x	
Cymbella tumida (Breb. ex Kutz.) V.H.	x	x	

Taxon	Fall ESF	Fall LEFR	Spring ESF
Diatoma tenue Ag.	x		
Diatoma vulgare Bory	**x**	**x**	**x**
Diploneis puella (Schum.) Cl.	x		
Diploneis sp.			x
Eunotia sp.			x
Fragilaria construens v. *pumila* Grun.	x		x
Fragilaria intermedia Grun.		x	
Fragilaria pinnata Ehr.	x		
Fragilaria strangulata (Zanon) Hust.		x	
Fragilaria vaucheriae (Kutz.) Peters.	x	x	**x**
Frustulia vulgaris (Thwaites) DeT.	x		x
Gomphonema abbreviatum Ag.	x	x	x
Gomphonema acuminatum Ehr.		x	x
Gomphonema angustatum (Kutz.) Rabh.	x	x	**x**
Gomphonema angustatum v. *intermedia* Grun.	x		
Gomphonema brasiliense Grun.	x		x
Gomphonema clevei Fricke	x	x	
Gomphonema grunowii Patr.	x	x	
Gomphonema intricatum Kutz.	x	x	x
Gomphonema olivaceum (Lyngb.) Kutz.	x	x	**x**
Gomphonema parvulum Kutz.	x	x	x
Gomphonema subclavatum (Grun.) Grun.			x
Gomphonema tenellum Kutz.			x
Gomphonema truncatum Ehr.	x	x	x
Gomphonema sp.			x
Gyrosigma acuminatum (Kutz.) Rabh.			x
Gyrosigma nodiferum (Grun.) G. West		x	
Gyrosigma scalproides Rabh. Cl.	x	x	x
Gyrosigma spencerii (Quek.) Griff. & Henfr.	x		
Hantzschia amphioxys (Rabh.) Patr.	x		x
Melosira ambigua (Grun.) O.F.Mull.	x		x
Melosira distans (Ehr.) Kutz.		x	x
Melosira distans v. *alpigena* Grun.	x	x	
Melosira granulata (Ehr.) Ralfs	x	x	x
Melosira varians Ag.	**x**	**x**	**x**
Meridion circulare (Grev.) Ag.			x
Navicula accomoda Hust.	x	x	x
Navicula anglica Ralfs			x
Navicula arvensis Hust.	x	x	
Navicula atomus (Kutz.) Grun.	x	x	x
Navicula bicephala Hust.	x		
Navicula biconica Patr.	x	x	x
Navicula canalis Patr.	x	x	
Navicula capitata Ehr.	x	x	x
Navicula capitata v. *hungarica* (Grun.) Ross		x	x
Navicula cincta (Ehr.) Ralfs		x	x
Navicula cf *cincta* v. *rostrata* Reim.	x	x	
Navicula clementis Grun.	x		
Navicula confervacea (Kutz.) Grun.	x		x
Navicula contenta v. *biceps* (Arn.) V.H.	x		x
Navicula contenta Reim.			x
Navicula cryptocephala Kutz.	x	x	x
Navicula cryptocephala v. *exilis* (Kutz.) Grun.	x	x	x
Navicula cryptocephala v. *veneta* (Kutz.) Rabh.	x	x	**x**

APPENDIX A. (*continued*)

Taxon	Fall ESF	Fall LEFR	Spring ESF
Navicula cuspidata (Kutz.) Kutz.	x		
Navicula decussis Ostr.	x		x
Navicula exigua v. *capitata* Patr.	x		
Navicula exigua Greg. ex Grun.			x
Navicula graciloides A. Meyer	**x**	**x**	
Navicula gregaria Donk.	x	x	**x**
Navicula grimmei Krasske	x		
Navicula gysingensis Hust.	x	x	
Navicula halophila (Grun.) Cl.			x
Navicula heufleri Grun.			x
Navicula hustedtii Krasske	x		
Navicula indifferens Hust.	x	x	
Navicula ingenua Hust.	x		
Navicula lanceolata (Ag.) Kutz.	x		x
Navicula latelongitudinalis Patr.	x		
Navicula lateropunctata Wallace	x		
Navicula luzonensis Hust.	x	x	**x**
Navicula menisculis Schum.	x	x	x
Navicula minima Grun.	x	x	x
Navicula mutica Kutz.	x	x	x
Navicula mutica v. *cohnii* (Hilse.) Grun.	x		x
Navicula mutica v. *stigma* Patr.	x		x
Navicula mutica v. *undulata* (Hilse.) Grun.			x
Navicula orchridana Hust.	x	**x**	
Navicula omissa Hust. emend. Schoeman	x		
Navicula paratunkae Peters	x	x	
Navicula paucivisitata patr.	x	x	
Navicula pelliculosa (Breb. ex Kutz.) Hilse.	**x**	x	x
Navicula placentula (Ehr.) Kutz.		x	
Navicula pupula Kutz.	x	x	x
Navicula pupula v. *capitata* Skv. & Meyer	x	x	
Navicula pupula v. *mutata* (Krasske) Hust.	**x**	x	
Navicula pupula v. *rectangularis* (Greg.) Grun.	x		
Navicula radiosa v. *parva* Wallace	x		
Navicula radiosa Kutz.			x
Navicula radiosa v. *tenella* (Breb. ex Kutz.) Grun.			**x**
Navicula rhynchocephala v. *germainii* (Wallace) Patr.	x	x	
Navicula salinarum v. *intermdia* (Grun.) Cl.	**x**	x	**x**
Navicula schroeteri v. *escambia* Patr.	x	x	
Navicula secreta v. *apiculata* Patr.	x	x	
Navicula seminulum Grun.	x	x	x
Navicula subhamulata Grun.	x		
Navicula symmetrica Patr.	**x**	x	x
Navicula tantula Hust.	x	x	
Navicula tenera Hust.	x	x	
Navicula tripunctata (O. F. Mull.) Bory	x	x	**x**
Navicula tripunctata v. *schizonemoides* (V.H.) Patr.	x	**x**	
Navicula viridula (Kutz.) Kutz. emend. V.H.	x	x	**x**
Navicula viridula v. *Linearis* Hust.	x	x	x
Navicula viridula v. *rostellata* (Kutz.?) Cl.	x		x
Navicula sp.			x
Nitzschia accomodata Hust.	x		x
Nitzschia acicularis W.Sm.	x	x	**x**

Taxon	Fall ESF	Fall LEFR	Spring ESF
Nitzschia admissa Hust.	x	x	
Nitzschia amphibia Grun.	x	x	x
Nitzschia angustata (W.Sm.) Grun.	x		
Nitzschia bacata Hust.	x	x	x
Nitzschia biacrula Hohn & Hellerman	x	x	
Nitzschia capitellata Hust.	x		
Nitzschia capitellata v. *siberica* Sk.	x		
Nitzschia clausii Hantz.	x	x	
Nitzschia communis v. *obtusa* Grun. in Cleve. et Grun.	x	x	
Nitzschia constricta v. *subconstricta* Cleve et Grun.	x		
Nitzschia diserta Hust.		x	
Nitzschia dissipata (Kutz.) Grun.	**x**	**x**	**x**
Nitzschia dissipata v. *media* (Hantz) Grun.	x	x	
Nitzschia fonticola Grun.	x	x	x
Nitzschia filiformis (W.Sm.) Hust.	x	x	x
Nitzschia frequens Hust.	x	x	
Nitzschia frustulum (Kutz.) Grun.	x	x	x
Nitzschia frustulum v. *perminuta* Grun.	**x**	**x**	**x**
Nitzschia frustulum v. *subsalina* Hust.	x	x	
Nitzschia gandersheimiensis Krasske			x
Nitzschia gracilis Hantz.	x		x
Nitzschia hungarica Grun.	x	x	x
Nitzschia kuetzingiana Hilse	**x**	**x**	x
Nitzschia linearis W.Sm.			x
Nitzschia mediastalisis Hohn & Hellerman	x		
Nitzschia palea (Kutz.) W.Sm.	**x**	**x**	**x**
Nitzschia paleacea Grun.			x
Nitzschia parvula Lewis		x	x
Nitzschia recta Grun.	x	x	
Nitzschia reversa W.Sm.	x		
Nitzschia romana Grun. in V.H.		x	
Nitzschia sicula v. *migrans* (Cleve) Hasle	x	x	
Nitzschia sigmoidea (Ehr.) W.Sm.		x	x
Nitzschia sinuata v. *tabellaria* Grun.	x	x	x
Nitzschia sociabilis Hust.	x	x	
Nitzschia subtilis (Kutz.) Grun.	x		
Nitzschia tarda Hust.	x		
Nitzschia tropica Hust.		x	
Nitzschia tryblionella v. *debilis* (Arnott) A. Meyer		x	x
Nitzschia tryblionella v. *levidensis* (W.Sm.) Grun. in Cleve et Grun.		x	
Nitzschia tryblionella v. *victoriae* Grun.	x		x
Nitzschia umbilicata Hust.	x		
Nitzschia wallacei Reim.	x		
Nitzschia sp.			x
Opephora martyi Herib.			x
Pinnularia borealis Ehr.			x
Pinnularia brebissonii (Kutz.) Rabh.			x
Pinnularia obscura Krasske	x		x
Pinnularia sp.			x
Rhoicosphenia curvata (Kutz.) Grun. ex Rabh.	x	x	**x**
Skeletonema potamos (Weber) Hasle	x	x	
Stauroneis anceps Ehr.			x
Stauroneis sp.			x
Stephanodiscus astrea (Ehr.) Grun. in Cleve et Grun.	x		x

APPENDIX A. (*continued*)

Taxon	Fall ESF	Fall LEFR	Spring ESF
Stephanodiscus astrea v. *minutula* (Kutz.) Grun.	x	x	
Stephanodiscus hantzschii Grun.	x		x
Stephanodiscus invisitatus Hohn and Hellerman	x		**x**
Stephanodiscus minutus Kutz.	x		x
Stephanodiscus sp.			x
Surirella alicula Hohn & Hellerman	x	x	
Surirella angustata Kutz.	x	x	x
Surirella minuta Breb.	x	x	x
Surirella ovalis Breb.			x
Surirella ovata Kutz.	x		**x**
Surirella ovata v. *pinnata* W.Sm.			x
Surirella suecica Grun in V.H.	x	x	
Surirella sp.			x
Synedra acus Kutz.			x
Synedra delicatissima v. *angustissima* Grun.	x		
Synedra fasciculata v. *truncata* (Greve.) Patr.	x	x	
Synedra pulchella Ralfs ex Kutz.			x
Synedra rumpens v. *familiaris* (Kutz.) Hust.	x	x	
Synedra rumpens Kutz.			x
Synedra rumpens v. *meneghiniana* Grun.		x	
Synedra socia Wallace		x	
Synedra tenera W.Sm.	x		
Synedra ulna (Nitz.) Ehr.	x		x
Synedra ulna v. *oxyrhynchus* Hust.		x	x
Thalassiosira fluviatilis Hust.	x		
Thalassiosira visurgus Hust.		x	
Cyanophyceae			
Anacystis marina (Hansy.) Drouet & Daily	x		
Anacystis montana (Light) Drouet & Daily	**x**	**x**	
Microcoleus vaginatus (Vaucher.) Gomont.	x		
Nostoc commune Vaucher.		x	
Porphyrosiphon splendidus (Greve.) Drouet	x		
Schizothrix calcicola (Ag.) Gomont.	**x**	**x**	**x**
Chrysophyceae			
Chrysophyte sp. No. 2 (small loricate)	x		
Chlorophyceaea			
Ankistrodesmus falcatus (Corda.) Ralfs	x		x
Chlamydomonas sp.			x
Cladophera glomerata (L.) Kutz.		**x**	
Closterium sp.		x	
Cosmarium sp.	x	x	
Oedogonium sp.	**x**		
Scenedesmus ecornis (Ralfs) Chodat	x	x	
Scenedesmus quadricauda (Turpin) Breb.	x		
Scenedesmus sp.			x
Stigeoclonium lubricum Kutz.	**x**	x	
Undetermined coccoid chlorophyte	x		
Euglenophyceae			
Euglena sp.	x		x
Phacus sp.	x		x
Trachelomonas sp.	x	x	x
Numbers of Taxa			
Bacillariphyceae (186 taxa)	167	129	140

Cyanophyceae (6 taxa)	5	3	1
Chlorophyceae (11 taxa)	7	5	3
Chrysophyceae (1 taxa)	1	0	0
Euglenophyceae (3 taxa)	3	1	3
Total (207 taxa)	183	138	147

APPENDIX B.

List of components and suppliers of key operational equipment employed in the Experimental Stream Facility.

Item	Supplier	Function
	Water Delivery	
Submersible pumps	WEMCO Corp., WEMCO-Hidrostal model E4K-S-EE3A6	Procure water from river
PVC pipe, Type 1, Grade 1	Hardware/plumbing suppier	Water distribution
	Flow Control	
Magnetic flowmeters	Foxboro Co., river water model 801H-WCT-A, sewage effluent model 8001-WCT-A	Automated flow control
Actuator operated control valves	POSCON Actuator, Overbook Control Co., model A300; control valve ITT Grinnell Corp. model MM-82 diaphragm valve)	
Process controller	Analog Devices, Inc., model μMAC 5000	
Personal computer	IBM Corp., model IBM PC-AT	
Data log/control software	CAMM Solutions, Inc., CAMM version 3.2	
	Water Quality Monitoring	
Platinum resistance temperature sensor and transmitter	Rosemount, Inc., sensor series 78S and transmitter model 444	Temperature monitor
Electrolyte conductivity sensor and transmitter	Foxboro Co., sensor model 871EC-EV3 and transmitter model 870EC-EV10D-P	Specific conductivity monitor
Combination pH/reference electrode and pION transmitter	TBI-Baily Controls, model 551 and transmitter Royce Instrument Corp., model 3750	pH monitor
Dissolved oxygen sensor and analyzer	Royce Instrument Corp., sensor model 64 and analyzer model 3000	Dissolved oxygen monitor

APPENDIX B. (*continued*)

Item	Supplier	Function
	Lighting	
Metal halide arc lamps (1000 Watt)	GTE Sylvania, model MS1000/C	Full spectrum daylight equivalent
Light sensors	LI-COR Inc., model LI-190SA	Monitor PAR
Suntracker	Paragon Electric Co., Inc.	Adjust daily photoperiods
	Chemical Feed System	
Tank mixer	Mixing Equipment Inc., Lightnin Series XJ	Mix stock tank test material
Diaphragm pump	PULSAfeeder Inc., Pulsa series 340	Meter out stock solutions to streams
Pressure switch	Dresser Industries, Ashcroft Series L	
	Stream Construction	
3/4" Plywood	Lumber supplier	
Polyvinyl film (10 mil)	Robeco Chemical Inc.	Line stream interior

Note: Additional details are available upon request.

LITERATURE CITED

1. National Research Council. 1981. Testing for Effects of Chemicals on Ecosystems. National Academy Press, Washington, D.C. 103p.
2. Touart, L.W. 1988. Aquatic mesocosm tests to support pesticide registrations. Hazard Evaluation Division, Technical Guidance Document, Office of Pesticide Programs, U.S. Environmental Protection Agencies, EPA 540/09-88-035. National Technical Information Service, Springfield, VA, 18p.
3. Woltering, D.M. and W.E. Bishop. 1989. Evaluating the environmental safety of detergent chemicals: a case study of cationic surfactants. pp. 345–389, in, Paustenbach, D.J., ed. *The Risk Assessment of Environmental and Human Health Hazards.* John Wiley, New York.
4. Lewis, M.A., M.J. Taylor, and R.J. Larson. 1986. Structural and functional response of natural phytoplankton and periphyton communities to a cationic surfactant with considerations on environmental fate. pp. 214–268, in, Cairns, J., Jr., ed. *Community Toxicity Testing,* ASTM STP 920. American Society for Testing and Materials, Philadelphia, Pennsylvania.
5. Lewis, M.A. and B.G. Hamm. 1986. Environmental modification of the photosynthetic response of lake plankton to surfactants and significance to a field-laboratory comparison. *Water Res.* 20: 1575–1582.
6. Hustedt, F. 1930. Bacillariophyta (Diatomeae). In, Pascher, A., ed., *Die Susswasser-Flora Mittleuropas.* Heft 10. Gustav Fisher, Jena. 466p.
7. Patrick, R. and C.W. Reimer. 1966. *The Diatoms of the United States.* Vol. I., Acad. Nat. Sci. Phila., Monograph 13. 688p.
8. Patrick, R. and C.W. Reimer. 1975. *The Diatoms of the United States.* Vol. II, Part I, Acad. Nat. Sci. Phila., Monograph 13. 213p.

9. Prescott, G.W. 1962. *Algae of the Western Great Lakes Area*. Wm. C. Brown and Co., Dubuque, Iowa. 977p.

10. Bourelly, P. 1966. *Les Algues D'eau Douce*. Vol. 1: Les Algues Vertes. N. Boubee and Cie, Paris, France. 511p.

11. Bourelly, P. 1968. *Les Algues D'eau Douce*. Vol. 2: Les Algues jaunes et brunes. Boubee and Cie, Paris, France. 438p.

12. Bourelly, P. 1970. *Les Algues D'eau Douce*. Vol. 3: Les Algues bleues et rouges. N. Boubee and Cie, Paris, France. 512p.

13. Drouet, F. and W. Daily. 1956. Revision of the coccoid myxophyceae. Butler University Botanical Studies 12. 222p.

14. Germain, H. 1981. *Flore des Diatomees. Diatomophyees eaus douces et saumatres du Massif Armericain et des Contrees voisines d'Europe occidental*. Boubee, Paris, France. 444p.

15. Beyer, W.H. 1981. *CRC Standard Mathematical Tables*, 26th Ed. CRC Press, Inc. Boca Raton, Florida.

16. Shannon, C. and W. Weiner. 1963. *The Mathematical Theory of Communication*. University of Illinois Press, Champaign, Illinois.

17. Stander, J.M. 1970. Diversity and similarity of benthic fauna off Oregon. M.S. Thesis, Oregon State University, Corvallis, Oregon. 58p.

18. Genter, R.B., D.S. Cherry, E.P. Smith, and J. Cairns, Jr. 1987. Algal-periphyton population and community changes from zinc stress in mesocosms. *Hydrobiologia* 153: 261–275.

19. Palumbo, T.E., P. Mulholland, and Elwood. 1987. Extraction with DMSO to simultaneously measure periphyton photosynthesis, chlorophyll, and ATP. *Limnol Oceanogr.* 32:464–471.

20. American Public Health Association, Water Pollution Control Federation and American Water Works Association. 1989. Standard Methods for the Examination of Water and Waste Water, 17th ed., Washington, D.C.

21. Wee, V. and J. Kennedy. 1982. Determination of trace metals and quaternary ammonium compounds in river water by liquid chromatography with condumetric detection. *Anal. Che.* 54: 161–163.

22. United States Environmental Protection Agency. Handbook for Chemical Analysis of Water and Wastes. Environmental Monitoring and Support Laboratory, Cincinnati, Ohio. EPA-600/4-79-020.

23. Smith, E.P., K.W. Pontasch, and J. Cairns, Jr. 1990. Community similarity and the analysis of multispecies environmental data: a unified statistical approach. *Water Res.* 24: 507–514.

24. Kimball, K.D. and S.A. Levin. 1985. Limitations of laboratory bioassays: the need for ecosystem level testing. *BioScience* 35: 165–171.

25. Belanger, S.E., J.B. Barnum, J.W. Bowling, D.M. Woltering, B.S. Schwabb, A.C. Palmisano, R.M. Ventullo, S. Sschermerhorn, and R.L. Lowe. 1990. Periphyton structure and function in response to consumer chemicals in experimental stream mesocosms. Presented at the Eleventh Annual Meeting of the Society of Environmental Toxicology and Chemistry, Arlington, Virginia, November 11–15.

26. Cuffney, T.F., D.D. Hart, K.C. Wolbach, J.B. Wallace, G.J. Lugthart and F.L. Smith-Cuffney. 1990. Assessment of community and ecosystem level effects in lotic environments: the role of mesocosm and field studies. Presented at the North American Benthological Society Technical Information Workshop: Experimental Ecosystems-Applications to Ecotoxicology, May 25, 1990, Blacksburg, Virginia, 40p.

27. Belanger, S.E., D.H. Davidson, J.L. Farris, D. Reed, and D.S. Cherry. Effects of cationic surfactant exposure to Asiatic clams in stream mesocosms. In press, *Environ. Toxicol. Chem.*

28. Weber, C.I. and B.H. McFarland. 1981. Effects of copper on the periphyton of a small calcareous stream. pp. 101–131, in Bates, J.M. and C.I. Weber, eds. *Ecological Assessments of Effluent Impacts on Communities of Indigenous Organisms.* ASTM STP 730. American Society for Testing and Materials, Philadelphia, Pennsylvania.

29. Cairns, J., Jr. 1977. Quantification of biological integrity, in, Ballentine, R.K. and L.J. Guarraia (eds.), *The Integrity of Water,* pp. 171–187. U.S. Environmental Protection Agency, Washington, D.C.

30. Johnson, B.E. and D.F. Millie. 1982. The estimation and application of confidence intervals for Stander's similarity index (SIMI) in algal assemblage comparisons. *Hydrobiologia* 89: 3–8.

31. Lewis, M.A. 1990. Are laboratory-derived toxicity data for freshwater algae worth the effort? *Review. Environ. Toxicol. Chem.,* 9: 1279–1284.

32. Kolkwitz, R. and M. Marsson. 1908. Okologie der pflanzlichen Saprobien. *Ber. Deut. Bot. Ges.* 26: 505–519.

33. Lowe, R.L. 1974. Environmental requirements and pollution tolerance of freshwater diatoms. USEPA, Cincinnati, Ohio. EPA-670/4-74-005.

34. Schwab, B.S., D.A. Maruscik, R.M. Ventullo, and A.C. Palmisano. 1992. Adaptation of periphytic communities in model streams to a quaternary ammonium surfactant. *Environ. Toxicol. Chem.* 11: 1169–1177.

CHAPTER 29

Case Histories:
Summary and Discussion

James H. Kennedy

Since the adoption of mesocosm testing by the U.S. EPA Office of Pesticide Programs as the final tier (tier IV), the role of outdoor surrogate ecosystems continues to grow in importance in risk assessment of chemicals on aquatic communities. "Risk" includes both toxicity and exposure elements.[2] Mesocosms are particularly effective in this regard as they include the exposure of pesticides to complex assemblages of organisms in the natural environment.

The chapters by Hill et al. (Chapter 24), Giddings et al. (Chapter 25), and Mayasich et al. (Chapter 26) describe the results of three different mesocosm studies that were among the first submitted to the U.S. EPA for use in the risk assessment of agricultural pesticides. These studies were performed under EPA guidelines and are similar in their study design. All three studies used similar size 0.1-surface acre (0.4 ha) earthen ponds. Construction (Hill et al. and Mayasich et al.) or refurbishment of ponds (Giddings et al.) used in these studies was usually completed less than a year before dosing began. Construction of the ponds used in the Hill et al. and Mayasich et al. studies were completed 10 months and 7 months, respectively, before the beginning of dosing. Ponds used by Giddings et al. were refurbished approximately 8 months before dosing began.

During the short interval between completion of construction or refurbishment and the beginning of testing, initial preparation of the test systems was controlled

0-87371-592-6/94/$0.00 + $.50

to facilitate the development of a biological community representative of those in established ponds. All three studies filled the mesocosms with water pumped from ponds with established biological communities. Additional sources of biological inoculum in all three studies included the transfer of sediment from established ponds. Hill et al. and Mayasich et al. planted macrophytes and added invertebrates captured in sweep net collections, to the newly constructed ponds. In addition, as required by current EPA criteria for mesocosm studies, a reproducing population of bluegill sunfish (*Lepomis macrochirus*) was stocked in the ponds.[3] Presumably these fish and their offspring are integrators of the system, and differences in numbers and size distribution between exposure levels provide requisite endpoints for risk management decisions. From a regulatory perspective, evaluation of fish is the one of the most important regulatory endpoints of a mesocosm study.

Variability of parameters within and among mesocosms can obscure our ability to detect differences in responses of the systems to pesticide treatment. Attempts were made in all three studies to control this variation. For example, Hill et al. and Mayasich et al. recirculated water among ponds (prior to pesticide application) to minimize the differences in water quality and plankton communities between mesocosms.

Due to the use patterns of the insecticides being tested, pyrethroids (Hill et al. and Mayasich et al.) and an organophosphate (Giddings et al.), test ponds received multiple applications of chemicals. Chemical application in the Giddings et al. study was via a simulated soil runoff while the other two studies were exposed to chemicals via both simulated spray drift and soil runoff. Spray and runoff dose levels were based on an estimated environmental concentration in the Hill et al. and Mayasich et al. study.

Examination of each of these chapters reveals that during the course of a mesocosm study, volumes of data were being collected on a wide spectrum of trophic levels and physicochemical variables. While direct and indirect effects were reported and discussed in all of these studies, the majority of the analysis relied on changes in density (total, subgroup, or individual) to measure effects. A variety of statistical tests were used in these studies. Analysis of Variance (ANOVA) followed by Dunnet's test (Mayasich et al.), *t*-tests (Hill et al.), and Williams' Test (Giddings et al.) were used in these studies to test hypothesis. In all of these studies, many types of samples taken among the ponds had a high variability which limits the power of inferential statistics to detect differences due to treatment effects. Giddings et al. calculated that in their study with Guthion a 30% reduction relative to the controls would be required to detect a significant difference with the Williams' Test.

In addition, Giddings used linear regression on counts of benthic invertebrates, emergent insects, and bluegill numbers/biomass. The choice of regression vs. ANOVA designs depends on the objectives of the study. The ANOVA design may be best if a specific environmental concentration of a chemical is of interest. The regression design, theoretically, has the ability to define the response of organism populations over a range of chemical concentrations. Each approach requires the careful choosing of test concentrations. The regression design requires a proper range of test concentrations for the populations of interest. Different components of the mesocosm ecosystem, however, may show widely differing responses to test chemicals, reflecting the relative sensitivities or differences in habitats that reduce or increase chemical exposure. It is also known that a mild dose can frequently increase populations even though a larger dose may decrease it, leading to a nonlinear response.[4] This may make it impossible to perform one mesocosm study and analyze all components by linear regression.

The ANOVA approach, on the other hand, requires the selection and exposure of the test system to a realistically expected environmental concentration (EEC) of pesticide that nontarget organisms may experience. The estimation of the expected environmental concentration can be enigmatic. Methodology for the estimation of exposure estimations is unclear, and validated or verified models are not always available.[2] If an acceptable EEC is not available, selection of an erroneous EEC may hamper the interpretation of mesocosm results.

If systems of sufficient size are available, then study designs that incorporate features of both ANOVA and regression techniques into the statistical evaluation of the study can be accommodated. No matter what procedures are used to measure effects of chemicals on ecosystems, users need to keep in mind that the inclusion of one of them in the data analyses does not preclude the use of sound ecological judgement in their interpretation of the results.[5]

A current problem that exists with the evaluation of mesocosm studies for registration of agricultural chemicals is that the endpoints are not clearly defined *a priori*. Variability is inherent in any biological system. Pratt and Bowers[6] advocate the use of coefficients of variation to select low variability metrics necessary for the best detection of effects of stressors on communities. Without clearly designated endpoints, it is difficult to formulate an experimental design that will minimize variation.

In the Hill et al. and Mayasich et al. studies it was concluded that the abundance of young bluegill obscured or complicated the evaluation of pesticides impacts on many invertebrate communities. It is well established that fish populations have both direct and indirect effects on ecosystem function. Liber et al. (Chapter 17 this volume) measured decreased variability among zooplankton populations in limnocorrals upon removal of fish. Arumugam and Geddes[7] determined that larval golden perch (*Macquaria ambigua*) predation significantly affected the densities of many zooplankton species and the overall zooplanktonic community structure. In separate limnocorral studies, Brabrand et al.[8] and Langeland et al.[9] similarly concluded that fish predation alters planktonic communities in eutrophic lakes and that the very presence of certain fish species may contribute to the eutrophication process. Deutsch et al.[8] stocked largemouth bass in pond mesocosms in order to control bluegill population growth, thereby potentially limiting among-system variability and provide a more natural surrogate system. Results from these studies indicate that fish can have considerable impact on the ecosystem. An evaluation should be made whether fish communities could be introduced that would enhance our ability to measure impacts on the aquatic ecosystem in these surrogate systems.

Basic knowledge needs to be gathered concerning test systems described in this section so they can be validated. For example, before we can provide risk assessments of pesticides on natural ecosystems using data collected from manmade ponds, we need to understand how the structural and functional components of these surrogate test systems relate to natural systems. In addition we need to validate the ability of these surrogate systems to accurately and consistently measure ecosystem responses to pesticides.

Community ordination techniques have been undergoing great development in recent years. The ultimate goal of ordination techniques is to elucidate those biological and environmental factors that may be important in determining the structure of the communities from which they were collected.[10] Correspondence Analysis, an ordination technique, was used by Lucassen and Leeuwangh (Chapter 27) in a field study using replicated small ponds to measure the effects of the organophosphorus pesticide, Dursban®, on the zooplankton community. Impacts of Dursban® on the zooplankton community as measured by correspondence analysis in this experiment are comparable with data reported in a number of studies completed in outdoor ponds and outdoor surrogate ecosystems. As pointed out by the authors, the acceptance of correspondence analysis for ordination of community data in this study must be considered tentative because of the low numbers of organisms and the lack of replication. However, the technique may prove useful as a more holistic view of ecosystem responses.

Belanger et al. (Chapter 28), in the final chapter of this section, used 11-m long experimental stream channels housed in an indoor facility to evaluate the fate and effect of surfactants on the periphyton community. Water for the streams is taken from a nearby river that has within its drainage basin few impacts that would affect water quality. A low percentage (10% of the water volume) of secondarily treated wastewater from a treatment plant effluent was added to all experimental stream. Wastewater was added because surfactants are introduced into the aquatic environment in a wastewater matrix. Taxonomic composition and chlorophyll *a* levels in periphyton communities were measured from unglazed tile substrates incubated in experimental streams. Primary production rates were determined using $^{14}CO_2$.

A surfactant was chosen for this study for which the biological responses in laboratory and field studies were well known prior to the start of the experimental stream study. Algal, periphytic community, and population responses measured in the experimental stream study were interpretable and generally consistent with this data. The authors did observe that a diversity of measures were necessary to understand community responses.

Although experimental streams, as discussed by Belanger et al., have been used for years to measure effects of chemicals on flowing water ecosystems, they have yet to be incorporated into the pesticide risk assessment process under FIFRA. Results of lentic water ecosystems may be able to provide insights into ecosystem level effects of chemicals that can be extrapolated to ponds, reservoirs, and lakes. How well the results measured in lentic mesocosms, which differ in flow regime, chemical exposure, and community structure, can be extrapolated to flowing freshwater ecosystems is a very difficult question. At present too few comparative ecosystem studies exist, even between mesocosms and natural lentic waters, making it difficult to extrapolate results from one system to another. The inclusion of lotic mesocosms may be justified in the FIFRA risk assessment process for those pesticides expected to impact flowing waters.

REFERENCES

1. Touart, L.W. 1988. Aquatic mesocosm tests to support pesticide registrations. EPA 540/09-88-035. U.S. Environmental Protection Agency, Office of Pesticide Programs, Washington, D.C.
2. Anonymous. 1992. *Improving Aquatic Risk Assessment under FIFRA Report of the Aquatic Effects Dialogue Group.* World Wildlife Fund, 98 pp.
3. Touart, L.W. 1987. Aquatic mesocosm testing to support pesticide registration. EPA 540/09-88-035. U.S. Environmental Protection Agency, Office of Pesticide Programs, Washington, D.C.
4. Smith, E.P., and D. Mercante. 1989. Statistical concerns in the design and analysis of multispecies microcosm and mesocosm experiments. *Tox. Assess.: Int. J.* 4:129–147.
5. Pontasch, K.W., E.P. Smith, and J. Cairns, Jr. 1989. Diversity indices, community comparison indices and canonical discriminant analysis: Interpreting the results of multispecies toxicity tests. *Wat. Res.* 23:1229–1238.
6. Pratt, J.R., and N.J. Bowers. 1992. Variability of community metrics: Detecting changes in structure and function. *Environ. Toxicol. Chem.,* 11:451–457.
7. Arumugam, P.T., and M.C. Geddes. 1986. An enclosure for experimental field studies with fish and zooplankton communities. *Hydrobiologia* 135:215–221.
8. Braband, Å, B. Faafeng, and J.P.M. Nilssen. 1987. Pelagic predators and interfering algae: stabilizing factors in temperate eutrophic lakes. *Archiv für Hydrobiologie* 110:533–552.
9. Langeland, A., J.I. Koksvik, Y. Olsen, and H. Reinertsen. 1987. Limnocorral experiments in a eutrophic lake — effects of fish on the planktonic and chemical conditions. *Polish Archives of Hydrobiology* 34:51–65.
10. Deutsch, W.G., E.C. Webber, D.R. Bayne, and C.W. Reed. 1992. Effects of largemouth bass stocking rate on fish populations in aquatic mesocosms used for pesticide research. *Environ. Toxicol. Chem.* 11:5–10.
11. Ludwig, J.A., and J.F. Reynolds. 1988. *Statistical Ecology. A Primer of Methods and Computing.* John Wiley & Sons, New York, NY, pp. 337.

SECTION VI

Ecosystem Analysis

INTRODUCTION

Historically, potential environmental effects of pesticides on aquatic ecosystems have been predicted from standardized laboratory toxicity tests performed on an array of single-species tests. Single-species tests, however, do not provide insight into ecosystem level parameters such as predator-prey interactions, energy transfer, and material cycling.[1] Nor do these types of tests provide realistic exposure levels for chemicals whose dispersion and degradation in natural environments may result in modification of direct and indirect toxicity effects in field situations. The U.S. EPA has recognized the importance of testing pesticides at the ecosystem level and has adopted the policy of evaluating agricultural chemical in replicated manmade ponds (mesocosms). These ponds are intended to serve as surrogates for natural systems.[2] While the use of mesocosms gives us the advantage of examining ecosystem level responses to agricultural chemicals, few ecosystem level analyses are performed on the data collected.

Studies submitted for pesticide registration tend to isolate species and other types of measured responses and then analyze the results as single-species responses with inferential statistics. Because of natural variability in the biological parameters measured, statistical differences are often difficult to detect unless treatment effects are large. More importantly, ecological significance does not always correlate with statistical significance. No integrative data analysis approaches are required for mesocosm data.[2]

The four chapters in this section discuss the types of indirect and cascade effects that can be observed on aquatic communities in mesocosms. Collectively they argue

0-87371-592-6/94/$0.00 + $.50
© 1994 by CRC Press, Inc.

for the inclusion of ecosystem level analyses in registration studies of agricultural chemicals. Although, they should be read in sequence, each chapter will stand on its own individual merit. All chapters in this section are based on research that has been performed at the University of Kansas, Nelson Environmental Study Area, over the last 10 years.

deNoyelles et al., University of Kansas Nelson Environmental Study Area (Chapter 30), present a general overview, based on information published in the scientific literature, of the types of direct and indirect effects that have been measured in mesocosm studies of pesticides. A long-term study on the effects of the herbicide atrazine, completed at the University of Kansas, is reviewed in detail.

Dewey and deNoyelles (Chapter 31) define ecosystem stability and develop the conceptual framework for the two chapters that follow. In addition, using the data from the atrazine study described in Chapter 30, they use a qualitative environmental correlation technique to measure the relative stability of the system.

The chapter by Johnson et al. (Chapter 32) describes a structural equation model, LInear Structural RELations (LISREL), that measures stability and provides a measure of the total, direct and indirect effects of all ecosystem components on all other components. Much of this chapter is devoted to providing a conceptual overview of LISREL and the explanation of the specialized terminology associated with this model. As such, it serves as a primer for the last chapter in this section by Huggins et al. (Chapter 33), in which LISREL is used to quantify the effects of an insecticide (atrazine) on mesocosm ecosystems.

CHAPTER **30**

Aquatic Mesocosms in Ecological Effects Testing: Detecting Direct and Indirect Effects of Pesticides

Frank deNoyelles, Jr., Sharon L. Dewey, Donald G. Huggins, and W. Dean Kettle

Abstract: Mesocosms are surrogate ecosystems which, in the form of experimental ponds or large enclosures in a pond or lake, can be used to simulate the responses of the corresponding natural aquatic ecosystems to an experimentally applied stress. A review of studies applying pesticides as the stress reveals the variety of direct and indirect effects that can develop in the multispecies/multicondition setting of the natural environment. Effects of the chemical appear through natural interactions among ecosystem components. Thirty-six pesticide studies including thirteen insecticides and eight herbicides demonstrate the cascade of effects that can occur particularly from the top of the food chain downward with insecticides and from the bottom of the food chain upward with herbicides. One study with the herbicide atrazine is examined in detail revealing differential direct effects, with decreased abundance and biomass of certain plants while others gained a degree of resistance. Some indirect effects developed as decreases in abundance and/or biomass of certain animals, including fish, tadpoles, and macroinvertebrates, possibly due to reductions in food, preferred microhabitat substrate, or refugia. Other indirect effects developed as increases in some animals including macroinvertebrates, possibly due to increases in food, better exploitation of altered microhabitats, or reduced predation. These effects of atrazine are compared to those revealed by single-species tests, and the advantages and disadvantages of both methods of ecological effects testing are considered.

0-87371-592-6/94/$0.00 + $.50
© 1994 by CRC Press, Inc.

INTRODUCTION

Various types of ecological effects tests have been performed on chemicals of environmental concern for many years. Most of these, following EPA guidelines, are single-species laboratory tests with various types of organisms exposed individually to individual chemicals under controlled laboratory conditions (e.g., References 1,2). With the completion of the EPA Technical Guidance Document, "Aquatic mesocosm tests to support pesticide registration",[3] a whole-ecosystem test was established for ecological effects testing of pesticides.

Mesocosms are surrogate ecosystems that can be controlled, manipulated, and replicated in the field to simulate the responses of natural ecosystems. The type of mesocosm outlined in the guidance document[3] is restricted to the lentic or standing water system as simulated by experimental ponds (e.g., References 4,5) or by enclosures of some flexible synthetic material within a larger body of water (e.g., References 5,6). Flowing water systems, though not considered in the guidance document or in this chapter, have been designed and operated as a series of channels (e.g., References 5,7), attempting to simulate the lotic environment. The mesocosm provides a testing arena where single species develop their responses to the perturbation in the multispecies/multicondition setting of the natural environment. The whole ecosystem response to disturbance can then develop, as the direct and subsequent indirect effects of the perturbation appear through natural interactions among the ecosystem components.

This chapter considers what the mesocosm test can provide to ecological effects testing for pesticides based on a body of literature on mesocosm studies with pesticides that have developed over the past two decades prior to the current registration studies. Thirty-six studies of twenty-one pesticides using a variety of types of mesocosms and procedures are reviewed. The types of effects that a surrogate ecosystem reveals are reported, particularly the indirect effects that single species separated from the natural environment are unable to reveal. One study with the herbicide atrazine is examined in detail to illustrate the variety of indirect effects that can be recorded when direct effects for this pesticide are limited to comparatively few types of organisms.

SINGLE-SPECIES VS. MESOCOSM EFFECTS TESTING

Single-species laboratory tests are the mainstay of ecological effects testing of potentially toxic chemicals. Many investigators (e.g., References 8 to 14) have considered what these tests are able to and unable to provide. It is anticipated that mesocosm testing included with single-species testing will provide some important information about pesticide effects that currently cannot be gained by using single-species tests alone.

Single-species tests are generally described as providing measures of direct effects of a chemical on individual organisms or species under controlled laboratory conditions. The acute and chronic toxicity of the chemical is determined for the test organisms under generally constant and optimal growth conditions. For particular measured responses of the test organisms, dose- and time-dependent relationships and estimates of threshold levels for effects can be determined. These tests also readily provide the necessary degree of replication for routine statistical measures of certainty for the effects recorded. In contrast, statistical evaluations of whole-ecosystem effects are more difficult due to the sometimes high degree of variation observed among replicated mesocosms. The standardized routines established over many years of experience coupled with the comparatively rapid, simple, and inexpensive nature of single-species tests makes them useful for assessing the ever-increasing number of xenobiotic chemicals of environmental concern.

The standardized routines of the single-species laboratory tests, which include controlled exposure and environmental conditions, have an advantage of providing a similar arena for evaluating and also comparing the direct effects of all chemicals. The advantages described above differ from those of the mesocosm arena and lead to the disadvantage of the lack of contact between the developing effects of the chemical and the natural environment. Species in the natural environment respond to an array of conditions, the response to any one affected by responses to some others. Here there is also an array of different species, the response of any one affecting responses of others. The whole ecosystem response to a stress, as the mesocosm can provide, develops as the direct and then indirect effects of the stress appear through natural interactions among the ecosystem components.

The natural environment and the environment as simulated by the mesocosm is heterogeneous and changing. At the outset this can produce temporal and spatial variation affecting fate and transport of the chemical and its bioavailability, results not produced by single-species tests. The mesocosm provides the setting for direct effects to develop into indirect or secondary effects on other organisms and on progressively higher levels of organization in the natural environment such as populations, communities, and ultimately the ecosystem. Single-species tests do not establish relationships between the direct effects they produce for particular organisms and changes in ecosystem structure and function as they develop through these levels of organization.

The fundamental difference between these two methods of effects testing is that when only single species are used for effects testing, the extent to which their responses can affect or be affected by other organisms and conditions is not included in the assessment. From some of the above comparisons it may be expected that a chemical will produce more effects in the natural environment than demonstrated in the laboratory even using a series of different test organisms from the natural community. However, with potential modification of the chemical by the environment and with the recognized adaptive potential of populations, communities, and ecosystems, effects in nature may sometimes be less than predicted from individual species responses.

It will be evident from the literature for mesocosm studies with pesticides that a whole ecosystem or valid surrogate exposed to a stress will inevitably produce responses that are more complex than those produced by single species in the laboratory. These complex responses are more difficult to quantify and evaluate statistically and also to compare with responses from other mesocosm studies using the same or different chemical for reasons discussed above. Some of the advantages of replicability and experimental strength of the single-species tests are becoming available to mesocosm testing from experimental designs of mesocosms in series with replicated reference (i.e., control) and treated systems and from more standardized study protocols now required by the EPA guidance document for pesticide registration studies.[3] However, the degree of replication in mesocosm experiments has been to some extent constrained by cost.

MESOCOSM STUDIES WITH PESTICIDES

For this review, mesocosms are defined as middle-sized multispecies experimental systems falling between microcosms of volumes less than 10 m^3 and the large, complex, real world macrocosm defined by Odum[15] as the natural pond or lake. Compared to the microcosm, we expect the mesocosm to function more like natural ecosystems or macrocosms function, by allowing natural colonization of most components, and including a more complete complement of the components of the macrocosm. Further comparisons and contrasts of mesocosm and microcosm experimental design and testing can be found in Buikema and Voshell.[5]

The mesocosm surrogates for the lentic or pond/lake ecosystem used for the pesticide studies summarized in Table 1 most commonly include excavated earthen ponds 1 to 2 m maximum depth, 0.01 to 0.08 ha surface area, and 100 to 1000 m^3 volume. Also used is the limnocorral, an openwater enclosure, and the littoral corral, a shoreline enclosure, of a dimension and volume at the smaller end of the range for pond mesocosms. Other types of excavated basins are occasionally used, including concrete or plastic lined ones with added sediments.

The 36 studies reviewed in Table 1 include testing of 13 insecticides and 8 herbicides. Studies differ greatly in experimental design with some being similar in design to the current designs established in the guidance document.[3] All share in common the exposure of a naturally colonized (except for fish) and maintained assemblage of most of the lentic ecosystem components. The intent of this review is to determine from these earlier studies the types of direct effects that were observed from such experimental ecosystems followed by the types of indirect effects that were observed to develop. Some discussion will be provided on the types of direct effects that led to particular indirect effects and the types of natural processes involved that the use of the mesocosm allows us to examine. Few studies provide rigorous statistical correlation between direct and indirect effects nor are other experimental methods incorporated in the field to test the relationships between direct and indirect effects. In many of the studies only selected components of the mesocosms were monitored, thus observing all possible effects of the chemical was not intended.

Insecticide Studies

All of the studies reviewed in Table 1 report observable direct effects of the chemical, and those that looked for indirect effects observed some. Insecticides generally produced direct effects on animals including zooplankton, insects, other macroinvertebrates, and fish. Fish have the fewest direct effects reported due in part to the fact that they were sometimes not included in the mesocosms, but also because these chemicals are basically intended to target insects, particularly at the lower concentrations generally used in such studies. Most studies in some way note the influence of other prevailing environmental conditions on the direct effects of the chemical, particularly in terms of effects on the fate of the chemical. The direct toxic effects of a chemical on particular organisms are sometimes described as being influenced by indirect effects such as, for predator species, altered prey abundances or, for prey species, altered grazing or predation pressure.[16,17]

Most indirect effects reported for the insecticides were increases in organisms thought to be released from grazing, predation, or competition pressure. Direct effects reducing the numbers of grazing rotifer and crustacean zooplankton commonly led to increases in phytoplankton abundance.[6,18-22] Reductions in predatory fish, insects, and zooplankton are sometimes ascribed as the cause for increases in particular prey animals.[18,19,21,23] Several studies suggest that reduced competition led to increases in certain animals, most notably the rotifer component of the grazing zooplankton with the decline of crustaceans,[18,19,21,24-29] the latter generally noted in the literature as having a competitive advantage over the former. Secondary declines of some predators resulted from direct effects that caused prey reduction.[16,18]

Some insecticides caused changes in certain physical/chemical conditions as indirect effects, though there are more cases of this with the herbicides. Oxygen declines severe enough to impact fish resulted from the death and decomposition of algae and macrophytes killed directly by the insecticide.[18,30] Oxygen increases sometimes accompanied increased plant growth resulting from reduced grazing.[31] Concentrations of other nutrients including C, N, and P sometimes increased following death and decomposition of plants or animals after insecticide addition.[31,32]

Herbicide Studies

The herbicides produced most of their effects on mesocosms initially as direct effects reducing plant production and biomass, though some were directly toxic to certain animals.[33-36] Indirect effects were equally common compared to insecticide studies and appeared through the mechanisms of reduced food for grazers, sometimes followed by reduced prey for predators, altered habitat structure for particular plants and animals, and reduced competition for unaffected plants.

Table 1. Direct and Indirect Effects of Pesticides Detected From Mesocosm Studies Conducted Prior to the Original Mesocosm Guidance Document (Touart, 1988) For Pesticide Registration Studies

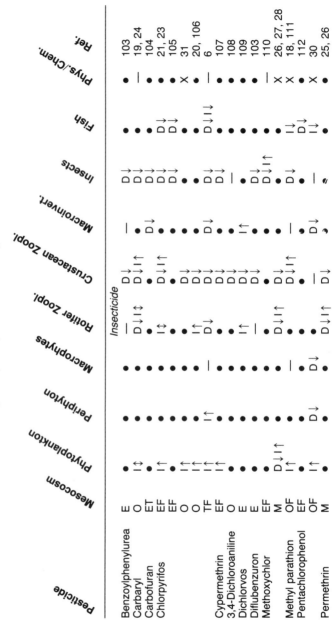

Pesticide	Mesocosm	Phytoplankton	Periphyton	Macrophytes	Rotifer Zoopl.	Crustacean Zoopl.	Macroinvert.	Insects	Fish	Phys./Chem.	Ref.
Insecticide											
Benzoylphenylurea	E	•	•	•	—	—	—	D	•	•	103
Carbaryl	O	I↔	•	•	D↓I↔	D↓I↔	•	D	•	—	19, 24
Carbofuran	ET	•	•	•	I↔	•	D	D	•	•	104
Chlorpyrifos	EF	I←	•	•	•	D↓I←	•	D	D→	•	21, 23
	EF	I←↓↔	•	•	I←	D→	•	D→	D→	•	105
	O	I↔	•	•	D	D→	•	D→	•	×	31
	O	I↔	•	•	•	D→	•	D→	D↓I↓	•	20, 106
	TF	•	•	—	•	D→	D→	D↓	D	—	6
Cypermethrin	EF	•	•	•	I←	D→	•	•	•	•	107
3,4-Dichloroaniline	O	•	I←	•	D↓	•	•	D↓	•	•	108
Dichlorvos	E	•	•	•	—	D↓	•	—	•	•	109
Diflubenzuron	E	•	•	•	D↓I←	D↓I↑	←	D↓I↑	•	•	103
Methoxychlor	EF	D↓I↑	•	—	•	D↓	•	D↓	•	×	110
Methyl parathion	M	I↑	•	•	•	•	•	D↓	I↓	×	26, 27, 28
Pentachlorophenol	OF	I←	D→	D→	•	I←	←	D→	D↓	•	18, 111
	EF	•	•	•	•	D↓	D→	•	I↓	×	112
Permethrin	OF	I←	•	•	D↓I←	I↑	?	•	•	•	30
	M	•	•	•	•	D↓	D↓	—	•	•	25, 26

Herbicide

Pesticide	Type	Ref.
Temephos	M / O	29 / 20, 106
Atrazine	NOF	36
	EF	37
	EF	41, 42
	EF	16, 38
	M	17, 43
Copper sulfate	EF	34, 35
2,4-D	EF	39
	E	113
Fenac	E	44
Prometone	NO	36
Propazine	NO	36
Simazine	NO	36
	EF	114
	EF	34, 35
	N	40
Sodium arsenite	OF	33

Note: Pesticides are listed by the common name for the active ingredient. Mesocosms, generally used in series, are of various types including experimental earthen ponds (E), farm or natural ponds (N), other pond basins variously constructed with concrete or plastic and lined with sediments (O), limnocorrals (M), and littoral corrals (T), the latter two being enclosures from surface to sediment within larger lentic ecosystems. Some were stocked with fish (F). Direct (D) and indirect (I) effects of a chemical are listed as increases (↑) or decreases (↓) in particular ecosystem components. For all studies only certain components were monitored even to the extent in some studies where indirect effects were not sought for most components. Components monitored but not observed to be affected are noted (—) and distinguished from others not monitored (●) thus, no effects for the latter could be recorded. The macroinvertebrate component does not include insects but does include macroinvertebrates such as oligochaetes, amphipods, arachnids, and mollusks. Some studies provide observations of altered physical/chemical conditions (X) due indirectly to the presence of the pesticide.

Food reduction effects are noted for grazers, including zooplankton,[29,37] insects,[16,38] and fish.[16,33,38] Altered habitat structures appeared to affect periphyton, with the loss of their macrophyte substrate,[16,38] and insects also requiring an epiphytic habitat.[16,38] The refugia provided by macrophytes for certain vertebrates such as fish, particularly juveniles, and tadpoles also appeared to produce indirect effects when reduced.[16] The herbicides that selectively reduced one plant type such as macrophytes generally caused increases in other plants[16,17,35,39-42] by reduced competition for nutrients and/or increases in nutrients accompanying decomposition. Most notably with atrazine, species within the same plant type as used in Table 1 can differ in their susceptability to the direct effects of the chemical, thus competitive releases can result in species shifts with herbicides as well.[16,17,32,41,43]

The direct effects of the herbicides on particular plants are expected in some cases to cause major changes in the physical/chemical environment as indirect effects. This is related to the role of plants in affecting physical conditions such as transparency and sediment cover and chemical conditions such as nutrient concentrations including oxygen.[16,40,44] The role of macrophytes in various pond physical/chemical processes has been reviewed and reported as significant.[45] Some of the impacts on physical/chemical processes caused by the loss of certain plants in mesocosms with herbicides such as atrazine and 2,4-D appear in Table 1. Turbidity increases have occurred as sediments became more exposed.[16] With turnover of macrophyte biomass generally low compared to that of the phytoplankton due to shorter generation time and greater grazing on the latter, macrophytes tend to lock up nutrients in the natural system. In some cases, then, macrophyte decomposition after herbicide addition can cause a sudden increase in nutrient availability, leading to algal increases and accompanying oxygen increases,[35,36,39] as well as the decrease in oxygen concentrations accompanying decomposition causing fish reductions.[34,35]

On a longer-term basis, macrophytes are considered to be a conduit for some nutrients from the sediments to the water column because of uptake from roots within the sediments and translocation to, and then some loss from, suspended leaves.[45] Therefore, macrophyte loss may, over a longer period of time, reduce nutrient availability in the water column. Also, a more indirect alteration of nutrients in the water column can occur either as increases or decreases accompanying increased turbidity. Increased turbidity can bring more nutrients into solution or, by adsorption to surfaces of particulates, remove some from solution. Mesocosm studies with pesticides have paid little attention to these nutrient relationships; thus, their influence is not documented in Table 1. Such relationships should be considered in the future, particularly with herbicide studies or in any study where macrophyte abundance differs between mesocosms or is being managed in the mesocosms, being entirely removed or allowed to become overgrown.

Conclusions

It is possible from this review to say that mesocosm studies with pesticides are able to assess direct effects on particular organisms, including the influence of other environmental conditions on the response. Such studies also have been able to allow direct effects to produce indirect effects through the natural network of interacting ecosystem components. In providing these, the mesocosm test reveals effects of a pesticide not observed in single-species tests.

The microcosm, though not being compared here, is a smaller (<10 m³) yet multispecies version of the mesocosm that can be operated with greater ease and perhaps greater replicability in the field. Because of its smaller size and usually fabricated basin of a glass aquarium or larger fiberglass tank, it is not expected that as many of the natural ecosystem components can be provided. Research continues with microcosms as described elsewhere in this volume and their use as ecosystem surrogates is still developing, as is also the case for mesocosms.

In the next section we use one pesticide mesocosm study extensively reported on in the literature to more specifically illustrate the array of direct and indirect effects possible from a mesocosm as identified generally in Table 1. For this pesticide, the herbicide atrazine, there is also information in the literature for ecological effects assessments from single-species tests that will be compared with the mesocosm assessments.

A POND MESOCOSM STUDY WITH AN HERBICIDE

Characteristics of the Herbicide

Plants are directly affected by the herbicide atrazine from the inhibition of photosynthesis. This occurs by the chemical occupying a specific binding site in the protein structure of the chloroplast, inactivating electron transport carrier proteins within photosystem II.[46,47] This action is rapid, following initial exposure of the plant, occurring within a few seconds for phytoplankton and a somewhat longer time for macrophytes, depending on translocation time.[48] Some plants, including algae and flowering plants, have been known to develop resistance through chance genetic alteration of the binding site.[16,37,49]

At the time of this study in the early 1980s, atrazine was one of the most heavily used pesticides in the U.S. (and still is) accounting for 10 to 14% of the annual pesticide application.[50,51] It was commonly detected in surface waters (e.g., References 16, 52 to 57) in regions with high corn and sorghum production, two of the few plants able to avoid most of the effects of atrazine by metabolizing the molecule before it reaches the binding site.[58] From these references, atrazine has been frequently reported as detected in pond and reservoir waters mostly at concentrations less than 4 μg/l with occasional levels up to 20 μg/l and adjacent to treated fields as high as 1 mg/l.

Following an initial exposure, atrazine remains in the water column beyond a month, with an observed half-life of a few months.[16,37,54,59] Though there is no evidence for biological magnification in the food chain, bioaccumulation in particular organisms, including vertebrates and invertebrates, directly from their surroundings has been observed in some cases[60-62] but not in others.[63]

Single-Species Aquatic Plant Tests

Single-species tests of aquatic plants exposed to atrazine have shown that most plants experience reductions in photosynthesis at concentrations above 10 µg/l. This includes both algae (e.g., References 64 to 69) and flowering plants (e.g., References 70 to 73). EC_{50} values are often between 30 and 100 µg/l and usually less than 500 µg/l.

Single-Species Aquatic Vertebrate Tests

Direct toxicity to aquatic vertebrates assessed by single-species tests, mostly observing the immediate acute effects, appears to be minimal at the concentrations found in contaminated habitats, including the 20 to 500 µg/l range used in the mesocosm study to be described here (e.g., References 63, 74 to 79). Less information exists for longer-term chronic effects using single-species tests, and long-term effects at concentrations above 200 µg/l and perhaps even lower cannot be ruled out.[63]

Single-Species Aquatic Invertebrate Tests

There is less information available for aquatic invertebrates, but what there is suggests that acute effects occur at concentrations similar to those cited for fish. As with fish, chronic effects on invertebrates cannot be ruled out even at concentrations somewhat lower than 200 µg/l.[63,75,78,80-83]

Mesocosm Test Design

From the characteristics of atrazine described above, including its occurrence and fate, mode of action, greater effects on plants than animals, and possibility of developed resistance in some plants, there is the likelihood of both direct and indirect effects developing in an exposed aquatic ecosystem. Select responses of 0.045-ha earthen pond mesocosms to a range of atrazine concentrations in one study will be examined here. Details of experimental methods and results not discussed here can be found in the literature.[4,16,37,38,55,67,84-91]

We use this detailed examination of a single mesocosm study to investigate whether direct effects predicted by single-species tests occurred in the mesocosm ecosystem and to what extent these effects were influenced by other conditions. We can then determine whether direct effects in the presence of other organisms and conditions led to indirect effects not demonstrated by single-species studies cited earlier from the literature. This provides one comparison of the range of effects detected for one chemical from single-species tests vs. whole-ecosystem mesocosm tests. Other comparisons are available from other studies reviewed in Table 1, if examined in greater detail.

The experimental design of this 3-year study (805 days) using ten 0.045-ha earthen pond mesocosms has been described.[16] This study examined the effects of 20, 100,

200, and 500 µg/l of atrazine in pairs of ponds, including 2 control ponds with no atrazine. The exposure was a single addition of atrazine at the beginning of each summer to the same ponds with 20-, 100-, and 200-µg/l treatments for 3 successive years (1981 to 1983), and the 500-µg/l treatment for 2 successive years (1982, 1983). Measured concentrations immediately after addition differed by <10% of nominal concentrations targeted. After 6 and 12 months, concentrations declined to about 70 and 25% of the original measured concentrations, respectively. Four species of fish were stocked in the ponds to represent different positions in the food chain.[16] These included *Ctenopharyngodon idella* (grass carp) a macrophyte grazer, *Dorsoma cepedianum* (gizzard shad) a filter-feeding omnivore, *Ictalurus punctatus* (channel catfish) a benthic-feeding omnivore, and *Lepomis macrochirus* (bluegill sunfish), a visual predator.

Direct Effects on Phytoplankton

From the single-species tests with algae cited earlier, some declines in phytoplankton production and biomass below control pond levels are expected at all of the atrazine concentrations. Such declines did occur within a few days of the first exposure to atrazine,[16] which was 21 June 1981 for 20- to 200-µg/l ponds and 30 May 1982 for 500-µg/l ponds. Some declines in pH and dissolved oxygen were noted for about 1 week[16] but, for the latter, not below 5 mg/l in middle of the water column, 1 h after sunrise. Thus, water chemistry effects on animals were unlikely. Other indirect effects of atrazine on physical/chemical conditions in the ponds are discussed below along with the organisms possibly impacted.

Within 1 month, phytoplankton production and biomass levels returned to control pond levels, exceeding them at times, in all treated ponds while decreases in atrazine concentrations were only slight.[16] Phytoplankton declines and returns in the 500-µg/l ponds after their first exposure to this concentration in 1982 can be seen in Figure 1a and 1c, and in greater detail, along with statistical evaluations of significant effects for this and other parameters, in deNoyelles et al.[16] In 1983 (Figure 1b and 1d), either the third (20 to 200 µg/l) or second (500 µg/l) annual exposure for the phytoplankton communities, no such declines are evident following the 25 May addition of atrazine.

It has been demonstrated in this study[16] by various means that species resistant (or tolerant, see Reference 16) to atrazine appeared in the phytoplankton communities exposed to all concentrations of atrazine. This resulted in the growth of algae in ponds with atrazine concentrations reported to severely inhibit single-species growth in the laboratory tests. Though production and biomass returned to control pond levels, even exceeding them at times, species composition was altered. Compositional changes are reported in detail for an earlier study on some of the same ponds.[37]

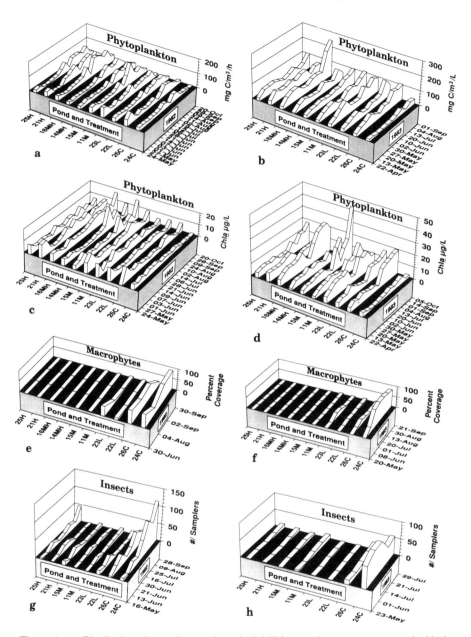

Figure 1.   Distribution of aquatic organisms in 0.045-ha pond mesocosms treated with the herbicide atrazine from 1981 to 1983 at concentrations of 20 μg/l (ponds 22L, 23L), 100 μg/l (ponds 11M, 15M), 200 μg/l (ponds 14MH, 16MH), and 500 μg/l (ponds 21H, 25H) along with two control ponds (ponds 24C, 26C) receiving no treatment (see text and Reference 16 for treatment details). Organism parameters recorded here for the second (a, c, e, g) and third (b, d, f, h) years of atrazine exposure (for method details see references cited) include phytoplankton production and biomass,[16] submersed macrophyte coverage of the pond bottom,[16,84,87] and total emerging insects.[16,38]

Animals Unaffected by Treatment

Animals more closely linked in the food chain to the phytoplankton base appear to be unaffected by atrazine treatment either directly or indirectly. These animals include grazing rotifer and crustacean zooplankton (Figure 2a) along with their predator, the planktonic insect *Chaoborus* (Figure 2b), and the filter feeding fish, gizzard shad (Figure 2h). Rotifers showed none of the increases in abundance common in other pesticide studies, but recall that these typically occurred only when crustacean zooplankton declined. Channel catfish (Figure 2h), though not linked to the phytoplankton as described for the organisms above, were also unaffected by treatment or showed slight increases in biomass, perhaps due to their omnivorous feeding behavior in the benthic zone (see Dewey and deNoyelles, Chapter 31). From the laboratory studies cited earlier, these animals would not be expected to be greatly inhibited by atrazine, if at all. However, other animals similarly unaffected according to single-species studies, but connected to a different food chain base, did decline in abundance as discussed below.

Direct Effects on Macrophytes

Based on the single-species laboratory tests with macrophytes cited earlier, these plants are expected, as are the algae, to be highly impacted in the mesocosms. During summers 1982 and 1983 the predicted impact on this particular compartment of the pond food chain is apparent for submersed macrophytes as a community (Figure 1e and 1f) and as particular species[16,84] and for emergent macrophytes.[16,84] Emergent macrophytes, primarily cattails (*Typha latifolia, T. angustifolia*), declined in abundance at the 100-μg/l treatment and above.[16] However, unlike the phytoplankton, macrophytes did not recuperate here or in a more recent study.[92]

There is no indication of resistant macrophyte species becoming established, though in the 20- and 100-μg/l ponds there was an increase in the abundance of species of *Chara* relative to the other macrophytes.[84] The lack of developed resistance for the macrophytes could be due to their longer generation time and lower number of individuals compared to the phytoplankton. Thus for macrophytes, a lower probability would exist of chance mutation introducing resistant individuals into the community that could then become rapidly established. Indirect effects of atrazine on other organisms, as either decreases or increases in abundance associated with macrophyate declines, are discussed next.

Declines of Macrophyte-Grazing Fish

Loss of macrophytes from a pond ecosystem could have variety of effects on the rest of the ecosystem (as reviewed by Carpenter and Lodge[45]) including effects associated with their multiple roles as a direct food source, as substrate or refugia for other organisms, and in nutrient cycling. Grass carp declined in individual biomass and biomass per pond (Figure 2h) at ⩾100 μg/l, likely due to the loss of

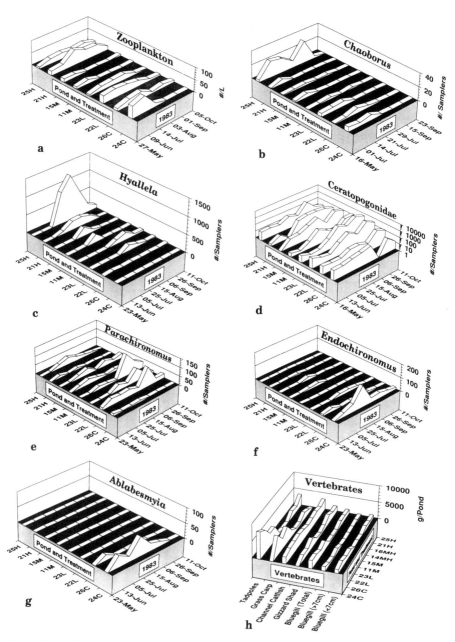

Figure 2. Distribution of aquatic organisms in pond mesocosms as described in Figure 1. Organism parameters recorded here for the third year of atrazine exposure include crustacean zooplankton,[16] the planktonic dipteran insect *Chaoborus punctipennis,*[16] other dipteran insects collected on plate samplers[55,87] including Ceratopogonidae, *Parachironomus* spp., *Endochironomus nigricans,* and *Ablabesmyia peleensis,* the amphipod *Hyallela azteca* also from plate samplers, and the vertebrates collected at the end of the study including tadpoles, grass carp, channel catfish, gizzard shad, and total bluegill and sizes <7 cm and ≥7 cm total length.

their macrophyte food source. Direct effects of atrazine on aquatic animals cannot be ruled out, due to the insufficient data from single-species tests particularly for species recorded in this study, including the grass carp. Grass carp did not decline in the 20-μg/l ponds, though macrophyte densities were low there as well. This was possibly due to higher macrophyte growth rates at this low treatment level than at the higher concentrations, as discussed below, still providing an adequate food supply.

Effects of Grazing on the Macrophyte Response

Grazing by grass carp affected the response of both submersed (discussed next) and emergent (see References 16, 84 for details) macrophytes, to similar degrees over the range of atrazine treatments. Increased loss of macrophytes resulted from the added pressure of grazing as evidenced by comparisons of macrophyte densities in the ponds to densities within fish exclosures occupying 5% of the area of each pond.[16] The maximum reductions of submersed macrophytes occurred with or without grass carp at ≥200 μg/l, but required both stresses for the similar maximum to occur at 100 μg/l (Figure 1e and 1f and Reference 16). At 100 μg/l, densities within exclosures were greater than on the outside and were about 40% of those in control and 20 μg/l pond exclosures. At 20 μg/l, densities within exclosures were not distinguishable from control pond densities within exclosures.[16] This demonstrates the influence of an accompanying stress on the pesticide stress, a condition rarely included in single-species studies but common in natural ecosystems.

Another perspective of the influence of grass carp grazing relative to the atrazine effects can be seen in comparing the two control ponds (Figure 1e and 1f) beginning in September 1982 when grass carp were lost from one pond (24C), probably due to transient vertebrate predation (snapping turtle). In 1983 a greater macrophyte coverage is evident in the control pond without grass carp, while the other control pond had a similar coverage to that of the 20-μg/l ponds (Figure 1f). From this comparison the influence of a macrophyte and/or grass carp difference in the absence of atrazine can be seen for other ecosystem components as discussed next.

Phytoplankton production and biomass differences between the two control ponds increased in 1983, the pond with less macrophytes (26C) increasing in phytoplankton (Figure 1a to 1d). This may indicate reduced nutrient competition for the phytoplankton in this pond. Increased nutrient turnover may also be involved resulting from grass carp processing vegetation and suspending sediments by their feeding behavior as suggested by others (e.g., References 93 to 96).

It is also evident from these control pond comparisons that by the third year, grass carp alone were eliminating macrophytes, but the nearly total elimination only occurred with atrazine present ≥100 μg/l. At the same time as in pond 24C (September 1982), grass carp were also lost and not replaced from a 500-μg/l pond adjacent to pond 24C, probably from the same cause. In this pond, macrophyte reduction was still nearly total, showing again that atrazine alone at this concentration was a sufficient stress for the maximum effect.

Comparisons of control and 500-μg/l treatment ponds with and without grass carp are also useful in analyzing an observed treatment effect on turbidity. Turbidity was greater in treated ponds in 1982 by a factor of 1.3 to 1.7 in 20-μg/l to 500-μg/l ponds compared to control ponds which were similar to each other with about 28 NTU. During 1983 the turbidity in the control pond with grass carp and reduced macrophytes increased slightly but remained different from the 20-μg/l to 500-μg/l ponds as in 1982. The other control pond with a dense coverage of macrophytes in 1983 (Figure 1f) decreased to a mean of 5 NTU, indicating further the influence of grass carp and macrophytes on turbidity. Because grass carp and the reduction of macrophyte cover have the same correlation with turbidity, it is difficult to separate their individual effects. The disruptive feeding behavior of grass carp, rumaging and rooting in the vegetation,[93-96] could conceivably account for increased turbidity in the control with grass carp compared to without grass carp. However, from the results of the 500-μg/l pond with no grass carp and with a herbicide-induced loss of macrophytes, we recorded the same elevated turbidity in 1983 as the 500-μg/l pond with grass carp. This implies that the loss of macrophytes in our ponds is a more important factor than grass carp behavior in causing turbidity increases.

Further comparisons of control ponds will be used below to evaluate other possible indirect effects of atrazine on animals associated with the macrophyte community.

Declines of Insects

Other animals declining in abundance with atrazine treatment did not feed directly on macrophyte biomass, but were associated with the macrophyte microhabitat. They were likely affected by macrophyte loss through some other dependency (e.g., epiphytic algal food and/or habitat). This conclusion of macrophytes as the main vector for atrazine effect on certain animal species was corroborated by control pond comparisons, where similar animal declines are seen during the second and third year in control pond 26C with less than half the macrophyte coverage of control pond 24C (Figure 1e and 1f). Total emerging insects (Figure 1g and 1h) in control ponds were dominated in numbers by Chironomidae, mostly nonpredators (herbivores, detritivores) in an epiphytic microhabitat. These insects declined with treatment and were also less abundant in the low-macrophyte control pond (26C) compared to the other control pond (24C) by the end of summer 1982 and 1983 (Figure 1g and 1h).

Insects with this response include *Endochironomus* (Figure 2f) and *Ablabesmyia* (Figure 2g) both within the dipteran Chironomidae. The decline of these insects with the macrophyte reduction in the low macrophyte control and all treated ponds is possibly associated with the loss of their epiphytic microhabitat. The former insect is a herbivore on the epiphytic algal community growing on the macrophytes and the latter is a predator on microcrustacea and Chironomidae such *Endochironomus*.[38,55] Other Chironomidae with this response include *Parakiefferiella* sp. and *Labrundinia* cf. *neopilosella*,[55] and others reported by Dewey[38] and discussed elsewhere in this volume.[87,102]

Declines of Tadpoles

Declines of tadpole biomass, primarily *Rana catesbeiana,* were evident (Figure 2h) by the end of the study at all atrazine concentrations. Loss of macrophyte feeding sites and egg deposition sites probably accounted for much of the decline in this vertebrate, the young of which feed primarily on periphyton. Frog populations colonized naturally in the ponds, so that modification of shoreline vegetation could affect egg deposition and survival. Loss of macrophytes could also affect refuge from predation by bluegill and channel catfish, though there is little evidence from the literature that these fish prey heavily on tadpoles. Reduction of the periphyton food resource by direct and indirect actions of atrazine was probably an important factor in the tadpole response. With the loss of both submersed and emergent macrophytes, there was an accompanying loss of periphyton, because their primary substrate in the water column was the macrophyte community.[16] There was also evidence of direct effects of atrazine on periphyton on the macrophytes that remained and on artificial structures in the ponds,[16,38] reducing the tadpole food resource even more.

Declines of Bluegill Sunfish

The predatory bluegill sunfish in ponds is in part supported by the phytoplankton-based food chain as younger fish (<3.1 cm, <1-year old) that feed on zooplankton, but then as they begin including a variety of macroinvertebrates as they get larger,[97,98] the macrophyte base also becomes supportive. Bluegill in experimental ponds similar in size to the ones used here have been shown to grow better and consume more prey at intermediate macrophyte densities,[97] approximate densities within our control pond with grass carp remaining and the 20-μg/l ponds prior to 1983.

Bluegill sunfish declined in total biomass at the end of the study in all ponds at or above 100 μg/l (Figure 2h and Reference 16). In an earlier 135-day study using some of the same ponds, adult bluegill from atrazine treated ponds (20 and 500 μg/l) at the end had fewer and less varieties of insect prey consistent with the findings of Crowder and Cooper[97] for low macrophyte ponds. With the reductions in emerging insects in our ponds (Figure 1g and 1h) and the shift to more sediment-dwelling macroinvertebrates discussed below, food availability likely declined for the bluegill, particularly the portion of the population >3.1 cm in length. These larger fish at the end of the study were $>75\%$ of the biomass in all but one pond where they were 60%. Foraging efficiency by bluegill for what macroinvertebrates were present in treated ponds may also have been reduced, particularly for bottom-feeding, as a result of increased turbidity caused, at least in part, by the herbicide.

Increases of Certain Macroinvertebrates

There remains one more group of animals affected by atrazine treatment and grass carp grazing, but showing increases in abundance at some treatment levels. The amphipod, *Hyallela azteca,* increased with treatment, being more abundant in

1983 in all treatment ponds compared to the controls (Figure 2c). Certain species of insects were more abundant at atrazine levels of 20 to 200 μg/l but only when compared to the high macrophyte control. Abundance declined for each in the 500-μg/l ponds but not below that of the low macrophyte control (26C). Insects with this distribution include species of the mayfly *Caenis* spp. and Chironomidae, including *Dicrotendipes simpsoni, Parachironomus* spp. (Figure 2e), and *Glyptotendipes* spp.[55] The Ceratopogonidae (Figure 2d) within the Diptera, including species of *Bezzia, Probezzia,* and *Palpomyi,* were the most numerically abundant insects in the ponds and had this same distribution except there were no consistent differences between the two control ponds.[55]

The amphipod and insects responding to macrophyte loss with increased abundance in both treated and control ponds may be favored by the loss of macrophytes to some degree but are then reduced by either direct or indirect effects of the highest atrazine levels (≥200 μg/l). In an experimental pond study with nutrient additions causing increased macrophyte densities, there was reported a shift from larger sediment-dwelling invertebrates to smaller more mobile forms that lived on the macrophytes.[98] A similar trend may be described by the invertebrate responses here, though in the reverse with atrazine/grass carp-induced macrophyte losses. Although we know with certainty only *Hyallela* to be a true benthic dweller, to varying degrees the insects with a similar response were able to successfully reside on the open pond bottom of the atrazine-treated ponds. Amphipods have been reported to be particularly vulnerable to bluegill predation in other experimental pond studies.[97,98] However, loss of cover in the form of macrophytes may have been compensated for by increased turbidity, creating another form of refuge from the highly visual bluegill predation in atrazine-treated ponds. Again there appears to be a series of possible indirect effects of atrazine leading to impact on organisms not shown in single-species tests.

CONCLUSIONS

It should be clear from the detailed examination of a whole ecosystem exposed to a pesticide stress that a complex "chain reaction" of effects can result. Examining other mesocosm studies with pesticides from Table 1 shows this as well, to varying degrees and in varying ways. Herbicides, with little direct effects on animals, tend to produce effects from the bottom of the food chain upward, and insecticides, with most direct effects on animals, from the top of the food chain downward. This is consistent with the well-established concept of cascading trophic effects in ecosystems, bottom up or top down (e.g., References 99 to 101). In addition, nontrophic (i.e., not food-related) indirect effects can manifest through the community, as seen in some of the atrazine-macrophyte-animal responses described above, where macrophytes represented refuge and/or habitat.

In the atrazine study examined, some species previously tested in single-species studies showed predicted direct responses to the chemical in the multispecies/multicondition setting of the natural mesocosm environment and then indirect effects of the chemical followed, produced through ensuing interactions among ecosystem components. This result shows the potential complementary relationship that can exist between single-species laboratory tests and natural ecosystem mesocosm tests. Other species showed greater or lesser effects in the mesocosm than predicted by single-species tests. This result shows the importance of supplementing single-species tests with ecosystem-level mesocosm tests that allow the full complement of natural interactions to produce responses.

In most of the studies reviewed, including the one detailed for atrazine, the pathways of interaction producing indirect effects from the direct ones are mostly speculative. More experimental evidence is required, for example as was done by using *in situ* exclosures to separate some of the effects of atrazine from those of the macrophyte grazer. There are also very few attempts to describe mathematically the pathways of effects, though with newly developing structural equation modeling techniques described elsewhere in this volume, this is being attempted for the atrazine mesocosm study described here.[55,87-89]

These pathway analyses may provide measures of the total response of the ecosystem, integrating the array of direct and indirect effects into one response measure as also discussed elsewhere in this volume.[89] Use of this and other mathematical models to analyze stability (e.g., Reference 89) within mesocosm ecosystems is a means of applying ecosystem stability theory to mesocosm effects testing (see Dewey and deNoyelles[102] for review). Various measures of stability may not only provide single measures of how ecosystem structure changes, which studies in the past have focused on, but also how ecosystem function changes and to what extent.

The ecosystem response to a pesticide begins with the responses of individual species or conditions. With the advent of mesocosm testing, we began to examine these responses in the natural setting as particular direct effects and then particular indirect effects, but still only as changes in particular components. In current registration studies we still tend to isolate species responses and analyze them as we do single-species test responses. We adopted mesocosm effects testing for pesticides from the broader use of mesocosms in basic aquatic ecology.[4] For pesticide ecological effects testing, we gained the immediate advantage of observing species-level responses in a more natural environment than the one of species isolated in the laboratory. In current registration studies we have not yet, however, gained the possible greater advantage of the surrogate ecosystem, which is to measure the integrated responses of the whole ecosystem as the response to stress to be extrapolated for regulatory purposes.

For the most complete ecological effects assessment both single-species and mesocosm tests must continue to be used in concert. The single-species laboratory tests can be used in a more practical and routine manner to characterize the direct effects of the ever increasing number of potentially toxic chemicals of all types entering the environment. The mesocosm test will refine our understanding of the pathways that direct effects follow in becoming indirect effects. This will help to identify the more important components of the whole ecosystem to examine as single components in the laboratory or to incorporate into the microcosm, which can serve as an intermediary between laboratory single-species and field natural community studies. Incorporating the mesocosm further into ecological effects testing requires further developments as suggested, but it also requires the willingness to continue taking the next step.

ACKNOWLEDGMENTS

For the atrazine mesocosm studies we appreciate the participation of C. Baker, E. Carney, L. Ferrington, C. Fromm, B. Heacock, B. Howard, M. Johnson, A. Kadoum, C-H. Lei, P. Liechti, T. Miller, M. Moffett, H. Montoya, D. Sinn, A. Trammel, and D. Weirick. We also appreciate the participation of the Kansas Biological Survey and D.P. Larson and F.S. Stay then from the EPA Corvallis Environmental Research Laboratory. The atrazine mesocosm studies were supported by grants from USEPA (R806641010 and CR808804010), Kansas Water Resources Research Institute (A-092-KAN and A-104-KAN), and University of Kansas (3488, 3709, and 3433). Although the research described in this article has been funded in part by the U.S. Environmental Protection Agency through contract to F. de-Noyelles, it has not been subjected to the agency's optional peer and policy review and, therefore, does not necessarily reflect the views of the agency, and no official endorsement should be inferred.

REFERENCES

1. U.S. Environmental Protection Agency. 1985. Methods for measuring the acute toxicity of effluents to freshwater and marine organisms. EPA-600/4-85/013. Environmental Monitoring and Support Laboratory, Cincinnati, OH, 216 pp.
2. U.S. Environmental Protection Agency. 1985. Short-term methods for estimating the chronic toxicity of effluents and receiving waters to freshwater organisms. EPA-600/4-85/014. Environmental Monitoring and Support Laboratory, Cincinnati, OH, 162 pp.
3. Touart, L.W. 1988. Hazard evaluation division, technical guidance document: aquatic mesocosm tests to support pesticide registration. EPA-540/09-88-035. U.S. Environmental Protection Agency. National Technical Information Service, Springfield, VA, 35 pp.
4. deNoyelles, F., Jr. and W.D. Kettle. 1985. Experimental ponds for evaluating bioassay predictions. In T.P. Boyle, ed., *Validation and Predictability of Laboratory Methods for Assessing the Fate and Effects of Contaminants in Aquatic Ecosystems.* ASTM STP 865. American Society for Testing and Materials, Philadelphia, PA, pp. 91–103.
5. Buikema, A.L., Jr. and J.R. Voshell, Jr. 1993. Toxicity studies using freshwater benthic invertebrates. In D.M. Rosenberg and V.H. Resh, eds., *Freshwater Biomonitoring and Benthic Macroinvertebrates.* Chapman and Hall, New York, NY, 344–398.

6. Siefert, R.E., S.J. Lozano, J.C. Brazner and M.L. Knuth. 1989. Littoral enclosures for aquatic field testing of pesticides: the effects of chlorpyrifos on a natural system. In J.R. Voshell, Jr. ed., *Using Mesocosms for Assessing the Aquatic Ecological Risk of Pesticides: Theory and Practice*. Miscellaneous Publication Number 75, Entomological Society of America, Lanham, MD, pp. 57–73.

7. Arthur, J.W., J.A. Zischke, K.N. Allen and R.O. Hermanutz. 1983. Effects of diazinon on macroinvertebrates and insect emergence in outdoor experimental channels. *Aq. Tox.* 4:283–301.

8. Cairns, J., Jr. 1983. Are single species toxicity tests alone adequate for estimating environmental hazard? *Hydrobiol.* 100:47–57.

9. Cairns, J., Jr. 1983. The case for simultaneous toxicity testing at different levels of biolgoical organization. In W.E. Bishop, R.D. Cardwell and B.B. Heidolph, eds., *Aquatic Toxicology and Hazard Assessment: Sixth Symposium*. ASTM STP 802, American Society for Testing and Materials, Philadelphia, PA, pp. 111–127.

10. Cairns, J., Jr. 1986. Multispecies toxicity testing: a new information base for hazard evaluation. *Curr. Prac. Environ. Sci. Eng.* 2:37–49.

11. Cairns, J., Jr. 1986. The myth of the most sensitive species: multispecies testing can provide valuable evidence for protecting the environment. *BioSci.* 36:670–672.

12. Chapman, G.A. 1983. Do organisms in laboratory toxicity tests respond like organisms in nature? In W.E. Bishop, R.D. Dardwell and B.B. Heidolphs, eds., *Aquatic Toxicology and Hazard Assessment: Sixth Symposium*. ASTM STP 802, American Society for Testing and Materials, Philadelphia, PA, pp. 315–327.

13. Kimball, K.D. and S.A. Levin. 1985. Limitations of laboratory bioassays: the need for ecosystem-level testing. *BioSci.* 35:165–171.

14. Mayer, F.L., Jr. and M.R. Ellersieck. 1988. Experiences with single-species tests for acute toxic effects on freshwater animals. *Ambio* 17:367–375.

15. Odum, E.P. 1984. The mesocosm. *BioSci.* 34:558–562.

16. deNoyelles, F., Jr., W.D. Kettle, C.H. Fromm, M.F. Moffett and S.L. Dewey. 1989. Use of experimental ponds to assess the effects of a pesticide on the aquatic environment. In J.R. Voshell, Jr., ed., *Using Mesocosms for Assessing the Aquatic Ecological Risk of Pesticides: Theory and Practice*. Miscellaneous Publication Number 75, Entomological Society of America, Lanham, MD, pp. 41–56.

17. Herman, D., N.K. Kaushik and K.R. Solomon. 1986. Impact of atrazine on periphyton in freshwater enclosures and some ecological consequences. *Can. J. Fish. Aquat. Sci.* 43:1917–1925.

18. Crossland, N.O. 1984. Fate and biological effects of methyl parathion in outdoor ponds and laboratory aquaria-II. Effects. *Ecotoxicol. Environ. Safety* 48:482–495.

19. Hanazato, T. and M. Yasuno. 1987. Effects of a carbamate insecticide, carbaryl, on the summer phyto- and zooplankton communities in ponds. *Environ. Pollut.* 48:145–149.

20. Hughes, D.N., M.G. Boyer, M.H. Papst and C.D. Fowle. 1980. Persistence of three organophosphorus insecticides in artificial ponds and some biological implications. *Arch. Environm. Contam. Toxicol.* 9:269–279.

21. Hurlbert, S.H., M.S. Mulla and H.R. Wilson. 1972. Effects of an organophosphorus insecticide on the phytoplankton, zooplankton, and insect populations of fresh-water ponds. *Ecol. Monogr.* 42:269–299.

22. Papst, M.H. and M.G. Boyer. 1980. Effects of two organophosphorus insecticides on the chlorophyll a and phaeopigment concentrations of standing ponds. *Hydrobiol.* 69:245–250.

23. Hurlbert, S.H., M.S. Mulla, J.O. Keith, W.E. Westlake and M.E. Dusch. 1970. Biological effects and persistence of Dursban® in freshwater ponds. *J. Econ. Entomol.* 63:43–52.

24. Hanazato, T. and M. Yasuno. 1990. Influence of time of application of an insecticide on recovery patterns of a zooplankton community in experimental ponds. *Arch. Environ. Contam. Toxicol.* 19:77–83.

25. Kaushik, N.K., G.L. Stephenson, K.R. Solomon and K.E. Day. 1985. Impact of permethrin on zooplankton communities in limnocorrals. *Can. J. Fish. Aquat. Sci.* 42:77–85.

26. Kaushik, N.K., K.R. Solomon, G.L. Stephenson and K.E. Day. 1986. Use of limnocorrals in evaluating the effects of pesticides on zooplankton communities. In J. Cairns, Jr., ed., *Community Toxicity Testing*, ASTM STP 920, American Society for Testing and Materials, Philadelphia, PA, pp. 269–290.

27. Solomon, K.R., G.L. Stephenson and N.K. Kaushik. 1989. Effects of methoxychlor on zooplankton in freshwater enclosures: influence of enclosure size and number of applications. *Environ. Toxicol. Chem.* 8:659–669.

28. Stephenson, G.L., N.K. Kaushik, K.R. Solomon and K. Day. 1986. Impact of methoxychlor on freshwater communities of plankton in limnocorrals. *Environ. Toxicol. Chem.* 5:587–603.

29. Yasuno, M., T. Hanazato, T., Iwakuma, K. Takamura, R. Ueno and N. Takamura. 1988. Effects of permethrin on phytoplankton and zooplankton in an enclosure ecosystem in a pond. *Hydrobiol.* 159:247–258.

30. Crossland, N.O. and C.J.M. Wolff. 1985. Fate and biological effects of pentachlorophenol in outdoor ponds. *Environ. Toxicol. Chem.* 4:73–86.

31. Butcher, J.E., M.G. Boyer and C.D. Fowle. 1977. Some changes in pond chemistry and photosynthetic activity following treatment with increasing concentrations of chlorpyrifos. *Bull. Environ. Contam. Toxicol.* 17:752–758.

32. Solomon, K.R., J.Y. Yoo, D. Lean, N.K. Kaushik, K.E. Day and G.L. Stephenson. 1986. Methoxychlor distribution, dissipation and effects in freshwater limnocorrals. *Environ. Toxicol. Chem.* 5:577–586.

33. Gilderhus, P.A. 1966. Some effects of sublethal concentrations of sodium arsenite on bluegills and the aquatic environment. *Trans. Am. Fish. Soc.* 95:289–296.

34. Tucker, C.S. and C.E. Boyd. 1978. Consequences of periodic applications of copper sulfate and simazine for phytoplankton control in catfish ponds. *Trans. Am. Fish. Soc.* 107:316–320.

35. Tucker, C.S., R.L. Busch and S.W. Lloyd. 1983. Effects of simazine treatment on channel catfish production and water quality in ponds. *J. Aquat. Plant Manage.* 21:7–11.

36. Walker, C.R. 1961. Simazine and other *s*-triazine compounds as aquatic herbicides in fish habitats. *Weeds* 12:134–139.

37. deNoyelles, F., Jr., W.D. Kettle and D.E. Sinn. 1982. The responses of plankton communities in experimental ponds to atrazine, the most heavily used pesticide in the United States. *Ecology* 63:1285–1293.

38. Dewey, S.L. 1986. Effects of the herbicide atrazine on aquatic insect community structure and emergence. *Ecology* 67:148–162.

39. Boyle, T.P. 1980. Effects of the aquatic herbicide 2,4-D DMA on the ecology of experimental ponds. *Environ. Pollut. (Ser. A)* 21:35–49.

40. Crawford, S.A. 1981. Successional events following simazine application. *Hydrobiol.* 77:217–223.

41. Gunkel, G. 1983. Investigations of the ecotoxicological effect of a herbicide in an aquatic model ecosystem. I. Sublethal and lethal effects. *Arch. Hydrobiol. Suppl.* 65:235–267.

42. Gunkel, G. 1984. Investigations of the ecotoxicological effects of a herbicide in an aquatic model ecosystem. II. Food chain significance and pesticide balance. *Arch. Hydrobiol. Suppl.* 69:130–168.

43. Hamilton, P.B. 1987. The impact of atrazine on lake periphyton communities, including carbon uptake dynamics using track autoradiography. *Environ. Pollut.* 46:83–103.

44. Simpson, R.L. and D. Pimentel. 1972. Ecological effects of the aquatic herbicide fenac on small ponds. *Search Agriculture* 2:1–59.

45. Carpenter, S.R. and D.M. Lodge. 1986. Effects of submersed macrophytes on ecosystem processes. *Aquat. Bot.* 26:341–370.

46. Moreland, D.E. 1980. Mechanisms of action of herbicides. *Annu. Rev. Plant Physiol.* 31:597–638.

47. Laasch, H., K. Pfister and W. Urbach. 1982. High- and low-affinity binding of photosystem II herbicides to isolated thylakoid membranes and intact algal cells. *Z. Naturforsch. Sect. C Biosci.* 37:620–631.

48. Arntzen, C.J., K. Pfister and K.E. Steinbeck. 1982. The mechanism of chloroplast triazine resistance: alternations in the site of herbicide action. In H.M. LeBaron and J. Gressel, eds., *Herbicide Resistance in Plants.* Wiley, New York, NY, pp. 185–214.

49. Erickson, J.M., M. Rahire and J-D. Rochaix. 1985. Herbicide resistance and cross-resistance: changes at three distinct sites in the herbicide-binding protein. *Science* 228:204–208.

50. Eickers, T.R., P.A. Andrilenas and T.W. Anderson. 1978. Farmer's use of pesticides in 1976. *Agricultural Economic Report No. 418,* Economic Research Service, U.S. Department of Agriculture, Washington, DC.

51. Delvo, H. and M. Hanthorn. 1983. Pesticides: outlook and situation. Economic Research Service, IOS-2, Washington, DC, 23 pp.

52. Butler, M.K. and J.A. Arruda. 1985. Pesticide monitoring in Kansas surface waters: 1973–1984. *Perspectives on Nonpoint Source Pollution,* EPA-440/5-85-001, U.S. Environmental Protection Agency, Washington, DC, pp. 196–200.

53. Frank, R., G.J. Sirons, R.L. Thomas and K. McMillan. 1979. Triazine residues in suspended soils (1974–1976) and water (1977) from the mouths of Canadian streams flowing into the Great Lakes. *J. Great Lakes Res.* 5:131–138.

54. Glotfelty, D.E., A.W. Taylor, A.R. Isensee, J. Jersey and S. Glenn. 1984. Atrazine and simazine movement to Wye River estuary. *J. Environ. Qual.* 13:115–121.

55. Huggins, D.G. 1990. Ecotoxic effects of atrazine on aquatic macroinvertebrates and its impact on ecosystem structure. Ph.D. thesis, University of Kansas, Lawrence, KS.

56. Leung, S.T., R.V. Bulkley and J.J. Richard. 1982. Pesticide accumulation in a new impoundment in Iowa. *Water Res. Bull.* 18:485–493.

57. Wu, T.L. 1980. Dissipation of the herbicides atrazine and alachlor in a Maryland corn field. *J. Environ. Qual.* 9:459–465.

58. Jensen, K.I.N. 1982. The roles of uptake, translocation, and metabolism in the differential intraspecific responses to herbicides. In H.M. LeBaron and J. Gressel, eds., *Herbicide Resistance in Plants.* Wiley, New York, NY, pp. 133–162.

59. Jones, T.W., W.M. Kemp, J.C. Stevenson and J.C. Means. 1982. Degradation of atrazine in estuarine water/sediment systems and soils. *J. Environ. Qual.* 11:632–638.

60. Gunkel, G.and B. Streit. 1980. Mechanisms of bioaccumulation of a herbicide atrazine (*s*-triazine) in a freshwater mollusc (*Ancylus fluviatilis* Mull.) and a fish (*Coregonus fera*). *Water Res.* 14:1573–1584.

61. Lynch, T.R., H.E. Johnson and W.J. Adams. 1982. The fate of atrazine and a hexachlorobiphenol isomer in naturally-derived model stream ecosystem. *Environ. Tox. Chem.* 1:179–192.

62. Metcalf, R.L. and J.R. Sanborn. 1975. Pesticides and environmental quality in Illinois. *Ill. Nat. Hist. Sur. Bull.* 377–436.

63. Macek, K.J., K.S. Buxton, S. Saunter, S. Gnilka and J. Dean. 1976. Chronic toxicity of atrazine to selected aquatic invertebrates and fishes. Ecological Research Series 600/3-76-047, U.S. Environmental Protection Agency, Washington, DC, 49 pp.

64. Bednarz, T. 1981. The effects of pesticides on the growth of green and blue-green algae cultures. *Acta Hydrobiol.* 23:155–172.

65. Brockway, D.L., P.D. Smith and F.E. Stancil. 1984. Fate and effects of atrazine in small aquatic microcosms. *Bull Environ. Contam. Toxicol.* 32:345–353.

66. Butler, G.L., T.R. Deason and J.C. O'Kelley. 1975. The effect of atrazine, 2,4-D, methoxychlor, carbaryl and diazinon on the growth of planktonic algae. *Br. Phycol. J.* 10:371–376.

67. Larsen, D.P., F. deNoyelles, Jr., F. Stay and T. Shiroyama. 1986. Comparisons of single species, microcosm, and experimental pond responses to exposure to atrazine. *Environ. Toxicol. Chem.* 5:179–190.

68. Plumley, F.G. and D.E. Davis. 1980. The effects of a photosynthetic inhibitor atrazine, on salt marsh edaphic algae, in culture, microecosystems, and in the field. *Estuaries* 3:271–277.

69. Stratton, G.W. 1984. Effects of the herbicide atrazine and its degradation products, alone and in combination, on phototrophic microorganisms. *Arch. Environ. Toxicol.* 13:35–42.

70. Forney, D.R. and D.E. Davis. 1981. Effects of low concentrations of herbicides on submersed aquatic plants. *Weed Sci.* 29:677–685.

71. Jones, T.W. and L. Winchell. 1984. Uptake and photosynthetic inhibition by atrazine and its degradation products on four species of submersed vascular plants. *J. Environ. Qual.* 13:243–247.

72. Jones, T.W., W.M. Kemp, P.S. Estes and J.C. Stevenson. 1986. Atrazine uptake, photosynthetic inhibition, and short-term recovery for the submersed vascular plant, *Potamogeton perfoliatus* L. *Arch. Environ. Contam. Toxicol.* 15:277–283.

73. Kemp, W.M., W.R. Boynton, J.J. Cunningham, J.C. Stevenson, T.W. Jones and J.C. Means. 1985. Effects of atrazine and linuron on photosynthesis and growth of the macrophytes, *Potamogeton perfoliatus* L. and *Myriophyllum spicatum* L. in an estuarine environment. *Mar. Environ. Res.* 16:255–280.

74. Birge, W.J., J.A. Black and D.M. Bruser. 1979. Toxicity of organic chemicals to embryo-larval stages of fish. EPA-560/11-79007, U.S. Environmental Protection Agency, Washington, DC, 60 pp.

75. Blackburn, R.A. 1988. The effects of single and joint toxicity of atrazine and alachlor on three non-target aquatic organisms. M.S. Thesis. University of Kansas, Lawrence, KS.

76. Cossarini-Dunier, M., A. Demal, J.L. Riviere and D. Lepot. 1988. Effects of oral doses of the herbicide atrazine on carp (*Cyprinus carpio*). *Ambio* 17:401–405.

77. Mayer, F.L., Jr., and M.R. Ellersieck. 1986. Manual of acute toxicity: interpretation and data base for 410 chemicals and 66 species of freshwater animals. Resource Publication 160, U.S. Department of Interior Fish and Wildlife Service, Washington, DC, 506 pp.

78. Streit, B. and H.M. Peter. 1978. Long-term effects of atrazine to selected freshwater invertebrates. *Arch. Hydrobiol. Suppl.* 55:62–77.

79. Ward, G.S. and L. Ballantine. 1985. Acute and chronic toxicity of atrazine to estuarine fauna. *Estuaries* 8:22–27.

80. Belluck, D.A. 1980. Pesticides in the aquatic environment. M.S. thesis. University of Illinois, Champaign-Urbana, IL.

81. Hartman, W.A. and D.B. Martin. 1985. Effects of four agricultural pesticides on *Daphnia pulex, Lemna minor* and *Potamogeton pectinatus*. *Bull Environ. Contam. Toxicol.* 35:646–651.

82. Pott, E. 1980. Die hemmung der futteranfnahme von *Daphnia pulex* — eine neue limnotoxikologische mebgrobe. *ZZ. Wasser Abwasser Forsch* 13:52–54.

83. Sanders, H.O. 1970. Toxicities of some herbicides to six species of freshwater crustaceans. *J. Water Pollut. Control Fed.* 42:1544–1550.

84. Carney, C.E. 1983. The effects of atrazine and grass carp on freshwater macrophyte communities. M.A. thesis. University of Kansas, Lawrence, KS.

85. deNoyelles, F., Jr. and W.D. Kettle. 1980. Herbicides in Kansas waters: evaluations of the effects of agricultural runoff and aquatic weed control on aquatic food chains. Contribution 219, Kansas Water Resources Research Institute, University of Kansas, Lawrence, KS, 40 pp.

86. deNoyelles, F., Jr. and W.D. Kettle. 1983. Site studies to determine the extent and potential impact of herbicide contamination in Kansas waters. Contribution 239, Kansas Water Resources Research Institute, University of Kansas, Lawrence, KS, 37 pp.

87. Huggins, D.G., M.L. Johnson and F. deNoyelles, Jr. The ecotoxic effects of atrazine on aquatic ecosystems: an assessment of direct and indirect effects using structural equation modeling, this volume, Ch. 33.

88. Johnson, M.L., D.G. Huggins and F. deNoyelles, Jr. 1991. Ecosystem modeling with LISREL: a new approach for measuring direct and indirect effects in ecosystem level ecotoxicological testing. *Ecological Applications* 1:383–398.

89. Johnson, M.L., D.G. Huggins and F. deNoyelles, Jr. Structural equation modeling and ecosystem analysis, this volume, Ch. 32.

90. Kettle, W.D., F. deNoyelles, Jr., B.D. Heacock and A.M. Kadoum. 1987. Diet and reproductive success of bluegill recovered from experimental ponds treated with atrazine. *Bull. Environ. Contam. Toxicol.* 38:47–52.

91. Stay, F.S., D.P. Larsen, A., Katko and C.M. Rohm. 1985. Effect of atrazine on community level responses in Taub microcosms. In T.P. Boyle, ed., *Validation and Predictability of Laboratory Methods for the Assessment of Fate and Effects of Environmental Contaminents in Aquatic Ecosystems,* ASTM STP 865, American Society for Testing and Materials, Philadelphia, PA, pp. 75–90.

92. Kettle, W.D. and F. deNoyelles, Jr. 1990. Determination of herbicide-induced alterations of aquatic habitats in Kansas. Contribution 272, Kansas Water Resources Research Institute, University of Kansas, Lawrence, KS, 53 pp.

93. Leslie, A.J., Jr., L.E. Nall and J.M. Van Dyke. 1983. Effects of vegetation controlled by grass carp on selected water-quality variables in four Florida lakes. *Trans. Am. Fish. Soc.* 112:777–787.

94. Prowse, G.A. 1971. Experimental criteria for studying grass carp feeding in relation to weed control. *Prog. Fish-Culturist* 33:128–131.

95. Richard, D.I., J.W. Small, Jr. and J.H. Osborn. 1984. Phytoplankton responses to reduction and elimination of submerged vegetation by herbicides and grass carp in four Florida lakes. *Aq. Bot.* 20:307–319.

96. Shireman, J.V., W.T. Haller, D.E. Colle, C.E. Watkins, D.F. DuRant and D.E. Canfield. 1983. Ecological impact of integrated chemical and biological aquatic weed control. Ecological Research Series 600/3-83-098, U.S. Environmental Protection Agency, Washington, DC.

97. Crowder, L.B. and W.E. Cooper. 1982. Habitat structural complexity and the interaction between bluegills and their prey. *Ecology* 63:1802–1813.

98. Hall, D.J., W.E. Cooper and E.E. Werner. 1970. An experimental approach to the production dynamics and structure of freshwater animal communities. *Limnol. Oceanogr.* 15:839–928.

99. Carpenter, S.R., J.F. Kitchell and J.R. Hodgson. 1985. Cascading trophic interactions and lake productivity. *BioSci.* 250:634–639.

100. Dewey, S.L. 1990. Cascading trophic interactions in experimental stream benthic communities. Ph.D. thesis. University of Kansas, Lawrence, KS.

101. Power, M.E., W.J. Matthews and A.J. Stewart. 1985. Grazing minnows, piscivorous bass, and stream algae: dynamics of a strong interaction. *Ecology* 66:1448–1456.

102. Dewey, S.L. and F. deNoyelles, Jr. On the use of ecosystem stability measurement in ecological effects testing, this volume, Ch. 31.

103. Ali, A. and M.L. Kok-Yokomi. 1989. Field studies on the impact of a new benzoylphenylurea insect growth regular (UC-84572) on selected aquatic nontarget invertebrates. *Bull. Environ. Contam. Toxicol.* 42:134–141.

104. Wayland, M. and D.A. Boag. 1990. Toxicity of carbofuran to selected macroinvertebrates in prairie ponds. *Bull. Environ. Contam. Toxicol.* 45:74–81.

105. Macek, K.J., D.F. Walsh, J.W. Hogan and D.D. Holtz. 1972. Toxicity of the insecticide dursban to fish and aquatic invertebrates in ponds. *Trans. Amer. Fish. Soc.* 101:420–427.

106. Papst, M.H. and M.G. Boyer. 1980. Effects of two organophosphorus insecticides on the chlorophyll *a* and pheopigment concentrations of standing ponds. *Hydrobiol.* 69:245–250.

107. Crossland, N.O. 1982. Aquatic toxicity of cypermethrin. II. Fate and biological effects in pond experiments. *Aquat. Toxicol.* 2:175–185.

108. Crossland, N.O. and J.M. Hillaby. 1985. Fate and effects of 3,4-dichloroaniline in the laboratory and in outdoor ponds: II. Chronic toxicity to *Daphnia* spp. and other invertebrates. *Environ. Toxicol. Chem.* 4:489–499.

109. Magovern, J.P. 1978. Changes in zooplankton populations after the addition of the organophosphonate pesticide, dylox, to man-made ponds. M.S. thesis. University of Oklahoma, Norman, OK.

110. Kennedy, H.D., L.L. Eller and D.F. Walsh. 1970. Chronic effects of methoxychlor on bluegills and aquatic invertebrates. Technical Paper No. 53, Bureau of Sport Fisheries and Wildlife, U.S. Department of the Interior, Washington, DC, 18 pp.

111. Crossland, N.O. and D. Bennett. 1984. Fate and biological effects of methyl parathion in outdoor ponds and laboratory aquaria. *Ecotoxicol. Environ. Safety* 8:471–481.

112. Robinson-Wilson, E.F., T.P. Boyle and J.D. Petty. 1983. Effects of increasing levels of primary production on pentachlorophenol residues in experimental pond ecosystems. In W.E. Bishop, R.D. Cardwell and B.B. Heidolph, eds., *Aquatic Toxicology and Hazard Assessment: Sixth Symposium,* ASTM STP 802, American Society for Testing and Materials, Philadelphia, PA, pp. 239–251.

113. Stephenson, M. and G.L. Mackie. 1986. Effects of 2,4-D treatment on natural benthic macroinvertebrate communities in replicate artificial ponds. *Aquat. Toxicol.* 9:243–251.

114. Mauck, W.L., F.L. Mayer, Jr. and D.D. Holtz. 1976. Simazine residue dynamics in small ponds. *Bull. Environ. Contam. Toxicol.* 16:1–8.

CHAPTER 31

On the Use of Ecosystem Stability Measurements in Ecological Effects Testing

Sharon L. Dewey and Frank deNoyelles, Jr.

Abstract: Current community and ecosystem stability theory is briefly reviewed, including a summary of general correlations between certain environmental parameters and stability, inferred from theoretical and empirical investigations. Two approaches are explored for the application of stability theory to experimental mesocosm studies: (1) use of mathematical models and computer simulations, and (2) use of environmental correlates. An example of the latter type of stability analysis is presented for mesocosm community responses to the herbicide atrazine. Atrazine was found to have a destabilizing effect on the overall mesocosm community, based on its effects on environmental parameters correlated with stability. In addition, a food web complexity measure, $\bar{l}$ (average number of trophic links per consumer species), was found to be significantly reduced in atrazine-treated ponds during the second year of a 3-year study. This provided a possible mechanism for the shift from trophic and habitat specialists to generalists. Inclusion of stability analysis in the evaluation of multi-species or ecosystem level toxicity tests is judged as potentially important for giving new insights into how toxins affect the functional mechanics of system integrity. Difficulties in meeting requirements for proper performance and interpretation of stability analyses in typical ecotoxicological mesocosm studies are discussed, with some recommendations for accommodating the requirements in experimental protocol.

INTRODUCTION

The classical approach to aquatic toxicity effects testing employs simple experimental systems to measure the direct and limited indirect effects on select species.[1] The advent of mesocosm experimentation in effects testing provided great potential for investigating a natural array of direct effects, as well as the myriad of possible indirect effects transmitted through naturally complex ecosystems (see deNoyelles et al., Chapter 30). One important and unexplored emergent quality of the complex biotic interactions that occur in mesocosms is system-level (community, ecosystem) stability. The stability of a natural system, or a surrogate natural system such as an aquatic mesocosm, is a measure of the system's ability to resist and/or recover from perturbation. As such, stability estimates are potentially valuable in mesocosm toxicology tests to predict the vulnerability of an aquatic community to given pesticide perturbations prior to application, and to assess the vulnerability of the altered system to future perturbations.

In this chapter we identify two basic approaches for the application of current stability theory to experimental aquatic mesocosm studies: use of mathematical ecosystem modeling and use of environmental correlates to estimate stability. Our objective is twofold: (1) to briefly review major concepts and predictions of early and current stability theory, and (2) to explore the potential for using the two approaches for stability analysis of aquatic mesocosms, with emphasis on use of environmental correlates.

STABILITY THEORY: AN HISTORIC VIEW

During the development of community and ecosystem stability theory, stability has been analyzed as an emergent quality of food web structure and function. As such, stability of a system depends on the nature of its component predator-prey and competitive (trophic exploitative) interactions controlling population densities. One of the earliest qualitative hypotheses linking stability to trophic structure viewed stability to depend on the number of alternative energy pathways in a food web;[2] the more possible pathways for energy to flow through the system, the more stable the system. Ecologically, it was thought that complex communities provided alternative prey for predators (allowing prey switching), thus dampening predator-prey oscillations and reducing the probability of extinctions due to overexploitation.[2-5] This intuitively plausible hypothesis initiated a 20-year period of theoretical and empirical work toward the development of the popular diversity-stability paradigm (e.g., References 3 to 7). The theory maintained that more biologically complex (i.e., more species-rich, or diverse, or interconnected) communities are more stable than simple ones because of their complexity. It was popularized by conservationists as scientific justification for preserving and promoting complex ecosystems such as the tropical rain forest.

However, much of the theoretical support for the paradigm was qualitative and anecdotal in nature, and empirical support often proved ambiguous under the scrutiny of later investigations.[23] Since the early 1970s the generality of the complexity-stability prediction has been systematically refuted by many formal analyses of mathematical model food webs and computer simulations (e.g., References 8 to 23) and (more tentatively) by a small body of empirical evidence (e.g., References 24 to 30). The principal conclusion common to these studies was that community/ecosystem stability is generally negatively, not positively, correlated with biological complexity, i.e., the more complex a community is, the less stable it is likely to be.

As evidence contradicting the diversity-stability theory amassed, Goodman[14] expressed concern for the cause of environmental conservation, left high and dry without the framework and guidance of the stability-diversity paradigm. However, the ecological explanation for the contradictory findings of current stability theory is as intuitively sensible as that for the original theory, and gives cause for even greater concern by conservationists for the fate of disturbed complex natural ecosystems: the more species and, particularly, the more interspecific interactions (i.e., connectance) a community has, the more pathways for the potentially destabilizing effects of a perturbation to be manifested directly and indirectly through the system. Thus, complex ecosystems such as the tropical rain forest may actually be more endangered than earlier thought: ''I would argue that the complex natural ecosystems currently everywhere under siege are less able to withstand our battering than are the relatively simple temperate and boreal systems'' (Reference 11, pp. 215–216). Finally, although the general prediction of current stability theory is that complex systems are less likely to be stable under a given perturbation than simple ones, there are several important exceptions to this trend. The current direction is for theoretical predictions to become much more system-specific than under the old paradigm.

CURRENT STABILITY THEORY

Definitions

The following definitions are drawn from Pimm's review of biological complexity and ecosystem stability (Reference 23, p. 322) and are adhered to in this discussion:

> Stability — A system is deemed stable if the variables (e.g., species population densities, species composition, trophic level density, biomass) all return to the initial equilibrium after being disturbed from it.

Local Stability — A system is locally stable if this return holds for small perturbations.

Global Stability — A system is globally stable if it returns to equilibrium from all possible perturbations.

Domain of Attraction — The set of all values of the variables from which the system returns to a particular equilibrium.

Resilience — How fast the variables return towards equilibrium following a perturbation. Resilience is not, therefore, defined for unstable systems.

Resistance — The degree to which a variable is changed, following a perturbation (the less the change, the higher the resistance).

Because the stability of a system can be defined as above for any relevant variable, it is possible for the system to be, for example, stable for biomass, but not for species composition and densities. We define the latter type of stability based on demographic criteria (i.e., numbers of individuals within populations, or other demographic units) as demographic stability, and the former type based on nondemographic criteria (e.g., biomass, productivity, product cycling) as nondemographic stability (see discussion below and Table 1). Which type of stability is measured depends on the focus of the particular study and the level of information about the system structure and function that the investigator or regulatory agency is willing to utilize. It is our opinion that taxonomic composition and community structure are particularly important qualities for understanding the functioning of natural ecosystems, and that consideration of these attributes should be included in

Table 1. Community and Ecosystem Parameters Associated With Stability

Species Richness (S) — The number of species

Species Diversity (e.g., H') — A combined index of S and the dominance distribution of species abundances in the community (evenness or equitability)

Complexity — S, H', and indices of interspecific predator-prey interactions in food webs:
    Connectance (C) — Number of actual trophic links (L) as a fraction of the number of possible links in a system
    Linkage (L) — Total number of actual trophic links or pathways in a system
    Average linkage ($\bar{l}$) — Average number of trophic links per consumer species

Mean Interaction Strength — The mean strength of interspecific interactions; a strong interaction is indicated when the removal of a predator or competitor causes a dramatic change in the population of associated prey/competitor[35,64]

Compartmentalization — Subsystems or compartments within a food web that are defined by differential connectance and interaction strengths, induced by food web dynamics or habitat quality

Total Biomass — Measures of total biomass of organisms at various levels of organization (species, trophic guild, community, etc.)

Product Cycling — The rate of turnover of various products, including nutrients, organic matter, other chemicals and minerals in the system

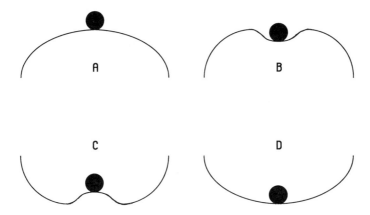

Figure 1. Conceptual model of community (or ecosystem) stability, where the community is a ball resting on a plane which defines the domain of attraction for the variables following a disturbance (see Definitions in text), such that the illustrated community has (A) low local, low global stability, (B) high local, low global stability, (C) low local, high global stability, and (D) high local, high global stability. (After Pimm[22] and Begon et al.[31].)

any experimental protocol aimed at protecting natural environments. Therefore, the emphasis of the following discussions is on demographic stability, referring to this type of stability unless otherwise stated.

A conceptual model devised by May[11,12] and further developed by Pimm[22] and Begon et al.[31] illustrates the concepts of local and global stability and some of the qualities of stability defined above. The model draws an analogy between the equilibrium condition of the community or ecosystem variables and a ball resting on a curved plane (Figure 1). The nature of the plane illustrates whether or not and to what degree that system is stable, should it be disturbed from the equilibrium point. For a system with low local and low global stability a perturbation large or small will send it permanently away from equilibrium (Figure 1 A). With high local and low global stability, the "ball" rests within the bowl of a volcano-like structure (Figure 1 B). A small perturbation relocates the system within the bowl, which represents the small local domain of attraction, and the system will tend to return to its original equilibrium condition. A large perturbation, however, will send it over the lip of the "volcano", permanently away from its original equilibrium point. Low local stability and high global stability (Figure 1 C) are represented by the inverse of the previous case. A small perturbation results in a new equilibrium condition, but limited in the degree of change by a large domain of attraction associated with high global stability (though to a nonpoint attractor and not to the original equilibrium). Both high local and high global stability are illustrated by a ball in a large bowl (Figure 1 D), where a large domain of attraction sends the system back to its original equilibrium state after a small or large perturbation. Also within this conceptual model, the slope of the walls defining the domain of attraction give us an idea of how resilient the system is;[13] a steep slope means the system will return quickly (i.e., high resilience), a shallow slope means it will return slowly (i.e., low resilience).

The above conceptual model demonstrates how a stable community behaves when dislodged from its equilibrium state by a perturbation. However, a stable community may also be resistant to an external force, and not change at all up to some threshold magnitude of disturbance. This type of stability has been called resistance stability,[32] and would be indicated in an ecotoxicological study by no change in community or ecosystem variables with pesticide application. More often in ecotoxicological studies we are concerned with "adjustment stability",[32] which is the community's ability to return to an equilibrium state after being disturbed from it (i.e., stability as defined here and by Pimm[23]). Therefore, the focus of this chapter is on adjustment stability.

Qualities of the Perturbation

Finally, there are some characteristics of the perturbation itself that need to be considered to properly assess the stability of a system. A perturbation is defined by its effect on the community or ecosystem as being a significant change to the variable of interest.[32] If the external force does not cause a specific change, then by definition it is not a perturbation (i.e., the system is resistant to the disturbance).[32] In terms of demographic responses, which most theoretical work addresses, a small perturbation is generally characterized by a small or moderate change to one or a few species population densities.[22,33] A large perturbation would include large changes in one to many species population densities or even species deletions.[22] Many factors affect the time required for recovery or for a system to settle to a new equilibrium after either type of disturbance. These include seasonal and ecological timing of the disturbance and the generation time of the species affected.[22,34] Connell and Sousa[32] stress the importance of allowing adequate time for populations to "turn over" or replace themselves in order to be able to make meaningful judgements about community stability. With respect to this need, Yodzis[34] provides a convenient rule of thumb for estimating the full recovery (or other response) time of a disturbed system, as equal to two times the maximum sum among all sums of generation lengths of single food chains containing the affected species: 2 max $\{T_c$ for all food chains, $c\}$, where $T_c = \sum_{i=1}^{s} g_i$, and g_i is the generation time of the ith species in a single affected trophic chain (c) consisting of S species.

An example of Yodzis' estimate might be a perturbation of the top-predator bluegill sunfish population in a pond mesocosm. Bluegill participate in two primary trophic pathways: bluegill-zooplankton-phytoplankton and bluegill-epiphytic insect-periphyton. A single chain in each pathway could consist of three or four species, depending upon whether it includes an intermediate zooplankton or insect predator. With only a little biological reasoning, we would conclude that the maximum sum of generation lengths will be a food chain consisting of bluegill-predatory insect-herbivorous insect-periphyton. Assuming an average bluegill generation time of 1 to 2 years, herbivorous insects 2 to 12 months, predatory insects 2 to 24 months,

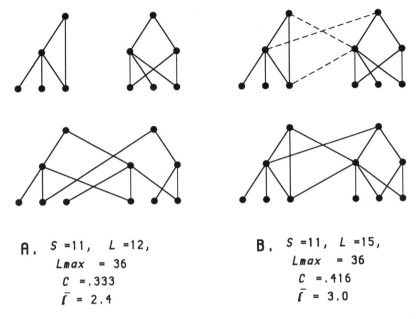

A. $S = 11$, $\quad L = 12$,
$\quad Lmax = 36$
$\quad C = .333$
$\quad \bar{l} = 2.4$

B. $S = 11$, $\quad L = 15$,
$\quad Lmax = 36$
$\quad C = .416$
$\quad \bar{l} = 3.0$

Figure 2. Compartmentalization and select measures of community complexity are illustrated in tri-level food webs, where S is species richness, L is actual trophic linkage, C is connectance $= L/Lmax$, and $\bar{l}$ is average number of trophic links per consumer species. Solid lines indicate strong and dashed lines weak trophic interactions. Each upper food web is identical to its paired lower web in the complexity parameters listed, but upper webs are compartmented, while lower webs are not. Compartmentalization can be achieved by (A) isolation of trophic interactions within dynamically or habitat-constrained compartments, or (B) weak trophic interactions between compartments relative to within compartments. (After Pimm[22].)

and most algae a few days during a summer season of the disturbance, the maximum Σg_i would be about 1.3 to 5 years through the bluegill-insect-insect-periphyton chain. Two times this sum is 2.6 to 10 years for full recovery or other equilibrium response. Certainly an understanding of the specific system and its food web would allow more refined estimates, but this crude calculation is useful at least to indicate that many if not most mesocosm toxicity tests probably allow insufficient time for the full response of the system to be assessed.

Environmental Correlates of Stability

There are several parameters of community and ecosystem structure and function that have been suggested to be associated with stability (Table 1). Most of these are self-explanatory. Figure 2 illustrates some of the lesser-known parameters associated with food web structure, particularly compartmentalization. Within each set of communities illustrated (Figure 2 A and 2 B) the upper and lower communities have identical species richness *(S)*, actual number of trophic links (linkage, *L*), connectance *(C)*, and average linkage for consumer species *(l)*. They differ within

Table 2. Correlations Between Community and Ecosystem Environmental Parameters and Demographic Stability

Environmental Parameter	Stability Correlation	Qualifications and Exceptions
Species Richness (S)	Negative[9-11,73]	Positive for donor-controlled systems and perturbations to predator-controlled herbivore and plant populations[7,23] Positive for biomass[74,75] and primary productivity[25] Negative for herbivore, carnivore productivity[25]
Species Diversity	Negative[10,11]	Positive for biomass[74-76]
Connectance (C)	Negative[8,10,11,21,33]	Positive for species deletions of plants, predator-controlled herbivores, or predators in donor-controlled systems[21,26,33] Positive for resistance to invasion[77] Positive for biomass[74]
Interaction Strength (Mean)	Negative[10,11]	
Compartments	Positive[10,11,29]	Positive for intermediate compartmentation[20,68] Most work on dynamically constrained not habitat-constrained compartments
Product Cycling Rate	Positive[78]	Greater cycling rates yield greater resilience[78]

Note: Correlations with nondemographic stability (e.g., biomass, productivity), resilience, and resistance are listed along with qualifications and exceptions for demographic stability. References that present evidence in support of the stability correlation are cited.

each set in that the upper communities are compartmented and the lower communities are not. The set of food webs on the right (Figure 2 B) further illustrates how compartments can still exist if the interactions between them are weak (dotted lines) relative to the interactions within them.

During the development of stability theory many investigators, both theoretical and applied, have contributed to what can be summarized as general correlations of community and ecosystem attributes with stability (Table 2). There are, however, some important exceptions and qualifications to these trends that should be noted (Table 2) which make the overall predictive relationship between system parameters and stability complex. This is particularly true for demographic stability. An important point about the community and ecosystem parameters (Tables 1, 2) is that it is feasible to measure most parameters, with the possible exception of mean interaction strength, which may require specific experimentation with food web components.[35] The fact that most parameters are reasonably measurable gives the ecologist a simple tool for applying stability analysis, at least qualitatively, to mesocosm experiments.

However, the inability to effectively assess interaction strength may not be a trivial exception. Yodzis[36] showed that using realistic instead of random interaction strengths among food web constituents in models describing real food webs made it more likely to find the communities stable. Paine[37] presented a strong argument for the importance of interaction strength in defining importance pathways that control stability. Without consideration of interaction strength, indices such as species richness, diversity, and connectance may be too general to be used with any precision for inferring the stability of a system, because a few species that strongly interact may have an overriding influence on stability.[24,37] He further argued for the importance of interaction strength in understanding and predicting trophic cascades (e.g., References 35, 38 to 41), a potentially significant mechanism for propagating indirect effects of perturbations. Progress is being made, however, towards the development of standard methods for determining interaction strength in natural communities (R. T. Paine, personal communication). Also towards this end, the authors are exploring the use of a type of path analysis as described by Johnson et al. (Reference 42, and Chapter 32) as a means of quantifying interaction strengths among species in naturally complex food webs.

APPLICATIONS

Mathematical Modeling Approach

One approach to applying stability analysis is to construct a mathematical model of the natural system in question. The models most commonly used include a series of equations, usually of a demographic nature, to describe the dynamics of the ecosystem components and coefficients of the interactions among components. Analysis of stability is then made mathematically from the interaction or community matrix and/or the results from computer simulations of the modeled system (e.g., References 43 to 51; see 52 for overview). Depending on the biological resolution of the study (e.g., to species, family, trophic guild), the model can consist of a few to very many equations and interaction coefficients, and can be relatively easy or difficult to analyze and fit to data.[53]

Some of the practical limitations of the mathematical modeling approach include difficulties associated with constructing the individual population growth or biomass accumulation equations of the dynamic models, and in particular determining interaction coefficients among components. Implicit in the interaction coefficient is a measure of interaction strength, which provides important information for environmental correlation stability analysis (see next section and discussion above). If species- or genus-level resolution is desired for the major taxonomic groups, the above limitations can be overwhelming and would make this approach impractical for standard application. A promising alternative to the conventional ecological modeling approach, which offers a solution to both of the above problems, is an empirical model (i.e., community and ecosystem components and interactions are described by dynamic equations generated from empirical response data) in a statistical package called LISREL (see Johnson et al.[42] and Chapter 32 for details).

One important limitation common to the mathematical models in this approach is that global stability cannot be explicitly calculated for realistically complex models. For the dynamic models this is because the linear approximations of the growth equations used to calculate local stability under small perturbations do not properly approximate the dynamics under large perturbations,[22] and analytic techniques for reasonably complex food webs with nonlinear functions are intractable. However, Pimm[22,33] proposed the analysis of local stability with the perturbation being a single-species deletion (species-deletion stability) as a reasonable approximation of a system's global stability.

Environmental Correlation Approach

The second approach to applying stability analysis is environmental correlation. This approach is qualitative and simply involves using the environmental correlates of stability (Tables 1 and 2) to estimate the relative stability of a natural system. Two immediately useful applications in aquatic mesocosm experiments would be to (1) compare the environmental correlates present in control mesocosms with those of treatments, and assess whether cumulative trends in the correlates enhance or reduce stability, and (2) include in the initial biological structure of the experimental ponds those correlates that enhance stability. The latter assumes that natural ponds have undergone some level of selection to incorporate stabilizing structure,[23] and that mesocosms are meant to simulate natural systems. In this section we present a brief example of the first of the above applications of the environmental correlation approach applied to mesocosm experiments with the herbicide atrazine.

The following analyses draw upon results from a 3-year (1981 to 83) mesocosm study that included three consecutive annual applications of atrazine,[54-56] and a 136-day (1979) atrazine study.[55,57] Results from both studies showed the herbicide atrazine to have direct effects on macrophytes and algae; an array of significant effects on benthic and pelagic animals was attributed to indirect responses to herbicide-induced changes in food (algae, macrophytes) and habitat (macrophyte) availability.

Species Richness and Diversity

Species richness and diversity of benthic (epiphytic) organisms declined with atrazine treatment at all treatment levels.[54-56] In order to assess what these results imply about the atrazine perturbation and mesocosm stability, we needed to examine the trophic structure and function of the ponds and the direct targets of the atrazine perturbation. This was necessary since demographic stability is negatively correlated with species richness in general (Table 2), but positively correlated for donor-controlled systems (i.e., systems where predators do not exert control over prey populations)[22] and for perturbations to predator-controlled herbivore or plant populations (Table 2). Reinterpretation of experimental results implied that the second exception (target of disturbance is predator-controlled plant populations) best fit this study. This seemed reasonable since the direct impact of the atrazine perturbation was on plants,[54-56,58] and herbivore indirect responses to atrazine implied strong interactions with their "prey",[55] suggesting herbivore control over plants. Therefore, stability was expected to be positively correlated with species richness in the case of the atrazine perturbation. The same line of reasoning applied to species diversity, which has the same correlation with stability (Table 2). Thus, we concluded from species richness and diversity data that atrazine had a destabilizing effect on the pond ecosystems. More specifically, the pond ecosystems resulting from atrazine treatment are expected to be less stable than controls.

Connectance

Since data on the actual trophic interactions in the mesocosm communities are incomplete, the following analysis was based on the maximum number of biologically reasonable predator-prey interactions in the largely epiphytic community examined by Dewey[54] and deNoyelles et al.[55] The partial food web (Figure 3) includes a top predator (bluegill sunfish), intermediate predators (predatory chironomid midges [Diptera]), and basal species (nonpredatory chironomid midges, mayflies, and caddis flies).[54] Since the maximum number of biologically reasonable trophic links was used for this analysis, and we did not have actual linkage data, proportional linkage or connectance *(C)* could not be calculated. However, total linkage *(L)* and the average number of trophic links per consumer species *($\bar{l}$)* in the maximized webs yielded information about the trophic complexity associated with species richness (Figure 3). In addition, $\bar{l}$ provided an interesting insight into the nature of the community structure observed in the benthic (epiphytic) insect communities.[54,56]

Despite the variation in average linkage *($\bar{l}$)* within treatments (Figure 3), log-transformed average linkage was significantly higher in control ponds (Figure 3 A) than in treatment ponds (Figure 3 B and C), which were not different from each other (SNK following nested ANOVA, $p = 0.048$). One implication of this result is that the average predator in the atrazine treatment ponds had far fewer prey to choose from than in the controls. This is consistent with reported results that atrazine treatment caused a shift to lower dominance of nonpredatory insects,[54] the potential prey in these systems. More generally, it is also consistent with Schoener's[59] results showing a positive correlation between species richness and mean number of prey species per predator for a wide array of food webs. One expectation that emerges from this trend is that trophic specialists might have difficulty surviving and generalists might prevail in the atrazine-altered systems. This expectation is consistent, for organisms with flexible feeding habits, with a body of optimal foraging theory that positively correlates degree of trophic specialization of a predator with available prey quality and quantity represented among prey species (e.g., References 60 to 67). From a population response perspective, it is reasonable to imagine that the same selective pressures that result in behavioral shifts from specialist to generalist feeding modes may also promote differential survival of generalists over obligate specialists. Finally, this specialist-generalist selection hypothesis (derived from food web structural changes with atrazine) provides a possible mechanism for what Huggins[56] described as an overall dominance shift in benthic invertebrate communities from trophic and habitat specialists to generalists with atrazine treatment during the third year (1983) of the study.

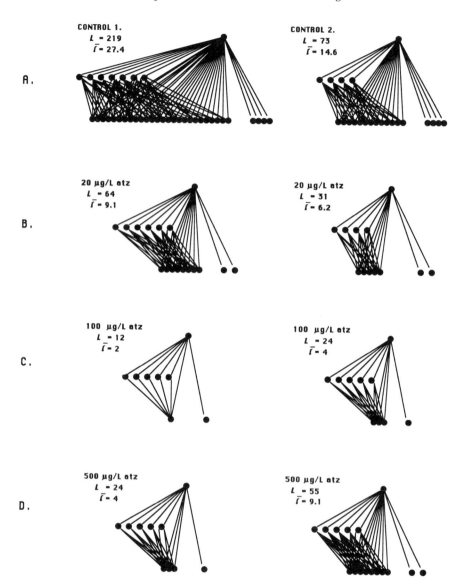

Figure 3. Stylized partial food webs derived from data from the 3-year atrazine (atz) meso-cosm study.[54,55] Top predator is the bluegill sunfish; intermediate predators are predatory chironomids, and basal species are chironomid (left) and other insect herbivores (right). Experimental replicate mesocosms include (A) Controls (0 μg/l atrazine, atz), (B) low, (C) medium, and (D) high atrazine concentrations. Complexity indices L and $\bar{I}$ are calculated for each food web, as defined in Figure 2 and the text. Webs are maximized such that L is the maximum number of biologically reasonable links.

Since the standardized proportional measure of complexity, connectance *(C)*, could not be calculated, the implications for stability were drawn from L and $\bar{l}$. Both parameters declined with atrazine exposure. Since these parameters are structurally linked and positively correlated with species richness,[59] these results also implied that atrazine had a destabilizing effect on the mesocosm community.

Compartmentalization

Pimm and Lawton[68] concluded that there was no compelling empirical evidence to show that natural food webs are divided into either dynamically or habitat-constrained compartments. An important omission in their survey and analysis of documented natural food webs, however, was consideration of trophic interaction strengths, which, if differentially distributed within and among guilds or habitats, can result in compartmentalization[35] (Figure 2 B). Although compartmentalization was not explicitly tested for, evidence for habitat-imposed food web compartments was strongly suggested by the ecological response data from both atrazine studies cited here. The compartments identified were communities associated with the macrophyte, benthic, and pelagic habitats of the mesocosms. Evidence for compartmentalization was drawn from the responses of organisms ecologically linked to these habitats.

The Macrophyte Compartment

Aquatic macrophytes and the biological community associated with the macrophyte habitat were the organisms most dramatically affected by atrazine. Atrazine directly reduced cattails and submerged macrophytes at all treatment levels.[54-56,69,70] Many species within the pond ecosystem that are tightly associated with the macrophyte habitat in turn were indirectly affected by atrazine:

- Abundances of tadpoles and nonpredatory, epiphytic insect species declined at even low treatment levels, presumably due to loss of food (epiphytic algae) and habitat (macrophytes)
- Grass carp biomass declined significantly in all treatments presumably due to loss of food (macrophytes)
- Bluegill biomass declined significantly with treatment at moderate to high treatment levels, presumably due to food (epiphytic insects and other invertebrates) and habitat (macrophyte) losses

The Benthic Compartment

Other habitat-imposed compartments, including the benthic (pond bottom) and pelagic compartments, were relatively unaffected by atrazine, or the effects appeared to be independent from trophic effects observed in the macrophyte compartment.

Channel catfish, an important benthic predator and scavenger, showed no response to atrazine treatment.[55] This was attributed to the broad diet of this species, consisting largely of components of the detritus-based benthic invertebrate food web, a trophic system expected to be relatively insensitive to the immediate actions of atrazine. Sampling methods used by Dewey[54] and Huggins[56] were not effective in collecting some important embenthic invertebrates such as gomphid and libellulid dragonfly nymphs. However, the highly active epibenthic amphipod, *Hyallela azteca,* occurred frequently enough in samples to allow statistical analysis.[56] Amphipod abundance increased with atrazine treatment. This result was attributed to the increased availability of suitable habitat (open pond bottom) caused by macrophyte decline.[56]

The Pelagic Compartment

The pelagic compartment was recognized primarily through the gizzard shad-zooplankton-phytoplankton interaction, with additional evidence from the planktonic predatory insect *Chaoborus*-zooplankton-phytoplankton interaction. The base of these food webs, the phytoplankton, was affected both demographically and nondemographically by the initial atrazine treatment in 1979 and 1981.[55,57,58] However, rapid replacement of sensitive phytoplankton species by tolerant species and the development of physiological resistance to the actions of atrazine by some species[58] allowed nondemographic measures (biomass, productivity) to return to normal levels within 3 weeks of atrazine addition. The altered phytoplankton community evidently provided an acceptable food resource for zooplankton grazers. Their numbers and species composition were unchanged after an initial slight decline in abundances corresponding to the temporary phytoplankton biomass depression.[55,57,58] No long-term effects were found for the pelagic predators on zooplankton, the gizzard shad,[55,56] and the dipteran insect, *Chaoborus punctipennis.*[54] The bluegill sunfish, which feeds in both the macrophyte and pelagic habitats (and occasionally benthic), was more closely linked to the macrophyte compartment in its indirect responses to atrazine, including reduced bluegill biomass.[55] Kettle et al.[71] found the diet of adult bluegill in atrazine-treated ponds to be significantly altered qualitatively (reduced prey insect species richness) and quantitatively (fewer prey individuals), signifying the importance of the trophic link between this species and the macrophyte compartment where its prey insects dwell. The above results taken together suggest that the interaction strengths between the pelagic and macrophyte compartments (through trophic links such as the bluegill sunfish) were weak compared to interaction strengths within each compartment.

Within-Compartment Stability

An interesting point about the results from the compartment analyses is the suggestion of differential within-compartment stability. No compartment showed overall demographic stability under the atrazine perturbation (at least for the recovery period between annual applications); however, the benthic and pelagic compartments showed a certain degree of nondemographic stability, while the macrophyte compartment did not. These results can, for the most part, be explained by differential biological compensatory abilities of the habitat-constrained compartments. The benthic compartment, which was indirectly affected by atrazine in the form of habitat alteration (increased benthic surface area), appeared to compensate through shifts in benthic prey species dominance[56] made possible by short generation lengths. Although no data are available on prey species biomass, biomass of the vertebrate benthic top-predator (channel catfish) was unaffected, implying that there may have been little net change in total prey biomass as well. The pelagic compartment experienced direct impact through atrazine's toxic effect on the phytoplankton food resource. The short generation time of the algae allowed compensation through the development of physiolgical resistance in some species and rapid abundance dominance shifts,[58] such that biomass of all trophic levels returned to previous levels within a few weeks of atrazine application.[55,58] The macrophyte compartment was also affected directly, through the toxic effect of atrazine on periphyton and macrophytes. However, the long generation time of the macrophyte species precluded any short-term selection for physiological resistance. Also, no macrophyte species were totally unaffected, such that a shift in dominance even through relatively rapid vegetative reproduction was not possible. In spite of the short generation time of periphyton, no physiological resistence developed in these species. Finally, the pelagic compartment, which consists of several trophic levels of short-lived organisms, also showed a tendency for demographic stability (resistance stability[32]) at higher trophic levels (microcrustacean grazers and predators, and gizzard shad).[55,58]

Overall Compartment Effects

The overall effect of atrazine on the above identified compartments was to nearly eliminate the macrophyte compartment, thus reducing the degree of compartmentalization in atrazine-treated ponds compared to controls. Because stability is positively correlated with compartmentalization (Table 2), this result also implies that treatment ponds were less stable than controls. In fact, results from all environmental correlates point to this conclusion. However, in order to interpret these results correctly we need to consider the recovery time required by the mesocosm community affected by atrazine, the time span of data collections, and the time span between atrazine applications. Given an estimated recovery time of 2.6 to 10 years (see above Qualities of the Perturbation), neither the data collection periods nor the time span between atrazine applications allowed sufficient time for the mesocosm communities to recover. Therefore, the lower inferred stability of the treatment ponds could imply two things: (1) a transient unstable condition of a system that would return to equilibrium given time, or (2) establishment of a potentially permanent new equilibrium condition that is generally less stable than the original pond community due to changes in the quality of environmental correlates.

DISCUSSION

In addition to the individual value of the two approaches for assessing stability of mesocosms, mathematical modeling, and environmental correlation, the analyses can be used in a complementary fashion. For example, Johnson et al. (Chapter 32) found the atrazine-treated mesocosms to be unstable using the LISREL analytic procedure. The above analyses based on environmental correlations support this result and offer some insight into the mechanisms of the instability. Also, insufficient sample size prevented Johnson et al. (Chapter 32) from calculating stability indices for control mesocosms separately from treatments, but environmental correlations analysis implies that controls were more stable than treatment mesocosms in face of the atrazine perturbation. Indeed, given sufficient recovery time, the mesocosms might have proven stable by returning to their original community structure.

Once we have established that there are reasonable means of quantitatively and qualitatively assessing the stability of mesocosms, we must determine how to best incorporate stability analysis into the objectives and methods of ecotoxicological studies and the agrichemical regulations process. Two general methodological applications are (1) to assess the stability of the undisturbed mesocosms with respect to the biological targets of the chemical perturbation and then test the prediction experimentally, and (2) to assess the stability of the altered mesocosms in the event of future perturbations.

One objective of the first application would be to allow us to assess the magnitude of the disturbance represented by a specific chemical in the context of whole-system integrity. For example, if undisturbed mesocosms were assessed as stable in the face of a perturbation, but experimental results showed dramatic long-term ecological effects, we might conclude that the chemical at applied dosages represented a larger perturbation than the global stability of the system could absorb (i.e., a catastrophic impact from the system's perspective). This could result in the case of key species deletions followed by development of a new community structure resistant to reinvasion by lost species. Atrazine, which directly caused several macrophyte species deletions, might fall into this category. More generally, incorporating mesocosm stability into toxicological studies as an endpoint for evaluation should give us new insights into the functional mechanics of how toxins affect community and ecosystem integrity.

One objective of the second application, assessment of disturbed mesocosms, would be to discover how a chemical affects the system's ability to recover from future perturbation. Of course, if the disturbed system is unacceptably altered, by society's environmental protection standards, from the original system in terms of what we desire to protect, its stability is not of much interest (e.g., a wolf-free tundra community might be more stable than one with wolves, but this is unimportant if it is wolves that we want to protect). Results from this type of application will be most useful if the mesocosm community has been altered in a way that falls within the limits of acceptability based on conservation criteria.

Finally, it is critical for either application that we allow sufficient recovery or other response time in order to determine if the mesocosms are stable.[34] For perturbations that affect long-lived organisms, however, this could mean years. And given the typical agrichemical application schedule of one to several times per year, long-term recovery would be virtually impossible in these cases. Connell and Sousa[32] discuss the temporal scale of perturbations and point out that the classical "concepts of stability or persistence refer to responses to discrete, punctuated disturbances that perturb abundances". Bender et al.[72] termed this type of disturbance a "PULSE" perturbation vs. a continual long-term disturbance, termed a "PRESS" perturbation. The latter includes species removals or deletions, and Yodzis[34] cited humanly produced toxins as an example of PRESS perturbations. Humanly produced toxins like agrichemicals can represent PRESS perturbations in two ways. Their application schedule may be such that the perturbation is repeated often enough to be relatively continual, and/or the half-life of the chemical may be long enough to extend the perturbation over a significant period of time. Yodzis[34] demonstrated by analyzing 100 randomly constructed plausible community matrices each from 16 published food webs[30] that the outcome of PRESS perturbations, in terms of community composition and trophic structure, cannot be predicted from the character of the original community. However, the data used by Yodzis contained a high degree of functional lumping of biological species into "trophospecies" (species with similar feeding habits).[30] There is some evidence that more functional or taxonomic refinement of the food web components (less lumping) reduces some of the indeterminacy of outcome of PRESS perturbation experiments.[34]

The above considerations taken together impose some serious restrictions on the application of demographic stability analysis to mesocosm ecotoxicological experiments. They also indicate changes in mesocosm experimental protocol that would improve interpretation of results. By definition[23] we need to know if a system can recover from a perturbation in order to know if it is stable. Use of stability analysis with current mesocosm experimental protocol (typically 1-year studies) would seem to be limited, then, to those cases where the half-life of the chemical is relatively short, affected species in the system have relatively short generation times, and the projected recovery time of the system is less than the time between applications of the chemical under investigation (i.e., a PULSE-type perturbation with respect to system response time and application schedule). For chemicals that represent PRESS-type perturbations (i.e., long chemical half-life and/or projected system response time is greater than the time between applications), mesocosm experiments must extend for relatively long periods of time, including a period after the perturbation has ceased (something approaching Yodzis' projection) in order to assess total effects. Finally, several lines of reasoning[34,37,72] argue for collecting ecological data at the finest taxonomic resolution possible or practical in order to improve interpretative and predictive abilities in experiments investigating the effects of perturbation on stability and system structure and function.

ACKNOWLEDGMENTS

We thank R. Graney, J. Kennedy, and J. Rodgers for inviting us to participate in the highly stimulating mesocosm symposium. We are grateful to Robert Holt and three anonymous reviewers for useful comments and criticisms in their careful reviews of the manuscript. We gratefully acknowledge financial and technical support from the Kansas Biological Survey and the Kansas Applied Remote Sensing Program, University of Kansas. The data used here were obtained in part from a project supported by the U.S. Environmental Protection Agency (grant number CR808804–02–0 to F. deNoyelles, Jr.).

REFERENCES

1. Cairnes, J., Jr. 1983. Are single species toxicity tests alone adequate for estimating environmental hazard? *Hydrobiologia* 100: 47–57.
2. Odum, E. P. 1953. *The Fundamentals of Ecology.* W. B. Saunders Company, Philadelphia, PA.
3. MacArthur, R. H. 1955. Fluctuations of animal populations, and a measure of community stability. *Ecology* 36: 533–536.
4. Hutchinson, G. E. 1959. Homage to Santa Rosalia or why are there so many kinds of animals? *American Naturalist* 93: 145–159.
5. Pimentel, D. 1961. Species diversity and insect population outbreaks. *Annals of the Entomological Society of America* 54: 76–86.
6. Elton, C. S. 1958. *The Ecology of Invasions.* John Wiley & Sons, Inc., New York, NY.
7. Hairston, N. G., J. D. Allan, R. K. Colwell, D. J. Futuyma, J. Howell, M. D. Lubin, J. Mathias and J. H. Vandermeer. 1968. The relationship between species diversity and stability: an experimental approach with protozoa and bacteria. *Ecology* 49: 1091–1101.
8. Gardner, M. R. and W. R. Ashby. 1970. Connectance of large dynamic (cybernetic) systems: critical values for stability. *Nature* 228: 784.
9. May, R. M. 1971. Stability in multi-species community models. *Mathematical Bioscience* 12: 59–79.
10. May, R. M. 1972. Will a large complex system be stable? *Nature* 238: 413–414.
11. May, R. M. 1974. *Stability and Complexity in Model Ecosystems.* Princeton University Press, Princeton, NJ.
12. May, R. M. 1975. Stability in ecosystems: some comments. In van Dobben, W. H. and R. H. Lowe-McConnel, eds., *Unifying Concepts in Ecology,* Dr. W. Junk B. V. Publishers, Pudoc, Wageningen, The Netherlands, pp. 161–168.
13. Maynard Smith, J. 1974. *Models in Ecology.* Cambridge University Press, New York, NY.
14. Goodman, D. 1975. The theory of diversity-stability relationships in ecology. *Quarterly Review of Biology* 50: 237–266.

15. Margalef, R. 1975. Diversity, stability and maturity in natural ecosystems. In van Dobben, W. H. and R. H. Lowe-McConnel, eds., *Unifying Concepts in Ecology,* Dr. W. Junk B. V. Publishers, Pudoc, Wageningen, The Netherlands, pp. 151–160.

16. Oriens, G. H. 1975. Diversity, stability and maturity in natural ecosystems. In van Dobben, W. H. and R. H. Lowe-McConnel, eds., *Unifying Concepts in Ecology,* Dr. W. Junk B. V. Publishers, Pudoc, Wageningen, The Netherlands, pp. 139–150.

17. Pimm, S. L. and J. H. Lawton. 1977. Number of trophic levels in ecological communities. *Nature* 286: 329–331.

18. Pimm, S. L. and J. H. Lawton. 1978. On feeding on more than one trophic level. *Nature* 275: 542–544.

19. Lawlor, L. R. 1978. A comment on randomly constructed model ecosystems. *American Naturalist* 112: 445–447.

20. Pimm, S. L. 1979a. The structure of food webs. *Theoretical Population Biology* 16: 144–158.

21. Pimm, S. L. 1979b. Complexity and stability: another look at MacArthur's original hypothesis. *OIKOS* 33: 351–357.

22. Pimm, S. L. 1982. *Food Webs.* Chapman and Hall, New York, NY.

23. Pimm, S. L. 1984. The complexity and stability of ecosystems, *Nature* 307: 321–326.

24. Paine, R. T. 1969. A note on trophic complexity and community stability. *American Naturalist* 103: 91–93.

25. Hurd, L. E., M. V. Mellinger, L. L. Wolf and S. J. McNaughton. 1971. Stability and diversity at three trophic levels in terrestrial successional ecosystems. *Science* 173: 1134–1136.

26. DeAngelis, D. L. 1975. Stability and connectance in food web models. *Ecology* 56: 238–243.

27. Odum, E. P. 1975. Diversity as a function of energy flow. In van Dobben, W. H. and R. H. Lowe-McConnel, eds., *Unifying Concepts in Ecology,* Dr. W. Junk B. V. Publishers, Pudoc, Wageningen, The Netherlands, pp. 11–14.

28. Whittaker, R. H. 1975. The design and stability of some plant communities. In van Dobben, W. H. and R. H. Lowe-McConnel, eds., *Unifying Concepts in Ecology,* Dr. W. Junk B. V. Publishers, Pudoc, Wageningen, The Netherlands, pp. 169–181.

29. Rejmanek, M. and P. Stary. 1979. Connectance in real biotic communities and critical values for stability of model ecosystems. *Nature* 280: 311–313.

30. Briand, F. 1983. Environmental control of food web structure. *Ecology* 64: 253–263.

31. Begon, M., J. L. Harper and C. R. Townsend. 1990. *Ecology: Individuals, Populations and Communities.* Blackwell Scientific Publications, Boston, MA.

32. Connell, J. H. and W. P. Sousa. 1983. On the evidence needed to judge ecological stability or persistence. *American Naturalist* 121: 789–824.

33. Pimm, S. L. 1980. Food web design and the effect of species deletion. *OIKOS* 35: 139–149.

34. Yodzis, P. 1988. The indeterminacy of ecological interactions as perceived through perturbation experiments. *Ecology* 69: 508–515.

35. Paine, R. T. 1980. Food webs: linkage, interaction strength and community infrastructure. *Journal of Animal Ecology* 49: 667–685.

36. Yodzis, P. 1981. The stability of real ecosystems, *Nature* 289: 674–676.

37. Paine, R. T. 1988. Food webs: road maps of interactions or grist for theoretical development? *Ecology* 69: 1648–1654.

38. Power, M. E., W. J. Matthews, and A. J. Stewart. 1985. Grazing minnows, piscivorous bass, and stream algae: dynamics of a strong interaction. *Ecology* 66: 1448–1456.

39. Carpenter, S. R., J. F. Kitchell and J. R. Hodgson. 1985. Cascading trophic interactions and lake productivity. *BioScience* 35: 634–639.

40. Dewey, S. L. 1990. Cascading trophic interactions in experimental stream benthic communities. Ph.D. thesis. University of Kansas, Lawrence, KS.

41. Power, M. E. 1990. Effects of fish in river food webs. *Science* 250: 811–814.

42. Johnson, M. L., D. G. Huggins and F. deNoyelles, Jr. 1991. Ecosystem modeling with LISREL: A new approach for measuring direct and indirect effects in ecosystem level ecotoxicological testing. *Ecological Applications* 1: 383–398.

43. McAllister, C. D. 1970. Zooplankton rations, phytoplankton mortality and the estimation of marine production. In Steele, J. H., ed., *Marine Food Chains*. University of California Press, Berkeley, CA.

44. McIntire, C. D. 1973. Periphyton dynamics in laboratory streams: a simulation model and its implications. *Ecological Monographs* 43: 399–419.

45. O'Neill, R. V. and J. M. Giddings. 1979. Population interactions and ecosystem function: phytoplankton competition and community production. In Innis, G. S. and R. V. O'Neill, eds., *Systems Analysis of Ecosystems*. International Cooperative Publishing House, Fairland, MD.

46. Collins, C. D. 1980. Formulation and validation of a mathematical model of phytoplankton growth. *Ecology* 61: 639–649.

47. Carpenter, S. R. and J. F. Kitchell. 1984. Plankton community structure and limnetic primary production. *American Naturalist* 124: 159–172.

48. Lane, P. A. and T. M. Collins. 1985. Food web models of a marine plankton community network: an experimental mesocosm approach. *Journal of Experimental Marine Biology and Ecology* 94: 41–70.

49. Loucks, O. L. 1985. Looking for surprise in managing stressed ecosystems. *BioScience* 35: 428–432.

50. Lane, P. A. 1986. Symmetry, change, perturbation, and observing mode in natural communities. *Ecology* 67: 223–239.

51. Pimm, S. L. and J. C. Rice. 1987. The dynamics of multispecies, multi-life-stage models of aquatic food webs. *Theoretical Population Biology* 32: 303–325.

52. Hall, C. A. S. and J. W. Day, Jr. 1977. *Ecosystem Modeling in Theory and Practice*. John Wiley & Sons, New York, NY.

53. Shoemaker, C. A. 1977. Mathematical construction of ecological models. Hall, C. A. S. and J. W. Day, Jr. *Ecosystem Modeling in Theory and Practice*. John Wiley & Sons, New York, NY, pp. 75–114.

54. Dewey, S. L. 1986. Effects of the herbicide atrazine on aquatic insect community structure and emergence. *Ecology* 67: 148–162.

55. deNoyelles, F., Jr., W. D. Kettle, C. H. Fromm, M. F. Moffett and S. L. Dewey. 1989. Use of experimental ponds to assess the effects of a pesticide on the aquatic environment. In Voshell, J. R., Jr., ed., *Using Mesocosms to Assess the Aquatic Ecological Risk of Pesticides: Theory and Practice*. Miscellaneous Publications of the Entomological Society of America 75: 41–56.

56. Huggins, D. G. 1990. Ecotoxic effects of atrazine on macroinvertebrates and its impact on ecosystem structure. Ph.D. thesis. University of Kansas, Lawrence, KS.

57. deNoyelles, F., Jr. and W. D. Kettle. 1980. Herbicides in Kansas waters: evaluations of the effects of agricultural runoff and aquatic weed control on aquatic food chains. Contribution 219. Kansas Water Resources Research Institute, University of Kansas, Lawrence, KS.

58. deNoyelles, F., Jr., W. D. Kettle and D. E. Sinn. 1982. The responses of plankton communities in experimental ponds to atrazine, the most heavily used pesticide in the United States. *Ecology* 63: 1285–1293.

59. Schoener, T. W. 1989. Food webs from the small to the large. *Ecology* 70: 1559–1589.

60. Emlen, J. M. 1966. The role of time and energy in food preference. *American Naturalist* 100: 611–617.

61. Emlen, J. M. 1968. Optimal choice in animals. *American Naturalist* 102: 385–390.

62. MacArthur, R. H. and E. R. Pianka. 1966. On optimal use of a patchy environment. *American Naturalist* 100: 603–609.

63. Schoener, T. W. 1971. Theory of feeding strategies. *Annual Review of Ecology and Systematics* 2: 369–404.

64. MacArthur, R. H. 1972. *Geographical Ecology: Patterns in the Distribution of Species.* Harper & Row, New York, NY.

65. Charnov, E. L. 1976. Optimal foraging: attack strategy of a mantid. *American Naturalist* 110: 141–151.

66. Charnov, E. L. 1976. Optimal foraging: the marginal value theorem. *Theoretical Population Biology* 9: 129–136.

67. Pianka, E. R. 1978. *Evolutionary Ecology,* 2nd edition. Harper & Row, New York, NY.

68. Pimm, S. L. and J. H. Lawton. 1980. Are food webs divided into compartments? *Journal of Animal Ecology* 49: 897–898.

69. Carney, C. E. 1983. The effects of atrazine and grass carp on freshwater macrophyte communities. M.A. thesis, University of Kansas, Lawrence, KS.

70. deNoyelles, F., Jr. and W. D. Kettle. 1983. Site studies to determine the extent and potential impact of herbicide contamination in Kansas waters. Contribution 239. Kansas Water Resources Research Institute, University of Kansas, Lawrence, KS.

71. Kettle, W. D., F. deNoyelles, Jr., B. D. Heacoch and A. M. Kadoum. 1987. Diet and reproductive success of bluegill recovered from experimental ponds treated with atrazine. *Bull. Environ. Contam. Toxicol.* 38: 47–52.

72. Bender, E. A., T. J. Case and M. E. Gilpin. 1984. Perturbation experiments in community ecology: theory and practice. *Ecology* 65: 1:13.

73. Watt, K. E. F. 1964. Comments on fluctuations of animal populations and measures of community stability. *Canadian Entomologist* 96: 1434–1442.

74. King, A. W. and S. L. Pimm. 1983. Complexity, diversity, and stability: a reconciliation of theoretical and empirical results. *American Naturalist* 122: 229–239.

75. McNaughton, S. J. 1977. Diversity and stability of ecological communities: a comment on the role of empiricism in ecology. *American Naturalist* 111: 515–525.

76. McNaughton, S. J. 1978. Stability and diversity of ecological communities. *Nature* 274: 251–253.

77. Post, W. M. and S. L. Pimm. 1983. Community assembly and food web stability. *Mathematical Biosciences* 64: 169–192.

78. DeAngelis, D. L. 1980. Energy flow, nutrient cycling, and ecosystem resilience. *Ecology* 61: 764–771.

CHAPTER 32

Structural Equation Modeling and Ecosystem Analysis

Michael L. Johnson, Donald G. Huggins, and Frank deNoyelles

Abstract: Structural equation modeling with LISREL was applied to the analysis of the effect of atrazine on a pond ecosystem. This ecosystem analysis is a path analytic approach that relies on modeling ecosystem structure and function. Ecosystems can be thought of as a set of compartments, each compartment being a structural component. These structural components were modeled as a series of concepts, each concept indicated by one or more actual variables. Functional relationships between compartments were represented by pathways connecting the compartments. LISREL provides a measure of ecosystem stability, and the total, direct, and indirect effects of all components on all other components. Three models were developed in which the vertebrate structural component of the ecosystem is decomposed into individual species components. A fourth model was developed that reorganizes the smallest compartment model. In general, as models became larger (more components), the ecosystem became more unstable. It was also found that as there was an increase in the number of same-sign direct and indirect effects, i.e., both are positive or both are negative, the ecosystem becomes less stable. Modeling with LISREL provides several insights into how chemical stresses can operate, and we discuss the role of LISREL in ecosystem modeling and ecotoxicological testing.

0-87371-592-6/94/$0.00 + $.50
© 1994 by CRC Press, Inc.

627

INTRODUCTION

From their initial use in ecotoxicological research and testing,[1,2] pond-scale mesocosms have proved to be valuable as experimental units for elucidating the fate and effects of chemicals in aquatic ecosystems. Mesocosms provide a majority of the features of natural aquatic ecosystems, and chemical treatments can be replicated to provide a reasonable database for analysis. However, a major problem in using mesocosms for ecotoxicological testing and research is finding an analytical methodology that can take full advantage of the experimental unit. This difficulty is a result not of any inherent problem with mesocosm experimental design, but rather with the complexity and our limited understanding of the ecosystem that is being investigated.

Nowhere is this difficulty more apparent than in the current EPA guidelines for mesocosm testing of agrichemicals for registration. The current analytical methodology often involves identifying a series of variables, e.g., the abundance of individual macroinvertebrate taxa, phytoplankton or zooplankton, monitoring those variables in control and treatment mesocosms, then testing with inferential statistics to determine if the treatments are different from the controls. Aside from the problem of identifying the correct or appropriate (or most informative, or even the most interesting) variables, the specific statistical methodology is the subject of considerable controversy. A Dunnett's t-test for comparing a series of treatments against a control can be performed to test the null hypothesis $H_0:\mu_T \geq \mu_C$ against the alternative hypothesis $H_1:\mu_T < \mu_C$, where μ_T and μ_C are the means of the treatment and control groups, respectively. However, if the data are highly variable, rejection of the null hypothesis can be extremely difficult, and consequently, another t-test has been developed[3] to test $H_0:\mu_T < b\mu_C$ against $H_T:\mu_1 \geq b\mu_C$, where $0 < b < 1$, to negate a presumptive effect. This test assumes that there is an effect of the chemical on the variable of interest, and the onus is on the investigator to reject the null hypothesis of a reduction in treatment levels compared to the control. The difficulty with this test is the same as with the previous test, i.e., the amount of variability commonly associated with biological variables. With coefficients of variation of 30 to 50%, it may be difficult to reject the null hypothesis with $b = 0.8$, the constant associated with a 20% effect level. At present, the controversy still rages over the proper approach.

However, the problem does not end there. For example, it is not clear how the experiment-wise error is affected by performing dozens of individual t-tests for each variable on each sample date, even if each individual t-test is adjusted to provide an error rate of $\alpha = 0.05$. Even if it becomes possible to decide on an appropriate statistical technique, that decision will not make it any easier to evaluate the effect of the chemical on the ecosystem because ecosystem level statistical testing is not performed. The hypotheses for either t-test assume that only direct toxic effects, or indirect negative effects (i.e., a reduction in the amount of the variable due to the presence of the chemical), are important. Indirect effects that result in treatment levels of a parameter above control levels apparently are not of interest. And yet, there are several examples where extremely high levels of some variable in treatment mesocosms compared to controls would be undesirable, for example, blooms of phytoplankton causing taste and odor problems and fish kills. In addition to eliminating the ability to detect a large number of critical indirect effects, the individual *t*-tests suffer from the fact that they are simply no more than a whole series of single parameter tests, and there is little attempt at integrating these effects. Ecosystem level testing becomes the subjective evaluation of the combined series of tests, and the criterion for judging the totality of the results, and therefore the effects of the chemical on the ecosystem are "best professional judgements". The pitfalls of such a subjective criterion are obvious. Consequently, it is clear that new analytical methodologies for evaluating ecosystem level effects in mesocosm testing are necessary.

The first question that must be addressed before deciding on a methodology for analyzing mesocosm data is what must the data provide, i.e., what does the experimenter wish to learn from the experiment? Usually, the answer at the broadest scale is what is (are) the effect(s) of the chemical on the ecosystem, i.e., what is the response of the ecosystem to exposure to the chemical? A second question must follow from the first. What is an appropriate measure of ecosystem response, where ecosystem response is defined as "its change of state of dynamics as a consequence of the stress"?[4] Changes in state or dynamics can be directly related to changes in ecosystem structure and function, and it is on these two features that any methodology must focus. We define ecosystem structure as all biotic or abiotic components which comprise or affect an ecosystem, and ecosystem function as the fundamental processes through which change in fate is affected among related structural components. These definitions are very close to the definitions of Matthews et al.[5] which were, in turn, based on the definitions of Odum.[6] Their[5] definition of structure and function is "measurement of abiotic or biotic characteristics at a point in time" and "measurement of any rate process of the ecosystem". Our definitions differ only in that we divorce the measurement from the actual components or rate processes, which is necessary for our modeling efforts (see below).

We propose that ecosystem stability, defined as the ability of the ecosystem to return to an equilibrium state once it has been disturbed, is the appropriate measure of ecosystem response because it incorporates both ecosystem structure and function into a measure of system dynamics. This definition of stability is the one commonly used in ecosystem analysis.[7-17] Because there may be multiple stable points for an ecosystem, we refer specifically to local or neighborhood stability, which is the behavior of the system when a perturbation moves the system a short distance from the equilibrium.

To measure ecosystem stability, it is necessary to identify indicators of the structure components of the ecosystem and designate the functional interactions among these components. Identification of structure and function relates to the investigator's knowledge of the ecosystem under study, and the measurement of stability is a problem of analysis. We believe that a structural equation modeling technique called LISREL (LInear Structural RELations) provides the necessary analytical methodology. LISREL is a modeling technique that allows the investigator to hypothesize a framework of causal interactions (hypothesis of structure and function) for the ecosystem of interest and use empirical data to test that hypothesis in a statistically rigorous manner.[18,19] The major outputs of LISREL modeling are (1) a chi-square goodness of fit test to determine if the causal framework of ecosystem components and functional pathways is consistent with the data, including the statistical significance of each pathway hypothesized to exist, (2) quantitative estimates of path coefficients in the units of the independent variable allowing the modeling to become predictive, (3) a matrix of the total effects of all components on all other components of the ecosystem, (4) stability properties of the system are provided so that it is possible to determine if the ecosystem, *as modeled,* is stable or unstable, and (5) the covariance structure among the error terms (see below) can be examined to determine if components are missing from the model.

In this chapter we (1) provide a brief introduction to LISREL, (2) discuss how the level of aggregation of variables within large ecosystem compartments influences the effects matrices and the stability index of the ecosystem, and (3) discuss the role of ecosystem level analysis with LISREL in ecotoxicological analyses in general, and specifically in the agrichemical registration process. We hope that we can begin to convince the reader that resolving the question of which *t*-test to use to analyze mesocosm data is ultimately not correct, or even important. These tests provide answers to "piece meal" questions that are of uncertain value in determining the response of an ecosystem to stress. Instead, what needs to be determined is what are the best variables to use and how the current mesocosm protocol can be used to take advantage of the potential benefits of modeling with LISREL.

HISTORICAL PERSPECTIVE

Structural equation modeling with LISREL is a path analytic technique, in that structural components of the ecosystem are identified, and causal (functional) relationships among those components are established. These systems are presented as path diagrams, where the structural components are connected by pathways representing ecosystem functional relationships. Several other modeling techniques have been developed utilizing the path analytic approach and we briefly review some of these. We offer LISREL as an alternative to other path analytic techniques that are currently used in ecosystem analysis such as food web theory and loop analysis, and input-output flow analysis. These techniques have had a substantial theoretical development over the last 20 years; however, for reasons we point out below, they have not been extended successfully to problems in ecotoxicology.

Path analysis commonly used by biologists was developed by Sewall Wright.[20] Paths originate with an independent variable which is then linked to a dependent variable (in the statistical sense). At the least, a functional relationship between the two variables is hypothesized, and it is often argued that these paths represent a cause and effect relationship.[21] Path coefficients are standardized regression coefficients calculated by least squares analysis and units of measure of the path coefficients are standard deviation units. LISREL differs from the path analysis developed by Wright in several aspects including: (1) paths link concepts (see below) not actual variables, (2) path coefficients are calculated by maximum likelihood estimation, and (3) path coefficients are in the units of measure of the concepts.

Food web analysis is a technique in which the nodes are species (or trophic or functional groups), and the paths connecting them are the biotic interactions, qualitatively determined by a +1 or −1. For example, predator-prey interactions are represented by two pathways connecting a predator and its prey, with the paths designated a +1 and a −1. The +1 represents the prey as food for the predator, and the −1 represents the prey removed from the population by the predators. The path diagrams are referred to digraphs (directed graphs[22,23]). Food web theory has been beneficial in helping to formulate appropriate questions concerning the fundamental processes that organize ecological communities[24] but has failed to be an operational theory.[24-28] In fact, Paine[25] maintains that the food web theory is nonpredictive, because food webs and connectance (a measure of the interactions among species present in the web, see below and Reference 29) cannot be defined operationally. Additionally, those interested in the effects of a potential toxicant on an ecosystem have found food web analysis to be too confining because of the inability to quantify effects and difficulties in modeling the abiotic components of the ecosystem.

Another potential modeling technique for mesocosm analysis is Odum's energy circuitry modeling (e.g., References 30, 31). With this technique, flows of energy, carbon, or nutrients through an ecosystem can be modeled using the language of electrical circuits. This technique appears to be flexible in that it does not appear to need a "common currency" (although most analyses convert all flows to energy), and it is possible to evaluate the effects of various disturbances in the system on all aspects of the system. However, because of the difficulty in obtaining measurements in the field for many of the parameters necessary for the modeling (e.g., rates of photosynthesis, respiration, and metabolism), this technique has received little use in ecosystem analysis in general and ecotoxicological analysis in particular.

Input-output flow analysis was developed for ecosystem modeling by Hannon[32] and has been used extensively since then (e.g., References 34 to 36; see Reference 19 for a discussion and comparison with LISREL). It has a well-developed theoretical framework developed mostly by Patten (e.g., References 37, 38) and his colleagues. Flow analysis relies on the concept of holons which are the basic particles within a whole system. Each holon has inputs and outputs (creaon and genon associations, respectively) of materials or energy, and, by investigation of these flows, patterns of causality can be established. Despite the rather complete theoretical development, this type of systems analysis has not enjoyed wide usage. The major difficulties in using food web analysis and flow analysis in ecotoxicological studies are measurement of the parameters (i.e., flows), and the lack of flexibility in modeling a number of ecosystem processes, e.g., age structuring in populations or responses of organisms to toxic substances. LISREL avoids these problems as all parameters can be measured in any appropriate and convenient manner, and the models can include any type of effect, including toxic effects. LISREL can readily accommodate any hierarchical level of analysis and can integrate processes that occur at the level of the individual, population, community, or ecosystem.

LISREL MODELING

Modeling with LISREL begins by separating ecosystem structure into concepts and their indicators. This separation can best be illustrated by an example. Suppose it is of interest to determine the effects of chemical "a" on bluegill sunfish. However, there are several aspects of bluegill sunfish biology that could be affected including survival, reproduction, growth, or some physiological parameter. Additionally, is the effect of chemical "a" to be determined by its concentration, loading, or exposure? Both chemical "a" and bluegill sunfish are concepts (structural components of the ecosystem). Chemical "a" is measured in μg/l making the variable, concentration, the indicator of the concept, chemical "a". To combine the parameters, survival, growth, and reproduction, the concept of bluegill sunfish is indicated by the variable biomass, measured in grams. In LISREL modeling, the functional relationship that exists is always between concepts and is represented by the pathway from one concept to the other (Figure 1a). Separating concepts from their indicators allows ecosystem structure and function to be hypothesized and tested without becoming preoccupied with the issue of measurement. The desired components of ecosystem structure can be established and then the indicator variables can be chosen. Additionally, if it is known how much of the variation in the indicator variable is due to actual measurement error, this amount of error can be accounted for by specifying that portion of the variance of the indicator variable to be error.[18] If the amount of measurement error is unknown, LISREL can estimate that error thereby providing a "quality control" aspect to be modeling.

Concepts are either exogenous or endogenous. Exogenous concepts (ξ in Figure 1a) are those that only influence other concepts and are never influenced by those concepts, and endogenous concepts (η in Figure 1a) are those that influence other concepts and are influenced, in turn, by those concepts. In general, exogenous concepts are external inputs to the ecosystem (corresponding to independent variables of a regression model), and endogenous concepts function within the ecosystem (corresponding to either independent or dependent variables of a regression model). We have argued previously[19] that the exogenous concepts are disturbance factors in the ecosystem, and they play no role in determining the stability of the ecosystem (see below). This point has important ramifications for evaluating the effects of exogenous concepts, such as chemical stresses, on the ecosystem.

The modeling technique involves the solution of three equations (Equation 1 to 3) involving four matrices, and the specification of four additional matrices. The basic structural equation is

$$\eta = \beta\eta + \Gamma\xi + \zeta \tag{1}$$

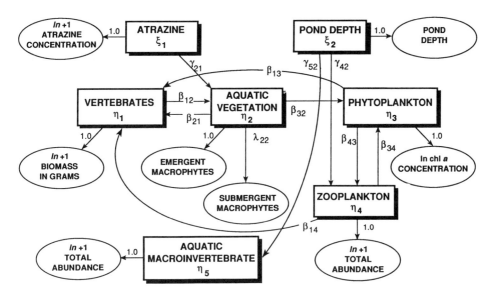

Figure 1a. Large compartment model of ecosystem structure and function. Each structural component (box) is a compartment (concept) indicated by an indicator variable(s) (circle). Exogenous concepts (ξ) are atrazine and pond depth, and endogenous concepts (η) are vertebrates, aquatic vegetation, phytoplankton, zooplankton, and aquatic macroinvertebrates. Functional pathways are represented by arrows, γ_{ij} connect exogenous to endogenous concepts, β_{ij} connect endogenous concepts to each other. Arrows connecting concepts to indicators are set at 1.0, with the exception of λ_{22}, meaning that the scale of measurement for the concept is the same as for the indicator. λ_{22}, as the second indicator of aquatic vegetation, must be estimated by the model. See the text for more details.

where $\eta = m \times 1$ vector of endogenous concepts, $\mathbf{B} = m \times m$ matrix of structural (path) coefficients, $\mathbf{\Gamma} = m \times n$ matrix of structural coefficients, $\xi = n \times 1$ vector of exogenous concepts, and $\zeta = m \times 1$ vector of errors in the conceptual model. Measurement Equation 2 links the endogenous concepts to their indicator variables, and measurement Equation 3 links the exogenous concepts to their indicators.

$$Y = \Lambda_Y \eta + \epsilon \tag{2}$$

$Y = p \times 1$ vector of endogenous indicators, $\Lambda_Y = p \times m$ matrix of structural coefficients, $\eta = m \times 1$ vector of endogenous concepts, and $\epsilon =$ vector of errors in the measurement model.

$$X = \Lambda_X \xi + \delta \tag{3}$$

$X = q \times 1$ vector of exogenous indicators, $\Lambda_x = q \times n$ matrix of structural coefficients, $\xi = n \times 1$ vector of exogenous concepts, and $\delta = q \times 1$ vector of errors in the measurement model. Four additional covariance matrices are necessary for the LISREL analysis: $\Phi = n \times n$ matrix of covariances among the exogenous concepts, ψ = m $\times$ m matrix of covariances among the error terms ζ, Θ_ϵ = p $\times$ p matrix of covariances among the errors (ϵ) in the endogenous measurement model, and Θ_δ = q $\times$ q matrix of covariances among errors (δ) of the exogenous measurement model.

The matrices Λ_Y and Λ_X contain the coefficients that relate the observed variables x_i and y_j to the concepts η_i and ξ_j. Specifying the elements on the diagonal to be 1.0 and all the off-diagonal elements to be 0 specifically links a single indicator to a single concept, and also specifies the scale of measurement for the concept.[18] Consequently, a unit of change in the indicator variable results in a unit of change of the concept. The choice of the indicators is determined by the context of the model and the appropriateness of the indicators can be determined indirectly by the statistical measures of the goodness of fit of that model to the data (see below). For example, a model that includes a predator-prey relationship between the concepts zooplankton and phytoplankton might use as appropriate indicators the abundances of zooplankton and phytoplankton. Inappropriate indicators might be species richness or species diversity measures. The model with abundances as indicators would be determined to be a "good" model, while the model with species diversity as indicators would be a "bad" model (see below).

The matrices ϵ and δ contain error variables for the measurement models and correspond to the error terms of regression models. Variances and covariances among the exogenous concepts are found in Φ, and should correspond closely to the variances and covariances among the indicator variables. The matrix ψ contains the variances and covariances among the error terms of the endogenous concepts. Terms along the diagonal are the error variances in the prediction of the endogenous concepts by the exogenous concepts.[18] Off-diagonal terms are the covariances among the error variables. Significant covariances indicate that a concept not included in the model is affecting two concepts in the model. Matrices Θ_ϵ and Θ_δ contain the error variances and covariances in measuring the endogenous and exogenous variables. Again, the covariance terms indicate that there may be a correlation between errors of two measurement equations, for either endogenous or exogenous variables, that is a result of the failure to include a variable that is a common cause of the two measured variables.[18] Terms along the diagonal represent error variance in the measurement of the indicator variables. If the amount of error is known, it can be specified in either the Θ_ϵ or Θ_δ matrices; if the measurement error is unknown, it can be estimated by LISREL (see Reference 18 Chapter 4 for a complete discussion of these matrices and the issues of measurement and reliability).

These equations and covariance matrices are used to specify the model (Figure 1a). The elements of **B** and Γ are the path coefficients β_{ij} and γ_{ij} for the model. The expanded form of the equations for the ecosystem hypothesized in Figure 1a is provided in Figure 1b. Note that only the pathways labeled in Figure 1a are included in Figure 1b. Zero values for parameters indicate pathways that are hypothesized to not be important (e.g., the pathway from phytoplankton to aquatic vegetation) or are zero by definition (e.g., β_{11} = β_{22} = 0). The model specified in Figure 1a represents a null hypothesis H_0 that is tested against H_1 = any other less restrictive model (a model involving fewer paths). The chi-square goodness of

Structural Equation

$$\eta = \beta\eta + \Gamma\xi + \zeta$$

$$
\begin{bmatrix} \eta_1 \\ \eta_2 \\ \eta_3 \\ \eta_4 \\ \eta_5 \end{bmatrix} =
\begin{bmatrix}
0 & \beta_{12} & \beta_{13} & \beta_{14} & 0 \\
\beta_{21} & 0 & 0 & 0 & 0 \\
0 & \beta_{32} & 0 & \beta_{34} & 0 \\
0 & 0 & \beta_{43} & 0 & 0 \\
0 & 0 & 0 & 0 & 0
\end{bmatrix} +
\begin{bmatrix}
0 & 0 \\
\gamma_{21} & 0 \\
0 & 0 \\
0 & \gamma_{42} \\
0 & \gamma_{52}
\end{bmatrix}
\begin{bmatrix} \xi_1 \\ \xi_2 \end{bmatrix} +
\begin{bmatrix} \zeta_1 \\ \zeta_2 \\ \zeta_3 \\ \zeta_4 \\ \zeta_5 \end{bmatrix}
$$

Measurement Model for Endogenous Concepts and Indicators

$$y = \Lambda_y\eta + \varepsilon$$

$$
\begin{bmatrix} y_1 \\ y_2 \\ y_3 \\ y_4 \\ y_5 \\ y_6 \end{bmatrix} =
\begin{bmatrix}
1.0 & 0 & 0 & 0 & 0 \\
0 & \lambda^y_{22} & 0 & 0 & 0 \\
0 & 1.0 & 0 & 0 & 0 \\
0 & 0 & 1.0 & 0 & 0 \\
0 & 0 & 0 & 1.0 & 0 \\
0 & 0 & 0 & 0 & 1.0
\end{bmatrix}
\begin{bmatrix} \eta_1 \\ \eta_2 \\ \eta_3 \\ \eta_4 \\ \eta_5 \\ \eta_6 \end{bmatrix} +
\begin{bmatrix} \varepsilon_1 \\ \varepsilon_2 \\ \varepsilon_3 \\ \varepsilon_4 \\ \varepsilon_5 \\ \varepsilon_6 \end{bmatrix}
$$

Measurement Model for Exogenous Concepts and Indicators

$$x = \Lambda_x\xi + \delta$$

$$
\begin{bmatrix} x_1 \\ x_2 \end{bmatrix} =
\begin{bmatrix} 1.0 & 0 \\ 0 & 1.0 \end{bmatrix}
\begin{bmatrix} \xi_1 \\ \xi_1 \end{bmatrix} +
\begin{bmatrix} \delta_1 \\ \delta_2 \end{bmatrix}
$$

$$\phi = \begin{bmatrix} \phi_{11} & \\ 0 & \phi_{22} \end{bmatrix}$$ Covariances among exogenous concepts

$$\psi = \begin{bmatrix}
\Psi_{11} & & & & \\
0 & \Psi_{22} & & & \\
0 & 0 & \Psi_{33} & & \\
0 & 0 & 0 & \Psi_{44} & \\
0 & 0 & 0 & 0 & \Psi_{55}
\end{bmatrix}$$ Covariances among errors in structural equation (ζ's)

$$\Theta_\varepsilon = \begin{bmatrix}
0 & & & & & \\
0 & \theta_{22} & & & & \\
\theta_{31} & 0 & \theta_{33} & & & \\
0 & 0 & 0 & 0 & & \\
0 & 0 & 0 & 0 & 0 & \\
0 & 0 & 0 & 0 & 0 & 0
\end{bmatrix}$$ Covariances among errors in endogenous measurement model (ε's)

$$\Theta_\delta = \begin{bmatrix} 0 & \\ 0 & 0 \end{bmatrix}$$ Covariances among errors in exogenous measurement model (δ's)

Figure 1b. The model from Figure 1a translated into the form of the equations used in the LISREL analysis. Also included are the parameters in the matrices ϕ, ψ, Θ_δ, and Θ_ε that are to be estimated by LISREL. The 0's along the diagonals of Θ_ε and Θ_δ indicate that these error variances have been estimated by the investigator and are already incorporated into the modeling process. Θ_{22} and Θ_{33} of Θ_ε are unknown and are to be estimated by the model. The 0's in the off-diagonal entries of Θ_δ and Θ_ε mean that there is no covariance expected among the error terms ε or δ. Θ_{31} indicates that there is a covariance expected between the ε's in the measurement equations for the vertebrate and phytoplankton concepts. This term was "suggested" by the modification indices in the LISREL output from a previous modeling effort. Again, see the text for more details.

fit statistic indicates whether the assigned or fitted parameter specifications of H_0 fit the data. For more details concerning model specification and development, see Hayduk[18] or Johnson et al.[19]

Data for the analysis consists of the variance/covariance matrix for the indicator variables. If the parameter values for the elements of the eight matrices are known (or can be estimated), it is possible to predict the variances and covariances among the original observed variables.[18] These predicted variances and covariances can be compared to the actual variances and covariances of the observed variables. This comparison between predicted and observed variances and covariances is the basis for the statistical test of H_0. The test statistic is a likelihood chi-square statistic for goodness of fit. The degrees of freedom for the test is the number of entries in the variance/covariance matrix of observed variables minus the number of parameters estimated by the model. A small chi-square value and a large p value indicates that the hypothesized ecosystem is a good fit to the data, i.e., the hypothesized ecosystem is a good representation of the true ecosystem. The individual path coefficients can be tested for significance by comparing the t-values for each coefficient (provided as part of the LISREL output) against a critical value for a specified level of significance from a normal probability table. A nonsignificant t-value indicates that the value of the path coefficient is not significantly different from zero, i.e., the path does not exist.

LISREL provides matrices of total effects, direct effects, and indirect effects of the exogenous concepts on the endogenous concepts and the endogenous concepts on each other. Additionally, LISREL provides standard errors for all effects. A sufficient condition for stability of the ecosystem, *as modeled,* is that the dominant eigenvalue of $\mathbf{BB}^T$ be less than one.[18,39]

AGGREGATION OF VARIABLES

Ecosystems can be modeled with either a few large compartments (concepts), or several small compartments, and it appears that there is no proper level of aggregation.[40] Gardner et al.[41] and Cale and Odell[42] have determined that, in general, ecosystem components can be aggregated into large compartments without introducing error into the analysis if their turnover times are comparable. However, there is little guidance for the disaggregation of large compartments into several smaller compartments. The problem that we explore here is how the stability value changes as the ecosystem is modeled with structural components at different levels of aggregation (large compartment vs. small compartment models). The data for this analysis comes from a 3-year mesocosm study of the effects of atrazine on pond ecosystems.[43-48] Details on the methods and experimental design are found in Huggins et al.[48] or Huggins.[47] Briefly, the study involved 8 identically constructed mesocosms: 2 control, and 2 each with 20 μg/l, 100 μg/l, or 500 μg/l atrazine. The ponds were initially exposed to atrazine in May 1981, and were periodically reexposed, keeping concentrations near treatment levels. Data used in this analysis are from 1983, the final year of the study. The variables used in these models are pond depth (measured in cm), atrazine concentration (in μg/l), aquatic vegetation (percent cover of submergent macrophytes and percent shoreline development of emergent macrophytes), phytoplankton (mg chl a/l), zooplankton (number per l), macroinvertebrates (total number from three artificial plate samplers), and all vertebrate variables were measured as biomass (in grams). All variables except pond depth were log transformed to equalize variances and normalize the distributions. The vertebrates' compartment was decomposed into smaller compartments (smaller meaning that fewer individual species are aggregated into a compartment). In a second set of analyses, the macroinvertebrate compartment was decomposed into smaller compartments in five separate models.[48] Species of vertebrates were tadpoles (*Rana* spp.), bluegill sunfish (*Lepomis macrochirus* Rafinesque), grass carp (*Ctenopharyngodon idella* Valenciennes), gizzard shad (*Dorosoma cepedianum* LeSueur), and channel catfish (*Ictalurus punctatus* Rafinesque). Channel catfish represented such a small portion of the total biomass that they were excluded from the individual species' model.

Pond depth and atrazine are modeled as exogenous concepts; aquatic vegetation, phytoplankton, zooplankton, macroinvertebrates, and vertebrates are endogenous concepts. Aquatic vegetation is indicated by both submergent and emergent macrophytes. We hypothesized several pathways of interaction between the concepts (Figure 1a; see Reference 48 for detailed explanations of hypothesized pathways).

The endogenous concepts represent the major biotic components of the ecosystem, and the exogenous concepts are the external disturbance factors.[19] The pathways represent the functional relationships among the structural components hypothesized to exist in the ecosystem. Model development was done with minimal use of the modification indices. Modification indices are provided as part of the LISREL output for all potential pathways that could be added to the model, and are the reductions in the likelihood chi-square value for the model that would occur if a specific pathway is added. In fact, in no models were pathways added as a result of modification indices. In several instances, pathways hypothesized to exist were "trimmed" from the models as a result of their nonsignificant t-values (Figure 2a: atrazine → vertebrates; aquatic vegetation → macroinvertebrates; Figure 2b: zooplankton → fish; Figure 2c: bluegill sunfish → aquatic macroinvertebrates; aquatic vegetation → grass carp). Modification indices primarily indicated that terms in the Θ_ϵ and ψ matrices could be added. These terms suggest that concepts are missing from the model, e.g., water chemistry variables or water temperature. These missing concepts (whatever they may be) are major structural components of the ecosystem, but we have not obtained data for them. They would be linked by functional pathways to other concepts (components) in the model, either established or themselves missing. Essentially, there are aspects of the ecosystem that we failed to include in our models that affect other parts of the ecosystem. An advantage to LISREL is that the modification indices can indicate that there are these missing concepts by "suggesting" the inclusion of parameters in the ψ, Θ_ϵ, and Θ_δ matrices. Unfortunately, LISREL cannot suggest what is missing, or how the functional

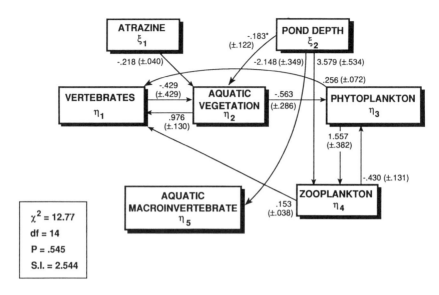

Figure 2a.   Large compartment model of Figure 1a, with the path coefficients and without the indicator variables. Standard errors for the path coefficients are in parentheses. The path coefficient λ_{22}, from pond depth to aquatic vegetation, is significant only at $p \approx 0.10$, but is retained in the model because it was hypothesized to exist and was significant at $p < 0.05$ in all other models.

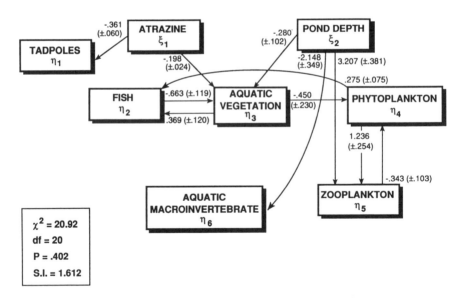

Figure 2b.   Intermediate compartment model with the vertebrate compartment decomposed into tadpoles and fish. The indicators for these concepts are the natural logarithm of biomass of tadpoles and fish, respectively. All other indicators remain the same as in Figure 1a. All path coefficients are significant at $p < 0.05$.

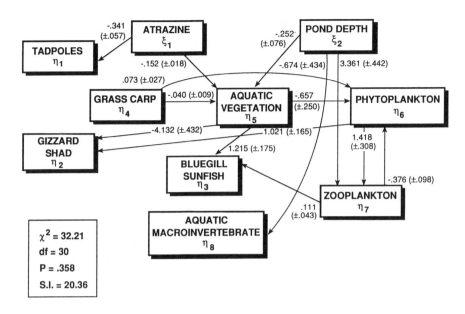

Figure 2c.   Small compartment model with the fish concept decomposed into individual species. Indicators of all fish species are the natural logarithm of biomass. All paths are significant at $p < 0.05$.

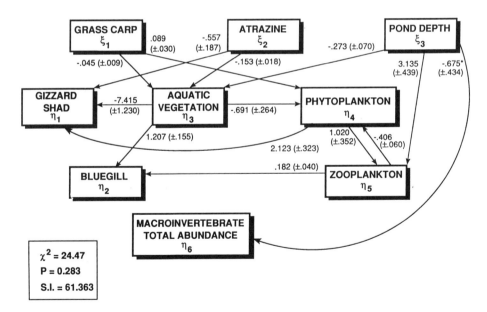

Figure 2d.   Reorganized model from Figure 2c. Tadpoles have been removed from the model and the grass carp concept has been redefined as an exogenous concept.

pathways between the established and missing concepts should be specific. However, we feel that even the suggestion of missing concepts is a significant advantage of LISREL over other analytical techniques, and often a reasonable candidate for the missing concept can be decided on by close examination of the covariance terms in the matrices.

Three ecosystem models were developed by decomposing the large compartment vertebrate model (Figure 2a) into successively smaller compartment models (Figures 2b, 2c), and a fourth model was developed by reorganizing the small compartment model (Figure 2d). The fourth model was developed to illustrate the difference between endogenous and exogenous concepts. Exogenous concepts are those that affect other concepts and are not affected by those concepts in return. These external disturbance factors are imposed upon the ecosystem from the outside. Figures 2a to 2c included only atrazine and pond depth as exogenous concepts. However, examining the small compartment model (Figure 2c) revealed that grass carp, designated as an endogenous concept, was not affected by any other concept. We reorganized the model by redefining grass carp to be an exogenous concept. In natural ecosystems, grass carp and aquatic vegetation are expected to be involved in a predator-prey relationship, much like zooplankton and phytoplankton. However, due to the short time frame of the experiment relative to the life expectancy of grass carp, it was not possible to detect an effect of aquatic vegetation on grass carp (grass carp were added during the second year of the experiment). Modeling with LISREL enabled us to detect the artificiality of the experimental system and properly assign grass carp as an exogenous concept.

The four models in Figures 2a to 2d all represent the same ecosystem from the same set of mesocosms, yet the stability indices vary from 1.612 (Figure 2b) to 63.363 (Figure 2d). Including models from Huggins et al.[48] results in stability index values as large as 164.43. These differences mean that the ecosystems, as modeled, change from slightly unstable to very unstable. The path coefficients for those paths that are common to all models, e.g., the pathway from aquatic vegetation to phytoplankton, also change (Table 1). However, in no instance for any path present in more than one model did the pathways change sign. In only one instance, the pathway from pond depth to aquatic macroinvertebrates, did the variation in magnitude exceed the overlap created by the standard errors. In only two instances are those pathways that are common to all models not consistently significant at $p <$ 0.05. The pathway from pond depth to aquatic vegetation in the large compartment model was significant at $p \approx 0.10$, and the pathway from pond depth to aquatic macroinvertebrates at $p \approx 0.12$. With the exception of these two pathways, the effects estimated by LISREL are consistent both in sign and magnitude. Both of these pathways involve pond depth, and the variation in the path coefficients is most probably the result of two factors. First, pond depth is a concept that is too broad and inclusive. For the pathway from pond depth to aquatic vegetation, this concept is light attenuation in the water column. However, other paths reflect other processes, such as the path from depth to aquatic macroinvertebrates reflecting changes in the area of the bottom (see Reference 48 for discussion). Additionally,

Table 1. Path Coefficients (± Standard Error) For Those Paths Common to All Models

Pathway	Path Coefficient			
	LC	IC	SC	RC
Atrazine → aquatic vegetation	-2.18 ± .040	-.198 ± .024	-.152 ± .018	-.153 ± .018
Pond depth → aquatic vegetation	-.183 ± .122[a]	-.280 ± .102	-.252 ± .076	-.273 ± .070
Pond depth → aquatic macroinvertebrates	-2.148 ± .349	-2.148 ± .349	-.674 ± .434	.675 ± .434[b]
Pond depth → zooplankton	3.579 ± .534	3.207 ± .381	3.361 ± .442	3.135 ± .439
Aquatic vegetation → phytoplankton	-.536 ± .286	-.450 ± .230	-.657 ± .250	-.691 ± .264
Phytoplankton → zooplankton	1.557 ± .382	1.236 ± .254	1.418 ± .308	1.020 ± .352
Zooplankton → phytoplankton	-.430 ± .131	-.353 ± .103	-.375 ± .098	-.406 ± .060

[a] $p \approx .10$.
[b] $p \approx .12$.

Note: LC = large compartment model, IC = intermediate compartment model, SC = small compartment model, RC = reorganized compartment model. Values are the number of units of change in the second concept with a unit change in the first concept.

Table 2. Numbers of Concepts, Pathways, and Connectance For the Eight Models Presented in This Chapter and Huggins et al. (Chapter 33)

Model	Source	Total Concepts	Endogenous Concepts	Total Pathways	Pathways per Concept	Endogenous Pathways	Endogenous Pathways per Endogenous Concept	Stability Index	Total Connectance	Endogenous Connectance
LC	Thisms	7	5	11	1.57	7	1.40	2.544	.367	.350
IC	Thisms	8	6	11	1.38	6	1.00	1.612	.262	.200
SC	Thisms	10	7	14	1.40	9	1.29	20.36	.194	.161
RC	Thisms	9	6	14	1.56	7	1.17	61.36	.292	.233
Specialists/ Generalists	Chapter 33	10	7	20	2.00	13	1.86	79.68	.317	.310
Habitat	Chapter 33	11	8	22	2.00	14	1.75	79.28	.275	.250
Trophic	Chapter 33	12	9	24	2.00	17	1.89	164.43	.242	.236
Species	Chapter 33	15	12	31	2.07	23	1.92	102.18	.185	.174

Note: Total connectance is the actual number of pathways divided by the total possible number of pathways for all concepts in the model. Endogenous connectance is only for the endogenous portion of the model. Abbreviations for the models are as in Table 1. The four models from Huggins et al. (Chapter 33) are decompositions of the macroinvertebrate concept as it appears in the reorganized model of Figure 2d. See text for more details.

it is necessary to consider the relationship among those terms in the model and those terms not in the model, but that are reflected in the covariance terms in Θ_ϵ and ψ. Off-diagonal terms are present in either or both matrices in all models, but as the models include larger numbers of specified concepts, more terms are present in the matrices. For example, an obvious concept for inclusion into an aquatic ecosystem model would be a water quality concept(s) indicated by the concentration of some nutrient or dissolved gas, e.g., oxygen or carbon dioxide. However, these variables were not regularly monitored in our study. In large compartment models, there is "internal compensation" within the compartments, such that increases and decreases in the constituent subcompartments resulting from the effects of other concepts cancel each other out. This results in the large compartment being relatively unconnected to the rest of the model. There is a reduction in the number of pathways per concept, and, consequently, a reduction in the importance of concepts not included in the model (Table 2). This explanation is consistent with our data (Table 2) as the total number of pathways per concept increases from 1.57 to 2.07 when the total number of concepts increases from 7 to 15. The same is true when considering only the endogenous portion of the ecosystem, as the number of endogenous pathways per endogenous concept increases from 1.40 to 1.92 and the number of endogenous concepts increases from 5 to 12. Missing concepts (e.g., water quality) become more critical in larger models (with smaller compartments). There is no internal compensation within smaller compartments, and both established and missing concepts would be connected to more concepts within those models.

Connectance is defined as the actual number of pathways divided by the total possible number of pathways and is often used as a measure of complexity.[11] We calculated connectance for the combined exogenous and endogenous portions of the ecosystem, and for the endogenous portion only. The connectance values were generally low, with total connectance ranging from 0.185 to 0.367, and endogenous connectance ranging from 0.161 to 0.350 (Table 2). A general hypothesis in community ecology is that increased connectance and increased numbers of species result in decreased stability.[7,11] For our model ecosystems, in which there is one nonbiological concept, the correlation between stability and total connectance is -0.333 ($p > 0.05$, Spearman's rho, n = 8) and the correlation between endogenous connectance and stability is 0.048 ($p > 0.05$, Spearman's rho, n = 8). These values suggest that there is no relationship between complexity and stability in our model ecosystems. These results also suggest that simply increasing the complexity of any ecosystem model by the decomposition of large compartments and increasing the number of pathways into and out of those smaller compartments does not cause an increase in the stability index. Increases in the stability index (increased instability) result from decompositions or additions of specific concepts with specific functional relationships within the ecosystem.

EFFECTS MATRICES AND STABILITY INDICES

The **B** and **Γ** matrices are the direct effects of the endogenous concepts on each other, or the exogenous concepts on the endogenous concepts, respectively. LISREL also provides a matrix of total effects, and, by subtraction, a matrix of indirect effects can be calculated. These effects are also measured in the metrics of the indicator variables. For example, in Figure 2a, there is no direct path from atrazine to the vertebrate compartment; however, there is a total effect of -0.131 of atrazine on vertebrates. This effect is entirely indirect and operates through the direct effect of atrazine on aquatic vegetation (γ_{21}) and the effects of aquatic vegetation and vertebrates on each other (β_{12} and β_{21}). Therefore, the addition of one unit of atrazine [(ln + 1) μg/l] to the mesocosms should result in the loss of 0.131 [(ln + 1) g] of biomass of vertebrates (remembering the transformations to equalize the variances). The measurement of total, direct, and indirect effects is potentially one of the most beneficial aspects of LISREL modeling, because only after examining both direct and indirect effects can the total effect of a chemical on an ecosystem be determined.

In the ecosystem modeled in Figure 2a there is an overall negative effect of atrazine on vertebrates and aquatic vegetation, and an overall positive total effect of atrazine on phytoplankton and zooplankton. The measurement of total effects points out the major problem with the current statistical methodology for analyzing mesocosm test data. First, there is a direct toxic effect of atrazine on aquatic vegetation (pathway γ_{21}), but for vertebrates, there is a negative indirect effect resulting from the ecological processes occurring within the ecosystem. The current analytical techniques cannot distinguish between these two very different types of effects. Second, because of the statistically significant direct negative effect of atrazine on aquatic vegetation, there are positive effects of atrazine on both phytoplankton and zooplankton. Because the *t*-tests, as they are currently performed, are one-tailed and designed to test only for reductions in variables relative to the controls, the positive total effects will not be distinguished as being effects at all. Finally, because the effects are measured in the metrics of the indicator variables, it is possible to determine exactly how much any concept (and its indicator variable) will be affected by the addition of any amount of a second concept. With LISREL, not only is it possible to determine if there are statisically significant interactions between concepts, a quantitative assessment of those interactions is provided.

The total, direct, and indirect effects for all models are presented in Appendix 1. Again, for all models that share the same concepts, the values of the total effects are similar (Table 3). There are no changes in sign, and the variation among all four models generally is smaller than for the direct effects (actual path coefficients).

In each model, there are about twice the number of indirect effects than direct effects. It is also true that in the overwhelming majority of cases in which there

Table 3. Total Effects of Concepts on Each Other For the Models in Figure 2

Pathway	Total Effect			
	LC	IC	SC	RC
Atrazine → aquatic vegetation	−0.162	−0.167	−0.152	−0.149
Atrazine → phytoplankton	0.055	0.053	0.065	0.069
Atrazine → zooplankton	0.085	0.065	0.092	0.071
Pond depth → aquatic vegetation	−0.165	−0.117	−0.252	−0.279
Pond depth → phytoplankton	−0.866	−0.735	−0.717	−0.781
Pond depth → zooplankton	2.231	2.229	2.344	2.132
Aquatic vegetation → phytoplankton	−0.369	−0.266	−0.428	−0.489
Aquatic vegetation → zooplankton	−0.390	−0.329	−0.608	−0.499
Zooplankton → phytoplankton	−0.255	−0.252	−0.245	−0.287
Phytoplankton → zooplankton	0.982	0.910	0.925	0.722

Note: Abbreviations and values are the same as in Table 1.

are both nonzero direct and indirect effects, the signs of the direct and indirect effects are opposite, i.e., when the direct effect of one concept on another is positive, the indirect effect is negative. This phenomenon of opposing signs may be an indication of negative feedback or homeorheic activity[4] within the ecosystem. It is also interesting that for the models in Huggins et al.[48] opposing signs for direct and indirect effects predominate. However, as the stability index increases, so does the number of instances of nonopposing signs. When all models from both papers are analyzed together, the correlation between the number of same-sign effects and the stability index is 0.982 (Spearman's rho, n = 8, $p < 0.05$), and the correlation between the stability index and the ratio of number of pathways with opposing signs to the number of same sign pathways is 0.964 (Spearman's rho, n = 8, $p < 0.05$). A preliminary conclusion is that as there is an increase in the number of same-sign direct and indirect effects, there is an increase in the stability index and the ecosystem tends to be less stable. While it is tempting to attribute the lack of stability to the number of same-sign effects in a cause-and-effect relationship, it remains only an interesting phenomenon that needs more investigation. It is not clear whether this relationship will be common to all mesocosm experiments, all aquatic ecosystems, or all ecosystems in general.

The most stable model is the intermediate compartment model with tadpoles separated from fish (SI = 1.612), and the least stable is the reorganized model with all species of fish modeled individually, but without tadpoles (SI = 63.36). However, the stability indices from the four models should not be viewed as contradictory information.[19] For example, the two models mentioned above indicate that the mesocosms will be able to retain fish, provided that we do not care what species are present. The biomass lost due to a reduction or loss of one species will be compensated for by an increase in biomass of a second species. It is even possible to determine which concept will be lost from the ecosystem, and the sensitivity of the stability index value to changes in the values of the path coefficients by performing projections with the effects matrices and sensitivity analyses (see below).

Rahel[49] recently investigated a similar problem by examining the stability of communities with three different hierarchical measures of stability. Rahel found that communities stable at one level of analysis, e.g., absolute abundances of the species,were not necessarily stable at other levels of analysis, e.g., abundance rankings of the species or species' presence and absence. He stressed that examining communities at different levels is actually a problem of the scale of resolution of the stability analysis. There is no reason why those scales should be contradictory, and examining the stability at different analytical scales could be helpful. Our problem is similar except that we used a single stability measure and several hierarchical taxonomic and functional categories.[43] However, our message is the same as Rahel's; the different measures of stability from the models with different taxonomic/functional categories should not be seen as contradictory or competing. Instead, they should be viewed as complementary.

ECOSYSTEM LEVEL MODELING AND ECOTOXICOLOGIC TESTING

From the preceding discussion, several points can be made concerning the use of LISREL in evaluating mesocosm tests. First, simply examining the structure of any LISREL model, with its exogenous and endogenous concepts, provides an insight into how a chemical could affect an ecosystem. Given that the stability of the ecosystem is determined by the interactions among the endogenous concepts, and that the exogenous concepts function as disturbance factors, it becomes clear that the stability of the ecosystem can be altered in two ways. There could be a retention of the current framework of ecosystem structure and functional pathways, but the external stress could change the value of the path coefficients within the ecosystem. This essentially amounts to a change in the time scaling of processes within the ecosystem. In some instances, the addition of the chemical stress may increase the rate at which certain processes occur leading to instability at a much faster rate than would naturally occur. To fully determine how and where within the ecosystem the collapse will occur requires a sensitivity analysis of the control and treatment ecosystems. This analysis provides an estimate of how much the stability index changes in response to changes in the path coefficients in the matrix of direct effects of endogenous variables on endogenous variables. This matrix of sensitivity values is the sensitivity structure of the matrix. A sensitivity analysis is critical in fully assessing the effects of the chemical stress on the ecosystem, especially for ecosystems that are naturally unstable, as pond mesocosm systems seem to be. A chemical stress can change the numerical values of the paths in the model while maintaining the sensitivity structure of the effects matrix, or it can alter the sensitivity structure of the matrix. By maintaining the sensitivity structure of the control ecosystem, a chemical stress could cause the ecosystem to "naturally" collapse. It will simply become unstable at a faster rate than the control. However, once the stress has been removed, the ecosystem should recover rapidly. By altering the sensitivity structure of the ecosystem, the stress could cause the ecosystem to "unnaturally" collapse, which might not recovery quickly. This aspect of recovery is referred to as resilience, defined as the inverse of the time required for an ecosystem to return to ''near-normal'',[4] and is often considered to be a function of the ecosystem alone. It is possible, however, that in some instances it may be a consequence of the interaction between the chemical stress and the ecosystem. We will address the whole issue of sensitivity analysis at a later time.

The second way an external stress could operate would be the induction or elimination of functional pathways. Of course, the extreme, but trivial, example of this effect would be a complete loss of a specific structural component resulting in the loss of all pathways into and out of that component. These two processes, alteration of path coefficients and induction/removal of pathways, are very different, and there are significant implications concerning the detection of these types of changes and their mitigation. For example, a stress that changes only the magnitude of certain path coefficients, but leaves the pathways unchanged, could be very difficult to detect, but it could be easier to mitigate these effects. Conversely, gain or loss of functional pathways might be easy to detect, but difficult to mitigate.

The two different modes of action of a chemical stress can be distinguished from each other with the proper mesocosm test design. A large number (six to ten) of control mesocosms and an equal number of treatment mesocosms could be used to resolve the mode of action. Unfortunately, our experimental design used only two control mesocosms, which are insufficient to provide good estimates of the path coefficients. What is not known at this time is whether a single set of controls used over a 1- or 2-year period will be adequate controls for comparison against treatment mesocosms over a period of several years, or whether a set of control mesocosms needs to be used each year for testing a specific chemical.

It appears that a single stability measure from a specific LISREL model (e.g., the large compartment model in Figure 2a) will not provide enough information about the ecosystem to be of use. The comparison of stability indices from a paired series of models, one set of models based on untreated, control mesocosms and the other from the treatment mesocosms, will be the most informative. However, a few points concerning those models are clear. Large compartment models will tend to be more stable and more buffered from the exogenous disturbances. Intuitively, it makes sense that a stress would be less likely to negatively impact all zooplankton as opposed to single species (or even finer decomposition of the concepts such as size or age classes). Therefore, the loss of stability due to a change of structure or function in a large compartment model would be seen as a serious problem. Small compartment models tend to be more unstable, because there is no opportunity for internal compensation within small compartments. However, the loss of small compartments would probably cause the stability index to change only slightly, and might not be viewed as seriously as the loss of large compartments (unless there is an additional reason why certain small compartments should not be lost from the ecosystem).

In our brief introduction to LISREL and its role in mesocosm studies, we have only touched upon the basics of the modeling procedure and its potential benefits in ecotoxicological research. We believe that structural equation modeling offers a viable solution to the problems of data analysis that is evidenced by the controversy surrounding the current statistical methodologies. LISREL offers not only the ability to measure the total effects of all components of an ecosystem on all other components, it provides true integration of these effects into a single ecosystem model (or hierarchical series of models). LISREL's utility is further demonstrated by its ability to generate predictions about how chemical stresses in the ecosystem can disrupt structure and function. We believe that with continued exploration of LISREL modeling, additional hypotheses and predictions can be generated that will provide significant contributions to ecotoxicological research.

ACKNOWLEDGMENTS

We extend our thanks to R. Graney, J. Rodgers, and J. Kennedy for the invitation to participate in the mesocosm symposium. Thanks also to Steve Bartell for bringing to our attention the phenomenon of the signs of the direct and indirect effects, as he edited a previous manuscript. Greg Howick provided helpful comments and suggestions on parts of this manuscript. Data used in this chapter were obtained from a project funded by the U.S. Environmental Protection Agency (grant number CR808804–02–0 to F. deNoyelles). Financial support was also provided by the Kansas Biological Survey.

REFERENCES

1. deNoyelles, F. and W. D. Kettle. 1985. Experimental ponds for evaluating bioassay predictions. In T. P. Boyle, ed. *Validation and Predictability of Laboratory Methods for Assessing the Fate and Effects of Contaminants in Aquatic Systems*. American Society for Testing and Materials ASTM STP 865, American Society for Testing and Materials, Philadelphia, PA, pp. 91–103.
2. deNoyelles, F., Jr., S. L. Dewey, D. G. Huggins and W. D. Kettle. 1993. Aquatic mesocosms in ecological effects testing: detecting direct and indirect effects of pesticides. In, R. Graney, J. Kennedy, and J. Rodgers, eds. *Utilization of Simulated Field Studies in Aquatic Risk Assessment*. Society of Environmental Toxicology and Chemistry.
3. Stunkard, C. 1992.
4. Kelly, J. R. and M. A. Harwell. 1989. Indicators of ecosystem response and recovery. In S. A. Levin, M. A. Harwell, J. R. Kelly and K. D. Kimball, eds. *Ecotoxicology: Problems and Approaches*. Springer-Verlag, New York, NY, pp. 9–35.
5. Matthews, R. A., A. L. Buikema Jr., J. Cairns Jr. and J. H. Rodgers. 1982. Biological monitoring. Part IIA-Receiving system functional methods, relationships and indices. *Water Research* 16: 129–139.
6. Odum, E. P. 1969. Relationship between structure and function in ecosystems. *Japanese Journal of Ecology* 12: 108–118.
7. May, R. M. 1973. *Stability and Complexity in Model Ecosystems*. Princeton University Press, Princeton, NJ.

8. Connell, J. H. and W. P. Sousa. 1983. On the evidence needed to judge ecological stability or persistence. *American Naturalist* 121: 789–824.

9. Waide, J. B. and J. R. Webster. 1976. Engineering systems analysis: applicability to ecosystems. In, B. C. Patten, ed. *Systems Analysis and Simulation in Ecology, Volume IV.* Academic Press, New York, NY. pp. 330–371.

10. Astor, P. H., B. C. Patten and G. N. Estberg. 1976. The sensitivity substructure of ecosystems. In, B. C. Patten, ed. *Systems Analysis and Simulation in Ecology, Volume IV.* Academic Press, New York, NY. pp. 390–429.

11. Pimm, S. L. 1982. *Food Webs.* Chapman and Hall, New York, NY.

12. Siljak, D. D. 1979. Structure and stability of model ecosystems. In, E. Halfon, ed. *Theoretical Systems Ecology.* Academic Press, New York, NY. pp. 151–181.

13. Harte, J. 1979. Ecosystem stability and the distribution of community matrix eigenvalues. In, E. Halfon, ed. *Theoretical Systems Ecology.* Academic Press, New York, NY. pp. 453–465.

14. Goh, B. S. 1979. Robust stability concepts for ecosystem models. In, E. Halfon, ed. *Theoretical Systems Ecology.* Academic Press, New York, NY. pp. 467–487.

15. Jeffries, C. Stability of holistic ecosystem models. In, E. Halfon, ed. *Theoretical Systems Ecology.* Academic Press, New York, NY. pp. 489–504.

16. DeAngelis, D. L., S. M. Bartell, and A. L. Brenkert. 1989. Effects of nutrient recycling and food-chain length on resilience. *American Naturalist* 134: 778–805.

17. DeAngelis, D. L., P. J. Mulholland, A. V. Palumbo, A. D. Steinman, M. A. Huston and J. W. Elwood. 1989. Nutrient dynamics and food-web stability. *Annual Review of Ecology and Systematics* 20: 71–95.

18. Hayduk, L. A. 1987. *Structural Equation Modeling with LISREL.* The Johns Hopkins University Press, Baltimore, MD.

19. Johnson, M. L., D. G. Huggins and F. deNoyelles, Jr. 1991. Ecosystem modeling with LISREL: A new approach for measuring direct and indirect effects in ecosystem level ecotoxicological testing. *Ecological Applications* 1: 383–398.

20. Wright, S. 1968. *Evolution and the Genetics of Populations. Vol. 1. Genetic and Biometrical Foundations.* University of Chicago Press, Chicago, IL.

21. Li, C. C. 1981. *Path Analysis — A Primer.* Boxwood, Pacific Grove, CA.

22. Levins, R. 1975. Problems of signed digraphs in ecological theory. In, S. A. Levin, ed. *Ecosystem Analysis and Prediction.* Society for Industrial and Applied Mathematics, Philadelphia, PA. pp. 264–277.

23. Levins, R. 1975. Evolution in communities near equilibrium. In, M. L. Cody and J. M. Diamond, eds. *Ecology and Evolution of Communities.* The Belknap Press of Harvard University Press. Cambridge, MS. pp. 16–50.

24. Pimm, S. L. and R. L. Kitching. 1988. Food web patterns: trivial flaws or the basis of an active research program. *Ecology* 69: 1669–1672.

25. Paine, R. T. 1988. Food webs: road maps of interactions or grist for theoretical development? *Ecology* 69: 1648–1654.

26. Hastings, A. 1988. Food web theory and stability. *Ecology* 69: 1665–1668.

27. Polis, G. A. 1991. Complex trophic interactions in deserts: an empirical critique of food-web theory. *American Naturalist* 138: 123–155.

28. Peters, R. H. 1988. Some general problems for ecology illustrated by food web theory. *Ecology* 69: 1673–1676.

29. Dewey, S. L. and F. deNoyelles, Jr. 1993. On the use of ecosystem stability measurements in ecological effects testing. In, R. Graney, J. Kennedy, and J. Rodgers, eds. *Utilization of Simulated Field Studies in Aquatic Risk Assessment.* Society of Environmental Toxicology and Chemistry.

30. Odum, H. T. 1976. Macroscopic minimodels of man and nature. In, B. C. Patten, ed. *Systems Analysis and Simulation in Ecology, Volume IV.* Academic Press, Inc. New York, NY. pp. 249–280.

31. Lugo, A. E., M. Sell and S. C. Snedaker. 1976. Mangrove ecosystem analysis. In, B. C. Patten, ed. *Systems Analysis and Simulation in Ecology, Volume IV.* Academic Press, New York, NY. pp. 113–145.

32. Hannon, B. 1973. The structure of ecosystems. *Journal of Theoretical Biology.* 41: 535–546.

33. Finn, J. T. 1976. Measures of ecosystem structure and function derived from an analysis of flows. *Journal of Theoretical Biology.* 56: 363–380.

34. Patten, B. C. and J. T. Finn. 1979. Systems approach to continental shelf ecosystems. In, E. Halfon, ed. *Theoretical Systems Ecology.* Academic Press, Inc. New York, NY. pp. 183–212.

35. Patten, B. C. and G. T. Auble. 1981. System theory of the ecological niche. *American Naturalist* 117: 893–922.

36. Barber, M. C., B. C. Patten and J. T. Finn. 1979. Review and evaluation of input-output flow analysis for ecological applications. In, J. H. Matis, B. C. Patten and G. C. White, eds. *Compartmental Analysis of Ecosystem Models.* International Co-operative Publishing House, Fairland, MS. pp. 43–72.

37. Patten, B. C., R. W. Bosserman, J. T. Finn and W. G. Cale. 1976. Propogation of cause in ecosystems. In, B. C. Patten, ed. *Systems Analysis and Simulation in Ecology, Volume IV.* Academic Press, Inc. New York, NY. pp. 457–579.

38. Patten, B. C. 1982. Environs: relativistic elementary particles for ecology. 119: 179–219.

39. Joreskog, K. G. and D. Sorbom. 1986. *LISREL VI. Analysis of Linear Structural Relationships by Maximum Likelihood, Instrument Variables, and Least Squares Techniques.* Scientific Software, Inc. Mooresville, IN.

40. Levin, S. A. 1989. Models in ecotoxicology: methodological aspects. In S. A. Levin, M. A. Harwell, J. R. Kelly and K. D. Kimball, eds. *Ecotoxicology: Approaches and Problems.* Springer-Verlag, New York, NY, pp. 213–220.

41. Gardner, R. H., W. G. Cale and R. V. O'Neill. 1982. Robust analysis of aggregation error. *Ecology* 63: 1771–1779.

42. Cale, W. G., Jr., and P. L. Odell. 1979. Concerning aggregation in ecosystem modeling. In, E. Halfon, ed. *Theoretical Systems Ecology.* Academic Press, New York, NY. pp. 55–77.

43. deNoyelles, F., W. D. Kettle and D. E. Sinn. 1982. The responses of plankton communities in experimental ponds to atrazine, the most heavily used pesticide in the United States. *Ecology* 63: 1285–1293.

44. Carney, E. C. 1983. Effects of atrazine and grass carp on freshwater communities. M.S. Thesis, University of Kansas, Lawrence, KS.

45. Dewey, S. L. 1986. Effects of the herbicide Atrazine on aquatic insect community structure and emergence. *Ecology* 67: 148–162.

46. Fromm, C. H. 1988. Effects of herbicide atrazine on eutrophic plankton communities. M.S. Thesis, University of Kansas, Lawrence, KS.

47. Huggins, D. G. 1990. Ecotoxic effects of atrazine on aquatic macroinvertebrates and its impact on ecosystem structure. Ph.D. Thesis, University of Kansas, Lawrence, KS.

48. Huggins, D. G., M. L. Johnson and F. deNoyelles, Jr. 1993. The ecotoxic effects of atrazine on aquatic ecosystems: an assessment of direct and indirect effects using structural equation modeling. In, R. Graney, J. Kennedy and J. Rodgers, eds. *Utilization of Simulated Field Studies in Aquatic Risk Assessment*. Society of Environmental Toxicology and Chemistry.

49. Rahel, F. J. 1990. The hierarchical nature of community persistence: a problem of scale. *American Naturalist* 136: 328–344.

APPENDIX

The following matrices are the effects matrices for the four models in Figure 2a to 2d. In each case, the upper set of matrices represent the effects of the exogenous concepts on the endogenous concepts. The lower set of matrices represent the effects of the endogenous concepts on each other. The names of the exogenous concepts are provided above the total effects matrices; refer to Figure 2 for the names of the endogenous concepts. The values e_{ij} of the effects matrices are read as the effect of the concept in the jth column on the concept in the ith row.

Large Compartment Model
Total Effects = Direct Effects + Indirect Effects

	Atrazine	Pond Depth								
η_1	−.131	0.041		.000	0.000		−.131	0.041		
η_2	−.162	−0.165		−.218	0.000		.056	−0.165	Effects of exogenous	
η_3	.055	−0.866	=	.000	0.000	+	.055	−0.866	concepts on endogenous	
η_4	.085	2.231		.000	3.579		.085	−1.348	concepts	
η_5	.000	−2.148		.000	−2.148		.000	0.000		

	η_1	η_2	η_3	η_4	η_5			η_1	η_2	η_3	η_4	η_5
η_1	−.257	.601	.220	.019	.000			.000	.976	0.256	.153	.000
η_2	−.318	−.257	−.094	−.008	.000			−.429	.000	0.000	.000	.000
η_3	.107	−.251	−.369	−.255	.000	=		.000	−.563	0.000	−.430	.000
η_4	.167	−.390	.982	−.396	.000			.000	.000	1.557	.000	.000
η_5	.000	.000	.000	.000	.000			.000	.000	0.000	.000	.000

	−.257	−.375	−.036	−.134	.000		
	.111	−.257	−.094	−.008	.000	Effects of	
+	.107	.312	−.369	.175	.000	endogenous	
	.167	−.390	−.575	−.396	.000	concepts on	
	.000	.000	.000	.000	.000	each other	

Total, Direct, and Indirect Effects for Intermediate Compartment Model

$$
\begin{array}{c}
 \\
\eta_1 \\
\eta_2 \\
\eta_3 \\
\eta_4 \\
\eta_5 \\
\eta_6
\end{array}
\overset{\text{Atrazine} \quad \text{Pond Depth}}{
\begin{bmatrix}
-.361 & .000 \\
-.047 & -.245 \\
-.167 & -.117 \\
.053 & -.735 \\
.065 & 2.299 \\
.000 & -2.148
\end{bmatrix}}
=
\begin{bmatrix}
-.361 & .000 \\
.000 & .000 \\
-.198 & -.280 \\
.000 & .000 \\
.000 & 3.207 \\
.000 & -2.148
\end{bmatrix}
+
\begin{bmatrix}
.000 & .000 \\
-.047 & -.245 \\
.031 & .063 \\
.053 & -.735 \\
.065 & -.908 \\
.000 & .000
\end{bmatrix}
$$

$$
\begin{array}{c}
\eta_1 \\
\eta_2 \\
\eta_3 \\
\eta_4 \\
\eta_5 \\
\eta_6
\end{array}
\overset{\eta_1 \quad\;\; \eta_2 \quad\;\; \eta_3 \quad\;\; \eta_4 \quad\;\; \eta_5 \quad\;\; \eta_6}{
\begin{bmatrix}
.000 & .000 & .000 & .000 & .000 & .000 \\
.000 & -.158 & .238 & .162 & -.056 & .000 \\
.000 & -.558 & -.158 & -.108 & .037 & .000 \\
.000 & .176 & -.266 & -.264 & -.252 & .000 \\
.000 & .218 & -.329 & .910 & -.312 & .000 \\
.000 & .000 & .000 & .000 & .000 & .000
\end{bmatrix}}
=
$$

$$
\begin{bmatrix}
.000 & .000 & .000 & .000 & .000 & .000 \\
.000 & .000 & .369 & .275 & .000 & .000 \\
.000 & -.663 & .000 & .000 & .000 & .000 \\
.000 & .000 & -.450 & .000 & -.343 & .000 \\
.000 & .000 & .000 & 1.236 & .000 & .000 \\
.000 & .000 & .000 & .000 & .000 & .000
\end{bmatrix}
+
\begin{bmatrix}
.000 & .000 & .000 & .000 & .000 & .000 \\
.000 & -.158 & -.131 & -.113 & -.056 & .000 \\
.000 & .105 & -.158 & -.108 & .037 & .000 \\
.000 & .176 & .184 & -.264 & .091 & .000 \\
.000 & .218 & -.329 & -.326 & -.312 & .000 \\
.000 & .000 & .000 & .000 & .000 & .000
\end{bmatrix}
$$

Total, Direct, and Indirect Effects for Small Compartment Model

$$
\begin{array}{c}
 \\
\eta_1 \\
\eta_2 \\
\eta_3 \\
\eta_4 \\
\eta_5 \\
\eta_6 \\
\eta_7 \\
\eta_8
\end{array}
\overset{\text{Atrazine} \quad \text{Pond Depth}}{
\begin{bmatrix}
-.341 & .000 \\
.695 & .308 \\
-.174 & -.045 \\
.000 & .000 \\
-.152 & -.252 \\
.065 & -.717 \\
.092 & 2.344 \\
.000 & -.674
\end{bmatrix}}
=
\begin{bmatrix}
-.341 & .000 \\
.000 & .000 \\
.000 & .000 \\
.000 & .000 \\
-.152 & -.252 \\
.000 & .000 \\
.000 & 3.361 \\
.000 & -.674
\end{bmatrix}
+
\begin{bmatrix}
.000 & .000 \\
.695 & .308 \\
-.174 & -.045 \\
.000 & .000 \\
.000 & .000 \\
.065 & -.717 \\
.092 & -1.017 \\
.000 & .000
\end{bmatrix}
$$

$$
\begin{array}{c}
\eta_1 \\
\eta_2 \\
\eta_3 \\
\eta_4 \\
\eta_5 \\
\eta_6 \\
\eta_7 \\
\eta_8
\end{array}
\overset{\eta_1 \quad\;\; \eta_2 \quad\;\; \eta_3 \quad\;\; \eta_4 \quad\;\; \eta_5 \quad\;\; \eta_6 \quad\;\; \eta_7 \quad\;\; \eta_8}{
\begin{bmatrix}
.000 & .000 & .000 & .000 & .000 & .000 & .000 & .000 \\
.000 & .000 & .000 & .229 & -4.569 & .666 & -.250 & .000 \\
.000 & .000 & .000 & -.038 & 1.147 & .103 & .073 & .000 \\
.000 & .000 & .000 & .000 & .000 & .000 & .000 & .000 \\
.000 & .000 & .000 & -.040 & .000 & .000 & .000 & .000 \\
.000 & .000 & .000 & .064 & -.428 & -.348 & -.245 & .000 \\
.000 & .000 & .000 & .091 & -.608 & .925 & -.348 & .000 \\
.000 & .000 & .000 & .000 & .000 & .000 & .000 & .000
\end{bmatrix}}
=
$$

$$\begin{bmatrix} .000 & .000 & .000 & .000 & .000 & .000 & .000 & .000 \\ .000 & .000 & .000 & .000 & -4.132 & 1.021 & .000 & .000 \\ .000 & .000 & .000 & .000 & 1.215 & .000 & .111 & .000 \\ .000 & .000 & .000 & .000 & .000 & .000 & .000 & .000 \\ .000 & .000 & .000 & -.040 & .000 & .000 & .000 & .000 \\ .000 & .000 & .000 & .073 & -.657 & .000 & -.376 & .000 \\ .000 & .000 & .000 & .000 & .000 & 1.418 & .000 & .000 \\ .000 & .000 & .000 & .000 & .000 & .000 & .000 & .000 \end{bmatrix} +$$

$$\begin{bmatrix} .000 & .000 & .000 & .000 & .000 & .000 & .000 & .000 \\ .000 & .000 & .000 & .229 & -.437 & -.355 & -.250 & .000 \\ .000 & .000 & .000 & -.038 & -.068 & .103 & -.038 & .000 \\ .000 & .000 & .000 & .000 & .000 & .000 & .000 & .000 \\ .000 & .000 & .000 & .000 & .000 & .000 & .000 & .000 \\ .000 & .000 & .000 & -.009 & .229 & -.348 & .131 & .000 \\ .000 & .000 & .000 & .091 & -.608 & -.493 & -.348 & .000 \\ .000 & .000 & .000 & .000 & .000 & .000 & .000 & .000 \end{bmatrix}$$

Total, Direct, and Indirect Effects for Reorganized Compartment Model

$$\begin{array}{c} \\ \eta_1 \\ \eta_2 \\ \eta_3 \\ \eta_4 \\ \eta_5 \\ \eta_6 \end{array} \begin{bmatrix} \text{Grass Carp} & \text{Atrazine} & \text{Pond Depth} \\ .515 & .715 & .397 \\ -.039 & -.171 & .099 \\ -.045 & -.153 & -.273 \\ .085 & .075 & -.766 \\ -.087 & -.076 & 2.353 \\ .000 & .000 & -.675 \end{bmatrix} = \begin{bmatrix} .000 & -.577 & .000 \\ .000 & .000 & .000 \\ -.045 & -.153 & -.273 \\ .089 & .000 & .000 \\ .000 & .000 & 3.135 \\ .000 & .000 & -.675 \end{bmatrix} + \begin{bmatrix} .515 & 1.292 & .397 \\ -.039 & -.171 & .099 \\ .000 & .000 & .000 \\ -.004 & .075 & -.766 \\ .087 & .076 & -.782 \\ .000 & .000 & .000 \end{bmatrix}$$

$$\begin{array}{c} \\ \eta_1 \\ \eta_2 \\ \eta_3 \\ \eta_4 \\ \eta_5 \\ \eta_6 \end{array} \begin{bmatrix} \eta_1 & \eta_2 & \eta_3 & \eta_4 & \eta_5 & \eta_6 \\ .000 & .000 & -8.453 & 1.501 & -.609 & .000 \\ .000 & .000 & 1.116 & .131 & .129 & .000 \\ .000 & .000 & .000 & .000 & .000 & .000 \\ .000 & .000 & -.489 & -.293 & -.287 & .000 \\ .000 & .000 & -.499 & .722 & -.293 & .000 \\ .000 & .000 & .000 & .000 & .000 & .000 \end{bmatrix} = \begin{bmatrix} .000 & .000 & -7.415 & 2.123 & .000 & .000 \\ .000 & .000 & 1.207 & .000 & .182 & .000 \\ .000 & .000 & .000 & .000 & .000 & .000 \\ .000 & .000 & -.691 & .000 & -.406 & .000 \\ .000 & .000 & .000 & 1.020 & .000 & .000 \\ .000 & .000 & .000 & .000 & .000 & .000 \end{bmatrix}$$

$$+ \begin{bmatrix} .000 & .000 & -1.038 & -.622 & -.609 & .000 \\ .000 & .000 & -.091 & .131 & -.053 & .000 \\ .000 & .000 & .000 & .000 & .000 & .000 \\ .000 & .000 & .202 & -.293 & .119 & .000 \\ .000 & .000 & -.499 & -.298 & -.293 & .000 \\ .000 & .000 & .000 & .000 & .000 & .000 \end{bmatrix}$$

CHAPTER 33

The Ecotoxic Effects of Atrazine on Aquatic Ecosystems: An Assessment of Direct and Indirect Effects Using Structural Equation Modeling

Donald G. Huggins, Michael L. Johnson, and Frank deNoyelles, Jr.

Abstract: The ecotoxic effects of atrazine (0, 20, 100, and 500 µg/l exposure rates) on experimental pond mesocosm ecosystems were evaluated after 805 days of continuous exposure. Analysis of data was done using LISREL, a structural equation modeling technique. A series of five ecosystem models were developed with LISREL with each model including a unique subset of macroinvertebrate concepts. All models identified both direct and indirect effects of the exogenous concepts (grass carp, pond depth, and atrazine) on all endogenous concepts (i.e., fish, aquatic vegetation, phytoplankton, zooplankton, and macroinvertebrate concepts). All models quantify the ecotoxic effects (direct and indirect) of atrazine on the various endogenous concepts (structural units of these ecosystems) by specifying the functional paths that connect them. Direct ecotoxic effects of atrazine were primarily limited to reductions in gizzard shad biomass and aquatic vegetation abundance (% cover). Most indirect effects associated with atrazine exposure were initiated by atrazine's direct effect on the aquatic plant community. Measures of the direct, indirect, and total effects of atrazine on affected endogenous concepts are provided and the ecological consequences of these effects are discussed.

INTRODUCTION

The primary source of information concerning the direct effects of atrazine on various aquatic organisms is through laboratory toxicity testing (see reviews by References 1, 2). Atrazine toxicity in aquatic plants varies greatly depending on the test species and response criteria used (e.g., References 3, 4). Generally, algal species display a wider range of EC_{50} values (8 to 1500 μg/l) than macrophyte species which range from 30 to 163 μg/l.

Acute toxicity data suggest that the toxicity of atrazine to aquatic vertebrates is generally low at concentrations normally found in contaminated environments.[5-12] For example, runoff studies have shown atrazine concentrations to range from 1 to 100 μg/l in receiving streams and lakes[13-16] while acute toxicity estimates (LC_{50}) were most often greater than 500 μg/l (e.g., Reference 10). Information on the acute toxicity of atrazine to invertebrates is more limited,[6,10,17-19] but again most reported lethal concentration values were greater (>700 μg/l) than ambient levels (1 to 100 μg/l) typically associated with aquatic habitats most frequently monitored (e.g., large rivers and lakes).

Of more concern is the paucity of chronic toxicity data, but the few chronic test results available suggest that direct atrazine toxicity may impact the aquatic animal community or "sensitive" species components of the community in moderate to highly contaminated water bodies.[10,20] Macek et al.[10] provided estimates for the highest no observed effect concentration (NOEC) and the lowest observed effect concentration (LOEC) levels of atrazine for several aquatic vertebrates and invertebrates using decreased survival, growth, and reproduction as indicators of toxic effects. The LOEC for invertebrates was observed to be at 230 μg/l for *Chironomus tentans* Fabricinus, 250 μg/l for *Daphnia magna* Straus, and 140 μg/l for *Gammarus fasciatus* Say. The most sensitive fishes tested were brook trout, *Salvelinus fontinalis* (Mitchell) (LOEC = 120 μg/l) and bluegill, *Lepomis macrochirus* Rafinesque (LOEC = 500 μg/l).

Belluck[20] found that the egg hatchability of the caddisfly *Triaenodes tardus* Milne was adversely affected at an atrazine concentration of 22 μg/l (EC_{50} value). Long-term chronic studies with several species of leeches and snails led Streit and Peter[21] to conclude that harmful effects to invertebrates occur at atrazine concentrations equivalent to 2% of the estimated LC_{50} value. Ecotoxic effects of atrazine on various ecosystem components (e.g., phytoplankton, zooplankton) and/or attributes (e.g., primary production) of freshwater mesocosms have been demonstrated by researchers in both the U.S. and Germany.[22-24] Experimental ponds dosed with various concentrations of atrazine demonstrated initial inhibition of photosynthesis within the phytoplankton community at concentrations as low as 20 μg/l.[22] Both Gunkel[23] and deNoyelles et al.[22] observed that phytoplankton communities exposed to atrazine levels greater than 100 μg/l for extended periods showed depressed growth and production. Subsequent successional changes led to the establishment of species of phytoplankton more resistant to atrazine. The zooplankton community was influenced concurrently by the reduction and recovery of the phytoplankton, suggesting an indirect effect of atrazine upon these primary consumers. Herman et al.[25] observed that a single dose of atrazine exposure regime of 100 μg/l inhibited production of periphyton within lake limnocorrals and altered antecedent species composition. In a later study utilizing the same lake and limnocorrals of Herman and co-workers,[25] but with multiple dosing with 100 μg/l of atrazine through the study period, produced similar algae community responses.[26] These researchers reported very similar taxa and community responses between their multiple atrazine applications study and single application studies. However, they did find phytoplankton community reductions were more gradual and the Chlorophyta community and primary production recovery took longer in the multiple dose study. On a smaller scale, laboratory microcosm experiments using constructed communities[27-29] recorded results similar to those in mesocosm studies. Similar recovery rates of net productivity (16 to 21 days) were found in most microcosm and mesocosm studies reviewed.

There is strong evidence from laboratory studies that partial photosynthetic recovery occurs in macrophytes continuously exposed for 3 weeks to ≤100 μg/l of atrazine when they are moved to atrazine-free water after exposure.[30,31] However, laboratory studies on the toxicity of atrazine to macrophytes may underestimate the true ecotoxic effects of this herbicide on naturally occurring macrophyte communities. In mesocosms, atrazine severely reduced the abundance of submergent macrophytes in ponds with nominal atrazine concentration of 100 and 500 μg/l.[32] Data from deNoyelles and Kettle[33] suggest that aquatic macrophytes in an experimental pond study (approximate 5-month exposure) were inhibited by 20 μg/l of atrazine, and this community did not develop resistance to concentrations as high as those tolerated by resistant phytoplankton (20 to 500 μg/l). It is possible that macrophyte communities and/or specific plant species may also be subject to indirect effects resulting from whole ecosystem exposure to atrazine.

The direct and indirect effects of atrazine upon benthic and fish communities within ecosystems (derived or natural) have not been well documented. Walker[34] concluded that the benthic fauna of fish ponds was negatively influenced temporarily at application concentrations between 500 and 2000 μg/l of atrazine. Leeches, gastropods, and certain insect larvae were eliminated or reduced in abundance 6 to 8 weeks after treatment. These field studies were not sufficiently sensitive to distinguish whether or not the observed effects resulted from the direct toxicity of atrazine at such high environmental concentrations or were caused indirectly by the changes observed in the plant community.

Dewey[35] concluded that the aquatic insect community could be perturbed by atrazine concentrations as low as 20 μg/l in experimental ponds. Declines in overall species richness and total abundance were noted in the open-water emerging insect populations of atrazine-treated ponds when compared to untreated ponds. She suggested that these responses could have been of an indirect nature because of concurrent macrophyte community losses. Kettle et al.[36] also found that atrazine concentrations as low as 20 µg/l in experimental ponds could have ecological consequences. They found effects on diet and reproductive success of bluegill recovered from ponds treated with atrazine and suggested the responses to be either direct and/or secondary due to effects of the herbicide on the ecosystem.

The ecotoxic effects of atrazine upon nontarget aquatic communities are likely to be of greater ecological significance than its direct toxicity. The direct and indirect effects of aquatic herbicides used to control aquatic plants has been reviewed by Brooker and Edwards,[37] and a substantial body of literature exists on the ecological impacts associated with intentional plant control and/or eliminations. Brooker and Edwards observed that the following indirect effects could occur when submergent plants are treated: (1) change in oxygen-carbon dioxide balance (reduced photosynthesis = reduced net utilization of carbon dioxide and decreased net oxygen production); (2) plant replacement; and (3) alterations in the abundance and kinds of aquatic fauna resulting from plant death and/or replacement.

Hurlbert's[38] review of secondary (indirect) effects of pesticides in aquatic ecosystems also suggests that a number of ecosystem components (e.g., macroinvertebrates, zooplankton, and fish) were affected by secondary consequences resulting from direct effects on other ecosystem variables. Reduced food, substrate, and habitat complexity were observed as probable causes of macroinvertebrate changes. He also noted that death of plants could create new food supply for decomposers and that certain macroinvertebrates could increase in abundance. Increased plant detritus was a viable food resource for some invertebrates (e.g., oligochaetes and chironomid larvae), which expanded their populations by two to ten times greater densities than control areas where herbicides did not cause plant reductions or eliminations. Perhaps his most important observation was that somewhere between "low" exposure concentrations and "high" concentrations, some organisms significantly *increased* in abundance when compared to control sites.

In general, it may be concluded that herbicide contamination (e.g., atrazine) can, through both direct and indirect effects, disrupt and impact all or most of the biotic components of aquatic systems. Current thought strongly suggests that indirect effects dominate in ecosystems (e.g., Reference 39). Dependent on the characteristics of the exposed ecosystem and the extent of contamination (concentration and duration), ecotoxic effects can theoretically vary greatly especially when indirect effects can induce both positive (e.g., increased species abundance) and negative (e.g., reduced taxon richness) responses within the ecosystem. Of ultimate interest is the assessment of the "overall" affect (all responses) of atrazine on various populations within an ecosystem or attributes interacting with, and dependent on, other ecosystem components or conditions. Atrazine could initiate a series of effects which elicit both negative and positive responses within a hierarchical grouping (e.g., species richness) creating a total effect that may or may not be observed and quantified by linear model statistical approaches often utilized in ecosystem studies. In many respects atrazine's ecotoxic effects are either unknown or unquantified in ponds and other aquatic ecosystems.

The primary objective of the research presented here was to utilize the data of Huggins[4] in an ecosystem-level approach to identify and quantify the direct and indirect effects of atrazine on experimental pond mesocosms continuously exposed to various atrazine concentrations for 805 days. Examination of these mesocosms during the final 6 months of exposure to atrazine provided long-term exposure data that was analyzed using a structural equation modeling technique called LISREL[40,41]

STUDY AREA

The data utilized in this study were obtained from a long-term study (June 1981 to October 1983) conducted at the University of Kansas Nelson Environmental Study Area (NESA) located 16 km north of Lawrence, KS in Jefferson County. A complement of 15 earthen ponds was constructed in 2 phases with a single series of 7 ponds and a larger reservoir constructed in November 1977 and the second pond series added in 1979 (Figure 1). The reservoir (0.33 ha, 3.0 m maximum depth) and the 14 experimental ponds (0.045 ha) have bermed perimeters to prevent contamination from surface runoff. The reservoir and all ponds may be independently filled or drained. Water for maintaining the reservoir and/or flushing the ponds was supplied by pumping groundwater from a well located in the Kansas River aquifer.

The large reservoir was filled with water immediately after construction and natural plankton and benthic communities were allowed to develop within it. No experimental manipulation of the reservoir is allowed. To initiate the experiment the experimental ponds were drained and then simultaneously gravity filled with water and plankton from the reservoir. This procedure allowed the experiment to begin with all ponds having similar water and plankton community characteristics (see Reference 22).

Experimental ponds are excavated and bermed square cells (20 × 20 m) with a 2:1 slope to the pond bottom (10.36 × 10.36 m) (Figure 2). Access piers for pond sampling were approximately 0.3 m above the water level (maximum pond depth = 2.1 m) and extended out from the west shore of each pond. These access piers permitted placement of biological samplers within the pond and general sampling procedures to be accomplished with a minimal amount of disturbance to the pond system.

EXPERIMENTAL DESIGN

Pond Manipulations

In May of 1981, ten experimental ponds were drained and filled from the reservoir. Since their construction, the ponds had developed similar though sparse macrophyte communities. Therefore, it was deemed unnecessary to actively seed

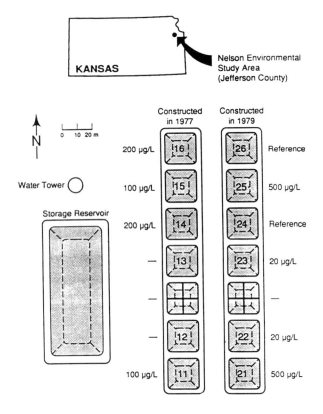

Figure 1.   The NESA experimental pond facility (mesocosm ponds) in 1983 showing location and treatment regime of ponds. Ony data from the reference 20-, 100-, and 500-μg/l pond were used in this study.

them with macrophyte propagules and no attempt was made to manipulate the macrophyte populations to provide between pond uniformity. After filling, each pond was stocked with 10 gizzard shad (*Dorosoma cepedianum* Lesueur), 50 bluegill (*Lepomis macrochirus* Rafinesque), and 20 channel catfish (*Ictalurus punctatus* Rafinesque). Mean total length values for the introduced gizzard shad, bluegill, and channel catfish were 15.2, 5.1, and 9.4 cm, respectively.

In June of 1981 atrazine concentrations of 20 or 100 μg/l were established in selected ponds using either reagent or commercial grade (CO-OP liquid, 41% active ingredient, EPA registration number 1990–81) atrazine. One pond received a 200 μg/l equivalent combined addition of atrazine. In May 1982, after no difference was detected between the water quality and biological pond responses to the different herbicide grades, all ponds were converted to reagent grade to reestablish original concentrations and to establish new higher treatment levels which were added to the study. The two 20-μg/l commercial grade ponds were converted to new 500-μg/l reagent grade ponds and the 100-μg/l commercial grade pond and the 200-μg/l combined treatment pond became the 200-μg/l reagent ponds. This provided

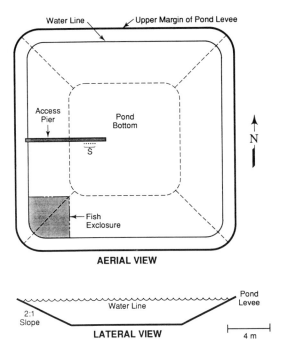

Figure 2. Diagrammatic views (aerial and lateral) of one of the experimental ponds used in this study. The relative placement of artificial substrate samplers (S) is indicated within the aerial view of the pond.

two replicate ponds each for reference, 20-, 100-, 200-, and 500-µg/l treatments in 1982 (Figure 2). In May and June of 1983, sufficient reagent grade atrazine was added to each pond to reestablish their 1982 concentrations.

In May 1982, 4 grass carp (*Ctenopharyngodon idella* Valenciennes) were added to each pond (stocking rate = 88 fish per ha) to control macrophyte abundance and to allow assessment of the interaction of this herbivorous fish and the herbicide on pond macrophytes. At the time of stocking, the total length of grass carp individuals stocked varied from 18 to 23 cm. Thus the 1982 fish community was comprised of four species. In order to assess the effects of atrazine alone upon the macrophytes, it became necessary to enclose a portion of each pond and exclude the grass carp from that area. This exclosure was created by fencing off approximately 7% of each pond using 2.5-cm mesh poultry fence. The exclosures were constructed in the southwest corners of all the ponds (Figure 2). This area was selected after considering the macrophyte sampling problems associated with dense stands of cattail, which tended to be less dense in the southwest corners.

Experimental Design

The final experiment design adapted to investigate the effects of atrazine on pond mesocosm ecosystems was a simple 2 × 4 design (200-µg/l ponds were excluded) consisting of replicated pond treatments that were sampled over three consecutive summer periods (1981 to 1983). However, this chapter will address only those data collected during the final study period, May 23 to October 11, 1983.

Pond volumes (depth) were equalized at approximately 2 m (maximum depth) with the annual (May) reestablishment of treatment-level concentrations of atrazine but were allowed to vary within each annual study period. Depth only recently has been recognized to influence seasonal atrazine concentrations and ecosystem structure.[4] Grass carp populations were lost in 2 of the 8 experimental ponds (a reference and 500-μg/l pond) used in this macroinvertebrate study, presumably in 1982. Data presented by Huggins[4] suggested that the loss of grass carp within the reference pond occurred between August 4 and September 2, 1982. Fish population losses recorded in 1983 were incorporated as zero biomass estimates for affected ponds in modeling efforts.

MATERIALS AND METHODS

Macroinvertebrates

The macroinvertebrate sampling program utilized multispaced, multiple plate artificial substrate samplers. This sampling method was thought to be minimally destructive or disruptive to the pond environment. The Fullner[42] modification of the Hester-Dendy multiplate sampler[43] was selected for use because of its wide acceptability in water quality studies[44] and availability. A full discussion of the macroinvertebrate method is given by Huggins.[4]

Macrophytes

The relative abundance of submergent macrophytes was measured periodically through most of the growing period in all years. A variation of the grapnel analysis technique[45] as used by Carney[32] was adopted for sampling the submerged macrophyte community. Both 1982 and 1983 submergent macrophyte communities were primarily composed of two species of *Potamogeton* L. and *Najas quadalupensis* (Spreng.) Magnus. In addition the macroalga *Chara* L. was included in this group because it functions more like an aquatic vascular plant than an alga.[46]

A second sampling method was developed to provide estimates of the emergent macrophyte (*Typha* spp.) populations found within each pond. Photointerpreted estimates of the percent of shoreline development with *Typha* spp. were made using photographs taken each summer for 1982 and 1983. A series of photographs (mostly 35-mm slides) were examined to obtain visual estimates of both the relative density and percent of shoreline occupied by *Typha* along each of the four pond sides including the fenced margins. A full description of methods utilized to estimate macrophyte abundance can be found in Carney[32] and Huggins.[4]

Fish

Ponds were drained during the latter part of October (1983) and the fish community in each pond was directly assessed. Fish were sorted to species in the field, preserved in 10% formalin, and returned to the laboratory for enumeraton and measurement of growth responses. All fish weight estimates were for preserved, wet weights and measured to the nearest gram.

Water Quality, Phytoplankton, and Zooplankton

Data for these ecosystem components were obtained from other researchers and studies.[3,35,47-49] The methods and materials utilized by these researchers can be found in the existing literature and were summarized by Huggins.[4] Typically, water for all analyses except *in situ* measurement of dissolved oxygen was collected by a column sampler operated from the end of the access pier between 0830 and 1000 hours. Samples were brought immediately into the laboratory for analysis. In general, most analytic methods followed methods described in *Standard Methods*.[50]

Phytoplankton and Zooplankton

Phytoplankton pigment analyses were performed on pond water samples according to Lorenzen[51] and Strickland and Parsons.[52] Corrected chlorophyll *a* estimates were taken to represent estimates of pond phytoplankton biomass. Zooplankton samples were collected between 2200 and 2400 hours with a 60-μm mesh, 94-cm long net with a diameter of 29 cm. For this study only Cladocera and Copepoda were enumerated and no nauplii were included in the counts. Counts were made in a Wild® counting wheel with a Zeiss compound microscope. Zooplankton were identified according to Edmondson.[53]

Pond Parameters

Pond depth recordings, which corresponded to water chemistry and plankton sampling events, represented maximum pond depths obtained from the end of the pier. As all mesocosms used in this study were of uniform construction, only pond depth and the associated changes in shoreline length and pond volume were considered as variables.

Analysis and Statistics

Relative abundance and/or biomass measures were utilized in assessing vertebrate, macrophyte, and macroinvertebrate responses with pond and treatment conditions. In addition, a number of derived macroinvertebrate variables representing various population and community attributes were analyzed for treatment and pond effects. All water quality variables were analyzed as originally measured except for dissolved oxygen values which were converted to percent saturation values.

Table 1. Macroinvertebrate Trophic and Habitat Classification System Developed For Use in This Study

Trophic Category	Dominant Food
Predators	Living animal tissue
Herbivores	Living plant (algae or macrophytes) tissue or fluids
Omnivores	Living plant and animal tissues
Detritivores	Decomposing fine or coarse particulate organic matter including some living or senescing algae cells

Habitat Category	Primary Habitat and Habit
Epibenthic	Occurring on but not in bottom substrates
Embenthic	Occurring in bottom substrates
Epibenthic/Embenthic[a]	Occurring on or in bottom substrates
Epiphytic	Occurring on or in plants
Epiphytic/Epibenthic[a]	Occurring on plants and bottom substrates

[a] These categories are composite groupings which reflect the diverse habitat requirements of those organisms comprising that category.

Note: Organisms were assigned to the trophic category that best reflected the major component (on a biomass or volume basis) of its diet. A single larva of the midge genus *Chaoborus* was classified as a predator with pelagic habits but was not included in any trophic or habitat guild categories.

Taxonomic unit (e.g, genus, species, family) data collected from the artificial substrate samples were reclassified into trophic (a functional property) and habitat categories for further analysis. Taxa were assigned to each of these categories by consulting available life history and ecological literature for each taxa or invertebrate group and/or examination of gut contents from study organisms. Four trophic and five habitat classification categories were developed for further analyses of invertebrate communities found in each pond (Table 1). The same approach was followed when establishing habitat categories; however, taxa that exhibited diverse habitat preferences were placed in composite categories. Abundance estimates for each trophic and habitat guild were used to represent these guild indicators in some of the ecosystem models developed in this paper.

The overall functional organization of invertebrate communities was further described by examining that portion of the community that was composed of "specialists" or "generalists" macroinvertebrates. Operational definitions for both groups (specialists, generalists) were formulated by merging specific trophic and habitat categories (Table 1). Specialists represented organisms that were epiphytic, epibenthic, or embenthic in occurrence and were predators or herbivores. Generalists were then considered to be only those invertebrates classified as epiphytic/epibenthic or epibenthic/embenthic organisms and were also characterized as omnivores or detritivores. Over 46% of the taxa (44 taxa) were not categorized as either generalists or specialists by the definitions used in this study because they were specialists (or generalists) in trophic requirements and generalists (or specialists) in habitat selection. This approach identified, in a conservative manner, the extremes in functional organization (specialists, generalists).

Originally, Huggins[4] analyzed these mesocosm data utilizing a wide array of graphic, clustering, and linear model analyses (e.g., Model I and II ANOVA, simple and multiple linear regression), including path analysis.[54] These results and models provided the fundamental knowledge base and model concepts used in the development of models from the structural equation modeling technique (LISREL) utilized in this paper. LISREL is a path analytical technique (e.g., References 55 to 57) that differs from classical path analysis by utilizing maximum likelihood estimation techniques to establish path coefficients rather than least squares regression techniques. This structural equation model approach incorporates measurement error into the modeling process by separating latent theoretical concepts from their measured indicator variable.[40,58,59] This and other modeling characteristics make this technique an extremely powerful method when dealing with causal modeling.[40] It must be remembered that LISREL is not a primary method for discovering causes, but a method applied to causal models formulated by the researcher based on knowledge and theoretical considerations. In cases in which the causal relations are uncertain, this method can be used to find the logical consequences of any particular hypothesis in regard to them. The following results are intended to illustrate the usefulness of this technique and quantify the existence of the effects of atrazine on mesocosm ecosystems.

Ecosystem models presented in this chapter are based on only two assumptions: (1) ecosystem structure is defined by perceived ecosystem concepts measured by one or more indicator variables, and (2) ecosystem function is represented by the regulatory links between concepts (see References 59, 60). Thus, structural elements of the models were joined by functional pathways determined to be significant "causal" linkages. Numerical analysis (e.g., determination of path coefficients, tests of significance) were utilized in these modeling efforts to quantify model concept relationships and to examine the effects of atrazine and other variables in the ecosystem. It is from the variable set used by Huggins[4] and his hypothesized ecosystem models that the following structural equation models were developed.

Some model concepts were identified that were characterized by multiple indicator variables.[40,59] Concepts often constitute theoretical constructs that are not directly measurable, but represent necessary aspects of the system. The relationship between emergent and submergent macrophytes and their combined usage as a total macrophyte variable exemplifies the existence of a concept (the macrophyte community). This concept exists despite the fact that emergent and submergent macrophyte variables apparently effect various ecosystem variables differently.[4]

The results of five structural equation analyses involving different macroinvertebrate variables or sets of variables are presented. All models and included paths (unless otherwise indicated) were determined to be statistically significant. Models were thought to reflect the direct and indirect impacts of atrazine on different macroinvertebrate attributes as well as indicate the general relationship of atrazine exposure on the whole pond ecosystem (see Reference 59 for explanation of tests of significances). Although not essential for numerical analysis, path diagrams and the ecosystem models they represent are useful for displaying the pattern of causal relations among the sets of concepts. Therefore, the models are presented as path diagrams with corresponding path coefficients. The direction of causal flow of each path was indicated utilizing an arrow-pointed line (path) from "cause" to "effect"

Table 2. Summary of Indicator Variables and Concepts Which Comprised the Structure of Various Ecosystem Models

Indicator Variable	Unit of Measure	Concept	Unit of Measure
Exogenous			
Grass carp biomass	Grams	Grass carp	Grams
Atrazine concentration (water column)	µg/l	Atrazine	µg/l
Maximum pond depth	Centimeters	Pond depth	Centimeters
Endogenous			
Gizzard shad biomass	Grams	Gizzard shad	Grams
Bluegill biomass	Grams	Bluegill	Grams
Emergent macrophyte shoreline development	Percent		
Submergent macrophyte cover	Percent	Aquatic vegetation	Percent
Chlorophyll *a*	µg/l	Phytoplankton	µg/l
Zooplankton abundance	Count	Zooplankton	Count
Macroinvertebrate total abundance	Count	Macroinvertebrates	Count
Embenthic macroinvertebrate abundance	Count	Embenthos	Count
Epibenthic macroinvertebrate abundance	Count	Epibenthos	Count
Epiphytic macroinvertebrate abundance	Count	Epiphytes	Count
Macroinvertebrate predator abundance	Count	Predators	Count
Macroinvertebrate detritivore abundance	Count	Detritivores	Count
Macroinvertebrate herbivore abundance	Count	Herbivores	Count
Macroinvertebrate omnivore abundance	Count	Omnivores	Count
Macroinvertebrate generalist abundance	Count	Generalists	Count
Macroinvertebrate specialist abundance	Count	Specialists	Count
Parachironomus spp. abundance	Count	*Parachironomus*	Count
Dicrotendipes cf. *simpsoni* abundance	Count	*Dicrotendipes*	Count
Caenis spp. abundance	Count	*Caenis*	Count
Glyptotendipes spp. abundance	Count	*Glyptotendipes*	Count
Endochironomus nigricans abundance	Count	*Endochironomus*	Count
Hyallela azteca abundance	Count	*Hylallea*	Count
Bezzia/Probezzia/Palpomyi morphogroup abundance	Count	*Bezzia*	Count

Note: All variable and concept values except pond depth values were utilized in these models as transform values (i.e., log of measurement value + 1).

concepts. Path coefficients indicate direction of change (sign) in the appropriate unit of measure for the effect concept (see Reference 59). Table 2 summarizes the relationship between indicator concepts, variables, and units of measure. A measure of ecosystem stability (stability index value = SI) was determined for all models and is listed with each ecosystem figure. Stability index values greater than one indicate unstable ecosystems (see Reference 59).

Concepts are separated into exogenous and endogenous concepts. Exogenous concepts are those that were known or found[59] to act only as causes and are never affected by other concepts in the model. Endogenous concepts can act as causes, and therefore affect other endogenous concepts, and can themselves be affected by both exogenous and other endogenous concepts. The endogenous concepts are considered to be the structural components of the ecosystem, and the pathways that connect these concepts represent the functional relationships between concepts. The exogenous concepts are considered to be the external disturbance factors of the ecosystem; they are aspects of the ecosystem that are controlled by factors outside the ecosystem (e.g., climate, geology, anthropogenic influences). In the models that are developed in this chapter, there are three exogenous concepts and five endogenous concepts common to all models. Grass carp, atrazine, and pond depth were determined to be the external disturbance factors. Grass carp is a somewhat unusual exogenous concept, as biological features usually are endogenous. However, because of the time frame of the experiment, no effects of any concept on grass carp were detected, which necessitated its inclusion as an exogenous concept (see Reference 59 for a more detailed explanation).

RESULTS

In identifying the ecotoxic effects of atrazine on macroinvertebrates and its impact on ecosystem structure, Huggins[4] examined, in detail, atrazine's relationship with 60 macroinvertebrate, 12 vertebrate, 6 macrophyte, 6 water quality, 3 plankton, and 2 experiment pond variables. The variable state for some of these mesocosm variables can be seen in graphical form in deNoyelles et al.[61] Overall, mesocosm variables of interest appeared to be responding directly to atrazine concentrations and/or other ecosystem variables which may or may not have been linked to atrazine exposure. The initial findings of Huggins[4] suggested that the effects of atrazine on macroinvertebrate variables, as identified by a time series of one-way ANOVAs and corresponding multiple range tests, varied both in time and space (e.g., change from positive to negative effects through time, or varied response on a date due to atrazine treatment levels, or both) (Table 3). These results strongly indicated that atrazine as an exogenous ecosystem variable could elicit an apparent negative effect (direct affect?) or atrazine exposure could lead to a positive effect which could only be determined as an indirect or secondary effect resulting from induced ecosystem change.

This hypothesis was supported by the simple and multiple regression models developed by Huggins.[4] Huggins, utilizing all available ecosystem attribute variables, constructed a series of regression models in an attempt to explain the role of atrazine in determining various ecosystem relationships. These models were considered only partially successful in clearly identifying both the direct and indirect

Table 3. Datewise Summary of Treatment Level Positive or Negative Ecotoxic Responses For Selected Macroinvertebrate Variables Measured as Abundances When Compared to Reference Conditions

Treatment (µg/l)	May 23			Jun 13			Jul 5			Jul 25		
	20	100	500	20	100	500	20	100	500	20	100	500
Caenis spp.	0	0	0	0	0	0	−	+	−	0	+	0
Endochironomus nigricans	0	0	0	−	−	−	0	0	0	0	0	0
Total abundance	−	−	−	−	−	−	0	0	0	0	+	+
Epiphytes	0	0	0	−	−	−	0	0	0	0	0	−
Detritivores	−	−	−	0	0	0	0	0	0	0	+	0
Generalists	−	−	−	−	−	−	0	0	0	0	+	+

Note: Nonsignificant ANOVA results ($p > 0.00625$) and/or Newman/Keuls range tests showing no specific treatment level difference when compared to reference conditions were recorded as zero responses (0). Abundance increases were noted as positive ecotoxic responses (+) and decreased abundance values were recorded as negative responses (−). Only the first four of eight datewise responses are shown (Huggins, 1990).

effects of atrazine on ecosystem components.[4] Regression models revealed relationships that clearly indicated the presence of many indirect effects resulting from exposure to atrazine and the effects of other factors on the variables of interest. For example, Huggins found that the model for the total abundance (transformed values) of the macroinvertebrate omnivore, *Hyallela azeteca,* included atrazine concentration as a positive independent variable and pond depth as a negative variable. It was assumed that atrazine did not provide some direct chemical stimulus that allowed *Hyallela azteca* populations to increase, but it was possible that pond depth (or pond volume) could be limiting this organism's abundance. The multiple regression model for bluegill abundance (log transformed values) included emergent macrophyte and phytoplankton abundance as positive independent variables, and submergent macrophyte abundance and turbidity were included as negative factors.[4] Again, this equation suggested strong indirect effects resulting from atrazine exposure were affecting this ecosystem attribute.

In an attempt to quantify both the direct and indirect effects of atrazine on various ecosystem endpoints, Huggins[4] employed a path analytic approach (path analysis) to modeling the structural component and functional relationships of these experimental ecosystems. Huggins[4] was able to identify the structure of these ecosystems and demonstrated the existence of significant pathways of direct and indirect effects, but limitations of this technique compromise its general usefulness in ecological studies. Path analysis[54] is limited to recursive models where causal flow in the model is limited to unidirectional pathways. That is, at a given point in time, a variable cannot be both a cause and an effect of another variable. This may not be realistic in ecosystem modeling. This and other path analysis assumptions[62] can severely limit its application to ecosystem studies. The recognition of these shortcomings lead to the search for and proposed use of a better ecosystem-level analytic approach — structural equation modeling.[40] The following results are based on the hypothesized ecosystem structure and function proposed by Huggins[4] and established by LISREL, the structural equation modeling technique proposed by Johnson et al.[40,59]

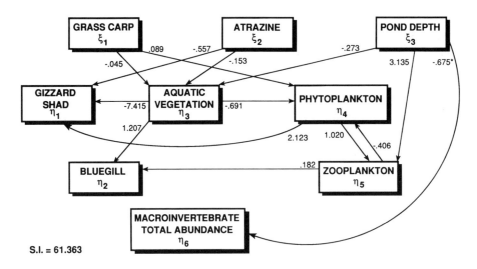

Figure 3. Macroinvertebrate Model. Ecosystem model and path coefficients that represent the amount of change in an "effect" concept resulting from the change of a single unit of measure change in the causal concept. All paths, as well as the model, were determined to be significant. The SI value (stability index value) for the model is also indicated.

Presented ecosystem models represent various "end point" aggregates of the macroinvertebrate concepts. As all models are similar in inclusion of exogenous concepts and their direct effects on certain endogenous concepts a simple model reflecting the macroinvertebrate, abundance response is presented first. Other discussions of models will not reiterate in any detail those model elements that were determined to be common model concept relations. Rather, the explanations of other ecosystem models will focus primarily on the examination of paths and concept relations to the new macroinvertebrate concepts.

Macroinvertebrate Abundance Model

The ecosystem model that includes the macroinvertebrate concept indicated by total abundance, consists of three exogenous and six endogenous concepts. In an attempt to present as comprehensive a model as available data allow and to represent all models in somewhat of a standardized format, many concepts are left in the models regardless of their relationship with the macroinvertebrate concept(s) of interest. Common to all models is the recognition and inclusion of the same three exogenous concepts (Figures 3 to 7). Pond depth, biomass of grass carp, and the concentration of atrazine in the water column were found to be of great importance in determining or controlling these ecosystems. All exogenous concepts appeared to function, in part, as habitat modifiers which either directly or indirectly "drove" the entire ecosystem.

Significant paths exist from all three exogenous concepts to aquatic vegetation (macrophytes) in all models. All path coefficients are negative, indicating that increased values of exogenous concepts suppress aquatic vegetation by an amount specificed by the path coefficient. The strong negative relationships between the herbicide (atrazine), herbivorous fish (grass carp), and aquatic vegetation most certainly reflect the presence of direct toxic or foraging effects, respectively, on macrophyte abundance (cover). The direct effects of these exogenous concepts on aquatic vegetation is not unexpected and is documented in the literature (e.g., References 33, 63, 64). In addition, increases in pond depth reduced aquatic vegetation directly. Since hydrologic regime is a principal factor in determining the distribution and composition of aquatic vegetation, it follows that fluctuations in this regime (e.g., charging water depth) could affect vegetation abundance (see Reference 65 for discussion of vegetation dynamics). It is hypothesized that depth, measured as the physical change in maximum pond depth, represented an associated change in pond surface area and light absorption. Thus, increased depth in concert with increases in surface area (or volume) and reduced light penetration at deeper depth were, in all probability, reducing aquatic vegetation cover. It has been suggested that in shallow lakes, increases in water depth may lead to more powerful waves which could prevent and/or reduce aquatic macrophyte abundance in regions of the lake susceptible to wave action and high turbulence (e.g., Reference 66).

Grass carp had a direct positive effect on phytoplankton (chlorophyll *a* values) in all models. It was hypothesized that the feces and macrophyte fragments generated by the feeding activity of these fish on aquatic vegetation were a source of nutrients that became available to the phytoplankton. Various authors have suggested that grass carp feeding could lead to nutrient enrichment and consequenty phytoplankton increases (e.g., References 67 to 70). Prowse[67,71] reported that when grass carp were used to control macrophytes in small ponds, algal blooms developed from nutrients released from fragmented vegetation and decaying feces. Oposzynski[72,73] found that aquatic ecosystems with grass carp displayed increased turbidity values, and high stockings of this fish resulted in increased eutrophication and primary production (planktonic) as macrophyte communities were reduced by grass carp. Stanley[74] was able to demonstrate in aquarium experiments that this fish increased the amount of nitrogen and phosphorus available for plant growth. Other studies failed to support these observations and suggested that grass carp additions to ponds did not promote chlorophyll concentrations and/or primary production increases.[63,75,76] Most of these authors did note that grass carp introduction lead to increased pond turbidity. The studies by Hestand and Carter,[77] Terrell,[78] and Leslie and co-workers[63] suggested that grass carp waste did not directly promote heavy phytoplankton blooms but did not indicate whether general increases in phytoplankton abundance were correlated with grass carp abundance. While the mechanism of the direct effects of grass carp on phytoplankton (chlorophyll *a*) remains unclear, we are able to demonstrate the presence of indirect effects of grass carp on phytoplankton through aquatic vegetation (Figures 3 to 7). Concurrently all exogenous concepts were indirectly promoting phytoplankton increases through the direct reduction of macrophytes (aquatic vegetation concept). These models recognized that increased aquatic vegetation was a cause of reduced chlorophyll *a* values (phyto-plankton concept). This negative relation could have been caused by shading and reduced light penetration resulting from extensive aquatic vegetation canopy. Habitat changes resulting from expanded aquatic vegetation cover and competitive inter-actions with these macrophytes and/or their ephiphytic algae communities (unmea-sured concept) may also have contributed to this negative relationship. Wetzel[68] concluded the inhibition of phytoplankton by submergent macrophytes could result from competition for nutrients, light, and other substances. He also noted that elevated pH values of the water within actively photosynthesizing submergent mac-rophyte beds reduced the rates of phytoplankton productivity. The production and excretion of inhibitory organic compounds by macrophytes that function in an antibiotic way on the growth of phytoplankton has been suggested by some studies (e.g., Reference 79). Regardless of the specific causes linking aquatic vegetation and phytoplankton concepts, it is obvious that atrazine was affecting chlorophyll *a* values indirectly.

The zooplankton concept (based on abundance) was found to be responding directly to the exogenous concept, pond depth. Zooplankton densities decrease with decreasing pond depth (and volume), and could be indicative of a general physical habitat shift from planktonic conditions favorable to zooplankton (e.g., certain Cladocera and Copepoda) to a shallow-water pond habitat less suitable for open-water microcrustacean populations but perhaps more advantageous to the devel-opment of littoral or benthic-dwelling microcrustaceans. The paths and path coef-ficients between phytoplankton and zooplankton concepts illustrate the interdepen-dent relationships existing between these planktonic primary producers and their consumers. As modeled, these pathways indicate that a single zooplankton indi-vidual (number per liter) can cause a decrease of 0.39 μg/l of chlorophyll *a*, which leads to an increase of 1.3 individuals in the number of microcrustaceans per liter. Thus it appears that the phytoplankton concept is controlling the zooplankton concept as measured by density and/or concentration.

Only two of four fish concepts (bluegill and gizzard shad) were successfully integrated into these models as endogenous concepts. The direct effect of atrazine probably is the result of chronic toxicity and is evident for both fish concepts. Increases in the concentrations of atrazine leads to reduced biomass of fish. The indirect effects of exogenous concepts on these fish are different and dependent, in part, on habitat preferences of affected fish species. Gizzard shad are pelagic "filter feeders";[80] thus, reductions in aquatic vegetation contributed to more open water pond conditions favorable to successful reproduction, feeding efficiencies, and survival of this fish, all of which contributed to biomass gains. Exogenous concepts acting as habitat modifiers or disturbances were creating open water habitat for this fish and its primary food resource (phytoplankton). Thus, exposure of this ecosystem to atrazine concentrations produced at least two secondary effects that directly contribute to increased gizzard shad biomass.

Similar secondary effect linkages (e.g., atrazine to aquatic vegetation to phy-toplankton to zooplankton) suggest that zooplankton abundance might be increased by any or all exogenous concepts which, in turn, lead to increases in bluegill biomass (Figures 3 to 7). Conversely, because of the presence of a positive path coefficient between the aquatic vegetation and bluegill concepts, these same exogenous concepts are indirectly suppressing bluegill biomass. Bluegill's affinity to vegetated habitats is well known with larval and juvenile fishes displaying a strong preference for limnetic habitats dominated by macrophyte growth (e.g., References 81, 82). In these mesocosms, increases in bluegill biomass are linked primarily to recruitment through reproduction and survival of juvenile bluegill.[4] The aquatic vegetation concept utilized in these models is measured by multiple indicators (emergent and submergent macrophyte abundance) while Huggins was only able to incorporate the measures as separate variables; a 1983 emergent macrophyte abundance variable displaying a positive relationship to bluegill abundance and a 1983 submergent macrophyte variable have a negative impact on abundance. The ramifications of these study differences remain unclear, but it is assumed that the positive relation between bluegill and aquatic vegetation concepts of these LISREL models reflects the overall positive effect of vegetation on bluegill young and thus total biomass. These findings are further supported by the dynamic trophic model and field data of Wiley et al.[83] that projects a positive linear increase in sunfish production with an increase percent in the plant cover.

The total abundance of macroinvertebrates (macroinvertebrate concept) is affected by a single causal concept, pond depth, such that increases in pond depth cause decreases in abundance of macroinvertebrates (Figures 3 to 7). These relationships reflect the general trend toward reduced macroinvertebrate (especially aquatic insects) diversity and numbers with increased depth.[84] While Brinkhurst discussed the characteristic change from "littoral" to "profundal" benthos communities as they relate to lakes, his general observations concerning reductions in benthos in deep water probably apply to smaller lakes and ponds. Pond depth as a concept (and those factors associated with depth changes) may simply represent those conditions that commonly characterize shallow and deep water environments. The literature generally suggests that shallower lakes support higher rates of secondary production (e.g., References 85 to 88). Additionally, in these experimental ponds, decreased pond depth proportionally decreases pond volume which may on the short-term tend to concentrate organisms and elevated population densities.

Habitat Guild Model

A habitat guild model was identified (Figure 4) which includes organisms classified as epibenthos, embenthos, and epiphytic macroinvertebrates. This model revealed that atrazine concentrations had no direct effect on the macroinvertebrate habitat guilds of these mesocosms, but indirect effects resulted in changes in all guild categories. Pond depth directly affected epibenthos and was the only exogenous concept that directly affects a habitat guild category. This relationship was negative, indicating that as mesocosm depth decreases epibenthos abundance increased. This response is most clearly relatable to those unmeasured factors associated with pond depth that enhance bottom habitat conditions. Certainly epibenthic

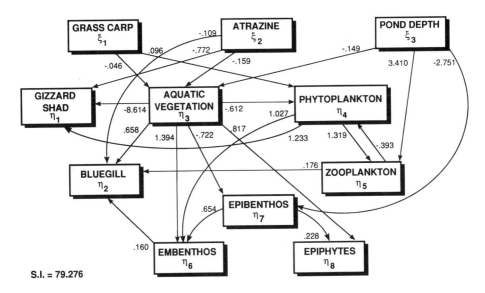

Figure 4. Habitat Guild Model. Ecosystem model and path coefficients that represent the amount of change in an "effect" concept resulting from the change of a single unit of measure change in the causal concept. All paths, as well as the model, were determined to be significant. The SI value (stability index value) for the model is also indicated.

organisms occurring on the substrate surface may best benefit from environment changes associated with decreases in pond depth. Shallow water may provide conditions favorable to an increase in food resource for epibenthos.[89] In addition, shallow water conditions promote bottom substrate heterogeneity that could support a more diverse and abundant epibentic community. Conversely, macroinvertebrates living either burrowed into the substrate or living on macrophytes were afforded some degree of independence from substrate surface conditions.

In general the abundance of epiphytic macroinvertebrates appears to be most influenced by the amount and morphological features of aquatic vegetation.[90-92] Aquatic vegetation provides these organisms with a complex multidimensional habitat that is structurally independent from bottom substrate area and condition. Aquatic vegetation directly affects all macroinvertebrate categories. Epibenthos was negatively affected as rooted macrophytes occupied more pond bottom surface while epiphytes responded directly to increases in their habitat resource. In addition, embenthos abundance was linked to increased aquatic vegetation which may be related to increased oxygenation of subsurface sediments as a result of oxygen production and gas leakage from roots and rhizomes. These rooted macrophytes probably help stabilize bottom substrates by reducing bottom currents and turbulence produced by wave action.

This model suggests that increases in epibenthos populations (abundance) caused concurrent increases in both epiphytes and embenthic populations. These relationships were most likely associated with interspecific interactions assumed to be

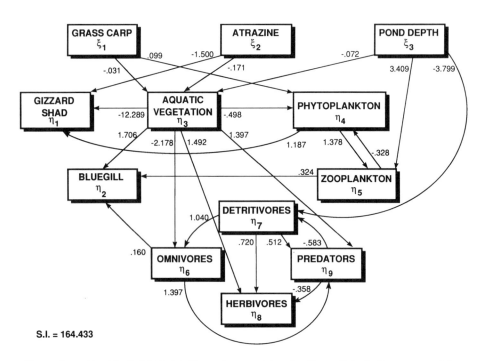

Figure 5. Trophic Guild Model. Ecosystem model and path coefficients that represent the amount of change in an "effect" concept resulting from the change of a single unit of measure change in the causal concept. All paths, as well as the model, were determined to be significant. The SI value (stability index value) for the model is also indicated.

present between habitat guild members. No specific mechanism(s) can be offered for these observed relations. This model supported the premise that bluegill are primarily dependent on embenthos as a food resource such that a significant positive pathway was found between these concepts (Figure 4). Kettle et al.[36] found that the embenthic mayfly, *Hexagenia* spp., was the largest and most frequently represented food item in the diet of bluegill from these same experimental ponds utilized in a prior study using 0, 20, and 500 µg/l atrazine treatments.

While all previous exogenous and endogenous concepts remain consistent in this model, one exception should be noted. A direct, negative pathway was established between atrazine and bluegill suggesting a toxic effect resulting from herbicide exposure in these mesocosms. However, this pathway is not present in all models (e.g., Figure 3 and 5).

Trophic Guild Model

A macroinvertebrate trophic model was produced that in many respects was very similar to the habitat model previously presented. The primary change, excluding macroinvertebrate relationships, was the loss of the direct atrazine effect on bluegill (Figure 5).

Only a single macroinvertebrate concept was influenced by an exogenous concept. Pond depth changes are inversely related to detritivore abundance. In these ponds, as pond depth decreases during the study period, detritivore abundance increases. Again, it is hypothesized that pond bottom conditions are the causal factors and are being represented by the pond depth concept. In addition, increasing pond depth may negatively affect depositional processes. Regardless of the exact cause(s), pond depth as a concept provides a measurable and inclusive factor that represents an important exogenous concept often unrelated to the original experimental design and atrazine treatments. The most logical relationships were the negative relations between detritivores, herbivores, and predators which were interpreted as prey/predator linkages. Similarly omnivore population increases were a significant source of prey and resulted in increased predator abundance. Increases in detritivore abundance are linked to increases in all other trophic concepts. Increased detritivore abundance resulted in increases in omnivores and herbivores which are most probably related to their processing of detritus to more available and different food resources either directly (e.g., reduced food particle size by feeding activities) or indirectly (e.g., nutrient release and uptake by algae). It is speculated that processed detritus resulted in the release of nutrients that promote benthic plant growth which was then utilized by some herbivores. Omnivores also benefited from detritivores as detrital materials were processed to a form more usable by omnivores. As no measure of detritus or associated benthic algal communities is available, the effects of these unmeasured concepts cannot be integrated into this or other models but are thought to be part of the cause(s) involved in the observed macroinvertebrate trophic interactions. The observed relationships among macroinvertebrate trophic groups suggested that these groups are truly interspecifically connected and were real concepts. Both omnivore and herbivore are affected by aquatic vegetation. Increases in aquatic vegetation resulted in increases in herbivore numbers and decreases in the abundance in omnivores. Herbivore abundance is positively influenced by the presence of macrophytes as a habitat, a direct food source, and as a substrate for epiphytic algae which is also utilized by herbivorous macroinvertebrates. The abundance of omnivores was negatively influenced by increased aquatic vegetation. The interpretation of the functional association between these concepts is complicated by the fact that the concept of aquatic vegetation is conversely a measure of the open water habitat of these mesocosms. It may be that aquatic vegetation directly suppresses omnivore numbers and/or that omnivores as a group respond positively to open water conditions. As pointed out by Huggins,[4] the scud *Hyallela azteca* was the most abundant organisms in this guild category, and its general scavenging habits and its rather broad habitat requirements[93] may allow this organism to more readily exploit an open pond system. However the relationship between the aquatic vegetation concept and omnivores are, no doubt, much more complex than what has been suggested and probably involves microclimate and habitat changes occurring within vegetation cover.

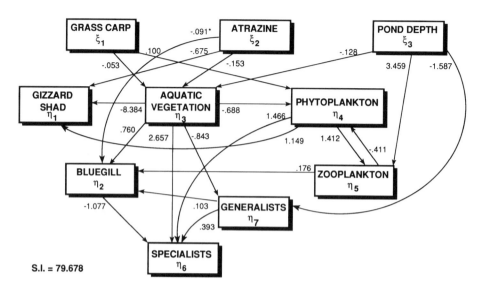

Figure 6.   Specialist/Generalist Model. Ecosystem model and path coefficients which represent the amount of change in an "effect" concept resulting from a single unit of measure change in the causal concept. All paths, as well as the model, were determined to be significant, except the path between atrazine and bluegill ($p \approx$ 0.06). The SI value (stability index value) for the model is also indicated.

Specialist/Generalist Model

Another model comprising hierarchical macroinvertebrate concepts is the specialist/generalist model (Figure 6). Interactions between exagenous concepts and most endogenous concepts representing the fish, plant, and plankton components of these ecosystems are similar to other models. While the path between atrazine and bluegill is not significant, its presence in one of the other models (Figure 4) suggests that this causal path might be of biological importance in helping explain bluegill biomass responses within these mesocosms. As with all other models, pond depth caused a negative effect on at least one of the macroinvertebrate concepts. The generalist concept is primarily composed of bottom-occurring organisms with generalized food habits (e.g., omnivores, detritivores), most of which have already been shown to be negatively influenced by pond depth (see Figures 4 and 5). This same concept acting through its negative relationship with aquatic vegetation and the presence of a direct negative effect of vegetation on generalists was shown to have a positive indirect effect on generalist abundance. Specialists, many of which are epiphytic organisms, did not respond directly to decreases in pond depth and the direct reduction in benthic habitat as shallow littoral areas are lost to the retreating shoreline. The specialist's dependence on plant communities (aquatic vegetation concept) is evident from the composition of organisms (e.g., herbivores, epiphytes) aggregated within this category and the positive cause/effect path between these concepts. However, several more complex indirect path linkages involving all concepts except gizzard shad were noted to affect the abundance of specialist. Most noteworthy is the existence of the pathways from atrazine and other exogenous concepts through aquatic vegetation which either directly, through this concept, or indirectly, through other linkages (e.g., phytoplankton, bluegill, or phytoplankton through zooplankton to bluegill), impose various specialist responses. In tracing the individual pathways or indirect effects of atrazine on specialists which always involve aquatic vegetation concept, it was observed that atrazine is responsible for both increases and decreases in the abundance of macroinvertebrate specialists. The simplest example of the indirect effects of atrazine on specialists is through its reduction of aquatic vegetation which has a positive causal relation with this macro-invertebrate concept. Conversely atrazine suppressed aquatic vegetation which, in turn, allowed increases in phytoplankton, a concept that directly caused specialist abundance to increase, presumably through increased food resources, to herbivores and other specialist organisms that benefits directly from algae biomass increases. Concurrently the phytoplankton increases resulting from atrazine exposure increased zooplankton densities which prompted increases in bluegill biomass that suppressed specialist numbers (Figure 6). As a result of these complex causal pathways, the total (or sum) effect of atrazine on specialists cannot be estimated from this model and path coefficients.

In addition to the complex interactions occurring between the specialist concept and the exogenous concepts, generalists are also directly and indirectly causing changes in specialists. The positive relation between generalists and bluegill (prey/predator interaction) and the negative effect of bluegill on specialists results in an indirect negative effect of generalists on specialists. The direct relationship between these two macroinvertebrate concepts is positive suggesting that generalists stimulated increases in specialists. This relationship is again considered the cumulative effect of numerous interspecific relationships linking these macroinvertebrate aggregate categories.

It was anticipated that if the specialist concepts truly represented organisms with very restricted or "narrow" environmental requirements (conditions), they should be highly connected to other ecosystem variables (highly defined pathways). Conversely, generalists would show limited connection to the entire system (defuse effects with limited or unrecognized effects) and be less vulnerable to stress effects through established pathways. Grassle and Grassle[94] suggested that the ability to survive under pollution stress often depends on opportunistic life histories (generalists). This model illustrated this general theory with specialists directly linked to four endogenous concepts but indirectly associated with all other model concepts except gizzard shad. Moreover, generalists were directly dependent on a single exogenous concept unrelated to atrazine exposure and aquatic vegetation, which was observed to be a "keystone" concept around which the ecosystem functioned. This situation would indicate the atrazine effects on generalists were limited and positive, being expressed only through suppressed aquatic vegetation cover.

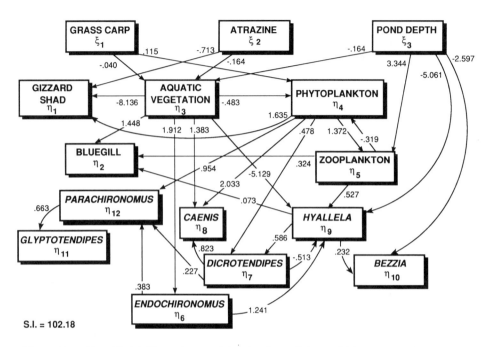

Figure 7. Taxa Model. Ecosystem model and path coefficients which represent the amount of change in an "effect" concept resulting from a single unit of measure change in the causal concept. All paths, as well as the model, were determined to be significant. The SI value (stability index value) for the model is also indicated.

Taxa Model

The final model presented is the taxa model (Figure 7) consisting of all previous nonmacroinvertebrate concepts linked to the seven species or taxa observed to comprise 82% of the macroinvertebrate abundance found to occur in these ponds during the 1983 study period.[4] Only general examination and explanation of the interactions observed in this model are provided. All nonmacroinvertebrate concept interactions were similar to previous models. This model failed to indicate a significant direct effect of atrazine on bluegill which was the prevalent condition noted in previous models. A full discussion of all indirect paths from atrazine to macroinvertebrate taxa is beyond the intent of this paper, but a few major points serve to illustrate the nature of atrazine exposure at the species or taxa level. First, there are no direct effects from atrazine to any taxa. This observation agrees with acute toxicity results reported in the literature for most concentrations tested. However, the indirect effects of atrazine are so numerous and inclusive as to effect all macroinvertebrate taxa. Typically all prior observations of effects, either direct or indirect, on larger macroinvertebrate aggregate concepts (Figures 3 to 6) were also observed within this model. The only exogenous concept to directly effect macroinvertebrate concepts (i.e., *Hyalella* and *Bezzia*) was pond depth. The direct negative effect of increased pond depth or reduced depth as was the general case in these ponds on these organisms was speculated to be related to general habitat requirements and bottom substrate conditions as most taxa were either epibenthic or embenthic in nature. This explanation seems flawed because these taxa were classified in this study and by Huggins[4] as either epiphytic/epibenthic omnivore *(Hyallela)* or embenthic predators *(Bezzia)* and neither the epiphyte or embenthos concepts were linked to pond depth. However, these aggregate macroinvertebrate concepts and their interactions are the result of all membership responses to an effect, and individual responses do vary according to each taxa's life history and environmental requirements. Thus, different aggregate concepts can lead to different observed interactions and interpretation of these responses as pointed out by Johnson et al.[59] Only the endogenous concepts of aquatic vegetation and phytoplankton directly affected macroinvertebrate taxa concepts (Figure 7). All path coefficients were positive except the coefficient between *Hyallela* and aquatic vegetation which has been previously discussed in relation to the omnivore response. All other macroinvertebrate concepts are detritivores or herbivores, which apparently benefit from increased phytoplankton and/or aquatic vegetation cover. In part, these effects were probably related to food resource interactions as algae or algal detritus became more abundant and available to epiphytic organisms (as planktonic and epiphytic algae) and to bottom-dwelling organisms as detrital "rain". Increases in two of the most abundant midge larvae *(Endochironomus* and *Dicrotendipes)* led to increases of the omnivore *Hyallela azteca* as it utilized these food resources as either living organisms or as they scavenged dead or dying individuals.[93,95,96] The direct relationships between the concepts of *Hyallela* and *Bezzia* and *Dicrotendipes,* plus that of *Parachironomus* and *Glyptotendipes,* are more involved and lend little to the recognition of the numerous indirect effects of atrazine on the ecosystem except to further support its subtle tendency to indirectly affect most recognized macroinvertebrate attributes. Unexplained are the many indirect paths that ultimately affected all taxa concepts which were shown to exist.

MODEL SUMMARY

It is clear that the preceding ecosystem models are biologically defensible and illustrate both the simple and complex pathways by which atrazine perturbed these ecosystems. Unknown is the effect(s) of atrazine on the structure of the ecosystems (alteration or elimination of pathways) due to the constraints of the experimental design (see Reference 59 for a discussion of these potential effects). Of concern is the ultimate effect (total effect) of atrazine on various ecosystem concepts as a result of differing direct and/or indirect effects of varying consequence (both positive and negative effects). LISREL provides matrices of direct, indirect, and total effects of the exogenous concepts on endogenous concepts and the endogenous concepts on each other. In support of these ecosystem models, a series of tables (matrix values) were produced that quantify the total, direct, and indirect effects associated with each model (Tables 4 to 7). These tables allow the direct examination and comparison of all effects of interest associated with each model. For example, Table 6 clearly shows that the atrazine had an overall (total) effect on all model concepts except macroinvertebrates (total abundance).

Common to all tables is a range of values for all five endogenous concept responses to exogenous causal factors (concepts). Aside from the actual effects noted in these models, it is of equal importance that the range of predictive effects values associated with the concepts common to all models are very consistent and display low variability (Tables 4 to 7). We measured the variation in total, direct, and indirect effects of the exogenous concepts on the endogenous concepts for those sets of concepts common to all models. The measure was calculated as the difference between the largest effect and the smallest effect, divided by the value of the largest effect. For example, the largest total effect of atrazine on phytoplankton is 0.075 for the Macroinvertebrate Model (Figure 3) which is based on total abundance and the smallest total effect is 0.059 (Trophic Model, Figure 5), giving a 21% variation for the total effect of atrazine on phytoplankton. As a percentage, total effects varied less than either direct or indirect effects associated with concepts affected by both types of paths (Table 6). Those total effects associated with atrazine exhibited the least variation across models (6 to 25%), and those associated with pond depth exhibited the most (6 to 100%). Grass carp exhibited intermediate levels of variation, with a difference between models of 12% for the total effects of grass carp on gizzard shad, and a difference between models of 72% for the total effects of grass carp on bluegill. These comparisons support our earlier observations that atrazine, as a concept, is behaving as a simple, well-defined causal factor. It is a toxic herbicide, and all indirect effects are the results of this toxicity. Both pond depth and grass carp are less well defined and encompass a broader range of causal mechanisms, e.g., changes in pond depth represent changes in light penetration, volume, and area of the bottom.

Also of interest is the assessment of the effect of atrazine exposure on ecosystem stability. Listed with each model (Figures 3 to 7) are the stability index values (SI) associated with each model. As explained by Johnson et al.,[59] these values are based on the endogenous relationships so the actual effect of atrazine on ecosystem stability can only be eluded to and could not be measured directly within the study. All models are unstable (≥ 1.0) with a general trend toward more unstable conditions as macroinvertebrate concepts were divided into smaller and smaller aggregate concepts (e.g., total abundance vs. taxa model SI values). However, model structure is not entirely related to model stability (Johnson et al., Chapter 32). Evaluation of Johnson et al.[59] findings revealed two pertinent factors concerning the assessment of atrazine impacts on these ecosystems both of which are related to the experiment design of the atrazine study that was the basis of this and the Johnson et al. paper.[59] It is clear that the inclusion of gizzard shad as an element in the experiment performed in these pond mesocosms was in error. This concept appears to be the most unstable concept in all models. In retrospect this is not surprising as this fish species is unsuited to a long-term existence in small vegetated environments, being by nature a pelagic organism. Secondly, the evaluation of ecosystem stability and structure must be compared in reference to a series of reference conditions and treatments.

In this experiment an inadequate number of reference conditions (ponds) prevent the comparison of the two types of impacts that are assumed to be related to any anthropogenic disturbance: (1) a temporal induced effect(s) resulting in a change in path coefficients but no change in pathways, and (2) an effect resulting in the loss or addition of causal paths. In addition the establishment of separate models with and without atrazine as an exogenous concept would allow a comparison of the change in SI values associated with different models.

Despite these shortcomings, an assessment of atrazine effects is possible in relation to all model concepts. In ecological risk assessments, the ultimate goal is to address total effects resulting from ecosystem exposure on perceived important ecosystem components (concepts). This LISREL analysis provided not only a framework of ecological structure and function but estimated all effects due to atrazine and other exogenous and endogenous concepts (variables) affecting these ecosystems (Tables 3 to 7).

DISCUSSION

In summary, the direct effects of atrazine are primarily limited to gizzard shad and aquatic vegetation concepts (Table 4) while indirect effects of this concept were evident in nearly all endogenous concepts (e.g., Table 5) except for macroinvertebrate total abundance. Only two of the five models presented included a direct path between atrazine and bluegill concepts. If we consider that the most complete laboratory study of atrazine toxicity on bluegill[10] estimated the LOEC for this species to be 500 μg/l (the highest nominal concentration use in this study), it is understandable that LISREL models varied in the inclusion/exclusion of this direct pathway. Macek and co-workers based their LOEC value (500 μg/l) on loss of equilibrium among bluegill continuously exposed to 500 μg/l atrazine for 28 days. In our mesocosm study, seldom did 500-μg/l treated ponds (nominal value) maintain this high a concentration for any length of time. In addition, the bluegill concept was based on bluegill biomass not equilibrium loss and it is unknown whether this chronic effect directly translates to biomass losses. We assumed that the exposure regime was at the upper threshold for a direct bluegill response; thus, model structure and data limitations created a situation where direct effects were only occasionally detectable. In effect, modeling inconsistencies were more a result of experimental design and models in general support Macek's general findings.[10] The large concept of macroinvertebrate total abundance apparently is so well buffered by internal composition (e.g., population responses and interactions) that atrazine exposure had no recognizable effect. The decomposition of the total macroinvertebrate abundance concept into smaller aggregate units and the accompanying atrazine effects on these concepts supported this contention. Table 5 values from LISREL calculations represent the quantified indirect effects values of each exogenous concept

Table 4. Direct Effects of Exogenous Concepts on Those Endogenous Concepts Comprising the Various Ecosystem Models (Figures 3 to 7)

Endogenous Concepts	Exogenous Concepts		
	Grass Carp	Atrazine	Pond Depth
Total Abundance Model			
Gizzard shad	0.000	−0.557 to −1.500	0.000
Bluegill	0.000	0.000 to −0.109	0.000
Aquatic vegetation	−0.031 to −0.053	−0.149 to −0.171	−0.072 to −0.279
Phytoplankton	0.089 to 0.115	0.000	0.000
Zooplankton	0.000	0.000	3.121 to 3.459
Macroinvertebrates	0.000	0.000	−0.675
Habitat Model			
Embenthos	0.000	0.000	0.000
Epibenthos	0.000	0.000	−2.751
Epiphytes	0.000	0.000	0.000
Trophic Model			
Omnivores	0.000	0.000	0.000
Detritivores	0.000	0.000	−3.799
Herbivores	0.000	0.000	0.000
Predators	0.000	0.000	0.000
Specialist/Generalist Model			
Specialists	0.000	0.000	0.000
Generalists	0.000	0.000	−1.587
Taxa Model			
Parachironomus	0.000	0.000	0.000
Dicrotendipes	0.000	0.000	0.000
Caenis	0.000	0.000	0.000
Glyptotendipes	0.000	0.000	0.000
Endochironomus	0.000	0.000	0.000
Hyallela	0.000	0.000	−5.061
Bezzia	0.000	0.000	−2.597

Note: All endogenous concepts are listed under the macroinvertebrate total abundance model with range values given for concepts that were common to all models (i.e., fish, aquatic vegetation, and plankton concepts). Only the macroinvertebrate concepts and values unique to each of the other models are presented under each model heading. Values for direct effects are in the metrics of the appropriate indicator variable.

to each endogenous concept. No indirect effects due to atrazine exposure were observed for aquatic vegetation, but negative effects were noted for bluegill, embenthos, herbivores, and specialists; all other concept relationships were positive. Other exogenous concepts were shown to have various effects on endogenous concepts (positive or negative effects) which explained much of the confusion experienced by Huggins[4] in his attempt to explain atrazine effects with standard linear modeling techniques.

While the recognition and quantification of direct and indirect effects are of an ecological interest, the main concern of ecosystem risk assessments is the ultimate impact of the total effects of the contaminate under evaluation in relation to the ecosystem. The total effect values of exogenous concepts (e.g., atrazine) on the endogenous concepts of each model were calculated by LISREL and are presented in Table 6. All endogenous concepts were affected by all exogenous concepts except for the largest concept of macroinvertebrates (total abundance) which appeared to be unaffected by grass carp and atrazine. In most models all three exogenous

Table 5. Indirect Effects of Exogenous Concepts on Those Endogenous Concepts Comprising the Various Ecosystem Models (Figures 3 to 7)

Endogenous Concepts	Exogenous Concepts		
	Grass Carp	Atrazine	Pond Depth
Total Abundance Model			
Gizzard shad	0.478 to 0.543	1.173 to 1.452	0.001 to 0.449
Bluegill	−0.011 to −0.053	−0.090 to −0.198	0.099 to 0.312
Aquatic vegetation	0.000	0.000	0.000
Phytoplankton	−0.021 to 0.020	0.055 to 0.075	−0.687 to −0.844
Zooplankton	0.087 to 0.128	0.071 to 0.085	−0.782 to −1.193
Macroinvertebrates	0.000	0.000	0.000
Habitat Model			
Embenthos	0.042	−0.081	−2.782
Epibenthos	0.033	0.115	0.107
Epiphytes	−0.030	0.104	−0.723
Trophic Model			
Omnivores	0.091	0.427	−2.578
Detritivores	0.022	0.052	1.157
Herbivores	−0.029	−0.095	−1.278
Predators	−0.001	−0.063	−2.357
Specialist/Generalist Model			
Specialists	0.019	−0.067	−2.322
Generalists	0.045	0.129	0.108
Taxa Model			
Parachironomus	0.086	−0.013	−1.174
Dicrotendipes	0.115	0.242	−1.758
Caenis	0.229	0.084	−3.069
Glyptotendipes	0.057	−0.008	−0.779
Endochironomus	−0.077	−0.314	−0.313
Hyallela	0.120	0.369	2.621
Bezzia	0.028	0.086	−0.567

Note: All endogenous concepts are listed under the macroinvertebrate total abundance model with range values given for concepts that were common to all models (i.e., fish, aquatic vegetation, and plankton concepts). Only the macroinvertebrate concepts and values unique to each of the other models are presented under each model heading. Values for indirect effects are in the metrics of the appropriate indicator variable.

concepts were causing co-occurring total effects that often elicited differing responses from the same endogenous concept. For example, bluegill were negatively impacted by increased grass carp biomass and atrazine concentration, while increased pond depth resulted in a positive total effect on this same concept. In other instances, the total effects of all exogenous concepts were similar in sign (e.g., total effects on aquatic vegetation were all negative). Therefore, it is not surprising that many standard statistic tests would fail to identify an atrazine effect on an ecosystem variable as most often this same variable was responding to other exogenous variables intentionally or unintentionally incorporated in mesocosm studies.

Values of the total effect of grass carp and atrazine on all macroinvertebrate concepts reflect the indirect effect values previously associated with these concepts, as no direct effects from these exogenous concepts were found. In general, this same situation existed in regard to the total effects observed for pond depth on macroinvertebrates except for the occurrence of direct depth effects on macroinvertebrate epibenthos, detritivores, generalists, and two of the seven taxa concepts.

Table 6. Total Effects of Exogenous Concepts on Those Endogenous Concepts Comprising the Various Ecosystem Models (Figures 3 to 7)

Endogenous Concepts	Exogenous Concepts		
	Grass Carp	Atrazine	Pond Depth
Total Abundance Model			
Gizzard shad	0.478 to 0.543	0.673 to 0.715	0.001 to 0.449
Bluegill	−0.011 to −0.039	−0.171 to −0.199	0.099 to 0.312
Aquatic vegetation	−0.031 to −0.053	−0.149 to −0.171	−0.072 to −0.279
Phytoplankton	0.079 to 0.094	0.059 to 0.075	−0.687 to −0.844
Zooplankton	0.087 to 0.128	0.071 to 0.094	2.266 to 2.402
Macroinvertebrates	0.000	0.000	−0.675
Habitat Model			
Embenthos	0.042	−0.081	−2.782
Epibenthos	0.033	0.115	−2.644
Epiphytes	−0.030	−0.104	−0.723
Trophic Model			
Omnivores	0.091	0.427	−2.578
Detritivores	0.022	0.052	−2.631
Herbivores	−0.029	−0.095	−1.278
Predators	−0.001	−0.063	−2.353
Specialist/Generalist Model			
Specialists	0.019	−0.067	−2.322
Generalists	0.045	0.129	−1.479
Taxa Model			
Parachironomus	0.086	−0.013	−1.174
Dicrotendipes	0.115	0.242	−1.758
Caenis	0.229	0.084	−3.069
Glyptotendipes	0.057	−0.008	−0.779
Endochironomus	−0.077	−0.314	−0.313
Hyallela	0.120	0.369	−2.440
Bezzia	0.028	0.086	−3.164

Note: All endogenous concepts are listed under the macroinvertebrate total abundance model with range values given for concepts that were common to all models (i.e., fish, aquatic vegetation, and plankton concepts). Only the macroinvertebrate concepts and values unique to each of the other models are presented under each model heading. Values for total effects are in the metrics of the appropriate indicator variable.

Thus any discussion of the total effects of atrazine on macroinvertebrate concepts would be an interaction of the results of indirect effects and will only be briefly summarized. While no total effect of atrazine was observed on the total abundance of macroinvertebrates, such was not the case with other macroinvertebrate concepts (Table 6). Both positive and negative total effects occurred within the macroinvertebrate concepts (guild categories) of all other models. These results, as with previous results, clearly support the fact that atrazine exposure can cause not only negative effects (e.g., reductions in abundance), but can cause ecological alterations of the ecosystem that results in positive effects on some ecosystem concepts. These positive (and negative) ecotoxic effects are most often associated with indirect effects which highly influence the observed total effects associated with nonplant concepts. The negative total effects of atrazine on most macroinvertebrate concepts in each model are readily explainable. Epiphytes, herbivores, and specialists are the principle concepts experiencing negative ecotoxic effects from atrazine exposure. In addition, all epiphytic herbivore taxa (i.e., *Endochironomus nigrocans,*

Glyptotendipes spp.) and the detritivorous larvae of *Parachironomus,* which are both epiphytic and epibenthic, were shown to be responding negatively to the total effects of atrazine exposure (Taxa Model). Common to all macroinvertebrate concepts affects by atrazine exposure is an ecological dependence on aquatic vegetation, either directly or indirectly. Conversely, atrazine through the same or similar pathways often causes increases in some macroinvertebrate concepts (e.g., *Hyallela* increased with decreased aquatic vegetation abundance caused by atrazine). Overall, the pattern of effects within the macroinvertebrate community appears to be one of loss and gain within populations and guilds such that the large scale measure of this community (macroinvertebrate total abundance concept) was not affected by atrazine. It is also clear that ecological effects resulting from ecosystem exposure to atrazine produce conditions favorable to macroinvertebrate generalists, while reducing numbers of specialists.

In summary, ecotoxic effects of atrazine on these ecosystems, as modeled, revealed consistent effects throughout the ecosystem and affected all but 1 of the 22 concepts included in the 5 models (Table 6). No model was or should be considered a better model as all models represent the same ecosystem, which was conceptually represented by differing model structural elements, thus sometimes differing functional pathways. A similar pattern of total effects is observed for atrazine and grass carp concepts on most endogenous concepts, except that atrazine causes more negative effects within the macroinvertebrate concepts (i.e., specialists, *Parachironomus* and *Glyptotendipes*). Atrazine's effect on nonmacroinvertebrate concepts were primarily limited to:

1. Direct toxic effects on both endogenous fish concepts
2. An indirect negative ecotoxic effect on bluegill
3. An indirect positive ecotoxic effect on gizzard shad and both plankton concepts

The pond depth concept was shown to function somewhat differently than other exogenous concepts, such that increases in depth results in increased fish and zooplankton populations and decreases in the plant and macroinvertebrate community, guild, and population concepts.

Lastly, LISREL produced a matrix of total effects of endogenous cause and effect relationships (Table 7). The interpretation and discussion of the numerous interactions represented in this matrix lie beyond the scope and purpose of this paper. However, a few observations concerning some of the relationships between endogenous concepts are offered in support of our previous findings. Gizzard shad has no effect on any other endogenous concept, and bluegill effects are limited to negative effects on macroinvertebrate herbivores and specialists, presumably due to predation by this insectivore. Plant concepts display similar relations with nonmacroinvertebrate concepts except for aquatic vegetation's negative effects on gizzard shad. The responses, both positive and negative, are thought to be a result of habitat changes associated with the loss or gain of macrophyte cover (abundance). Herbivore and predator populations with discrete habitat requirements, along with the herbivore and epiphytic larvae of *Glyptotendipes* and *Endochironomus,* were

Table 7. Total Effects of Endogenous Concepts on Those Endogenous Effects Concepts Comprising the Various Ecosystem Models Presented (Figures 3 to 7)

	Giz. Shad	Bluegill	Aq. Vegetation	Phytoplankton	Zooplankton
Gizzard shad	0.000	0.000	-8.341 to -12.696	0.727 to 1.501	-0.268 to -0.609
Bluegill	0.000	0.000	0.565 to 1.154	0.120 to 0.328	0.111 to 0.224
Aquatic vegetation	0.000	0.000	0.000	0.000	0.000
Phytoplankton	0.000	0.000	-0.336 to -0.489	-0.293 to -0.368	-0.226 to -0.291
Zooplankton	0.000	0.000	-0.472 to -0.615	0.722 to 0.954	-0.293 to -0.368
Macroinvertebrates	0.000	0.000	0.000	0.000	0.000
Embenthos	0.000	0.000	0.508	0.677	-0.266
Epibenthos	0.000	0.000	-0.722	0.000	0.000
Epiphytes	0.000	0.000	0.653	0.000	0.000
Predators	0.000	0.000	0.365	0.129	-0.042
Detritivores	0.000	0.000	-0.306	0.124	-0.041
Herbivores	0.000	-0.505	0.558	-0.115	-0.126
Omnivores	0.000	0.000	-2.496	0.108	-0.036
Generalist	0.000	0.000	-0.843	0.000	0.000
Specialists	0.000	-1.077	1.078	0.748	-0.501
Parachironomus	0.000	0.000	0.077	0.773	-0.193
Dicrotendipes	0.000	0.000	-1.475	0.482	0.084
Caenis	0.000	0.000	-0.513	1.811	-0.382
Glyptotendipes	0.000	0.000	0.051	0.513	-0.128
Endochironomus	0.000	0.000	1.912	0.000	0.000
Hyallela	0.000	0.000	-2.243	0.256	0.324
Bezzia	0.000	0.000	-0.521	0.059	0.075

	Macro	Emben	Epib	Epip	Pred	Detri	Herb	Omni	Gen	Spec
Gizzard shad	0.000	0.000	0.000	0.000	0.000	0.000	0.000	0.000	0.000	0.000
Bluegill	0.000	0.000	0.000	0.000	0.137	0.110	0.000	-0.064	0.000	0.000
Aquatic vegetation	0.000	0.000	0.000	0.000	0.000	0.000	0.000	0.000	0.000	0.000
Phytoplankton	0.000	0.000	0.000	0.000	0.000	0.000	0.000	0.000	0.000	0.000
Zooplankton	0.000	0.000	0.000	0.000	0.000	0.000	0.000	0.000	0.000	0.000
Macroinvertebrates	0.000	0.000	NA	NA	NA	NA	NA	NA	NA	NA
Embenthos	NA	0.000	0.654	0.000	NA	NA	NA	NA	NA	NA
Epibenthos	NA	0.000	0.228	0.000	NA	NA	NA	NA	NA	NA
Epiphytes	NA	0.000	NA	0.000	NA	NA	NA	NA	NA	NA
Predators	NA	NA	NA	NA	-0.401	0.688	0.000	-0.141	NA	NA
Detritivores	NA	NA	NA	NA	-0.386	-0.338	0.000	-0.135	NA	NA
Herbivores	NA	NA	NA	NA	-0.482	0.213	0.000	-0.250	NA	NA

	Parac	Dicro	Caenls	Glyto	Endo	Hyal	Bezzia	Omnivores	Generalist	Specialists	Parachironomus	Dicrotendipes	Caenis	Glyptotendipes	Endochironomus	Hyallela	Bezzia
Gizzard shad	0.000	0.000	0.000	0.000	0.000	0.000	0.000	NA	NA	0.232	NA	NA	NA	0.000	−0.338	0.580	0.000
Bluegill	0.000	−0.029	0.000	0.000	0.070	0.056	0.000	0.000	0.000	NA	NA	NA	NA	NA	NA	NA	0.000
Aquatic vegetation	0.000	0.000	0.000	0.000	0.000	0.000	0.000	0.000	0.283	NA	NA	NA	NA	NA	NA	NA	0.000
Phytoplankton	0.000	0.000	0.000	0.000	0.000	0.000	0.000	NA	NA	NA	NA	NA	NA	NA	NA	NA	0.000
Zooplankton	0.000	0.000	0.000	0.000	0.000	0.000	0.000	NA	NA	NA	NA	NA	NA	NA	NA	NA	0.000
Macroinvertebrates	NA	NA	NA	NA	NA	NA	NA	NA	NA	NA	NA	NA	NA	NA	NA	NA	NA
Embenthos	NA	NA	NA	NA	NA	NA	NA	NA	NA	NA	NA	NA	NA	NA	NA	NA	NA
Epibenthos	NA	NA	NA	NA	NA	NA	NA	NA	NA	NA	NA	NA	NA	NA	NA	NA	NA
Epiphytes	NA	NA	NA	NA	NA	NA	NA	NA	NA	NA	NA	NA	NA	NA	NA	NA	NA
Predators	NA	NA	NA	NA	NA	NA	NA	NA	NA	NA	NA	NA	NA	NA	NA	NA	NA
Detritivores	NA	NA	NA	NA	NA	NA	NA	NA	NA	NA	NA	NA	NA	NA	NA	NA	NA
Herbivores	NA	NA	NA	NA	NA	NA	NA	NA	NA	NA	NA	NA	NA	NA	NA	NA	NA
Omnivores	NA	NA	NA	NA	NA	NA	NA	NA	NA	NA	NA	NA	NA	NA	NA	NA	NA
Generalist	NA	NA	NA	NA	NA	NA	NA	NA	NA	NA	NA	NA	NA	NA	NA	NA	NA
Specialists	NA	NA	NA	NA	NA	NA	NA	NA	NA	NA	NA	NA	NA	NA	NA	NA	NA
Parachironomus	0.000	0.175	0.000	0.000	0.510	0.102	0.000	NA	NA	NA	NA	NA	NA	NA	NA	NA	NA
Dicrotendipes	0.000	−0.231	0.000	0.000	0.559	0.451	0.000	NA	NA	NA	NA	NA	NA	NA	NA	NA	NA
Caenis	0.000	0.632	0.000	0.000	0.460	0.371	0.000	NA	NA	NA	NA	NA	NA	NA	NA	NA	NA
Glyptotendipes	0.663	0.116	0.000	0.000	0.338	0.068	0.000	NA	NA	NA	NA	NA	NA	NA	NA	NA	NA
Endochironomus	0.000	0.000	0.000	0.000	0.000	0.000	0.000	NA	NA	NA	NA	NA	NA	NA	NA	NA	NA
Hyallela	0.000	−0.395	0.000	0.000	0.954	−0.231	0.000	NA	NA	NA	NA	NA	NA	NA	NA	NA	NA
Bezzia	0.000	−0.092	0.000	0.000	0.221	0.178	0.000	NA	NA	NA	NA	NA	NA	NA	NA	NA	NA

Note: The range of total effect values identified between concepts occurring in all models are presented along with the single total effect value associated with the macroinvertebrate concept(s) incorporated into individual models. Nonapplicability (NA) occurs when concepts do not co-occur in the same model.

positively related to increased aquatic vegetation. All other macroinvertebrate concepts except *Parachironomus,* a chironomid genus whose detritivorous larvae are both epiphytic and epibenthic, display increasing abundance with increasing vegetation cover. However, as a whole, the total abundance of macroinvertebrates (macroinvertebrate concept) was not affected by aquatic vegetation (Tables 4 to 7). Phytoplankton as a causal concept displayed positive effects on both fish concepts and zooplankton. These relationships of cause/effect relations were easily conceptualized.

Except for herbivores, phytoplankton exerted either no effect or a positive effect on all invertebrate concepts. The reason(s) for the negative effect of phytoplankton biomass on herbivorous macroinvertebrates are not clear and like some other interactions shown in this table (e.g., zooplankton's negative impact on gizzard shad and all macroinvertebrate trophic categories) may represent the effects of interceding concepts (unmeasured or unidentified) which occurred within the path between these concepts. While no measures of detritus, epipelion, epiphytic algae, or nonplankton microcrustacean populations were obtained in this study, it is hypothesized that they constitute important concepts. It is assumed that missing and undefined concepts that are not included in these models, in part, compromise the biological interpretation of some of the interactions among concepts. Despite the assumed impact of missing concepts on model interpretation, most endogenous interactions represent logical cause/effect relationships. Most macroinvertebrate concepts are observed to be functionally related to other concepts as one would suspect. Increasing predators cause reductions in all other trophic categories. Of interest is the limited effects of macroinvertebrate concepts on the rest of the ecosystem. These effects are restricted to interactions with bluegill and most were positive. Crowder and Cooper[97] demonstrated that bluegill populations within ponds having various macrophyte cover reduced *Hyallela* and other populations of large-sized organisms when compared to control ponds without fish. While the bluegill concept did not cause a reduction in *Hyallela, Hyallela* increases were related to increased bluegill biomass (Table 7).

Bluegill also exerted a negative effect on itself, a phenomena which was observed in a few invertebrate endogenous concepts (i.e., zooplankton, predators, detritivores, *Dicrotendipes,* and *Hyallela*). These interactions are attributed to density dependent factors associated with each aggregate group or population. Typically, intraspecific regulatory interactions were not found in the endogenous concepts utilized in these models (15 of 22 concepts displayed total effects equal to zero).

In general, the relationships observed between endogenous concepts and presented as total effects values support the premise that the models are predictive for the effects of atrazine and are realistic and representative of those ecosystems examined in this study. We believe this information and other data produced by LISREL can be utilized in sensitivity analysis which, in turn, can be used to determine which endogenous concepts are most sensitive to change. It is anticipated that our future research with LISREL can provide more predictive information as to the impact of both exogenous and endogenous concepts on ecosystems (see Johnson et al., Chapter 32).

ACKNOWLEDGMENTS

We would like to thank the organizers (R. Graney, J. Kennedy, and J. Rodgers) of this SETAC sponsored symposium for the opportunity to participate in this endeavor. We also wish to acknowledge Paul Liechti who reviewed this manuscript and shared his knowledge on aquatic ecology throughout this research effort. Much of the data used in this chapter were obtained from a cooperative project funded by the U. S. Environmental Protection Agency (grant number CR 808804–02–0 to F. deNoyelles, University of Kansas). Additional research initiatives and funding were provided by the Kansas Biological Survey as part of the Ecotoxicology Program.

REFERENCES

1. Eisler, R. 1989. Atrazine hazards to fish, wildlife, and invertebrates: a synoptic review. U.S. Fish Wild. Serv. Biol. Rept. 85(1.18).
2. Kansas Department of Health and Environment. 1989. Atrazine in Kansas. Div. Environ., Topeka, KS., March 1989.
3. Larsen, D. P., F. deNoyelles, Jr., F. Stay and T. Shirogama. 1986. Comparison of single species, microcosm, and experimental pond responses to exposure to atrazine. *Environ. Toxicol. Chem.*, 5: 179–190.
4. Huggins, D. G. 1990. Ecotoxic effects of atrazine on aquatic macroinvertebrates and its impact on ecosystem structure. Ph.D. dissertation, Univ. Kansas, Lawrence, KS.
5. Hiltibrand, R. C. 1967. Effects of some herbicides on fertilized fish eggs and fry. *Trans. Am. Fish. Soc.* 96: 414–416.
6. FWPCA. 1968. *Water Quality Criteria*. Rep. Natl. Tech. Adm. Comm. to Secr. Int., Fed. Water Poll. Control Adm., U.S. Dept. Interior.
7. Bathe, J. F., L. Ullmann and K. Sachsse. 1972. Toxizitatsbestimmung von Pflanzen-schutzmitteln an Fishen. Schr. – R. Ver. Wass. – Boden – Lufthyg., Berlin-Dahlem, H. 37: 241–256.
8. Tooby, T. E., P. A. Hursey and J. S. Alabaster. 1975. The acute toxicity of 102 pesticides and miscellaneous substances to fish. *Chem. Indust.* 12: 523–526.
9. Gunkel, G. and H. Kausch. 1976. Die akute toxizitat von Arazin (s-triazin) auf sandfelchen (*Coregonus fer* Jurine) in Hunger. *Arch. Hydrobiol. Suppl.* 48: 207–234.
10. Macek, K. J., S. Buxton, S. Sauter, S. Gnilka and J. Dean. 1976. Chronic toxicity of atrazine to selected aquatic invertebrates and fishes. EPA/600/3–76–047. U.S. Environ. Protect. Agency, Environ. Res. Lab., Duluth, MN.
11. Birge, W. J., J. A. Black and D. M. Bruser. 1979. Toxicity of organic chemicals to embryo-larval stages of fish. EPA-560/11–79–007, U.S. Environ. Protect. Agency, Washington, D.C.
12. Birge, W. J., J. A. Black and R. A. Kuehne. 1980. Effects of organic compounds on amphibian reproduction. Univ. Kentucky, Water Resourc. Res. Inst., Lexington, KY, Rpt. 121.
13. Wu, T. L. 1980. Dissipation of the herbicides atrazine and alachlor in a Maryland corn field. *J. Environ. Qual.* 9: 459–465.

14. Baker, D. B. 1981. The concentration and transport of pesticides in northwest Ohio Rivers — 1981. U.S. Corps of Eng., Buffalo, Lake Erie, Wastewater Management Study, Tech. Publ. 20.

15. Leung, S. T., R. V. Bulkley and J. J. Richard. 1982. Pesticide accumulation in a new impoundment in Iowa. *Water Res. Bull.*, 18: 485–493.

16. Glotfelty, D. E., A. W. Taylor, A. R. Isensee, J. Jersey and S. Glenn. 1984. Atrazine and simazine movement in Wye River estuary. *J. Environ. Qual.* 13: 115–121.

17. Sanders, H. O. 1970. Toxicities of some herbicides to six species of freshwater crustaceans. *J. Wat. Pollut. Cont. Fed.* 42: 1544–1550.

18. Pott, E. 1980. Die hemmung der futteranfnahme von *Daphnia pulex* — eine neue limnotoxikologische mebgrobe. *ZZ. Wasser Abwasser Forsch* 13: 52–54.

19. Hartman, W. A. and D. B. Martin. 1985. Effects of four agricultural pesticides on *Daphnia pulex, Lemna minor* and *Potamogeton pectinatus. Bull. Environ. Contam. Toxicol.* 35: 646–651.

20. Belluck, D. A. 1980. Pesticides in the aquatic environment. MS. thesis. Univ. Illinois, Urbana-Champaign, IL.

21. Streit, B. and H. M. Peter. 1978. Long-term effects of atrazine to selected freshwater invertebrates. *Arch. Hydrobiol. Suppl.* 55: 62–77.

22. deNoyelles, F., Jr., W. D. Kettle and D. E. Sinn. 1982. The responses of plankton communities in experimental ponds to atrazine, the most heavily used pesticide in the United States. *Ecol.* 63: 1285–1293.

23. Gunkel, G. 1983. Investigations of the ecotoxicological effects of a herbicide in a aquatic model ecosystem. I. Sublethal and lethal effects. *Arch. Hydrobiol. Suppl.* 65: 235–267.

24. Lay, J. P., A. Muller, L. Peichl, W. Klein and F. Korte. 1984. Long-term effects of the herbicides atrazine and dischlorbenil upon the phytoplankton density and physico-chemical conditions in compartments of a freshwater pond. *Chemosphere* 13: 821–832.

25. Herman, D., N. K. Kaushik and K. R. Solomon. 1986. Impact of atrazine on periphyton in freshwater enclosures and some ecological consequences. *Can. J. Fish. Aquat. Sci.* 43: 1917–1925.

26. Hamilton, P. B., G. S. Jackson, N. K. Kaushik, K. R. Solomon and G. L. Stephenson. 1988. The impact of two applications of atrazine on the plankton communities of *in situ* enclosures. *Aquatic Toxicol.*, 13: 123–140.

27. Brockway, D. L., P. D. Smith and F. E. Stancil. 1984. Fate and effects of atrazine in small aquatic microcosms. *Bull. Environ. Contam. Toxicol.* 32: 345–353.

28. Hamala, J. A. and H. P. Kollig. 1985. The effects of atrazine on periphyton communities in controlled laboratory ecosystems. *Chemosphere* 14: 1391–1408.

29. Stay, F. S., D. P. Larsen, A. Katko and C. M. Rohn. 1985. Effects of atrazine on community level responses in Taub microcosms. In T. P. Boyle ed. *Validation and Predictability of Laboratory Methods for the Assessment of Fate and Effects of Environmental Contaminants in Aquatic Ecosystems.* ASTM STP 865, Am. Soc. Test. Materials, Philadelphia, PA, pp. 75–90.

30. Kemp, W. M., W. R. Boynton, J. J. Cunningham, J. C. Stevenson, T. W. Jones and J. C. Means. 1985. Effects of atrazine and linuron on photosynthesis and growth of the macrophytes, *Potamogeton perfoliatus* L. and *Myriophylum spicatum* L. in an estuarine environment. *Mar. Environ. Res.* 16: 255–280.

31. Jones, T. W., W. M. Kemp, P. S. Estes and J. C. Stevenson. 1986. Atrazine uptake, phytosynthetic inhibition, and short-term recovery for the submerged vascular plant, *Potamogeton perfoliatus* L. *Arch. Environ. Contam. Toxicol.* 15: 277–283.

32. Carney, C. E. 1983. The effects of atrazine and grass carp on freshwater macrophyte communities. MA thesis. Univ. Kansas, Lawrence, KS.

33. deNoyelles, F., Jr. and W. D. Kettle. 1983. Site studies to determine the extent and potential impact of herbicides in Kansas waters. Office Water Res. and Tech. Rept. Washington, D.C.

34. Walker, C. R. 1964. Toxicological effects of herbicides on the fish environment. Proc. Ann. Air. Wat. Pollut. Conf. 8: 17–34.

35. Dewey, S. L. 1983. The effects of the herbicide, atrazine, on aquatic insect community structure and emergence. MS thesis. Univ. Kansas, Lawrence, KS. 105 pp.

36. Kettle, W. D., F. deNoyelles, Jr., B. D. Heacock and A. M. Kadoum. 1987. Diet and reproductive access of Bluegill recovered from experimental ponds treated with atrazine. *Bull. Environ. Contam. Toxicol.* 38: 47–52.

37. Brooker, M. P. and R. W. Edwards. 1975. Aquatic herbicides and the control of water weeds. *Water Res.* 9: 1–15.

38. Hurlbert, S. H. 1975. Secondary effects of pesticides on aquatic ecosystems. *Residue Rev.* 58: 81–148.

39. Higashi, M. and B. C. Patten. 1989. Dominance of indirect causality in ecosystem. *Am. Nat.* 133: 288–302.

40. Johnson, M. L., D. G. Huggins and F. deNoyelles, Jr. 1991. Ecosystem modeling with LISREL: A new approach for measuring direct and indirect effects in ecosystem level ecotoxicological testing. *Ecol. Appli.* 1: 383–398.

41. Hayduk, L. A. 1987. Structural equation modeling with LISREL. John Hopkins Univ. Press, Baltimore, MD.

42. Fullner, F. W. 1971. A comparison of macroinvertebrates collected by basket and modified multiple-plate samplers. *J. Water Poll. Control Fed.* 43: 484–499.

43. Hester, F. E. and J. S. Dendy. 1962. A multiple-plate sampler for aquatic macroin-vertebrates. *Trans. Amer. Fish. Soc.* 91: 420–421.

44. Weber, C. I. (ed.). 1973. Biological field and laboratory methods for measuring the quality of surface waters and effluents. EPA-670/4–73–001, U.S. Environ. Protect. Agency, Environ. Monit. Series.

45. Wood, R. D. 1975. *Hydrobotanical Methods.* University Park Press, Baltimore, MD.

46. Hutchison, G. E. 1975. *A Treatise on Limnology.* Vol. 3-*Limnological Botany.* John Wiley & Sons, Inc., New York, NY.

47. Kettle, W. D. 1982. Description and analysis of toxicant-induced responses of aquatic communities in replicated experimental ponds. Ph.D. dissertation, Univ. Kansas, Lawrence, KS.

48. deNoyelles, F., Jr. and D. Kettle. 1985. Experimental ponds for evaluating bioassay predictions. In T. Boyle, ed. *Validation and Predictability of Laboratory Methods for Assessing the Fate and Effects of Contaminants in Aquatic Ecosystems.* ASTM STP 865, Am. Soc. Test. Materials, Philadelphia, PA, pp. 91–103.

49. Fromm, C. H. 1988. Effects of herbicide atrazine on eutrophic plankton communities. MS thesis, Univ. Kansas, Lawrence, KS.

50. APHA, AWWA and WPCF. 1981. *Standard Methods for the Examination of Water and Wastewater.* 15th ed. Amer. Publ. Health Assoc., Washington, D.C.

51. Lorenzen, C. J. 1967. Determination of chlorophyll and pheo-pigments: spectropho-tometeric equations. *Limnol. Oceanogr.* 12: 343.

52. Strickland, J. D. H. and T. R. Parsons. 1968. *A Practical Manual of Sea Water Analysis.* Bull. Fish. Res. Bd. Canada, 167.

53. Edmondson, W. T. (ed). 1959. *Fresh-water Biology,* 2nd ed. John Wiley & Sons, Inc., New York, NY.

54. Wright, S. 1921. Correlation and causation. *J. Ag. Res.* 20: 557–585.

55. Wright, S. 1968. *Evaluation and the Genetics of Populations.* Vol. 1. *Genetic and Biometrical Foundations.* Univ. Chicago Press, Chicago, IL.

56. Li, C. C. 1981. *Path Analysis — A Primer.* Boxwood, Pacific Grove, CA.

57. Crespi, B. J. 1990. Measuring the effect of natural selection on phenotypic interaction systems. *Am. Naturalist* 135: 32–47.

58. Herting, J. R. 1985. Multiple indicators using LISREL. In H. M. Blalock, Jr. ed. *Causal Models in the Social Sciences.* Aldine Publ. Co., New York, NY, pp. 263–319.

59. Johnson, M. L., D. G. Huggins and F. deNoyelles, Jr. 1993. Structural equation modeling and ecosystem analysis, this volume, Ch. 32.

60. Odum, E. P. 1962. Relationship between structure and function in ecosystems. *Jap. J. Ecology* 12: 108–118.

61. deNoyelles, F., Jr., S. L. Dewey, D. G. Huggins, and W. D. Kettle. 1993. Aquatic mesocosms in ecological effects testing: detecting direct and indirect effects of pesticides, this volume, Ch. 30.

62. Pedhazur, E. J. 1982. *Multiple Regression in Behavioral Research,* 2nd ed. Holt, Rinehart, and Winston, New York, NY.

63. Leslie, A. J., Jr., L. E. Nall and J. M. Van Dyche. 1983. Effects of vegetation control by grass carp on selected water quality variables in four Florida Lakes. *Trans. Am. Fish. Soc.* 112: 777–787.

64. Carney, C. E. and F. deNoyelles, Jr. 1986. Grass Carp as a potential control agent for cattails. *Trans. Kansas Academy Sci.* 89: 86–89.

65. Kantrud, H. A., J. B. Miller and A. G. van der Valk. 1989. Vegetation of wetlands of the prairie pothole region. In A. van der Valk (ed.) *Northern Prairie Wetlands,* Iowa State Univ. Press, Ames, IA. pp. 132–187.

66. Wallsten, M. and P. Forsgren. 1989. The effects of increased water level on aquatic macrophytes. *J. Aq. Plant. Management* 27: 32–37.

67. Prowse, G. A. 1971. Experimental criteria for studying grass carp feeling in relation to weed control. *Prog. Fish-Culturist* 33: 128–131.

68. Wetzel, R. G. 1975. *Limnology.* W. B. Saunders Co., Philadelphia, PA.

69. Buck, D. H., R. J. Baur and C. R. Rose. 1975. Comparison of the effects of grass carp and the herbicide diuron in densely vegetated ponds containing golden shiners and bluegills. *Prog. Fish — Culturist* 37: 185–190.

70. Richard, D. I., J. W. Small, Jr. and J. H. Osborn. 1984. Phytoplankton responses to reduction and elimination of submerged vegetation by herbicides and grass carp in four Florida lakes. *Aq. Botany* 20: 307–319.

71. Prowse, G. A. 1969. The role of cultured pond fish in the control of eutrophication in lakes and dams. *Internationale Vereinigung Fur Theoretische und Angewandte Limnologie Verhandlungen* 17: 714–718.

72. Opuszynski, K. 1972. Use of phytophagous fish to control aquatic plants. *Aquaculture* 1: 61–74.

73. Opuszynski, K. 1979. Weed control and fish production, 103–138 pp. In Proceeding of the grass carp conference, J. V. Shireman (ed.). Univ. Florida, Gainesville, FL.

74. Stanley, J. G. 1974. Nitrogen and phosphorus balance of grass carp, *Ctenophryngodon idella*, fed elodes, *Egeria densa*. *Trans. Am. Fish. Soc.* 103: 587–593.

75. Rottman, R. W. and R. C. Anderson. 1978. Limnological and ecological effects of grass carp in ponds. Proc. Ann. Conf. Southeast Assoc. Game and Fish Commissioners 30: 24–39.

76. Mitzner, L. 1979. Research and management of grass carp in Iowa. In J. V. Shireman ed. Proceedings of the grass carp conference. Univ. Florida, Gainesville, FL. pp. 103–138.

77. Hestand, R. S. and C. C. Carter. 1978. Comparative effects of grass carp and selected herbicides on macrophytes and phytoplankton communities. *J. Aq. Plant Management* 16: 43–50.

78. Terrell, T. T. 1982. Responses of the phytoplankton community to the introduction of grass carp into some Georgia ponds. *J. Freshwater Ecol.* 1: 395–406.

79. Kogan, S. I. and G. A. Chinnova. 1972. Relations between *Ceratophyllum demersum* (L.) and some blue-green algae. *J. Hydrobiol.* (USSR, Translation Ser.) 8: 14–25.

80. Cross, F. B. 1967. *Handbook of Fishes of Kansas*. Univ. Kansas Mus. Nat. Hist., Misc. Publ. 45.

81. Engel, S. 1985. Aquatic community interactions of submerged macrophytes. Dept. Nat. Res., Madison, WI., Tech. Bull. 156.

82. Conrow, R., A. V. Zale and R. W. Gregory. 1990. Distribution and abundances of early life stages of fishes in a Florida lake dominated by aquatic macrophytes. *Trans. Am. Fish. Soc.* 119: 521–528.

83. Wiley, M. J., R. W. Gordan, S. W. Waite and T. Powless. 1984. The relationship between aquatic macrophytes and sport fish production in Illinois ponds: a simple model. *North Am. J. Fish. Management* 4: 111–119.

84. Brinkhurst, R. O. 1974. *The Benthos of Lakes*. St. Martin's Press, New York, NY.

85. Johnson, M. G. 1974. Production and productivity. In R. O. Brinkhurst, ed. *The Benthos of Lakes*. St. Martin's Press, New York, NY.

86. Zytkowicz, R. 1976. Producton of macrobenthos in Lake Tynwald. *Acta. Univ. Nicolai Copernici Nauk. Matem.-Przyrod* 38: 75–97.

87. Matuszek, J. E. 1978. Empirical predictions of fish yields of large North American lakes. *Trans. Am. Fish. Soc.* 107: 385–394.

88. Brylinsky, M. 1980. Estimating the productivity of lakes and reservoirs. In E. D. LeCren and R. H. Lowe-McConnell, eds. *The Functioning of Freshwater Ecosystems*. IBP 22, Cambridge Univ. Press, Cambridge, U.K.

89. Murkin, H. R. 1989. The basis for food chains in prairie wetlands. In A. van der Valk (ed.) *Northern Prairie Wetlands*. Iowa State Univ. Press, Ames, IA. pp. 316–338.

90. Krecker, F. H. 1939. A comparative study of the animal population of certain submerged aquatic plants. *Ecology* 20: 553–562.

91. Dvorak, J. and E. P. H. Best. 1982. Macro-invertebrate communities associated with the macrophytes of Lake Vechten — structural and functional relationships. *Hydrobiologia* 95: 115–126.

92. Reid, F. A. 1982. Wetland invertebrates in relation to hydrology and water chemistry. In Proceedings of water impoundments for wildlife: A habitat management workshop. U. S. Dept. Agriculture, Gen. Tech. Rept. NC-100, pp. 72–79.

93. Pennak, R. W. 1989. Fresh-water invertebrates of the United States, 3rd ed. John Wiley and Sons, Inc., New York, NY.

94. Grassle, J. F. and J. P. Grassle. 1974. Opportunistic life histories and genetic systems in marine benthic polychaetes. *J. Mar. Res.* 32: 253–284.

95. Hargrave, B. T. 1970. The utilization of benthic microflora by *Hyallela azteca* (Amphipoda). *J. Anim. Ecol.* 39: 427–437.

96. Mathias, J. A. 1971. Energy flow and secondary production of amphipods *Hyallela azteca* and *Crangonyx richmondensis occidentalis* in Marion Lake, British Columbia. *J. Fish. Res. Board Can.* 18: 711–726.

97. Crowder, L. B. and W. E. Cooper. 1982. Habitat structural complexity and the interaction between bluegills and their prey. *Ecology* 63: 1802–1813.

CHAPTER 34

Ecosystem Analysis:
Summary and Discussion

James H. Kennedy

The transition from evaluating the potential risk from chemical exposure via single species toxicity tests to multispecies tests is still evolving and the subject of much discussion, particularly in regard to regulatory decisions.[3] Analyses of data collected in most outdoor microcosm and mesocosms studies, completed for regulatory or other purposes, has been relatively simple, using regression or hypothesis testing designs. The underlying premise of all the chapters presented in this section is that for regulatory agencies to receive maximum information from complex surrogate ecosystems requires that they be evaluated in a more integrated, holistic fashion, rather than as a profusion of single-species tests. One of the most overwhelming tasks associated with assessing the impact of pesticides is that of measuring the indirect effects that result from the direct effects of pesticides on the biota.[4]

The four chapters in this section are authored and coauthored by scientists associated with the University of Kansas, Nelson Environmental Study Area. Each of the chapters is interlinked in that they argue for a more holistic approach be taken in the analysis of the data being generated by EPA registration studies. In addition, a common data set taken from a multiyear aquatic mesocosm project performed at the University of Kansas, Nelson Environmental Study Area, on the

0-87371-592-6/94/$0.00 + $.50
© 1994 by CRC Press, Inc.

effects of the herbicide atrazine is used to illustrate concepts and techniques presented in each paper.

The first paper in this section by deNoyelles et al. (Chapter 30) provides a summary of the types of direct and indirect effects that have been observed in aquatic mesocosm studies using pesticides as stressors on the aquatic community. They concluded from this literature that most of the secondary trophic level effects reported were those that resulted from the release of competition (i.e., enhanced rotifer populations when cladoceran populations are directly effected). Other indirect trophic effects reported included direct effects on predators and then an increase in their prey populations. In general deNoyelles concluded that herbicides, with little direct effects on animals, influenced food chains bottom upward, while insecticides with direct effects on animals impacted food chains from the top downward. Also discussed were nontrophic indirect effects, such as macrophyte loss, a direct effect, causing changes in community structure through the loss of habitat or refugia. The authors point out that in the studies reviewed pathways connecting direct effects to indirect effects are mostly speculative. Few experiments have been performed to test these relationships. In addition these pathways were rarely described mathematically.

deNoyelles et al. discussed in detail the direct and indirect effects revealed in a 3-year mesocosm study with atrazine, a herbicide, that are modeled in the chapters following this one. The herbicide's direct effects were primarily on plants. Most of the macrophytes declined in abundance during the study. Changes in components of this community caused a cascade of indirect effects on other communities in the systems. Some phytoplankton declined in number or were eliminated and subsequently caused a population decline in the herbivorous copepods. Phytoplankton species tolerant to the herbicide increased in abundance due to reduced competition and grazing pressures, causing a change in the species assemblage. Population changes were noted for those organisms that depend on the macrophytes or phytoplankton as a substrate, habitat, or food. A comparison is made between the results from single-species tests performed with atrazine in the laboratory under controlled conditions to responses revealed from the mesocosm test. They conclude that while these types of tests can provide useful information for protecting populations, the endpoints measured in the laboratory are modified in the natural environment by competition, predation, availability of food, and habitat and provide only a partial picture.

deNoyelles and Dewey (Chapter 31) make the point that mesocosm studies need to evolve and use statistical methods other than inferential statistics to measure direct and indirect effects of pesticides on aquatic ecosystems. They feel, as do the other authors in this section, that more comprehensive methods need to be explored to measure secondary effects. Procedures that integrate ecosystem responses should be adopted. They propose that the concept of ecosystem stability, defined as the ability of the ecosystem to return to its original state once it has been disturbed, may be an appropriate measure of ecosystem response because it incorporates both ecosystem structure and function into measures of system dynamics. The authors discuss how ecosystem stability can be evaluated through the compartmentalization and connectance of food webs. An example of this method is discussed using structural measurements taken from the atrazine study described in Chapter 30. This measure of ecosystem stability diagrammed species recorded at each level of atrazine exposure in terms of predator-prey relationships in the food web. Functional relationships including the total number of actual trophic level links and the average number of trophic links per consumer species can be determined from these diagrams. Despite the variation in the average linkage (1 bar) within the treatments, the average linkage was statistically significantly higher in control ponds compared to the treatment ponds. These results are consistent with the ecosystem changes noted in Chapter 30 and provided a measure of the stability of atrazine treated ponds.

The last two papers in this section by Johnson et al. (Chapter 32) and Huggins et al. (Chapter 33) use LInear Structural RELations (LISREL), a structural equation modeling technique, to quantify the direct and indirect ecotoxic effects of atrazine on an aquatic community. Johnson et al. provide an introduction to LISREL and the basics of the modeling procedure. LISREL provides a measure of ecosystem stability and the total, direct, and indirect effects of all ecosystem components on all other components. Community and ecosystem components and interactions are described by dynamic equations generated from empirical response data, providing insight into how the chemical stresses the ecosystem. Huggins et al. using LISREL techniques was able to measure and quantify direct and indirect effects of atrazine on the aquatic community. In summary, the effects of atrazine on the community relationships over a 3-year period as measured by LISREL techniques in surrogate ecosystems have been realistic and the models predictive.

Dewey and deNoyelles (Chapter 31) point out that one of the key components in measuring community stability is to allow adequate time for populations to "Turnover" or replace themselves. They provide estimates that it might take from 2.6 to 10 years for this to occur. Registration studies currently are performed over a short period of time (spring through late fall). Whether biological responses measured over a single 9-month period in newly constructed ponds, usually less than 1-year old, stocked with a single species of fish are adequate for evaluation by stability measurements and LISREL techniques remains to be determined.

Questions addressing stability, interaction strengths, and connectance are very interesting and might prove useful for providing a holistic analysis. However, as pointed out by Mount,[5] toxicology tests and methods of analysis must be decisive to be of value to regulators. Before we embrace new measures of ecosystem responses as regulatory requirements we need to clearly understand the assumptions and limitations of these tests. In addition, we need to understand what the endpoints of these tests are telling us. Whatever methods are ultimately adopted for measuring trophic level interactions, if they are to be used for regulatory purposes, it will be important that they have probability attached to them so that we can extrapolate with definable error.[5]

In order to provide risk assessments of pesticides on natural ecosystems using data collected from surrogate manmade ponds, we need to understand how the structural and functional components of the test systems relate to natural systems. Our concerns for protection of ecosystems are on a continent or global scale, not on a small lake or stream segment.[3] Ultimately, the actual value in mesocosm testing for the risk assessment of pesticides is not the ability to measure effects in surrogate ecosystems, but rather their potential to predict responses in natural ecosystems. Before these types of extrapolations can be made, we must evaluate the validity of mesocosms to predict ecosystem level responses. Modeling approaches, like those discussed in this chapter, may ultimately prove to be useful in extrapolating from mesocosms to natural ecosystems.

The mesocosm studies being performed under the FIFRA statute are measuring hundreds of biotic and abiotic aquatic parameters and are producing voluminous data sets. We have an inadequate understanding, however, of the biological processes that occur in many ecosystems.[6] A systematic synthesis of this existing information will aid mesocosm testing to reach its full potential as a tool for prediction of impact (as opposed to just effects assessment). Information needed to **begin** answering questions concerning system functioning, the development, and refinement of ecosystem level evaluation tools and extrapolation is contained in these data sets. Unfortunately, time constraints imposed for completion of these megalithic field studies, the mandatory data analyses prescribed by the regulatory agency, in addition to the proprietary nature of these data have left an untapped resource for information concerning ecosystem responses. Ecologists, toxicologists, and the chemical industry should collaborate to evaluate this wealth of data from the standpoint of ecosystem response. Without a clear-cut, systematic synthesis of the existing information, mesocosm testing will never reach its full potential as a tool for prediction of impacts (as opposed to just effects assessment). The most useful information may ultimately come about from innovative procedures that cross traditional disciplinary boundaries[7] and lead to a more defensible regulatory tool that incorporates ecosystem level analyses in the ecological risk assessment.

It is unlikely that the analysis of mesocosm data will rely on one method as the panacea for the risk assessment of agricultural chemicals. Rather, assessment of this type of data will include a synthesis of techniques that embrace biological concepts, multivariate statistics, modeling, and experimental approaches to the analysis of effects and extrapolation to real world situations. As Cairns[8] succinctly stated: "If environmental toxicology is to come of age, it must begin to ask more searching questions, develop broader hypotheses involving natural systems, and develop models that are validated in landscapes, not laboratories."

REFERENCES

1. Odum, E. P. 1984. The mesocosm. *BioScience* 34: 558–562.
2. Touart, L. W. 1988. Aquatic mesocosm testing to support pesticide registration. EPA 540109–88–035. U.S. Environmental Protection Agency, Office of Pesticide Programs, Washington, D.C.
3. A.E.D.G 1992. Improving aquatic risk assessment under FIFRA report of the aquatic effects dialogue group. World Wildlife Fund, 98 pp.
4. Wallace, J. B.. G. J. Lugthart, T. F. Cuffney, and G. A. Schurr. 1989. The impact of repeated insecticidal treatments on drift and benthos of a headwater stream. *Hydrobiologia* 179: 135–147.
5. Mount, D. I. 1985. Scientific problems in using multispecies toxicity tests for regulatory purposes. In J. Cairns, Jr., ed., *Multispecies Toxicity Testing,* Jr. Pergamon Press, Elmsford, NY, pp. 13–18.
6. Golley, F. B. 1985. What Ecologists Expect from Industry? In J. Cairns, Jr., ed., *Multispecies Toxicity Testing.* SETAC Special Publications Series, Pergamon Press, New York, NY, pp. 27–35.
7. Maciorowski, A. F. 1988. Populations and communities: Linking toxicology and ecology in a new synthesis. *Environ. Toxicol Chem.,* 7: 677–678.
8. Cairns, J., Jr. 1992. Paradigms flossed: The coming of age of environmental toxicology. *Environ. Toxicol. Chem.* 11: 285–287.

Index

INDEX

ABIB, see Augmented balanced incomplete block design
Ablabesmyia spp., 192, 592
ABS, see Artificial benthic substrates
Absorption, 204, 219–220
Abundance patterns, 188, 191
Acanthocyclops sp., 530
Achnanthes lanceolata, 551, 555
Achnanthes minutissium, 439, 551, 554
Acricotopus sp., 196
Actinomycetes, 419, 434
Adjustment stability, 610, see also Stability, theory
Adsorptive matrix technique, 416
AEDG, see Aquatic Effects Dialog Group
Aeslinida, 453
AFST, see Aquatic Field Studies Team
Aggregation of variables, 636–642, see also Structural equation modeling
Aircraft, fixed-wing, 205, 206, 207–209, 213–214, see also Spray drift
Alachlor, 50
Algae, see also Blue-green algae; Green algae
  community structure in TMAC stream mesocosm, 541–543, 545, 546–548, 557
  cyflurin-treated pond mesocosms/concrete tank microcosms, 345
  lambda-cyhalothrin toxicity, 423, 441, 461
Alisma plantago-aquatica, 312, 315
Alkalinity
  cyflurin-treated mesocosms/microcosms, 347, 349, 377, 379
  guthion mesocosm study, 471
  lambda-cyhalothrin mesocosm study, 434
  lauryl trimethyl ammonium chloride stream mesocosms, 543–544
  sulprofos-treated ponds and tanks, 326, 327
  tralomethrin mesocosm study, 501
  untreated pond mesocosms, 111
Alona rustica, 351, 364, 387
4-Amino-3-methyl-6-phenyl-1,2,4-triazin-5(4H)-one, see Goltix
Ammonia, 111, 377, 378–379
Amphibians, 581–583
Amphipoda
  atrazine effect, 593–594, 618
  colonization in pond mesocosms, 114, 116

cyflurin-treated mesocosms/microcosms, 350, 352, 361, 382–383
Amphora veneta, 551, 555
Anabaena, 287
Ancylidae, 449
Anisoptera
  cyflurin-treated mesocosms/microcosms, 383, 385, 391
  lambda-cyhalothrin toxicity, 426, 453, 461
Ankyra ancora, 311
Annelida, 114
Anodonta imbecilia, 449
ANOVA, see One-way analysis of variance
Anureopsis spp., 525
Aphanizomenon flosaquae, 436
Aquatic Effects Dialog Group (AEDG), 203, 296–297
Aquatic Field Studies Team (AFST), 19
Aquatic Mesocosm Tests to Support Pesticide Registrations, 11
Artificial benthic substrates (ABS), 326, 329, 330, see alsoArtifical substrates; Macroinvertebrate artificial substrates
Artificial substrates, see also Artificial benthic substrates; Macroinvertebrate artificial substrates
  cyflurin-treated mesocosms/microcosms, 345, 352–353
  lambda-cyhalothrin study, 424–426, see also individual entries
  untreated pond mesocosms, 109–110, 113, 116
Asplanchna sp., 525, 530, 531
Asterococcus sp., 281
Atrazine, see also Mesocosm, data and effects predictions for aquatic ecosystems; Stability, theory; Structural equation modeling
  bluegill sunfish decline, 593
  characteristics, 585
  direct/indirect effects, 583, 588
  herbicide effects, 583, 584
  insect decline, 592
  LISREL modeling of mesocosm, 633–644
  macroinvertebrate increases, 593–594
  macrophyte-grazing fish decline, 589, 591, 592
  mesocosm test design, 586–587

resistance, 589, 590
single-species aquatic tests, 586
stability analysis, 614–615
tadpole decline, 593
toxicity in pond mesocosms, 243–244,
    246, 247
Augmented balanced incomplete block (ABIB)
    design, 161–162
Autotrophic index, 423, 440
Azinphos methyl, see Guthion
A.I. chlorpyrifos, see Dursban

Bacillariophyceae, 112, 281, 282
Bacillariophyta, 429, 436, 560–564
Bacteria, 419, 434
Baetidae
    cyflurin toxicity, 350, 352, 358, 363, 385
    lambda-cyhalothrin toxicity, 426, 447,
        450–453, 461
    tralomethrin toxicity, 505, 507, 508
Baythroid, see Cyflurin
BCF, see Bioconcentration factor
Belostomatidae, 454
Benthic communities, 52, 53
Benthic compartment, 617–619
Benthic feeding, see Feeding behavior
Benthic organisms
    atrazine toxicity, 655
    cyflurin-treated mesocosms/microcosms,
        345, 352, 358–360, 365, 381–384
    Goltix-treated ponds, 308
    guthion toxicity, 472, see also
        Macroinvertebrates
    sulprofos-treated ponds and tanks, 329, 330,
        332
Benzoylphenylurea, 582
Berosus sp., 383
Bezzia spp., 594, 677
Bioaccumulation of insecticides, 49–51
Bioavailability
    cyflurin, 356, 390
    Dursban, 522
    toxicants in pond mesocosms, 245, 247
    tralomethrin, 509
Biocoenosis, 308–310, 519
Bioconcentration factor (BCF), 22
Biomass, see also individual entries
    bluegill sunfish
        guthion mesocosm study, 478, 479
        lambda-cyhalothrin toxicity study,
            456, 457
        tralomethrin mesocosm study, 509, 511

fish and macroinvertebrate abundance
    model, 669
periphyton in lambda-cyhalothrin mesocosm
    study, 440
phytoplankton
    atrazine effects, 587
    lambda-cyhalothrin mesocosm study, 420,
        434–436, 439
    zooplankton in Dursban pond mesocosm
        study, 520, 524, 531
Biometrics, 110–111, 116–122, 124–125
Bioturbation, 245–246
Biovolume, 263, 277, 279–281
Blooms, 288
Bluegill sunfish, see *Lepomia machrochirus*
Blue-green algae, 287, 288, see also Algae
Body size, 250–252
Bolstar 6L, see Sulprofos
Bonverroni multiple comparison procedure,
    135, 140
Booms, see Sprayers, boom
Bosmina longirostris, 265–267
Bosmina spp.
    cyflurin-treated mesocosms/microcosms, 364
    statistical analysis of herbicides in
        enclosures
        abundance data, 134, 137
        regression analysis, 143–144, 146, 148
        RM-ANOVA, 140, 141
Brachionus sp., 351, 360, 388, 525, 530
Bremerhaven Cassions, 52, 53
Bridge spanner, 341, 342, 357, 501, 503, see
    also Spray drift; Sprayers, ground
    boom
Brook trout, see *Salvelinus fontinalis*
Brown-Forsythe ANOVA, 135, 140, see also
    One-way analysis of variance

Caddisfly, see *Triaenodes tardus*
Caenidae
    cyflurin-treated mesocosms/microcosms,
        350, 352, 361, 363, 385
    lambda-cyhalothrin toxicity, 426, 447, 450,
        452, 461
    tralomethrin toxicity, 505, 507, 509
Caenis spp.
    atrazine effect, 594
    cyflurin-treated mesocosms/microcosms,
        352, 359, 365, 381
    lambda-cyhalothrin toxicity, 452
    tralomethrin toxicity, 507
Calcium, 284, 544

Calcium carbonate, 111
Callibeatis floridanus, 352, 358, 365,
    381, 385
Callibeatis sp., 451–453
Callitriche sp., 312
Calothrix spp., 439
CAMM computer program, 539–540
Canonical correspondence analysis (CCA),
    521, 525–527, 531
Carbaryl, 249, 582
Carbofuran, 582
Carbon, 581, 582, see also individual entries
Carp, 389
Cattails, see *Typha* spp.
Causal models, 663–665, see also Atrazine
CC, see Coefficient of Community
CCA, see Canonical correspondence analysis
CEPEX system, 52, 53
Ceratopogonidae
    atrazine effect, 594
    cyflurin-treated mesocosms/microcosms,
        355, 363
    guthion toxicity, 487, 489, 490, 493
    lambda-cyhalothrin toxicity, 455, 461
Ceriodaphnia lacustris, 265–267, 286
Chaetogaster sp., 449
Channel catfish, see *Ictalurus punctatus*
Channel flow, 218, 219, see also Runoff
Chaoboridae, 350, 353, 363, 385, 455
Chaoborus crystallinus, 531
Chaoborus punctipennis, 618
Chaoborus sp.
    atrazine effects, 589, 590
    colonization in untreated pond mesocosms,
        113
    cyflurin-treated mesocosms/microcosms,
        355, 362
Chara spp., 246, 471, 519, 589, 660
Chemical fate
    cyflurin, 346–348, 356–358, 376–379,
        388–389
    Dursban, 520
    Goltix, 313–316, 318
    guthion, 474–477
    laboratory vs. mesocosm predictions, 242,
        258–259
    tralomethrin, 504
Chironomidae
    atrazine effect, 592, 594
    colonization in untreated pond mesocosms,
        115, 116, 122
    cyflurin-treated mesocosms/microcosms,
        350, 355, 363, 364, 366, 381

emergence
    chronology, 186–188, 194
    species richness, 182–186, 189, 191–193
    endosulfan effects, 42
    guthion toxicity, 486–488, 493
    lambda-cyhalothrin toxicity, 426, 447, 455
    sulprofos-treated ponds and tanks, 329, 330,
        333, 334
    tralomethrin toxicity, 509, 510
Chironominae
    cyflurin-treated mesocosms/microcosms,
        350, 355, 361–362, 364, 366,
        382, 385
    lambda-cyhalothrin toxicity, 447, 449, 453,
        461
    tralomethrin toxicity, 509, 510
Chironomini
    emergence, 183–185
    guthion toxicity, 487, 488, 493
    phenology patterns, 187, 189, 190, 195,
        200
    species richness, 192
Chironomus spp., 192, 195, 654
Chlamydomonas sp., 389
Chlorella vulgaris, 281
Chloride, 543–544
2-Chloro-4-ethylamino-6 isopropylamino-1,3,5
    triazine, see Atrazine
Chlorophenol
    degradation and toxicity relation in
        limnocorrals, 285, 286
    residue analysis from limnocorrals,
        261–262, 264–265, 268–270, see also
        DIATOX
Chlorophyceae, 112, 564, 565
Chlorophyll *a*
    atrazine-treated mesocosm, 661
    cyflurin-treated mesocosms/microcosms,
        345, 349, 351, 377, 379
    DIATOX-treated limnocorrals, 262,
        277, 278
    Goltix-treated ponds, 307, 317
    lambda-cyhalothrin mesocosm study, 420,
        423, 437, 439
    lauryl trimethyl ammonium chloride stream
        mesocosms, 543–547, 557
    macroinvertebrate abundance model,
        668–669
    phytoplankton in untreated pond systems,
        110, 112, 117
    sulprofos-treated ponds and tanks, 326, 327
Chlorophyta

DIATOX impact on abundance, 280–282, 287
Goltix effect, 310, 311
Lambda-cyhalothrin, 435, 436, 439
Chlorpyrifos, 582
Chromagaster ovalis, 275
Chroococcus sp., 281, 288–289
Chrysophyceae, 112, 564, 565
Chrysophyta, 280–282, 287, 436
Chydorus sp., 364, 387, 522, 528
Cladocera
  atrazine-treated mesocosm, 661
  colonization in untreated pond mesocosms, 113
  cyflurin-treated mesocosms/microcosms, 350–353, 360, 363–364, 386–387
  DIATOX impact, 265, 270, 273, 288
  Dursban toxicity, 520, 522, 524, 527–529, 531
  esfenvalerate effect in pond mesocosms, 248, 249
  Goltix-treated ponds, 308, 309
  guthion toxicity, 480–481, 483, 493
  lambda-cyhalothrin toxicity, 446
  macroinvertebrate abundance model, 669
  statistical analysis of herbicides in enclosures
    abundance data, 134, 138
    regression analysis, 143, 145, 146, 148, 150, 151
    RM-ANOVA, 141, 142
  sulprofos-treated ponds and tanks, 328, 329, 333, 335
  tralomethrin toxicity, 505, 506
Cladotanytarsus spp., 191, 196
Clay slurries, 325, 327, 328, see also Runoff; Spray drift
Cocconeis placentula, 550, 551, 553
Coefficient of Community (CC), 182, 186, 193
Coefficient of variation (CV), 95–99, 102–103, 163–165
Coenagrionidae
  cyflurin-treated mesocosms/microcosms, 350, 354, 363, 366, 391
  lambda-cyhalothrin toxicity, 447, 454
Coleoptera
  colonization in untreated pond mesocosms, 115, 116
  cyflurin-treated mesocosms/microcosms, 383
  guthion toxicity, 485
  lambda-cyhalothrin toxicity, 426, 447, 455, 461

Colonization, see also Emergence
  atrazine-treated pond mesocosms, 246
  cyflurin-treated mesocosms/microcosms, 352, 358–360, 365, 366, 381–384
  pond mesocosms, 112–116
    pesticide testing, 122–123
    time table, 106
Community
  alteration and pond testing of pesticides, 39–41, 246–247
  cyflurin-treated mesocosms/microcosms, 366
  DIATOX impact, 275, 277, 279–281, 287
  Goltix-treated ponds, 310, 312, 314
  guthion mesocosm study, 481–484
  lauryl trimethyl ammonium chloride stream mesocosm, 541–543, 557
  parameters and stability relation, 608, 609, 612
  predator-prey relationships and pesticides, 51
Community similarity analysis (SIMI), 546, 548, 551, 558
Compartmentalization, see also Stability, theory; Structural equation modeling
  LISREL modeling, 636–642
  stability relation, 608, 611–612, 617–619
Complexity, 608
Concentration-response
  pond mesocosms, 242
  theory
    conclusions, 152–153
    general data characteristics, 137–139
    linear regression, 142–147, 148
    materials and methods, 131–134
    nonlinear regression, 147, 149–151
    overview, 129–131
    RM-ANOVA, 138, 140–142
    statistical analysis, 134–137
Concepts, see also Endogenous concepts; Exogenous concepts; Structural equation modeling
  depth of pond, 670, 673
  LISREL modeling, 632, 637, 640, 643–644, 665
Conductivity
  Dursban pond mesocosm study, 519, 522, 523
  Goltix-treated ponds, 307
  guthion mesocosm study, 471
  lambda-cyhalothrin mesocosm study, 422, 427, 434
  lauryl trimethyl ammonium chloride stream mesocosms, 539, 543–544, 546, 549

sulprofos-treated ponds and tanks, 326, 327, 331
Conjugatophyceae, 311
Connectance, see also Stability, theory; Structural equation modeling
   atrazine mesocosm study, 615, 617
   food web theory, 631
   LISREL modeling, 640–642
   stability relation, 608, 611, 612, 615–617
Conochilus unicornis, 275, 276, 286
Copepoda
   atrazine-treated mesocosm, 661
   colonization in untreated pond mesocosms, 113
   cyflurin-treated mesocosms/microcosms, 350, 363, 364
   DIATOX impact, 265, 270, 273
   Dursban toxicity, 520, 524–525, 529, 530
   esfenvalerate effect in pond mesocosms, 248, 250
   guthion toxicity, 480–482, 493
   lambda-cyhalothrin toxicity, 446
   macroinvertebrate abundance model, 669
   statistical analysis of herbicides in enclosures
      abundance data, 134, 138
      regression analysis, 145–146, 148, 150, 151
      RM-ANOVA, 141
   sulprofos-treated ponds and tanks, 328, 329, 333
   tralomethrin toxicity, 505
Copper sulfate, 583
Coregonus lavaretus maraenus, 390
Corixidae, 454
Corn, 585
Corpulence factor, 307, 308
Cotton pest control, see also Pyrethroid, simulations in aquatic mesocosms
   pesticide application rate, 234, 236
   pesticide runoff, 223–225
   pyrethroid insecticides, 203–204, 206, 213–214
Covariance matrices, 633–635, see also Structural equation modeling
Covariance structure, 630, see also Structural equation modeling
Cricotopus sp., 191, 196
Crop canopy, 205, 207–210, see also Pyrethroid, simulations in aquatic mesocosms; Runoff; Spray drift
Cropping practices, 218, see also Runoff; Spray drift

Cross-contamination, 414–415
Crosswind, see Wind
Crustacea
   atrazine effects, 589, 590
   colonization in untreated pond mesocosms, 114
   lambda-cyhalothrin toxicity, 445–448, 461
   tralomethrin toxicity, 505, 507
Cryptochironomus sp., 192
Cryptophyta, 311, 436
Ctenopharyngodon idella
   atrazine
      effects, 587, 589, 591, 617, 678, 681
      establishment, 659, 660
   macroinvertebrate abundance model, 668
   vertebrate compartment in LISREL modeling, 637–640
Culicidae, 455
CV, see Coefficient of variation
Cyanophyceae, 112, 564, 565
Cyanophyta
   DIATOX impact, 277, 281, 282, 287, 288
   Goltix-treated ponds, 311
   lambda-cyhalothrin toxicity, 435
Cyclops vernalis, 531
Cyclostella pseudostelligera, 550, 552
Cyflurin
   application, 376
   biological effects
      fish, 349–350, 359
      microinvertebrates, 352–355, 358–365
      phytoplankton/periphyton, 351
      zooplankton, 351–361
   bluegill effect on invertebrates, 381–388, 390–391
   chemical fate, 346–348, 356–358, 376–378, 388–389
   conclusion, 391–392
   impact on bluegill, 379–381, 389–390
   impacts of fish, 362–366
   methods and materials, 338–346, 375
   other factors, 366–367
   overview, 337–338, 374–375
   physical and chemical parameters, 347–349, 378–379
   sample collection and analysis, 376–377
   statistical analysis, 377
Cyfluthrin, 203, see also Pyrethroid, simulations in aquatic mesocosms
Cypermethrin, see also Pyrethroid, simulations in aquatic mesocosms
   application and spray drift, 203, 213–214, 215, 216

direct/indirect effects, 582
runoff entry, 223–226
testing in aquatic mesocosms, 227–229
toxicity to rainbow trout, 389

2,4-D, 583, 584
Dactylothece sp., 281
Damselfly, see Coenagrionidae
Daphnia longispina, 522, 528
Daphnia magna, 243, 317, 492, 654
Daphnia pulex, 521, 527–528
Daphnia spp.
  cyflurin-treated mesocosms/microcosms,
    364, 366
  DIATOX impact, 265, 266, 285, 286, 288
  fenvalerate impact, 389
  survival and refugia relation, 377
DDT, 222, 223
Degradation of insecticides, 204, 313
Deltamethrin, 203, 207, see also Pyrethroid,
    simulations in aquatic mesocosms
Demographic stability, 608, 612, 619, 621,
    see also Stability, theory
Dero sp., 449
Desaminometamitron, 314–316, 318, see also
    Goltix, mesocosm/microcosm ponds
Desmidaceae, 112
Detergents, see Lauryl trimethyl ammonium
    chloride (TMAC)
Detritivores, 662, 673, 674, 677, see also
    Structural equation modeling
Diamesa spp., 196, 200
Diaphanosoma brachyurum, 351, 353, 364,
    387
Diaptomus spp., 530
Diatoma vulgare, 551
Diatoms, 281, 311, 542, 545
DIATOX
  chlorophenol residues, 264–265, 268–270,
    285
  dissolved oxygen, chlorophyll *a*, Secchi
    disk, 275, 277, 278
  materials and methods, 259–264
  overview, 257–259
  phytoplankton, 277, 279–282, 287–289
  water chemistry, 281, 283, 284
  zooplankton
    abundance, 265–268, 270, 272–276,
      284–285
    community structure, 275
    variability, 270, 271, 286–287
DIC, see Dissolve inorganic carbon

3,4-Dichloroaniline, 582
Dichlorvos, 582
Dicrotendipes spp., 192, 594, 677
Difflugia sp., 442, 443
Diflubenzuron, 582
Dinaophyceae, 281, 282
Dinobryon bavaricum, 436
Dinobryon divergens, 281
Dinophyceae, 112
Diptera
  atrazine effects, 592, 594
  colonization in untreated pond mesocosms,
    115, 116
  cyflurin-treated mesocosms/microcosms,
    350, 355, 384
  guthion toxicity, 485–487, 493
  lambda-cyhalothrin toxicity, 426, 447, 448,
    455–456, 461
Direct effects, see Effects matrices
Dissolved inorganic carbon (DIC), 284
Dissolved organic carbon (DOC), 284, 345,
    348, 349
Dissolved organic material (DOM), 245
Dissolved oxygen (DO)
  atrazine-treated mesocosm, 587, 661
  cyflurin-treated mesocosms/microcosms,
    345, 349, 377, 379
  DIATOX-treated limnocorrals, 263, 275,
    278, 287, 288
  Dursban pond mesocosm study, 519,
    522, 523
  Goltix-treated ponds, 307, 317
  guthion mesocosm study, 471
  herbicide/insecticide-treated pond systems,
    245
  insecticide indirect effects, 581, 582
  lambda-cyhalothrin mesocosm study, 422,
    427, 433–434, 437
  lauryl trimethyl ammonium chloride stream
    mesocosms, 539, 543–544
  sulprofos-treated ponds and tanks, 326,
    327, 331
  tralomethrin mesocosm study, 501
  untreated pond mesocosms, 111
Disturbance, see Perturbation
Disturbance factors, 632, 665, see also
    Structural equation modeling
Diversity index, see Shannon-Weaver index
DO, see Dissolved oxygen
DOC, see Dissolved organic carbon
DOM, see Dissolved organic material
Domain of attraction, 608, 609, see also
    Stability, theory

Dominant eigenvalue, 636
Dorosoma cepedianum
  atrazine testing, 587, 589, 590, 618,
    658, 678
  biomass and macroinvertebrate abundance
    model, 669
  vertebrate compartment in LISREL
    modeling, 637–640
Dosage
  DIATOX in limnocorrals, 260–261
  Goltix, 306
  herbicide/insecticide toxicity relation,
    249–251
  levels in mesososm testing, 3, 36, 245–246
  optimal design of aquatic field studies,
    166–170
Dragonfly, see Libellulidae
Drift, see Spray drift
Drift deposition cards, see also Spray drift
  endosulfan testing, 39, 40
  insecticide spray drift evaluation, 207,
    213–215
Droplet size and insecticide application, 204,
    205, 208, 228, 230, see also
    Pyrethroid, simulations in aquatic
    mesocosms
Dunnett's test, see also Student's *t*-test
  multiple range, 90, 346, 349, 350, 365,
    377, 491
  t-test, 333–335, 628
Dursban
  bioassay with *Daphnia pulex*, 527
  cladocera, 522, 524
  correspondence analysis, 525–527
  discussion, 527–531
  materials and methods, 518–521
  ostracoda/copecoda, 524–525
  overview, 518
  physicochemical characteristics, 521–523
  rotifers, 525
  similarity ratio, 527
  zooplankton biomass, 525
Dytiscidae, 426, 455

Earthen ponds, mesocosm, see Sulprofos
Ecological endpoints, see Endpoints
Ecosystems, see also Structural equation
      modeling
  analysis
    introduction, 575–576
    summary and discussion, 693–697

aquatic, chemical toxicity, see Mesocosm,
      data and effects predictions for aquatic
      ecosystems
  guthion responses, 493
  mesocosms as surrogates, 578
  parameters and stability relation, 608, 609,
    612, see also Stability theory
  stability in farm ponds, 44
  structure and function in atrazine-treated
    mesocosms, 663
  structure in LISREL modeling, 629, 632
Ecotoxicology, comparative, 55
EEC, see Expected environmental
      concentration
Effect size, 74–77, 80, 81, 83
Effects matrices, 643–645, 650–652
Egg hatchability, 654
Einfeldia sp., 196
Ekman dredge, 472
Electron capture, 415–416
Eleocharis sp., see Spike rush
Elmidae, 455
Elodea nuttallii, 519
Emergence, see also Emergence traps
  Chironomidae taxa in mesocosm test ponds,
    182–186, 189–190
  chronology in mesocosm test ponds,
    186–188
  guthion mesocosm study, 485–491
  phytoplankton in Goltix-treated ponds,
    312–313, 318
Emergence traps, see also Emergence
  cyflurin-treated mesocosms/microcosms,
    345, 354–355, 363–364, 376, 384–385
  guthion mesocosm study, 472
  lambda-cyhalothrin study, 424, 450,
    454–456
  tralomethrin mesocosm study, 501, 509, 510
Emibenthos macroinvertebrates, 662,
    670–671, see also Macroinvertebrates
Enclosures, aquatic, 158–159, 202, 227, see
      also Concentration-response, theory;
      Field studies, optimal design of aquatic;
      Pyrethroid, simulations in aquatic
      mesocosms
Endochironomus nigricans, 191, 192,
    196, 682
Endochironomus sp., 592
Endogenous concepts, see also Concepts;
      Exogenous concepts; Structural
      equation modeling
  habitat guild model, 672

LISREL modeling, 632, 634, 637, 643, 645
  atrazine-treated mesocosms, 664, 665,
    678, 679–686
  macroinvertebrate abundance model, 667
  specialist/generalist model, 674, 675
  taxa model, 677
Endogenous indicators, 633, 635, see also
    Indicators; Structural equation modeling
Endosulfan
  estuarine field studies, 51
  farm pond testing
    materials and methods, 37–40
    results and discussion, 40–44, 67
  runoff entry, 222
Endpoints, 26–28, 62, 64–65, 463
Energy circuitry model, 631
Energy flow, 606–607, see also Stability,
    theory
Entry routes, pyrethroid insecticides, 204, see
    also Pyrethroid, simulations in aquatic
    mesocosms
Environment, 397–398, 579, 584
Environmental Protection Agency (EPA)
  aquatic mesocosms
    conduct of study, 19–20
    final review, 20–23, 62–63
    overview, 17–18
    protocol review, 19
  inverted *t*-test, see Inverted *t*-test
EOT index, 110, 120
Ephemeroptera
  colonization in untreated pond mesocosms,
    110, 114, 115
  cyflurin-treated mesocosms/microcosms, 385
  guthion toxicity, 485, 486, 489, 490, 493
  lambda-cyhalothrin toxicity, 426, 447, 448,
    461
  sulprofos-treated ponds and tanks, 329, 334
  tralomethrin toxicity, 505
Epibenthos macroinvertebrates, 662,
    670–671, see also Macroinvertebrates
Epimerization, 428, 430
Epiphytes, 662, 670–671, 674, 682, see also
    Macroinvertebrates; Structural equation
    modeling
Equilibrium, 609, 610, 629, see also Stability,
    theory
Erosion equation, 224
Erthemis simplicolis, 454
ESCORT, see Metsulfuron methyl
Esfenvalerate, see also Mesocosm, data and
    effects predictions for aquatic
    ecosystems

dose response in fish, 168–170
life history changes in pond mesocosms,
    248–251
toxicity in pond mesocosms, 243–244,
    246–248
Estuarine system contaminantion, 49–51
Eta-square, 73, 74, 82
Ethylan, 222
Eucyclops agilis, 113
Euglena acus, 289
Euglenophyceae, 564, 565
Euglenophyta, 281, 282, 311, 436
Eukiefferiella sp., 200
Euorthocladius sp., 200
EXAMS, see Exposure Analysis Modeling
    Systems
Exogenous concepts, see also Endogenous
    concepts; Structural equation modeling
  habitat guild model, 672
  LISREL modeling, 632, 637, 639, 643,
    645, see also Concepts
    atrazine-treated mesocosms, 664, 665,
      678, 679–686
  macroinvertebrate abundance model,
    667–669
  specialist/generalist model, 674, 675
  taxa model, 676
  trophic guild model, 673
Exogenous indicators, 633, 635, see also
    Indicators; Structural equation modeling
Expected entry rates, 202–203
Expected environmental concentration (EEC)
  ANOVA techniques in mesocosm design,
    158
  compartmentalization, 617–619
  connectance, 615–617
  EPA pesticide registration, 18, 64
  guthion, 470, 493
  hexazinone and metsulfuron methyl,
    132–133
  insecticide dosage relation, 202–203
  optimal design of aquatic field studies,
    166–167
  regression analysis approach, 90
  species richness and diversity, 614–615
Experimentwise error rate, 135
Exposure, 12
Exposure Analysis Modeling Systems
    (EXAMS), 12
Exposure-response, 1–2, 48, 492, 493
Exuviae
  cyflurin toxicity, 345, 354–355, 363–364,
    376, 385–386

guthion mesocosm study, 472

Farm ponds, see Pond systems
Fathead minnow, 168
Federal Insecticide, Fungicide and Rodenticide
    Act (FIFRA), 7, 8, 13, 25, see also
    Risk assessment
Feeding behavior, see also Predation
    benthic, 245–246
    bluegill in cyflurin-treated mesocosms/
        microcosms, 380, 389, 390
    environmental stability relation, 615–619
Fenac, 583
Fenthion, 50
Fenvalerate, 51, 203, 220, 221, 389, see also
    Pyrethroid, simulations in aquatic
    mesocosms
Field exposure, 397–398
Field studies, see also individual entries
    actual vs. simulated
        farm pond concept, 37–44, 66, see also
            Pond systems
        mesocosm concept, 36–37, 66
        overview, 35–36
    EPA pesticide registration, 18
    marine and estuarine systems
        conclusions, 55–56
        exposure comparison, 54–55
        literature review of marine ecosystem
            pesticide levels, 49–53
        overview, 47–48
    optimal designs of aquatic
        choice of replicate allocation, 158–162
        conclusion, 170–171
        nonconstant variances, 162–166
        number of doses to test, 166–170
        overview, 157–158
    site-specific vs. simulated, 2–3
FIFRA, see Federal Insecticide, Fungicide and
    Rodenticide Act
Filinia longiseta, 351, 356, 360, 388
Filinia sp., 525
Finfish, 14, 28, 62, see also individual entries
Fire ants, 49
Fish, see also individual entries
    atrazine-treated mesocosm, 661
    guthion mesocosm study, 472
    herbicide effects, 581, 583
    insecticide indirect/direct effects, 581, 582
    planktonivorous, 284, 286, 296
Fisher's Least Significant Differences (LSD)
    test, 245

Food chain, 374, 594, 610, see also Atrazine;
    Cyflurin; Food web; Stability, theory
Food supply, 460, 584, 656
Food web, see also Food chain; Stability,
    theory; Structural equation modeling
    analysis, 630, 631
    atrazine mesocosm study, 615–619
    stability theory relation, 606, 611–613
Fragilaria vaucheria, 551
FSCBG drift model, 217, see also Pyrethroid,
    simulations in aquatic mesocosms
Fungi, 419, 434

Gambusia sp., 456
Gastropoda
    colonization in untreated pond mesocosms,
        115, 116
    cyflurin-treated mesocosms/microcosms,
        350, 351, 353, 354, 363, 383
    sulprofos-treated ponds and tanks, 334
Gas-liquid chromatography (GLC), 415–417
Generation time, 250–252
Gerridae, 447, 454, 455, 461
Gizzard shad, see *Dorsoma cepedianum*
GLC, see Gas-liquid chromatography
Gloeocystis planktonica, 436
Global stability, 608, 609, 613–614, see also
    Stability, theory
Glyptotendipes spp., 192, 594, 677, 683
Goeldichironomus sp., 196
Goltix
    discussion, 317–318
    material and methods, 304–308
    overview, 303–304
    results
        chemical fate and mass balance, 313–316
        fish, 307, 308
        macrophytes, 312–313, 315
        phytoplankton, 309–311, 313, 314
        water quality, 317
        zooplankton and benthic, 308–309
Gomphidae, 453
Gomphonema angustatum, 551, 555
Gomphonema olivaceum, 551
Grabtolebris testudinaria, 522
Grass carp, see *Ctenopharyngodon idella*
Grazing pressure, 288, 289, 581, 582,
    591–592
Green algae, 287, see also Algae
Gross erosion, 224
Growth, 249, 379, 380, 381, 389, see also
    Lepomis machrochirus

Guidelines for Registering Pesticides in the
   United States, 9–10
Guthion
   discussion, 491–494
   methods, 471–475
   overview, 469–470
   pond mesocosm, 158–159
   results
      concentration effect, 476–477
      fish, 477–480
      macroinvertebrates, 485–491
      zooplankton, 480–484
Guttipelopia sp., 196
Gyrinidae, 455

Habitat, 245–248, 581, 656, 662
Habitat guild model, 670–672, 680–682, see
   also Structural equation modeling
Habitat switching, 390
Habitat-imposed compartments, see
   Compartmentalization
Haliplidae, 447, 455, 461
Haptageniidae, 505, 508
Hardness
   cyflurin-treated mesocosms/microcosms,
      347, 349, 377
   sulprofos-treated ponds and tanks, 326, 327,
      331
   tralomethrin mesocosm study, 501
Hazard, 12
Helisoma sp., 449
Hemiptera
   colonization in untreated pond mesocosms,
      114, 116
   guthion toxicity, 486
   lambda-cyhalothrin toxicity, 426, 447, 454,
      461
Herbicides, see also individual entries
   aquatic enclosures, 131–133, see also
      Concentration-response theory, aquatic
      enclosures
   field studies, 50
   mesocosm testing, 580–582
   triazine, see individual entries
   specialist/generalist model, 674
   taxa model, 677
   total effects of atrazine, 682
   trophic guild model, 662, 673
Herpobenthos, 447
Hexagenia spp., 672
Hexarthra mira, 351, 360
Hexazinone

aquatic enclosures, 131–133, see also
   Concentration-response, theory
linear regression analysis, 147, 148, 151
repeated measures ANOVA analysis, 135,
   138, 140
High pressure liquid chromatography (HPLC),
   475, 544
Holopedium spp., statistical analysis
   abundance data, 134, 137
   regression analysis, 144, 146, 148, 150, 151
   RM-ANOVA, 141, 142
Homeorheic activity, 644
Homogeneity of variance, 134, 135
HPLC, see High pressure liquid
   chromatography
HSD, see Tukey's honestly significant
   difference
Hyallela azteca
   atrazine effect, 593–594, 618, 666
   esfenvalerate effect, 168, 170
   taxa model, 677
   trophic guild model, 673
Hybrid tests, 29–31, 65
Hydrobaenus spp., 191, 195, 196, 200
Hydrocarcina, 382, 391, 450, 461
Hydrolysis stability, 313
Hydrophilidae, 455, 461
Hydrosoil
   lambda-cyhalothrin mesocosm study, 407
   microbial population assessment,
      419–420, 434
   residue analysis, 417–419, 421, 429–431
   tralomethrin mesocosm study, 501
Hymenoptera, 486
Hypalon barriers, 407, 408
Hypothesis tests, 29–30, 65, see also One-way
   analysis of variance; Regression
   analysis

Ictalurus punctatus
   atrazine testing, 587, 589, 590, 618, 658
   vertebrate compartment in LISREL
      modeling, 637–640
Indicator variables, 634, 664–665, see also
   Structural equation modeling
Indicators, 632, 634, see also Structural
   equation modeling
Indirect effects, see Effects matrices, LISREL
   modeling
Input-output flow analysis model, 630, 631
Insecta, 114, see also individual entries
Insecticides

foliar applications, see Pyrethroid,
    simulations in aquatic mesocosms
mesocosm testing, 580–582
organochlorine, 49–50, 51
organophosphorus, see Dursban; Guthion
Insects, see also Emergence; Emergence traps
    atrazine effects, 587, 588, 592, 593,
        655–656
    cyflurin-treated mesocosms/microcosms,
        345, 350, 352–356, 362–364, 380
    emergent and bluegill impact, 390
    herbicide effects, 581, 583
    insecticide direct effects, 581, 582
Internal compensation, 642, 646
Interpretational endpoints, see Endpoints
Invertebrates
    atrazine toxicity, 506, 585, 654
    cyflurin-treated mesocosms/microcosms,
        390–391
Inverted *t*-test (EPA), see also Dunnett's test;
    Student's *t*-test
    procedure, 93–95
    Student's *t*-test comparison, 95–99,
        100, 102
Ions, major, 284
Ischnura posita, 454

Karate, see Lambda-cyhalothrin
Kellicotia spp., 275
Kepone, 49–50
Keratella cochlearis
    DIATOX impact, 265–267, 275–276, 286
    cyflurin-treated mesocosms/microcosms,
        351, 360
    lambda-cyhalothrin toxicity, 443, 445
    statistical analysis of herbicides in
        enclosures
        abundance data, 134, 137
        regression analysis, 144, 146, 148,
            150, 151
        RM-ANOVA, 140, 141
Keratella sp., 525, 530
Keystone predators, 364, 374, 391, see also
    Lepomis machrochirus

Labeling of pesticides, 13
Laboratory exposure, field comparison, 1–2,
    242, 258–259, 397–398
Labrundinia spp., 191, 196, 592
Lagerheimia sp., 281
Lake St. George, 259, see also DIATOX
Lambda-cyhalothrin, see also individual entries

biological assessments
    bluegill sunfish, 425–426, 456–461
    filamentous algae and macrophytes, 423
    macroinvertebrates, 424–426, 447–456
    microbial, 419–420, 427, 434
    periphyton, 422–423, 439–441
    phytoplankton, 420, 422, 434–439
    zooplankton, 424, 441–447
biological establishment, 409–410
description of pond, 405–409
discussion and conclusions, 462–465
overview, 404–405
pesticide treatment, 410–415
physicochemical determinations, 417–419,
    426, 433–434
residue analysis, 415–417, 421, 428–433
sampling, 415, 416
spray drift/runoff simulations, 226–229,
    232, 234
statistical analysis, 426
Largemouth bass, see *Micropterus salmoides*
Larsia sp., 192
Lauryl trimethyl ammonium chloride (TMAC)
    community primary productivity, 554, 556
    discussion, 554–559
    dominant taxa, 550–555, 560–565
    equipment, suppliers, 565–566
    facility and site, 537–441
    materials and methods, 541–545
    overview, 536
    periphyton community structure, 545–551
Lecane luna, 275
Leeches, 655
Legislation, 5, see also Environmental
    Protection Agency; Risk assessment
Lepidoptera, 485
Lepomia machrochirus
    atrazine
        environmental stability, 617, 618
        establishment in mesocosm, 658
        total effects, 678, 681
        toxicity, 587, 593, 654
    biomass and macroinvertebrate abundance
        model, 669, 670
    cyflurin toxicity
        benthic colonization, 381–383
        effects, 379–381, 389–390
        emergence, 384–385
        exuviae, 385
        pond systems, 339, 349–350, 362–366
        zooplankton, 386–388
esfenvalerate, 168, 169, 243, 245

food chain perturbations and stability, 610–611
guthion toxicity, 471, 472, 477–479, 493
habitat guild model, 672
lambda-cyhalothrin toxicity, 410, 425–426, 461
pesticide effects in pond mesocosms, 250, 251
predation, 296
specialist/generalist model, 674, 675
sulprofos toxicity, 323, 330–331, 333–335
tralomethrin toxicity, 500, 509, 511
trophic guild model, 672–673
vertebrate compartment in LISREL modeling, 637–640
Leptoceridae, 447, 455, 461
Levene's test, 134, 135, 138
Libellulidae, 350, 354, 361, 363, 366, 453
Life cycle studies and risk assessment, 17
Life history, 248–252
Light stability, 313
Lighting, artificial, 539, 540
Limit of determination (LOD), 417, 418, 428, 429, 432
Limnocorrals, 2, 202, 227, 580, see also DIATOX; Pyrethroid, simulations in aquatic mesocosms
Limnodribus hoffmeisteri, 449
Limnophyses sp., 196
Linear model analysis, 662, see also Atrazine
Linear regression analysis
  guthion mesocosm study, 475, 480, 485, 491–492
  herbicide effect in aquatic enclosures, 135–136, 142–147, 148
Linear Structural Relations (LISREL), see Structural equation modeling, ecosystem analysis
Linkages, see also Structural equation modeling
  atrazine mescosm study, 615–616, 618
  specialist/generalist model, 675
  stability relation, 608, 611
LISREL, see Structural equation modeling, ecosystem analysis
Littoral corrals, 3, 580
Littoral zone, 211, 212, see also Pyrethroid, simulations in aquatic mesocosms
Loading
  cyflurin in microcosm/mesocosm systems, 340, 341, 357
  Goltix artificial pond systems, 306

pesticide and Environmental Protection Agency guidelines, 21
  sulprofos in earthen pond/tank systems, 324–325
Local stability, 608, 609, 614, see also Stability, theory
LOD, see Limit of determination
LOEC, see Lowest observable effect concentration
Lotic systems, 3
Lowest observable effect concentration (LOEC), 135, 142, 654, 679
LSD, see Fisher's Least Significant Differences test
Ludwigia uruguayensis, 410, 441
Lumber industry, 258–259
Lymnaeidae, 449
Lyngbya limnetica, 436, 439

Macrocosms, 580, see also Mesocosms; Microcosms
Macroinvertebrate abundance model, 667–670, 680–682
Macroinvertebrate artificial substates (MAS), see also Artificial benthic substrates; Artificial substrates
  cyflurin-treated mesocosms/microcosms, 345, 352, 353, 376, 384
  tralomethrin mesocosm study, 501, 505, 508, 509
Macroinvertebrates
  atrazine-treated mesocosm, 593–594, 660, 662
  biometrics in pond mescosms, 110, 120–122
  colonization in untreated pond mesocosms, 108–110, 113–116, 123
  compartment in LISREL modeling, 637–640
  cyflurin-treated mesocosms/microcosms, 350, 352–356, 361–362, 364–365
  guthion toxicity, 472, 485–491
  herbicide effects, 581, 583
  insecticide direct effects, 581, 582
  lambda-cyhalothrin toxicity, 409–410, 424–427, 447–456, 461
Macrophytes, see also individual entries
  atrazine effects, 587–589, 591–593, 655, 660
  compartment and stability, 617–619
  cyflurin-treated mesocosms/microcosms, 348, 356–357, 389
  Goltix-treated ponds, 312–313, 315, 317

guthion mesocosm study, 471, 486
habitat guild model, 671
herbicide/insecticide effect in pond systems,
    245–248
lambda-cyhalothrin toxicity, 409–410, 423,
    441, 460–461
large ponds systems, 305–306, see also
    Goltix, mesocosm/microcosm ponds
macroinvertebrate abundance model,
    668, 670
refugia role, 584
tralomethrin mesocosm study, 500
trophic guild model, 673
untreated pond mesocosms, 112
Macrothrix rosea, 353, 387
Magnesium ions, 284
Malathion, 50
Mallomonas caudata, 281
Marine systems, 48, 49, 51
MAS, see Macroinvertebrate artificial substrate
Mass spectrometry (MS), 417
Mathematical models, see also Runoff; Spray
        drift; Stability, theory
    pesticide runoff, 224, 226–227
    spray drift deposition, 215, 217
    stability, 613–614
MAX models, 72–77
Maximum estimated entry rate (MEER),
        87, 89
    foliar-applied pyrethroid insecticides, 217
    spray drift/runoff entry of pesticides, 231,
        233–237
Maximum estimated environmental
        concentration (MEEC), 27
Maximum likelihood estimation, 631
Mean interaction strength, 608, 612–613, 618
Measurement error, 632, 635, see also
        Structural equation modeling
Measurement model, 633, see also Structural
        equation modeling
MEEC, see Maximum estimated environmental
        concentration
MEER, see Maximum estimated entry rate
Megaloptera, 485
Melosira varian, 550, 551
MERL systems, pesticide studies, 52
Mesocosms
    atrazine testing design, 586–587, see also
        Atrazine
    chemical testing evaluation, 2–3
    data and effects predictions for aquatic
        ecosystems

methods and materials, 243–245
    overview, 241–242
    results and discussion, 245–251
    summary and conclusions, 251–252
definition, 580
design and statistical analysis, 69–70
    ANOVA vs. regression approach, 89–90
    appendix, 102–103
    conclusions, 100
    criteria, 86–89
    discussion, 99–100
    multiple comparison test use, 90–91
    overview, 85–86
    student vs. inverted *t*-test, 91–99
farm ponds comparison, see Field studies,
        actual vs. simulated
foliar insecticide application simulations, see
        Pyrethroid, simulations in aquatic
        mesocosms
Goltix testing, see Goltix
management and treatment overview,
        177–178
Office of Pesticide Programs guidelines, 8,
        see also Risk assessment (Office of
        Pesticide Programs)
pesticide studies design, 51–53
pond, see Cyflurin; Pond systems
regulatory endpoints in tests
    conclusions, 31–32
    discussion, 30–31
    endpoint descriptions, 25–28
    experimental design, 29–30
    overview, 25–26
single-species laboratory test comparison,
        578–580
Mesocyclops edax, 272, 274
Metabolism, 422, 437
Metamitron, pond systems, see Goltix
Meteorology, 205, 207–211, 213, 216, see
        also Weather
Methoxychlor, 207, 222, 582, see also
        Pyrethroid, simulations in aquatic
        mesocosms
Methyl parathion, 582
Metsulfuron methyl, 131–133, see also
        Concentration-response, theory
Microcosms, 2, 580, 585, see also Cyflurin;
        Goltix
Microcystis aeruginosa, 558
Micropterus salmoides, 463
MIN models, 72–77
Mirex, 49, 51

Modification indices, 637
Moina micrura, 531
Mollusca, 447, 449, 461
Monoalkyl ammonium compounds, see Lauryl
    trimethyl ammonium chloride (TMAC)
Monostyla spp., 275, 388
Monte Carlo permutation test, 521, 527
Mortality, 8, 14, 28, 308, 379, 389
Mosquito, 50
Mougeotia sp., 439
MS, see Mass spectrometry
Multiple comparison tests, 90–91, 161
Municipal wastewater treatment plants
    (WWTP), 536, see also Lauryl
    trimethyl ammonium chloride (TMAC)
MUSLE, see Universal Soil Loss Equation

Naididae, 350, 353, 360, 449
Najas spp., 246, 471, 660
Nanocladius sp., 192
Naucoridae, 454
Nauplii, 505, 506, 520, 524
Navicula cryptocephala, 551
Navicula graciloids, 550, 551, 553, 559
Navicula gregaria, 551
Navicula luzonensis, 551
Navicula pelliculosa, 558
Navicula pupula, 552
Navicula radiosa, 551
Navicula salinarum, 550, 551, 553
Navicula symmetrica, 550, 552
Navicula viridula, 551
Naviculaceae, 545
Nekton, 447
Nematoda, 114
Neurotoxins, 375
Neuston, 447
Nitrates, 111, 284, 347–349, 377, see also
    Nitrites; Nitrogen
Nitrites, 111, 284, 377, see also Nitrates;
    Nitrogen
Nitrogen, see also Nitrates; Nitrites; Particulate
    nitrogen
    cyflurin-treated mesocosms/microcosms,
        347, 349, 377
    Goltix-treated ponds, 307, 317
    guthion mesocosm study, 471
    insecticide direct effects, 581, 582
    lauryl trimethyl ammonium chloride stream
        mesocosms, 543–544
Nitzchiaceae, 545
Nitzschia acicularis, 551, 554

Nitzschia dissipata, 554
Nitzschia frustulum, 550, 551, 553, 555, 557
Nitzschia kutzingiana, 550, 553
Nitzschia palea, 439, 550, 551
No Observable Concentration Effect (NOEC)
    atrazine in aquatic ecosystems, 654
    bluegill sunfish in lambda-cyhalothrin-
        treated mesocosms, 459–460
    DIATOX impact in limnocorrals, 260, 285,
        286, see also DIATOX
    Environmental Protection Agency pesticide
        registration, 18
    regression analysis, 89
NOEC, see No Observable Concentration
    Effect
Nondemographic stability, 608, 619, see also
    Stability, theory
Nonlinear regression analysis, 136–137, 147,
    149–151
Nostoc spp., 439
Notonectidae, 447, 454, 455, 461
Nutrients, 111, 584

Observational endpoints, see Endpoints
Odonata
    colonization in untreated pond mesocosms,
        110, 114–116
    cyflurin-treated mesocosms/microcosms,
        350, 354, 361, 366, 380
    guthion toxicity, 485, 486, 489, 491
    lambda-cyhalothrin toxicity, 453–454, 461
OECD, see Organization for Economic
    Cooperation and Development
Oecetis inconspicua, 455
Oedogonium sp., 439
Office of Pesticide Programs (OPP), see Risk
    assessment (Office of Pesticide
    Programs)
Offsets, 207–211, 213, 215, see also Spray
    drift
Oligochaeta
    colonization in pond mesocosms, 114, 116
    cyflurin-treated mesocosms/microcosms,
        350, 353, 361, 391
    lambda-cyhalothrin toxicity, 447, 449–450,
        461
Oliveridia sp., 200
Omnibus F-test, 72, 75–76
Omnivores, 662, 673, 674, 677, see also
    Structural equation modeling
Oncorhynchus mykiss, 304, 307, 308, 371,
    389

One-way analysis of variance (ANOVA)
  atrazine-treated mesocosms, 665
  bluegill survival and esfenvalerate toxicity,
    250–251
  cyflurin-treated mesocosms/microcosms,
    346, 377
  lambda-cyhalothrin mesocosm study, 426
  lauryl trimethyl ammonium chloride stream
    mesocosms, 545
  mesocosm design, 158
  regression analysis comparison in mesocosm
    studies, 89–90, 100
  repeated measures (RM-ANOVA)
    DIATOX impact in limnocorrals, 260,
      264, 273, 274, 279, 282, see also
      DIATOX
    herbicide effect in aquatic enclosures,
      135, 138, 140–142
  replicates for analysis, 88
  tralomethrin mesocosm study, 503
Oocystis sp., 281, 436
Organic carbon, 501, see also Carbon;
    Dissolved organic carbon; Particulate
    organic carbon; Total organic carbon
Organization for Economic Cooperation and
    Development (OECD), 317
Organochloride pesticides, 220
Organophosphorus insecticide, see Dursban;
    Guthion
Orthocladiinae
  emergence, 183–185, 192
  phenology patterns, 187, 189, 190, 195
  species richness, 192, 193
  tralomethrin toxicity, 509, 510
Orthocladius sp., 195, 200
Oscillatoria amphibia, 436, 439
Oscillatoria spp., 277, 288
Osmoregularity, 375
Ostracoda
  colonization in untreated pond mesocosms,
    113
  Dursban toxicity, 520, 524–525, 529, 530
  lambda-cyhalothrin toxicity, 446
Overcrowding, 460
Overland flow, 218, 219, 224
Oxyethria sp., 455

Palpomyi spp., 594
Pantala sp., 454
Parachironomus spp., 594, 677, 683
Parakiefferiella sp., 592
Paramerina smithae, 191, 196

Particle detachment, 218, 219
Particulate nitrogen (PN), 281, 283, see also
    Nitrates; Nitrites; Nitrogen
Particulate organic carbon (POC), see also
    individual entries
  cyflurin-treated mesocosms/microcosms,
    345, 348, 349, 377
  DIATOX-treated limnocorrals, 281, 283
Particulate organic material (POM), 245
Path analysis, 630, 663, 666, see also
    Structural equation modeling,
    ecosystem analysis
Path coefficients, see also Atrazine; Structural
    equation modeling
  macroinvertebrate abundance model,
    668, 669
  structural equation modeling using LISREL,
    630, 631, 634, 636–641, 645, 646
  atrazine-treated mesocosms, 663
  taxa model, 677
PCP, see Pentachlorophenol
Pelagic compartment, 618, 619, see also
    Stability, theory; Structural equation
    modeling
Pentachlorophenol (PCP), 51, 247, 259, 582
Perca flavescens, 261, 263–265, 268, 285,
    see also DIATOX
Peridinium, 117
Periphyton
  atrazine toxicity, 593, 655
  cyflurin-treated mesocosms/microcosms,
    377, 379
  herbicide effects, 583, 584
  lambda-cyhalothrin toxicity, 345, 422–423,
    439–441
  lauryl trimethyl ammonium chloride stream
    mesocosm, 541–543
Permethrin, see also Pyrethroid, simulations in
    aquatic mesocosms
  application and spray drift, 203, 207, 213,
    215, 216
  direct/indirect effects, 582
  field studies, 50
  fish feeding behavior, 390
  runoff entry, 220, 221, 223
  zooplankton growth in pond mesocosms,
    249
Permutation test, 308
Perturbations, 609–611, see also Stability
    theory
Pesticide Assessment Guidelines, 10, 36
Pesticide Run-off Simulator program, 88, see
    also Runoff

Pesticides, see also individual entries
  application rate and frequency, 231,
    233–235, 249, see also Pyrethroid,
    simulations in aquatic mesocosms
  direct/indirect effects
    atrazine mesocosm testing, 585–594, see
      also Atrazine
    conclusions, 594–596
    mesocosms, 580–585
    overview, 578
    single-species vs. mesocosm, 578–580
  Environmental Protection Agency, see
    Environmental Protection Agency,
    aquatic mesocosms
  Federal Insecticide, Fungicide and
    Rodenticide Act, 8, 11
  field studies of effects in literature, 49–51
  Office of Pesticide Programs guidelines,
    9–11
  registration, pond ecosystems relation, 338
  risk assessment process, 11–13
    atrazine effects, 587
    cyflurin-treated mesocosms/microcosms,
      345, 347, 349, 377, 379
    Dursban pond mesocosm study, 519,
      522, 523
    Goltix-treated ponds, 307, 317
    guthion mesocosm study, 471, 476
    lambda-cyhalothrin mesocosm study, 422,
      427, 434
    lauryl trimethyl ammonium chloride stream
      mesocosms, 539, 543–544
    macroinvertebrate abundance model, 669
    sulprofos-treated ponds and tanks, 326,
      327, 331
    untreated pond mesocosms, 111
Phaeophytin *a*, 326, 420, 437, 439, 543
Phantom midge, see *Chaoborus*
Phenology patterns, 187–190
Phormidum sp., 439
Phosphate, 112, 347, 349, 581, 582
Phosphorus, 307, 317, 471, 543–544
Photodegradation, 285
Photolysis, 259, 286
Photoperiod, 539, 540
Photosynthesis, see also Chlorophyll *a*;
    Phaeophytin *a*
  cyflurin-treated mesocosms/microcosms,
    345, 377, 379
  inhibition by atrazine, 585, 654, 656
  lauryl trimethyl ammonium chloride stream
    mesocosms, 543

phytoplankton and lambda-cyhalothrin
  toxicity, 420, 434, 437, 438
Physa sp., 449
Physidae, 449
Phytoflagellates, 436
Phytoplankton, see also individual entries
  atrazine toxicity, 249–250, 587–588,
    655, 661
  biometrics in pond mescosms, 110,
    117, 118
  colonization in pond mesocosms, 112, 123
  compartment in LISREL modeling, 637–640
  cyflurin-treated mesocosms/microcosms,
    345, 350, 351, 377
  DIATOX toxicity in limnocorrals, 262–263,
    277, 279–281
  Goltix-treated ponds, 309–314, 317–318
  herbicide response in aquatic enclosures,
    132–133, 140, see also Concentration-
    response, theory
  lambda-cyhalothrin toxicity, 420, 422,
    434–439
  macroinvertebrate abundance model,
    668, 669
  specialist/generalist model, 675
  statistical analysis of herbicides in
    enclosures, 136
  total effect of atrazine, 678
  untreated pond mesocosm, 108, 109
Planariidae, 449
Planktivores, 364, 390
Planktonic communities, 53
Planorbidae, 449
Plants, see also individual entries
  atrazine toxicity, 586, 614, 654, 655–656
  endosulfan effects in farm ponds, 42, 43
  herbicide direct effects, 581, 583, 584
  insecticide direct effects, 581, 582
Ploesoma spp., statistical analysis of herbicide
    effects
  abundance data, 134, 137
  regression analysis, 143–145, 148
  RM-ANOVA, 141
Plot size, 220–223
PN, see Particulate nitrogen
POC, see Particulate organic carbon
Point estimate tests, 29, 30, 65
Pollution stress, 675, see also Stress
Polyarthra remata
  cyflurin-treated mesocosms/microcosms,
    351, 357, 360, 364, 388
  lambda-cyhalothrin toxicity, 443, 445

Polyarthra spp.
  DIATOX impact, 265–268, 275, 285, 286
  Dursban toxicity, 525, 530
Polypedilum sp., 191, 196
POM, see Particulate organic material
Pond depth concept, see also Structural
      equation modeling
  macroinvertebrate abundance model,
      668, 670
  specialist/generalist model, 674
  taxa model, 676–677
  total effects of atrazine, 681
  trophic guild model, 673
Pond systems, see also Dursban; Pyrethroid,
      simulations in aquatic mesocosms;
      Sediment transfer
  ecological development and biometry
    biometrics, 116–122, 124–125
    colonization, 112–116, 122–124
    environmental characteristics, 111–112
    methods, 108–111
    overview, 105–107
    site, 107–108
  Environmental Protection Agency pesticide
      registration, 18, 20
  foliar insecticide applications, 209, 211,
      215, 216
  materials and methods, 37–40
  mesocosms, 580
  models of freshwater mescosms, 54
  pesticide application and spray drift/runoff,
      223–225, 227
  results and discussion, 40–44
  test system, 2
Potamogeton spp., 312, 471, 500, 660
Potassium, 294, 544
Potato, 213, 215
Potthastia sp., 200
Predation, see also *Lepomis machrochirus*
  atrazine mesocosm study and environmental
      stability, 615, 618
  bluegill
    cyflurin-treated mesocosms/microcosms,
      380
    invertebrates, 390–391
    lambda-cyhalothrin toxicity study,
      454, 460
  cyflurin-treated mesocosms/microcosms,
      364–366
  DIATOX-treated limnocorrals, 284–285
  grass shrimp and Mirex exposure, 51
  herbicide effects, 581

insecticide indirect effects, 581, 582
  sulprofos-treated ponds and tanks, 334
Predator-prey interactions, 631, see also
      Predation
Premixing, 230
Press perturbation, 621, see also Perturbations
Prey depletion, 390
Probezzia spp., 594
Procladius spp,, 191, 192, 195, 196
Product cycling, 608, 612
Productivity, see also individual entries
  chemical toxicity relation, 245–248
  cyflurin-treated mesocosms/microcosms, 385
  lauryl trimethyl ammonium chloride stream
      mesocosms, 543, 554, 556, 557, 559
  phytoplankton and lambda-cyhalothrin
      toxicity, 420, 437, 438
  pond mesocosms, 120
Prometone, 583
Propazine, 583
Proportional means tests
  discussion, 81
  environmental hypothesis, 78–80, 107
  minimal acceptable designs, 72–77
  overview, 71–72
  technical notes, 82–83
Protozoa, 442, 461
PRZM model, 217, 226
Psectrocladius sp., 196
Psectroctadium vernalis, 191
Pseudochironomini
  emergence, 183–186
  guthion toxicity, 487, 489, 493
  phenology patterns, 187, 189, 190, 195
Pulse perturbation, 621, see also Perturbations
Pyrethroid, simulations in aquatic mesocosms,
      see also Runoff; Spray drift
  applications
    methods, 228–232
    overview, 227–228
    rates and frequencies, 231, 233–235
  entry routes, 204
  overview, 201–203
  review, 203–204
  runoff entry
    agricultural in-use determinations,
      223–225
    conclusions, 227, 237
    factors influencing, 218–219
    literature review, 219–223
    mathematical models, 224, 226–227
  spray drift deposition

agricultural in-use determinations,
213–216
conclusions, 217, 236
factors influencing, 205
literature review, 205–213
mathematical models, 215, 217
Pyrethroid pesticides, see Cyflurin
Pyrrhophyta, 436

Quality assurance, 23
Quotient method, 12–13

R factor, see Rainfall erosivity
Rainbow trout, see *Oncorrhynchus mykiss*
Rainfall, 220–222, 225, 226
Rainfall erosivity (R factor), 224
Rana catesbeiana, 459, 593, 637–640, see
also Tadpoles
Randomized complete block design (RCBD),
260, 264
Rannunculus aquatilis, 312, 315
RCBD, see Randomized complete block
design
Recovery time, 610, 619, 620, 645, 655, see
also Stability, theory
Refugias, see also Macrophytes
concrete tank microcosms, 339
cyflurin-treated mesocosms/microcosms,
377, 389
herbicide effects, 584
lambda-cyhalothrin mesocosm study,
460–461
Regression analysis, 89–90, 100, 158, see also
One-way variance of analysis
herbicide effects in enclosures, 144–146,
148, 150
Regression models, 665–666, see also
Atrazine
Regulatory endpoints, see Endpoints
Relative humidity, 205, 208, 216
Repeated measures one-way analysis of
variance, see One-way analysis of
variance, repeated measures
Replicates
atrazine mesocosm study, 658–660
guthion mesocosm study, 476–477
lambda-cyhalothrin mesocosm study,
410–411, 463
mesocosm studies, 88, 93, 95–99
optimal allocation in design of aquatic field
studies, 158–162
Reproduction, 349, 459, 477–478, see also
Lepomis machrochirus

Residue analysis, see also Hydrosoil; Water
quality
cyflurin-treated mesocosms/microcosms,
342, 344–346, 347–349
Dursban pond mesocosm study, 519,
521, 528
Environmental Protection Agency pesticide
registration, 19, 21–22, 63
Goltix-treated ponds, 308, 314–316
guthion mesocosm study, 474–475,
476–477
lambda-cyhalothrin mesocosm study,
415–417, 428–429, 431–433
sulprofos-treated ponds and tanks, 326
tralomethrin mesocosm study, 501
Resilience, 608–610, 645, see also Stability,
theory
Resistance, see also Atrazine; Stability, theory
atrazine, 587, 589, 655
definition, 608, 610, 619
Respiration, community
cyflurin-treated mesocosms/microcosms,
345, 377, 379
phytoplankton and lambda-cyhalothrin
toxicity, 420, 437, 438
Restricted Use classification, 13, 28
Rhaphidiopsis curvata, 436
Rhoicosphenia curvata, 551, 555
Rill flow, 218, 219, 224, see also Runoff
Risk assessment (Office of Pesticide
Programs)
ecological-based regulatory decisions, 13–14
mesocosms vs. actual field study, 11, 61–62
overview, 7–8
pesticide regulation and guidelines history in
field tests, 8–11
process, 11–13
Rotifera
atrazine tolerance, 589, 590
colonization in untreated pond mesocosms,
113, 117
cyflurin-treated mesocosms/microcosms,
350, 351, 355, 360, 363, 388
DIATOX impact, 265–268, 270–271,
275–276, 285, 286
Dursban toxicity, 524, 525, 529, 530
esfenvalerate effect in pond mesocosms,
248, 249
guthion toxicity, 480–481, 484, 493
lambda-cyhalothrin toxicity, 442–445, 461
statistical analysis of herbicides in
enclosures

abundance data, 134, 138
   regression analysis, 144–146, 148, 150
   RM-ANOVA, 141
   sulprofos-treated ponds and tanks, 328
Runoff, see also Spray drift
   agricultural in-use determinations, 223–225
   atrazine toxicity assessment, 654
   conclusions of study, 227
   cyflurin
      concrete and pond microcosms, 339–341,
         346, 348, 358
      mesocosms/microcosms, 376–378
   factors influencing, 218–219
   guthion mesocosm study, 473–474
   lambda-cyhalothrin mesocosm simulation,
      412–414, 428, 431, 432
   literature review, 219–223
   mathematical models, 224, 226–227
   overview, 217–218
   pesticide application
      application rate and frequency, 231,
         233–235
      farm ponds, 39, 40
      sulprofos-treated ponds and tanks, 324, 325,
         333
      tralomethrin mesocosm study, 501, 503

Salvelinus fontinalis, 654
Sampling, see also individual entries
   atrazine-treated mesocosm, 660–665
   Dursban pond mesocosm study, 520–521
   guthion mesocosm study, 471–472
   lambda-cyhalothrin mesocosm study,
      415, 416
   tralomethrin mesocosm study, 500–502
Sapstain, 259
SAS, see Statistical Analysis System
Scaling, see also individual entries
   concrete tank microcosms/pond mesocosms,
      339, 358
   Dursban pond mesocosm study, 531
   introduction, 301–302
   pesticide testing, 462, 464
   research needs, 399–400
   summary and discussion, 397–400
Scapholeberis mucronata, 275
Scenedesmus subspicatus, 317, 318
Scenesecmus quadricauda, 436
Schizothrix calcicola, 551
Schroderia setigera, 436
Scout, see Tralomethrin
SCS, see Soil Conservation Service

Season, 610, see also Stability, theory
Secchi depth, 112, 263, 277, 278
Secondary effect, 288, 289, see also Effects
   matrices; Stability, theory; Structural
   equation modeling
Sediment transfer
   abundance patterns, 188, 191, 196–197
   conclusion, 197
   discussion, 188–191
   emergence chronology, 186–187, 194–195
   methods and materials, 180–182
   overview, 179–180
   phenology patterns, 187–190, 195–196
   species richness and emergence
      composition, 182–186, 191–193, 296
Seeding, mesocosms, see Sediment transfer
Selenastrum capricornatum, 558
Sensitivity structure, 645
Sewage effluents, 538, see also Lauryl
   trimethyl ammonium chloride
Shannon-Weaver index
   DIATOX-treated limnocorrals, 263, 264,
      275, 277, 279
   Goltix-treated ponds, 310
   lauryl trimethyl ammonium chloride stream
      mesocosms, 543, 548, 550
Shrimp, grass, 51, see also Predation
Simazine, 583
Simocephalus sp., 522, 528
Simulator for Water Resources in Rural Basins
   (SWRRB) model, 12, 88, 224, 226,
   463, 501
Single-species laboratory tests, 1, 578–580
Snails, 330, 491, 493
Socioeconomic endpoints, see Endpoints
Sodium, 544
Sodium arsenate, 583
Sodium-potassium transport, 375
Soil, 218–222, see also Hydrosoil
Soil Conservation Service (SCS), 224
Soil-water slurry, 414, 445, 452
Solubility, pesticides, 204, 219–220
Soluble reactive phosphorus (SRP), 284, see
   also Phosphorus
Sorbents, 245
Sorghum, 585
Sparganium erectum, 312
Special Reviews, 14, 29, 62, see also Risk
   assessment (Office of Pesticide
   Programs)
Specialists, 682
Specialist/generalist model, 674–675, 680–682

Species diversity, 608, 612, 614–615
Species richness, see also Taxa richness
  atrazine effect, 656
  mesocosm test ponds, 182–186, 192–193
  stability relation, 608, 611–612, 614–615
Spike rush, 42, 43
Spirogyra sp., 435
Spray drift, see also Runoff
  cyflurin application, 339, 341, 346, 347,
    357, 376–377
  guthion mesocosm study, 473–474
  lambda-cyhalothrin mesocosm study
    simulation, 411–414, 428, 431, 432
  pesticide application rate and frequency,
    231, 233–235
  pyrethroid insecticides
    agricultural in-use determinations,
      213–216
    conclusions, 217, 236
    factors influencing deposition, 205
    literature review, 205–213
    mathematical models, 215, 217
    overview, 204
  sulprofos-treated ponds and tanks, 324, 327
  tralomethrin mesocosm study, 501, 503
Spray Drift Task Force, 203, 217, 464
Sprayers, ground boom, see also Spray drift
  Dursban application, 519
  foliar-applied pyrethroid pesticides, 205,
    213, 228–230, 232
  lambda-cyhalothrin application, 411
SRP, see Soluble reactive phosphorus
Stability, see also Structural equation modeling
  definition, 607, 629
  index, LISREL modeling, 640, 642–646,
    665, 678
  theory
    definitions, 607–610
    discussion, 620–621
    environmental correlates, 611–619
    historical background, 606–607
    mathematical modeling approach,
      613–614
    perturbation qualities, 610–611
Stander's index, 308
Statistical Analysis System (SAS), 245
Stephanodiscus invisitatus, 551
Stream mesocosms, see Lauryl trimethyl
    ammonium chloride
Stress, 380, 389, 645–646
Strombidium delicatissiumum, 442
Structural equation modeling

ecosystem analysis
  aggregation of variables, 636–642
  appendix/effects matrices, 650–652
  ecosystem level modeling and
    ecotoxicological testing, 645–647
  effects matrices and stability indices,
    643–645
  historical background, 630–632
  LISREL modeling, 632–636
  overview, 628–630
ecotoxic effects of atrazine
  discussion, 679–686
  experimental design, 657–660
  habitat guild model, 670–672
  macroinvertebrate abundance model,
    667–670
  materials and methods, 660–665
  model summary, 677–679
  overview, 654–657
  results, 665–667
  specialist/generalist model, 674–675
  study area, 657–659
  taxa model, 676–677
  trophic guild model, 672–673
Student's *t*-test, see also Dunnett's test;
    Inverted *t* test
  EPA inverted *t*-test comparison, 95–99,
    100, 102
  lambda-cyhalothrin mesocosm study, 426
  procedure, 91–93
Stylaria, 449
Sulfate, 284, 543–544
Sulfonylurea herbicides, see Metsulfuron
    methyl
Sulprofos
  discussion, 331–336
  methods, 322–326
  overview, 321–322
  results, 326–335
Surface films, 230
Surfactants, see Lauryl trimethyl ammonium
    chloride (TMAC)
Surirella spp., 558
Swath displacement, 207
Swimming behavior, 380
SWRRB, see Simulator for Water Resources
    in Rural Basins model
Sympotthastia sp., 200
Synchaeta spp., 525, 530
Synechocystis sp., 281
Synedra rumpens, 439
Syrphidae, 455

Tabanidae, 455
Tadpoles, 456–459, 593, 617, 637–640
Tagging, fish, 19
Tanks, fiberglass, as mesososm, see Sulprofos
Tanypodinae
  cyflurin-treated mesocosms/microcosms, 350, 353, 355, 361, 363, 382, 385
  emergence, 183–185
  lambda-cyhalothrin toxicity, 447, 454, 455, 461
  phenology patterns, 187 189, 190, 195
  species richness, 192
  tralomethrin toxicity, 509, 510
Tanytarsini
  emergence, 183–185, 192
  guthion toxicity, 487, 489, 493
  phenology patterns, 187–190, 195
  species richness, 193
Tanytarsus spp., 191–193, 196
Target swath, 207, 209, 210
Taxa model, 676–677, 680–682
Taxa richness, see also Species richness
  cyflurin-treated mesocosms/microcosms, 384–386
  lauryl trimethyl ammonium chloride stream mesocosms, 546–548, 558
  untreated pond mesocosms, 110, 117–122
Technical Guidance Document, 19, 20, 29, 64–65
TEEAM, see Terrestrial Ecosystem Exposure Assessment Model
Temperature
  cyflurin-treated mesocosms/microcosms, 345
  DIATOX-treated limnocorrals, 263
  Dursban pond mesocosm study, 519, 523
  Goltix-treated ponds, 307, 317
  guthion mesocosm study, 471
  insecticide spray drift relation, 205, 208, 216
  lambda-cyhalothrin mesocosm study, 433–434
  lauryl trimethyl ammonium chloride stream mesocosms, 539, 543–544, 546, 549
  sulprofos-treated ponds and tanks, 326, 327
  tralomethrin mesocosm study, 501
Terbufos, 50
Terrain, 205
Terrestrial Ecosystem Exposure Assessment Model (TEEAM), 217
Tetrachlorophenol (TeCP), see DIATOX
2,3,4,6-Tetrachlorophenol, see DIATOX

Thermoionic specific detection (TSD), 134, 137
Thiodan 3EC, see Endosulfan
Three point diel oxygen method, 345, 501
Tidal exchange, 48
Time of day, 261, see also DIATOX
Tipulidae, 455
TKN, see Total Kjeldahl nitrogen
TOC, see Total organic carbon
Tolerance, 587, 589, 655, see also Atrazine
Topsoil, 106, 108, 123
Total biomass, 608, 612, see also Biomass; Stability, theory
Total effects, 630, 677–679, see also Effects matrices
Total Kjeldahl nitrogen (TKN), 284, see also Nitrogen
Total organic carbon (TOC), see also individual entries
  cyflurin-treated mesocosms/microcosms, 345, 348, 349, 377, 378
  Dursban pond mesocosm study, 519, 521–522
  lauryl trimethyl ammonium chloride stream mesocosms, 543–544
  sulprofos-treated ponds and tanks, 326, 327
Total phosphorus-filtered (TPF), 283, 284
Total suspended solids (TSS)
  cyflurin-treated mesocosms/microcosms, 347, 349, 377
  guthion mesocosm study, 471
  sulprofos-treated ponds and tanks, 326, 327, 333
  tralomethrin mesocosm study, 501
Toxaphene, 222
Toxins, 621, see also Perturbations
TOXSTAT computer program, 475
TPF, see Total phosphorus-filtered
Trachelomonas hispada, 289
Tralomethrin, see also Pyrethroid, simulations in aquatic mesocosms
  materials and methods, 499–504
  overview, 497–499
  results and discussion, 504–514
  spray drift/runoff simulations, 203
Transport, insecticides, 218, 219, see also Pyrethroid, simulations in aquatic mesocosms
Triaenodes tardus, 654
Triazine herbicides, see Hexazinone
Trichocerca sp., Dursban toxicity, 525
Trichoptera

cyflurin-treated mesocosms/microcosms, 350, 352, 363, 385
  guthion toxicity, 485
  lambda-cyhalothrin toxicity, 447, 448, 455, 461
Trophic guild model, 672–673, 680–682, see also Structural equation modeling
Trophic interaction theory, 285, see also DIATOX
Trophic structure, see also Atrazine; Stability, theory
  atrazine/esfenvalerate effect in pond mesocosms, 245–248, 252
  atrazine mesocosm study, 614–616, 662
  stability theory relation, 606, 610, 611
Trophospecies, 621
Trout, rainbow, see *Oncorrhynchus mykiss*
TSD, see Thermoionic specific detection
TSS, see Total suspended solids
Tubificidae, 449
Tukey's honestly significant difference (HSD), 503
Tukey's test, 90
Turbellaria, 449, 461
Turbidity
  atrazine-treated pond mesocosms, 592–594
  cyflurin-treated mesocosms/microcosms, 347–349, 379, 389
  herbicide effects, 584
  lambda-cyhalothrin mesocosm study, 434, 456
  macroinvertebrate abundance model, 668
Type I/II errors, 95, 100, 107, 135, 174
Typha latifolia, 312
Typha spp., 589, 617, 660, see also Atrazine
Typhloplanidae, 449

Universal Soil Loss Equation (USLE), 224, 227
USLE, see Universal Soil Loss Equation

Variances, nonconstant, 162–166
Variance/covariance matrix, 636, see also Structural equation modeling
Vegetation, 213
Veliidae, 447, 454, 455, 461
VELPAR, see Hexazinone
Vertebrates, 375, 506, 585, 637–640
Volatilization, 204, 357

Water circulation systems
  guthion mesocosm study, 473–474

lambda-cyhalothrin mesocosm study, 407, 409
  cyflurin study, 339, 340
Water fleas, see *Daphnia magna*
Water quality, see also individual entries
  atrazine-treated mesocosm, 661
  cyflurin-treated mesocosms/microcosms, 345, 347, 349
  DIATOX-treated limnocorrals, 262–263, 281–284
  Goltix-treated ponds, 317
  guthion mesocosm study, 471
  lambda-cyhalothrin mesocosm study, 416–417, 428–429
  lauryl trimethyl ammonium chloride stream mesocosms, 539, 543–544
  tralomethrin mesocosm study, 501
Water temperature, see Temperature
Weather, see also Meteorology
  Goltix-treated ponds, 317
  lauryl trimethyl ammonium chloride stream mesocosms, 546
  pesticide runoff, 218, 220–222, 225, 226
Weed beds, 409
Weight, 379–382
Weight-of-evidence, 25–26, 31
  review, 12, 13, 62
Welch's *t*-test, 135, 140, 141
Whitefish, see *Coregonus lavaretus maraenus*
Williams' test, 470, 480, 485, 487, 489, 491
Wind, 205, 207–211, 213, 216, see also Spray drift
WWTP, see Municipal wastewater treatment plants

Yellow perch, see *Perca flavescens*
Young-of-the-year (YOY), 250, 251, 375
YOY, see Young-of-the-year

Zannichellia palustris, 312
Zooplankton, see also individual entries
  atrazine
    esfenvalerate toxicity, 243–244, 246, 247–251
    environmental stability, 618
    toxicity, 655, 661
  biometrics in untreated pond mescosms, 110, 117, 119, 120
  bluegill predation, 391
  colonization in untreated pond mesocosms, 113, 123
  compartment in LISREL modeling, 637–640

cyflurin-treated mesocosms/microcosms, 345, 362, 377, 386–388
DIATOX impact, 262, 265–268, 270–275, 285
Dursban toxicity, 520, 522, 524, 527–530
Goltix-treated ponds, 308, 310, 317
guthion toxicity, 471–472, 480–484
herbicide effects, 581, 583
    aquatic enclosures, 132–134, 136, 137, see also Concentration-response, theory
insecticide direct effects, 581, 582
lambda-cyhalothrin toxicity, 424, 441–447, 461
macroinvertebrate abundance model, 669

specialist/generalist model, 675
statistical analysis
    regression analysis, 138–139, 142–145, 150, 151
    RM-ANOVA, 140–142
sulprofos-treated ponds and tanks, 328, 329, 330, 331, 334
tralomethrin toxicity, 505
untreated pond mesocosm, 108, 109
Zygoptera
    cyflurin-treated mesocosms/microcosms, 383, 385, 391
    lambda-cyhalothrin toxicity, 426, 447, 454, 461